MAKING BOURBON

O.F.C. 113. A.
GRAIN ELEVATOR
FREE WAREHOUSE.
BONDED WAREHOUSE No 2. A.
CARLISLE BONDED WAREHOUSE No 2. B.

MAKING BOURBON

A Geographical History of Distilling in Nineteenth-Century Kentucky

KARL RAITZ

Cartographic Design and Production by Dick Gilbreath,
Gyula Pauer Center for Cartography and GIS

Paperback edition 2023
Copyright © 2020 by The University Press of Kentucky

Scholarly publisher for the Commonwealth,
serving Bellarmine University, Berea College, Centre
College of Kentucky, Eastern Kentucky University,
The Filson Historical Society, Georgetown College,
Kentucky Historical Society, Kentucky State University,
Morehead State University, Murray State University,
Northern Kentucky University, Spalding University,
Transylvania University, University of Kentucky,
University of Louisville, University of Pikeville,
and Western Kentucky University.

Editorial and Sales Offices: The University Press of Kentucky
663 South Limestone Street, Lexington, Kentucky 40508-4008
www.kentuckypress.com

Unless otherwise noted, photographs are from the author's collection.

Library of Congress Cataloging-in-Publication Data
Names: Raitz, Karl B., author.
Title: Making bourbon : a geographical history of distilling in
   nineteenth-century Kentucky / Karl Raitz.
Description: Lexington : University Press of Kentucky, 2020. | Includes
   bibliographical references and index.
Identifiers: LCCN 2019045519 | ISBN 9780813178752 (hardcover) | ISBN
   9780813178776 (pdf) | ISBN 9780813178783 (epub)
Subjects: LCSH: Bourbon whiskey—Kentucky—History. |
   Distilleries—Kentucky—History. | Distilling
   industries—Kentucky—History.
Classification: LCC TP605 .R36 2020 | DDC 338.4/766352—dc23
LC record available at https://lccn.loc.gov/2019045519
ISBN 978-0-8131-9701-2 (pbk. : alk. paper)

This book is printed on acid-free paper meeting
the requirements of the American National Standard
for Permanence in Paper for Printed Library Materials.

Manufactured in the United States of America

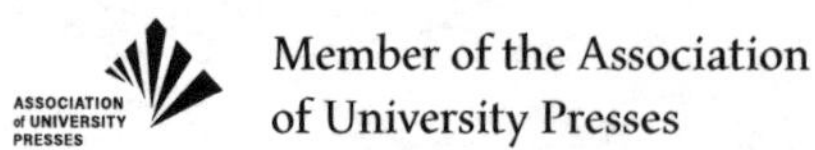

Member of the Association
of University Presses

Contents

Illustrations

Maps

Photographs, Drawings, and Art

Tables and Graphs

Introduction

Kentucky Bourbon in Time and Place

I was born into a culture that then, as now, treasured place and geography more than time and history.
—Edmunds V. Bunkše

The manufacture of whisky is one of the most extensive and valuable interests of the entire Blue Grass Region. Indeed, the blue grass seems to have a beneficial effect on whisky, as it has on everything else that comes in reach of it.
—William H. Perrin

In the heroic age our forefathers invented self-government, the Constitution, and bourbon, and on the way to them they invented rye. Our political institutions were shaped by our whiskeys. . . . They are distilled not only from our native grains but from our native vigor, suavity, generosity, peacefulness, and love of accord. Whoever goes looking for us will find us there.
—Bernard De Voto

Kentucky distillers have produced alcohol spirits for more than two centuries. The frontier craft began by utilizing simple equipment and traditional techniques to convert Indian corn into a clear spirit sometimes called white whiskey.[1] When barreled and aged, this raw spirit was transformed into mellow red bourbon. Some distillers adopted mechanization and became commercial businesses during the nineteenth century, despite being reproved by temperance advocates and subjected to increasing government taxation and regulation. A national recession in the 1870s destabilized capital markets. Grain production faltered. The demand for bourbon was unpredictable. Overproduction and marketing problems dogged the industry from the 1880s into the 1910s, and few distillers survived the national prohibition experiment that began in 1920 and continued until its repeal in 1933. During World War II, the few remaining Kentucky distillers converted their factories to industrial alcohol production. By the late 1940s, distillers had resumed making whiskey. Those distillers that successfully negotiated some eight decades of difficult market conditions not only maintained whiskey production; by the end of the twentieth century, they were leading an industrial revival. In 2018 the thirty-three member distilleries of the Kentucky Distillers' Association produced nearly 1.7 million barrels of bourbon, and there were 7.5 million barrels stored in aging warehouses. Kentucky distillers currently account for about 20,000 jobs with an annual payroll that exceeds $1 billion. In addition to being in demand nationally, bourbon's market now extends worldwide.[2]

To see a contemporary distillery from a distance is not to understand how it works; how it relates to its site, employees, or neighboring properties; or how it evolved to become a tangible artifact. While the distillery is a largely anonymous place to the casual observer, its image and its product have come to represent the industry, and the state of Kentucky, to the world.

The place and landscape of distilling have always been important elements in promoting and advertising Kentucky bourbon. Nineteenth-century advertisements and bottle labels prominently featured the name of the county of origin and often included real or fanciful images of operational distilleries. Bourbon producers have long revered traditional distilling techniques. Distillers followed proven recipes and branded their whiskeys as "Old," as in Old Times and Old Log Cabin. Or their brands alluded to historical places and personages, such as Rolling Fork and Evan Williams. While many manufacturers of consumer goods tout their products as "new and improved," contemporary bourbon advertising campaigns continue to emphasize tradition and heritage. Perhaps more than any other industry in America, the Kentucky bourbon whiskey business has purposely maintained its traditions and methods. Bourbon's heritage has its own folklore and founding narrative. Its heritage is iconic; it is tangible; it is the essence of the industry's legendry. And heritage is marketable. Present-day distilleries welcome tourists with guided tours, visitors' centers, and museums. While marketing to tourists is a comparatively new phenomenon in the bourbon distilling business, the industry image presented to visitors is infused with tradition. Indeed, many distilleries operate, at least partially, in old buildings, a venerable legacy of the past. Distilling's most valued images, those depicted in museums, on tours, and in advertising, are taken from the nineteenth and early twentieth centuries—the era when this frontier craft became a modern industry.

The general historical outline of Kentucky distilling has been well documented, and there is no need to duplicate those studies.[3] Rather, the purpose here is to examine the geographic history of the transformative nineteenth century, when distilling changed from artisanal craft to large-scale industry, and to examine how distillers created the signature distilling landscape that remains at the core of the contemporary industry's identity.

Distilling has long been associated with farming and milling. It also has supply linkages to the timber and cooperage industries, which provided staves and barrels; the copper mining and smithing industries, which contributed still equipment; and the pottery and glass industries, which provided the stoneware jugs and bottles in which the product was sold in the retail market. Raw materials and supplies arrived at distilleries by road or by railroad, and perhaps by river in some locations; the finished product was transported to customers first by packhorse trains, then by wagons and river flatboats, and later by steam-powered conveyances. All these things—the farms and mills, the lumberyards and cooperage shops, and the turnpikes, railroads, and steamboats—contributed elements to the distilling landscape.

As a manufacturing process, distilling is simple in concept and complex in execution. Distilling is a serial or stepwise operation in which raw materials are processed to make raw alcohol spirits that are further transformed by special aging techniques before eventually yielding a potable and salable product. Distilling begins with enzymes in malted barley that convert the starch in a slurry of water and milled grain into sugar. Cooking the mixture produces a mash, which the distiller ferments; adding yeast converts the sugar into alcohol. When fermenting is complete, the product is referred to as distiller's beer. The liquid then passes through a two-stage distilling process. The first stage produces a low wine, which is then redistilled in the second stage into a high wine, often in a traditional copper pot still or doubler. Complexity arises because the distiller must work with precision and speed, using varied recipes. Corn mash requires a different cooking temperature than rye or wheat mash. Malted barley enzymes lose their effectiveness if the mash is too hot. Yeast requires oxygen to work effectively and may die if the mash pH is too acidic or too alkaline or if the mash temperature is too high. The selective movement of raw materials and the precise control of the distilling process in an industrial distillery require a dense rookery of conveyors, vats, pipes, pumps, valves, gauges, and filters. This assemblage demonstrates that the modern distilling process is intricate and demands exacting attention to detail.

Yeast was (and distillers might argue it remains) the key to making quality whiskey, and some yeast recipes have been proprietary family secrets for generations. In the early nineteenth century Lexington porter and ale brewer John Coleman made and marketed beer, but he also made yeast of "preeminent superiority" that he sold to "scientific and experienced" distillers.[4] In the 1830s yeast makers in Scotland and Ireland learned their profession through apprenticeships, and yeast makers were often distillers as well.[5] Richard Cummins (or Cummings), for example, was born in County Carlow, Ireland, southwest of Dublin, in 1830. He began a four-year yeast-making apprenticeship at age fourteen. Cummins migrated to America in 1848 during Ireland's Great Famine. After holding several distilling jobs, he moved to Raywick in Marion County, Kentucky, in about 1868. He built and operated the Coon Hollow Distillery near New Hope and distilled Old Cummins straight bourbon whiskey. Some say that his was the first post–Civil War distillery to mash and distill with machinery.[6]

Many nineteenth-century Kentucky distillers claimed Scottish and Irish heritage. Some migrated to America in the 1830s and 1840s. The exceptional hardships leveled on Ireland during the Great Famine of 1845 to 1849 produced a diaspora of more than 1 million to foreign lands; many ended up in America. These famine-motivated migrants likely included a cross section of the general population, including, no doubt, distillers and others experienced in the distilling business. Less well documented were distillers prompted to migrate by an Irish temperance crusade lasting from 1839 to 1842. Father Theobald Mathew of Cork, along with several other Catholic clergy, led a movement to convince thousands of people to pledge total abstinence from alcoholic beverages. Many public houses and distilleries closed, and the consumption of spirits fell by half. It is possible that this transformative movement led some Irish distillers to move to America to resume their trade, but in the absence of definitive biographical information, the issue remains open to conjecture.[7] A determinist might argue that whiskey distilling was established in central Kentucky because of its "limestone water." But such a simplistic explanation discounts the importance of an experienced, knowledgeable cadre of craftspeople, many of them with Old World traditions, whose proficiencies attracted like-minded people who contributed to the development of the industry, among them distillers, millers, coopers, coppersmiths, blacksmiths, mechanics, farmers, merchants, and financiers.

The distilling process has undergone comprehensive change over the last two centuries, prompted in part by technical and procedural innovations. Kentucky's frontier distillers were happy to use Indian corn of several different varieties as their primary distilling grain. Contemporary distillers prefer non–genetically modified or organic grains if they can obtain them, although more than 85 percent of America's commercially grown corn crop is now genetically modified. But as distilling technology and business practices became more sophisticated, legal policies became more burdensome. Those engaged in distilling, or those whose businesses were linked to distilling, had to adjust to volatile market geographies while defending themselves from the political intrigues of temperance enthusiasts and adapting to the grasping authority of tax collectors.

Elements of Geographic History

This analysis of geographic history centers on three concepts that merit some elaboration: making, landscape, and historical ecology.

Making

People make things based on what they know. They refine or modify things based on what they learn. Knowing and learning may spring from demonstrations or tutorials, but they can also be the product of self-learning or self-discovery.[8] *Making* is a comprehensive expression that encompasses specific types of creative processes and haptic skills such as crafting, building, constructing, fashioning, producing, and manufacturing. Those who study landscapes recognize that built environments have many contributors. Some are local, unlettered, and vernacular; vernacular making may yield a useful structure that follows an age-old functional form. Other contributors are ingenious "mechanics" whose

Kentucky artist Paul Sawyier's painting of the Old Crow Distillery, ca. 1870. The Old Crow Distillery operated on a wide terrace along Glenns Creek in northern Woodford County. The large white gable-roofed building housed the mash tubs, fermenting barrels, and still. Smoke from the steam engine issues from the tall brick chimney. A cooper's shed to the left is flanked by two piles of stave wood; the red brick building in the upper left is likely the storage warehouse. The small cabin between the distiller's house and the warehouse may be the government storekeeper's office. The site represents the medium-sized industrial works that replaced the limited-production distilleries operated by farmer-millers from the early 1800s to the 1860s. (Courtesy of William Coffey, Paul Sawyier Galleries)

primary concern is making structures and machines that perform a function or task in a comprehensible, useful, and efficient way. Some contributors may be experienced professionals, trained in formal design and design history. They may adopt ideas from formal catalogs of building principles that reference grand historic design themes that are weighty with prescribed meaning. Formal, considered design and construction may produce a functional structure, but this type of making may also declare the builder's concern for art and beauty, elegance and refinement. Each contributor possesses a different personal repertoire of interactive knowledge—customs, habits, values—about how to make and use things. Importantly, every person who participates in making a landscape element—be it a clearing in the forest, a building, or a road—becomes a part of that element because it captures a bit of what they knew, what they learned, and, perhaps, what they imagined.[9] Over time, different forms of knowledge intertwine to yield practices that people employ to make practical and functional material landscapes.[10] And each form or

Paul Sawyier's painting of the Old Taylor Distillery, ca. 1910. Millers and distillers were building works on the lower reaches of Glenns Creek near Millville by the early nineteenth century. E. H. Taylor Jr. acquired a distillery here in 1882 and, over several years, rebuilt it, achieving a grand industrial scale. Stonemasons built the turreted castle-form still house (seen in the distance in the upper right) of Tyrone limestone, which weathered to a brilliant white. Warehouse B (in the foreground), at four stories tall and 530 feet long, was thought to be the largest whiskey aging warehouse of its type in the world. Workers installed a sunken garden, next to the still house, in 1906, and the following year the Kentucky Highlands Railroad completed a spur to the site, providing high-capacity rail access for large-scale distilling (mashing capacity reached 500 bushels of grain per day). The railroad track traces a light gray arc across the center of the painting. In addition to delivering grain and hauling away whiskey, the connection allowed Taylor to hire laborers who commuted from Frankfort, and it allowed the distillery to entertain train-borne visitors. (Courtesy of William Coffey, Paul Sawyier Galleries)

sphere of knowledge has equal value or equal potential to contribute to the creation of functional structures and infrastructures.

Posing analytical questions about the making and maintenance of a built environment requires one to recognize that material landscapes are not simply visual and tactile products of an architect, engineer, or building contractor; rather, they are the consequence of unseen creative processes that involve multiple, complex, intersecting circuits of knowledge possessed by people of many social stations and cultural persuasions.[11] Awareness of these knowledge circuits requires what anthropologist Tim Ingold terms "knowing from the inside," or knowledge gained from studying *with* people, which is fundamentally different from the study *of* people.[12] Applying this principle to the study of landscape suggests that the creative process people engage in to make artifacts may become subsumed in the study of the artifacts themselves. An analysis of factories and farms might produce a compendium of industrial buildings, rail yards, farm fields, and barns but yield little appreciation of the inventive processes and social and economic priorities that gave rise to these structures.[13] Insight into the modification of an old factory building might be aided by knowing the owner's name or the year the building was completed. But an appreciation of the cultural practices that produced the building and its new addition might require one to follow the process of modernization and the manner in which the owner evaluated inventions and established relationships with financial institutions and insurers. If an invention were deemed workable, affordable, and insurable, it might be applied to the industrial process, thereby requiring a factory alteration. If such an invention is a dynamic, ongoing process, however, one should take care in judging a building, and a landscape in the larger sense, as "finished." Reality is fluid, and landscapes are always in a state of becoming, continually unfolding along paths of creation, expansion, stasis, decay, and regeneration.[14] Understanding how landscape making works requires that we consider how people bring to bear the knowledge and skills that make up the time-specific technologies that prescribe favored forms and processes. It is imperative that we attempt to position ourselves on the inside looking out. Limited insight into landscape making is gained by acceding to an external perspective couched in tabulating a landscape's visual elements or focusing primarily on enumerating contingent historical economic, political, or social conditions.[15]

This study of the creation and historical development of Kentucky's bourbon-producing landscape employs this "knowing from the inside" perspective to the extent that resources allow. Numerics, drawings, maps, and photographs provide objective information that documents the growth and expansion of the distilling industry, but they also suggest questions about management objectives, labor profiles and practices, and the invisible influence of legal requirements that might direct or regulate industrial activity. Period documents such as business ledgers and personal daybooks and diaries convey selected objective facts, but they can also provide insights into an individual's subjective day-to-day decision making. Details are essential if we are to appreciate a people's inherent and acquired knowledge and understand how they applied information to make things work, a process that might be termed "management strategy" in contemporary idiom. Records outlining plans and priorities can convey what people sought to accomplish and amplify our understanding of how they dealt with the obstacles they encountered. To be concise, we wish to learn how people made their lives and, in the process, made their landscape.[16]

Landscape

People are necessarily situated in place and time, and their occupation of place through time yields landscape. *Landscape* can therefore be defined as "the tangible, visible, impress of human activity on the surface of the earth."[17] Landscape is both a physical entity and a register of contemporary and historical knowledge; it is the physical and built environment and a way of knowing the world. Humans imprint their activities on the land. Sometimes they do this in obvious ways, such as logging a forest, cultivating a field, or building a town. Or the imprint may be more subtle, such as when a political authority erects a sign to indicate the presence of a prohibitory ordinance or an administrative boundary.[18] Through learned behavior and practices, or culture, people are provided with a storehouse of concepts and tools for utilizing and modifying environments. Dwelling with and in a landscape may involve adapting to and exploiting the natural world while at the same time applying to it structures, technologies, organizations, and other cultural elements and practices that permit and promote making a living.[19]

Landscapes are inherently spatial; they occupy space and exhibit geographic patterns and variability. How we observe and describe landscape is not scale bound; that is, landscapes can be small, intimate, and personal—family or community scale—or they can be expansive, dynamic, and collective—regional scale. People alter landscapes in response to cultural change or adjustment. Landscapes are also modified by natural processes. One of our objectives is to sort through the complex natural and cultural processes that shape landscape in order to understand how particular landscapes were formed and maintained or changed.[20]

We conduct our lives in and on a palimpsest of the overlapping remains of landscapes created by past generations. We may add our own signature atop the historical artifacts if the cost of removing or altering them would be prohibitive; the implication here is that landscapes have inertia—in this case, fiscal inertia. If we revere the past as heritage, and if old landscape elements have symbolic value, we may wish to preserve or reconstruct them, thereby lending them tangible inertia.[21] There is, therefore, a reciprocity between people and the landscapes with which they live. As Winston Churchill famously noted in a 1943 address to the British Parliament: "We shape our buildings, and afterwards, our buildings shape us."[22]

Landscape is something that we see and sense, even though it is often constructed and modified by unseen people and practices. But it is also a way of seeing and sensing that is predicated on our cultural context.[23] When viewing landscapes, we must remember that our observations of artifacts constructed by others, during eras different from our own, are biased by our cultural values and traditions and likely are not altogether appreciative of the values and tenets of others.[24] Thinking analytically about landscapes from a geographic historical perspective might begin with a reconnaissance of visual landscape signatures: land use, material structures, technologies, transport routes, and boundaries. Documenting a landscape's geology, geomorphology, hydrology, vegetative cover, and soil types can provide a context for examining how people engaged the physical environment to draw a living from it.[25]

Finally, thoughtful engagement with landscape requires that we attempt to understand how it is valued and what it means to its occupants and observers. Landscape meaning can have multiple manifestations in times past and at present. Meaning can be reflected in what people write about the landscape, such as in personal memoirs or family records, or how it is depicted in paintings, photographs, or other images. A landscape element such as a childhood home, for example, may connote prideful ancestry, pleasant early-life experiences, and endearing personal histories—identity, in short—for a small cadre of former residents. Passersby, in contrast, may ignore the home entirely or see it as a symbolic display of privilege and wealth or exploitation and marginalization. Necessarily, there is a distinction between making and using something, and a structure or practice may carry different meanings for those who interact with it. A prideful nineteenth-century distiller might have boasted of his work's high milling and mashing capacity or its outsized and expensive copper pot still, whereas the laborers who milled the grain, cleaned the still, and moved the filled barrels might have regarded the distillery as a place where they performed demanding work for a low wage. As generations pass, the associations between landscape making and meaning may be recast. For some people, landscapes, or segments thereof, may fade into obscurity and anonymity; others may embrace those same landscapes, perhaps selectively, as their heritage and deem them worthy of formal remembrance, recognition, and preservation.[26]

Historical Ecology

People live in geographic places where others have lived before. In the process, they create and re-create material landscapes.[27] Thus, landscape carries both transient and enduring records of the lives and works of past generations.[28] Time is, unavoidably, an important dimension of all landscape; put another way, all landscapes are historical. The adjective *historical* refers to diachronic study through time.[29] The places where people live and the landscapes they create, through time, are multidimensional, comprising physical environments and cultural constructions. *Ecology*, as the term is used here, refers to the study of the varied and complex relationships among humans and the relationships between human beings and the physical environments they inhabit and their cultural constructions—whatever their form or content. An ecological perspective does not impose an empirical separation between humans and nature.[30] Historical ecology, then, is a broad-based investigation of the physical and human elements of present-day landscapes through a diachronic reconstruction of the conjunctions of people, places, and processes through time, with attention to the order, timing, and contingency of influencing or causal events. This perspective is akin to viewing a subject in high-relief detail with binocular vision, as opposed to flat, two-dimensional monocular sight. To undertake such a study requires that we re-create, to the extent possible, the experiences and everyday activity patterns of the people who make or frame landscapes, both past and present, and whatever their gender, heritage, or vocation—enslaved peoples, laborers, farmers, mechanics, merchants, industrialists, bankers, and legislators among them.

Kentucky's contemporary distilling industry is grounded in traditions and techniques that were more than two centuries in the making. During this time, production methods and technical capabilities developed and changed, as did demographic structures, cultural practices, societal values, and political organizations. The industry's relationship to nature has not remained static either. Cultural practices can change, sometimes in an evolutionary manner, when people refine and elaborate technologies incrementally with new inventions or by borrowing and improving ideas provided by others. Traditional practices can also change abruptly when an event demonstrates the effectiveness of a new idea or artifact, such as an improved crop variety that produces higher yields while resisting diseases and pests. Both types of change often result in landscape adjustment and alteration. We must also acknowledge that although members of a social group may share a large body of conventional understandings about how the world works, they may not participate in the making of cultural practices equally or from the same perspective. Some may favor using a resource in a certain way, while others object to that use or wish to use the resource differently. Some may deploy various types of leverage to use land and erect and utilize structures that those with lesser influence would not attempt. Cultural practice is neither constant nor consistent. And perceived contradictions in cultural customs may lead to change.[31]

People living in different historical eras and possessing different cultural practices develop distinctive cultural landscapes within the context of their physical world, landscapes that can exhibit both a historical and a contemporary character. Artifact construction must take place within or upon a physical environment that offers certain opportunities—a humid or arid climate, fertile or sour soils, steep rocky slopes or low-gradient alluvial floodplains. Environmental change can be gradual and evolutionary or fitful and abrupt. A forest fire can kill susceptible tree species, allowing their gradual replacement by other vegetation types. Catastrophic floods can erode new channels across floodplains and alter soil profiles. A physical environment is not an undifferentiated terrain or a neutral background on which people act; nor are environments readily separable from human action.

Many landscape elements exhibit a tangible presence that can be empirically reviewed by observation, photography, or inventory and otherwise objectively assessed. A corollary, however, is that landscapes can also be shaped by unseen economic influences, social relations, and political policies. America's frontier settlers produced staple crops for consumption by family and neighbors. New technology in the form of improved tools and plant varieties permitted those farmers to produce surpluses that could be shipped to distant markets if prices permitted. Governments sought to enhance the production and flow of domestic farm products by leveling tariffs on inexpensive imported goods; alternatively, governments placed controls on that same domestic production in order to extract taxes. Plant genetics, market prices, tariffs, and production controls may not be signatures that are readily visible on the landscape, but they can influence production patterns. And if one knows what to look for, such factors may leave behind subtle signs on the landscape. With some study, the casual observer can distinguish a field of wheat infested by the Hessian fly from a field planted in a resistant seed variety, just as one can differentiate between water- and steam-powered gristmills. Understanding the historical ecology of landscape requires an appreciation of its totality, of the complex conjunctions that lend structure, function, and presence to what people build and how they use land.[32]

A Synthesis of Making, Landscape, Historical Ecology, and Distilling

Distillers and their cognate businesses make consumer products, of course. But the production process also creates a material landscape of distinctive structures and land uses. America's physical landscape is constructed of common materials such as stone, brick, wood, concrete, iron, and steel, which are transformed into structures with foundations, walls, girders, pipes, and so on. Many industrial structures may be familiar to us in profile, but the manner of their design, assembly, and function remains obscure. Yet it is a structure's purpose that suggests appropriate materials and forms and thereby contributes definitive character to the landscapes that people occupy—be they employees or passersby.

To better understand the connections between the distilling industry and its landscape, we may employ two contrasting but complementary perspectives: the local or vernacular, and the national or formal. The first perspective recognizes that building landscape is a local process and represents the intersection of local practices, materials, economies, social networks, and politics. Each locality has its own cultural center, and its landscape can be examined as the product of local circuits of knowledge and practice. Within this context, local practice is tempered to varying degrees by regional, national, or even international influences such as migration, the development of transportation networks and markets, and the flow of raw materials. Attention to local scale permits a detailed examination of personal economic strategies; the social relations among laborers, farmers, and distillers; and the workings of various financial and political institutions. Knowing how distillers established their businesses, the nature of their labor forces, and their business relationships with grain and barrel suppliers, bankers, wholesalers, and other businesses can provide an appraisal of how distillers functioned and how those functions became materialized in landscape.[33] At the local level, a landscape is made human.

A national perspective on landscape creation self-referentially configures landscape elements according to national or regional trends or models, especially through the development of widely applicable technologies, the creation of a national marketing system for commodities and financing, and the increasing reach of the national government, particularly its regulatory and production control policies. Those local distilling businesses that succeeded financially often increased in size and became large industrial works.[34] Such change was enabled, in part, by new technologies whose application was formalized by science, engineering, and the economies of a national marketplace. The Kentucky distiller's primary product—bourbon whiskey—commanded a profitable price, or distilling stopped. Increased production also compelled distillers to develop business relationships with financiers and insurers. Overproduction reduced prices and profits and implied that other producers were engaged in creating a competitive market. The product was also susceptible to governmental taxation and regulation. Those distillers that competed with other producers, embraced new technologies to enhance production, and acquiesced to governmental regulation created landscape elements that resembled those of their competitors. And all distillers were subject to the controls imposed by regional and national commodity and financial markets. As a result, the distilling process and the landscape it created began to adhere to regional and national norms.

At the local scale, individual distillers operated at different sites that required idiosyncratic adaptations to those places. They constructed buildings with the materials at hand, using whatever carpentry and masonry skills they possessed or could hire. Each distillery was different, its character a reflection of its place and its builder. But at the regional and national scale, each distiller was subject to advances in technology, refinements in business procedures, and increasing government oversight, all of which tended to normalize the distilling process and its landscape.[35]

The overall objective here is to understand how local-scale industry developed through the insights and innovations of individuals with different origins and life experiences. These individuals included Irish and Scottish immigrants, Maryland and Pennsylvania millers and farmers, and native-born Kentuckians. Nineteenth-century distilling did not emerge as a refined manufacturing process in the complex and chaotic locales of early American industrial development such as New Jersey's Delaware Valley, the upper reaches of the Ohio and Monongahela River Valleys in western Pennsylvania, or the large East Coast cities of Philadelphia, New York, and Boston. Rather, distilling began as a small-scale enterprise, usually in concert with farming or grain milling; only gradually did it develop into an integrated commercial business operating at an industrial scale. Distilling experience gained in distant locales certainly provided an advantage to those early settlers who wished to carry on the business, but the distilling practices that subsequently blossomed in Kentucky were a product of that specific place and the circumstances presented by the environmental, cultural, social, and political milieu that developed there.

Kentucky's distilling landscape shares some common elements with other industries that process agricultural products, such as flour milling and livestock feeding. Yet distilling is distinct, in that it applies traditional knowledge of elementary chemistry and commodity characteristics to produce a potable product whose potential market extends as far as information and transportation permit. At the same time, distillers have had to defend themselves from those seeking to capture their market or to control or even prohibit production and consumption. At each phase of industrial development, Kentucky's whiskey makers created or adopted utilitarian structures to house their activities and shelter their raw materials and products. When long-standing technology was refined or when new or larger equipment was developed, existing distilleries might be moved or torn down and replaced. Alternatively, many distillers adapted older structures to new purposes and added new buildings to the distillery site. The result, though functionally pragmatic, often presented a rather disorderly appearance.

American distillers did not spawn ancillary industries such as iron and copper metalworking, cooperage, or glass manufacturing. Those industries emerged independently, and their standardized products were accepted by distillers who adapted them to their own needs. A distillery might require copper kettles, brass valves, iron pipes, leather belts, and reciprocating iron and brass steam engines, but these products were not initially custom designed and built for distillers. Rather, distillers initially built their own equipment from bulk material such as copper sheeting or used "off-the-shelf" products. Gradually, as distilling grew into an industrial business independent of farming but dependent on farmers' products, distillers began to develop and test ways to improve product quality and production efficiency. They attracted supporting industrial innovations that led to the development of dedicated milling equipment and grain storage facilities. Increasingly, farmer-distillers could disconnect from their dependence on neighborhood mills and coopers and purchase ready-made, custom-built copper stills; specialized valves, pumps, and fittings; cooperage machinery; and patented warehouse storage rack systems. New glass bottle-making technology permitted a radical change in the packaging and sale of distillery products and underwrote a specialty bottle-making industry that allowed distillers to refine their marketing and distribution strategies. Innovations of another type—improvements in botanical crop varieties—increased grain yields but also resulted in grains with superior milling and distilling characteristics.[36]

All industrial landscapes are composed of intricately layered and interrelated elements, but Kentucky's nineteenth-century distilling landscape was especially complex. Rural distilleries stood beside springs or creeks and processed grain from surrounding farms. Urban distilleries drew water from rivers or wells and patronized rail lines that delivered their grain and shipped their product. The distillery workers who transformed grains into whiskey may have been enslaved people prior to 1863 or the sons of local farmers. Skilled coopers and coppersmiths found work supplying barrels and still equipment. The land carried the imprint of these and many related activities.

The distilling landscape can be secretive and proprietary, on the one hand, and subject to disclosure owing to governmental regulation and oversight, on the other. Industrial innovations produced new equipment that distillers installed and operated according to their own traditional procedures. But beginning in the 1860s, federal taxing regulations exerted increasing control over the configuration and functioning of the distilling land-scape. Regulation enclosed distillery buildings with invisible legal boundaries and ceded operational control to federal appointees. The distillery landscape, then, with the exception of a few signature structures, is partially opaque to the casual observer. But, when it is examined closely, one finds a complex production and marketing system with ties to agriculture, industry, business, government, and law and local, regional, and even national cultural mores.

It is likely that most nineteenth-century distillers were not aware of all the historical contingencies that affected the development of their industry. Similarly, we recognize that we have not discovered all the people and events that influenced the construction of Kentucky's traditional distilling landscape. Nevertheless, we can construct a narrative that demonstrates that landscapes, however ordinary they seem, are linked to an expansive web of cultural influences and contingencies that entrain from local to regional to national and even international scales. It is also evident that contemporary distillers are keenly aware of their forebearers' contributions to the development of the state's signature industry and maintain an appreciation of that heritage as it is manifest in landscape preservation and product marketing.

I

Making Kentucky's Distilling Landscape

1

Heritage and Process

My aim is to diffuse new light on everything that relates to the formation of spirituous liquors that may be obtained from grains. Most arts and trades are practiced without principles, perhaps from a want of the means of information. For the advantage of the distillers of whiskey, I will collect and offer them the means of obtaining from a given quantity of grain, the greatest possible quantity of spirit, purer and cheaper than by the usual methods.
—Anthony Boucherie

Building on Heritage

American spirits distilling grew from European and colonial traditions. New England distillers, among them John Hancock and Samuel Adams, made rum out of fermented Caribbean molasses to supply local tipplers with spirituous drink, to stock merchant and whaling ships with the makings for grog, and to serve as currency in the Triangular Trade.[1] In the Middle Colonies, German settlers in southeastern Pennsylvania's Piedmont and the adjacent counties in Maryland distilled spirits from rye, as did George Washington in eastern Virginia. Protestant Scots-Irish from Ulster country and Catholic Irish from Ireland brought a thousand-year whiskey-making tradition to western Pennsylvania and the Ohio River Valley in the trans-Appalachian West.[2] In 1794, upon finding themselves subject to federal enforcement of Alexander Hamilton's excise tax on the whiskey they made, some 2,000 farmer-distillers in western Pennsylvania migrated west, many of them to Kentucky.[3] These Whiskey Rebellion refugees further reinforced frontier Kentucky's rapidly developing distilling economy.

Pioneers and innovators are often rewarded with citations as the "first" to accomplish something in the legends and lore that make up a founding myth. Evan Williams, Elijah Craig, and other eighteenth-century Kentucky distillers have been so acknowledged, but absent definitive documents, one can only admit of general recognition. It is sufficient to know that distilling was commonplace in 1780s Kentucky.[4] Second-generation distillers were providentially positioned to see their accomplishments recorded as literacy rates increased, personal circulation and travel reduced isolation, and communication improved with the establishment of local newspapers. Accordingly, the contributions of distiller James C. Crow are more fully understood and illustrate how a talented individual influenced the development of quality whiskey. "To him, more than any other man, is due the international reputation that Kentucky enjoys, and the vast distilling interests of the country are largely the result of his discoveries," pronounced the *New York Sun* in 1897.[5] Crow was born in Dirleton, Scotland, in 1779. A "Presbyterian of the John Knox type," he graduated from the College of Medicine and Surgery in Edinburgh in 1822 as a physician and chemist.[6] He migrated to America and, after spending a short time in Philadelphia, moved to Kentucky in 1823, where he entered the distilling business on Glenns (also spelled Glens or Glenn's) Creek, near Millville in Woodford County. Distilling at the time was rarely directed by rules; nor was it exacting in terms of quantity and proportion of ingredients, critical temperatures, or precise timing. Though his still house had a very limited capacity, producing only two to two and a half gallons per day, by embracing chemistry as a hobby and experimentation as a practice, Crow applied scientific methods and advanced instrumentation to his distilling practice. He was eventually able to consistently produce a superior corn spirit and, in the process, changed whiskey distilling from a folk craft to a semblance of a science. Brisk sales soon followed, with Henry Clay, Andrew Jackson, William Henry Harrison, and Daniel Webster among his better-known customers.[7]

Kentucky's frontier folk, whatever their heritage, often engaged in small-scale farm distilling, producing enough corn spirits for themselves and perhaps some to sell to neighbors or ship to southern markets if they had ready access to one of the Ohio River's navigable tributaries. Kentucky's first *commercial* distillery may have been in operation in Louisville, at the Falls of the Ohio, in 1783. Many of the skilled coopers who built the wooden tubs and barrels used to distill and store whiskey had served apprenticeships and worked in Virginia, Maryland, and Ohio, as well as in Belgium, France, and Germany. Distillers who wished to increase the scale of operations beyond what their family members could accomplish hired itinerant and immigrant laborers or pressed enslaved African Americans to provide the workforce.

Process: A Distilling Primer

Grain starch is the only ingredient used in the manufacture of whiskey. Millers initiated the distilling process by milling corn and rye or wheat to the consistency of a fine meal. Barley has a high "diastatic power"—that is, malted barley contains enzymes that can convert up to 2,000 times the weight of starch into glucose.[8] Maltsters prepared barley through a wetting process that caused the grain to sprout; thereafter, the malted barley was dried and milled. Distilling begins with the conversion of grain sugars and starches into grape sugar, which then undergoes vinous fermentation. To make a distillable mash, distillers traditionally mixed milled corn and rye or wheat with boiling water in large wooden tubs. After cooking, the mixture was cooled to about 152°F, at which point the distiller added barley malt. When the starch conversion process was completed, the mixture or mash was cooled, and yeasts began fermenting the sugars into alcohol.[9] Yeast fermentation, a form of combustion, lasted seventy-two to ninety-six hours. The fermented mash, or distiller's beer, contained alcohol and congeners such as methanol, fusel alcohols, aldehydes, and tannins, which provided the distilled whiskey with some of its taste and aroma. This process produced a sweet mash; to make sour-mash whiskey, distillers added to the fermenting vat stillage residue, or "back set," from a previous batch.

In Kentucky, fine whiskeys were traditionally made in alembics, or onion-shaped copper pot stills; these were later augmented by the vertical column still. In the 1880s commercial-scale column stills were three to four feet in diameter and thirty to seventy-five feet tall. Both still types were comparatively simple mechanical devices whose operation was based on the differential boiling points of water and alcohol. Pot and column stills removed alcohol from fermented beer by heating the beer to about 173.1°F, or 38.9°F lower than the boiling point of water. At this critical temperature, the alcohol "goes over," or is vaporized. Since alcohol vapor is lighter than water vapor, it separated and left the water vapor behind. The alcohol vapor was then captured and passed through a coiled copper pipe, or worm, where it condensed into liquid ethyl alcohol, also known as drinking alcohol or spirits.[10] This condensed spirit was termed low wine or singlings, and it was 90 to 130 proof, or 45 to 65 percent alcohol.[11] The low wine was then passed into a second still or doubler, commonly made of copper and traditionally heated by a wood or coal fire, where it underwent a second distillation. The alcohol vapor produced then passed through a second worm and emerged as high wine or whiskey of roughly 150 proof, or 75 percent alcohol. By adding distilled or local water, the distiller further reduced the proof to 100 to 110. The mixture of water and spent grain, known variously as grain stillage, wash, or slop, was left behind and required disposal. The contemporary distilling process is little changed, in principle, from that employed in the nineteenth century. The final step in whiskey production is barreling and aging. Nineteenth-century American whiskey barrels held 48 gallons; standard barrel volume was increased to 53 gallons, or 200 liters, sometime during the twentieth century. Laborers moved the filled barrels to warehouses for aging for a period of two years or more. Fine whiskeys may be aged for a decade or longer.

If the spirit so produced was distilled from a fermented mash of not less than 51 percent corn and stored at no more than 125 proof in new, charred oak barrels, it can legally be identified as "bourbon whiskey." Whiskeys stored in appropriate containers and aged for two years or more are identified as "straight bourbon whiskey."[12] As this standard implies, all bourbon is whiskey, but not all whiskey is bourbon. Nineteenth-century Kentucky distillers sold bourbon whiskey to wholesalers and retailers in full-size barrels or in smaller containers such as half barrels or kegs. Saloon keepers served their customers whiskey from full-size barrels tapped with a spigot. Some distillers sold their whiskey at distillery "quart houses" to customers who brought their own stoneware jugs.

Kentucky: The Center of Bourbon Whiskey Distilling

How does one explain Kentucky's traditional association with whiskey distilling? Many of the state's eighteenth-century Anglo settlers had a modicum of distilling experience, so the industry did enjoy the advantage of an early start and historical inertia. But folk distilling was also widely practiced elsewhere, from the Carolinas and Tennessee north and west to Indiana and Illinois, well into the nineteenth century. One might consider the extraordinary concentration of whiskey distilling in Kentucky to be a matter of rational economic geography that can be readily explained, as it is for other businesses and industries, by an analysis of production costs. Industrial processes incur costs to purchase, move, and process raw materials; to train and pay a labor force; and to distribute a finished product. If, for example, the location of bourbon distilling is dependent on the availability of large quantities of distilling grains, then, according to best economic practices, the industry should be located in the most productive parts of the Corn Belt—Indiana, Illinois, and Iowa—to ensure availability of grain and low transportation costs. It should not be centered in Kentucky. Yet, in 2014, Kentucky distillers used more than 12 million bushels of corn and 4 million bushels of other grains, and farms outside Kentucky supplied about 50 percent of the corn and 80 percent of the wheat and rye processed by the state's distilleries.[13] Very little rye or barley was grown in Kentucky during the nineteenth century. In the 1860s and 1870s many distillers bought malted barley from Canadian suppliers, and the majority of contemporary malting barley production takes place in North Dakota and Saskatchewan.

Moreover, distilling is not concentrated in Kentucky because of superior transport access or freight rate advantages. Distilling has a very low weight-loss ratio; that is, the weight of raw materials used in distilling is very similar to the weight of the finished product. Transportation costs for finished products are usually higher than the costs of shipping raw materials, in part because those materials can be handled in bulk by specialized equipment. In addition, finished products may require special handling because they are more perishable than the raw materials. Whether bourbon was shipped in barrels, kegs, or jugs during the nineteenth century or in glass bottles thereafter, transport costs included losses due to breakage and pilferage. Distilleries are most favorably located near their customers, yet the consumption of spirituous beverages in Kentucky is comparatively low, and 48 of the state's 120 counties are nominally dry.

Distilled bourbon must be aged in American white oak (*Quercus alba*) barrels, often for four years or more, to assure a quality product. Indeed, the costs of cooperage and warehousing, insurance, and product loss through evaporation can be higher than the costs of raw materials, labor, and overhead combined. And the expenses associated with aging or warehousing would be similar in other states, so Kentucky does not enjoy an advantage there in transforming raw whiskey into aged bourbon.[14] Given such anomalies in raw material quality and availability and the costs of manufacturing and transport, why is bourbon whiskey production centered in Kentucky? Kentucky distiller John Atherton acknowledged these inconsistencies when he testified before a US congressional committee in 1888 concerning internal revenue regulations. "The distillers located in the best grain country (the Middle West and central and northern Great Plains) and with the best railroad facilities would drive other distillers (in Kentucky) out of existence," he said, without the protection of the internal revenue law for fine whiskey production.[15]

Perhaps Kentucky possesses optimal environmental conditions that abet distilling. Corn and wheat thrive in Kentucky's climate and soils, yet rye and barley do not. Accepted lore suggests that early distillers touted the state's "pure limestone water" as a favorable factor if not the primary locational criterion for the founding and building of Kentucky's distilling industry. Kentucky's groundwater, which was said to be "pure" and iron free, supposedly lent the finished product a particular taste. Distillers in other states likely vigorously debated this presumption. The Great Valley is floored with limestone for much of its length from Maryland and Virginia to eastern Tennessee, as are many of the fertile farming valleys of eastern Pennsylvania. And what about Tennessee? The famed Nashville Basin is also a limestone plain and a near mirror image of the Bluegrass, to which it is geologically related. If limestone water was the secret to whiskey production, it seems likely that distillers would have favored these other places as well. Meanwhile, central Illinois was covered in many feet of glacial till, yet Peoria became one of the nation's major nineteenth-century whiskey distilling centers. By the 1840s, distilleries operated in most Kentucky counties, regardless of their geological provenance. Limestone is scarce in Kentucky's Appalachian counties and is incidental on the Ohio River floodplain downstream from Louisville. Two Kentucky regions, the Greater Bluegrass and the Pennyroyal, are deeply underlain by limestones that husband groundwater and are the source of productive springs and wells. Yet the most extensive development of distilling occurred in the Bluegrass but not in the Pennyroyal.

Can one argue that distilling succeeded in Kentucky because of the social and political attitudes of the public and government officials? Kentucky had an active temperance movement during the nineteenth century. In 1834 the Kentucky Legislative Temperance Society was formed, with Governor John Breathitt (Democrat) as president and Lieutenant Governor James Morehead (Republican) as one of its five vice presidents.[16] Well-attended Temperance Society meetings were often held in towns with large operating distilleries.[17] Temperance activist Carrie Nation was born in Garrard County, Kentucky, in 1846, although she did most of her temperance work in Missouri, Kansas, and Texas.

Perhaps the most compelling explanation for Kentucky's emergence as the nation's principal bourbon whiskey producer is that, from the earliest decades, Kentucky distillers sought to make high-quality whiskey, an inclination that became an operative requirement by the last third of the nineteenth century. Neighboring states were major whiskey production centers—Illinois produced more whiskey than any other state. But Kentucky distillers argued that those Corn Belt whiskeys were poorly made and insufficiently aged; the end product, which was made tolerable only by rectifying or compounding the whiskey with flavorings, was deemed suitable only for a nondiscriminating market. Kentucky distillers, in contrast, purposely cultivated their reputation for producing a first-class product and expanded their sales through clever marketing and advertising.[18]

Kentucky Distilling

Craft to Commercial Enterprise

The rapidity of improvements in the western parts of the United States, is a matter of some consideration to the distillers of the Atlantic states. They have already made considerable progress in the art of distillation, and the vast quantities of grain which are produced by their fertile lands, beyond the necessary consumption, cannot be so well disposed of in any way as in pork and whiskey. Hence we already find Kentucky whiskey in our sea ports.
—Harrison Hall

A Potable Product and Transactional Currency

The bourbon distilling industry has long been associated with Kentucky, although whiskey distilling was common throughout the Atlantic Seaboard during the colonial period. Log cabin distilleries operated by migrants from Maryland, Pennsylvania, and Northern Ireland produced unaged white whiskey on Bluegrass farms and in the town settlements of central Kentucky by 1775.[1] Whiskey distilling was a craft, a vernacular folkway that was perhaps the first widespread industry in Kentucky that processed local resources—water and grain—into a product that could be readily sold. Local customers patronized distillers, but the product was also exported by overland trails to river towns such as Louisville and Cincinnati, where it was sold or loaded onto riverboats for delivery to markets downriver. As corn and wheat production moved into the trans-Appalachian West, whiskey distilling followed, and by 1850, Cincinnati had become the largest whiskey market in the world.[2]

Distillers used whiskey to barter with farmers; a gallon of whiskey for a bushel of corn was a common equivalence. In lieu of coin money or specie, frontier whiskey often served as currency. Wherever farmers produced grain, it was milled, and any milled grain that was not directly consumed as flour or meal was often distilled into whiskey. Milling and distilling developed in concert, both geographically and historically, and these frontier businesses moved away from the economically limiting barter-exchange system to the medium of debits and credits, which became the basis for the transition from a subsistence to a market economy in communities where these businesses prospered.[3]

In central Kentucky, whiskey was used in barter-exchange transactions, primarily in rural areas beyond the ready reach of merchants and bankers in commercial towns. In larger settlements, whiskey was part of a cash economy and was not commonly used in lieu of cash. John Moylan operated a general store on Main Street in Lexington in the early 1790s. He offered for sale a wide range of goods: linen and calico; groceries, glassware, and gun locks; hand tools, harnesses, and hymnbooks; and tobacco and whiskey. In 1792 and 1793 Moylan periodically purchased whiskey from Nathaniel Ashby, James Dunn, William Kearney, John McCall, Payton Short, William Trotter, and Robert Wallace for resale to customers. He bought whiskey in amounts ranging from five to more than sixty gallons, for which he paid three shillings six pence per gallon. His whiskey suppliers—the record is unclear whether these men were also distillers—maintained accounts at the store for sundry articles such as hardware and hand tools, including a cooper's adz, cloth, groceries, sugar, and salt. Moylan sold the whiskey at retail to his customers, sometimes in multigallon lots, for one shilling six pence per quart, or a sales margin of about 25 percent over his cost. One entry in his ledger specified that the spirit for sale was "Bourbon." Most of Moylan's customers settled their accounts in cash, including the whiskey suppliers. Moylan did not receive whiskey in lieu of cash or as part of barter arrangements; he bought spirits for resale in the same manner that he purchased locally produced nails, wood, meal, flour, and vegetables.[4]

Industrial Imprints

During the nineteenth century, distilling changed from a vernacular artisanal craft to a commercial industry, and this resulted in pronounced alterations in operations and landscape. While the craft distillers' landscape was never static, the adoption of industrial techniques and technology led distillers to radically alter their traditional procedures. Some of those alterations were nuanced, and others were conspicuous, but they all reflected distillers' preferences for how the work should be done.[5] Distilleries, and kindred businesses such as farming and milling, adopted new technologies and business practices: construction materials such as portland cement and Pittsburgh iron; steam, electrical, and internal combustion power sources; farm machinery such as the reaper and the corn planter; industrial machinery that facilitated grain handling and automated glass blowing; the automobile, truck, and tractor to move people, materials, and farm implements; and insurance to mitigate the risk of fires and accidents. Between 1810 and 1850 the number of distilleries in America declined by 90 percent, but the quantity of spirits produced more than doubled.[6] All this was reflected in the changing distilling landscape.[7]

The industrial transformation of Kentucky distilling took place between 1820 and 1860 and was enabled primarily by the adoption of mechanization. Some aspects of agricultural production in the trans-Appalachian West were also mechanized.[8] During these four decades, farmers, mechanics, and artisans transformed the production and handling of grains and livestock into commercial-scale farming, milling, and butchering businesses.

Kentucky Distillery Census, 1810–2018

Year	Number of Distilleries
1810	2,000 (estimated; Fayette County had 139)
1840	889
1864	166
1870	141
1876	225
1880	242
1894	170
1919	183 (on the eve of Prohibition)
1920–1933	6 (licensed during Prohibition)
2012	10 (craft and industrial scale)
2015	31
2018	68 (craft and industrial scale)

Compiled from various sources.

Home food production, processing, and preservation were present in most communities from the time of their initial settlement, especially in the Middle West and adjacent Ohio Valley states. Manufacturing new tools and implements required skill, so the successful application of new technology was contingent on the development of new proficiencies. The industrialization of farming, milling, and distilling therefore proceeded in concert with technical advances in manufacturing.[9] Farmers made harnesses and built wagons; cleared land and marketed sawtimber; invented and improved plows, harrows, drills, and reapers; and often became first-generation manufacturers. Inventors patented hundreds of new farm tools and implements.[10] Reapers replaced sickles, scythes, and rakes; threshers replaced forks and flails. Instead of planting grain by hand-broadcasting, farmers used drills and planters. Cutting grain with one of Cyrus McCormick's horse-drawn reapers gave farmers a five-to-one production advantage over a hand-swung scythe and cradle. By the end of the Civil War, farmers were planting and cultivating corn with machinery, and production increased three- to fourfold over the yield obtained with wooden plows and hoes.[11] After harvest, farmers could shell corn off the cob by machine rather than by hand. Importantly, farmers created the industries that processed their products: grain milling, cattle and hog butchering, lumber milling, brewing, and distilling.[12] Over the century from 1830 to 1930, mechanized grain production provided an advantage of roughly thirty-two to one over traditional manual labor practices. And transforming distilling from vernacular craft to mechanized industry was contingent on sizable increases in grain production.[13]

Commercial, industrial-scale distilleries gradually replaced vernacular farm stills. This process was abetted by new specialized machines that were efficient and productive but also required more space that offered protection from the elements. Distilleries could no longer function in simple lean-tos or single-room sheds; they required larger spaces often configured as a series of rooms, each housing a different function. These compound distillery buildings sometimes appeared to be cobbled together rather than purpose-built. Larger copper pot stills and new vertical column stills increased the volume of whiskey production, but they also changed the distillery's size and floor plan. Barrel storage warehouses evolved from low wooden sheds to large-capacity multistory timber and brick structures. Some industrial distilleries added dedicated cooperage shops. Stationary steam engines required fuel, careful tending, and regular maintenance. Increased distilling capacity required a reliable supply of raw materials, necessitating the construction of grain bins and corncribs. The need for ready access by horse- and mule-drawn wagons obligated the distiller to give careful consideration to building placement and road access.

Although craft and early industrial bourbon distilling left distinct imprints on the landscape, the manufacturing process did not ravish the land. Unlike the heavy industrial development in southwestern Pennsylvania, where coal mines, coke ovens, and steel mills created a pervasive "black country," Kentucky's small rural distilleries fit glove-like into the countryside. Antebellum industrial stills often operated in simple log or frame buildings that resembled barns or sheds. Adjacent sheds sheltered hogs or cattle that were fed on the distiller's spent grains, or slop. Woodcutters and sawyers harvested trees from nearby woodlands to provide fuel, and coopers obtained wood from adjacent lumberyards. Neighboring farm fields produced distilling grains. And loaded wagons moved along country roads, bringing raw materials to distilleries and taking barreled whiskey to markets or river landings for shipment. Kentucky's distillers often encouraged or underwrote turnpike construction, and after the state's first railroad began operation in the 1830s, distillers lobbied rail companies to build lines into their counties.

As a new rail network developed after the 1850s, some distillers moved their works to cities, where they had direct access to the railroad and other factories in light industrial districts. Steam power enabled distillers to significantly increase their scale of operations, which in turn stimulated a change in traditional business relationships with suppliers and customers. As distillers refashioned their industry during the nineteenth century, craft production gradually transformed into science-based manufacturing; the still moved from a log cabin into a post-and-plank shed or barn and, eventually, into a large, industrial-scale brick-and-stone works. As factory distilling increased in volume, the imprint on the land was amplified. Related industries such as gristmills, lumberyards, and cooperage shops all contributed to this new industrial footprint, as did toll roads and railroad yards and sidings. Some distillers prospered and built great houses for themselves and stores and churches for their workers; others went bankrupt and lost their holdings to banks, lenders, or other distillers.

Locational Patterns in Kentucky Distilling

In 1810 more than 2,000 distilleries operated in Kentucky, although the county-by-county

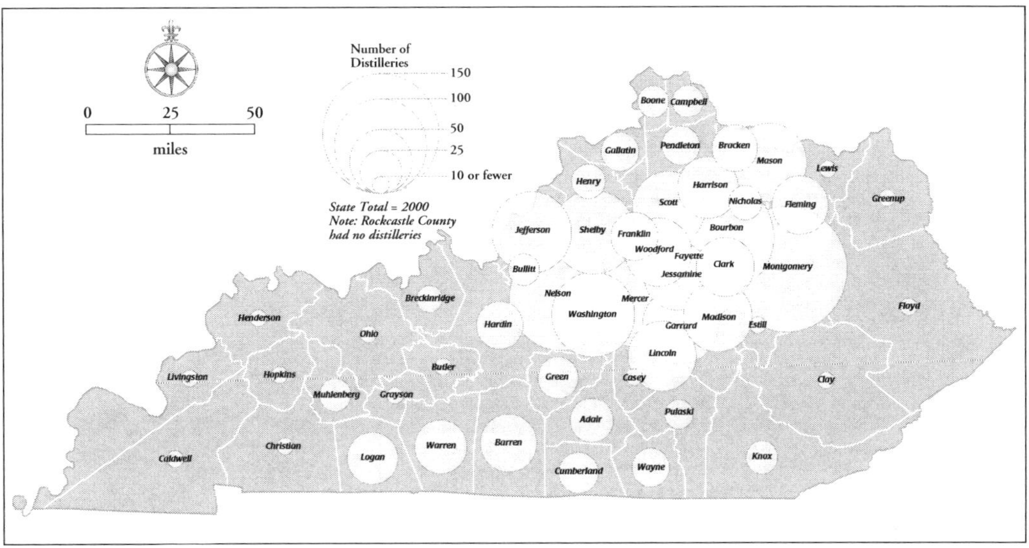

Kentucky's distilleries, 1810. (From US Census, 1810, "A Series of Tables of the Special Branches of American Manufacturers," 122–23)

distribution was profoundly asymmetrical. This pattern was related to opportunities offered by benign topography and benevolent geology, but it was likely also influenced by transport route quality and access to distilling grains and markets for surplus production.

Whiskey issued from twenty-five or fewer distilleries in each of the eastern Appalachian Mountain counties—at least those enumerated by census marshals. Correspondingly, few distillers could be found in the Western Coal Field or the far western counties bordering the Mississippi River. Each of the eastern Pennyroyal counties, such as Barren, Warren, and Hardin, had twenty or more distilleries in operation. The largest number of distilleries was clustered in the Greater Bluegrass region, with 100 or more operating in several counties. More than 150 distilleries produced spirits in both Fayette and Montgomery Counties.[14]

The early stages of the transition from farmer-distiller craft production to commercial-scale production are reflected in distillery production capacity. Again, the overall pattern of concentration in the Bluegrass region obtained—the largest-capacity distilleries extended from Boone County on the Ohio River in the north to Lincoln County in the south, and from Mason County in the east to Jefferson County in the west. By comparison, the mountain and Western Coal Field distilleries were small in both number and capacity. And some important anomalies appeared. For instance, Harrison County had only about fifty distilleries in 1810, but total production approached 3,000 gallons per distillery. At the same time, nearby Montgomery County vied with Fayette County for the largest number of distilleries, but overall production in Montgomery lagged well behind, indicating small works with limited capacity. Most dramatic, however, was the production pattern in the

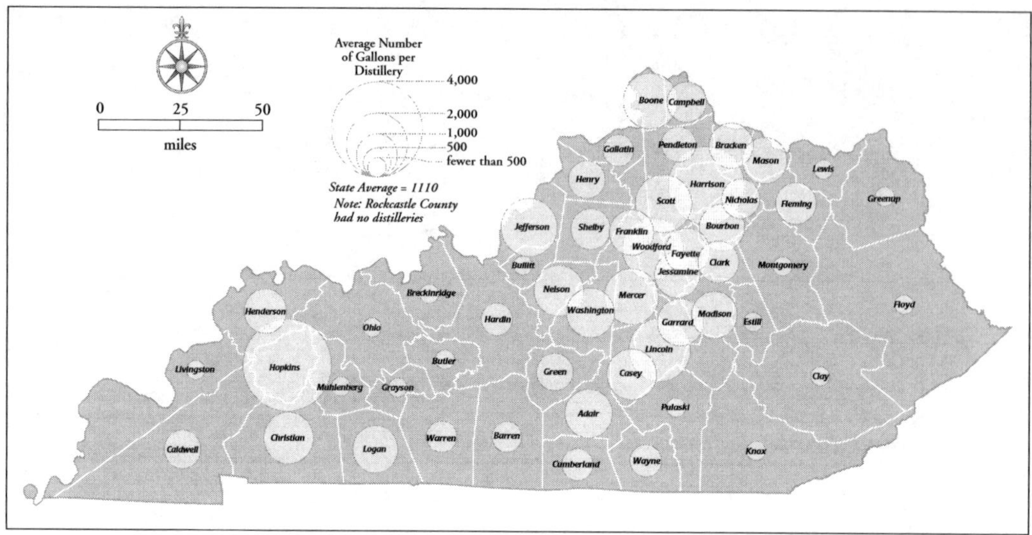

Distillery production, 1810. (From US Census, 1810, "A Series of Tables of the Special Branches of American Manufacturers," 122–23)

Western Pennyroyal counties of Logan, Christian, and Hopkins. Fewer than ten distilleries operated in Hopkins County, yet each distillery produced nearly 4,000 gallons annually.

By 1830, the era of industrial distilling was under way, although many traditional distilleries continued to operate. Industrial-scale distillers supplied whiskey to local and regional markets, despite the limitations of primitive transportation. And their working capital was checked by taxation imposed by a federal government anxious to pay its bills. Although the distilling process remained little changed until the mid-nineteenth century, successful distillers expanded their facilities and extended their geographic reach across the countryside to obtain grain, cooperage wood, and new and specialized mechanical equipment. Accordingly, farm fields, hardwood forests, distillery structures, mud tracks or stone-surfaced roads, and ancillary businesses became important elements in the distilling landscape.

By 1840, the total number of distilleries operating in the state had declined to 889. Although information is incomplete or missing for some counties, the overall pattern established by 1810 was still evident in 1840, with some noticeable exceptions. The southeastern mountain counties retained their distilleries, or new works were established, especially in Perry and Harlan Counties. The eastern Pennyroyal counties maintained or increased the number of working distilleries, and new works appeared in the far western Jackson Purchase counties. In the Greater Bluegrass, Fayette County lost a significant number of distilleries, as did Mercer and Franklin. By comparison, Harrison County's distilleries increased in number, as did those in the western Bluegrass counties of Nelson, Marion, Spencer, and Washington. After the Civil War, this latter group of counties became one

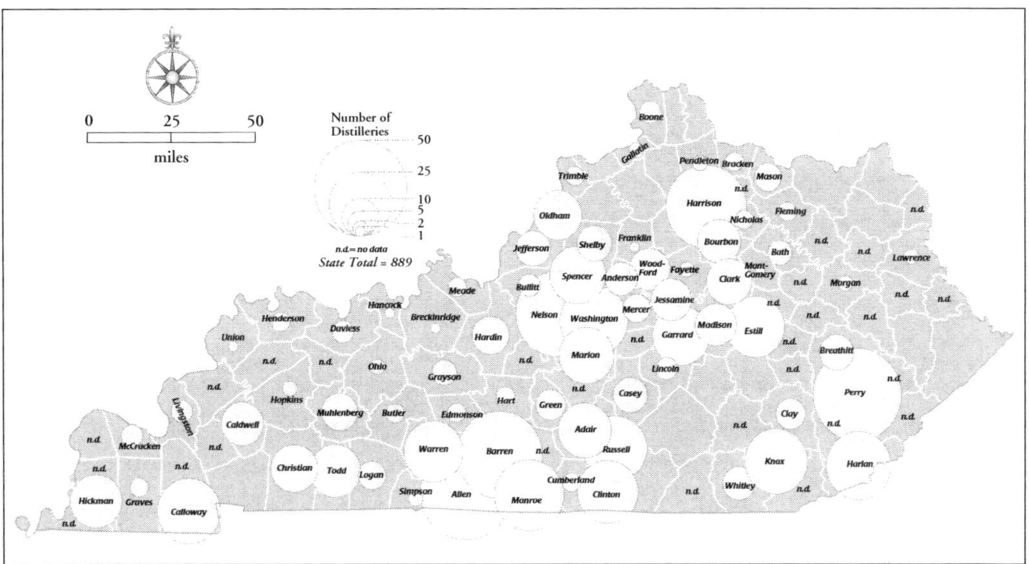

Kentucky's distilleries, 1840. (From US Census of Agriculture, 1840)

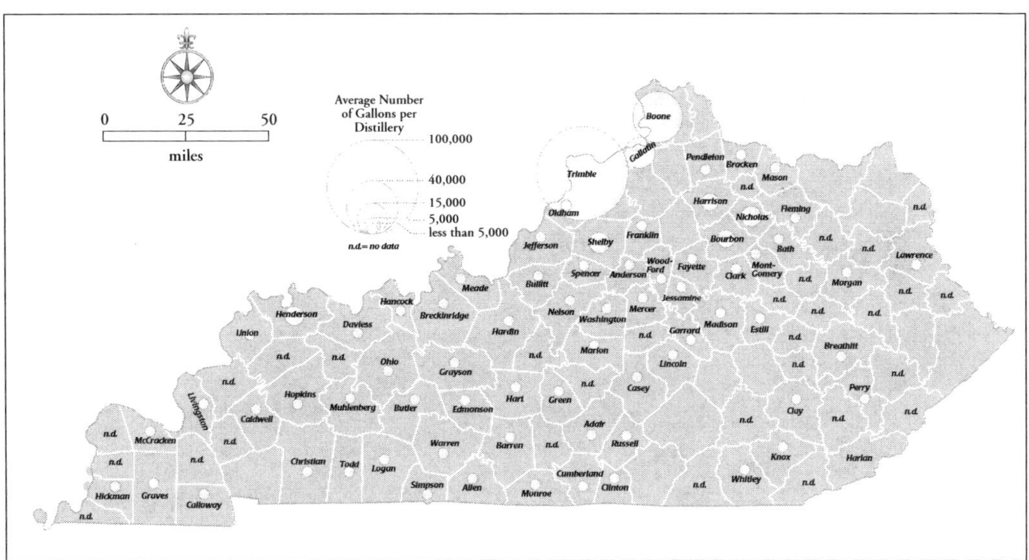

Distillery production, 1840. (From US Census of Agriculture, 1840)

of the core distilling areas; they survived Prohibition in the 1920s and 1930s and thereafter emerged as a major distilling center.

The impact of industrialization and the rapid improvements in river transportation enabled by the steamboat after 1820 was reflected in the distillery production capacity in 1840. Very few of the state's distilleries could muster an annual production of 5,000 gallons or more. The inland counties of Harrison, Nicholas, Bourbon, and Shelby managed to reach this threshold, likely because they had access to navigable streams such as the Licking River and its larger tributaries or because they were served by the state's first railroad line. The most dramatic production increases, however, occurred in selected Ohio River counties. Each of Henderson County's distilleries produced about 5,000 gallons; those in Gallatin exceeded 15,000 gallons, and Boone County distilleries produced about 40,000 gallons each. Two Trimble County distilleries averaged 100,000 gallons a year. By a considerable margin, the dominant works was William and John Snyder's distillery at Petersburg in Boone County. The Snyders increased production to 164,000 gallons in 1850 and surpassed 1 million gallons in 1860, becoming the exemplar of how the adoption of industrial-scale technology and the studied use of high-capacity river transportation could boost production. Riverine distilleries also enjoyed an unusual transport advantage during the winter: they could haul grain and whiskey on the frozen river and avoid difficult or impassable roads. Teamsters hauled sleds loaded with whiskey from William Appleton's Petersburg distillery about one mile up the frozen Ohio River to the railroad freight depot in Lawrenceburg, Indiana, in early February 1872. Sled teams made several subsequent trips across the ice, and one returned with a load of distillery malt (Lawrenceburg was also a significant distilling center).[15] Proximity to navigable water was not simply a convenience or an expedient for conveyance. Some distillers and their customers recognized that while awaiting the seasonal rise in water level to accommodate safe river navigation, barreled whiskey acquired some age, and some thought the motion of travel improved its taste.[16]

Prior to the 1840s, farmer-miller craft distilleries were generally distributed across the state. By 1894, the locus of industrial factory distilling had concentrated in the Inner or Outer Bluegrass, the eastern and central Pennyroyal, the Ohio River floodplain in the Western Coal Field near Owensboro and Henderson, and Louisville–Jefferson County. Industrial distillery sites were clustered at special nodes with linkages to local, regional, and national grain and coal supplies and to distillery materials such as copper stills, cooperage, and specialized equipment and building supplies. Marketing distilled products became increasingly regional and national.

Anderson, Harrison, and Madison Counties had the largest number of distillery works in the Inner Bluegrass, while the early production center in Nelson and adjoining counties in the Outer Bluegrass continued to operate. Urban industrial distilleries at Owensboro (Daviess County) and Louisville (Jefferson County) sought sites with river and rail access and drew their labor force from the cities' comparatively large working-class populations, advantages that helped these places become major whiskey production centers.

Creating a Distillers' Landscape

Observing a modern bourbon distillery is a useful exercise if one wants to learn how whiskey is produced, but it is not likely to provide a comprehensive appreciation of all the elements that constitute the historical and contemporary distilling landscape, nor will it

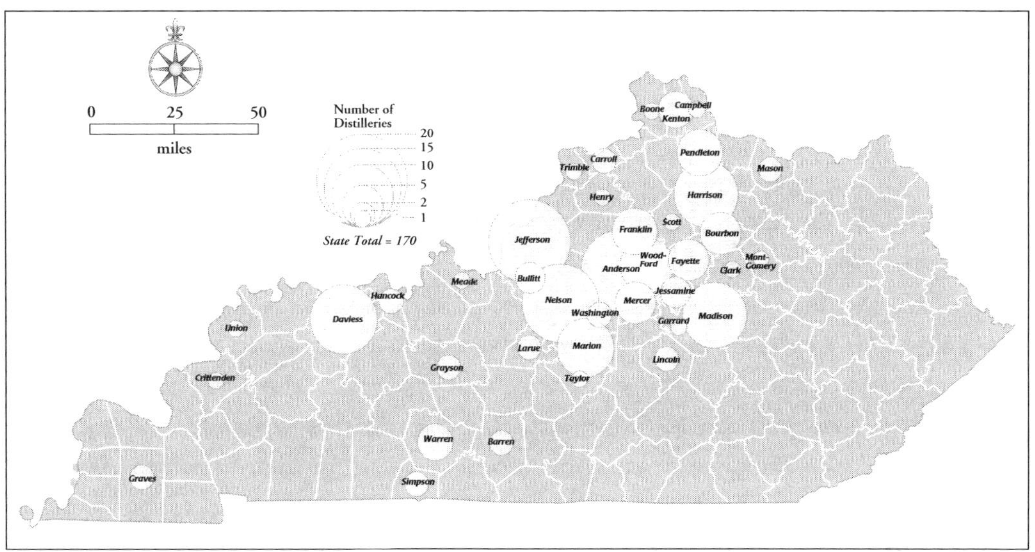

Kentucky's distilleries, 1894. (Compiled from *Sanborn's Surveys of the Whiskey Warehouses of Kentucky and Tennessee* [New York: Sanborn-Perris Map Company, 1894]; US Census of Agriculture, 1894)

assist one in understanding how distilling evolved from a traditional agrarian industry into a large-scale modern commercial enterprise. This transition is of particular interest because while distilling could be urbanized to take advantage of transportation facilities, associated industrial suppliers, and a concentrated labor force, whiskey production could also continue in the rural countryside where it originated. Urban distilleries often blended in with adjacent industrial plants, and their landscape was not immediately evident or distinguishable unless one paused to study the structures, watch the flow of commodities into and out of the buildings, and experience the pungent aroma of fermenting mash. Close observation of any distillery, be it rural or urban, will confirm that distillery structures are largely functional; they have an essential identity. Like most industrial architecture, early distilleries were designed to house or shelter the processing of raw materials into a consumer product, not to attract attention or declare values or status. In short, their design or architecture could be thought of as vernacular industrial, comprising simple frame buildings or sheds whose sole purpose was to accommodate various distilling functions and shelter them from the elements.

After the Civil War, as distilling became increasingly industrialized and construction of a regional railroad system proceeded, large distilleries appeared, often housed in buildings designed by engineers or architects. Distillery design became considered and rational, a factory-like transformation into a formal industrial process, and the principles of aesthetics gradually joined with the principles of engineered efficiency, productivity, and safety. Eventually, some modern distilleries, especially their interiors, were designed or redesigned to be showplaces, with large, gleaming copper pot stills or steel column stills and polished and painted fittings and piping. From the modern distillery emerged products packaged in distinctively shaped, lustrous glass bottles provided by an associated industry; these products were identified by colorful labels and brand names that, when advertised, elicited recognition almost anywhere in the country.

What is not apparent in the modernization of distilling are the changing relationships among technology, economy, and cultural and social relations that enable the evolution of the production process, shape the structures erected to house its technology, and market and distribute the finished product. Importantly, one must recognize that industrial landscapes are closely tied to the people who build and work in them. The appreciation of construction, especially of historical buildings and other industrial structures and facilities, is often attributed primarily to the vision or work of architects or engineers. While such recognition can be informative, it can also focus our attention primarily on the technical and the aesthetic. A richer interpretation of landscape extends to the identification of historical precursors that can alert us to important innovations and suggest a timeline of landscape change or adjustment. Further, although architects and engineers play a key role in the creation of industrial landscapes, it is the common person, the wage laborer or salaried employee, who builds the structures and brings the entire operation into production. Knowing something about the laborers and their social and cultural worlds may expand and enrich interpretations of how industrial places work and how they relate to people's lives as well as the places they occupy. Limiting one's perspective on landscape construction to concerns about economic geography gives short shrift to the complex interweaving of social experiences and cultural influences that constitute the industry's historical ecology and more fully disclose how whiskey is made.

Making Connections

Support linkages to kindred industries such as the making of machine tools allowed manufacturing processes to convert to machine-based production.[17] Distribution connections related to grain processing included the manufacture and dispersal of food products such as bread and cereals, as well as brewed beer and distilled spirits. Food processing created a demand for cooperage, which was the product of the timber cutting, lumber milling, and woodworking industries. Lumbering required investment in planning mills, which produced building materials, barrel staves, and dimensional wood for boxes, furniture, wagons, carriages, and railroad ties. Agricultural-industrial development in the Ohio Valley and the Middle West was partly stimulated by the availability of productive land and the regional demand for local and imported manufactured products. The industrialization of the Ohio Valley and the Middle West included traditional heavy industries such as iron and steel manufacturing, as well as large-scale metal fabrication and the specialized development of steam-powered machines such as riverboats, locomotives, and stationary power plants. There also developed medium-scale industries such as grain and lumber milling, butchering and tanning, and brewing and distilling.[18] In addition to its primary reaper works in Chicago, for example, Cyrus McCormick's farm machinery company established a factory for the manufacture of the Kentucky Harvester in Louisville in 1856, suggesting that there was a demand for such machines by the state's grain producers.[19] Cyrus's cousin J. B. McCormick of Woodford County, Kentucky, became a sales agent for McCormick reapers in 1845; he sold machines across the Ohio Valley, thereby underwriting increases in grain production.[20]

Nineteenth-century industry in America was also closely related to technology, cultural tradition, and social structure. When moving water was the primary industrial power source, manufacturers chose locations at stream rapids and falls, where, with the addition of dams and millraces, hydraulic head pressure was obtainable. And because such advantageous sites were locationally restricted, industries tended to cluster at specific places along streams that offered power capability. Likewise, productive agriculture required fertile soils, so the most desirable land, despite being more expensive, was settled, cleared, and planted before marginal land. The requisites for overland transport routes were shortest distance, lowest gradient, and driest ground. Water transport would thrive if harbors were sheltered from storms, deep-water dockage was available, and the waterway was free of obstructions. Raw materials, be they potable water for distilling, clay for bricks, iron for tools, or copper for containers, were rarely ubiquitous; they were often found in specific places that were not always advantageously located near transport routes or population centers. Until the development of a systematic network of stone-paved turnpikes, beginning in the second and third decades of the nineteenth century, hauling bulky materials such as grain and lumber was slow and exceedingly expensive. After grains were converted into flour, the transport problem was exacerbated; corn, wheat, rye, and barley were comparatively durable if kept dry and free of vermin, but once they were milled into flour or meal, the product was perishable and required special containers and expedited handling.

Most industrial activity, past or present, utilizes extant technology to erect the buildings that house the machines and tools required to make useful products. Industrial structures usually reside at important geographic focal points, such as sites that can supply raw material, a labor force, a transportation node, or a market. A successful, long-lived industry may build structures at one focal point—because of the limited extent of a mineral deposit, for example—or it may operate at multiple sites over a broad area—because the raw materials are widely available and the markets for the product are abundant. During the nineteenth century, the production of fine ceramics in America tended to be localized at high-quality clay deposits such as those at Trenton, New Jersey, and East Liverpool, Ohio. Wagon and carriage making, in contrast, was common and widespread across the Northeast and Middle West in part because hardwood lumber and iron were widely available and the market was ubiquitous.[21]

Each industrial type has specific raw material requirements and may utilize dedicated technologies that require different kinds of infrastructure, be it buildings for manufacturing and storage space or facilities for by-product disposal. In the aggregate, industrial structures contribute to a distinctive landscape signature or imprint. For those working in the industry or living nearby, the industrial landscape becomes central to their sense of place. Eventually, though, an industry may adopt new technologies and change its infrastructure, or it may move and abandon the work site. The structures left behind also constitute a landscape, perhaps relict, that may continue to represent the inhabitants' industrial heritage.

Kentucky's distilling industry contributed to the process of industrial development and landscape creation in the Ohio Valley and Middle West. Migrants established distilleries before Kentucky achieved statehood in 1792, and the vocation evolved from subsistence-scale farms making corn whiskey and fruit brandy into modern, twenty-first-century commercial businesses that produce primarily bourbon at an industrial scale. Industrial distilling developed irregularly, as change was contingent on each distiller's interests and financial wherewithal. Each increment of change was associated with a distinctive ensemble of landscape elements—structures and their placement relative to operant technology, water as a power source or transport medium, overland transport routes, raw material supplies, labor availability, and regulatory and market considerations.

Commercial industrial distilling was and is a complex process with scientific, technological, economic, social, and political dimensions. And the landscape formed by this multifaceted historical ecology is, necessarily, multidimensional. Of the more than 2,000 distilleries in operation during the nineteenth century, many were dismantled. Others burned. A few exist as ruins, and only a small number remain in production. Interpreting and understanding the industry's historical ecology requires familiarity with locational decision making, applicable technologies, inventions and patents, state and federal regulations and revenue policies, and business linkages with associated industries.

3

Kentucky's Distilling Environment

If it form the one landscape that we, the inconsistent ones,
Are consistently homesick for, this is chiefly
Because it dissolves in water. Mark these rounded slopes
With their surface fragrance of thyme and, beneath,
A secret system of caves and conduits; hear the springs
That spurt out everywhere with a chuckle. . . .
. . . what I hear is the murmur
Of underground streams, what I see is a limestone landscape.
—W. H. Auden

Conjunctions—An Ecological Context

The historical ecology of nineteenth-century whiskey making reveals a complex process that operated at the conjunction of several contributing or controlling natural, cultural, and social elements, among them the physical environment, agriculture, technology, invention and innovation, transportation, workforce, federal and state laws, and social constraints. Each of these elements was related to the others in complex ways, such as when inventions permitted increases in transport speed and reliability or when temperance-influenced legislation forced distilleries to close, costing laborers their jobs, farmers their grain sale contracts, and the federal government its tax revenue. To understand distilling, and the landscape it created, one must recognize how the totality of these various conjunctions lent that landscape its structure, function, and presence through time. This chapter explores the historical ecology of the conjunction of distilling and environment and the landscape features that extended from that relationship.

Humans relate with their natural environment in distinctive ways and, at specific locations and times, in ways that repeat if they are found to be useful or meaningful.[1] Understanding these relationships is difficult because it requires sorting through what historian-geographer William Cronon calls a crowded and disordered chronological and spatial reality.[2] One might begin by looking for general geographic patterns in human behavior, the regular or repeating forms of land use and settlement that people establish to live and make a living. Agriculture, forestry, and mining are primary activities that yield useful raw materials and products and, in the process, create patterns of fields and farmsteads, cutover forests and logging trails, and opencast pits from which fuel and minerals are drawn. Products so obtained can be processed by manufacturing or some secondary economic activity that is necessarily attached to primary work by a labor force, supply lines, and financial transactions. Corporate actions may create patterns of bonded and immigrant workers, roads and river ports, or barter and bank transactions. And all this takes place upon and within a physical environment that lends context but may also present opportunities as well as impediments. Objective numerics and temporal context are useful in revealing and outlining the patterns embedded in such activities. But reading the narratives people wrote about their experiences and their purposes for action is indispensable if we are to appreciate how their behaviors and decisions created and modified the places in which they lived.

The technologies deployed in contemporary agriculture—labor-saving machinery, hybrid and disease-resistant seed, chemicals and fertilizers, irrigation, weather forecasting—greatly extend the contemporary farmer's ability to raise crops and livestock in superlative environments, while also facilitating the extension of production into less hospitable places. But while these products of science and engineering allow farmers to increase yields and product quality and reduce the impact of environmental hazards, they can also obscure one's view of the agricultural past. Eighteenth- and nineteenth-century Kentucky farmers lived not so much *on* the land as *in* the land. Following the furrow cut by a single-moldboard plow pulled by a mule put the farmer in intimate contact with every clod, every stone, every crease and irregularity in a field. Once a field's trees and briers were cleared and its sod was turned, the dirt often stocked sufficient nutrients to produce plentifully. But the soil was also a storehouse of seeds dropped over decades by grasses, forbs, and other native plants, and working a field with a hand plow and harrow brought them to the surface, where they flourished and, if left untended, returned the field to sod.

Lore and experience taught farmers that the sky was not innocent. Some clouds signaled fair weather; others produced hail and torrential rain. Clear skies for days on end could bring drought and heat sufficient to kill crops and livestock. In the early nineteenth century, the intrepid forecaster could do little better than summon weather lore, which could be enlightening or misleading. "Red sky at night, sailors' delight" has some basis in fact. "Rain before seven, clear by eleven" rhymes rather well but has little empirical validity. A successful harvest provided seed for the following season's planting, grain that the miller's burr stones ground into meal and flour, feed for livestock during the winter, and, if one's granary was built with ill-fitting logs, food for the neighborhood's vermin. Livestock might thrive or perish for no obvious reason. And while losing hogs or cattle to disease was costly, the loss of a draft horse or mule could be catastrophic. A farmer could work a garden with hand and hoe, but working a farm required a plow and wagon pulled by draft animals.

Appreciating the intimate relationship between distillers and their physical world requires a basic familiarity with the vocabulary of nature: climate, geology, geomorphology, hydrology, soils, topography. These environmental elements are all coinfluences—for some distillers, codeterminants—of distilling's geographic patterns and processes.

In 1881 C. E. Bowman, Kentucky's commissioner of agriculture, made the bold assertion that the superior quality of Kentucky's bourbon whiskeys was attributable to the state's soil and water, not the artful distilling techniques employed by the state's distillers. Kentucky's "corn, as do its wheat and barley, and oats and grasses, partakes of the elements of the soil upon which it is raised, and of the water that percolates through them," he said. "The whiskies made in no other States have or can come in competition with [Kentucky's whiskey] . . . *because* the corn of which they are made and the water used in their distillation lack the essential qualities and ingredients possessed by Kentucky corn and Kentucky water. Herein lies the reason of it. Kentucky distillers possess *no art or skill or superior knowledge* above other distillers."[3] Bowman's rather maladroit and deterministic premise likely earned sharp reproval from nineteenth-century distillers. An effective rebuttal would be enhanced by a nuanced understanding of the industry's evolving historical ecology and the multiplicity of environmental and human influences that intersected to produce a superior bourbon whiskey. Corn and water were key elements in the distiller's progress toward "fine" whiskey, but they were not the sole determining factors.

Water: Hydrology and Distilling

Kentucky's frontier settlers were engaged in the production, consumption, and sale of whiskey by the 1780s. In 1810 census enumerators counted some 2,000 stills in the state;

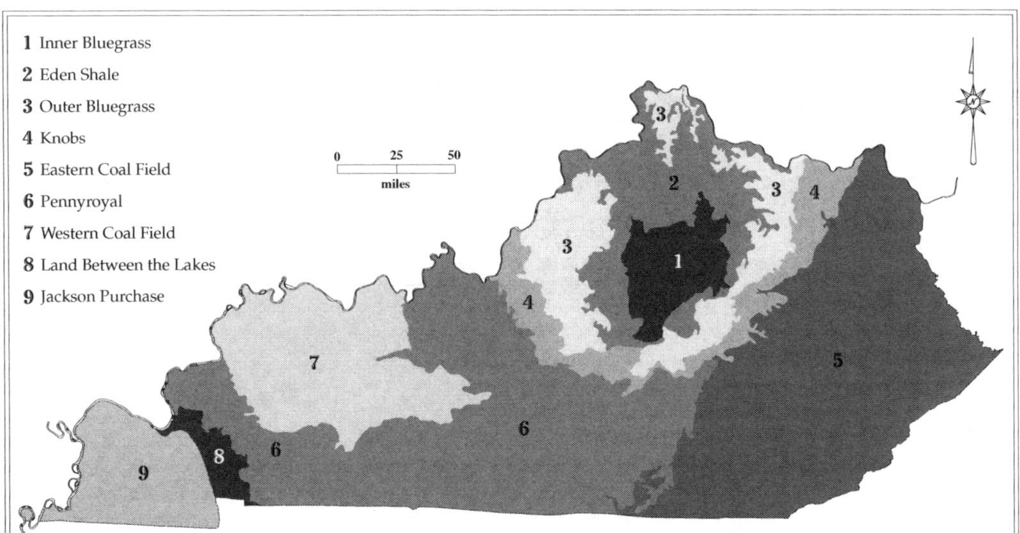

Kentucky's physiographic regions.

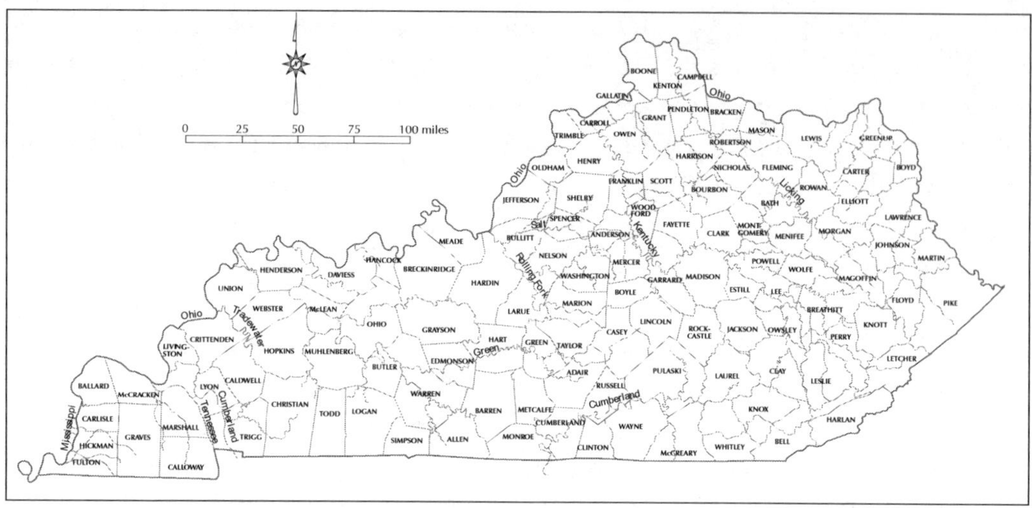

Kentucky's counties and major streams.

many more likely escaped official notice. That Kentucky whiskey's singular character is derived from "pure limestone water" is long-standing lore, but pioneer whiskey making was likely based more on pragmatic production requirements than on romantic notions verging on environmental determinism. The frontier distiller required a source of sugar, be it from fruit or from the starch in grains or potatoes, and a source of clean, fresh, cool water. Farmers provided the source of sugar, and water was readily available in most places because central Kentucky's Bluegrass country is underlain by thick beds of water-bearing Ordovician-age limestone and shale. The alluvial bottomland on the Ohio River's flood-plain terraces—such as those at Maysville, Covington, Carrollton, and Louisville—was formed of Pleistocene glacial outwash sands and gravels and proved to be a productive aquifer when tapped by wells.[4] Precipitation, more than forty inches annually, falls as soft water and runs off the surface by way of an extensive stream network.

Because water is an effective solvent, it dissolves many of the minerals and gases with which it comes in contact, meaning that, from a scientific perspective, pure water rarely occurs in nature. Rain acquires carbon dioxide as it falls through the atmosphere and forms a dilute form of carbonic acid. A substantial volume of rain percolates down into the soil and bedrock, where it can dissolve limestone, absorb minerals, and enter a groundwater system that is as intricate as the surface drainage network. In Kentucky's limestone country, groundwater feeds thousands of springs, which were favored sites for early settlements. Pioneers built Fort Harrod near Harrodsburg Spring, and Georgetown grew up beside Royal Spring, as did Elijah Craig's distillery sometime around 1786. Royal Spring's flow rate is 400,000 gallons or more per day.[5] Early settlers preferred to site their farmsteads at perennial or "never failing" springs that supplied water for their households and livestock. Farms and village settlements on creeks and rivers could draw water directly from those watercourses. Ohio River towns drew water from the river and nearby springs. In time, wells bored into the Ohio floodplain's water-bearing gravels would yield enough fresh water to support municipal and industrial consumers.

How pioneer settlers came to regard limestone groundwater as "pure" resides dimly in yesteryear's lore, yet it is a staple of Kentucky's distilling origination narrative. Chemically "pure" water can be made by distillation to remove soluble salts. But how could an eighteenth-century farmer-distiller analyze the chemistry of spring or creek water other than by its appearance, taste, and smell? Perhaps if the water was clear rather than cloudy and tasted fresh rather than saline or sulfurous—as it did at Bullitt's Lick, Lower Blue Licks, Mayslick, and Drennon Spring—it was deemed suitable for drinking and distilling. Certainly, the region's expansive distribution of springs and surface streams enabled distilling, and sensory water "testing" sufficed until scientific chemical analysis became available. "Pure" water was mother's milk to early distillers, but they also valued flow volume and reliability. Discovery of a new spring at the Frankfort Distillery near the Forks of Elkhorn east of Frankfort was deemed of sufficient importance to warrant a front-page announcement in the city's newspaper. While blasting rock to create a pool for water storage, workers unexpectedly exposed "a stream of water as large as a man's leg, clear and pure and with a steady temperature of 56 degrees." The new spring was deemed "one of the finest in the State," and its "water is from a limestone formation and is absolutely pure. It flows with some force and shows no signs of diminution." The notice concluded: "As the water in Kentucky is what makes Kentucky whiskey different from any other whiskey the spring is invaluable."[6]

During the 1870s E. H. Taylor Jr. operated three Frankfort distilleries: O.F.C. (Old Fire Copper), Carlisle, and J. S. Taylor. He published a publicity booklet describing the works in which he extolled the merits of the water issuing from a spring on the property. Citing the recently published *Map of Kentucky* by John H. Procter, director of the Kentucky Geological Survey,[7] Taylor embraced the science represented by the map, noting that it "shows . . . a small section of the state made famous by its fine sour-mash whiskeys, the rare-birds-eye limestone of the lowest stratum of the Lower Silurian formation." He continued: "The O.F.C. and other brands of the E. H. Taylor, Jr., Co., are produced upon the depressed apex of this stratum, thus securing the best limestone drainage it can possibly afford. The result in fine whiskey is no doubt largely due to the water that, percolating through the limestone, becomes impregnated with its properties, and imparts them to the spirit during the process of manufacture. Opinions and assertions are debatable—a geological fact, stereotyped in and reflected from the earth's crystallized strata, is as solid and immovable as the everlasting hills."[8] Taylor's booklet cited "distinguished chemists," including Professor Barnum of Louisville, who stated, "The water is of wonderful purity, and of peculiar adaptedness for the manufacture of whiskey. Your water contains very appreciable quantities of phosphate of lime, which would have the same affect in promoting the growth of the yeast plant that a dressing of bone phosphate would on a wheat field."[9] An engraving of the spring shows it issuing from a hillside a few feet above the level of the Kentucky River.[10]

By 1877, the O.F.C. Distillery had passed to George T. Stagg and others, who conducted business as Stagg, Hume & Company. In 1894 the distillery's water supply still came from Taylor's spring, carried to the works by a three-and-a-half-inch pipe. Insurance maps published that year, however, show that the O.F.C. Distillery used two force pumps to draw water directly from the Kentucky River via two pipes of different diameters—one eight inches, the other three and a half. Moreover, the pipe drawing water from the spring seemed to be spliced directly into the pipe of the same diameter drawing river water, so it appears that the springwater was mixed with river water before it entered the distillery.[11]

Other distillers held different opinions about the importance of a water source for whiskey distillation. Captain Luther Van Hook, superintendent at the H. E. Shawhan Distillery near Cynthiana, in Harrison County, believed that "sulphur water" made the best whiskey. The water kept the mash tubs "sweet," in his experience, and he ventured an opinion that the famous Blue Licks water made the best whiskey.[12] Since prehistoric times, the Lower Blue Licks Spring near the Licking River in northern Nicholas County had attracted bison and other animals to the site, where they licked mineral accumulations in the soil. Indians hunted there, and early pioneer settlers established a salt-boiling operation nearby. By the mid-nineteenth century, the spring had become the focus of a large spa, where visitors came to drink and bathe in the "medicinal waters." Dr. Robert Peter, professor of medical chemistry at the Kentucky School of Medicine and Transylvania University, published a paper on Blue Licks water in 1850. His analysis revealed that the water did indeed contain considerable amounts of sodium chloride, potassium, calcium, magnesium, and other minerals, and its "rotten egg" smell indicated the presence of hydrogen sulfide.[13] While it is unlikely that many other distillers shared Van Hook's opinion that water with a high sulfur and high sodium content was desirable for making bourbon, his contention suggests that there was broad latitude in acceptable water quality.

As long as distillers accessed their stills and markets by overland tracks or trails, a distillery's location near a perennial spring or stream was economically viable. Transport by packhorse or wagon supported rather than countermanded the decision to site a distillery by a spring or creek. But, with the construction of hard-surfaced turnpikes and railroad networks in the mid to late nineteenth century, many distillers in central Kentucky abandoned their springs and relocated close to the railroad tracks. If "pure" limestone water was the critical locational element in the geography of distilling, then either "pure" water could be found wherever there were railroad tracks, or distillers could use any "clean" water source that was convenient to the transportation network, and water quality became a secondary consideration. Questions of appropriate distillery location were more

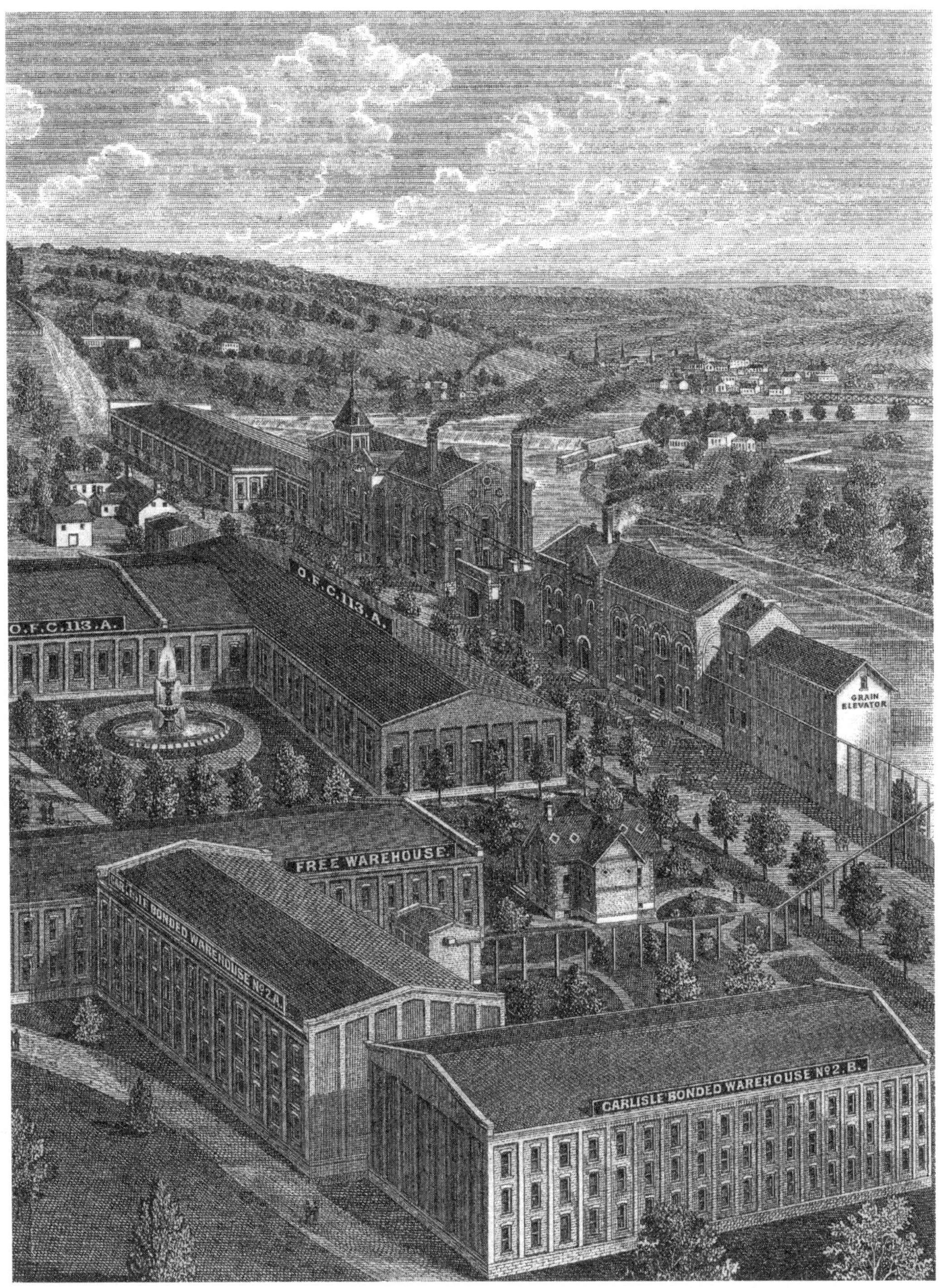

Engraving of E. H. Taylor's original O.F.C. and Carlisle Distilleries, 1869. (From E. H. Taylor Jr., *Description of the O.F.C., Carlisle, and J. S. Taylor Distilleries* [Frankfort, KY: E. H. Taylor Jr. Co., ca. 1886], plate 1)

complex for the miller-distiller. Many Bluegrass springs had flow rates of less than fifteen gallons per minute; this was sufficient for a farm household and a small distillery but not for industrial-scale milling and distilling, unless large storage tanks or catchment basins were available. A miller's primary consideration in mill location may have been sufficient water flow from a creek or spring to turn a mill wheel; supplying water to a nearby distillery was important but secondary.

Although distillers concentrated their production in the Bluegrass limestone country, nineteenth-century whiskey production extended across the state into nonlimestone topography. After the Civil War, commercial distilling increased in Daviess County in the Western Coal Field physiographic province. Daviess County's bedrock geology and surface geomorphology comprise two distinct sections: sandstone, shale, and coal in the hilly upland in the south, and deep, nearly level alluvial plains in the north. Pleistocene glacial meltwater pooled there, creating a large periglacial lake that, when drained, left behind an extensive floodplain dotted with small hills. The floodplain lay atop 100 to 150 feet of alluvial gravels, sands, silts, and clays, closely akin to the alluvial plain in west Louisville.[14] Soils were deep, but given the lack of gradient, natural drainage was sluggish at best. Much of the land between the Green and Ohio Rivers was forested swamp when pioneer settlers arrived. Distillers erecting new works in the late 1860s preferred accessible locations near the Ohio River's south bank, rather than the low-lying swamps to the south. In the early 1870s J. W. M. Fields bought a distillery in Daviess County, which he dismantled and then reassembled about three miles east of Owensboro. The new location was likely along the railroad tracks at Pleasant Valley Road, about half a mile from the Ohio River.[15] Distilling began there in 1873. In 1882 Fields drilled a 1,000-foot well to obtain groundwater to enhance the distillery's water supply. Driller's notes logged several small coal seams and "impure" limestone, and at 765 feet the well struck white sandstone, which produced a weak brine. At 900 feet the well penetrated 70 feet of "coal-black" slate (likely shale). As the drill descended, "the water was growing less and less fit for the contemplated use." As noted by the Kentucky Geological Survey, the drill log confirmed that the distillery stood atop the Western Coal Field's sandstone, shale, and coal, making further drilling futile.[16] Ohio River water was available, of course, but pumps and pipes were required to bring it to the distillery.

Modern industrial distilleries use water in three distinct ways: mixing, dilution, and cooling. Potable water is mixed with milled grain to make fermentable mash at a rate of about twenty gallons per bushel. Ideally, mash water should contain calcium sulfate to enhance the rapid conversion of starch to sugar, but it should also be low in iron, bicarbonate, and dissolved solids. The most desirable water lacks organic matter, bacteria, and fungi.[17] Bourbon whiskey is distilled at no more than 80 percent alcohol, or 160 proof. Potable water is used to dilute barreled whiskey to reduce the alcohol content to no more than 62.5

percent, or 125 proof. Whiskey may be diluted with water again when it is bottled, bringing the final proof to 80 or higher. Water is also used to cool fermenting vats. Groundwater drawn from wells is used extensively for cooling because of its consistent temperature, which typically ranges from 56 to 59°F.[18]

The chemicals found in surface water and groundwater that are of greatest concern to distillers are iron, calcium carbonate, and dissolved inorganic solids. Iron is found in most types of rocks and soils and can dissolve in water; levels greater than 0.3 parts per million (ppm) result in reddish brown stains on laundry. The US Public Health Service recommends that the iron content in drinking water not exceed 0.3 ppm.[19] Mildly acidic rainwater or groundwater dissolves limestone, producing calcium carbonate and hard, mineralized water. Water containing calcium carbonate tends to reduce strong acids. Calcium concentrations higher than 120 ppm are considered very hard water, although elevated levels of calcium and magnesium carbonates can provide nutrients for yeast. Counts of 150 to 200 ppm, though, are cause for concern in the operation of steam boilers and water heaters. Groundwater can also dissolve some naturally occurring inorganic minerals, metals, and salts from bedrock, including potassium, chloride, arsenic, chromium, lead, and mercury. The US Public Health Service recommends that domestic and industrial water contain no more than 500 ppm dissolved inorganic solids.[20]

Working on behalf of the Kentucky Geological Survey and the US Geological Survey, several geologists conducted a study of industrial water sources in forty-three Bluegrass counties in the early 1950s. The study included a complete chemical analysis of forty-nine industrial groundwater sources, including springs and wells used by several distilleries. Chemical analysis of spring and well water at a number of Bluegrass distillery sites found a range of mineral and iron contents, and such water was not consistently "pure." Most distilleries drew their well water from Ordovician or Silurian limestones, and their readings for iron, bicarbonate, and dissolved solids varied significantly. Only a small number of them used water that contained less than 0.3 ppm iron.[21]

In the Inner Bluegrass, the Benson Creek Distillery in Franklin County, west of Frankfort, used about 144,000 gallons of water per day, obtained from a large spring and from Benson Creek. The springwater contained 0.39 ppm iron, 245 ppm bicarbonate, and 302 ppm dissolved solids.[22] Several other distilleries used water with comparable chemical compositions. Two distilleries used water that was exceptionally high in iron. The Carstairs Brothers Distillery at Lair, in Harrison County, pumped water from the Licking River and from one well. The well water, which came from limestone bedrock, contained 6.6 ppm iron, 316 ppm bicarbonate, and 345 ppm dissolved solids.[23] At Camp Nelson in Jessamine County, the Kentucky River Distillery drew water from two wells and from the Kentucky River. The sampled well was sixty feet deep and pumped water from Ordovician limestone; its water contained 11.0 ppm iron, 306 ppm bicarbonate, and 307 ppm dissolved solids.[24] In the Outer Bluegrass at Deatsville, in Nelson County, the rail-side T. W. Samuels Distillery drew water from two wells. The sampled well contained 3.3 ppm iron, 224 ppm bicarbonate, and 216 ppm dissolved solids.[25] Most of these distilleries used water that they, in effect, distilled by passing it through a commercial boiler.

A US Geological Survey analysis of groundwater drawn from wells throughout central Louisville's industrial and residential areas from the early 1940s through 1955 revealed that "ground water in the Louisville area is very hard, very high in iron content, and generally of the calcium bicarbonate or calcium sulfate type."[26] The study included water drawn from seven distillery wells. Most samples contained concentrations of iron exceeding the 0.3 ppm standard, with values ranging from 0.23 to 5.6 ppm. Dissolved solids in these samples ranged from a low of 526 ppm to a high of 1,270 ppm. Carbonate concentrations were exceptionally high; the lowest sample recorded was 327 ppm, and the highest was 619 ppm. Many concentrations exceeded 400 ppm.[27]

From 1937 to 1940 Louisville distilleries were the area's largest consumers of groundwater, averaging more than 6 million gallons per day, or about 16 percent of the daily industrial water use. More than 88 percent of the total groundwater withdrawal came from wells pumping from sand and gravel aquifers; the remainder came from wells in limestone formations.[28] During World War II, industrial groundwater use more than doubled, with a peak withdrawal rate of 17.2 million gallons per day in 1943, as distillers manufactured industrial alcohol to support the war effort.

The water-quality narrative was still in play in 1959 when Martin Mayer interviewed Edgar Bronfman of Seagram, one of Louisville's largest distilleries at the time. Bronfman, it seems, deemed the limestone water myth worse than nonsense. "All of the famous old country distilleries used pond water," Bronfman said, "rain water most of it, the softest water imaginable."[29] Samuel Pollack, chief chemist for Schenley's Distillery, was equally skeptical. "I've read volumes and volumes on water. But from a scientific point of view I have not reached a formula," he said, "so many grains of hardness, so much magnesium and so forth. Our water supply is analyzed twice a year, and it varies all over the lot." Pollack even wrote a report on the relationship between water and whiskey quality. "'Despite all the allegations in writing on the effects of water on the bouquet of whiskey,' he noted, 'we must rely on fifty-four years of practical experience. Irrespective of any chemical or bacteriological analysis, there is no effect of water on whiskey.'" In conclusion, "The only general rule is, if it is good enough for drinking purposes as water, it's good enough for distiller's purposes."[30] One's belief (or not) in the pure limestone water narrative seems to depend on one's perspective and motivation. Small nineteenth-century countryside distillers embraced the narrative as part of their tradition and heritage. Modern industrial distillers disdained the myth; their beliefs were informed by science and empirical investigations, and their distilleries' locations were based on access to transport and labor, rather than propinquity to a spring presumed to issue "pure," unchanging water.[31]

Climate and Weather

Climatologists catalog Kentucky's climate as humid subtropical. Average temperatures from November through March exceed 32°F, so winters are comparatively short and mild. Prevailing westerly winds bring a succession of high- and low-pressure weather systems. Low-pressure systems produce clouds and precipitation in all seasons. High-pressure systems bring cool or cold, dry air from the north and northwest, often accompanied by clear skies. Summers are generally hot and humid, with a growing season that varies from 170 to 200 days. Precipitation ranges from 40 to 50 inches per year and is often associated with low-pressure systems that bring warm, moist air masses into contact with cool, dry air along cold fronts. About half of the annual precipitation falls from April to September or during the growing season, which is optimal for agriculture. Summer droughts occur from time to time, as do periods of heavier-than-normal rainfall.

Weather records suggest that, with the exception of some long-term variations, Kentucky's contemporary weather is very similar to that of a century ago. In 1839, the second year of a two-year drought in central Kentucky, the corn crop was severely damaged. The heat turned pasture grasses to crisps and depleted the water sources used for livestock and domestic purposes. Hundreds of farmers drove their cattle miles to reach streams, hoping to find a sustained trickle or slack-water pools, no matter how stagnant. They hauled household drinking water five miles or more.[32]

In any era, weather imposes uncertainty and risk on farmers. Nineteenth-century farmers had only limited options as to how they adjusted to weather variability. Before the development of disease-resistant crop varieties, hot and humid summers brought the likelihood of debilitating diseases such as rust and smut in grain crops. Heavy rains caused severe soil erosion before the use of winter cover crops, contour plowing, and sodded waterways. Farm life was fraught with weather-related risk, and each week brought new apprehensions. Insights into the vagaries of weather events and crop disease are offered by John Jones, a middling farmer and pre–Civil War slave owner from southern Bourbon County. Jones's land lay near North Middletown, in the Inner Bluegrass. He was active in his community and founded the North Middletown & Cane Ridge Turnpike Road Company. In the 1860s and 1870s he worked his farm with freedmen and immigrant laborers and used mules and oxen for motive power to produce grain, clover, feeder cattle, and hogs. On July 28, 1866, he observed, "The rain of yesterday has washed the ground greatly damaged the grain materially & washed away a good deal of fencing." Four years later, in 1870, he wrote:

[Saturday, June 11.] Wheat begins to turn, and it will not be long before harvest is upon us again.

[Sunday, June 12.] Last night it began to rain & I think continued all night

& it looks this morning if it would rain all day. The ground is thoroughly wet & warm sun will lift the corn out of the clods.

[Monday, June 13.] The wheat does not look well. The rust is already upon it. Thinning the corn. The ground is thoroughly soaked.

[Friday, July 1.] Commenced cutting white wheat. I noticed the clouds rising pretty much in all directions & soon a deluge of water rushed in every direction. The plowed land is washing in gulleys & hundreds of cart loads has been swept from the surface. But I am truly thankful it is no worst for I really felt afraid of a perfect tornado.

[Saturday, July 2.] The fields are most terribly washed but thankful I am that it is not worst. The heat is intense.[33]

The farmers' intimate connection to the physical environment required anticipation and planning and a complex integration of diverse operations. If extraordinary weather events threatened agricultural production, they also passed risk and uncertainty directly into the distilling industry and related supply businesses, affecting teamsters, coopers, lumber mills, hardware retailers, wholesale brokers, and banks.[34]

Seasonality

Ohio Valley farmers established a corn, wheat, and livestock economy in the six decades between 1800 and the beginning of the Civil War.[35] In Kentucky, farmers also produced tobacco and hemp. Grain production required concerted attention for five months of the year. Corn planted by early May required two or three cultivations before being "laid by" in June, by which time the plants had grown too large to permit further cultivation. The corn crop did not require workers again until harvest in September or October, when the extended process of cutting, shocking, husking, and cribbing began. The total time expended on wheat could be less than one month, primarily at spring planting and again at harvest, which required one to two weeks of intensive work during midsummer. Maintaining a herd of feeder cattle or hogs required daily attention for feeding, but these tasks were often considered ongoing chores rather than concerted work. Farmers raising tobacco followed a twelve-month calendar from seed selection and field preparation in the late winter to barn curing, stripping, and marketing in the late fall and early winter. Hemp production required labor for field preparation and spring planting, but the most intensive work occurred in the fall, which involved cutting, shocking, retting, breaking, cleaning, and storage.

Depending on the scale and type of farming practiced in a given area, rural laborers could rely on steady employment for half the year, perhaps more on a tobacco farm. The remainder of the year, rural laborers (or mechanics) found itinerant work cutting and splitting timber for shingles, fence rails, construction lumber, and fuelwood, as well as building and repairing fences and farm buildings. Day laborers also performed other jobs such as butchering hogs, salting meat, and digging potatoes.[36] Off-farm work might be available with turnpike contractors, gristmills, or distilleries. The nineteenth-century distilling season began in November or December and extended to May or June. Still-house temperatures in the summer and early fall months were generally too warm to be suitable for effective malting and fermentation.[37]

Kentucky farmers preferred to plant winter wheat rather than spring wheat, in part because that crop permitted adjustments for seasonality. Wheat planted in the fall would sprout, grow four to six inches, and then lay dormant over the winter. In early spring it rejuvenated and matured quickly and was usually ready for harvest in June or July, ahead of the strong midsummer thunderstorms that brought high winds, driving rain, and drifts of hail that could break plant stems or flatten them on the ground. The winter wheat planting and harvest cycle also fit with the farmers' labor demand because the wheat was planted after the corn harvest in the fall and was harvested after corn cultivation stopped.

John Jones organized his seasonal farm activity systematically. Related tasks followed a logical sequence, such as sharpening plows in January or February and plowing fields in March or butchering hogs and salting meat in November and marketing hams in May. Unrelated tasks, such as clearing woodland, were performed as time and weather permitted. Jones adjusted all tasks to seasonal environmental changes. Managing the corn crop required year-round attention. Hired hands shucked corn from stored stalks in February. They planted corn in March and April and then continued to work on shucking and shelling the preceding season's crop through May. Corn cultivation also began in May. Jones "laid by" his corn in June. In August he sold corn harvested the previous year. Corn harvest began in September and continued into October with cutting, shocking, and husking. Hauling, husking, and cribbing corn continued in November, and in December the hands built or repaired a split-rail pen to hold the newly harvested crop. This extended period of corn-related tasks is a reminder that all these jobs were accomplished by manual labor, and prolific yields meant that processing corn required months of effort. Distillers who wished to buy corn for their winter distilling season would have to be cognizant of this schedule and sequence their purchases accordingly—perhaps buying corn the previous year and storing it to ensure an adequate supply.

Favorable Environments: Milling, Distilling, and Agriculture

Frontier farmers and millers erected their mills and distilleries at advantageous sites utilizing local materials.

Topography and Site Choice

Desirable site characteristics included a perennial spring or stream, abundant timber for fuel, productive farmland as a source of grains, a local labor force, and waterway or road access. Millers built gristmills and companion distilleries at two general types of sites: creek-side and upland. During the frontier decades, the preferred mill and distillery sites lay along large-volume creeks, ideally with access to a navigable river. Creek-side mills were often capable of a substantial output and remained operational until the mid-nineteenth century or later. Creek valleys in stream-dissected country with 200 to 400 feet of local relief were heavily wooded and difficult to access. Nevertheless, some creek-side sites were large enough to host several dams, millraces, and mills along the valley bottom. Rock ledge springs feeding the creeks were often used for water power, in addition to dammed pools on the main creek channel.

Lower Howard's Creek in Clark County offered several potential creek-side industrial sites. Rising on the limestone upland south of Winchester, Lower Howard's Creek flowed south to the Kentucky River, cutting a gorge through 400 feet of horizontally bedded limestone to river level. Mill construction and operation along Lower Howard's Creek began in the late 1700s, shortly after the founding of Boonesborough across the Kentucky River and a short distance upstream. The creek was not only one of Kentucky's first industrial centers but also one of the first large-scale industrial sites in the trans-Appalachian West, hosting fifteen gristmills, four mills, and sawmills, as well as cooper shops, leather tanneries, homes, and related outbuildings.[38] Stonemasons opened quarries on several sections of the Lower Howard's Creek gorge to retrieve blocks of Oregon limestone, or Kentucky River marble, and Tyrone or bird's-eye limestone, which they fashioned into building stones used by masons to erect industrial buildings and houses.[39] Importantly, early mills and distilleries and their associated dams and millraces were often designed and built by skilled millwrights, carpenters, and stonemasons, assisted by common laborers. These artisan builders, some of them both millwrights and millers by profession, possessed the special tools required to work wood and stone and were known in their communities by reputation and through advertisements in local newspapers.[40] Laborers moved barrels of flour and whiskey produced at Lower Howard's Creek down the valley to a landing on the Kentucky River, where the barrels were loaded onto flatboats for the extended trip down the Kentucky to the Ohio River and then, perhaps, on to New Orleans and southern markets. Some mills and distilleries on Lower Howard's Creek stopped production during the Civil War. Others continued to operate until the late 1800s.

Millwrights and stonemasons developed steep-gradient, creek-side mill and distillery sites on several other Kentucky River tributaries. At the mouth of Boone Creek in southern Fayette County, Eli Cleveland had established a landing on the Kentucky River by 1780, and within three decades, four mills and two distilleries were operating along the creek's lower section. In eastern Anderson County, Cedar Brook Run flows through a valley into the Kentucky River at Tyrone; millers and distillers developed works at several sites along the valley's lower section.

The second type of preferred site for mill and distillery development was the rolling upland sections of small streams in both the Inner and Outer Bluegrass country. In 1788 Bourbon County resident Isaac Ruddell petitioned the Virginia General Assembly for permission to build a dam on Hinkston Creek, near its junction with Stoner Creek, to impound water to power a gristmill and a sawmill.[41] By 1790, carpenters had completed a bridge across Hinkston Creek, and five more mills stood at the site. This cluster of mills attracted laborers and other businesses, and the settlement became known as Ruddles Mills.[42] By 1800, sixty-six mills operated in Bourbon County, many of them on Hinkston and Stoner Creeks.[43] Distillers built works at Ruddles Mills in the early nineteenth century, and some mills were converted into distilleries in the late 1860s and 1870s. Outer Bluegrass trunk streams in Nelson County such as Beech Fork and Rolling Fork, and their tributaries such as Knob Creek, accommodated numerous mills and distilleries.

Some creek-side industrial properties were limited to their immediate sites, while others included farmland on the adjacent uplands. Both types of developments were attractive real estate investments. In April 1826 Samuel Livingston of Jessamine County offered for sale two such developments on Jessamine Creek. The lower section included "one first rate *Merchant Mill, Horse Mill, Saw Mill,* and *Distillery,* with three comfortable *Dwelling Houses,* and other necessary buildings, such as Stables, Coopers' shops, etc."[44] Although this site was comparatively remote, even by present-day standards, the mill's twin overshot mill wheels turned a rubbing burr stone, and the mill was outfitted with elevators and sieves, apparently following Oliver Evans's widely replicated mill design.[45] The one-year-old stone distillery operated on steam power and had a production capacity of five barrels of whiskey per day. "The mashing is done by steam upon the third floor and cooled off on the second, and run on the first, except the slap-mash, which is mashed on the first floor." Livingston's advertisement lauded the distillery's design, which "exceeds any distillery that has ever been erected in the state."[46] The property included 153 acres of woodland that required clearing and more than 80 acres of cleared land. Livingston's second or upper "establishment" was half a mile upstream and included another merchant mill powered by an eighteen-foot waterwheel, a sawmill, and frame and stone houses. Livingston extolled the site's physical qualities. "These seats are well calculated for water works of any kind," he claimed; "the fall [is] great and the stream is good."[47]

Farmers, millers, and distillers also developed upland sites at or near large perennial springs, often surrounded by quality farmland but not close to any large creek or river. The scale of upland mills and distilleries was likely smaller than that of early creek-side operations of the same period, and these upland sites were tethered to overland wagon tracks and packhorse trails. The oldest distillery in Bourbon County at such a location may have been the Jacob Spears Distillery on Route 1836 near Ewalts Crossroads. Spears built a stone house, a distillery, and a warehouse near an all-weather spring on lands overlain by deep Maury-McAfee-Lowell soils, the Inner Bluegrass region's most productive soil association. In Washington County in the Outer Bluegrass, a 500-acre farm four miles from Springfield was advertised for sale in 1839. The upland farm "is well adapted for a stock farm being well watered," the ad stated, "and is so arranged as to throw each field to water—It is also well adapted for a distillery, having now an establishment for that purpose. It is well timbered, and all the land is good and lays well."[48] Farmers and millers viewed their upland properties as both agricultural and industrial enterprises and advertised them as such. Those offering land for sale believed that a distillery or a mill increased a farm property's desirability, making it worthy of consideration by prospective farmer-miller-distillers. In June 1839 the *Franklin Farmer* carried an advertisement by Mr. B. Dougherty, who stated, "I have for sale, in Franklin county, a Farm containing about 200 acres situated about five miles from Frankfort on the Georgetown Turnpike road. Also, an Engine, Mill and Distillery. Apply on the premises."[49] Most such ads mentioned road access—an increasingly important amenity as mills and distilleries increased in size and production capacity. By 1850, Kentucky's residents included 742 millers and 224 millwrights.[50]

Soils and Agricultural Potential

Although distilleries operated in most sections of the state by 1810, the most attractive locations for large-scale rural distilling operations were the Bluegrass country and the Ohio River floodplain, where the soils were fertile and farms productive. Alfisols are the predominant soil order found throughout the Ohio Valley; the suborder Udalf is found across central and western Kentucky and is composed of several distinct soil series. Alfisols occur in temperate, humid climates—or humid subtropical, in the climatologists' idiom—and they supported hardwood forest at some time during their development. Depending on their bedrock parent material, Alfisols are comparatively high in calcium, potassium, and magnesium and tend to have excellent native fertility.

The Bluegrass region comprises three distinct physiographic zones or subregions: the Inner Bluegrass, the Eden Shale Hills, and the Outer Bluegrass. Each subregion has a distinct bedrock geology and surface geomorphology, and the boundaries between them are visually demarcated by narrow transitions in bedrock and topography. Two major escarpments delineate the Bluegrass from bordering regions. The Pottsville Escarpment bounds the region on the east at the edge of the Appalachian Plateau and Eastern Coalfield, and Muldraugh Hill marks the region's southern and western boundary with the Pennyroyal.

With the exception of the deeply stream-dissected tributary valleys near the Kentucky River, the Inner Bluegrass surface is a gently rolling upland that formed on phosphate-rich Ordovician-age limestones. The soils on this surface are residual, having developed through weathering and erosion; they vary from shallow to deep and tend to be very fertile. Precipitation runs off the surface in low-gradient streams or percolates through the soil into the soluble bedrock, where it erodes rock surfaces and fissures, often enlarging cracks into small passages and caves that may feed surface springs, creating a karst topography. The surface may be dimpled by saucer-shaped sinkholes, some hundreds of feet wide; surface streams may disappear into swallow holes, flow underground, and then reemerge miles away.

The annular Eden Shale Hills subregion forms an asymmetrical circle around the Inner Bluegrass. The foundation bedrock is thinly bedded limestone, siltstone, and calcareous shale. The shale readily erodes, creating a sharply corrugated topographic surface of narrow ridges connected to narrow valleys by high-angle slopes. Eden soils are thin. The slopes are heavily eroded, and their surfaces are often littered with stones. When cleared of the beech–black oak–hickory woodland that favors this area, farmers find the soils comparatively infertile. The only sections suitable for field agriculture are the broader ridgetops and valley bottoms.

The Outer Bluegrass subregion is also annular and asymmetrical and lies in a broad arc around the Eden Shale Hills—wide on the east and west, narrow on the north and south. The Outer Bluegrass surface is a gently rolling upland with generally deep, fertile soils that developed atop comparatively young Ordovician limestones. The eastern section of the Outer Bluegrass is traversed by the Licking and Kentucky Rivers and their tributaries. To the west, the Salt River, Beech Fork, and Rolling Fork flow in looping meanders across the low-gradient surface toward the Ohio River.

The outer perimeter of the Outer Bluegrass is set off from the fringing escarpments by the Knobs, a nearly horseshoe-shaped subregion that forms a three-quarter circle on the east, south, and west. Only a few miles in width, the Knobs topography is punctuated by conical hills or knobs that are capped by erosion-resistant conglomerate rocks and rise 300 to 400 feet above the surrounding countryside. A native of the Middle West's mildly rilled glacial till plains might mistake these prominences for minor mountains. The Knobs' geologic base material is shale, and the interstitial areas between the hills are also underlain with shale and sandstone. The topographic surface is deeply dissected by streams, and the bedrock yields soils of modest fertility.

River floodplain and terrace soils are largely depositional or transported, rather than residual, and are composed of alluvial silts and sands with some fine-textured windblown loess. Although floodplains are often nearly flat and poorly drained, their deep soils are fertile and very productive when cleared of trees and ditched or artificially drained.[51]

To establish productive agriculture, eighteenth- and nineteenth-century Kentucky farmers had to adapt their practices to environments that often differed in complex ways over short distances.[52] A few gentlemen farmers studied farming practices systematically and published their findings and recommendations in journals such as *American Agriculturalist, Southern Farmer,* and *Southern Planter.* The contemplative farmer could, they argued, improve crop varieties through seed selection, cross-breed livestock to increase weight-gain potential, or rotate crops to improve yields. Many middling farmers—individuals with sufficient acreage and expertise to produce crops and livestock for self-sufficiency as well as for sale in local and regional markets—understood, at least in a general sense, the rudimentary differences in soil capability from one place to another.[53] Neighbors exchanged information about their successes and failures and the experience they gained through trial-and-error farming. Nevertheless, in the era before scientific agriculture, folk knowledge prevailed, and farming success did not always correlate well with soil fertility and benign weather.

As recently as the 1870s and 1880s, there was no direct relationship between crop yields and soil potential—the highest yields were not always produced on the best soils. Drought, plant diseases, and insect pests reduced yields, as did farmers' limited understanding of optimal crop production techniques. Regardless of the type of soil on their land, farmers were not yet fully capable of using good soils to full advantage. Kentucky commissioner of agriculture C. E. Bowman was critical of poor farm husbandry strategies across the state as late as 1880.[54] Yet distillers and millers depended on neighborhood farmers to produce ample amounts of grains and sell the surplus to them. Those farmers who were the first to develop successful husbandry practices and refine them through experimentation were likely farming the best soils. Distillers and millers, therefore, had an interest in conducting their operations on or adjacent to land with the best farming potential, however they made that assessment.

The relationship among environment, agricultural production, and the evolving geography of distilling begins with soils and their potential for grain production. Three county clusters in the Greater Bluegrass region, where distilling became a notable industry, serve as illustrations.

Bourbon and Harrison Counties occupy the northeast corner of the Inner Bluegrass physiographic region. The stream-dissected Eden Shale Hills region encroaches on the north and southeast. The Maury-McAfee-Faywood soil association is highly desirable for agriculture and is representative of large sections of the Inner Bluegrass portion of these two counties.[55] The carbonate Lexington limestone parent material contains phosphate, some of which is imparted to deep, loamy, fertile soils.[56] The land surface is a rolling upland with broad karst sinkholes and supports intensive agriculture. The Eden Shale Hills region offers only limited agricultural potential.

The northwest portion of the Inner Bluegrass—represented by Anderson, Franklin, and Woodford Counties—is bisected by the Kentucky River palisades, a 300-foot-deep gorge with nearly vertical limestone cliffs. Land within four to five miles of the palisades is deeply dissected by tributary streams. The rolling upland surface is underlain by Lexington limestone and has a soil profile that closely resembles that of Bourbon and Harrison Counties. The surface is pocked with sinkholes in some areas, but the loamy soils are well suited for intensive agriculture and have been productive in the past. The western half of Anderson County and the northern half of Franklin County are in the Eden Shale belt, where field agriculture has been confined largely to the valley bottoms.[57]

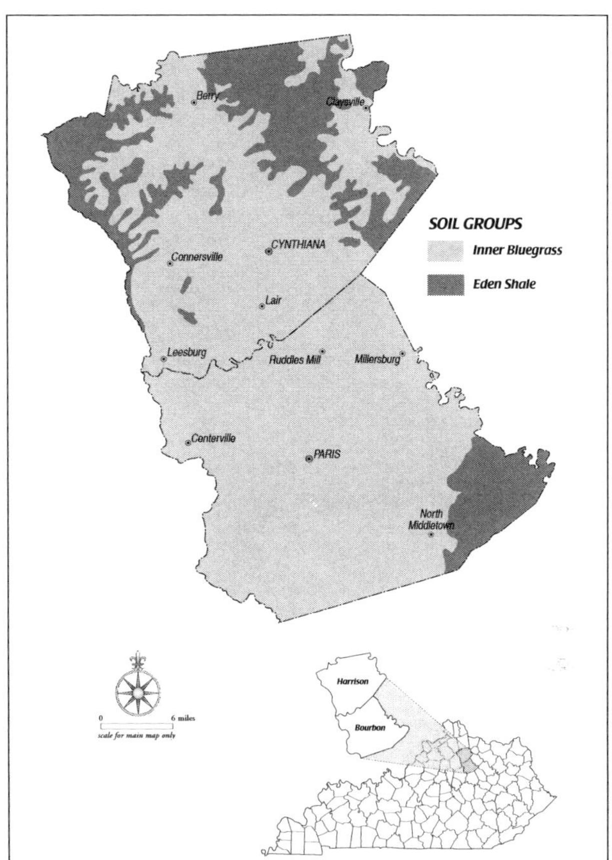

Physiographic boundaries and soil associations: Bourbon and Harrison Counties.

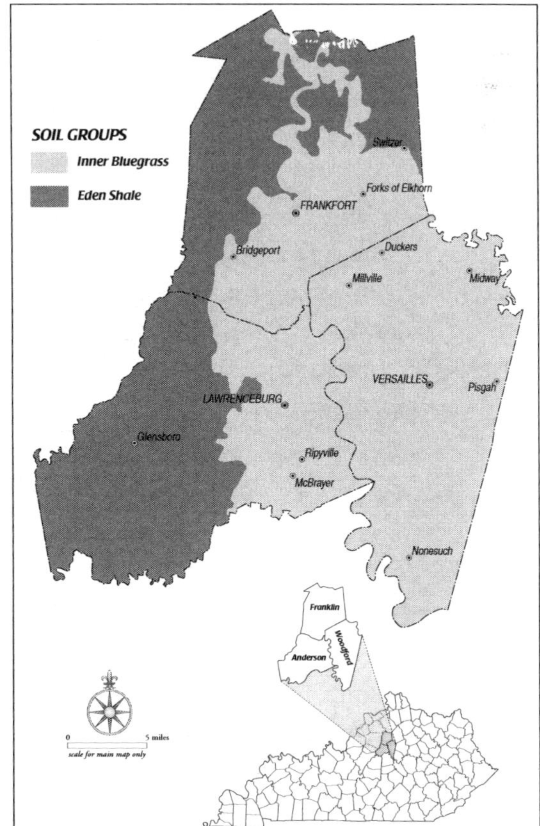

Physiographic boundaries and soil associations: Anderson, Franklin, and Woodford Counties.

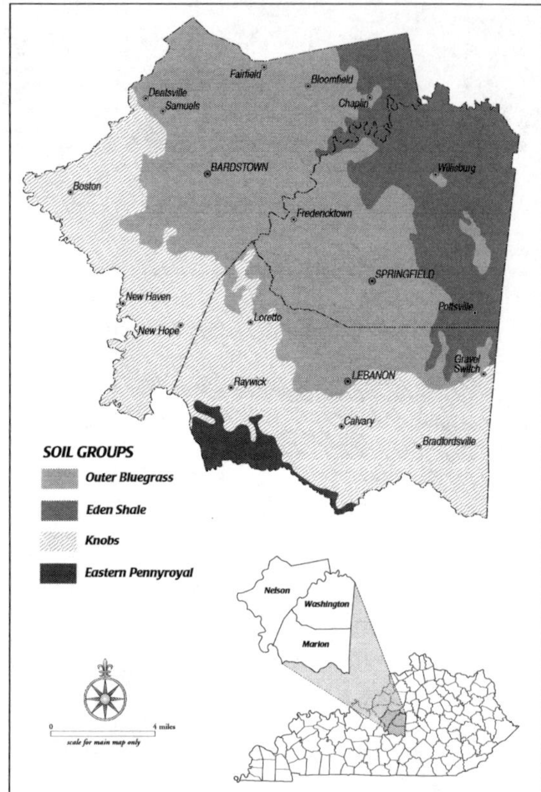

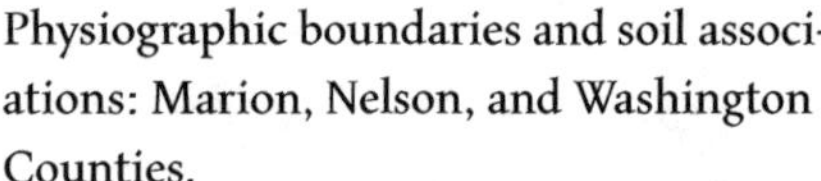

Physiographic boundaries and soil associations: Marion, Nelson, and Washington Counties.

The western section of the Outer Bluegrass region is represented by Marion, Nelson, and Washington Counties. Karst and underground drainage here is not as well developed as in some parts of the Inner Bluegrass. The surface drains west and north by way of the Chaplin River, Beech Fork, and Rolling Fork and their extensive tributary networks. Streams issue from the Eden Shale Hills to the east and cross the limestones of the Outer Bluegrass before entering the shales and sandstones of the Knobs region to the west. The Eden Shale Hills occupy the eastern half of Washington County and a small section of northeastern Nelson County. The topography and soils here are similar to those found in the Eden Shale sections of Anderson and Franklin Counties.

The Outer Bluegrass section is underlain by two different limestone formations. Grant Lake limestone, which is interbedded with shale, occurs in a small section of northern Marion County and significant portions of Washington and eastern Nelson Counties. A belt of Laurel dolomite runs northwest to southeast across the central section of Nelson County, with an extension into Washington County. Soils on the Grant Lake limestone are represented by the Lowell-Fairmount association; these soils occur on rolling land and are deep, finely textured, and of moderate fertility. The Pembroke association, weathered out of Laurel dolomite, has deep, loamy, fertile soils. The land surface is nearly level to rolling and invites intensive agriculture.

The Knobs region covers the western quarter of Nelson County and the southern half of Marion County. Shale and sandstone parent materials yield soils represented by the Rockcastle-Colyer association. Rockcastle soils are found on heavily rolling to steep slopes; they are thin, do not readily retain moisture, and are only moderately fertile. Lower slopes will entertain agriculture, but the steeper land is best suited to woodland.[58]

The Environmental Conjunction

Appreciation of distilling's historical ecology is enhanced by knowledge of how people engaged with their physical environments through time. We want to understand an environment's attributes in an objective sense but also uncover how people understood the environments in which they lived and how they interpreted an environment's qualities in order to draw their livelihoods from it. Part of that understanding includes the realization that the physical world is not static; it is always changing, in part because people modify it, and in part because natural processes are subject to variation and perturbation. Cutting trees to clear farmland created fields but exposed the ground to erosion and changed the infiltration rate of precipitation into the water table; as a result, springs dried up and stream flow became problematic in dry periods. Rivers might achieve their maximum flow in a given season of the year, but they could go dry or overflow their banks episodically in response to other environmental processes.

Eighteenth- and nineteenth-century farmers, millers, and distillers were bearers of traditional knowledge and practices that they used to inhabit their physical environments. Drawing from tradition, their use of the environment might have been tentative until they realized that some tactics worked better than others, at which point they could direct their efforts through informed decisions and choices. Working within an environment over time allowed people to refine and reform their perceptions and knowledge of how an environmental system worked and how they might adjust their behaviors to achieve more productive ends; perhaps they learned that as they adjusted, the environment also changed. The selection of mill or distillery sites was heavily influenced by people's perceptions and their understanding of the environment available to them. Farmers, millers, and distillers accomplished much of their work by manual labor, and they repeated their day-to-day and season-to-season activities if they proved effective. Yet their activities were often circumscribed by conditions and situations that they could not fully address until tools were developed that permitted them some degree of dominion over their environment.[59]

4

Distilling Grain, Feeding Livestock

The general use of Indian corn in distilling, has furnished a field for investigating the nature of that grain: many distillers find it yet impossible to bend the nature of that flinty grain to the production of spirit. I am convinced that a proportion of that grain worked with rye is highly favorable to the production of spirit.
—Michael Krafft

Conjunctions—Grain and Livestock Production

Kentucky's nineteenth-century distilleries produced bourbon whiskey from a fermented mash of corn, small grains such as rye or wheat, and malted barley. A common whiskey mash bill, or recipe, called for 30 percent small grains and 70 percent corn.[1] Until reliable turnpikes allowed grain transport over significant distances, Kentucky distillers were acutely aware of the importance of locating their distilleries at sites that had access to large volumes of quality grains. Central Kentucky grain farmers were often livestock producers as well. From the distiller's perspective, raising livestock was both oppositional and complementary: oppositional because livestock were often fed corn (along with hay, grass, and other grains) and were therefore competing with distillers for limited grain supplies, but complementary because hogs and cattle could be fed a distillery's spent grains, known as "slop," which provided distillers with a supplementary source of income and disposed of this otherwise nuisance by-product.

Grains

Corn and wheat were widely produced in the nineteenth century, but not always in sufficient amounts for home use, livestock feeding, and distilling. Some farmers produced rye and barley, but in limited quantities. The editor of the *Cynthiana Democrat* offered the following observation on the relationship between grain and distilling in 1869: "Farm lands are renting from $5 to $10 per acre. A great amount of sod land is being broken up; farmers are preparing to plant a large crop of corn. The distilling interests here amply justify the planting of a great breadth of land in this grain. They are but in their infancy and are as sure as the demand for whisky, which is as sure as the demand for bread."[2] Farmers and distillers alike deemed access to a reliable grain supply as important as access to a dependable spring or stream, and as distilling capacity increased, the expansive bluegrass pastures that had sustained the region's early cattle industry began to yield to the grain farmers' plows and harrows.

Whether grain furnished flour for home consumption or meal for distilling, its use was predicated on proper milling. And although whole ear or shelled corn might be fed to hogs, cracked or milled grain, if available, was often fed to other livestock. Water-powered gristmills operated in four steps: grinding the grain between burr stones, followed by cooling, sifting, and sacking the flour. Most early mills used a labor-intensive gravity-flow system to move grain through the mill, which meant that someone had to carry or hoist the grain by hand to the top floor. Cooling, sifting, and sacking were also accomplished by hand. In the 1780s Oliver Evans developed the first continuous-flow automated flour mill in Delaware. The ingeniously designed Evans mill used bucket elevators, augers, conveyors, and chutes to move grain to the top of the mill, where it descended through millstones and sifters. An elevator then returned the flour back to the top of the mill, and it made its way down through automatic cooling and sifting devices before exiting the mill in sacks or barrels on the ground floor. Evans also developed a steam engine that could power a mill, eliminating the need to site mills beside perennial streams on gradients sufficient to impel a waterwheel.[3]

Grains high in starch content are the preferred whiskey distilling ingredients, and because corn has more starch than wheat, rye, or barley, it produces more fermentable sugar per dollar of grain. The nonstarch components of distillery grain include concentrated nutrients with a comparatively high protein content; millers often sold this material to farmers as livestock feed.[4]

Corn or Maize

Indian corn (*Zea mays*) was Kentucky's frontier crop of choice. John F. D. Smyth, a British soldier who traveled extensively along America's Eastern Seaboard and in the trans-Appalachian West in the early 1780s, was impressed by the versatility of Indian corn. Farmers cured the leaves for horse feed, and the tassels, stalks, and husks made fine fodder for cattle. Smyth remarked that the grain "supports the inhabitants themselves, both white and black, besides feeding their horses, and fattening their hogs." More impressive, he thought, was corn's productivity when properly planted and cultivated. "A bushel of corn will plant near twenty acres; and on the richest lands twenty acres will produce two hundred and fifty barrels or one thousand two hundred and fifty bushels. A most astonishing increase indeed!"[5] In 1793 frontier land speculator George Imlay published a glowing report evaluating Kentucky's fecund Bluegrass and Green River country. Newly arrived settlers, he said, equipped with only basic tools, could in their first year clear three acres of land, plant corn, and expect yields of thirty bushels per acre. Given the extraordinary productivity of crops and livestock, a farmer who cleared and improved just two to three acres of land each year would be able to repay his costs in three years.[6] By the 1880s, Kentucky's commissioner of the Bureau of Agriculture declared corn to be "our universal crop."[7]

Preceding statehood in 1792, Kentucky settlers pursued land claims based on private surveys or Virginia military warrants. According to Virginia law, settlers could claim title to land if they surveyed and improved it by building a house and planting a crop. The most valued claims lay in the Bluegrass country, where the land may have been partly or wholly covered in grass, woods, or cane (*Arundinaria gigantean*) and may have included a perennial spring.[8] Arduous labor transformed raw land into working farms. Farm families cleared woodland or canebrakes, built houses and stables, split rails for fences, and planted gardens and crops—corn was usually the first crop planted.[9] Small single-family properties of 50 to 200 acres might have 1 to 3 acres of corn under cultivation by the second or third year of settlement. The fertile limestone soils produced generous yields. Even desultory planting and weeding of a 1-acre cornfield could produce 50 to 80 bushels. Established farmers might, in rare instances, produce more than 100 bushels per acre.[10] Corn became a staple crop; farmers fed it to livestock, and farm families consumed it at every meal. Within a few years, Bluegrass farms were producing surplus corn well in excess of farm and household needs. Among farmstead structures, the corncrib was arguably second in importance only to the house. Town sites were often surveyed according to an old English plan that included small half-acre residential and business lots surrounded by 10-acre out-lots on which many owners planted corn.[11]

European settlers adopted several corn varieties grown by Native Americans. Indians in the Northeast grew a flint corn (*Zea mays* var. *indurate*) with thin ears and eight to ten rows of hard kernels. Mid-Atlantic Indians raised several varieties, including flint corns and one with thick ears and fourteen to thirty rows of soft kernels.[12] Varieties of Indian corn grown in the South included southern big and small white flint, yellow Peruvian, and Virginia white gourdseed.[13] Anglo settlers from Virginia, Maryland, and the Carolinas who moved to Kentucky and the Ohio Valley often referred to their corn as Virginia gourdseed. Because corn produces seed by open pollination, accidental crosses between flint and gourdseed varieties were common and produced a dent-type corn (*Zea mays* var. *indentata*) that was commonly grown in Tennessee and Kentucky, the country's top two corn-producing states by the 1820s.[14] Farmers practicing artificial seed selection led to the further development of improved corn varieties. Baden corn was celebrated for its high yields; Maryland farmers developed the variety, and central Kentucky farmers were raising it by the mid-1830s.[15] A farmer in Boone County, Indiana, developed a deep-kernel dent variety in about 1874 that became known as Boone County white; it was favored for making meal used in bread and for distilling.

Corn Production and Consumption

By the early nineteenth century, corn was in demand as a food grain (cornmeal) and as a feed grain for livestock (ear corn or cracked kernels). It was also the basis for the distilling of spirits. Corn became an indispensable ingredient in the southern diet and was eaten

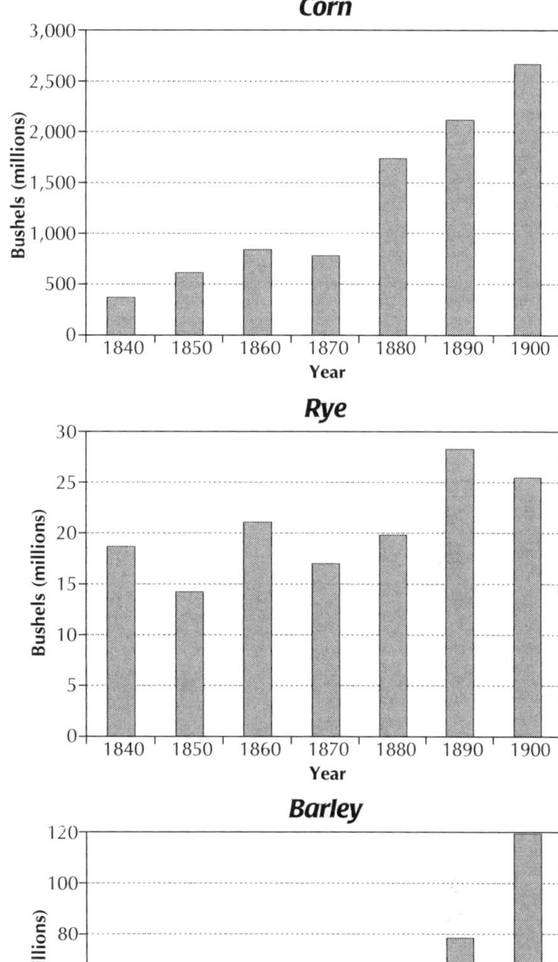

Primary distilling grains produced in the United States, 1840–1900. (Compiled from US Census of Agriculture, 1840–1900)

as meal, mush, and hominy; fresh ears were boiled or roasted. Markets priced wheat at 75 cents to $1 per bushel in 1820; rye was 50 cents, and corn, 37 to 50 cents. Wheat and rye flours were luxuries.[16] Millers sold wheat flour for $3 per hundred pounds.[17]

As a distilling grain, corn was the cheapest source of starch and the primary ingredient in most of the distilled spirits manufactured in the nineteenth century.[18] Milling corn to the consistency of coarse meal released starch. In the 1870s Ohio Valley starch manufacturers and distillers found that corn grown south of the Ohio River could be left in the field until it was fully ripe and dry, yielding more starch than corn produced in the Middle

West with its shorter growing season. Knowing this, distillers were often willing to pay a premium of a few cents per bushel for corn grown in Kentucky and Tennessee.[19]

By the mid-nineteenth century, corn was the chief national agricultural product, accounting for nearly twice the gross revenue of any other crop. Farmers planted about 31 million acres of improved farmland in corn, and the average yield was eighteen bushels per acre. National corn production reached more than 592 million bushels. Kentucky ranked second, producing about 59 million bushels of corn, or nearly 10 percent of the total national production. Malt beverage and spirituous liquor producers consumed some 11 million bushels, or less than 2 percent of the national corn supply, in 1850.[20] Although corn remained a staple food grain in the South, farmers increasingly raised it as a feed grain for livestock. Farmers marketed some corn for cash, but much of the crop provided the foundation for the region's livestock economy.[21] By midcentury, the center of corn production had moved north and west into what became known as the Corn Belt.[22]

Husbandry Practices

Before the market for corn could expand beyond the household, production had to increase substantially. Selection of superior-yielding seed varieties increased production, but prior to the development of herbicides, improved husbandry practices were the most effective way to increase yields. Grain crops planted by broadcasting seed shaded the ground during the growing season and provided some weed control. But weeds were the bane of open-field row crops such as corn. Hoeing weeds from large cornfields was laborious and impractical for small farm families. By planting corn in rows forty inches apart, with seed hills spaced at forty-inch intervals, farmers using hand plows harnessed to a horse or a mule could till their fields along and across the rows. Cross-cultivation became an effective technique for weed control and an important part of the effort to increase production. Mechanical horse-drawn corn planters appeared in the 1830s. Several patented improvements followed, including a machine that planted seeds in rows at exact intervals; this eliminated hand planting while permitting cross-cultivation between plants. Manufacturers introduced new horse-drawn cultivators in the 1830s, enabling farmers in the rolling uplands of Kentucky and the Middle Western states to mechanize corn cultivation. Two-wheel cultivators appeared in the 1850s and were continually refined by patented improvements through the 1870s.[23]

While the mechanization of corn planting and cultivation proceeded apace, harvesting remained an exercise in protracted manual labor. The primary factor that limited corn production in the mid to late nineteenth century was not the total acreage planted or the yield per acre but the number of workers available for hire at harvest time.[24] Farmers cut corn by hand when the ears and stalks were ripe. They gathered the individual stalks from several hills and tied them together into bundles, or shocks.[25] The shocks stood in the field for a month to fully cure before the stalks could be used as animal feed or fodder. When the stalks had cured, the ears were pulled and husked by hand, a task that required mo-tile skill and endurance. Farmhands then loaded the ears onto wagons and hauled them to the farmstead for storage in cribs. Frontier farmers harvested corn when the ears were fully ripe—about 20 percent moisture—and stored them in simple unchinked log cribs. Wooden planks later replaced logs in corncrib construction. Sawn planks permitted tight-er joints, which helped repel squirrels, raccoons, and rats.[26] Whole ears of corn could be fed to hogs. A common custom among Kentucky farmers was to place hogs in large pens next to the cornfields and throw the picked corn directly into the pens.[27] Ear corn had to be shelled before it could be ground into meal for home consumption or sale to distillers.

Whether corn was destined for human consumption, animal feed, or distilling, nine-teenth-century corn production must be evaluated in light of its manual labor require-ments. Labor efficiency readily improved small grain production from 1850 to 1890, when horse-drawn reapers and steam-powered threshing machines mechanized planting and harvesting, roughly in parallel with the distilling industry's mechanical transformation. Mechanical corn shockers were available by the 1890s, but they required handwork and did not eliminate the steps of removing and husking the ears. Reliable mechanical corn pickers finally began to replace hand harvesting in the 1920s.[28]

Bearing in mind that distillers had to compete with other buyers in the corn market, it is important to assess whether production was sufficient to meet the on-farm demand for livestock feed and the general market demand for cornmeal for human consumption. Two measures can provide a basis for judging the volume of Kentucky's nineteenth-centu-ry corn production.[29] The first measure considers the proportion of farmland dedicated to corn. Nineteenth-century farmland could generally be assessed as either improved—land planted in crops and pasture grasses—or unimproved—woodland or land that was oth-erwise unused for crop or livestock production. The ratio of improved and unimproved farmland acreage has never been stable, and this was especially true during the nineteenth century, when farmers were clearing fields of trees and rocks, draining lowlands, and oth-erwise improving their farms amidst the disruptions caused by the Civil War and finan-cial panics. Nonetheless, it is reasonable to estimate a yield of 7.5 bushels of corn per acre of improved farmland when assessing the amount of land dedicated to corn. A second measure examines the relationship between corn production and corn availability for live-stock feed. The production of 18.5 bushels per head of livestock is a reasonable estimate of sufficient corn grown for the maintenance of feeder livestock. Recognizing that crop production differs from year to year, given the variables of weather, insects, plant diseases, and other contingencies, we will assume for measurement purposes that farmers typically practiced a four-year crop rotation cycle, that 20 percent of all farmland was unimproved, and that the average corn yield was eighteen bushels per acre. Bourbon County farmer John Jones exemplified the process of crop rotation. He planted a given field in corn the first year, in a winter grain crop or spring-planted oats the second year, and in timothy for

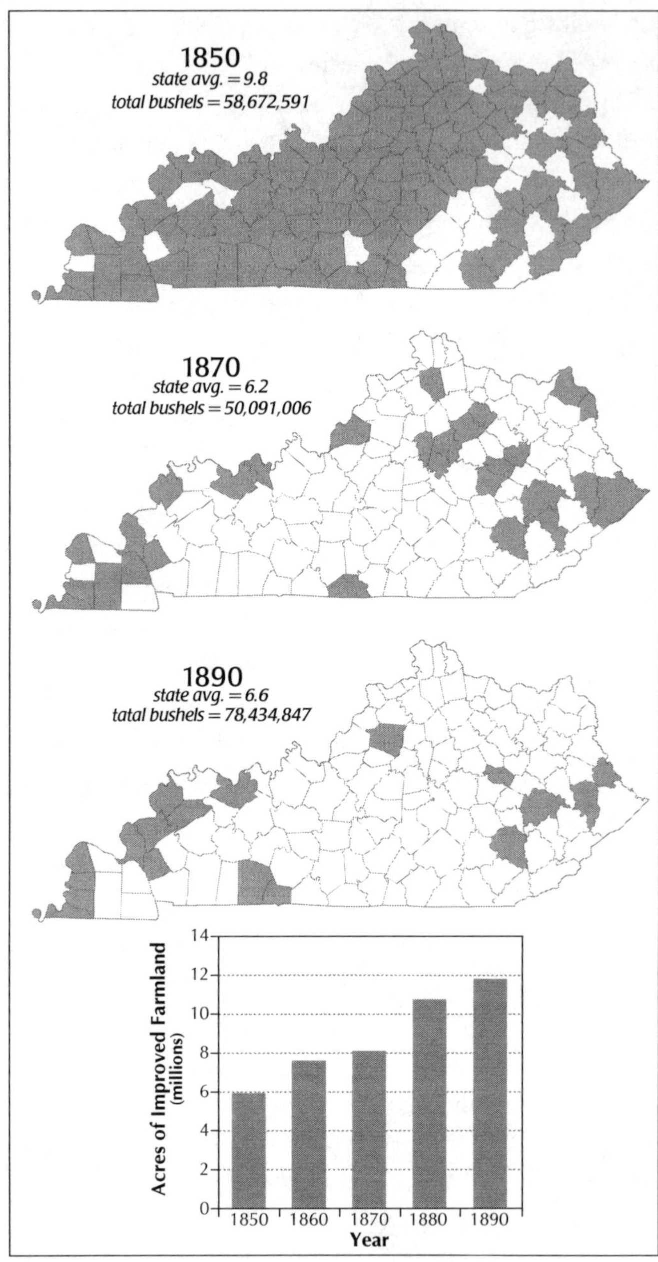

Corn production and improved farmland, 1850, 1870, and 1890. The maps show those counties producing more than 7.5 bushels of corn per acre of improved land. (Compiled from US Census of Agriculture, selected years)

hay the third year; the timothy crop wintered over for pasture grazing the fourth year. In the fifth year, the cycle began again by plowing under the timothy as green manure and planting corn.[30]

In 1850 corn production throughout the Bluegrass and Pennyroyal regions met or exceeded the threshold of 7.5 bushels per acre of improved farmland. But from 1870 to 1890, the state average fell below the threshold of 7.5 bushels, with only a few contiguous counties on the Ohio River floodplain and the Mississippi River counties of the Jackson Purchase maintaining threshold levels. Two factors contributed to this apparent anomaly. First, although total state corn production fell between 1850 and 1870, likely related to difficulties associated with the Civil War, in the two decades from 1870 to 1890, production grew from 50 million bushels to more than 78 million bushels, or an increase of roughly 40 percent. Second, the total amount of improved farmland roughly doubled from 1850 to 1890, so even though production per acre remained below the threshold, total production was robust.

Farmers producing feeder cattle and hogs for commercial markets, whether local or distant, often fed the animals corn as a primary supplement to pasture grass and other feeds. Using the figure of 18.5 bushels of corn per head as a threshold, we can assess the potential for producing sufficient grain to meet the on-farm demand for stock feed. In 1840 the state average was 12.9 bushels of corn per head, well below the threshold. Only five Inner Bluegrass counties exceeded that level of production. From 1840 to 1850 the total number of feeder livestock—cattle and hogs—increased to about 3.5 million head; that number declined by roughly 1 million head by 1870, before increasing again to about 3 million in 1890. Corn production expanded in the fertile Bluegrass and Ohio River counties by 1850, and in 1870 total state production reached 19.7 bushels per head, exceeding the 18.5-bushel threshold for the first time. Farmers in the Pennyroyal and the Ohio River counties west of Louisville to the Cumberland River also increased production levels. In 1890 state production averaged 25.8 bushels per head of livestock, and all but twelve counties in the southeastern corner of the state exceeded the 18.5-bushel threshold.[31] Distillers who did not grow their own grain, especially those who built their works in towns and cities, were compelled to compete directly with livestock feeders for sufficient grain to operate. The Ashland Distillery in Lexington regularly advertised for grain. "Farmers will please understand," one ad said, "we are *always* in the market [for corn], and willing to pay the highest cash price. We also want a few thousand bushels of No. 1 Rye."[32]

In the 1880s and 1890s Kentucky corn production ranged from 72 million to 78 million bushels per year. Distillers used 2.3 percent of the state's total corn production in 1880 and 7.7 percent in 1890. One should regard these numbers as illustrating only the *direction* of relative change, because as distillers mechanized their operations and increased their production capacity in the 1870s, they rapidly drew down local grain supplies. In Bourbon County, for example, distillers purchased all their grain from within the county until 1880–1881. Thereafter, most of their grain requirements had to be met by farmers elsewhere in Kentucky or even by out-of-state sources. Those distillers who were unable to obtain "imported" corn were forced to curtail operations.[33] By the 1880s, corn cost 40 to 45 cents per bushel, yet it remained relatively cheap, considering that rye often sold for 75 to 80 cents. Assuming adequate access to transportation and discounting interest on loans, moderately priced grain permitted large, efficient distilleries—those that produced more than four gallons of distilled spirits per bushel of grain—to profit if their whiskey sold for at least 25 cents per gallon.

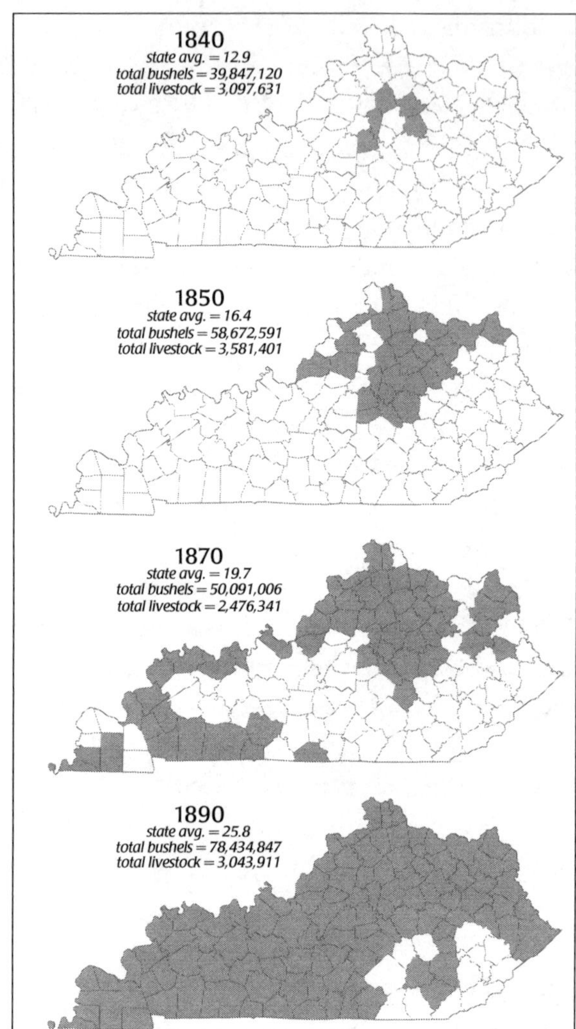

Corn production and feeder livestock, 1840, 1850, 1870, and 1890. The maps show those counties producing more than 18.5 bushels of corn per head of cattle and hogs. (Compiled from US Census of Agriculture, selected years)

Kentucky's Corn Production and Distillery Consumption, 1880 and 1890

Year	Corn Produced (Bushels)	Corn Consumed by Distilleries (Bushels)	Percentage Consumed by Distilleries
1880*	72,852,263	1,685,440	2.3
1890	78,434,847	6,075,330	7.7

Sources: US Internal Revenue Service, *Report of the Commissioner of Internal Revenue, 1879 and 1890* (Washington, DC: GPO, 1879, 1890); US Census of Agriculture, 1880 and 1890.
*Because the IRS report for 1880 was not available, 1879 figures were substituted.

By the twentieth century's first decade, the expanding national railroad network allowed Kentucky's industrial distillers to obtain grain shipments from farms more than 1,000 miles away. Farmers on the northern Great Plains raised quality corn, wheat, rye, and barley, and some began to cater to the distilling industry. At least one Kentucky distiller, Lynn Stambaugh, purchased farmland in North Dakota to experiment with corn production for distilling purposes. When Mr. Stambaugh visited Jamestown, North Dakota, during the harvest season in 1911, the state's largest newspaper took note:

> That North Dakota grown corn has a special value for distilling purposes is the statement of Lynn H. Stambaugh, representative of a large whisky distilling company in Kentucky, who was in this city Tuesday. Mr. Stambaugh is on a tour of the state with other representatives of his distilling company to get first-hand information with regards to actual corn raising conditions in this state. He was also encouraging the raising of corn here for distilling purposes in localities that would raise a crop especially adapted to this business and suggesting the planting of special varieties for the distilling trade. The firm of Stambaugh & Sons own a large farm near Bentley in Adams County, where experiments in corn raising with special attention paid to the value of the product for distilling purposes are going on constantly.[34]

Stambaugh may have known that large-scale bonanza farming had been under way in eastern North Dakota since the 1870s and that railroads and other infrastructure were in place to support grain farming and long-distance shipping.

Wheat

As a food grain, wheat (*Triticum aestivum*) provided sustenance. French settlers in the Wabash River Valley near Vincennes produced wheat surpluses that they shipped downriver to New Orleans by 1750.[35] Colonial farmers found that although wheat grew well in southeastern Pennsylvania and upstate New York, the grain did not thrive in the South, in part because the climate was favorable to blight, mildew, and a variety of other plant diseases. The region also harbored insects that attacked the plants. As settlement expanded west, the demand for wheat flour followed, and wheat acreage increased rapidly on the Ohio Valley frontier. Corn was more widely planted initially, and it had long been a primary staple in the frontier diet, but wheat commanded a much higher price by the 1840s. A farmer received an estimated 70 cents per bushel for wheat, compared with about 12 cents for corn, although prices varied substantially from place to place.[36] The amount of wheat ground into flour and for human consumption was estimated at more than four bushels per capita per year in the 1840s, and total consumption approached total production in several states. Given the demand for flour, wheat was often too expensive to be used as a primary distilling grain, other than in limited production.

Wheat cultivation west of the Appalachians increased in direct relationship to the extension and expansion of settlement, and grain milling and wheat farming tended to develop in tandem. "The miller's absence meant for the farm family a diet centered about home-pounded cornbread; his presence made possible white wheat bread and also furnished a market for flour."[37] Wheat varieties or cultivars planted in Kentucky and the "middle states" included Kentucky white bearded (also called silver bearded), Virginia white May, blue stem, old red chaff, Indiana, and Spanish Talavera, although many farmers preferred the red varieties Mediterranean and Lancaster.[38] Bakers favored soft wheat varieties for making flat bread, crackers, and pastries, and millers preferred soft wheats because they were easier to mill. Hard wheat had a higher gluten content and was used to make general-purpose flour and bread. The steel roller mill, invented in 1878, readily ground the hard wheat varieties and produced quality flour.

HUSBANDRY

Farmers who repeatedly planted wheat on the same fields soon exhausted the soil's residual fertility, a problem readily addressed by diligent crop rotation. Some farmers actively sought to improve their grain production through plant selection. Careful experimentation produced new varieties that improved yields and demonstrated resistance to pests and diseases, while also allowing production in less-than-ideal environments.[39] Yields var-

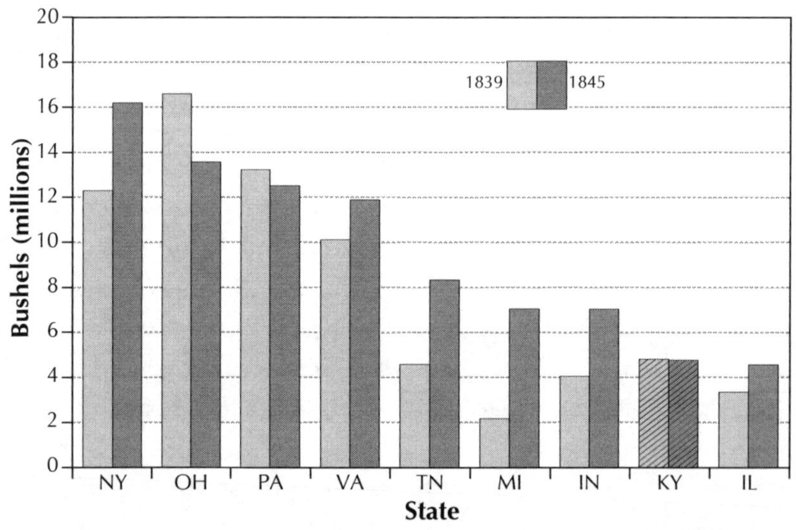

Wheat production in the United States, 1839 and 1845. Only those states producing more than 2 million bushels are included. Figures for 1845 are estimates. (Compiled from *National Magazine* 2, no. 11 [April 1846]: 1033)

ied from ten to forty bushels or more per acre, depending on the variety planted, husbandry practices, and the whims of weather.[40]

Wheat yields also varied year to year based on the plants' vulnerability to insect pests and disease. The northeastern states experienced widespread crop failures in 1836 and 1837, whereas the 1838 wheat crop produced excellent yields.[41] The mechanization of grain production proceeded apace. Grain drills available in the 1870s placed seed and fertilizer such as bone dust in the same furrow. Some farmers reported that the addition of minerals increased yields by fivefold or more.[42]

MARKETING

By the 1840s, wheat remained a subsistence crop in some areas, but Ohio Valley farmers who produced small surpluses began to sell the grain to millers and general markets as a cash crop. Initially, wheat sales were largely local; the cost of transporting wheat any significant distance to a commercial market could approach 40 percent of its potential value. Corn remained a basic staple crop in the Ohio Valley, where the average annual consumption of corn equaled that of wheat; corn consumption exceeded wheat consumption by two or three times in the South.[43] Certainly the corn-wheat price differential did not favor using wheat as a primary distilling grain, although some distillers used wheat as a secondary grain in place of rye, largely because it imparted a mellower flavor to whiskey; this was an important consideration when most whiskeys were given limited time to age before being sold.

Nineteenth-century rural communities were often served by small custom mills that ground farmers' grains in return for a portion of the grain as payment. Farmers loaded sacks of wheat onto packhorses or mules for delivery to the mill for grinding. As roadways improved and wheat acreage increased, merchant millers specialized in buying quantities of sacked grain delivered to the mill by wagon. The millers then shipped the grain or flour to wholesale and retail merchants. This system of shipping sacked wheat and other grains from farms to mills, then to warehouses and riverboats, and finally to wholesalers and retailers was widespread across the Ohio Valley and Middle West prior to the development of reliable railroads.[44] But the supply area from which merchant millers drew their grain was restricted by costs, which in turn were predicated on production patterns and the extent and quality of the transport net.

Rye

Commonly grown as a winter- or spring-planted crop, rye (*Secale cereale*) does well in cool climates. It was grown widely in the colonies from New England south to the Potomac River on those lands where wheat struggled. Colonists grew rye before they began to produce wheat, and rye flour and cornmeal were their primary sources of breadstuffs.[45] By the end of the Revolutionary War, farmers in western Pennsylvania were planting rye on the rough and stony uplands, where it became a staple grain. Surplus grain was not readily transportable to distant markets, so rye became the principal ingredient in farm whiskey distilling.[46] When the national government exacted a tax on whiskey to retire Revolutionary War debts, Pennsylvania farmer-distillers rebelled; many removed to the Ohio River Valley and Kentucky.

Rye production in the 1840s centered in New York and Pennsylvania, but acreage devoted to the crop declined by one-third during that decade because of less demand by distillers, "to which a large part of the crop is applied."[47] In western New York, "rye, where it is grown, was a medium crop. Since the extinguishment of the distillery fires, the cultivation of rye is not much of an object, except on the poorest sandy soils, where, if the crop is small, its quality renders it more suitable for bread, either alone, or mixed with corn."[48] In 1850, of the 14,188,600 bushels of rye produced in the United States, 2,144,000 bushels, or 15 percent of the total, were consumed in the manufacture of malt and spirituous liquors.[49] Rye acreage continued to decline in the eastern states as the crop accompanied the settlement frontier into the upper Middle West and the eastern Great Plains.[50]

HUSBANDRY

Rye varieties were less well developed than wheat varieties; the most widespread were winter, spring, black, white, and common.[51] By the 1830s, Kentucky farmers were planting rye in the fall on the stubble of harvested grain fields. The growing rye formed a dense mat that shaded and protected the soil from winter erosion. Some farmers turned cattle onto their rye fields to graze during the winter months, weather permitting. In early spring the rye revived from dormancy and continued to grow. When the field was dry enough, the rye was plowed under to provide a green manure fertilizer before planting the field in corn or other crops.[52]

Malting Barley

Barley (*Hordeum vulgare*) has a wider ecological range than any other grain.[53] Malting barley is best grown in cool, humid climates, although it tolerates higher temperatures and semiarid conditions. Dry, sunny weather in mid to late summer lowers the risk of mold or smut, which gives the grain a musty smell that makes it unacceptable for malting. Production of a quality crop requires well-drained loam soils; barley will not tolerate heavy clay, which means that much of the South is not well suited to barley production. Barley was one of the first grains planted by New England colonists, and by 1611, the grain was under cultivation in pioneer settlements in Virginia, New Netherland, Nova Scotia, and Quebec. By 1648, barley production was widespread across the Middle and North Atlantic colonies, and the crop's importance among grains was exceeded only by corn and wheat.[54] From the beginning, colonists grew barley primarily to make malt for brewing rather than for use as a food or feed grain.[55]

In 1850 total US barley production was 5,167,000 bushels, 3,780,000 of which, or more than 73 percent, was consumed in the production of malt and spirituous liquors; much of the remainder was used as livestock feed.[56] In 1844 the Kentucky barley crop was estimated at 14,000 bushels, compared with 3.9 million bushels of wheat and 2.3 million bushels of rye.[57] From 1840 to 1850, New York's Mohawk Valley and Finger Lakes regions led the nation in barley production. In 1860 New York's production was surpassed by that of Central Valley farmers in California. In 1880 California and New York remained the leading barley-producing states, accounting for 46 percent of total US production. Kentucky's barley crop had increased to 20,000 acres producing 486,000 bushels, or 1.1 percent of the total national production. East of the Rocky Mountains, the increase in barley acreage from 1850 to 1880 was attributed to better prices, as the demand for malted beverages increased, and to a decrease in wheat production because of predation by the Hessian fly.[58] From New York, barley production moved into the upper Middle West and central and northern Great Plains, north of the developing Corn Belt, where it flourished.[59] By 1900, the Red River Valley of northwestern Minnesota and the eastern Dakotas, the high plains of western Kansas and Nebraska, and California's Central Valley had become the nation's three primary barley-producing areas.[60] Yields ranged from thirty to fifty bushels per acre or more.

Husbandry and Malting

In the 1850s European and American farmers planted several barley varieties that were preferred by maltsters because they were disease resistant and produced a full kernel. As concerns about quality and production efficiency increased in the nineteenth century's latter decades, maltsters came to prefer the six-row Manchuria and Oderbrucker varieties because of their superior malting performance. Between 1870 and Prohibition in 1920, the distilling and brewing industry's requirements for malting barley became more exacting. Malting-quality barley commanded a premium price, but three-fourths of the crop typically failed to meet malting standards. Contemporary maltsters select barley that has plump, mature kernels with fully developed proteins and starches, is free from disease or weather damage, and is clean and carefully threshed.[61]

The malting process converts barley into a distilling grain. Maltsters steep selected clean, disease-free grain in cold water and spread it on a malting floor to germinate. Chemical changes make the grain's starch more soluble and readily extractable by water. The sprouting process is stopped by floor and kiln drying. The sprouts are removed, and the dried grain kernels are the malt.[62] Distiller's malt is generally germinated at relatively high moisture levels and dried in low heat to approximately 6 percent moisture; malts processed in this manner produce comparatively high enzyme activity.[63] Modern malting requirements are stringent, and the regional limitations on growing malting-quality barley are more restrictive than those applicable to barley produced for flour or animal feed.[64]

Although malted barley is required for whiskey distilling, nineteenth-century Kentucky distillers often could not obtain locally produced malt. Some distillers were obliged to use malted rye—or perhaps oats or wheat—if acceptable malted barley was not available.[65] As regional railroad networks expanded into the Middle West and the Great Plains after the Civil War, Cincinnati and Louisville brokers could purchase malted barley from distant suppliers. Because malt weighs less than unmalted barley, maltsters in the Dakotas and the Prairie Provinces of Canada could ship malt to Ohio Valley brokers at comparatively low cost. By the mid-1880s, eighteen malting sales offices and three malt kilns operated in Cincinnati, in addition to the malt manufactured by the city's brewers for their own use. Cincinnati was a primary malt market for Kentucky's distillers.[66]

In the 1850s grain accounted for roughly 80 percent of the cost of distilling a gallon of whiskey.[67] In the second half of the nineteenth century the increased distilling capacity of a rapidly growing number of industrial distilleries pushed the demand for distilling grains beyond the capacity of local farms to supply. Nonetheless, distillers were abetted in obtaining grain supplies by a rapidly expanded national railroad network and the development of long-haul grain shipping.

Livestock

Livestock were ubiquitous on Kentucky's nineteenth-century farms, where they provided meat (feeder cattle and hogs), milk (dairy cattle), wool (sheep), and motive power (horses, mules, and oxen). Farm distillers discovered early on that feeding their spent grains to cattle and hogs led to significant weight gains from a by-product that otherwise would have been discarded. As whiskey production volume increased, some distillers bought feeder cattle and hogs from local farmers to fill their feedlots; others contracted with farmers to feed their stock on a per-animal or per-week basis.

Hogs

Bird Smith, writing in the *Franklin Farmer* in 1838, reported that hogs raised in New England, where they were confined to barns and fed vegetables rather than being allowed to run in open pastures, produced pork that brought 10 cents per pound in Boston markets. And confinement of the animals allowed farmers to collect the manure, which was worth $2 to $3 per load, or 25 percent of the total value of pork produced.[68] Hogs produced in Kentucky could not match this return: pork prices in Louisville were only 5 cents per pound, and manure could not be collected because the hogs were not confined. Smith concluded that an extensive piggery could not be profitable in Kentucky unless it was connected with a distillery. Hogs confined in pens at a distillery could be fed distillery slop during the distilling season and put out to graze during the late spring and summer.

In the 1830s Kentucky's hog breeders marketed their surplus hogs to local butchers but also drove them overland to markets in the Carolinas. Farmers believed that English hog breeds—Bedfords, Berkshires, and Byfields—were best suited for feeding, given their tendency to gain weight rapidly. They were also deemed best for home consumption. The English breeds were not desirable as droving stock on backcountry trails, though, given their weight and inability to travel on foot.[69]

In 1840 the Society of Shakers at Pleasant Hill, in Mercer County, offered Berkshire and Lancashire hogs for public sale. The hogs were crossbred with stock recently imported from England and China. An advertisement placed by William Curd in Fayette County announced that he was standing at stud three purebred Berkshire boars and an Irish Grazier boar. A $10 fee would "insure a sow with pig."[70] Irish immigrants imported the Irish Grazier during the 1830s and 1840s. In the 1870s and 1880s farmers across central Kentucky were actively importing and breeding blooded stock. Distiller T. W. Samuels of Deatsville, in Nelson County, imported and bred English Berkshire hogs. He advertised widely and sold breeding stock to farmers in central Kentucky, southern Indiana, Maryland, and Louisiana.[71]

Feeding hogs at distilleries offered an advantage to hog farmers, thanks to a convenient overlap in the seasonality of feeding and distilling. Distilling activity began in November and reached peak production during the winter and early spring months. Given the dormancy of pasture grass and the limited availability of grain crops in the winter, farmers' demand for livestock feed was highest during the time when, by happy coincidence, the supply of distillery slop peaked. As distillers industrialized and increased their scale of operations after the Civil War, their demand for feeder stock, especially hogs, also increased. When a distillery on Elkhorn Creek in Scott County was put up for sale in 1870, the advertisement described the property as including "a Blacksmith Shop; Stone-House; Two new Corn Cribs that will hold Two Thousand barrels of corn; Hog Pens, capable of holding One Thousand Hogs; and situated on one of the best lots in the county, for slopping Cattle and Mules."[72] Although the phrase "best lots in the county" may have referred to the distillery property itself, it may have applied as well to the exceptionally productive surrounding farmland as a possible source of both distilling grain and feeder stock.

Cattle

Early settlers moving from Maryland, Virginia, and the Carolinas to claim land in Kentucky's Bluegrass region often drove their livestock on foot while carrying their belongings on pack animals. By the late 1700s, central Kentucky farms were producing feeder cattle for home consumption and local and regional markets. The cattle driven west by the pioneers were "scrub" or common cattle, not purebreds. Farmers grazed their cattle on the region's natural woodland pastures and expansive stands of cane.[73] By 1800, some central Kentucky cattle farmers were also driving their herds to grasslands in west-central Ohio for summer pasturing. Following Virginia custom, these farmers grew corn, cut the cornstalks, and stacked them in shocks to cure. They turned their two-year-old cattle into the cornfields to feed on the ear corn and stalks over the winter. The following spring and summer, the cattle grazed on grass pastures. After fall harvest, the cattle were again fed corn until February, when they were driven to distant markets. Some farmers deemed Kentucky's climate better suited to cattle production than that of England or any other part of America; they also judged corn fodder to be a superior winter feed.[74] Cattle breeding and feeding were sufficiently extensive in central Kentucky that by 1800, the area had become the first of four major cattle-feeding areas in the Ohio Valley. The typical Bluegrass farm might have thirty to fifty cattle, horses, hogs, sheep, and a still.[75] By the turn of the nineteenth century, Kentucky drovers were conducting regular cattle drives over the Wilderness Road and through the Cumberland Gap to their primary markets in the cotton plantation country of the Carolina Coastal Plain and the Piedmont. The Scioto Valley and Miami Valley of Ohio and the Wabash Valley of north-central Indiana were all important cattle-feeding centers by 1830.

Driven cattle—also termed "stocker" or "thin" cattle—were produced not on prime farmland but on marginal lands in Kentucky's Knobs and "barrens" by 1820.[76] These common cattle breeds were adaptable to frontier conditions and long drives, but their slow rate of weight gain made them less desirable as feeder cattle on the specialized stock farms of the Bluegrass and the Pennyroyal.[77] Common cattle might not be ready for market until they were five years old; improved breeds could reach market weight in three and a half years.[78]

Prior to 1817, the first improved feeder cattle breeds were brought to Kentucky by Matthew Patton, who migrated from a farm on the South Branch of the Potomac in Virginia. Lewis Sanders imported English shorthorn and longhorn cattle in a transaction that became known as the importation of 1817. The Sanders imports were the basis for improved breeds for several years thereafter. Also in 1817, Lexington's Henry Clay imported Herefordshire (Hereford) cattle. Importations of Durham shorthorn cattle continued into the 1830s. Shorthorns were deemed good travelers and were favored by farmers who drove their herds to eastern markets. By the 1840s, the Patton stock had disappeared from Kentucky, succeeded by Sanders's 1817 shorthorn stock. Clay's Hereford stock had also disappeared, giving way to the red, white, and roan shorthorn Durhams, which were preferred for their early maturity, rate of weight gain, and milk production, although some farmers believed they did not travel well because of their heavier weight.[79] Cattlemen also imported and promoted Devon cattle from southwestern England—Cornwall, Devon, Dorset, and Somerset—and, during the 1870s, dairy cattle from the Channel Island of Alderney.[80] In the 1870s high-quality shorthorn cows sold for 5 cents per pound, or $50 for a 1,000-pound animal; at the same time, quality Inner Bluegrass farmland sold for $20 to $30 per acre.[81]

Complementarity of Distilling and Livestock Feeding

By the mid-nineteenth century, the cattle-feeding industry was a long-standing tradition in central Kentucky. Distillery hog and cattle feedlots were part of that infrastructure and augmented the feeding industry's further development.

Distilling and livestock feeding continued as tandem businesses into the twentieth century. The Taylor & Williams Distillery near Fairfield, in Nelson County, was a medium-sized works with a cattle shed and a hay barn. A slop pipe likely carried spent grains to a nearby hog or cattle feedlot enclosed by a perimeter wood-slat fence.

The farming and distilling businesses shared a high level of complementarity in terms of the seasonality of grain and slop production and consumption and labor use and availability. The agricultural production cycle ranged from four to six months for corn and small grains; it took six months to a year to produce fat hogs and three to five years to breed and fatten cattle. At the beginning of a year's work cycle, farmers had to either own breeding animals, feed, seed, and tools or have access to cash and credit to buy them.[82] Farmers were the distillers' primary source of both distilling grains and feeder stock. But farmers and distillers sometimes required access to financing and often drew credit from the same banks and other financial sources. Distillers also had to make periodic federal tax payments on their stored product after 1868, and if they did not have cash on hand, they borrowed the funds from the same financial institutions that served farmers. In this way, farmers and distillers were engaged in noncomplementary businesses, in that they competed for access to the same sources of financing. Despite these tensions, grains and feeder stock continued to move from area farms to distilleries.

After the Civil War, many of those distilleries with railroad access industrialized and increased their production of slop, thereby increasing their demand for feedlot cattle and hogs. The railroad also made possible the long-distance marketing of fattened stock. By the 1890s, stock feeding in the distilling counties had become a complex and productive business. Distillers contracted with local farmers for feeder stock and fed stock owned by absentee brokers. Anderson County distilleries, for example, fed 4,990 head of cattle and 1,150 hogs in 1892.[83] Distillers who fed livestock preferred to feed those breeds that attained the most rapid weight gain, thereby reinforcing the transition from the casual production of common-breed stocker cattle to the focused management of purebred or crossbred stock.

5

Distillery Configurations

Distillation was for a long time confined to farmers, who only carried on the work during the winter season, and men of small capital, who being obliged to make quick sales, were more attentive to quantity, than the quality of the spirit distilled.
—Harrison Hall

Conjunctions—Technology and Distilling

Kentucky's frontier settlers planted corn, and cornmeal became their dietary staple. But shelling ear corn and converting shelled corn into meal by hand was burdensome at best. Millers were among the state's frontier settlers, and the miller's skills and equipment were important to easing life's labors and providing a source of food if a farmer's crops failed. Eighteenth-century farmers likely would have agreed that "a grist mill was . . . a greater necessity than a store, a courthouse, or a professional physician."[1] Distillers also considered gristmills high-value neighbors. Following folk traditions, eighteenth-century farmer-distillers took their corn and grains to the nearest mill to be ground into mashing meal. As distillers grew in number and increased their processing capacity, they bought corn from neighbors if surpluses were available. And some millers became distillers if they were proficient and if local corn and grain production were sufficient.

Colonial farm stills were elementary: small, portable copper pots holding little more than a gallon of a fermenting mash of ground corn or rye, or larger three- to five-gallon cylindrical copper tanks set atop a stone-lined fire pit. A few wooden barrels provided fermenting containers and storage for the finished product; the whiskey was clear in color and, by contemporary standards, harsh in taste. Family and neighbors consumed the spirits produced, although limited amounts might have been marketed further afield. By the late 1700s, some distillers operated as part of a localized milling center erected beside a perennial stream. In 1801 farmer and hat manufacturer Montgomery Bell owned such a works on the upper reaches of a creek in Jessamine County about five miles north of the Kentucky River. The complex included a distillery, a three-story merchant flour mill, a gristmill, a sawmill, and an equipped blacksmith shop. The large distillery building, measuring thirty-four by sixty feet, included stills and boilers. Bell had planted fifty acres of his farm in winter wheat and wished to rent the land and the mills either separately or together. Also available for hire on the premises were "fifteen negroes; Men, Women, and

Boys."[2] These were likely enslaved people belonging to Mr. Bell who worked the farm and operated the mills.

By 1810, mill and distillery combinations such as Bell's operated on several creeks that emptied into the Kentucky River and were the focus of local trails connecting to river landings and county seats. Such clusters also represented a craft or artisan stage of industrial development for mills and distilleries. That is, distillers began to consciously improve their product by hiring experienced employees and purchasing better equipment. They catered to customer preferences, considered the quality of the ingredients they used, and situated themselves in a positive marketing position by recognizing competitors' products and trying to best them. Whether distilleries operated in rural isolation or in a mill-distillery cluster, they increased productivity and quality by adopting innovations and inventions in distilling and related processes. These mill and distillery centers also benefited from a form of geographic symbiosis or local integration, whereby adjacent millers and distillers maintained a seller-buyer relationship.

Early Kentucky distillers—some of them with distilling experience in Ireland, Scotland, and England and others from Maryland and Pennsylvania—located primarily in rural areas at waterpower sites or on the edges of emerging urban centers at river landings such as Maysville, Covington, Louisville, Owensboro, and Frankfort. Some distillers set up business at important inland crossroads such as Lexington. While rural Kentucky distillers could obtain grains from their own farms or from neighbors, urban distillers lacked local grain sources. For their part, farmers wished to minimize transportation costs by growing grain as close to city distillers and markets as possible, and urban fringe grain growers began to distort the traditional rural grain production pattern.

In Scotland and England, eighteenth-century distilling was primarily an urban industry, with most large distilleries located in or on the outskirts of Edinburgh and London. Scottish and English brewers and distillers practiced the long-standing tradition of feeding spent grains to hogs and cattle housed in large barns. Distillers bought livestock at the onset of the distilling season and sold them to slaughterhouses after feeding them for fourteen to sixteen weeks. Because distillers sold their fattened livestock at lower prices than farmers could, livestock markets were chronically depressed, forcing farmers to sell their unfattened stock to distilleries rather than feeding and marketing the animals themselves.[3] Hogs were the exception to distillers' market dominance, however, because "country-fed" pork was deemed superior to "distillery" pork. In some markets, the price of country-fed pork was nearly double that of distillery pork.[4] Kentucky distillers developed similar distilling and livestock feeding relationships, although the process evolved over several decades before becoming fully organized by the mid-nineteenth century.

First Generation: Farm Stills, 1790s–1830s

Frontier farm stills, whether legal or not, were modest in size and output and were housed

Illicit Distillation, by A. W. Thompson, depicts an illegal still operating at night. Its component parts would have been similar to small-scale farm stills used in nineteenth-century Kentucky. An onion-shaped pot still, or alembic, presumably made of copper, sits atop a stone fire box. Two men tend the fire, while another carries a small wooden keg to a sled used to move heavy objects through hilly terrain. The lean-to roof on the right may cover the mash barrels. (*Harper's Weekly* 11, no. 571 [1867])

in informal, idiosyncratic buildings; a small log barn or simple lean-to shed often sufficed. A farmer could produce whiskey with a few small wooden vessels in which to combine water and grain meal and ferment the mixture with yeast; a copper pot or kettle served to boil off the alcohol, and a coiled copper condensing pipe reduced the vapor to a liquid that, if one had the fortitude, was ready to consume.[5]

Farm structures and artisan craft buildings such as gristmills, sawmills, blacksmith shops, cooperages, barns, and sheds were shaped by common knowledge—part traditional, part practiced or professional.[6] When built of hewn timbers and sawn lumber, most of these structures conformed to basic forms and proportions that were repeated from one building to the next. Early distillers utilized these same flexible building forms that could be adapted, with minimal modification, to suit the site and scale of a given operation. Such structures were functional and required a minimal investment of time and effort. Some early Kentucky mills were basic vernacular structures; others that were built by accomplished millwrights and stonemasons exhibited clever engineering and aesthetic qualities. Millraces, dams, and foundations were often made of dimensional cut stone rather than rough-faced field rock. A mill's superstructure might be stone or, more commonly, heavy timber post-and-beam framing covered with vertical planks or horizontal clapboard siding.[7] A distillery building often consisted of an internal timber frame that followed a basic geometry of squares and rectangles, topped with a gable or simple shed roof. First-generation distillery builders demonstrated little concern for artistry, never mind the picturesque or antiquarian values so popular with late-nineteenth-century Victorians. Traces of these concerns appeared in some second-generation buildings.[8] Farm stills were comparatively isolated and consumed limited amounts of water, grain, and cooperage while producing small amounts of distilled whiskey—initially a nonaged product. Later, whiskey would be aged in charred oak barrels. Whiskey markets were often local and, over time, became regional.

Power Source: Waterpower

The transition from self-sufficient farm stills to commercial-industrial distilling was a gradual but relatively short process, compressed into four or fewer decades. Eighteenth- and early-nineteenth-century mills and distilleries followed structural forms that were dictated in part by the power source available. Water-powered millstones ground the grain; manual labor was required to dip the water, lift the pails, and mash the ingredients into a fermentable slurry. Distilling required grain ground to a gritty or granular texture, not the exceptionally fine texture of milled and bolted flour. The miller adjusted the millstones to produce a chopped or coarse grind—a medium texture for milled rye, and a fine texture for flint-type corn to ensure thorough scalding.[9]

Falling water powered most of Kentucky's first-generation mills and distilleries, although small steam engines were available in some places by the 1830s. Operating under waterwheel or steam engine power required a building that was sturdy enough to hold the combined weight of the power machinery, the filled mashing and fermenting vessels, and the charged still.[10]

SITE SELECTION

The primary concern in distillery site selection was to enhance distilling operations. The basic requirements were ready access to moving water for power, potable water for mashing, and a general water source for cooling, along with an elevation change sufficient to permit the use of gravity to move grains and liquids. Selecting a site amidst productive grain farms was important but secondary. If a distillery was not sited to take advantage of elevation, water would have to be moved manually. Hauling water to a distillery and carrying or pumping water by hand constituted a significant portion of the expense of operating a distillery and therefore limited a distillery's size and output.[11]

In part, builders adapted a still's plan to its site, be it on a level river floodplain, on a low-angle hillside, or in a narrow, steep-sided valley. Leveling a large site with hand tools was impractical. A site near a spring on a south-facing hillside offered the advantages of a cool water source, an elevation change, and shelter from northerly winds during the winter distilling season. If a water-powered gristmill operated in conjunction with a distillery, a waterfall of at least six feet was deemed necessary to create the gravity power required to turn a vertical overshot waterwheel and drive the millstones and pumps.[12] If only a stream with a low gradient or a limited flow was available, millers could capture that lower power potential by building a small mill with an impact-driven undershot waterwheel.[13]

Heat and Fuel

Early distilleries heated water and distilled alcohol with stone, brick, or iron fireboxes; wood was the predominant fuel. Because wood was heavy and awkward to move, the distiller required a timber source in close proximity and a wood yard for drying and storage. Distillers preferred to use spring or creek water at a temperature of 60°F or lower to cool hot liquids and as a medium for mashing grain meal. By late spring, the surface water temperature of creeks or rivers had warmed, effectively limiting the distilling season to the period from late October to the end of May.

Equipment

Some distillers made their own still equipment using basic tools. Others purchased equipment made locally or manufactured regionally by coopers, coppersmiths, and blacksmiths according to studied and practiced craft techniques.[14] The still apparatus, therefore, could take many forms, with elements made from iron, copper, or wood. Fermented mash could be distilled by direct heat—a fire directly under the still vessel—or by indirect heat supplied by a small iron firebox with a flue to direct hot air to the bottom of the still. Operational efficiency required low fuel consumption and high output of alcohol per unit of grain distilled. The most efficient copper still vessel had a broad concave bottom that was heated directly and short sides to maximize the volume of mash exposed to the fire.[15] To attain these properties, skilled coppersmiths fabricated onion-shaped pot stills, or alembics. The pot still had to be scalded clean after each batch. Ideally, the coiled copper condensation pipe, or worm, had a tapered diameter and was round in cross section. The section of coil that passed through the cooling water vessel, or flake stand, might measure up to forty feet long. In the 1820s Scottish Lowland distillers began to use vertical continuous-flow column stills, following a design patented by John Haig and Robert Stein in 1826 and later modified by numerous refinements.[16] Column stills appeared in Kentucky when craft distilling began its transition to industrial-scale production.

Local coopers supplied mash tubs and fermenting vats made of wooden staves bound by wooden or iron strap hoops. The vats resembled the large hogsheads used to store and ship tobacco. Part of the distillery's interior space had to be shaped and sized to permit the proper positioning of the tubs and vats as they were filled with milled grain and, in the appropriate sequence, hot and cold water. Tubs and vats also had to be spaced to accommodate the workers moving between them as they hand-stirred the water and grain mixture with wood-handled mash rakes and cleaned the vessels between batches. Fermentation produces carbon dioxide, so an enclosed fermentation room required venting by sidewall or clerestory windows. Some distillers placed their fermentation vats along a wall outside the distillery building under a shed roof.[17] The alcohol fermentation cycle varied from two to four days, and some distillers used at least one vat for each day of the cycle. Depending on scale of operation and the rate of production, a distiller might be conducting mashing, fermenting, and distilling operations simultaneously. Distillery structures had to be large enough to accommodate this attenuated production cycle, and increased production required additions to existing buildings or new structures altogether.

Warehouses

Barrel aging warehouses varied in dimension, material, and design. Distillers operating small works might store filled barrels in an existing barn or outbuilding. Larger production favored discrete warehouse buildings at the distillery site so that additional units could be built if necessary—arranged in a row on a level upland site, or arranged along a floodplain at a riverside site. Some distillers built unheated and unvented warehouses that were fully enclosed except for access doors. Others installed side windows to admit light and provide some measure of temperature control in the summer. The monitor roof with clerestory windows and vents was a common configuration for American industrial buildings in the pre-electric nineteenth century, and several distilleries adapted this configuration for their warehouses to provide both light and ventilation.

Additions

Although farm distillery buildings were comparatively small, they were often accretive, as distillers added or attached buildings to accommodate new technology or replace structures that had burned. Those farm stills that reached commercial scale often replaced old, undersized structures and adopted formal building plans such as those used in general industrial buildings and manufacturing works. Instead of basic multipurpose buildings, dedicated industrial buildings were purpose-built to house appropriate machinery and equipment and were laid out in a rational manner.[18] Similarly, millers often upgraded their gristmills to utilize elements of Oliver Evans's conveyor and gravity feed system and converted from waterwheel to steam power. If old mill buildings could not be economically retrofitted, a miller would construct new buildings of two or more stories. These new structures were subjected to mounting demands by insurance companies that mills and distilleries employ materials and designs that promoted fire containment.

Many commercial-scale distillers converted to steam engine power as technology and financing permitted. Installing a small steam engine in an existing distillery might require a new room or an attached shed with an external iron or brick chimney, in part to conserve interior space for distillery functions, and in part to prevent a boiler explosion and ensuing fire from consuming the entire works. Wood was the predominant engine fuel, and the transition from wood to coal required advances in transportation. Some distillers sited their new works adjacent to navigable rivers, where they could receive direct coal shipments. Inland distillers often awaited the railroad before converting from wood to coal.

Discharge

Distilleries produced spent grains in volumes proportional to the amount of grain processed and the quantity of water used for mashing. Feeding cattle and hogs on distillery slop was sometimes the most profitable part of the distilling business, and it was preferable to dumping the waste. Livestock feeding pens and barns were usually located downslope and away from the distillery works. A hillside distillery site might provide the elevation required to move slop by gravity from the still to the feeding pens. Alternatively, elevating the still floor or the slop holding tanks allowed the distiller to use gravity to direct the slop into feeding troughs. Efficient distilleries could feed up to thirty hogs for every five bushels of grain processed each day.[19]

Summary: Early Distillery Function and Form

Late-eighteenth- and early-nineteenth-century whiskey distilling in Kentucky largely followed long-standing traditional practices. Adoption of new techniques and technology required confidence and foresight, as well as a willingness to accept a high but unknowable degree of risk. Failure could mean losing a season's production and the money invested in purchasing grain and livestock. Some farmer-distillers avoided risk by adhering to traditional practices. Others embraced new technology and were early adopters. As businesses, distilleries operated within a larger and changing geographic context. Transport infrastructure expanded with the development and adoption of steam power, and turnpike construction created durable roads that reduced transport times and costs. As individuals, distillers understandably demonstrated different levels of experience and business acumen. But all operated within the context of volatile external influences—be it unpredictable weather truncating their grain supplies or a capricious economy closing off the market for their product. Generalizing such variations into meaningful periods or eras that reflect broad trends is not straightforward, but it can serve as a pragmatic investigative strategy to illustrate, in a general way, the process whereby the nineteenth-century distilling landscape was constructed and maintained or decommissioned and erased.

Early craft distillers followed vernacular traditions and developed their skills by apprenticing with experienced distillers and conducting trial-and-error experimentation. Literate distillers may have encountered books on distilling techniques, such as those by Anthony Boucherie, Harrison Hall, and Michael Krafft.[20] Some distillers became aware of new science-based distilling techniques and adopted instruments such as the hydrometer to measure specific gravity, but progress toward the scientific application of chemistry was ploddingly slow. Distillers took pride in using traditional methods to produce a traditional product. Nevertheless, some distillers began to adopt technology-assisted techniques, and as they did so, the distilling landscape changed. Distillers added new buildings and remodeled or replaced others. Transport by packhorse and wagon along modest overland trails limited the distillers' source of grain to local producers. But if their grain requirements exceeded the capacity of their own farms, they could take advantage of new turnpikes to extend their supply lines beyond their local neighborhoods. The comparatively high value of their whiskey allowed them to seek out markets at even greater distances if they could gain access to river ports.

In the aggregate, distillers who wished to industrialize were reliant on inventions in kindred or complementary manufacturing businesses that could be adapted to distilling or that increased the productivity of other industries such as ceramics and glass, metal fabrication, farm machinery, and power production.[21] For their part, many distillers engaged in a more or less continuous invention process and routinely patented machinery and developed procedures that increased production capacity and product quality. Such interdependencies suggest that early-nineteenth-century distillers could remain traditionalists if they wished, but they could also adopt new technology and techniques if they were interested and financially able.

Second Generation: Industrialization, 1830s–1880s

Collectively, the transition from craft to industrial distilling was not an abrupt revolution in structure and function but a gradual process of adjustment. For individual distilleries, gradual change was punctuated by the adoption of steam engines, the installation of new equipment, or the construction of railroads. As distillers added steam engines and hardware designed to move grains and liquids, distillery design changed to accommodate the size and shape of this new equipment and the weight of outsized vats. Internal space requirements and the flow of materials now directed building design, rather than vernacular habit. Industrial distilleries, therefore, were usually robust, purpose-built structures that bore only a casual resemblance to generic livestock barns and utility buildings. A building's size and shape had to be appropriate to the equipment it housed.[22] The further expansion of production often necessitated the appending of sheds or lean-tos or the construction of entirely new buildings.

Distillery Structures

Increasing distilling capacity required ready access to sustaining volumes of appropriately processed raw materials. Early commercial distillers therefore often paired their works with flour mills or gristmills. The mill-distillery structure might be designed and erected by a distiller, a millwright, or an experienced carpenter who selected horizontal plan dimensions and vertical cross-section configurations based on a repertoire of traditional forms, pragmatic experience, and astute reasoning.[23] As each new distillery began production, it became a demonstration model for builders and distillers, illustrating how effectively that structure served its intended purpose. The distiller remained an artisan, employing long-standing practices, but this tradition also gave way to applied science. The transition to mechanized, high-capacity distilling also required changes in complementary businesses, such as marketing, and access to capital; developing professional relationships with wholesalers and bankers became increasingly important.

The nomenclature linked to the term "still" depended on context, but it generally referred, individually and collectively, to the pot or column still fixture and to the assembly of tubs, vats, valves, coils, and pipes that moved the grain and liquid. These mechanicals and the structure that housed them constituted the "distillery." Early distilleries were single-room buildings that accommodated all the equipment and activity. Over time, as new technologies developed, as operational scale increased, and as governmental regulations were put into place, the distillery structure was significantly enlarged and subdivided into separate rooms with different functions: mashing, fermenting, distilling, doubling, and so on. The distillery so modified became an outsized compound building. Distillers added powerhouses, storage warehouses, bottling plants, cisterns, and livestock pens, all tied together with in-ground or overhead pipes, walkways, and roads. The various elements operated together like a factory or manufacturing works. Distillery employees likely used specific building names or numbers, while referring to the collective as "the distillery."

During the early decades of industrialization, distilleries did not adopt standard factory buildings, as did the textile mills of New England. Unlike the ranks of identical textile looms that were powered by water or steam and run off of precisely arranged pulley-and-shaft millwork, distilleries functioned in several different configurations with nonstandard equipment that was homemade or supplied by a variety of artisans and manufacturers. By the late nineteenth century, large-production industrial distilleries were housed in expansive brick buildings, all of which bore some resemblance to one another. But, given that each distiller practiced somewhat different manufacturing techniques, each distillery was configured somewhat differently, and each may have operated custom distilling equipment built to individual specifications.[24] Although distilleries were engaged in product manufacturing, and their aggregated buildings were often described as "works," distilleries were rarely referred to as factories. The term "factory" implied the production of standardized goods, and most distillers, then as now, did not aspire to make a standardized product. On the contrary, each distiller wished to portray his whiskey as different, if not unique, and as handmade, not the product of machines.

Distillers often moved into and out of the business. A farm-distillery might be abandoned and salvaged or improved in small increments. A mill-distillery owner might sell the works to another operator who invested additional money in equipment and buildings. As much landscape adjustment occurred through the buying and selling of distilleries as by the gradual improvements made by a single owner. In June 1839 B. Dougherty placed a property-for-sale notice in the *Franklin Farmer,* published in Frankfort: "I have for sale, in Franklin County, a Farm containing about 200 acres, situated about five miles from Frankfort on the Georgetown Turnpike road. Also, an Engine, Mill and Distillery, Wool Machinery, etc. which will be sold with the Farm . . . or separate. Apply on the premises."[25] B. Dougherty may have built this distillery or may have been its second or third owner. One sees here the beginnings of industrial-scale distilling, not in a town or at a river landing but in the countryside on a farm, albeit one served by a turnpike. The term "engine" in the ad implies a steam engine, which may have powered both the mill and the distillery, and though we cannot be certain, the mill was likely outfitted with continuous-process elevating, grinding, and bolting equipment, which, by that time, was in general use.

In addition to specialized structures and equipment, industrial distilling required adjustments in social arrangements. The Elliot household provides an example. George Elliot operated a farm, a mill, and a distillery near New Haven in southern Nelson County. His immediate family included his wife and four children; four employees also lived in the house as boarders. Matthew Graves worked as a farmer laborer; John Willitt, son of a farmer, worked in the distillery; Joseph Lang, born in Bavaria, was a cooper; and William James worked as a wood chopper. James was likely clearing Elliot's farm of trees, and the wood he cut had various potential uses, including fuel for a mill and distillery steam engine as well as for household stoves. Farmers traditionally boarded hired hands, as did turnpike contractors; rural distillers were often obliged to follow suit. Before mechanized personal

George W. Elliot Household, Nelson County, Kentucky, 1880

Name	Age	Relationship	Occupation	Place of Birth
George W. Elliot	40	Husband	Farmer-miller-distiller	Kentucky
Amanda Elliot	39	Wife	Keeping house	Kentucky
William Elliot	14	Son	Works on farm	Kentucky
Lisa Bell Elliot	11	Daughter	At home	Kentucky
John W. Biles	21	Stepson	Works on farm	Kentucky
Joseph K. Biles	18	Stepson	Works on farm	Kentucky
Matthew Graves*	20	Boarder	Works on farm	Kentucky
John Willitt	21	Boarder	Works in distillery	Kentucky
Joseph Lang	42	Boarder	Cooper	Bavaria
William James	21	Boarder	Wood chopper	Kentucky

Source: US Census of Population Manuscripts.
*Name may be misspelled.

transportation, employing farmhands or other laborers who lived more than a short walking distance from the works was impractical. And the wages paid to young farm boys were too small to permit them to maintain horses of their own.

Industrial-Scale Rural Distilleries

In 1840 Kentucky's distilleries were distributed across the state, with large clusters in the Bluegrass and eastern Pennyroyal regions, a small cluster in the southeastern mountains, and a thin distribution in the Western Coal Field and Jackson Purchase regions. The first distilleries to achieve industrial-scale production were along the Ohio River in Boone, Gallatin, and Trimble Counties, with a smaller cluster at creek-side mill sites in the Licking River drainage in Harrison, Nicholas, and Bourbon Counties.

Maintaining industrial-scale production required access to a reliable transportation network, be it water, road, or rail, to sustain high-volume grain and fuel deliveries and provide product marketing egress. Mechanized equipment minimized manual labor, but larger production volumes and machinery maintenance requirements likely increased overall employment. The distilling process now aggregated discrete functions into an interconnected ensemble of high-capacity grain storage and milling, multiple mashing and fermenting vats, and one or more high-capacity stills, all integrated by elevators and gravity flow or pumps and pipes. Steam engines provided heat and horsepower and required a continuous supply of fuel, be it wood or coal.

Operations were no longer confined to a single modestly sized structure; the distilling process had expanded to occupy clusters of integrated buildings and second and third stories. Vertical expansion was necessary to accommodate gravity-flow grain-handling equipment and the new vertical column still—twenty to forty feet tall—if distillers chose to replace or supplement their traditional pot stills. Distillers operated fewer machines that were one-off purpose-built—perhaps made onsite by employees. Instead, they purchased standardized equipment from local and regional manufacturers. Building design and configuration nevertheless remained idiosyncratic, in part because distillers had to adapt structures to the verities of the site—spring location, creek frontage, slope, and road or rail access. Industrial processing expanded a distillery's footprint. Distillery fires were commonplace, and the greater size and number of distillery structures potentially at risk piqued insurance companies' interest in "fireproof" construction, spurring the erection of new buildings with brick walls and iron doors and window shutters.

Power Source: Stationary Steam Engines

Many eighteenth- and early-nineteenth-century gristmills, flour mills, and sawmills in America were powered by direct-drive waterwheels.[26] Mill wheels and windmills were the prime movers, in that they converted natural power—moving water or air—directly into mechanical power. Stationary steam engines were in very limited use during the eighteenth century, but they pumped water from mines in New Jersey, powered a cotton factory in Philadelphia, and, by the early 1790s, were powering a small number of sawmills and gristmills in the trans-Appalachian West.

The conversion from water to steam power was the key enabling factor in the industrialization of distilling. Steam power was initially applied to distilling and other manufacturing activities directly. That is, stationary steam engines converted heat and reciprocating motion into rotational or turning motion that directly operated belts, pulleys, shafts, and gears that mechanics connected to grain mills, mash mixing equipment, liquid pumps, and other machinery. In the late nineteenth century steam engines operated manufacturing works indirectly by powering electricity generators, which in turn ran electric motors that operated the same mechanical systems.

Philadelphia inventor Oliver Evans opened a steam engine manufacturing shop in Pittsburgh in 1811, and within two decades, Pittsburgh and the downstream Ohio River cities of Cincinnati and Louisville had become centers for steam engine production.[27] By 1821, Louisville's Prentice and Bakewell foundry and other mechanical shops were building steamboat and stationary steam engines in such quantity that they were among the city's leading exports.[28] A large steam mill on the Cincinnati riverfront powered a flour mill, a fulling mill, and a distillery in 1826.[29] Pittsburgh, Cincinnati, and Louisville manufacturing firms also built small engines that powered Louisiana sugarcane mills by the 1830s.[30]

Skilled blacksmiths and machinists using hand tools built a limited number of basic stationary steam engines for mill applications. According to the US Treasury, the earliest report of a stationary steam engine adapted for use in a distillery and for corn milling occurred in 1837.[31] Large-scale engine production was under way in New York by the mid-1840s, but the production of efficient and effective stationary steam engines in quantity awaited the development of machine tools and interchangeable parts, which became generally available in the 1850s.[32] Steam engine production in the Middle West and Ohio Valley industrial cities was well established by the 1850s, and as the number of engines in use increased, the variety of applications broadened rapidly. Stationary engines tended to operate at low rotating speeds and moderate pressures of 30 to 60 pounds per square inch—less than half the pressure used on steam-powered riverboats; these engines were simpler to build and maintain. Whether refitting waterwheel mills or constructing new mills, potential customers were primarily concerned that the engines be effective and efficient. Engine effectiveness was a measure of reliability, frequency and difficulty of maintenance, and the overall cost of operation. Efficient engines delivered a high ratio of power per unit of fuel burned.

Louisville attorney W. S. Pilcher posted a notice in the *Louisville Daily Democrat* in September 1852, offering for sale a four-and-a-half-acre parcel of "limestone land" at Fisherville, about fifteen miles east of Louisville in Jefferson County. The property bordered Floyds Fork and included "a dwelling house, storehouse, other necessary buildings for a family; a large corn mill and distillery house, 3 stories high, 60 feet long by 30 feet wide, pens for 400 hogs, and a corn-crib nearly of the same size as the distillery building. It has a steam engine, and all distillery machinery complete."[33] Other village businesses included a second corn mill and distillery, a corn mill and sawmill, mechanic shops, retail stores, a Mason's Lodge, and, ironically, a Temperance Hall. Importantly, the Taylorsville and Louisville Turnpike passed through Fisherville, and a triweekly mail coach served the village. The turnpike not only provided access to area grain suppliers and an outlet for marketing barreled whiskey but also permitted the delivery of heavy machinery such as steam engines.

Between the end of the Civil War and the great financial panic of 1873—and for several years thereafter—Kentucky's industries increased in scale, many converting from handcraft to machine-assisted manufacturing. Many factories, especially in the riverfront cities of Covington-Newport, Louisville, and Owensboro, converted to steam power, and new works were purpose-built to operate on steam. Such was the case at Louisville's Grainger & Company's Phoenix Foundry, which had been in business since 1833 making waterwheels and distillery and mill machinery. By 1868, the foundry had expanded into steam boiler and engine manufacturing.[34] Master mechanics also established machine shop businesses in county seats, thereby supporting those small-town and rural businesses that wished to convert from waterpower to steam. In 1883 the Macdonagh brothers, who had acquired proficiency during apprenticeships in England, opened a machine shop in Paris in Bourbon County, offering repairs on steam engines and all forms of mill and distillery machinery.[35]

Power Sources and Workforces of Selected Kentucky Industries, 1870

Industry	Number of Establishments	Number of Waterwheels	Number of Steam Engines	Number of Workers	Number of Workers per Establishment (Average)
Cooperages	95	0	2	383	4.0
Flourmills and gristmills	696	321	314	1,686	2.4
Distilleries	141	6	95	1,033	7.3
Lumber mills	562	101	378	2,497	4.4
Steam engine and boiler manufacturers	10	0	9	2,238	223.8
Tin, copper, and sheet iron works	127	0	3	531	4.2
State totals	5,390	459	1,147	30,636	5.7

Source: US Census of Manufacturing, 1870.

Rural flour mills and gristmills continued to operate on a limited, seasonal basis, each employing only a few workers. Some mills converted from water to steam power, but by 1870, more than half of the state's 696 mills were still powered by waterwheels. By comparison, more than twice as many steam engines (1,147) as waterwheels (459) powered Kentucky's expanding catalog of urban and transport-oriented industries. And mechanization proceeded apace in those businesses that adopted steam power. The firms that made distillers' machinery, along with ten works in the state that built steam engines and boilers, were the most technically advanced; they were largely steam powered and employed an average of more than 220 workers per factory. Kentucky's steam industry was, consequently, the first industry to achieve large-scale mechanized operations. Of the other industry types with a total of more than 1,000 employees, only bagging (eleven establishments; two steam engines; and 1,228 employees, with an average of 112 per factory) and pig iron foundries (nineteen establishments; twenty-eight steam engines; and 1,565 employees, with an average of 82 per factory) approached or exceeded 100 employees per plant. Distillers and other distilling-related industries operated at a more modest scale. Ninety-five of the state's 141 distilleries were steam powered by 1870, and they employed, on average, seven people. Cooperage remained a handcraft industry; only two of ninety-five establishments were steam powered, and they employed an average of four workers.

By 1880, the number of steam engine and boiler manufacturers in Kentucky had increased to fourteen. Skilled workers were paid $2.40 per ten-hour day; unskilled workers received $1.25. Construction and repair of steam engines and boilers used in steamboats involved more exacting work than that for stationary land-based engines, so riverine industrial centers required skilled workers and acquired new machine tools as they became available. River cities, consequently, became important centers for the manufacture and repair of steam engines and boilers. The steam engine manufacturing business became a pioneer industry, in the sense that early firms established the initial nucleus of skilled labor and manufacturing facilities. When the mills and distilleries in Louisville's rural hinterland began to adopt small stationary steam engines, the pioneer factories were already in place and expanded to produce not just steam engines but also a range of generic and proprietary machinery and hardware.[36]

Steam engines required fuel, of course, and the early engines ran primarily on wood.[37] Dry white oak, for example, produced about 6,750 BTUs per pound of wood. But the logistics of ensuring a reliable wood supply were awkward at best. Steam-powered distilleries employed woodcutters or purchased cordwood from area farmers, who cut trees during the winter months to clear their land. Periodic wood delivery required that teamsters contend with difficult roads during the fall through spring distilling season, a contingency that likely prompted attentive distillers to stockpile firewood when they could.

Conversion from wood to coal became possible when Ohio Valley mines opened and shipped coal downriver via barges. Bituminous coal produced roughly 12,000 BTUs per pound, or about twice the heat value of dry, high-quality wood. Overland coal hauling was simplified and transport costs were lowered when rail companies extended their lines from river city ports into the countryside. In the 1880s distillers along the Ohio River fueled their steam engines with bituminous coal from southwestern Pennsylvania, whereas distillers with access to interior railroads used Kentucky coal.[38] Rail lines released millers and distillers from the geographic imperative of stream-generated waterpower and gave them the enviable option of vacating their isolated rural waterpower sites and moving to towns, where turnpikes and railroads offered ready access to affordable coal and grain as well as market connections.[39]

As an alternative fuel source, some distillers purchased well-dried cob corn and stored it in slat-sided cribs before shelling and milling. When thoroughly dry and shelled of kernels, corncobs were a good source of heat energy, producing between 6,700 and 7,400 BTUs per pound, or approximately 56 to 62 percent of the energy derived from an equal volume of coal and slightly more heat value than white oak. In 1869 James Stone, at the Elkhorn Distillery in Scott County, mixed corncobs with coal and may have burned cobs exclusively for a time if good-quality coal was unavailable.[40]

Distillers likely made a gradual transition to steam power, as suggested by the August 1870 US marshal's auction of the John R. Lair Distillery at Lair Station, south of Cynthiana in Harrison County. The distillery stood on a low bluff adjacent to the Licking River near a railroad station. The property inventory included the following: "1 Frame Distillery Building about 120 by 25 feet; 1 Frame Warehouse Building about 50 by 25 feet; 1 Portable Engine, complete, about 20 Horse Power; 1 Corn Mill; 1 Corn Sheller; 2 Mash Kettles; 4 Large Copper Stills; 58 Fermenting Tubs; 5 Beer Heaters; 1 Beer Pump; 1 Water Pump; Shafting; Belting; Hose; Pipe & etc."[41] That Lair's steam engine was portable suggests that it had been retrofitted into an existing distillery. The miscellaneous collection of pumps, shafts, belting, and pipe suggests that waterpower equipment had been replaced, that the conversion to steam was not complete, or that Lair had purchased too much equipment.

Not all distillers were early adopters of steam technology. When the Scott County commissioner announced the sale of the Shawhan & Atkins Distillery in August 1870, the notice described the large, year-old (1869) distillery as having a "capacity of Four Hundred Bushel per day. In addition to the latest Improvements and Conveniences, it has complete Water Power to Run all the Machinery."[42] These "latest Improvements" did not include a steam engine, and powering a new distillery with water seems puzzling. Perhaps the economies of location outweighed the impulse to modernize. Shawhan and Atkins built their distillery on North Elkhorn Creek at Lemmon's (also spelled Lemons) Mill Pike, the site of a mill, a mill dam, and a bridge. Roads to mills were often well traveled and of higher quality than farm roads. Moreover, had the distillery been powered by a coal-fired steam engine, the nearest railroad line on which coal might have been delivered was the Cincinnati & Southern Railroad, four miles away in Georgetown. The Frankfort & Cincinnati Railroad between Georgetown and Paris, known as the "Bourbon Railroad," would run three miles to the north, but it would not be completed until 1890. Shawhan and Atkins would have encountered similar logistical difficulties had they wished to fire an engine with wood. By 1870, there was little woodland on Inner Bluegrass farms that owners were willing to cut down and feed to a steam engine, thereby obliging distillers to haul wood considerable distances.[43]

Symbiotic Industries

Factories in the Manufacturing Belt of the 1880s and later operated according to self-discovered economic efficiencies that produced distinctive connections between raw materials, manufacturing, and finished products and their respective locational patterns. Agriculture- and forestry-related industries were the core of the American Industrial Revolution and demonstrated a fundamental geography of symbiotic relationships—the tendency for dissimilar industries to locate near one another for mutual advantage.[44] Cooperage factories, for example, were located at or near distilleries in part because it was more expensive to ship finished barrels than to ship the raw lumber or staves. Flour mills and distilleries were also symbiotic industries; one used a product produced by the other. By-products also attracted associated industries. Cattle and hogs consumed the slop produced by distilleries, leading to another geographic linkage: slaughterhouses and meatpacking plants preferred to operate in towns and cities where they had ready access to livestock produced in the countryside or at urban distilleries, and most slaughterhouses adjoined a stockyard. By opportune coincidence, urban slaughterhouses and packing plants were also proximal to customer markets, an advantage to the sale of freshly processed meat, which was more perishable than the live animal. That relationship changed with the introduction of the railroad refrigeration car in about 1870. Business propinquity extended to tanneries, which were traditionally located close to slaughterhouses to reduce the costs of shipping low-value unprocessed hides long distances.

Third Generation: Locational Refinement, Competition, and Attrition, 1880s–1920

By 1894, Kentucky's distilling capacity had become concentrated in three areas that generally corresponded to the three different modes of distilling that had emerged (see the map "Kentucky's Distilleries, 1894" in chapter 2). Small-scale rural distillers operated in an arc of Inner and Outer Bluegrass counties from Pendleton and Harrison in the northeast to Madison and Marion in the south. Large-scale rural distillers concentrated in the traditional centers of Anderson and Nelson Counties. And large-scale urban distilleries operated along the Ohio River in Kenton, Jefferson, and Daviess Counties and on the Kentucky River in Franklin County.

The end of the 1890s marked the end of the turnpike–toll road era. Responding to vehement public protests, turnpike companies closed their tollhouses and sold their pikes to county and city governments to be operated as public roads.[45] Free roads reduced transport costs for some rural distillers and grain farmers, but steam power and the rapid construction of overland rail lines and riverine navigational improvements ceded the advantage to those places with rail stations and river port connections.

Railroads became more efficient during the 1880s, consolidating lines and adding spurs that greatly extended access avenues. The American railroad network was national by 1871, with more than 60,000 miles of track; by 1890, the interlaced rail net had accrued more than 167,000 miles.[46] Some distillers moved their works trackside. Railroad companies accommodated new large-capacity distilleries with dedicated switches and sidings. And the spreading rail net permitted Kentucky distillers to market their product nationally and internationally, with seaport connections through New York, Philadelphia, and Baltimore.

Industrial-scale whiskey production was achieved so rapidly that by the late 1880s, industry leaders believed that small producers wishing to open new distilleries could no longer compete with large established works. In testimony before the House of Representatives' Committee on Manufacturers in 1888, John Atherton, a Larue County distiller, noted that Kentucky had only about 150 small distillers who mashed up to sixty bushels of grain a day and produced three or fewer barrels of whiskey. Small distillers continued to operate in conjunction with farms and mills, and distilling was not their sole business. They sold their product locally, often at the distillery itself, or regionally. Larger distillers who mashed 100 bushels per day or more were usually professional operatives with no other occupation. They obtained their grain from regional markets and marketed their whiskey to national and international markets. Atherton drew a geographic distinction between traditional rural distillers and what he called the "Ohio River distilleries" that manufactured whiskey in high volume for sale to rectifiers, who added flavorings and sold the unaged product at low prices. He claimed that the Ohio River distilleries—by which he apparently meant those operating in northern Kentucky and southwestern Ohio—were not engaged in making "fine Kentucky whiskey." He thereby informed legislators that commercial distilling was actually two distinct industries: one group of traditional Kentucky distillers who made "fine whiskey," and a second group concerned with large-volume production, expedience, and immediate sales at low prices. Kentucky distillers defined "fine whiskey" as that "made for ripening" or aging.[47]

W. H. Thomas, a wholesale whiskey dealer and former distiller from Louisville, saw the relationship between small traditional distillers and industrial-scale distilling quite differently. Small distillers were a competitive threat, he thought, and the federal internal revenue tax should be maintained and vigorously enforced to reduce their numbers, thereby allowing industrial distillers to avoid unnecessary competition. In parsimonious testimony before the same House Committee on Manufacturers that Atherton addressed, Thomas complained that small traditional distilleries would benefit unfairly if the internal revenue tax of 90 cents per gallon of whiskey were repealed, and the resulting competition would ruin every Kentuckian operating a large distillery. Large distillers invested a great deal of money in industrial distilling and, as he saw it, "if there was no tax upon the product there would be distilleries upon every farm in the United States that had a stream running through it." Small-scale whiskey production required little capital, and "any farmer in my State of Kentucky could make whiskey on a capital [investment] of $100." Substantial overproduction and falling prices would no doubt result, according to Thomas.[48]

Rectified Whiskey

Distillers of fine spirits were confounded not just by large-volume producers but also by a cadre of secondary liquor manufacturers, or rectifiers, who compounded or blended whiskeys and cut them with exotic ingredients to produce a cheap but profitable product for uncritical mass consumption.[49] Rectified whiskey was made from an undistinguished, mass-produced spirit manufactured by large distilleries such as those in St. Louis, Chicago, Cincinnati, and Peoria. A few Kentucky distillers produced bourbon, rye whiskey, and alcohol spirits for sale to large rectifying distilleries in northern Kentucky at Covington. By the 1880s, rectified whiskey accounted for two-thirds to three-fourths of all spirits made in the country. Distillers used a high ratio of corn to rye and malted barley in making these whiskeys, as corn was cheaper than rye. Rectifiers then blended this product with flavorings and perhaps some aged whiskeys made in Kentucky and elsewhere. Rectified whiskeys usually contained "essential oils" and other additives such as sugar syrup sweeteners and herbs and spices. Iodine, caramel, or burnt sugar was often added to give color to clear alcohol spirits, and prune juice softened young whiskeys, giving them "the appearance of older goods."[50] Rectified whiskeys were not intended to be aged.

To aid rectifiers in creating specific whiskey flavors, compounder Charles Cross offered for sale a general "bourbon extract" and "essence," as well as five other bourbon essences that he identified as "Cynthiana," "Harrison County," "Kentucky," "Paris," and "Sour Mash."[51] A rectifier might make forty-five gallons of "bourbon whiskey" from cheap alcohol spirits by adding one ounce of bead oil (a soap made with castor oil), "bourbon essence," and six ounces of brown sugar.[52] Bourbon essence might be formulated in various ways, but it often included rectified fusel oil (primarily amyl alcohol to provide flavor complexity), amyl acetate (for a scent of apples), pelargonic (a rancid scent), orris (a perfume derived from iris plants), and acetic ether (a fruit-like scent).[53] During Atherton's testimony before the House of Representatives, Henry Smith of Wisconsin asked him whether the cheap rectified whiskeys made principally by large-scale Middle West manufacturers had any redeeming qualities. Atherton responded in censorious fashion: "I have not been in the habit of drinking cheap whisky and I could not tell you that. You may know more about it than I do."[54]

Fine Whiskey

Rectified whiskey producers had little direct business interaction with the distillers of fine whiskeys produced in Kentucky, Maryland, Pennsylvania, and Tennessee. The product that distillers referred to as "fine" was commonly known as "Kentucky whiskey," but it was also called "bourbon whiskey," "family whiskey," or "table whiskey." Most of Kentucky's industrial-scale whiskey distillers aged their product for several years in charred white oak barrels, a process that produced a mellow, caramel-flavored red whiskey that was ready for drinking. And because fine whiskey was not cut with additives or blended with other whiskeys, it was also referred to as "straight whiskey." This type of whiskey was usually sold under a brand name and priced according to its age. Fine whiskey distillers reckoned the age of their product in terms of summers. A whiskey that had been in a barrel for three summers was said to be three years old and was ready for drinking. Most fine whiskeys, though, remained in barrel storage for four to seven years, improving in quality each year.[55] In 1908 the *Frankfort Weekly News and Roundabout* reported that straight Kentucky bourbon whiskey was in high demand because "the United States Government has begun to enforce strictly the proper branding of liquors under the last pure food law. This law requires all liquor to be labeled according to what it really is, and the rectifiers and those who have been adulterating whiskey have been shoved into the background."[56]

Surging Production and Repercussions

By the 1880s, many distillers had converted their operations from water to steam power, thereby increasing the size of their operations and producing whiskey more efficiently and more economically. The volume of spirits produced and stored in aging warehouses increased dramatically. Some traditional rural distilleries radically altered their landscape imprint by adopting steam power and expanding their old still houses or building new ones. Distillers also put several large-capacity works into operation in the 1870s and 1880s, financed in part by investors seeking to supply the growing national and foreign markets.[57] In 1879 Congress passed the Carlisle Revenue Act, which amended the internal revenue tax code by extending from one year to three years the amount of time distillers could store whiskey in bonded warehouses before they had to pay production taxes.[58] In 1890 newly distilled whiskey sold for 35 to 50 cents per gallon, whereas whiskey aged for at least three years sold for five times that price. An unintended but salutary effect of delaying tax payments was that storing whiskey became as lucrative as making it.

In 1871 US warehouses contained 6.7 million gallons of spirits, with aging whiskey accounting for a large proportion of the total amount in storage. By 1879, the number of gallons in storage had nearly tripled to 19.2 million; it tripled again in 1881, to 67.4 million gallons. The following year, the volume of spirits in storage increased by more than 20 percent, cresting at 84.6 million gallons in 1882. Volume subsequently decreased until the 1890s, when it surged again, reaching 152.1 million gallons in 1896. Kentucky's share of national spirits production ranged from about 50 percent to more than 70 percent between 1878 and 1900. In 1878 Kentucky distillers stored half of the nation's spirits, or 7.5 million gallons. By 1882, Kentucky's stored spirits had increased eightfold, to 59.6 million gallons; by 1893, Kentucky had 87.3 million gallons in storage—nearly the same amount stored in the entire nation in 1890. The American distilling industry had successfully marketed whiskey abroad, but foreign market demand began to fall off in the 1880s, leaving distillers with the capacity to produce four times what the home market could consume.

Curtailing production was not a simple matter for a single distillery, never mind an

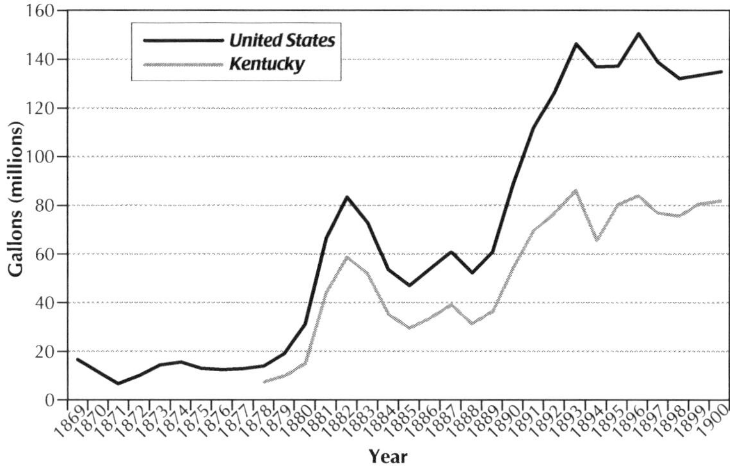

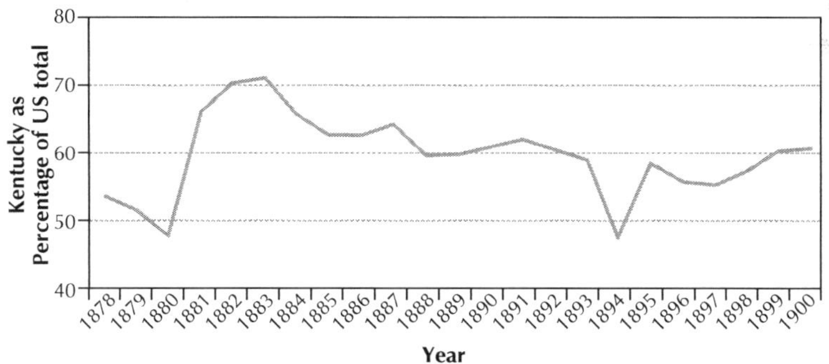

Distilled spirits in warehouse storage in the United States and Kentucky, 1869–1900. Spirits include aged and rectified whiskeys, rum, gin, brandy, etc. Note that prior to 1878, the annual reports of the internal revenue commissioner but did not list amounts stored in warehouses by state. (Compiled from Annual Report, Secretary of the Treasury on the State of the Finances for 1880, 1889; Annual Report of the Commissioner of Internal Revenue, 1879, 1883, 1892, 1898, 1900, 1901)

entire industry. Many distillers held contracts with farmers and marketers for grain deliveries; disrupting production threatened a cascade of other financial discomfitures for coopers, copper still fabricators, iron hoop and stave producers, teamsters, and railroads. Most distillers fed slop to cattle or hogs, and slowing or stopping production required thinning or dispersing these herds. For some distillers, the best solution was to organize into "pools" whose members agreed to voluntarily restrict production, a strategy that proved difficult to administer and enforce.[59] But perhaps the most extensive modifications to the distilling landscape started in 1887 with the formation of the Whiskey Trust or, as it was later known, the Distillers' and Cattle-Feeders' Trust. Fashioned after the Standard Oil

Trust and the Sugar Trust, it appointed a board of trustees to coordinate operations, and distillers that joined the trust relinquished control of their distilleries to the board.[60] Over the following decade, the trust acquired eighty-six distilleries nationwide through voluntary membership or outright purchase. By 1899, the Kentucky Distillers and Warehouse Company, a subsidiary of the trust operated by C. H. Stoll, had consolidated fifty-seven Kentucky distilleries, about 90 percent of those in operation, and closed many of them. The works that closed were either salvaged or left to disintegrate.[61]

Power Source: Electricity Supplements Steam

Specific information about the distilling industry's conversion from water and steam power to electrical power is scanty at best, but national- and state-level reports provide a general view of how the changeover progressed. In the nineteenth century's last three decades, Kentucky's small- and large-scale industrial plants, including distilleries, increased their use of steam power fivefold—from 31,928 horsepower in 1870 to 162,829 horsepower in 1905.[62] Initially, manufacturers used steam engines as prime movers to turn shafts and operate pulleys directly, but by the 1870s, steam engines were increasingly used to direct power via compressed air or water pressure or to generate electricity.[63] Industries continued to use steam for heating, but as the technology of electricity matured, they also deployed steam engines to run dynamos to produce direct current or alternators to produce alternating current.

Kentucky's output in electric motor production increased from 100 horsepower in 1890 to 3,415 in 1900 and to 10,690 in 1905.[64] In 1905, 784 US businesses manufactured electrical equipment, including 92 in Ohio and 34 in Indiana. Only three electrical equipment factories operated in Kentucky in 1905. Two steam engines and sixty-seven electric motors powered their operations, and 133 employees produced products valued at about $170,000.[65]

An important innovation, the steam-powered central electrical generating station, allowed cities to direct electrical power through transmission lines to industrial, commercial, and domestic customers. The Westinghouse Company installed Kentucky's first commercial alternating-current power transmission system at Louisville in 1892, only two years after the first such system in America became operational in Portland, Oregon.[66] Central generating stations gave industries the flexibility to erect manufacturing plants without concern for the logistics of fuel delivery. Industries served by new power transmission lines had much greater latitude in plant site selection; locations in proximity to raw materials, markets, large urban labor pools, other industries that supplied parts, and transport hubs, as well as building sites on cheap land, were now viable options.

Urban Distilleries

During the nineteenth century, five of Kentucky's largest cities became distilling centers. Distilleries operated in several Kentucky cities prior to the Civil War. Thereafter, railroads permitted distilleries to go much further afield to obtain grain and cooperage lumber and to extend their markets across the nation.

Urban Population Growth, 1800–1900

Year	Population				
	Louisville	Covington	Lexington	Owensboro	Frankfort
1800	359	n/a	1,795	n/a	628
1810	1,357	n/a	4,326	n/a	1,099
1820	4,012	n/a	5,270	n/a	1,679
1830	10,341	743	6,026	229	1,682
1840	21,210	2,026	6,997	n/a	1,917
1850	43,194	9,408	8,159	1,215	3,308
1860	68,033	16,471	9,321	2,308	3,702
1870	100,753	24,505	14,801	3,437	5,396
1880	123,755	29,720	16,656	6,231	6,958
1890	161,129	37,321	21,567	9,837	7,892
1900	204,731	42,938	26,369	13,189	9,487

Source: US Census of Population.
n/a, not available.

Lexington was the state's largest city in 1800. In 1815 the *Niles Weekly Register* opined that Lexington "promises to be the great inland city of the western world," noting that it was "situated in the centre of an extensive plain of the richest land," referring to the fertile and coveted Inner Bluegrass country, which attracted early land speculators and settlers.[67] Lexington became the commercial center for surrounding farms, processing their hemp in rope works and bagging factories, milling their grain, and processing their wool. Merchants stocked their stores with goods imported from East Coast cities via the Ohio River and the Maysville Road, Kentucky's frontier highway. The Middle Fork of the Elkhorn, or Town Branch, as it was later known, became the city's industrial center and the site of early distilleries. As Ohio River steamboats supplemented and then displaced flatboats, commercial development in Kentucky's river port cities escalated, stimulating rapid population growth.

Louisville's population exceeded Lexington's in 1830; by 1850, Covington's population had also surpassed Lexington's. For the remainder of the nineteenth century, Louisville added at least 20,000 to 30,000 people each decade. Between 1880 and 1890, Louisville's population increased some 38,000, nearly twice Lexington's total population in 1890. Owensboro's nineteenth-century growth was less dramatic, but its population nearly doubled between 1870 and 1880; by 1890, Owensboro was larger than Lexington had been only three decades earlier. The Kentucky River flows northwest across the Greater Bluegrass country to join the Ohio River at Carrollton. Frankfort's site on the Kentucky River is sixty-five miles upstream from that riverine junction. In addition to being the state capital, Frankfort became a nineteenth-century sawmill center, processing logs floated downriver from eastern Kentucky's mixed mesophytic forest. The city's population growth was stimulated by this river-related commerce, but it also became a major distilling center, processing primarily locally produced grains.

The basic requirements for urban and rural distilleries were similar: convenient access to a fresh water source and grain supply. For rural distillers, access to an unfailing cold-water spring or creek was paramount. The primary locational criterion for large urban distilleries was ready access to a railroad or a river port. Urban distillers were therefore obliged to drill wells or draw water from the Ohio or Kentucky River—hardly an endorsement of the timeworn "pure limestone water" narrative. Both rural and urban distillers fed their spent grain by-product to livestock. But operating cattle and hog pens in or near a city was different from feeding livestock in the countryside, where odors were carried away on a breeze and manure disposal was a straightforward matter of fertilizing adjacent farm fields.

Louisville

The Ohio River's 981-mile course from its head at Pittsburgh to its confluence with the Mississippi River at Cairo, Illinois, follows a deeply incised valley through the western uplands of the Allegheny Plateau past Cincinnati and Louisville to Tell City, Indiana. There, the river emerges from its picturesque vale to flow in great looping meanders across a broad alluvial plain on its way southwest to the Mississippi. The river's natural channel includes shoals, nearshore islands, and deep channel sections, but the most dramatic feature, and its only geologic obstruction, is the Falls of the Ohio. About 606 miles below Pittsburgh, the Ohio cascades across the exceptionally competent Jeffersonville limestone—coral rich and, at 390 million years old, Devonian in age—and drops twenty-six feet over a distance of two miles. The rapids are especially vigorous in low water, which, given the Ohio Valley's precipitation pattern, can persist for nine months out of the year. Boat captains moving downstream might attempt to run the rapids by one of three chutes. Those with better judgment landed above the falls and off-loaded cargo and passengers for portage to another landing below the obstruction. Overland trails, including the western branch of the Wilderness Road, coalesced at the falls; given the imperative of riverine and overland travel, frontier settlements developed at this geographic nexus.[68]

The two-mile rapids were less a focal point than a corridor that invited development along both the south and north banks. From the first years of European settlement in the Ohio Valley, five villages developed near the falls. Pioneers founded Louisville at a site above the falls in 1779; Clarksville materialized on the north bank, below the falls, four years later. Jeffersonville was established above the falls and across the river from Louisville in 1802; Shippingport below the falls and west of Louisville in 1806; and Portland, also below the falls, in 1814. Commercial flatboats carrying Ohio Valley goods to

New Orleans were active from the first years of settlement, prompting the US Congress to declare Louisville a port of entry in 1789.[69] By 1800, Louisville had 359 inhabitants. The first steam-powered boat arrived there in 1814, and by 1819, nearly seventy steamboats operated on the Ohio. As railroads developed in the mid-nineteenth century, they focused at established river landings near the falls, initially linking inland towns to the waterway. By 1882, more than 5,000 miles of railroad track linked the falls corridor to the national rail net.[70] Engineers and laborers completed the two-mile-long Louisville and Portland Canal, bypassing the falls, in 1830.[71] The following year, 406 steamboats passed through the canal. Six years later, in 1837, more than 1,500 boats negotiated the lock, or a fourfold increase in traffic. This count did not include the boats that landed above and below the falls without passing through the canal.[72]

Ohio Valley industrialization was abetted, in part, by the availability of coal to fire steam engines. Beginning in 1814 and for some eight decades thereafter, coal mined in southwestern Pennsylvania fueled Louisville's industrial development. The first coal barges from the Monongahela Valley arrived in 1814. In the mid-1850s steamboat-powered tows delivered coal, supplemented in the 1870s by the railroads.[73] By the mid-nineteenth century, manufacturing and related commercial businesses clustered in the five settlements near the falls, especially at Louisville. Factories processed farm products and prepared salted beef and pork, smoked bacon, lard, flour, tobacco, and hemp rope and bagging for shipment. Maltsters produced malted barley and rye for Kentucky distillers and brewers.[74] Wholesalers' warehouses stored and routed produce and products for shipment via river or rail. By the 1880s, Louisville's manufactured and processed goods were available throughout the South and Middle West, with deliveries extending as far as the new railroad towns springing up on the Great Plains. Kentucky whiskey, whether measured in volume or value, was a major export.

Paradoxically, before the 1860s, little of the whiskey shipped through Louisville's river port was distilled in the city. A few pioneer distillers such as Evan Williams operated works in Louisville and Jefferson County in the late eighteenth century, but their production was small and intermittent. New England investors led by Rhode Island resident James D'Wolf Jr. built the large Hope Distillery near the Ohio River in Portland in 1816. The steam-powered works produced 1,200 gallons of whiskey per day but closed in 1820.[75]

Two firms, Schrodt & Laval Company and N. D. Smith & Company, built distilleries in Louisville in 1857. The Smith Distillery burned in 1858 and was not rebuilt. The Schrodt & Laval Distillery operated near the Ohio River on Second Street, between Washington and Water Streets, approximately where the KFC Yum Center currently stands. The distillery had an operating capacity of 60 barrels per day, or 12,000 barrels during a 200-day operating season, but the steam-powered works made industrial alcohol for lamp fuel and chemical manufacturing, not whiskey.[76]

By the eve of the Civil War, only one small distillery operated in Louisville. After the

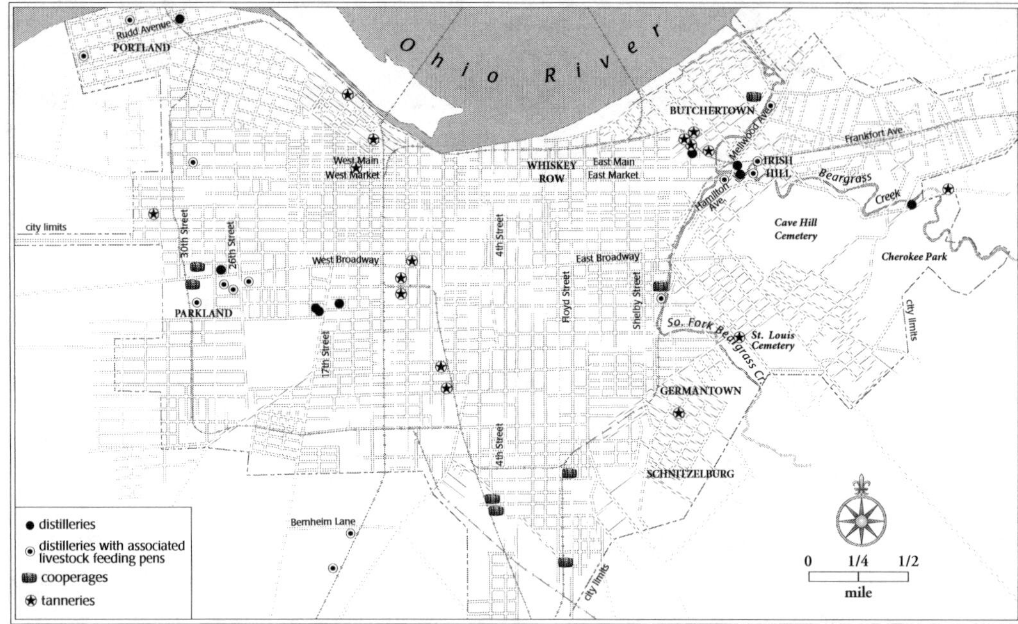

Distilleries, cooperages, and tanneries, Louisville, 1892–1905. (Compiled from *Insurance Maps of Louisville, Kentucky*, vols. 1–3 [New York: Sanborn-Perris Map Company, 1892]; *Insurance Maps of Louisville, Kentucky*, vol. 4 [New York: Sanborn Map Company, 1905]; *Caron's City Directory of the City of Louisville*, vol. 2 [Louisville: Bradley & Gilbert, 1895])

war, urban industrial distilling expanded rapidly in new large-scale plants at sites with railroad connections; direct river access had become a secondary consideration. The new industrial distilleries appeared on the city's periphery, at level sites that fronted creeks or allowed the drilling of deep wells. As the city industrialized after the Civil War, it also suburbanized as factories and residential neighborhoods expanded into the surrounding farmland. From Louisville's original site on the Ohio's south bank, suburban expansion extended up- and downriver and south across the nearly level glacial outwash plain. Property subdivision was laissez-faire, driven primarily by landowners and speculators, although lot sales were often tepid and desultory. The city established a horse-drawn trolley car system in 1865, and by 1889, the network was electrified. Speculators platted small-lot subdivisions along trolley lines to attract working-class residents, many of them Irish and German immigrants, seeking to become homeowners.[77] During this same postwar period, railroad companies extended tracks south of the river, creating rail corridors attached on the south to a circumferential belt line. High-density industrial development took place in the city's old riverfront section as well as along the new railroad corridors.[78]

Louisville's 620 manufacturing establishments in 1860 nearly doubled in number to 1,191 by 1880. Factories turned out durable and nondurable goods: tobacco products; wooden furniture; farm machinery, especially steel plows; iron pipes, boilers, and stoves;

cotton and woolen goods; and glass, brick, and tile products. On the banks of the Ohio, boatyards assembled steamboats.[79] In terms of monetary value of operations and products, pork packing was Louisville's largest industry; whiskey distilling ranked second. From one distillery in 1860 the industry grew to six in 1870; by 1880, thirty-six large-capacity distilleries operated across the city. One works covered twenty-five acres and produced some 8 million gallons of fine whiskey annually.[80] At least five distillers from elsewhere in Kentucky sold their works and relocated to Louisville: John G. Mattingly and George W. Beall moved from Marion County in 1878; Isaac and Bernard Bernheim moved from McCracken County in about 1888; and John G. Roach moved from Green County to Louisville, where he built a new distillery in around 1893.[81] The addition of thirty large distilleries in ten years to Louisville's industrial base, given the city's population of 100,000 in 1870, would have been an admirable achievement if only local and state capital had been available, but capital markets were national or even international, and distillers with access to funds provided by investors in eastern cities and Europe expanded their operations at an astonishing rate.

The new high-capacity industrial distilleries bore limited resemblance to long-established farm and mill distilleries. Louisville's postbellum distilleries were housed in multistory brick buildings and warehouses with dedicated rail sidings. During the 1880s and 1890s, the city's collective grain mashing capacity was roughly 35,500 bushels a day, or more than 100 railroad boxcars of grain. A few distilleries had mashing capacities of less than 400 bushels per day; eleven mashed more than 1,000 bushels per day.[82]

Certain complementary industries functioned as distilleries' input suppliers or output consumers.[83] Farmers supplied corn and rye, and coopers furnished oak barrels. In 1870 fifteen Louisville cooperages employed 129 coopers.[84] Coopers obtained staves from Kentucky and Indiana lumberyards and shipped their barrels to local and state distilleries; therefore, most cooperages were positioned along rail corridors.

Distillers fed their spent grain to livestock. James D'Wolf's distillery at Portland fed slop to cattle in 1816. In the 1880s and 1890s fourteen Louisville distilleries operated cattle feedlots, and others sold or gave slop to area farmers. Meat processors and distillery feedlots were complementary businesses. Butchertown, an east Louisville neighborhood near Beargrass Creek, was a small nineteenth-century slaughterhouse and animal products processing center at the terminus of the Frankfort and Lexington Turnpike. Prior to the development of a regional railroad network, farmers drove their livestock overland to Butchertown for slaughter and processing. The Bourbon Stockyards started there in 1836 as a place to consolidate livestock for feeding and watering prior to sale and slaughter. Fresh meat was perishable, so it had to be either sold promptly at local urban markets or salted or smoked and packed in barrels for storage or shipment elsewhere. Operators usually sited stockyards, slaughterhouses, and packing plants in close proximity. Many of Butchertown's shops were family-operated businesses that could process a limited volume of livestock, but these small operations began to fade during the 1880s and 1890s. Commercial meatpackers built several large processing plants in Louisville, motivated in part by the ease of livestock shipment by rail, the availability of refrigerated railcars to haul processed carcasses, and increased livestock feeding at the city's distilleries.[85]

Just as slaughterhouses were reliant on livestock producers, tanneries were reliant on slaughterhouses and butcher shops to supply hides. Awkward overland transport motivated tanners to locate on good roads, near hide suppliers, and close to ready markets for tanned hides.[86] Six small tanneries operated in Louisville in 1860, and there were eleven in 1870; by 1880, seventeen large industrial tanneries operated in the city.[87] Enabled by steam-powered mechanization, railroad transport, and increased livestock production in the state, the tanning industry flourished, and Louisville became a major leather manufacturing center. The tanneries, in turn, supported boot and shoe, belt, and harness manufacturers. By the 1880s, as the national rail net matured, Louisville tanneries processed more than 150,000 hides annually, many shipped in from as far away as Texas and Colorado.[88]

Nineteenth-century Kentucky tanners followed traditional Old World practices, which required only two raw materials to process animal hides: lime and tannic acid.[89] When crushed limestone was burned in kilns at high temperatures, it yielded calcium oxide, which, when hydrated, produced the caustic lime water tanners used to scrub hair from hides. Tanneries extracted tannic acid from hemlock and oak bark. Upon receiving bark shipments by wagon or rail, tanners stored the bark in covered sheds to dry thoroughly before pulverizing it in a bark mill. The bark powder was then soaked in warm water to extract tannic acid. The traditional tannery was a simple operation: an open tanyard, a rude cluster of shallow wooden tanning vats sunk into the ground, a small mill with bark-grinding stones turned by a horse or powered by steam, and an open shed or beam house for working hides.[90] Nineteenth-century tanneries were noisome places, but they were often surrounded by working-class residential neighborhoods.

By the late 1860s, Louisville's tanning industry had become industrialized and was supplied by the railroad. The tanneries proper were arranged in clusters of large four-story, factory-type brick buildings with vat rooms, curry and finish rooms, bark mill, beam house, hide house, and engine room. Wooden frame buildings contained bark storage and leeching rooms. The industrial tannery closely resembled a brick whiskey aging warehouse, although the tanneries were outfitted with more windows for interior illumination. The Conrad, Ohio Falls, Phoenix, and Shuff tanneries were clustered between Eleventh and Twelfth Streets, adjacent to the Louisville & Nashville Railroad shops, and together they occupied one and a half city blocks.[91] Factory clustering reduced procurement, processing, and distribution costs and made attractive destinations for railroad sidings. But clustering also increased the risk of catastrophic fire. In August 1890 fire destroyed a Kentucky Distilling Company warehouse containing 25,000 barrels of whiskey, the distillery, and cattle sheds, at an estimated loss of $800,000. The adjoining Conrad Seller slaughter-house and pork packing plant was also destroyed, at a loss of about $50,000.[92]

Many of Louisville's industrial distilleries, including J. T. S. Brown & Sons and the House of Weller, opened branch management offices, distilleries, and warehouses on East and West Main Street in the 1870s. That cluster soon became the armature of the city's distilling business and acquired the nickname Whiskey Row. The structures on the north (river) side of the 100 and 200 blocks of West Main Street stood atop the slope leading down to the Ohio River. The Main Street entrances opened into each building's first floor. Washington Street ran parallel to Main one block north and was sufficiently downslope

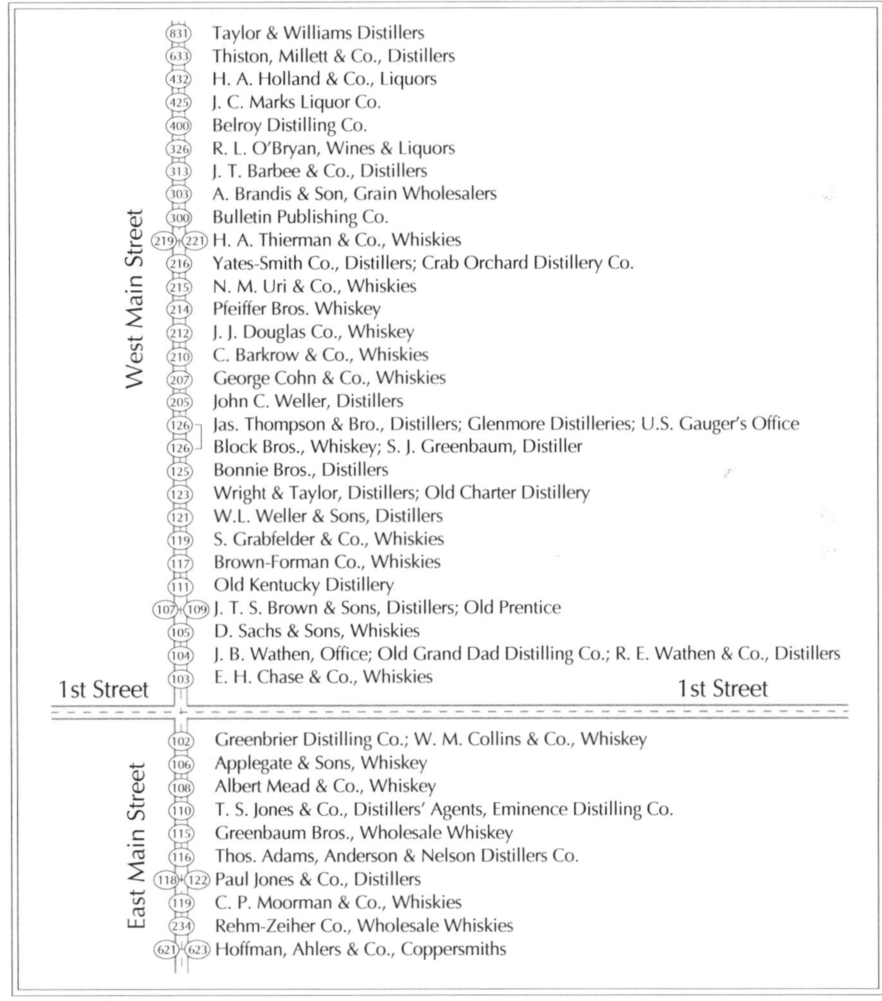

Louisville's Whiskey Row, 1909. Actual distances between addresses are not accurately depicted. (Compiled from *Caron's Annual Directory of the City of Louisville*, vol. 39 [Louisville: Bradley & Gilbert, 1909], 2046–54)

to provide access to basement entrances at the rear. The large Belknap Hardware & Manufacturing Company occupied three contiguous blocks behind the Whiskey Row buildings in the 100 blocks of East and West Main Street. River-frontage rail sidings served the Belknap complex and may have provided transport access to the whiskey wholesalers and warehouses on Main Street as well.[93]

Some rural distillers opened offices and shipping warehouses on Louisville's Main Street or on adjacent streets. By 1909, forty-six distillers and related businesses operated offices, distilleries, or warehouses on the Main Street corridor from the 600 block on the east side to the 800 block on the west. This cluster included related businesses such as Hoffman, Ahlers & Company, a coppersmith that became the Vendome works; A. Brandis & Son, a grain wholesaler and distilling grain supplier; Bulletin Publishing Company, which published *Wine & Spirit Bulletin;* and the gauger's office of the US Treasury's Bureau of Internal Revenue.[94] Several buildings in this district, including grand Romanesque Revival, Beaux Arts, and Chicago-style structures with stone, brick, and cast-iron façades, remain intact today. The contemporary district is undergoing rejuvenation as whiskey-related businesses open new craft distilleries and tasting rooms.

Covington

Covington's site was surveyed and platted in 1815 on an expansive alluvial basin inside the right angle formed by the confluence of the Licking and Ohio Rivers. The Licking's watershed extends from the Ohio's south bank 300 miles southeast to its headwaters in eastern Kentucky's Appalachian Plateau. The Covington basin, directly across the Ohio River from Cincinnati, was the terminus for early buffalo trails and roads, including the Covington-Lexington Turnpike, incorporated in 1834, and the Covington & Lexington Railroad, which was completed to Paris with connections to Lexington in 1854.

Industrial distilling in northern Kentucky, and manufacturing agglomeration more generally, was abetted by the opportune coincidence of navigable rivers, broad floodplains, and highly productive agricultural hinterlands—not by abundant calcium-rich limestone springwater. The Ohio River represents a topographic seam between two different but complementary regions: Kentucky's Bluegrass region to the south and Ohio's Miami Valley to the north. Though physically and culturally distinct, these two regions were both important frontier farming areas; by the 1830s, they both served as incubators for the development of specialized corn and livestock agriculture that spread across the Middle West and came to characterize the Corn Belt.[95] Nineteenth-century farmers in southwestern Ohio and the Kentucky Bluegrass country produced large surpluses of hogs, cattle, and grain, especially corn. Drovers directed livestock herds to Covington and Cincinnati slaughterhouses, and corn moved by wagon and railcar to Covington distilleries.

Four tributaries form a confluence with the Ohio River at or near Covington and Cincinnati. The Great and Little Miami Rivers, the Licking River, and the smaller Mill

Creek flow into the Ohio just east of the great meander curve, where the trunk stream's valley changes course. The Ohio's general direction of flow pivots ninety degrees from northwest to southwest at a point marked by an Ohio town appropriately named North Bend. Mill Creek and the Great and Little Miami Rivers flow out of southwestern Ohio to enter the Ohio River immediately up- and downstream from Cincinnati. The Licking River flows into the Ohio opposite present-day Cincinnati's business district. In cross-section width, each tributary stream's valley differs dramatically from the Ohio Valley.[96] Just downstream from Cincinnati, the Ohio Valley—bank to bank—is about 4,500 feet wide. The valley of the Great Miami west of Cincinnati is about 5,000 feet wide near its confluence with the Ohio. Upstream from Cincinnati, the Little Miami's valley is 8,500 feet wide near the Ohio. Mill Creek is at present a small stream running in a concrete conduit for much of its lower reach before trickling into the Ohio. Its valley is outsized and about two miles wide. For the last mile or more above its confluence with the Ohio, the Licking's valley is 9,000 feet wide—and it is much wider than that where the bluffs recede to accommodate the riverside towns of Covington, Newport, Bellevue, and Dayton. Six miles upstream at DeCoursey, the Licking's valley is still 4,500 feet wide, the same width as the Ohio River Valley.

The unusual stream configuration at Covington was brought about by Pleistocene glaciation when the northward flow of the Old Licking and Old Kentucky Rivers was blocked during the Illinoian glacial advance about 300,000 years ago, and the Ohio drainage was reconfigured during the Wisconsin glacial epoch, about 15,000 years ago.[97] Vast quantities of glacial meltwater runoff widened old river channels and deposited significant volumes of alluvial material—silt, sand, and gravel—in their beds. At Covington, the depth of alluvial deposits reaches 150 feet. The present-day Licking and Ohio River channels occupy only a small part of the old alluvial surface at Covington, allowing expansive urban development across the valley floor.[98]

The rivers' generous floodplains offered square miles of nearly level land, and the river nexus was a favored place for urban and industrial development. Steel mills, stockyards, slaughterhouses, packing plants, tanneries, flour mills, lumber and coal yards, and industrial distilleries all required large expanses of level land and were avid consumers of the Ohio and Licking floodplain flats, a rare resource in the otherwise deeply stream-dissected country near the Ohio River. These industries also required access to Ohio and Licking River shipping.[99] Nineteen distilleries operated in the Covington area during the nineteenth century; all but two were established after the Civil War and operated as high-volume industrial-scale works with connections to valley railroad lines and riverfront docks.[100]

Although transport access and level land were the primary attractions for distillery development at the confluence of the Licking and Ohio Rivers, fresh water could also be obtained in quantity from both surface and ground sources at this node, although nineteenth-century water quality was problematic at best. Water for industrial and residential use could be drawn from different sources at Covington: from the Licking and Ohio Rivers, from wells sunk into the alluvial river flats, and from deep bedrock. During the second half of the nineteenth century, when distilling flourished in the area, river water contained silt and a heavy pollution load and had to be treated before use. The Pleistocene alluvial material was a valuable aquifer with moderate to large amounts of fair- to excellent-quality water. Water pumped from wells drilled into the Ordovician limestone and shale bedrock below the city was meager in quantity and generally too salty and sulfurous for use. The water obtained from several area wells had an iron content too high for use in beverage manufacturing or food processing.[101]

The Cincinnati firm Walsh, Brooks & Kellogg operated a rectifying distillery and built a second rectifying-redistilling operation beside the Roebling Bridge on the Ohio River in Covington in about 1859. Irish immigrant James Walsh and other business partners reorganized the company and renamed it the James Walsh Company. The Covington distillery drew water from six bored wells and processed it at a private waterworks.[102] A second rectifying distillery, the New England Distillery Company, operated a small works in a crowded industrial area near the Covington railroad station and the Kentucky Central Railroad freight warehouse. A lumberyard, livery stables, and carriage factory operated nearby. The New England Distillery used water from two wells for cooling and condensing.[103] The Old '76 Distillery on the Licking River south of the Ohio River drilled a 1,200-foot well to the St. Peter sandstone formation; the water obtained was characterized as salty and sulfurous, a variation on the "Blue Lick" water used for cooling, but it was so corrosive to pipes that the well was abandoned.[104] "Blue Lick" is the term applied to the sulfurous, briny waters that occurred in northern Kentucky. The "blue" mineral water found at Lower Blue Licks on the central Licking River west of Flemingsburg was thought to have medicinal qualities and was the primary attraction at the nineteenth-century resort that grew up at that site.

In the 1880s at least three distilleries operated along the Licking River as part of an intensive rail and river industrial development that extended south of Covington for more than three miles.[105] Northernmost in this alignment was the Dorsel & Wulftange works, which produced bourbon and rye whiskey at Rickey Street in the shadow of a large steel rolling mill. The distillery used city water and received whole ear corn that was stored in a crib and shelled onsite. Dorsel & Wulftange did not operate a livestock feeding lot; slop was collected in a large tub that stood beside the street, suggesting that the by-product was sold, given away, or hauled to the Licking River and dumped.

The B. K. Reynolds or Kenton County Distillery operated beside the Licking River about one and a half miles south of the courthouse. Despite having no direct railroad connection, this was a high-production plant; the distillery mashed 800 bushels per day, and the works included a corncrib and a sheller. Cattle and pigs were fed on slop piped to large pens—effectively, livestock latrines—beside the Licking River. Runoff from the feedlot most likely flowed directly downslope and into the river.

The Willow Run or Milldale Distillery operated about three miles south of the Covington courthouse. A coal-fired steam plant powered the works, which mashed about 400 bushels of grain per day. Willow Run had access to railroad tracks and stored grain shipments in an onsite elevator. Seven small warehouses stored the barreled whiskey. The Willow Run works used well water in its operation. Given the heavy pollution load carried by the Licking and Ohio Rivers from industrial and residential contamination, nineteenth-century distilleries were reluctant to sully their product by using river water.[106]

LEXINGTON

Lexington and Fayette County are positioned near the center of the Inner Bluegrass Ordovician limestone plain. The land surface is 900 to 1,000 feet above sea level, and surface water drains primarily northwest toward the Kentucky River by way of the Elkhorn Creek watershed's North, Middle, and South Forks. Precipitation runoff also drains by way of underground channels whose direction and volume of flow are not yet fully understood, but this process is evident in hillside springs and valley swallet holes that capture surface streams and direct them underground. In the early 1790s Lexington's surveyors drew up a street plat on which Main Street paralleled Middle Fork, a tributary of South Elkhorn Creek that was later renamed Town Branch. Some perennial springs, such as McConnell Springs, appeared on the surface and flowed aboveground for a short distance before disappearing into a swallet hole. Other springs flowed into Town Branch. In dry weather the branch was sluggish and meandering, but with heavy rains it became a torrent. Despite flow variability and the potential hazard of flooding, several industrial works—tanyards and water-powered mills of various types—operated along or near Town Branch by 1802. A gunpowder mill was likely in operation at McConnell Springs near the branch by 1813.[107] Shallow wells and water pumped through underground log pipes supplied water to distilleries. By 1810, the city's 4,300 residents obtained their domestic water largely from wells.[108]

Some suspected that Lexington's cholera epidemics of 1833 and 1849 were caused by contaminated drinking water, and after another outbreak in 1873, most shallow wells were abandoned and water usage from public wells was discontinued, except for industrial purposes. People built concrete cisterns to collect rainwater, but precipitation was seasonal; it was sufficient from December to July but often scarce in late summer and fall.[109] Dependence on rainfall left Lexingtonians and their businesses and industries susceptible to water shortages. Although the cholera epidemics had undermined public confidence in the quality of water drawn from shallow wells, many continued to believe that the area's springwater remained pure.[110]

Robert Peter, a physician and a professor of chemistry at Transylvania University, was concerned about Lexington's water quality, and he reported on the water drawn from two wells bored through limestone rock. In 1852 a druggist drilled a well to a depth of forty feet at Cheapside on the courthouse square. The well yielded water that contained "hard masses of iron pyrites, and the water was such a strong chalybeate, sparkling with carbonic acid gas, depositing oxide of iron on exposure to the air, and containing quite a variety of saline ingredients, that it soon became quite popular as mineral water." Upon exposure to the air, though, the water's odor became offensive. Near the racetrack, about one mile east of the courthouse, another well was bored ninety feet through limestone; that water contained carbonate of soda, nitrates and other salts, and fatty organic matter that smelled like soap. Dr. Peter deemed the water to be impure and found that every well bored in the Town Branch watershed contained nitrate salts and hydrogen sulfide. The source of the groundwater's contamination, he concluded, was drainage from backyard privies, bathhouses, and laundries.[111]

As the basis for a land claim, Kentucky pioneer settler William McConnell built a station between Old Frankfort Pike and Town Branch near the site of a large "sinking spring," later named McConnell Springs. The spring was near the Headley Distillery, also known as the Henry Clay Distillery, built in 1858 by the Headley & Farra Company.[112] The distillery burned in about 1873, and the Blue Grass Pork House packing plant operated at the site from 1875 to 1879. The James E. Pepper Company purchased the packing plant and in 1880 erected a new brick and stone industrial-scale distillery with corncribs and shed space for 500 feeder cattle. Wedged between Town Branch and Manchester Street, the forty-eight-acre site had access to creek water, the road network, and the railroad. Powered by two steam engines, the distillery consumed 250 bushels of coal and mashed some 500 bushels of grain per day. The Pepper Company elected not to use water from Town Branch or from a bored well on the distillery site; instead, it contracted with John and Laura Wilson, the owners of Wilson's Spring (also known as McConnell Springs), to pipe springwater to the distillery to support operations. Thereafter, an iron pipe carried Wilson's Spring water approximately 3,000 feet to the Pepper Distillery.

In about 1866 Turner, Clay & Company established the Ashland Distillery some 2,500 feet east of the Henry Clay–James Pepper Distillery. The eleven-acre site was on Manchester Street, near Cox Street, Town Branch, and the railroad tracks; it had once been a lead works and thereafter a pork packing plant.[113] William Tarr & Company purchased the distillery in 1871 and operated it until it burned in 1879, whereupon the company rebuilt the distillery on the same site. The distillery's three steam engines burned 160 bushels of coal per day, likely delivered via the Louisville, Cincinnati & Lexington Railroad. The distillery mashed between 420 and 1,000 bushels of corn, rye, and malted barley per day and fed the slop, supplemented with hay, to cattle tethered in a large pen (measuring one-third of an acre) beside Town Branch. The works fed about 500 cattle to market weight each year. Two warehouses beside Manchester Street stored the aging whiskey. A cooper shop operated just to the north (downstream), near the Red River Lumber Yard and a railroad siding.

Rather than draw water from Town Branch or a bored well, the company leased springwater from the nearby Ater Spring, using some 200,000 gallons per day.[114]

The managers at neither the Pepper nor the Tarr distillery chose to use Town Branch water. One must conclude that this decision was based on what they knew and what they believed.[115] The distillers likely knew that Town Branch's flow was seasonally variable and therefore unreliable—although the highest seasonal flow would have occurred during much of the distilling season from December to May. Or they may have recognized that Town Branch was heavily polluted—not only by surface runoff from city streets but also by the city's garbage and effluent from privies, livery stables, tanyards, and other industries. Or they may have read Dr. Peter's reports on Lexington's well water.[116]

Four large distilleries operated in rural Fayette County during the second half of the nineteenth century.[117] One began as a five-story brick and stone cotton spinning and weaving mill purchased by Stoll, Hamilton & Company in 1880. The company converted the steam-powered mill into a distillery that stood adjacent to Sandersville Road, three miles north of Lexington; it was located near two springs, Boiling Spring and Hillenmeyer Spring, whose water quality was deemed excellent.[118] A 40,000-bushel crib supplied 300 bushels of corn and 150 bushels of rye and malted barley for each day's mashing. An extended trough delivered slop directly to cattle pens situated downstream from the distillery. The works also fed 1,000 hogs. The Cincinnati & Southern Railroad track was about 2,000 feet distant, and four wagon teams moved grain from track to distillery and whiskey from distillery to warehouse.[119]

FRANKFORT

At Frankfort, in Franklin County, the irregular splice between the Middle Ordovician limestones of the Inner Bluegrass and the Upper Ordovician limestones and shales of the Eden Shale Hills is on exhibit along the Kentucky River. South of Frankfort, the river meanders through a narrow, palisaded gorge. The gorge widens to accommodate elevated river terraces in only a few places, such as at Tyrone in Anderson County east of Lawrenceburg. Tributary creeks have dissected steep valleys into the adjoining uplands, giving the river corridor a rugged, ribbed character; it is difficult to cross, even at points where tributary valleys on the east correspond with valleys across the river on the west side. Near Frankfort, the Kentucky River exits the gorge and begins a transition into the Eden Shale Hills. The river flows across the competent Lexington limestone there but readily erodes laterally into the compliant shale, widening the valley.[120] The hillsides rake back at low angles and the valley widens, providing ready access to the river along several tributary creek valleys.

The river has changed course here over the millennia, leaving behind cutoff meander loop terraces with nearly level floors. One of these loops is horseshoe shaped, with its open end attached to the Kentucky River on the southwest. The meander's northwest arm is the site of Frankfort's first settlement at Leestown and the point where a buffalo track and Indian trail, called Alanant-O-Wamiowee, crossed the river. The meander's southeast arm was the site of the first capitol, early residential development, and a small business district.

Before engineers completed river improvements and dam construction on the lower river at and below Frankfort, numerous shoals and rapids hampered river navigation; even small craft had to be carried or pulled through some sections during low water.[121] Construction crews completed the first five dams between the river's confluence with the Ohio at Carrollton and Lawrenceburg by 1842. Thereafter, steamboats could readily traverse the slack-water pools between dams in most seasons of the year.

During the nineteenth century, people and freight wagons moving overland toward Frankfort from the east could gain access to the river by following a broad, tapering ridge down to river level. This was the route likely followed by English traveler Fortesque Cuming in 1807, and it is similar to that followed in the 1880s by the Frankfort and Georgetown Turnpike and, today, by US Highways 60, 421, and 460.[122] This was the most direct route west from the Forks of Elkhorn, where the North and South Forks of Elkhorn Creek joined about three and a half miles from Frankfort. The creeks flowed northwest from the Lexington area, and during the nineteenth century, they were intensively developed by millers.[123] Approaching from the west, wagons could descend the low-gradient valley cut through the hills by Big Benson Creek to its confluence with the Kentucky River. Railroads used the Big Benson Creek valley as a route west toward Louisville. The Lexington & Ohio Railroad gave Frankfort its first rail connection to Lexington in 1830, but engineers and gandy dancers did not complete a reliable, high-capacity railroad connection to Louisville and the Ohio River until 1851.

Small distilleries operated in the county by the early nineteenth century. A distillery site adjacent to the river at Leestown was developed by a series of owner-distillers. Edmond H. Taylor Jr. modernized the distillery in 1869 and operated it for a time as the Old Fire Copper (O.F.C.) Distillery. Taylor added the Carlisle Distillery on adjacent land in about 1880. When the O.F.C. Distillery burned in 1882, Taylor had it rebuilt to resemble a two-story brick and stone Victorian mansion with a three-story tower and Greek Revival and Romanesque architectural details. Beer from copper-lined fermenting vats traveled by copper pipe to a copper beer well, where it was heated before arriving at the copper stills and doublers (see the engraving in chapter 3).

Taylor regarded Franklin County as a choice distilling site because its "waters, climate, and special facilities have long caused it to be known as the almost exclusive locality for the manufacture of pure, old-style, sour-mash whiskey."[124] The O.F.C. Distillery drew water from a riverside spring, and a 4,600-foot-long pipe brought water from a reservoir to the east. A coal-fired steam engine powered the works, which mashed 500 bushels of grain per day. An elevated, gravity-flow slop pipe carried spent grains to distant cattle and hog feedlots. O.F.C. whiskey was barrel-aged in four warehouses with an aggregate storage capacity of 35,680 barrels. The adjacent Carlisle Distillery stood beside the river and end to end with the O.F.C. works. The Carlisle had a large riverfront grain elevator, mashed 500 bushels per day, and aged whiskey in three brick warehouses with a combined capacity of 28,760 barrels. The Leestown Turnpike passed through the distillery property, but by 1886, neither the Carlisle nor the O.F.C. was served by a railroad.

W. A. Gaines & Company, financed by investors from Kentucky and New York, built the industrial-scale Hermitage Distillery in 1868 on a terrace some fifty-feet above the Kentucky River in south Frankfort. By 1886, the large wooden distillery mashed more than 900 bushels per day using water from two wells and the river, apparently not sharing E. H. Taylor's passion for "pure spring water." The Hermitage aged whiskey in eight brick warehouses with a combined storage capacity of 50,300 barrels. The Hermitage management also operated a large dedicated cooper shop to provide barrels, presumably using staves and barrel heads supplied by local sawmills, including one located just south of the distillery. Wood and coal fueled the distillery's steam engine. A large shed adjacent to the Hermitage's aging warehouses sheltered slop-fed cattle. The distillery had no railroad connection and depended on wagons and river barges. Barge deliveries arrived at a river landing and reached the distillery via a steep track up the bluff face. Management compensated for irregular grain deliveries by building two large wooden "corn houses" adjacent to the distillery.[125]

Laborers and artisans completed construction of Kentucky's new limestone, granite, and marble state capitol building in 1910 on farm property south of the river. The distance between the capitol grounds and the Hermitage Distillery was less than half a mile.[126] Four other small distilleries operated in Frankfort, and four large industrial distilleries operated in rural Franklin County, including the Baker Brothers works, which became the Old Grand-Dad Distillery in 1940.[127]

OWENSBORO

West of Louisville, distilling was a minor antebellum activity practiced by a small number of farmer-distillers in environmental circumstances profoundly different from those in central Kentucky's Bluegrass region. Only five distilleries in Owensboro and Daviess County predated the Civil War. The presence of distilling there was based not on the availability of "pure" limestone springwater but on the experience of a few early settlers whose family migration paths included origins in Maryland and Pennsylvania and a presence in Washington, Marion, and Nelson Counties in the Outer Bluegrass before moving to western Kentucky. The Ohio River or shallow wells provided water for distilling.[128] Land near the Ohio River above Owensboro was lightly settled by 1810, but the low-lying swamplands to the south and west were slow to develop. By 1850, those lands were still largely undeveloped or occupied by squatters.[129]

On the eve of Daviess County's great postbellum distillery boom, its farmland inventory included 115,000 acres of improved land, countered by 107,226 acres of woodland—that is, roughly 48 percent of the farmland acreage produced wood but no crops.[130] By contrast, in the Inner Bluegrass, farmland in Bourbon County included 155,304 acres of improved land, while only 17,569 acres, or 9.9 percent, remained in woodland. A portion of the Daviess County woodland was hilly uplands. The Ohio and Green River alluvial plain was largely old-growth swamp forest that included extensive stands of bald cypress (*Taxodiom distichum*), swamp white oak (*Quercus bicolor*), and pin oak (*Quercus palustris*). Some sections near the Ohio River drained well, but low-lying areas along Panther Creek and the Green River to the south readily retained standing water.[131]

Creating agricultural land out of waterlogged swamp required cutting the timber and organizing extensive land-draining projects. Land reclamation was under way by the 1900s; engineers used steam dredges to straighten and channelize streams and dig a system of tributary surface drainage ditches. By the 1920s, Daviess and Henderson Counties accounted for about two-thirds of Kentucky's artificial drainage systems.[132] In 1920, 913 of Daviess County's 3,808 farms (23.9 percent) had drainage systems, including 205 miles of open ditch. Prior to drainage, the county's timber and cutover land totaled 14,079 acres; ditch drainage reduced timberland to 9,402 acres, a 67 percent decrease.[133]

In 1880, prior to the drainage programs, Daviess County farmers grew 53,321 acres of corn, 1,460 acres of rye, and 151 acres of barley. Corn production was sufficient for the US Census to include Daviess and surrounding counties in the Corn Belt.[134] But rye acreage was inadequate to sustain more than a dozen high-capacity distilleries, and barley production was insufficient. A few early distilleries operated their own barley malting shops, but they soon closed. An 1888 US Treasury report summarized the industry's grain supply logistics: a "large amount of corn received from the adjacent country [was] loaded on flatboats and barges and towed in for consumption by the eight distilleries here."[135]

Although accurate tabulation is difficult, more than thirty distilleries operated in Daviess County as farm and industrial works between 1810 and 1900. Fourteen distilleries were in operation in 1899, ten of which occupied large industrial sites.[136] During the nineteenth century, some distilleries were shut down and abandoned; others burned. New distilleries were, in some instances, rebuilt on the sites of previous works. Distillers built at least two distilleries near the Green River in the southwestern part of the county, and more than twenty distilleries operated on the Ohio River's south bank, several on adjoining riverfront property. Only five distilleries operated in Owensboro itself or within walking distance of the city's residential areas.[137] Most distilleries in and near Owensboro were built between 1869 and 1895 as industrial works financed by local distillers and businesspeople; by distillers from Nelson, Marion, and Washington Counties; and by investors from Louisville and New York.[138]

The Owensboro and Daviess County economies were closely tied to river transport, and most postbellum distilleries fronted the Ohio at river landings. Owensboro was dependent on road and riverboat transport until 1884, when the north-south Owensboro & Russellville rail line opened, connecting Owensboro to Central City and Russellville. The Louisville, St. Louis & Texas Railroad completed an east-west track from Louisville to Owensboro, paralleling the Ohio River, in 1888. The following year, it completed a line west from Owensboro to the Green River, Henderson, and Evansville, Indiana.[139] Most riverfront distilleries were already in operation in 1889; by the 1890s, they could have obtained dedicated rail spurs, but several continued to operate as river distilleries.

The Owensboro riverfront distilleries were configured differently from rural Bluegrass distilleries or large industrial distilleries in Louisville and Jefferson County, where farmers' wagons or railcars could be counted on to deliver grain or coal on a regular schedule—daily, if needed. Barges served Owensboro's riverfront distilleries, and they moved at the whim of river conditions. Low water, floods, and winter ice impeded deliveries.[140] At each distillery river landing, barges arrived with grain and coal, which were off-loaded and elevated by steam-powered conveyors or trams to the distillery's corncrib, granary, or coal storage building. Filled whiskey barrels returned via the same tram to riverboats for shipment to market. Outsized storage buildings allowed the riverfront distilleries to accommodate irregular delivery schedules, and tramways reduced the awkwardness of moving raw materials and finished products up and down the steep riverbank.[141] Shippers delivered corn "on the ear," which distillers shelled and milled into mashing meal. Corncobs supplemented coal as a fuel source. Coal could have been obtained from several sources, including a local commercial mine that opened in 1905. Most Daviess County distilleries maintained feeding yards where they fed spent grains to hogs or cattle, yet a complementary livestock butchering and meatpacking industry did not develop until the 1890s, when four slaughterhouses were established on a small creek adjacent to the Daviess County Distillery northwest of Owensboro.

The authors of *History of Daviess County* and *An Illustrated Historical Atlas Map of Daviess County* conjectured that distillers came to the Owensboro area because grain was plentiful. "There is no necessity for sending long distances for grain to be converted by the distiller's art into a clear, sparkling, appetizing beverage, but the distilleries, at their very doors, have command of a cheap and abundant supply—inexhaustible as long as the hills yield their harvests and the valleys do not fail of their crops."[142] But, as US Census tabulations confirm, this assertion was doubtful at best.

History of Daviess County cited another advantage enjoyed by Daviess County distillers: "An additional item is the cheapness of transportation. The Ohio River furnishes shipping facilities of an unequalled character, and the railroad running south from Owensboro is always ready to carry freight on reasonable terms. It is mainly for these reasons that the large distilling business of the County has been carried on so successfully, while it has languished and died in other sections of the country."[143] In truth, the volume of local grain production was problematic, railroads did not arrive until the late 1880s, and distillers in Daviess County struggled with taxes and overproduction just as they did elsewhere in the state.

We see in Daviess County a variation on the origin narrative or the folk explanation of why distilling flourished in Kentucky. In the Bluegrass country, the origin narrative centered on the region's pure limestone water. In Daviess County, the narrative made no mention of water; rather, it emphasized grain availability and cheap transportation. *History of Daviess County* was a promotional essay, funded in part by patrons or subscribers, not a critical scholarly work. But therein lies its value as a window into traditional thinking and understanding.[144] The *Illustrated Historical Atlas Map of Daviess County* included lithograph views of farmsteads and distilleries and, comparable to other historical atlases of the time, was accurate and richly informative in its pictorial depiction of natural features, routeways, railroads, mills and manufacturers, farm properties, schools and churches, blacksmith shops, and villages.[145] The narrative, in contrast, adopted the tone and tactics of boosterism, which was commonplace in Victorian America's business-oriented towns and cities.[146]

Configuring Distilleries, Creating Narratives

A timeline for the century from the 1790s to the 1890s would include most of the events, inventions, and innovations associated with the industrialization of Kentucky's distilling enterprise. Distillers began the century following craft traditions in the operation of their farms, mills, and distilleries; they ended it as professional industrialists and adopters of scientific methods and studied business practices. Distilleries that had operated in hand-built vernacular structures became, through the adoption of proprietary technology, modern industrial structures. The most advanced works were carefully designed and planned to operate efficiently and at high capacity. Other distillers retrofitted their works with new equipment and otherwise reconfigured their facilities to increase production while also complying with federal law. Steam power was the most important enabling technological advance, but a host of other inventions in kindred industries were important elements in the century-long transition to industrial standing.

The distilling industry's origination narrative was also nurtured during the nineteenth century. The popular belief that fine whiskey distilling flourished in Kentucky because of its unique and abundant pure limestone springwater gained in volume if not in acceptance through the century. For myths to be accepted and perpetuated, they must be plausible and believable. People may have been unaware of the scientific reports written by early-nineteenth-century scientists about the quality of groundwater in some places and the foul condition of the water flowing in surface streams in others. For some distilling operations, the availability of pure water—a subjective judgment unproved by dispassionate science—was an important element in the decision to locate a distillery in one place rather than another. But the myth does not account for any personal predisposition toward the distilling industry based on family tradition, experience, or apprenticeship. Nor does it abide the consideration of hard-nosed business decision making, such as when a farmer contemplated the possible advantages of transporting ear corn or corn whiskey long distances over a wagon-swallowing mud track. Founding myths often have constituent elements, or collections of interrelated legends, that serve as popular or folk explanations of origins and cause, of historical precedent, of geographic determination. Kentucky's distilling legend is augmented by myths that explain the origin of the term "bourbon," identify the state's first distiller or the location of the first distillery, and provide the rationale for the first charred barrel. While such myths may not be proved by objective analysis, they are plausible and believable. And they contribute to the body of lore that underwrites the industry's heritage and self-image.

6

Technology's Tools

In the early operations of the grain distillery, much fuel, labour, and time, was employed, in effecting comparatively a small quantity of work. The object of the improvements and machinery adapted by different persons, has been to lessen the expenditures of one or more of these.
—Harrison Hall

Conjunctions—Inventions, Innovations, and Mechanization

Traditional craft or farm distilling was an important element in a Kentucky farmer's interest in acquiring agricultural assets—land, buildings, livestock, and tools—sufficient to ensure the long-term continuity of the family unit.[1] Early farm stills, like other outbuildings, were functional in design, simple in construction, and organic or idiosyncratic in form. Small, single-pen barns constructed of round logs or sheds with hewn or sawn frames housed most of the equipment, although in some cases, the still sat atop a small stone fire pit covered only by an angled plank and riven-shingle roof set atop four corner posts.

Nearby log or plank cribs and granaries provided storage. Hot summer weather obliged open-air distilleries to limit their operations to a late fall–to–early spring distilling season. Operators lacked the machinery necessary to control their environment, other than feeding farm-cut wood to a fire or carrying cool water from a nearby spring or creek. Small operations using traditional techniques could be managed by a small labor force comprising family members and neighborhood farmers; the latter may have assisted in the grain milling and mashing process as part of an informal labor exchange system.

Increased distillery capacity and profitability, and the transition from craft to industrial-scale production, were not simply matters of expanding the works by installing a larger still, adding mash tubs, and hiring more employees. Distilling, as with most nineteenth-century manufacturing processes, could enhance efficiency while improving product quality by developing dedicated expertise. Innovative types of stills and new technologies produced by complementary industries such as grain processing, malt manufacturing, and lumber planing and jointing all contributed to the distilling industry's mechanization. The change from craft production—distillers were wont to describe their product as "handmade"—to industrial production was expensive. If a distiller's income or

Old log distillery. (*Harper's New Monthly Magazine* 15 [June–November 1857]: 732)

credit was sufficient to bear the cost, he might upgrade his works, but not all distillers were predisposed to adopt modern equipment or novel and nontraditional processes. Some distillers were also retail merchants or engaged in other businesses that provided a primary source of income they could draw on to build state-of-the-art distilleries, which they operated as avocations. Other distillers upgraded their works when they partnered with outside investors who joined the business and funded equipment and operating expenses.

Inventions

Greek alchemists recorded experiments in distillation as early as AD 200.[2] In America, colonial distilling began as a local craft enterprise based largely on traditional techniques. Change was built on new ideas and equipment: innovations and inventions. An invention is the creation of a new, original device or process. In an industrial economy, invention tends to be an irregular but ongoing process to which many individuals contribute. Engineer Robert Henry Thurston summarized invention's accretive character in 1891: "I propose to call attention to the fact that [the steam engine's] history illustrates the very important truth: '*Great inventions are never, and great discoveries are seldom, the work of one mind.*' Every great invention is really either an aggregation of minor inventions, or the final step of a progression. It is not a creation, but a *growth*."[3] Through time, industrial processes have been refined and refocused by inventions and innovations. Industry has become increasingly grounded in science and engineering; production has been enhanced, output increased, and products improved, all of which has enhanced sales.

Invention is different from discovery, in that discovery involves an awareness of or revelation about something that already exists in nature. Recognizing that alcohol has a lower specific gravity than water is a discovery, and discoveries cannot be patented. A hydrometer developed to measure that difference is an invention, and inventions can be patented. An innovation is the application of an invention, an inventive process, or a discovery in a new and creative way. The lag time between an invention and its acceptance is variable and depends on perceived need and application by potential adopters.[4] The acceptance of innovations over time tends to stimulate the invention process, leading to an increase in the number of patents. Patenting a complex machine, provided it has a potential market, is likely to stimulate a chain of secondary or follow-on improvements that may refine an operation or increase its durability and speed. First-generation industrial machines were typically crude and unreliable, although they did demonstrate the merits of fundamentally modifying the manner of making or doing something. Follow-on inventions were often incremental improvements developed through trial and error.[5]

Complementary inventions are those made in cognate or related fields that may be directly applicable to distilling because they improve some aspect of the production process. During the two decades from 1840 to 1860, important mechanical inventions in steam-powered railroads and boats, farm equipment, and manufacturing in general occurred more or less simultaneously with biological innovations such as improved grain varieties and selective breeding to enhance weight gain in livestock.[6] Developing and adopting new technologies permitted manufacturing, as a general process, to move from handcraft or vernacular work to proprietary mechanical production. Technological improvements also permitted an increase in operational scale from local to national or even international. As inventions stimulated growth in industrial capacity, the demand for raw materials and energy or fuel also increased, prompting improvements in transportation so that industries could obtain these commodities from greater distances and at lower cost.

The geography of industrial invention does not necessarily correlate with centers of production; rather, inventions tend to emerge in areas that have a long-standing industrial focus and a sizable cadre of mechanically inclined people. From 1790 to 1847, the US Patent Office granted 180 patents related to distilling. Only six of those patents were issued to Kentucky residents, even though the state's distilling industry was vigorous and increased its productivity during those six decades.[7] During the first half of the nineteenth century, America's industrial core remained in the Northeast, although some industrial products were being produced by new companies in the trans-Appalachian West. A catalog of the industrial products exhibited at the Great National Fair in Washington, DC, in 1846, for example, listed forty types of industrial products from Massachusetts and New York and more than ninety types from Pennsylvania. Kentucky displayed three products: water-rotted hemp, mustard, and handkerchiefs.[8]

Although there was an initial increase in the number of operational stills in Kentucky as settlement extended across the state from 1790 to the 1840s, the number of stills had dropped rapidly by the mid-nineteenth century. Most early stills were traditional handcraft operations of modest scale working in conjunction with farms and mills, not industrial works. After 1850, the number of stills continued to decline, but the remaining distilleries moved toward industrial production. They adopted new inventions and innovations made possible by improvements in still technology and by secondary advances in complementary industries such as steam power, machine tools, metal fabrication, and glassmaking. In addition to mechanical inventions, innovations in finance, insurance, and marketing all played a role in the self-elimination of traditional farmer-distillers in favor of more specialized miller-distillers and singularly focused merchant-distillers.[9] The conversion to industrial distilling coincided with an increase in the number of distilling-related patents awarded to Kentucky and Ohio Valley inventors.

Patents

Robert Thurston's observation that inventions are never the work of one mind is illustrated in the refinement of the still and the distilling process in Europe and America. Prior to the sixteenth century, whiskey distilling was a domestic practice in Celtic Ireland and Scotland. By the 1750s, spirits production was sufficient to supply local markets and export substantial amounts to England. By the 1820s, with the development of dedicated technologies, distilling in Ireland and Scotland had evolved from a peasant craft to a factory industry.[10]

THE POT STILL AND CYLINDRICAL STILL

In about 1818 English inventor Henry Tritton patented a still design that enclosed the still and condensers in sealed vacuum chambers, following principles developed by James Watt. The vacuum, achieved with an air pump, allowed the fermented wash liquid (termed fermented mash or distiller's beer in America) to be boiled at a lower temperature. Even though Tritton's still was fire heated, he enclosed the still chamber in a water-filled vessel that functioned as a double boiler. When placed on an open flame, the mash at the bottom of a standard still often scorched; enclosing the still in water prevented the temperature of the mash from exceeding 212°F. Tritton's still thereby improved the quality of the spirits produced by eliminating what he called *empyreuma,* the peculiar odor produced by scorching the organic oils in the fermented mash. Tritton's design received accolades, but it was neither widely adopted nor commercially successful, perhaps because the vacuum system and the water-filled chambers were costly to construct.[11]

During the first two decades of the nineteenth century, several French, English, and

Scottish inventors refined the distilling process by moving away from the traditional onion-shaped pot still and designing cylindrical stills with multiple distilling chambers in which the mash was heated to evaporation temperature by rising steam rather than the direct heat of an open fire. Early in the nineteenth century, French inventor Edward Adam developed a multichambered horizontal cylindrical still. The design's operational principles were refined by Laurent Solimani, M. Baglioni, M. Collier Blumenthal, and several others.[12] In 1822 English inventor Anthony Perrier patented an improved still in which the mash passed continuously through a circular labyrinth of copper tubes arranged along a horizontal chamber or boiler that was fire-heated from below.[13]

French distiller Robert Winter patented a two-chambered still in 1824. Though mounted atop brick fireboxes, both chambers were housed inside larger water-bath containment vessels that provided indirect heat. The first chamber's temperature was maintained at 170°F, the second at 140°F. The mash moved through the first chamber by way of small pipes. The liquid heated quickly and uniformly to evaporation temperature. The condensed liquid then entered the second chamber and passed through an arrangement of annular copper cylinders that created uniform evaporation of the remaining alcohol. The condensed spirit was said to be high strength or proof and free of *empyreumatic* contaminants.[14]

The Column Still

In 1828 Robert Stein of Scotland refined Perrier's still by creating a continuous-flow, steam-charged horizontal cylinder divided into eight chambers. Fermented mash containers were preheated by the circulation of spirit vapors and spent mash through the container jackets. The hot mash was then injected under pressure through small pipes into the still's chambers, creating a mist. Steam introduced into the chambers quickly and thoroughly vaporized the hot mash. The vapor then passed into a spiral-wound pipe, or worm, where it condensed into a liquid. Stein's still produced more alcohol with less fuel than pot stills, and it became the first commercially successful column still in the United Kingdom.[15]

Aeneas Coffey was born in France of Irish parents. He entered the Irish Excise Service as a tax collector on home products such as whiskey and was eventually promoted to inspector general of excise in Ireland. He resigned in 1824 and, in one of the great ironies of the age, established his own distillery. Six years later, Coffey patented an improved continuous-flow column still based in part on Stein's 1828 design.[16] The Coffey still had vertically stacked chambers divided by perforated copper plates. Fermented mash was introduced at the top of the still column; steam entered from the bottom, heating the plates and mash droplets, thereby evaporating the alcohol. The column still produced up to ten times the volume of traditional pot stills and resulted in higher-proof spirits, albeit absent many of the congeners that gave Irish whiskey its distinctive flavor. Distillers accustomed to the taste of whiskey produced by traditional copper pot stills vigorously resisted the Coffey

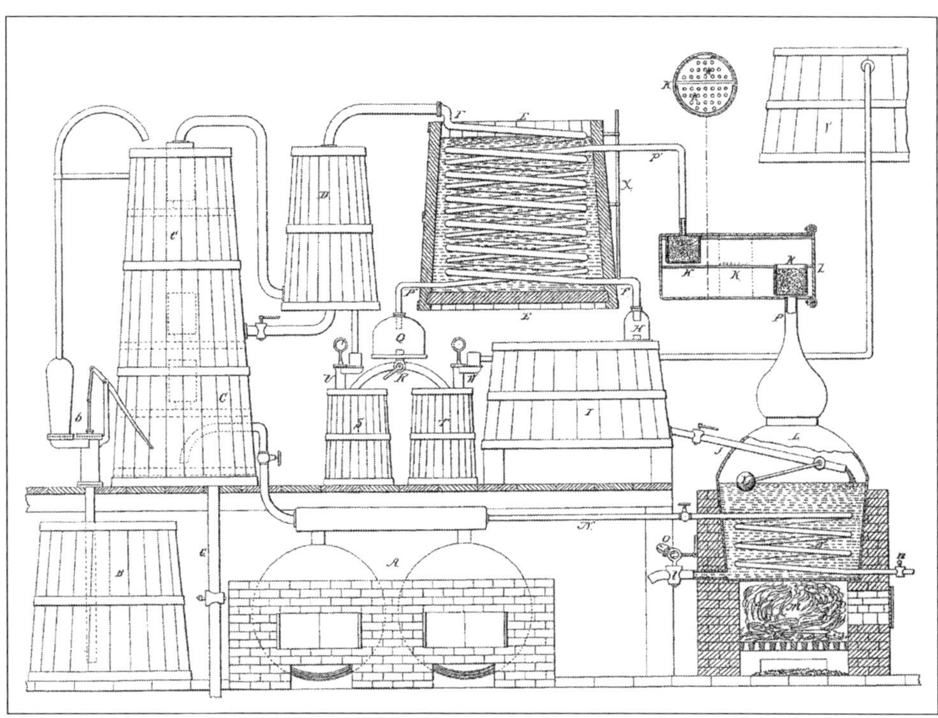

William and George Robson's improved alcohol still (patent no. 92,477, granted July 13, 1869). (W. and G. W. Robson, "Alcohol Still," US Patent and Trademark Office, http://www.uspto.gov)

still and referred to its product as the "silent spirit."[17] Scottish distillers adopted the Coffey still for large-volume commercial production, and by the 1840s, it was broadly accepted.[18]

In America, distilling inventions and patents increased after the Civil War. Patents for stills rarely represented full or radical departures from traditional designs or sequences; rather, they offered incremental improvements on established devices. Two patents are illustrative: one was obtained by George Robson of Cincinnati, Ohio, and Melvin Hughes of Paris, Kentucky, in March 1869; the second was obtained by William Robson and George Robson of Cincinnati, Ohio, in July 1869. The Robson and Hughes patent featured a still immersed in a steam-fed water jacket and a tank in which mash or low wine was preheated before entering the second distillation stage or doubler. This design eliminated the risk of scorching the mash by direct heat.[19]

The still patented by William and George Robson had a more elaborate design.[20] It incorporated two brick furnaces or heating units, labeled A and M in the patent drawing, for continuous double distilling. Furnace A heated a cylindrical steam boiler N. Fermented mash was pumped from vat B up into column still C. A pipe from the steam boiler introduced steam into the bottom of the still. Condensed alcohol vapor exited the top of the still through a pipe to a second still or doubler D. Vapor from the doubler then passed through pipe F into the flake stand, a vessel containing cool water, where the copper pipe coiled into a worm condenser. Spent mash exited the bottom of the still through pipe C, and low wine flowed from the flake stand into holding tank I. Pipe J emptied the low wine into the second still L, a traditional copper pot still heated directly by a masonry firebox. The low wine in the pot still was also indirectly heated by the coiled steam pipe N from the steam boiler. Vapor rose to the top of the pot still and entered the horizontal drum K, which contained optional, removable colanders (depicted as small square boxes) into which flavoring elements such as juniper berries, coriander seeds, and peach pits could be added. The technique of flavoring alcohol was used by whiskey rectifiers and would not have been employed by distillers of straight bourbon whiskey. Vapor, flavored or not, exited the drum by pipe P and entered the flake stand to condense into high wine, the clear, high-proof alcohol that was the still's final product. From the flake stand, the liquid drained into storage tank T. The inventors claimed that their still accomplished double distillation using both a steam-heated column still and a fire-heated copper pot still.

John Peden lived in Lawrenceburg in Anderson County, one of Kentucky's primary nineteenth-century commercial distilling centers. In 1886 the US Patent Office awarded Peden a patent for his improved column still. The patent was co-assigned to the Bourbon Copper and Brass Works of Cincinnati, which manufactured fire hydrants and copper and brass goods for distilleries, breweries, and steamboats.[21] Peden's column still was a refinement of the continuous-flow stills patented by Stein and others and used by several distilleries in the Scottish Lowlands in the 1820s. It was similar in design to the still patented by Coffey in 1831.[22]

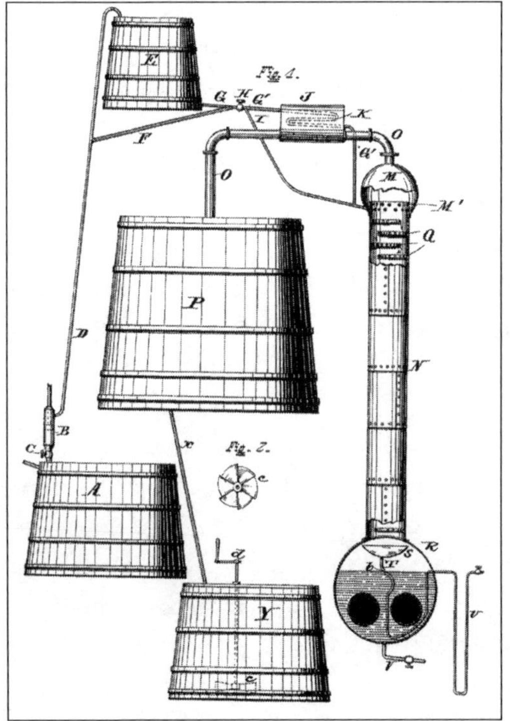

John Peden's distilling process and apparatus (patent no. 349,449, granted September 21, 1886). (J. C. Peden, "Process of and Apparatus for Distilling," US Patent and Trademark Office, http://www.uspto.gov)

Peden's patent drawing depicts the distillery elements in one plane, although when the individual units were built and installed, they would be variously positioned in horizontal and vertical dimensions. The drawing illustrates the flow of once-distilled or singles alcohol from tank A through pump B and pipe D into elevated tank E. The singles liquid exited tank E through pipe G. The fluid then flowed either into a coil in drum J, where the fluid was heated, or through pipe I into the top shelf of the tubular column still M. The column still was a shell (N) made of copper sections bolted together. The shell could be any height and any diameter a distiller required to achieve the desired volume of production. The upper four-fifths of the still column was fitted with metal pans Q, which were smaller in diameter than the still tube. The pans were perforated and had upturned rims. Liquid falling from the top hit the top pan and flowed out through the perforations in a fine spray or mist. The pans were bolted to the exterior wall in an alternating pattern so that outflow from the pan above fell into the pan immediately below, in sequence, down through the column. Maintaining the fine spray through the column permitted ready separation of the remaining alcohol.

A steam doubler at the bottom of the column housed in vessel R included a two-flue boiler (black circles seen in cross section) with a discharge pipe V. Catch basin S at the bottom of the column was drained by pipe T. Steam from the boiler rose vertically through the column, increasing the temperature of the descending singles mist to the point of evaporation. The alcohol vapor exited the top of the column through pipe O, passed through preheater drum K, and flowed into condenser tank P, which contained the copper worm. The liquid then drained through pipe X into tank Y for testing and temporary holding. Peden claimed that his column still effectively increased the proof of the alcohol produced. And, importantly, the heavier impurities and refuse containing fusel oil were caught in basin S at the bottom of the column and removed before they could be volatilized and mixed into the alcohol condensate passing up the column.

Distillers who wanted to adopt new distilling equipment had to consider whether it would fit into their existing buildings. If not, the distillery would have to be enlarged or torn down and replaced by a new structure of appropriate dimensions. Traditional pot stills, even the high-capacity renditions, were comparatively low profile, but the new vertical column stills required taller buildings, and their higher output led to the construction of additional aging warehouses.

Many of Kentucky's industrial-scale distilleries installed large column stills in the 1870s and 1880s to increase production; distillers rarely claimed that the transformation also increased product quality. For instance, J. G. Mattingly installed and perfected a continuous-flow columnar or "Coffee [*sic*]" still in his West End Distillery in Louisville in 1888, with the goal of increasing output.[23]

Nearly two decades earlier, on March 4, 1869, the *Cynthiana News* announced that Mr. T. J. Megibben and his brother had just completed construction of a new distillery near

Lair's Station in rural southern Harrison County. "It is said to be the best in Kentucky," the newspaper declared. "Its capacity is about equal to fifteen hundred bushels per day, but we understand that not more than four hundred will be mashed."[24] The distillery building was seventy-nine feet long, forty feet wide, and three stories tall. The J. & E. Greenwald Company of Cincinnati installed the steam engine and related machinery. The Robson Company of Cincinnati (likely run by the same William and George Robson whose still designs were patented that year) built and installed the copper still equipment. "The De-Bus Company manufactured the mashing tubs which were said to be some of the largest in the world." The distillery also introduced "corn screens" that removed all corn dust and chaff from the mash, allowing the Megibben brothers to "make a pure article of Kentucky whiskey."[25] The Megibben Distillery was something of a showcase for new inventions. The distillery building and equipment were outsized; the three-story height accommodated the tall column still and elevated milling equipment. Megibben may have operated a dedicated cooperage, but he also purchased whiskey barrels from the Andrew J. Oots & Sons cooperage in Lexington.[26]

Some small-volume distillers chose to continue to use traditional open-fire pot distilling and defended their choice of technologies. The Jacob Laval & Son Crystal Spring Distillery in Louisville made a sweet-mash "Pure Copper Whisky" in the 1860s; owner T. H. Sherley's advertisement in the *Louisville Daily Express* declared, "No steam used in the process of distillation."[27]

The Grain Elevator

Moving grain from farm to mill to distillery, be it corn, rye, wheat, or barley, had traditionally been a labor-intensive process performed by hand with scoop shovels. Farmers thrashed grain with hand flails or other devices and then scooped the loose kernels into sacks, which they loaded onto pack animals or wagons for delivery to a mill or distillery. In the late 1780s prolific Philadelphia inventor Oliver Evans designed a semiautomated flour mill employing a waterwheel to turn bucket elevators and conveyors that moved grain from delivery wagons to the mill's upper floor. From there, the grain descended through the milling stones and screens and reached the ground floor as sacked flour—all by gravity flow. Although Evans's elevator system was slow to be adopted, it was eventually used extensively in flour mills; distillers also installed elevating equipment to lift grain to a mill located on the distillery's top floor. The T. J. Megibben Distillery in Harrison County may have had a grain elevator and an upper-floor mill. High-capacity column stills and bucket elevators also invited the construction of grain bins and corncribs at distilleries, as occurred at most distilleries operating in Owensboro in the last quarter of the nineteenth century.

The Grain Dryer

The distilling process culminated in a product, bourbon whiskey, and in a by-product, variously called spent grains, stillage, or slop. Distillers had traditionally fed slop to cattle or hogs or disposed of it by dumping it into creeks, rivers, or sinkholes. Hot liquid slop was awkward to handle. Distillers ran pipes from slop storage tanks to their feedlots. Slop that was sold to farmers for livestock feed had to be transported from the distillery in barrels or special tank wagons. Even though the material cost to farmers was usually low—sometimes free—slop was heavy, expensive to move, and noisome. Drying or dehydrating the slop would reduce its weight and increase its storage life and transportability, allowing it to be shipped dry in sacks.

From 1790 to 1873 the US Patent Office issued 125 patents for grain and grain stillage dryers.[28] Inventors used various designs consisting of vertical or horizontal kilns or drums that circulated heated air or steam through flues or pipes onto the grain stillage.[29] Early dryers were intended primarily to dry high-moisture grain to prepare it for storage, but most operated poorly because of condensation and plugging. Direct-heat dryers also scorched the grain and presented a fire risk.[30]

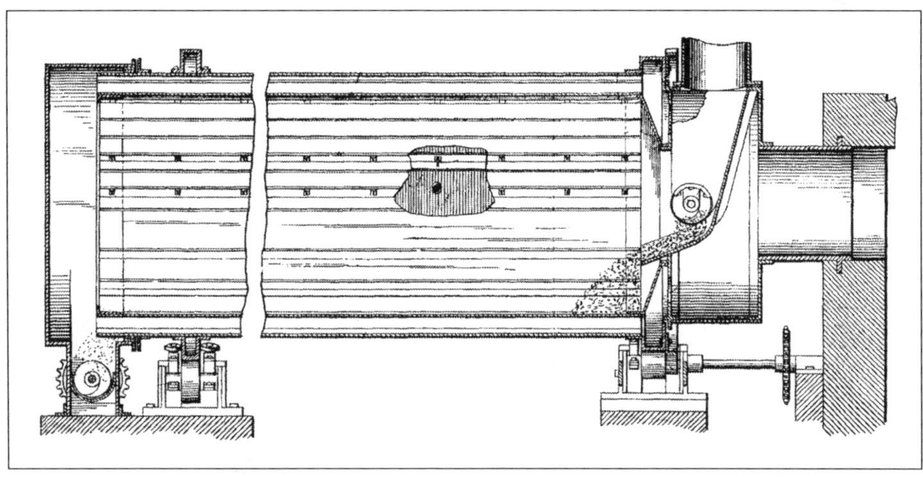

John E. Turney's rotary dryer (patent no. 774,859, granted November 15, 1904). (John E. Turney, "Rotary Drier," US Patent and Trademark Office, http://www.uspto.gov)

Inventor and English subject John H. Turney established the Turney Drier Company in Louisville in August 1899. He patented a horizontal rotating cylinder grain dryer in 1903.[31] The following year, working with Charles E. Geiger, W. K. Koop, and G. W. Fisk, Turney patented an improved version of the dryer.[32] This was an innovative departure from other dryer types. The dryer body was a double-walled horizontal cylinder, set at a slight incline and supported by a wheeled cradle in which it rotated. Flues carrying heated air extended longitudinally through the cylinder's interior. An operator introduced the wet material into one end. Vapor escaped the cylinder via a discharge tube, and the dried material exited the opposite end into a conveyor. As the cylinder rotated, heated flights projecting from the cylinder wall mixed the material uniformly, exposing the entire volume to indirect radiant heat. Because the Turney dryer conducted heat through sealed flues, the hot air or gas did not come into direct contact with the drying material, thereby reducing or eliminating the risk of scorching and fire. The dried material moved by gravity through the inclined cylinder in a continuous-flow process, and workers bagged the dried grain solids for shipment to livestock feeders. In 1905 Turney extended his dryer business interests through a new association with the Henry Vogt Machine Company in Louisville, which manufactured dryers and roller and filter presses.[33]

As distilleries adopted drying equipment, it became necessary to add drying rooms to house the associated gear. The fully assembled Turney rotary dryer was more than thirty feet long and ten feet wide, and it required connections to a steam-heating boiler; additional space was needed for weighing the dried grain and sewing shipping bags. The entire operation required a large enclosed room that had to be built onto an existing works.[34]

The Whiskey Aging Warehouse

The manufacture of fine Kentucky whiskey was not complete until it had been aged or ripened in specially made white oak barrels. Traditional nineteenth-century aging warehouses were single-level frame buildings. Building larger, multifloor storage warehouses with load-bearing walls permitted the distiller to store barrels in a manner that reduced space requirements, conserved labor, and decreased the risk of damage and contamination. But such structures also required innovative ways to move full barrels from ground level to the upper floors. Two types of inventions—hoists and storage racks— addressed these issues; when perfected, they radically changed whiskey warehousing methods and dramatically altered the distilling landscape.

The Hoist

Large industrial whiskey distilleries were capable of high-volume output and filled dozens of barrels each day during the distilling season. Distillers, by and large, stored their filled barrels near the distillery. Prior to the 1870s, most warehouses were one-story sheds with open floor plans, commonly termed "stack" warehouses. Moving full whiskey barrels into and out of these warehouses was labor-intensive. Workers laid the barrels on their sides, rolled them into the building, and lined them up on the floor in rows. Wooden chocks held each row in position. Workers placed wooden planks on top of the bottom row of barrels to support a second row; a third and fourth tier could be added, if needed. This storage method was fraught with problems. Storing 400- to 500-pound barrels on top of one another exerted considerable pressure on the bottom barrels, possibly causing them to leak or collapse. Whiskey evaporating from white oak barrels introduced alcohol vapor into the air, which invited the growth of a black mold (*Baudoinia compniacensis*). Air could not readily circulate through the vertical stack of barrels, so the mold might lend a musty odor

Stack-style brick warehouse (formerly Hillenmeyer's Nursery), Sandersville Road, Fayette County.

to the warehouse and possibly even the whiskey. Because the oldest whiskey in a warehouse was contained in the first barrels laid down, removing the oldest barrels from the bottom of a four-high stack was arduous and dangerous.[35] Prior to the 1870s, some distillers built two-story warehouses into hillsides, with entrance doors on two levels. Others had two or three levels connected by internal ramps. Both configurations required workers to roll barrels up two floors or more. But most warehouses were configured as horizontal rather than vertical spaces, making them extravagant consumers of expensive distillery property.

Hoisting devices have an ancient heritage, but the need to move minerals out of mine shafts and raw materials and finished products through multistory factory buildings in the industrial era stimulated systematic experimentation and development. Numerous hoist or elevator designs appeared in the late eighteenth and early nineteenth centuries. A common lifting device, likely of considerable antiquity, consisted of a pair of stout hooks crafted from flat metal bar stock. The hooks were placed on opposite ends of a barrel, resting on the flat stave surface just above the head. A clevis or other closure device attached the hooks to a strong rope. To gain mechanical advantage, the rope passed through a series of pulleys commonly referred to as a block and tackle. When hitched to a mule or draft horse, the device lifted the barrel to the desired level in the warehouse.[36]

Steam and hydraulics were used to power large-capacity industrial elevators by the 1850s. Hand-operated traction elevators had practical application in small factories and warehouses. These units were operated by ropes, pulleys, and counterweights that provided mechanical lifting power. Henry Reedy of Cincinnati, Ohio, patented one such hoisting machine in 1868 that was suitable for use in distillery warehouses.[37] Reedy's hoist consisted of a square base frame that formed a hatchway. Vertical stanchions attached to the base frame and supported a top frame, to which were mounted a grooved pulley and a gear mechanism. An operator pulled on a rope that turned the pulley and the gear and windlass arrangement. A pulley-guided lift rope passed under the lift platform and attached to two 150-pound counterweights that moved vertically between the stanchions. A chain-operated lever and gear assembly provided a brake that could hold the platform in any position. Reedy claimed that his hoist was capable of lifting 3,000 pounds without slippage, a capacity sufficient to move six or seven full whiskey barrels. Such hoists could be used to fill multifloor warehouses, but still lacking was an efficient method of stacking the barrels on each floor.

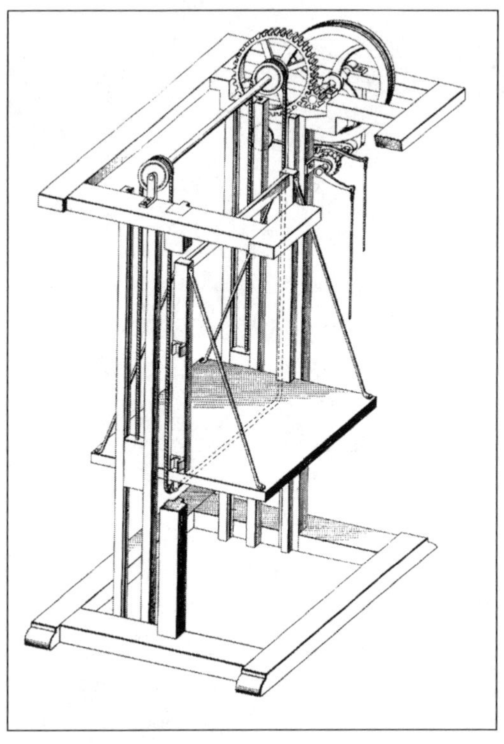

Henry Reedy's hoisting machine (patent no. 4,273, reissued February 21, 1871). (H. J. Reedy, "Hoisting Machine," US Patent and Trademark Office, http://www.uspto.gov)

The Storage Rack

Alsatian distiller Frederick Stitzel migrated to America in about 1848 and established a distillery in west Louisville sometime around 1870.[38] In 1879 Stitzel received a patent for "Improvement in Racks for Tiering Barrels."[39] The Stitzel rack design was inspired. Deceptively simple in layout and material requirements, the rack was modular; it could be built and used singly or aligned in horizontal rows and stacked vertically. Carpenters set four stout wooden posts (E) vertically into horizontal base supports (D) with mortise and tenon joints (H). They set horizontal wooden rails (F) into the vertical posts with square notches. Horizontal cross bolts and nuts (J) passed through each post-and-rail junction, tying the open rack together. At each end, the rack's vertical support posts were spaced far enough apart to allow barrels (G) to be placed on the horizontal rails and rolled into position. Stitzel's rack permitted air circulation between barrels, and workers could readily add or remove barrels from the rack's bottom level without moving the barrels above. The rack became known as the "patent rack" and was widely adopted by distillers; a version of this rack system is still being used to construct entire multistory warehouses consisting of a single unified rack with access passageways.[40]

Builders commonly configured warehouses with one rack, three barrels high, per floor or story, and these structures often rose six stories or more. Distillers developed a stout loading platform or table to address the problem of lifting heavy barrels into each

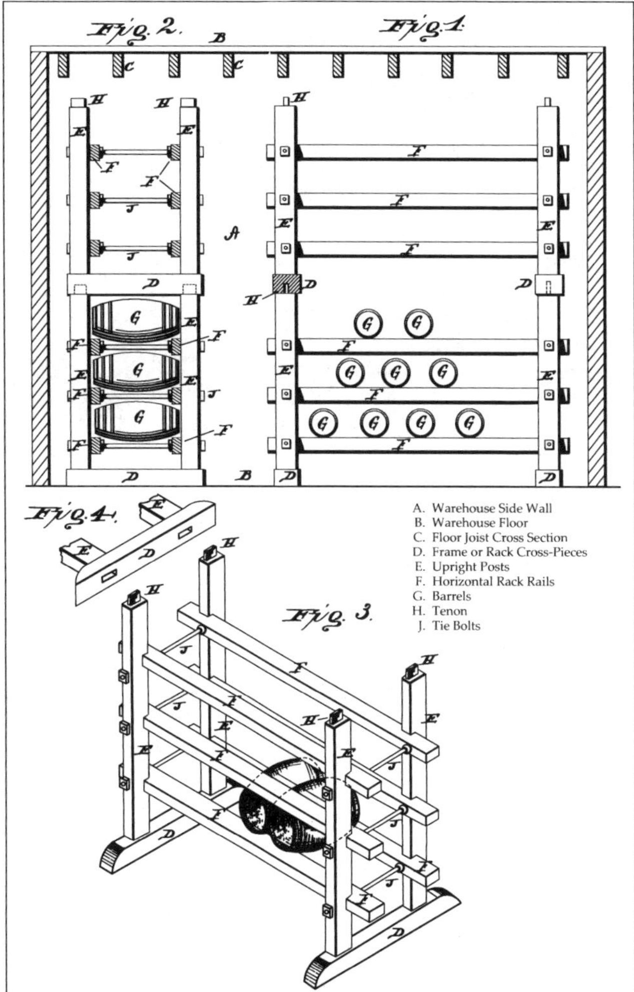

Frederick Stitzel's bourbon barrel racking system (patent no. 9,175, reissued April 27, 1880). (Frederick Stitzel, "Rack for Tiering Barrels," US Patent and Trademark Office, http://www.uspto.gov)

rack's second and third levels. With the table laid on its side, laborers could roll barrels up a plank ramp to the second tier. Standing the table on end and resetting the ramp's planks provided access to the third tier. Development of a wheeled hydraulic lift substantially reduced the effort required to place barrels in racks. New warehouses are similarly configured. They are built from the inside out, and the interior rack system forms the entire structure's supporting core. When the rack structure is completed to the desired height and width, exterior cladding is attached like a curtain wall.

Maker's Mark warehouse at Loretto, Kentucky, under construction.

The Monitor Roof

A third warehouse innovation, the clerestory-windowed monitor roof, admitted light and improved air circulation. Builders addressed airflow problems in musty warehouses by raising a linear section of the roof along the ridge beam to form a monitor roofline. When fit with vent panels and windows, the monitor roof ventilated the building and admitted light to the upper floors. This warehouse format was used at several Kentucky distilleries and was very similar in design and function to early-nineteenth-century engineering shops in Europe, New England textile mills, and Middle Atlantic industrial buildings.[41]

Synthesis: Mechanization and Landscape, 1830s–1890s

Inventions and innovations do not stand alone; they have social, economic, political, and geographic implications. New processes and equipment, individually or in the aggregate, foster landscape change—sometimes subtle, sometimes bold. The inventions that distillers adopted to increase production or enhance product quality often developed incrementally, as a sequence of refinements. And each element in the invention-adoption sequence had the potential to require adjustments in the whiskey production process and, unavoidably, the built environment or landscape where those processes took place. Distillers increased the horizontal and vertical dimensions of their still houses to accommodate larger equipment and their warehouses to increase storage. They often spliced new rooms onto existing structures or erected new stand-alone buildings to house specialized equipment. As distillery industrialization proceeded, the distillery proper and its attendant buildings came to resemble a modern manufacturing works with specialized structures and transport umbilicals that delivered raw materials and removed finished products.

The Brick and Iron Industrial Distillery

New technology requires accommodation. The conversion from animate to inanimate power increased productivity and changed the scale of operations, thereby requiring high-capacity infrastructure and high-volume transport access. Whereas the structures used by farm craft distillers varied in size and form, industrial distilleries became increasingly standardized, following the practice under way in other industrial sectors such as textile mills, metal fabrication plants, and machine tool manufacturers.[42] Commercially manufactured building materials replaced farm-harvested logs and local stone. Distilleries transformed into large brick and iron buildings that were often linear in form and combined new and existing structures in one functioning unit. Large buildings required long-span joists and load-carrying frameworks. When Covington, Kentucky, distiller N. J. Walsh decided to tear down an old distillery in Paris and replace it with a new works in 1897, the building plans described "a brick structure four stories in height, 225 feet long and 154 feet wide. The front arches will be of iron and the floors cement. . . . The new distillery will use a process by which more whiskey can be distilled from a given amount of grain than is gotten by the process now in use."[43] Although industrial distilleries enclosed expansive interior spaces, they were differentiated internally into rooms or sections with different names and functions: granaries and cribs, milling and mashing rooms, and sections housing fermenting vats and the still. Pipelines separated product from by-product.

Whereas the small farm distillery may have obtained raw materials locally and otherwise functioned within the farm production unit or larger rural neighborhood, the industrial distillery was increasingly dependent on important linkages to other processes that might not be obvious at first. Somehow, potable water, steam engine equipment and fuel, distilling grains, copper kettles, oak cooperage, and roadways and railroad tracks had to be combined with business knowledge to produce a profitable product. Increased production obliged distillers to obtain more grain, perhaps outstripping the yield of their own farms and that of their neighbors'. Hauling grain and fuelwood to a distillery required willing teamsters, healthy draft animals, sturdy wagons, and all-weather roads. Some riverfront distilleries at steamboat landings acquired railroad sidings and yards, giving them highly valued water-to-rail connections. If railroad companies did not build lines on both sides of a river, distilleries on the "wrong" side were compelled to compensate. West of Cincinnati, the Ohio & Mississippi Railroad ran along the north side of the Ohio River to Lawrenceburg, Indiana, the site of several nineteenth-century distilleries. South of the river, at Petersburg in Boone County, Kentucky, William Appleton shipped whiskey from his distillery across the river by corn boat to the freight depot in Lawrenceburg.[44] Rural distilleries sought connections to cross-country rail lines by building sidings and spurs. When clustered together in cities, distilleries integrated into industrial districts tethered to rail lines.

Steam Power

Steam power made possible the development of the modern industrial distillery. Adopting the stationary steam engine required distillers to adopt related mechanical systems as well, thereby creating an engineered distillery; the use of engineering principles entailed adjusting one's language and reasoning to include concepts such as accuracy, precision, and tolerances. Resilient brick, which was ubiquitous but of variable quality, became the favored building material. Builders used fire-retardant iron for girders, beams, doors, and window shutters. Distilling with steam power required large, dedicated engine rooms—a powerhouse—and a reliable fuel source, which distillers secured by stockpiling wood or coal. Engine rooms required concrete engine mounts and brick or concrete floors. Smoke exited the firebox through large-diameter metal chimneys, which were in common use by the 1870s. Brick chimneys were more expensive to build but were installed at large-capacity distilleries, especially those operating in urban areas. Steam engines also powered liquid pumps and grain-handling equipment, and they piped steam-heated water to tanks and mash tubs. The high-capacity distillery incorporated large mash tubs and fermenting tanks, some exceeding 10,000 gallons in capacity; when filled, one tank could weigh in excess of 80,000 pounds. Large distilleries might use ten or more such tanks. Because basic timber or concrete floors could not support so much weight, supplementary concrete footers had to be installed for each tank. The adoption of steam power also affected distilleries' operating costs. Craft farm distillers often used their own or locally produced grain, fuel cut from their own woodlots, and barrels they made themselves or obtained from local coopers. The industrial whiskey distillery acquired external suppliers, and expenditures were sizable, with the cost of grain, fuel, and cooperage being primary concerns.[45]

Distillers had been subject to federal taxation, episodically, since 1791. After the Civil War, taxation and regulation increased to pay war debts and underwrite the federal government's operating costs. Accelerating tax rates paralleled, to some extent, increased industrial production, and distillers needed considerable business acumen to understand the requirements. Consider the arcane wording of the federal excise tax law enacted in 1813:

> And for every boiler, however constructed, employed for the purpose of generating steam in those distilleries where wooden or other vessels are used instead of metal stills, and the action of steam is substituted to the immediate application of fire to the materials from which the spirituous liquors are distilled, for a license for the employment thereof, double the amount on each gallon of the capacity of the said boiler including the head thereof, which would be payable for the said license if granted for the same term and for the employment on the same materials of a still or stills to the contents of which, being the materials from whence the spirituous liquors are drawn, an immediate application of fire during the process of distillation is made.[46]

This curiously convoluted statute specified that the fee to license an industrial steam-based still was double the fee imposed on a traditional fire-based distillery—per gallon of capacity—perhaps because steam systems were more efficient and therefore more productive. Legislators also may have wished to protect traditional craft distillers by handicapping those inclined to adopt industrial machinery.

Some distilling operations were not readily changed by new nineteenth-century technologies. The annual climatic cycle required distillers to maintain a seasonal production schedule. The optimal temperature for yeast growth is about 80°F, and yeast cells die at temperatures above 140°F. Prior to the arrival of Willis Carrier's air-conditioning technology in 1902, distillers could most readily control the temperatures required in the fermenting and distilling process by operating during the coolest months of the year.

Changing Production, Changing Landscape

Kentucky's distilling industry was historically centered at discrete nodes of production with connections to raw materials and markets. As production capacity increased, distilleries were obligated to extend their reach for raw materials beyond the local to regional or even national suppliers. Industrialization also led to the redistribution of distilleries from scattered rural farm sites to transport-oriented locations, some of them in urban areas. For distillers who wanted to send or receive shipments of grain, fuel, lumber, barreled whiskey, and other materials in volume, their best option in the last third of the nineteenth century was to select a site beside a navigable river, a hard-surfaced turnpike, or a railroad line. Relocation to sites in or adjacent to cities also provided distillers with direct access to banks and insurers. Early commercial stills were often family ventures that were either self-financed or operated with funds obtained from local or regional sources. By the 1880s, corporations with access to national and international financial markets owned most of Kentucky's large industrial distilleries.[47] When considered in the aggregate, commercial distilling created an industrial landscape composed of the distillery and its associated works, grain fields and bins, woodlots and sawmills, cooper's shops and wood yards, livestock barns and feedlots, and hardware manufacturers and suppliers.

Many of the processes and structures involved in the twenty-first-century distilling industry were developed during the period from 1830 to 1910, an era of revolutionary, often problematic change for Kentucky distilling. The turgid term "golden age" is sometimes used to emphasize the accomplishments of a specific period. Such epochs are largely fictional because it is common to imagine that cultural achievement and social life conformed utterly to some ideal. In truth, the development of industrial distilling was uneven and fraught with doubt, risk, compromise, and failure. Collectively, progressive economic and social accomplishments were difficult to achieve, and individual initiatives were often tempered by risky decisions, chance, and bad luck.[48] The process of nineteenth-century industrialization was never stable, never predictable; it was always changing and adjusting. Yet the retrospective view of that era is often one of idyllic nostalgia; it is idealized as a time when larger-than-life actors created the foundation of a modern industry. Some individuals achieved admirable business success, but their progress was neither linear nor unproblematic. The developing distilling industry was subject to innumerable influences, both negative and positive, whether they arose from the environment, family and community, business and finance, labor and technology, or government and public opinion. Distilling's industrialization was not an inexorable sequence of grand accomplishments but an episodic period of intersecting events. Some were random—a crop lost to bad weather or an unanticipated invention; some could be anticipated with a degree of actuarial probability—loss of facilities by fire or other catastrophes; and some were certain—government-imposed taxes. Even though the people involved only partially understood or perceived the larger social and economic milieu in which they operated, they managed to construct out of these idiosyncratic events one of Kentucky's quintessential industries and one of its most distinctive landscapes. It is this corporate accomplishment by so many participants from various social, economic, and political stations that makes this such a compelling story.

7

Complementary Industries

There is perhaps no part of the apparatus of the distillery on which more attention should be bestowed by the distiller, than that of the construction of his stills. It forms in itself the sole dictator of the establishment. In the United States, the coppersmith becomes the sole agent and artist, he is the distiller's alma mater, in short he regulates the whole destiny of the manufacturer.
—Michael Krafft

Craftwork and Innovation

Nineteenth-century industrial distilling required the support of several complementary industries, among them cooperage, coppersmithing, glass manufacturing, and grain processing. Each of these industries was based on long-standing practical technologies applied in a context of confounding problems and constraints. Raw materials were "raw" in the sense that they had to be taken or harvested directly from the environment within a geographic radius limited by resource distribution and transportation costs. Industrial productivity and product quality were often dependent on the experience and skill level of individual craftspeople and were very difficult to improve. Innovation in process and mechanical production tended to be gradual and incremental. Key improvements often occurred not in the primary industry but in related or supporting industries, such as when a nail-making machine was patented in 1795 and refined in the early nineteenth century. What followed was a protracted transformation of the practices used in the construction of wooden buildings.[1] High-capacity industrial distilling required innovation and mechanization, but technological advances in associated or complementary industries such as woodworking and metal fabrication were also essential. This chapter examines four such industries and reviews how they worked, how they changed, and how those changes affected whiskey distilling.

Cooperage

Terra-cotta amphorae, elongated clay jars with handles and cone-shaped bottoms, were used to transport liquids such as olive oil and wine throughout the Mediterranean region in Phoenician times. The winemakers of Georgia's Caucasus Mountains used barrel-shaped terra-cotta containers to store and ship wine some six millennia BC. The Celts may have invented wooden casks or barrels, likely starting with tapered wooden buckets, which were subsequently adopted by the Gauls and the Romans by about AD 300.[2]

The safe and economical storage and shipment of bulk liquids and solids have long been concerns of producers, traders, and merchants. Ancient container technology was practical and elementary—people used baskets, bags, boxes, crates, and jars to store and transport a wide variety of goods. These common containers often lacked durability and were awkward to handle if they were large and heavily loaded. Celtic coopers made wooden casks or barrels out of staves they bent and planed to a convex cross section; they bound the round ends, or heads, together with wood or iron strapping. When properly made, Celtic barrels were strong, versatile, and easily moved.

Barrel making was physically demanding. The work required specialized tools and years of practice, in large part because the barrel's convex shape was difficult to fashion, especially if one required a tight seal between staves. Yet the shape proved advantageous because it allowed a barrel to be moved on flat or angled surfaces, such as riverboat gangplanks, or pivoted to change direction. The convex barrel was much easier to roll than a cylindrical container, and it was near friction-free compared with the movement of boxes or crates, which required a dead lift or a skid. An empty or lightweight barrel could be moved by tipping it and rolling it on its top or bottom rim. A heavy filled barrel was best moved by laying it on its side, resting on its narrow, wheel-like circumferential surface.

The modern wooden whiskey barrel's anatomy is straightforward. The body is composed of twenty-four to thirty-six staves—some four to five inches wide, and others narrower, measuring two to three inches—assembled alternately into a jig-like retaining hoop. Staves may be soaked in water or steamed to facilitate bending. In cross section, each stave is planed into an elliptical shape with chamfered or angled edges. Coopers notch, or croze, staves about one and a half inches from each end. When the staves are assembled, these notches form a circumferential groove into which the head or end cover fits. Circular barrel heads are made out of wood sections of varying widths jointed or doweled together with a beveled outer edge. When assembled with the staves, the head's beveled outer edge rests in the groove. Six to eight wooden or iron bands or hoops hold the staves in place. Traditionally, coopers made the wooden bands, and blacksmiths and iron foundries made the iron bands and retaining rivets. The cooper cuts a round bunghole in the center of one of the wide staves to allow filling and emptying. The hole is filled with a tight-fitting plug, often made of yellow poplar (*Liriodendron tulipifera*). If filled barrels are stored on their sides, the bung stave is positioned on top.

Coopering and joinery remained exclusively skilled manual trades in the early 1800s. In Cincinnati, lumber mills began to mechanize by 1815, producing staves for cooperage. But skilled coopers made and finished barrel staves, heads, and hoops with hand tools until the various steps were mechanized, beginning in the 1840s. Many Kentucky coopers learned their trade in Europe or by apprenticing to European coopers who had migrated to America.[3] William Dorsey operated a cooperage on Main Street in Lexington at the turn of the nineteenth century. The "help wanted" ad he placed in the *Kentucky Gazette* in December 1801 suggests that the European method of developing skilled craftsmen through apprenticeships was also practiced in Kentucky. Dorsey's ad stated: "Wanted Immediately, Two or Three JOURNEYMEN COOPERS, to whom good wages will be given—Also Two or Three APPRENTICES to the above business." Dorsey's ad also included an appeal for 8,000 to 10,000 staves.[4]

Barrels intended to hold liquids—naval stores such as turpentine, brine containing pickles or sauerkraut, wine, whiskey, and potable water—were deemed "tight" and required skilled, exacting work to shape the staves, joint the heads, and assemble (or "raise," in cooperage idiom) the barrels. Tight barrels were made predominantly of oak. Solids such as sugar, flour, meal, tobacco, peanuts, smoked and salted meats, metal goods such as hardware and nails, and cement were stored and shipped in "slack" barrels. A range of woods could be utilized to make slack barrels; species of elm, pine, gum, beech, basswood, and maple were most commonly used.[5]

From the first decades of European settlement through the mid-nineteenth century, America's material culture and economy were enabled by wood. It was the most important source of thermal energy. Wood was the primary construction material, whether used to build houses, barns, mills, bridges, fences, or ships. Craft workers converted wood into furniture, wagons, carriages, farm implements, hand tools, and, of course, containers. Conifer tree sap was the basis for a large naval stores industry, and wood could be burned to produce potash or processed and distilled into methanol.[6] Frontier farmers gauged a soil's productivity potential by the species of trees it supported.

America's eastern old-growth forest extended from the Atlantic littoral to the eastern margin of the Great Plains. Trees were a nuisance to farmers, as heavy labor was required to clear the land before fields could be planted. Yet the demand for dimensional lumber fostered timber cutting and a sawmill industry that produced rough lumber, beams and boards, and shingles and scantling. Simple wooden tubs and slack barrels could be made from cut logs with basic hand tools such as axes, augers, and froes. Tight barrels required seasoned, precisely cut wood produced by planing mills; oak was preferable, in part because of its low permeability.[7] From the 1870s through the 1910s, Kentucky's rapidly developing timber industry harvested old-growth forest, especially in the eastern and southern counties, where a wide variety of hardwood tree species grew—predominated by white oak (*Quercus alba*). This venerated tree adapted to a broad range of environmental conditions and grew vigorously in valleys and on north-facing slopes. White oak was more widely distributed and more plentiful than any of the region's other tree species. In some areas, oaks accounted for 17 percent of all forest growth—more than twice the amount of hickory, three times that of maple, and seventeen times that of black walnut.[8]

White oak lumber mills had multiple suitors. The timber was favored not only for general building, tools, furniture, and cooperage but also for wagon and steamboat construction. Cincinnati was one of the nation's manufacturing centers, and wagon factories there utilized large volumes of oak and other hardwoods. During the first half of the nineteenth century, two large boatyards operated in Jeffersonville, Indiana, across the Ohio River from Louisville; one opened in 1819 and the other in 1834. Both built boats through the early 1880s, but by that time, they were compelled to purchase their white oak from West Virginia lumber mills, 550 miles away, because the local supply was nearly exhausted. Downstream, below Evansville, Indiana, oak stands were still plentiful, and mills there supplied boatyards at Paducah, Kentucky, and Cairo, Illinois.[9]

Timber cutters and "bolters" judged standing white oak trees' suitability for barrel wood. They placed the highest value on trees that were at least seventy-five years old, had large-diameter trunks, and were branchless from butt or base to twenty feet or more.[10] A superior oak had "two cuts of 'queen's pipe,' five cuts of whiskies and short 'splayed stock' clear to the limbs."[11] "Queen's pipe" was straight-grained wood, without imperfections, and yielded stave stock sixty-six inches long; "whiskies" were staves with small flaws but suitable for whiskey barrels; and "splayed stock," obtained from the upper trunk, was used for small cone-shaped vessels.

From the mid-1830s until the 1870s, lumbering in eastern and southern Kentucky was conducted almost entirely by itinerant laborers and farmers clearing their own land and floating logs by creek and river to downstream sawmills. Working with basic hand tools such as the broadax, froe, and maul, farmers also produced rough dimensional lumber, laths, fence rails, and stave billets.[12] Farmers worked their timber stands seasonally in a complementary farming–timber cutting cycle. Tree cutting began in late fall, after crop harvest, and continued through winter until field work and planting commenced again in March. Winter rains and early spring snowmelt filled streams and carried cut logs to downstream sawmills. Timber cut close to the Ohio River was often sawn by sawmill boats.[13] Most timber was processed into dimensional lumber used for a wide range of products, including planking, wheel spokes, and tool handles. Rough-cut wood was transformed into shingles, fence posts and rails, railroad ties, and cordwood for fuel. Some farmers produced cooperage staves from their own woodland; others worked seasonally for lumber companies.[14]

The production of wood for staves usually began on farms along the lower reaches of accessible creek bottoms; as the closest stands of quality white oak were cut, the industry gradually moved upstream toward the drainage basin's headwaters. Sawyers cut oak logs into stave-length bolts or billets. Bolters used a froe to hand-split the billets into staves, which were then roughly dressed with a drawknife. "Riven" staves so produced were acceptable for slack barrel construction. Farmers hauled their raw stave stock to the nearest railroad station, where they stacked the wood in large piles to await shipment to in-state cooperages or international customers.[15] By the 1870s, stave-cutting crews were forgoing their hand tools in favor of steam-powered band and barrel saws and other mechanized equipment.[16]

In 1872 Lexington distillers William Tarr and James Thomas, together with Cynthiana distiller Thomas Megibben, incorporated the Kentucky Union Railroad. They planned to build a line into southeastern Kentucky to assure their distilleries a reliable supply of coal and staves. The rail line was completed from Winchester east to Clay City in 1885; an extension west from Winchester to Lexington opened in 1890. The rail line gave eastern Kentucky lumberyards an outlet that did not depend on high water in the Kentucky River, and by 1900, Clay City had become one of the largest stave shipment centers in the nation.[17]

By the 1880s, Kentucky and Indiana lumber mills were the primary suppliers of barrel staves to the American cooperage market.[18] From the 1880s through the 1910s, the extension of railroads into Appalachian Kentucky, Tennessee, and West Virginia supported the construction of stave mills in those regions, which not only supplied American barrel makers but also produced enough staves—1 million pieces annually—to export to whisky distilleries and breweries in Austria, Canada, Croatia, Ireland, and Scotland.[19] In 1900, 1,280 Kentucky sawmills cut more than 392 million board feet of white oak lumber.[20] Sawmills clustered along the rivers that drained the Appalachians, especially the Big Sandy, Little Sandy, Licking, and Kentucky. Seven saw and planing mills operated in Frankfort on the Kentucky River; most were clustered near the mouth of Benson Creek.

In 1883 and 1884 the commerce moving on the sixty-five miles of the lower Kentucky River between Frankfort and Carrollton included 130,000 sawlogs and 500,000 staves worth $40 per thousand. In July 1884, two years after completion of the four locks on the river's lower section, annual commerce was valued at more than $5 million; three years later, in 1887, the value reached $10.8 million.[21] Cooperage materials cut by Kentucky sawmills in 1900 included more than 63 million staves and 3.5 million sets of barrel heads. In total value, Kentucky sawmills produced more than $21 million worth of cooperage materials—hoops, staves, and headings. Although cooperage wood accounted for only about 6 percent of all forest products, staves, heads, and hoops were more highly valued than basic dimensional lumber.[22] The market demand for tight cooperage closely mirrored distillery whiskey production, and stave mills operated at capacity when whiskey production surged.

Wood imparts flavor to a barrel's contents; more than 400 flavor congeners, including esters, carbonyls, lactones, and phenols, can be present in barrel-aged whiskey.[23] American white oak can taste of vanilla and coconut, but thorough drying can reduce the intensity. White oak's cell structure evolved to resist invasion by pathogens, but it also blocks or slows the passage of liquids. Stave wood is usually air-dried in outdoor stacks for one to two years, or until the moisture content is reduced to 12 to 14 percent. Nineteenth-century cooperage shops were often surrounded by small lumberyards, where wood was stacked under open-sided sheds or left uncovered. Though initially sawed or split into straight lengths, barrel staves must be bowed. To facilitate bending, sawmills cut the oak trunk wood to standard lengths, or bolts, followed by quarter-sawing to achieve a desired grain angle that resisted cracking or breaking.[24] Coopers also preferred to use quarter-sawn staves because they thought the sawing technique reduced evaporation from the barrels.[25]

Barrel stave edges are doubly curved. Because barrels are wider across the center than at the ends, each stave has opposing curves; the broad center section tapers to narrow, equal-width ends. When viewed from the side, the stave bows from the ends to the center. A stave's side edges are flat but tapered to match the angle of the barrel's radius. Coopers traditionally cut these edges with a special jointer plane or a mechanical stave jointer.[26] Assuming that each stave's side edges are correctly tapered, the press fit provided by the iron retaining hoops should reduce leakage. Filled whiskey barrels are "professional drinkers," in the sense that the wood absorbs liquid; in addition, both alcohol and water evaporate through the wood—the ambient air temperature determines which one vacates first—and liquid may leak through the stave joints. This loss through evaporation or leakage is termed "outage."[27] In the 1880s even well-constructed oak barrels could lose ten to twelve gallons during a three-year aging period.

The cooperage industry brought together raw materials—white oak and iron for whiskey barrels—and special tools used to make barrels, kegs, tubs, and hogsheads for a broad market. Accomplished coopers possessed a selection of axes, adzes, froes, planes, shaves, chamfer knifes, hammers, and hoop drivers. Woodworking tool manufacturers that made specialized hand tools for coopers and carriage and wagon makers began to produce specialized steam-powered machines that coopers adopted to modernize their factories. Planing and mortising machines were available by about 1830.[28] By 1900, toolmakers had invented, patented, and offered for sale more than 380 specialized cooperage machines. Some were powered by hand cranks or foot pedals; others turned by pulley and belt systems driven by steam engines. Straight, quarter-sawn oak slats that artisan coopers laboriously but expertly transformed with hand tools into curved and chamfered staves could be precisely shaped by machines at rates of more than 1,000 per hour. Some machines made barrel heads with grooved edges and curved metal hoops of differing diameters. Other machines lashed staves together with power windlasses and drove hoops into position for riveting.[29]

Although cooper-crafted containers were generally in high demand, cooper shops, unlike iron foundries and textile mills, did not necessarily operate in centralized manufacturing centers. The cooper's stave material was widely available. Power requirements were minimal before mechanization, and thereafter, steam power could be adapted to a broad range of locations and building sites. The cooper's customer base included the general public as well as manufacturers whose goods required wooden containers, whether meat markets, flour mills, nail and bolt foundries, or distilleries. In 1850, 1,027 coopers worked in Kentucky in a variety of situations.[30]

Coopers working in small independent shops as owners or employees could supply diverse markets. George Sibert of Nelson County was one such cooper. He was twenty-seven years old in 1850, and his household included his wife and young son, as well as two boarders who were also coopers and likely Sibert's employees. Small-scale cooperages like Sibert's made extensive use of specialized hand tools and were slow to mechanize, although these coopers were proficient and productive. They may have purchased raw lumber and worked it into staves, heads, and hoops before assembling the parts into finished barrels. Cooper shops often operated on a made-to-order basis, and work slowed when distilleries shut down for the summer or started production late in response to slow markets. Alternatively, coopers might produce slack barrels, although this diversification involved different raw materials, suppliers, and customers.[31]

George Sibert Household, Nelson County, Kentucky, 1850

Name	Age	Relationship	Occupation	Place of Birth
George Sibert	27	Father	Cooper	Kentucky
Mary K. Sibert	22	Mother		Kentucky
John W. Sibert	1	Son		Kentucky
Dillard Cottingham	25	Boarder	Cooper	Kentucky
George W. Johnson	47	Boarder	Cooper	Ocean*

Source: US Census of Population Manuscripts.
*He was likely born on a ship to parents migrating across the Atlantic in 1803.

A second type of cooperage shop was a larger commercial business that operated in an urban area and had a dozen or more employees. These urban coopers might make staves and heads or buy finished staves, headings, bungs, and hoops; after assembly, they sold the barrels to a variety of manufacturing, retail, wholesale, and private customers, including distilleries. The large-scale commercial cooperage likely adopted specialized machinery as soon as it became available, thereby enabling substantial increases in productivity. A completely mechanized cooperage shop might operate fifty or more different machines, each one performing a specific operation but integrated into an assembly line.[32]

A third type of cooperage was dedicated to one distillery or a group of distilleries. These coopers might be distillery employees or operate the cooperage under contract. The C. B. Cook & Company Distillery in Harrison County, for example, operated a cooper shop with eight coopers as regular employees.[33] During the 1880s the Hermitage Distillery in Frankfort employed sixteen coopers. Each day the riverside works mashed about 1,000 bushels of grain, producing sufficient sour-mash bourbon to require forty "iron bound barrels."[34] Distillery cooperages often adopted mechanization early on and were capable of high-volume production. This type of cooperage likely purchased stave and head stock, perhaps preformed, and worked in tandem with a distillery blacksmith who produced iron barrel hoops and rivets. Given that whiskey barrels were never used twice, high-capacity distilleries required staves, barrel heads, and hoop iron in large quantities. Some distillers preferred to buy their hoop iron directly from foundries in Pittsburgh, Youngstown, and Cincinnati or from the iron-producing areas of Michigan, Minnesota, and Wisconsin. Certain distillers regarded Pittsburgh hoop iron as the best quality available.[35]

A dedicated cooperage might also include a sawmill. The Loretto Distilling Company built a high-capacity, post-Prohibition distillery in western Marion County in 1936. The

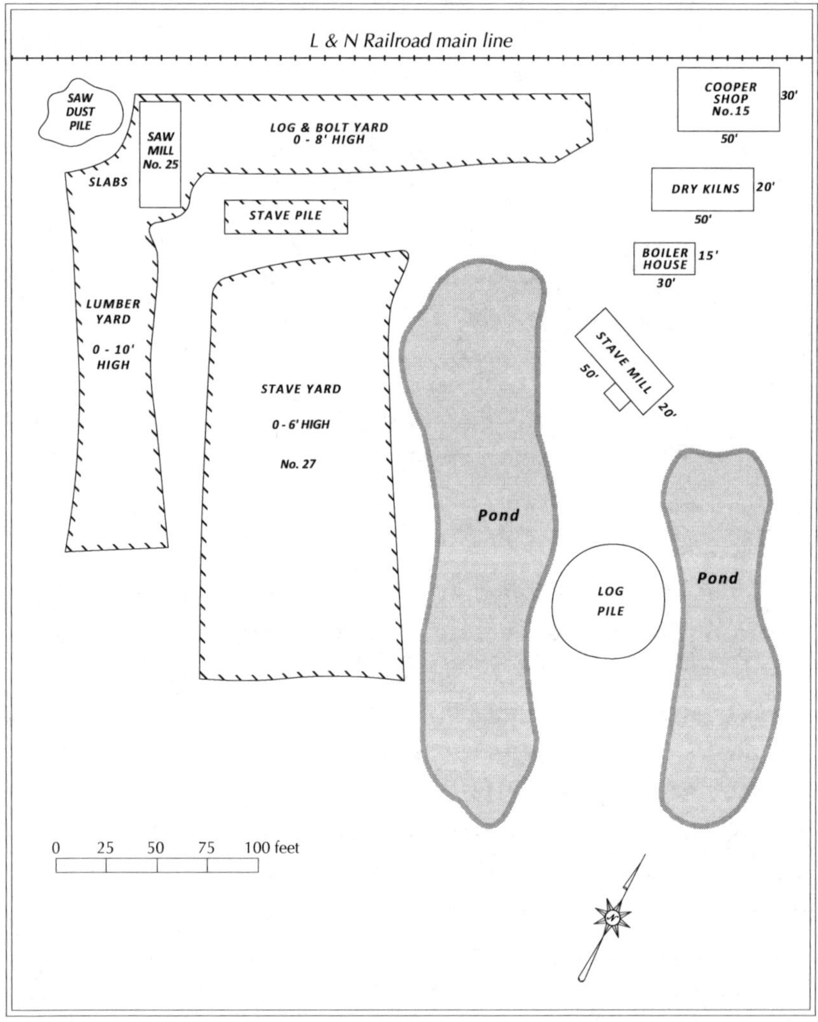

Dedicated sawmill and cooperage shop, Loretto Distilling Company, Marion County. Note that the distillery was built on the Louisville & Nashville Railroad, and the cooperage was adjacent to the distillery in Building No. 15. (Kentucky Inspection Bureau, Report No. 1828, 1946, Oscar Getz Museum of Whiskey History, Bardstown, KY)

distillery works stood beside the Louisville & Nashville Railroad tracks. In the adjacent wood yard, a sawmill cut logs into lumber and stave stock; after drying, a stave mill processed the raw wood into finished staves for the cooper shop.

In the Ohio Valley, Cincinnati was a major woodworking center; it had fifty-six cooperage establishments employing 1,270 people.[36] Three cooperages and five distilleries operated across the Ohio River in Covington. Owensboro had eight distilleries and three cooperages. In Evansville, five cooperages employing 425 people used wood largely from southern Indiana to produce barrels for Kentucky and Illinois flour mills and distilleries.[37] Louisville had twenty-three cooperages in 1886, producing $950,000 worth of products annually.[38] Kentucky's seventeen-county Fifth Whiskey District extended from Owen County in the north to Adair County in the south. In 1887 the district's distillers produced more than 35 million gallons of whiskey requiring some 165,000 barrels.[39] Industrial distilleries that were planning to increase their mashing and distilling capacity had to ensure not only that their grain and coal suppliers could fill their orders but also that their cooperage shops had sufficient white oak staves on hand and skilled coopers capable of supplying additional barrels. The owners of Anderson County Distillery No. 418 at Tyrone, Kentucky, installed new machinery in 1899 and increased their mashing capacity to 4,000 bushels of grain per day, making theirs "the largest 'mash tub' distillery in the world" and creating a substantial market for tight white oak barrels.[40] In 1903 many Kentucky distillers shut down some two months early, ending their distilling season in March because they were unable to procure enough barrels to continue operations.[41]

Andrew J. Oots and his sons operated an urban cooperage at 239 West Main Street in Lexington from the 1880s through the 1920s. The family resided in an adjacent building.[42] The Oots cooper shop manufactured both tight and slack barrels, including full-size forty-eight-gallon whisky barrels, forty-two-gallon tierce barrels used for salt pork and sausage, half and quarter barrels, five- and ten-gallon kegs, and tubs of various sizes.[43]

At the Oots cooperage, production and sales followed a seasonal pattern. Slack barrel manufacturing predominated in the summer and fall months, with peak production in July. The timing likely correlated with the harvest of grains, fruits, and vegetables. Tight barrel production began in late September, reached 400 barrels per week in December, and increased progressively to 600 barrels per week in April—the barrel production schedule mirroring the annual distilling season. Tight and slack barrel production overlapped from September to mid-November, with the shop producing 100 to 200 barrels of different types each week.

Mr. Oots's three sons, W. Edgar, Homer Franklin, and Robert, worked full time in the cooperage in the 1880s. They used hand tools to shape staves and cut and assemble heads, making four to five barrels per day. In addition, they delivered barrels to distilleries. John Randolph, who also worked in the shop, was apparently a skilled and experienced cooper; he made six barrels each day. Mr. Oots employed other coopers from Ireland and Italy

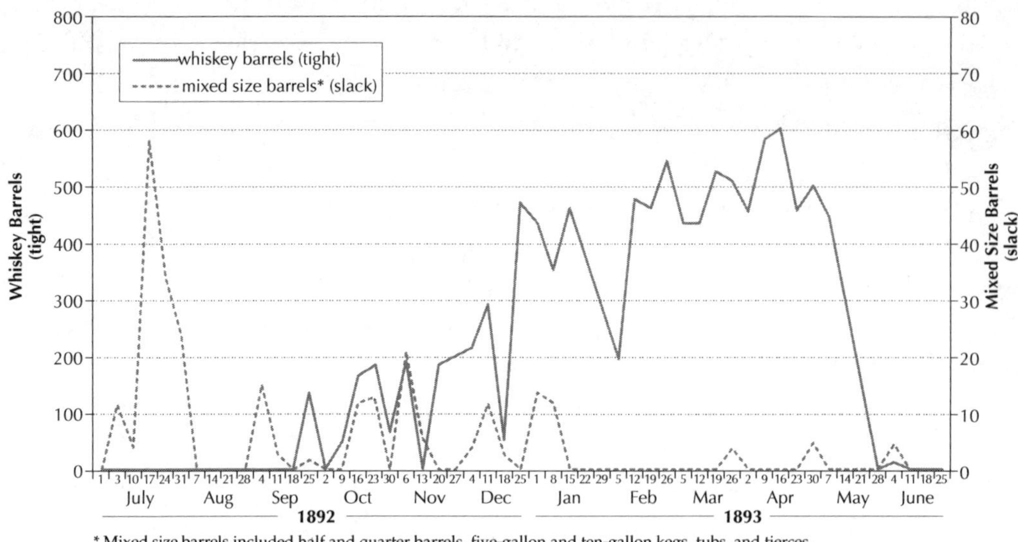

Cooperage sales of Andrew J. Oots & Sons, Lexington, July 1, 1892–June 30, 1893. (Compiled from Andrew J. Oots Cooper Shop Records, 1880–1926, Special Collections, University of Kentucky)

and, possibly, two brothers from Hungary. In the early 1880s there were five to twenty-two coopers working in the Oots shop during any given week. Seven coopers could produce 600 barrels per month, if sales warranted.

Although Mr. Oots did not sell any barrels during some weeks in the summer and early fall, his coopers worked steadily making and stockpiling them. The coopers also repaired barrels, often traveling to works as distant as the McBrayer Distillery at Mt. Sterling, in Montgomery County, to make repairs onsite. From January 1 to June 30, 1881, Oots's coopers visited William Tarr's distillery in Lexington on twenty-seven different days and made more than $300 in repairs, including installing two slop cisterns, replacing barrel staves and heads, repairing yeast and mash tubs, and installing new tubs.

Mr. Oots purchased stave wood, heads, bungs, and hoops from twenty-seven different suppliers across Kentucky and in several adjacent states, and he bought surplus stock from other distilleries. Many of his sources were likely independent farmers or sawmill operators in the watersheds of the Kentucky, Licking, and Big Sandy Rivers in eastern Kentucky. Other suppliers were clearing and farming land in the Knobs of south-central Kentucky, on the rim of the Outer Bluegrass. Multiple raw materials dealers were a prerequisite for operating an efficient and profitable cooperage, as not all dealers had equal or timely access to usable logs. Reputable sawmills and lumberyards graded their lumber by quality and moisture content and declined to sell green wood that had not been thoroughly air-dried. Competition among dealers moderated prices. And, given the exhaustive nature of timber cutting, the lumber business had to be mobile; uninterrupted access to un-cut forest by road and rail was paramount. Limited access to trees or transport, or failure to compete effectively with other lumber operations, placed some dealers on the edge of insolvency, which in turn placed cooperages in financial jeopardy unless they could obtain wood from other suppliers.

Andrew Oots's cooperage shop in central Lexington occupied the back half of a lot that extended from West Main Street to Water Street. The shop opened onto the Water Street railroad tracks, adjacent to three warehouses on lots to the east.[44] Oots usually had stave wood shipped to his cooperage via railcar—about 7,400 staves per carload. Although the cooperage operated on the building's first and second floors—connected by an ele-vator—the shortage of floor space suggests that Oots may have used one or more of the adjacent warehouse buildings to store his lumber and protect it from the weather. Quality oak staves cost $40 per thousand, and barrel heads cost about 30 cents for a set of two. Shipping costs in the early 1890s varied with distance and commodity, but it commonly cost $23 to $29.50 per carload of staves delivered to Lexington and more than $57 for one carload of barrel heads.

Other suppliers were large, commercial-scale saw and planing mill operations. Oots bought more than 5,000 sets of heads from the H. Alfrey Company of Memphis, Tennes-see, from 1891 to 1894. Little Sandy Cooperage Company operated near the Ohio River in Greenup, Kentucky, and sold 94,000 staves to Oots in 1892, 116,000 in 1893, and 64,000 in 1895. In 1908 Little Sandy Cooperage purchased the Greenup Cooperage Company's plant for $6,000, and the consolidated works continued to operate in Greenup.[45] Oots also ob-tained wood from the Southern Lumber Company in Arkansas.

Stave supplier H. L. Weisahau likely had a special relationship with the Oots coo-perage. Mr. Weisahau sold four freight cars of staves—25,250—to Mr. Oots on August 31, 1894. A week later, on September 7, Oots paid Weisahau $577.50 for the staves and $90 for freight costs and bought him a "suit of clothes" for $22. During the period for which records exist, Oots's largest annual purchase of staves—207,000—occurred in 1892. The F. S. Ashbrook Company of Cynthiana, an active distillery with a dedicated cooperage shop, sold heads and staves to Oots in the mid-1890s in what may have been a sale of sur-plus stock.[46] Oots also sold barrel heads and iron hoops to other distillers. In August 1895 Thomas Ripy, owner of the Ripy Distillery in Anderson County, bought forty-four bales of iron hoops, fifty sets of heads, and twenty-five barrels from the Oots cooperage.

Oots obtained iron barrel hoops from three different commercial suppliers. In 1891 he bought 4,216 pounds of iron hoops from the J. Weller Company in Covington, Kentucky. That same year, he bought 22,870 pounds of iron hoop—about 1,430 six-hoop sets—for $434.47 from the Youngstown Rolling Mill in Ohio. He made no additional purchases from the Weller Company or the Youngstown Mill thereafter. The Painter & Sons Rolling

Mill in Allegheny County, Pennsylvania, was the cooperage's largest supplier, selling more than 61,000 pounds of hoop iron to Oots in 1891, 59,000 pounds in 1892, and nearly 61,000 pounds in 1893.

The ledgers kept by Andrew J. Oots & Sons cooperage for the years 1891 to 1895 listed barrel and mash tub purchases by twenty-seven central Kentucky distilleries. Over the five-year period, Oots sold to distillers 33,960 barrels, or an average of about 6,800 per year. His largest customers were Wiley Searcy's Old Joe Distillery in Anderson County, Christian Stege's Glenn Creek Distillery in Woodford County, and the Bourbon County Distillery (also known by various other names, including the Howard & Bowen Distillery).[47]

Mr. Oots paid his coopers 55 to 80 cents each for the barrels they made in the 1890s; he charged customers $1.95 to $2.35 for whiskey barrels, $1.30 to $1.60 for half barrels, and $1.75 to $2.15 for mash tubs. He charged only 75 cents for ten-gallon kegs. In the late 1890s Oots sold "sets" of beer barrel stock to the R. B. Moore Company and G. F. Hasty & Sons (600 sets for $417.50). His cooperage also made oval tubs and quarter and half barrels that he sold to the Lexington Brewing Company in 1898.

Oots shipped large barrel orders to customers via over-the-road wagons and railroad freight cars. He also made local deliveries in and around Lexington. In June and July 1893, for example, Oots charged $54.48 to haul 1,363 barrels from his cooperage on West Main Street to the J. E. Pepper Distillery on Manchester Street. Delivery charges to distilleries beyond Lexington included turnpike toll fees, which were based on the size of the wagon and the width of its wheel rims.

Local businesses and individuals also bought small barrels and kegs from the Oots cooperage. August Belmont Jr. of New York was a notable customer. Belmont established the Kentucky Nursery Stud Thoroughbred Farm north of Lexington in the 1890s. On August 28 and again on November 29, 1895, Oots sold $8 water barrels to Belmont's farm. The Kentucky Nursery Stud Farm later produced Man o' War, winner of twenty of twenty-one stakes races and the horse that many regard as the best Thoroughbred to ever race in America.

The distilleries that were Oots's primary customers were subject to a capricious business environment. Drought-related grain shortfalls reduced production in some years. The market demand for whiskey fluctuated from year to year. And the national economy could be volatile—the financial panic of 1893 destabilized many businesses. In the 1880s and 1890s distillers cycled into and out of the cooperage market. Some distillers sold their businesses to others who operated the works under different names or ceased operations altogether, reducing overall production. Such volatilities had a dramatic effect on Oots's customer list. In 1881, for example, the William Tarr Distillery, a large, 400-bushel-per-day operation in Lexington, bought more than $12,000 worth of barrels from the Oots cooperage. Tarr's last transaction with Oots was in 1894, and Tarr went bankrupt three years later.[48] Recurrent hints of widespread financial distress among distillers appeared in Oots's ledgers from the 1890s. Instead of paying cash for cooperage upon delivery, some

Cooperage Sold to Distilleries by Andrew J. Oots & Sons, 1891–1895

Distillery	Registered Distillery Number	County	1891 Barrels	1891 Mash Tubs	1892 Barrels	1892 Mash Tubs	1893 Barrels	1894 Barrels	1895 Barrels
T. H. Bond	236	Anderson						1,378	140
Bourbon County	102	Bourbon			985		1,449	2,029	
Commonwealth	112	Anderson	62		1,365		304	30	1,656
Isaac H. Davidson	148	Fayette	13		30		22		
William J. Frazier	50	Woodford	730	16	787	28	701		
W. W. Green	220	Elliott							20
S. J. Greenbaum	9	Woodford	7						925
J. A. Grimes	117	Bourbon	13		108		115		
Nat Harris	46	Fayette	1,043		881				
Lexington Club	86	Jessamine						656	812
Silas Maggard	107	Elliott							10
J. K. Marshall		Scott	12						
J. W. Masters	802	Madison						76	
John H. McBrayer	17	Montgomery			310				
T. J. Megibben Co.	77	Bourbon			442		2,182		75
F. P. Merritt	22	Clark	47	16	35	6	65	12	10
Edward Murphy & Co.	400	Anderson					104	56	
Peacock		Bourbon			155		1,633		
James E. Pepper	5	Fayette					346		
Thomas B. Ripy	112	Anderson	280						25
Wiley Searcy	45	Anderson	1,107		1,764		1,159		
Zack Sellers	49	Woodford	742	30	404	18	125		
J. J. Skaggs	294	Lawrence							11
Christian Stege	57	Woodford	2,210		1,296		1,829	147	
William Tarr & Co.	1	Fayette			228		46		
Waterfill & Frazier	41	Anderson			796	20			
Wigglesworth Brothers	10	Harrison						230	
Total			6,266	62	9,586	72	9,734	4,690	3,684

Sources: Compiled from Andrew J. Oots Cooper Shop Records 1880–1926, 1880–1910 (bulk dates), 1 FM-49, Special Collections, University of Kentucky; Chester Zoeller, *Bourbon in Kentucky: A History of Distilleries in Kentucky,* 2d ed. (Louisville, KY: Butler Books, 2010).

Note: Oots charged about $2.25 retail per barrel in 1891, with discounts for large orders. His labor cost per barrel varied from 55 to 90 cents. In addition to "tight" barrels for whiskey, water, and other liquids, Oots manufactured "slack" barrels and kegs in various sizes for pickles, lard, sausage, corn, etc. No mash tubs were sold in 1893, 1894, or 1895.

distillers began to request payment on thirty- to ninety-day notes, including the McBrayer Distillery in Anderson County in January 1893, the S. J. Greenbaum Distillery at Midway in January 1895, and Commonwealth Distilling Company in Anderson County in July 1895. As distilleries suspended production or closed in the mid to late 1890s, Andrew Oots increasingly turned to producing mixed-size tight and slack barrels and tubs for the general commercial market, used for the storage of comestibles and hardware.

Coppersmithing

Copper, rather than aluminum, brass, iron, pewter, or tin, has long been the preferred material for making stills, condensation coils or worms, and distillery pipes. Copper has several qualities that make it superior for use in distilling. The metal is malleable and can be readily worked into complex shapes without heavy dies and presses. Copper conducts and distributes heat evenly across a surface but also effectively extracts heat from vapor, thereby promoting condensation. The metal resists corrosion and is, to an extent, antimicrobial. When used in a whiskey still, copper is thought to react with the sulfur compounds produced by yeast fermentation, such as dimethyl trisulfide. These sulfur compounds may produce unpleasant odors, and exposing heated, fermented mash to a copper still pot creates copper sulfate, which adheres to the copper vessel but can be removed by post-fermentation cleaning.[49]

Obtaining refined and fabricated copper sheets, bars, and wire stock for use by American coppersmiths has been a complicated process rife with difficulties, including the discovery of and access to sizable ore deposits, the development of suitable refining technologies, the resolution of transportation issues, and international colonial and post-colonial trade policies and regulations. The principal sources of copper for the general world market during the first third of the nineteenth century were England, Ireland, Cuba, and Chile. England exported more than 1,400 tons to the United States in 1844, much of it purchased by East Coast metal fabricators and shipbuilders in port cities such as Boston and Baltimore.[50]

Copper was first commercially mined in colonial America at Simsbury, Connecticut, in 1709. Thereafter, a modest amount of North American copper mining began to occur in the uplands of eastern Canada, at various discrete sites on the Piedmont, and in the Appalachians from Maine south to Tennessee and Alabama. Small sulfide copper deposits with assays of 1 to 5 percent copper could be found in veins in certain limestone strata and as bodies of ore in some volcanic rock formations.[51] Mines were operating in eastern Pennsylvania by 1700 and in New Jersey by 1712. The copper ore was mined, concentrated, and shipped to England as part of a British colonial policy to send American raw materials to England, where they would be used in manufacturing, shipbuilding, or military ordnance, for example, or turned into finished goods for export back to America.[52] The Piedmont copper deposits were exhausted by the beginning of the Revolutionary War.[53]

During the 1790s metal brokers and merchants in Philadelphia and Baltimore imported from England sheet copper as well as crocks, kettles, rivets, and stills ranging in size from 30 to 250 gallons. American smiths lacked the technology to refine raw copper ore, and in the early 1800s they resorted to fabricating copper by resmelting the metal hull cladding salvaged from ships. The metal was then rolled into sheets for use in manufacturing. Paul Revere operated such a copper rolling mill in Canton, Massachusetts, in

1801. After the Revolutionary War, American copper miners exploited small ore bodies in Vermont (1809), Maine (1870s), Maryland (until the 1850s), southeastern Tennessee (by the 1850s), and northern Alabama (1870s). But with few exceptions, mining operations were ephemeral and production trifling.

The largest Appalachian sulfide copper ore discovery was made on the Toccoa-Ocoee River drainage of southeastern Tennessee at Copper Hill and Ducktown in 1843. After basic refining, processors shipped much of the ore to Baltimore. Early in the nineteenth century, Baltimore refiners adopted a Welsh concentration process, and the city became an important processing and rolling mill center, producing quantities of copper sheeting that metallurgists referred to as "red metal." The industry grew rapidly, and Baltimore became the most important red-metal refining center on the Atlantic coast. By the eve of the Civil War, it was one of the world's largest copper production centers.[54]

The extensive native copper deposits in Michigan's Keweenaw Peninsula on Lake Superior were commercially mined by the mid-1840s; by the 1870s, Michigan accounted for more than 90 percent of America's copper production. The Keweenaw copper was pure, requiring only concentration, and it was much cheaper to produce than the sulfide copper refined at Baltimore. By 1870, the Baltimore smelters had closed.[55]

As practiced in America, sulfide copper refining was costly not only monetarily but also in terms of the appalling environmental degradation it caused.[56] The development of an electrolytic copper refining process in Wales in 1869 revolutionized the practice. American refiners adopted the technology and opened a new commercial refining plant in Newark, New Jersey, in 1883. By 1905, the United States had nine large copper refineries in operation, all of them employing high-capacity, low-voltage, direct-current electric dynamos. The largest refinery had a capacity of 350 tons of copper per day, and a significant proportion of the nation's early direct-current electricity production was employed in electrochemical manufacturing, especially copper refining.[57] The new electrolytic refining technology developed in tandem with the discovery of vast copper oxide deposits in Arizona, New Mexico, Montana, and Utah. The nation's rapidly expanding transcontinental railroad system provided an economical means of delivering refined ingots to fabrication and manufacturing centers in the Middle West and on the East Coast. By 1900, fabricated copper sheets and ingot stock were widely available at reasonable prices.

Nineteenth-century coppersmithing was largely a traditional craft that required significant handwork. Many smiths worked with hand tools that had been in use for generations or even centuries. Historically, coppersmithing in Kentucky followed two different forms: folk or traditional craft work, and professional work performed by an experienced "mechanic."[58] Traditional smiths were largely self-taught craftspeople who formed simple containers out of flat copper sheets that they rolled into cylinders and brazed or soldered together (brazing joins two metal parts with a third molten metal filler at 840°F or higher, whereas soldering joins metals at less than 840°F). To make a basic still cooker, the smith simply cut a shape out of flat sheet of copper with hand shears, rolled the cut material into a cone, and bonded the seam. He then fitted the cone to the top of a similarly fashioned copper cylinder, attached a flat copper circle to the bottom, and soldered or brazed all the seams. Pipes could be made of flat sheet material that was similarly rolled and soldered and then bent or coiled to form a condenser.

Professional or mechanical coppersmithing was a demanding occupation that combined practical knowledge of metallurgy, an appreciation of the art of design, and an understanding of applied geometry, as well as the tactile skills to accurately fabricate an item and the kinetic finesse to properly assemble and finish it.[59] One gained such advanced skills through extended training or apprenticeship. The work's complexity resided not in the tools—a forge and a small selection of hand tools sufficed—but in the careful geometric calculations and measurements required to create the precise patterns that allowed the coppersmith to cut and form flat sheets into complex shapes. The smith also had to master the heat ranges and melting points of different copper, brass, and zinc alloys and the chemistry of fluxes when soldering and brazing. The shaping techniques essential for coppersmithing were not readily mechanized, except for simple forms.

Although coppersmithing and coopering were both important complementary trades to the distilling industry, they were fundamentally different in organization and location. Skilled professional coppersmiths commanded high wages and usually made a variety of practical household utensils such as kettles, pots, and pans. They also made hardware—pipes, pumps, scoops, and large specialized tanks and stills. When E. H. Taylor Jr. rebuilt his O.F.C. Distillery in Frankfort in the 1880s, his construction crew lined the fermenting tanks with cement covered with copper sheeting. This impermeable surface, in Taylor's view, avoided the problem of acid saturation, which was common in wooden fermenting vats, and allowed thorough cleanup between batches.[60] Coppersmiths also supplied the brewing, marine, and railroad industries with various copper containers, boilers, pipes, and brass fittings.

Unlike coopers, skilled coppersmiths tended to work in large urban centers, not in small towns or rural industrial works. In the decade from 1860 to 1870, Nelson County, one of Kentucky's major distilling centers, had only one resident coppersmith (see the table "Distilling-Related Employment, Nelson County, 1850–1920," in chapter 11). Most coppersmiths worked in larger towns and cities such as Louisville, Cincinnati, Lexington, and Owensboro.

The Fishel & Gallatine copper- and tinsmith shop in Lexington opened on Main Street in the early nineteenth century. The shop made stills, kettles, and other copper and tin vessels and advertised that it was a place "where those who pleased to favor them with orders, may depend on their being strictly executed." The smiths announced that they had "received a fresh supply of thick copper" and also wished to purchase old copper and pewter.[61] Most nineteenth-century coppersmiths bought scrap copper, brass, pewter,

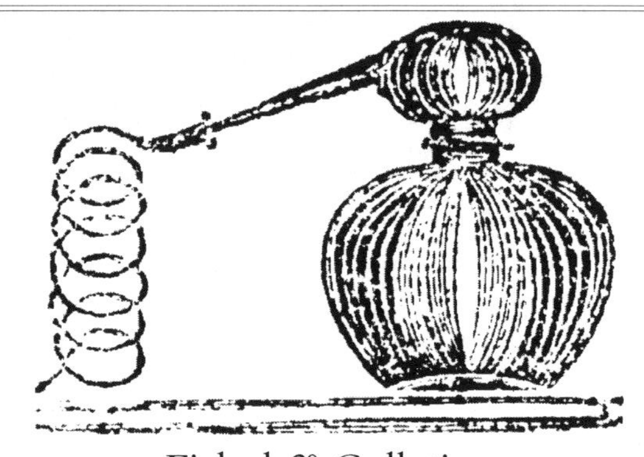

Advertisement for Fishel & Gallatine, Copper and Tin Smiths, Lexington. (*Kentucky Gazette*, February 27, 1809)

and tin, suggesting that these metals were in short supply. Copper stock was expensive to buy and transport, but most accomplished smiths possessed the necessary metallurgical knowledge and equipment to repair old objects and to process scrap into sheets, bars, and wire material for reuse. Later in the century, illegal stills became another source of copper. When US marshals and revenue agents raided illegal moonshine operations, they demolished the equipment and confiscated the copper, which they sold to coppersmiths, who might, ironically, use it to make another still.[62] Brokers bought and sold copper by weight rather than by dimension.[63] In 1810, after Fishel and Gallatine liquidated their partnership, Mr. Fishel advised customers that his shop had received a "large assortment of COPPER & TIN, and has engaged from the Eastward, some of the first workmen in his line of business, from which circumstance he can with full confidence assure his friends and the public, that any work done by him will be executed in a *superior manner*, to any done in this *State heretofore*."[64] Teamsters made regular trips by road and river from central Kentucky to Philadelphia and Baltimore to obtain goods for sale, including sheet copper. Carriage costs over that distance were estimated at $65 per ton, and a fully loaded Conestoga-type freight wagon carried five tons.[65] The cost of one wagonload of freight so transported would be roughly $6,500 in present-day currency.

By the 1850s, more than twenty metal fabrication shops operated in Louisville, employing more than eighty smiths. Skilled smiths were in demand, as were shop boys or untrained assistants, who earned $5 to $7 per week. Louisville smiths produced whiskey stills, condensing coils, and related distillery equipment in significant amounts. They also fabricated equipment for brewers, steamboat builders, hotels, and hardware manufacturers.[66] Candlemakers, such as Procter & Gamble in Cincinnati, converted tallow produced by slaughterhouses and butcher shops into a stearic fatty acid that they distilled into candle material. Candlemakers initially obtained their large copper stills from Philadelphia, but by the 1850s, Louisville coppersmiths were able to fill their orders with acid-resistant equipment.[67]

Louisville's metal fabrication shops tended to cluster together. They were situated to take advantage of access to the Ohio River and primary roads, as well as for the convenience of having like businesses nearby as a source of raw materials, tools, or warehouse space. Importantly, proximity allowed metalsmiths to interact directly. The spread of innovative ideas is enhanced by homophily, or occupation similarity and workplace propinquity. With clustered businesses, metalworkers were much more likely to discuss common problems and possible solutions with one another, and they could observe the features and quality of a competitor's product if their shops were proximal.[68] Most of Louisville's metal fabrication shops operated along Main Street from Third Street to Ninth Street or somewhere nearby.[69]

A Cincinnati copper and brass works, Hoffman, Ahlers & Company, established a successful branch shop in Louisville in 1879. A related family business, Hoffman & Company, opened a short-lived copper works at 725–729 East Main Street in the early 1900s. By about 1900, Hoffman, Ahlers & Company had become the Vendome Copper and Brass Works, operating at 625–627 East Main Street.[70] Vendome manufactured still equipment for several industrial distilleries in the 1890s and 1910s, including Glenmore Distillery in Owensboro, E. H. Taylor Jr. & Sons near Frankfort, and J. T. S. Brown & Sons, which operated distilleries in several locations, including the Old Prentice Distillery near Lawrenceburg. Vendome became one of the nation's premier red-metal still manufacturers; it remains so today and serves a worldwide market.[71]

Glassworks and Bottle Making

The demand for glass whiskey containers developed in two stages. Traditionally, when barrel aging was complete and the whiskey was ready to market, distillers sold it in full barrels to retailers and wholesalers. The marketing process was straightforward: whiskey was moved in its original storage barrel by way of shipping docks, wagons, and perhaps railcars to stores and saloons, where it was sold by the glass, flask, or jug.[72] In the early 1850s, the Farley & Taylor general store in Richmond, Kentucky, dispensed barreled whiskey in handblown flasks made to order by the Kentucky Glass Works in Louisville.[73] Some distillers sold whiskey directly to customers who brought their own containers for filling at the distillery's "quart house." By the 1860s, saloon keepers often bought whiskey by the barrel and diluted it with water or other additives before selling it to customers in small or short-capacity takeout bottles referred to as jojo, picnic, or shoofly flasks. As the sales of "mobile" whiskey increased, so did the demand for flasks and bottles.[74] Handblown glass was expensive, though, and distillers did not begin bottling and selling whiskey in cases until automation allowed glass manufacturers to reduce their prices. The innovation of marketing Kentucky whiskey in glass is attributed to George Garvin Brown, a Louisville wholesaler. In 1870 Brown began to dispense Old Forrester (later renamed Old Forester) bourbon in handblown bottles to doctors and druggists, who marketed the product as medicinal whiskey. In 1886 the J. E. Pepper Distillery in Lexington began to sell Old Pepper whiskey in handblown embossed bottles.[75]

Manufacturing handblown glass is an expensive and complex craft that requires significant skill and physical endurance, along with a broad range of raw materials—pure glass sand, soda ash, nitrate of soda, lime and limestone, arsenic, manganese, litharge or lead oxide, and pearl ash or potash—which glassworks obtained from national and international sources.[76] Nineteenth-century American glassmakers built single-purpose factories that manufactured glass containers for the food processing and patent medicine industries. Jars and bottles were individually handblown into clay, wood, brass, copper, or cast-iron molds. After the 1820s, glass containers could be shaped in chilled iron presses in a process developed by the New England Glass Company of Cambridge, Massachusetts.[77] Even though glass was blown in the same molds, no two containers were exactly the same, and nonstandard shapes and sizes required hand filling. Three decades would pass between Brown's initial sale of bourbon in handblown glass bottles and machine-made bottle manufacturing that produced standard shapes, dimensions, and volumes that were also suitable for automated filling machines.

Early American industrial glassmakers concentrated in New Jersey and New England, but by 1870, glass companies were operating new works in western Pennsylvania near Pittsburgh and in West Virginia's northern panhandle at Wheeling.[78] Glassmaking was energy-intensive, and the cost of fuel was the most important influence on the location of nineteenth-century American glass plants. Proximity to large markets was also a requisite. Packing and transporting glass bottles—given the high freight rates for glass—could account for 20 percent of total production costs.[79]

From 1860 to 1880 the cost of the fuel used to make glass amounted to about one-third of the value of all glass produced in the United States.[80] East Coast glassworks were initially fueled with wood and then by hard coal, once the anthracite fields of eastern Pennsylvania opened. Glassworks in western Pennsylvania and West Virginia were fueled by soft or bituminous coal. By 1880, national glass production was centered in the Pittsburgh area, where 61 of the nation's 211 firms operated, fueled primarily by cheap, locally mined bituminous coal. When natural gas became available, it was the fuel of choice, partly because of its high heat value and partly because it could be readily transported by pipeline. A glass plant at Rochester, Pennsylvania, on the Ohio River, twenty-five miles downstream from Pittsburgh, became the first to be fueled by natural gas in 1875. The number of works using gas increased steadily thereafter.

In northwestern Ohio, wells drilled into what later became known as the eastern rim of the Trenton limestone gas field began to produce natural gas and oil in about 1860. Gas-field towns such as Findlay, Fostoria, Tiffin, and Toledo offered free or cheap gas to industries to induce them to relocate or build new plants in the region. A well drilled near Marion, Indiana, in 1886 began to produce natural gas from the main body of the Trenton formation. Building on the gas bonanza that followed, more than a dozen Indiana towns, including Gas City, offered free gas to the glass factories operating in their areas. More than 160 new glass factories opened in the Indiana gas country between 1880 and 1900, many of them specializing in flask manufacturing.[81] Yet the ready supply of cheap fuel did not change the process of glass production, which remained traditional; in 1900 all bottles made in the United States were gathered, blown, and finished by hand by skilled workers following century-old techniques.[82]

Edward Libbey moved his New England Glass Company—one of the largest glass manufacturers in the nation at the time—from Cambridge, Massachusetts, to Toledo, Ohio, in 1888 to take advantage of the region's cheap fuel.[83] Libbey also benefited when miners found high-quality glass sand twelve miles west of Toledo at Silica, Ohio.[84] To enhance his Toledo labor force, Libby recruited experienced glassworkers from West Virginia, including Michael J. Owens, a young Irish immigrant employed at a glassworks near Wheeling.[85] Owens was both an accomplished glassblower and a mechanical prodigy. In 1895 he patented a semiautomatic machine for blowing lightbulbs. The machine's multiple molds rotated around a central hub, and it was actuated by a complicated assembly of driving pulleys, shafts, and gears. The assembly eliminated several manual preparation steps that had previously required hand, foot, or lung power.[86] By 1905, the Libby Glass Company of Toledo had licensed some 250 of Owens's semiautomatic glassblowing machines to other glassworks across the Middle West.

By the late 1890s, Owens had extended his experimental work, focusing on fully automated narrow-neck bottle production.[87] In 1903 he organized the Owens Bottle Machine Company, and the following year he patented a large, multiarmed, fully automatic bottle-blowing machine, a mechanical marvel with more than 10,000 parts.[88] The continuous-operation machine rotated above a tank of molten glass. The machine pulled the liquid glass into molds by suction, and air pressure conformed the glass to the molds. The Owens machine could produce 14,000 identical bottles in two twelve-hour shifts, and even though the process required skilled mechanics and machine operators, the per-bottle cost was less than 4 percent that of handblown bottles. Owens's machine-made bottles were suitable for automated filling by high-speed machines, and they ensured that the contents' weight and volume were standardized. Equal container size permitted Congress to issue packaging specifications as part of the Pure Food and Drug Act of 1906.[89]

Glass manufacturing was one of the last major American industries to mechanize, and until the industry adopted the Owens machine, glassmaking was a rudimentary, labor-intensive process. Furnaces generated oppressive heat, and the work was physically demanding in the extreme. Boys (sometimes regarded as apprentices) performed much of glassmaking's difficult work, and few American industries had a greater demand for child labor.[90] Owens licensed his glassblowing machines to other manufacturers, and by 1917, some 200 were in operation. In the aggregate, the Owens machines were equal in capacity to 10,000 manual glassblowers and accounted for half the nation's glass container production.[91] Whiskey and milk bottles required thicker bottoms, so the Owens machine had to run at a slower speed to make them than it did to produce soda or beer bottles.[92] Nevertheless, the machine dominated glass bottle manufacturing, thereby ending a millennium of melting pot–blowpipe bottle making.[93]

William Smith and Edward Levis founded the Illinois Glass Company in Alton, Illinois, in 1873 with the objective of producing handblown bottles for whiskey, medicines, and soda.[94] The company became a primary source of bottles sold to Ohio Valley distilleries, including the U.S. Bottlers' Supply Company in Louisville, which specialized in supplying bottles, corks, caps, and labels to area whiskey distillers.[95] In 1911 Illinois Glass began installing the Owens automatic bottling machines.[96]

Kentucky's nineteenth-century glass manufacturing capacity was modest, with one glass manufacturing firm in operation in 1860 and five in 1880, or about 2 percent of the nation's 211 firms. The state's glassworks were located primarily in Louisville and roughly matched the number of glassworks in West Virginia, although the output of West Virginia's plants had twice the value of Kentucky's.[97] Louisville's first glass manufacturing plant opened in 1850, when the Kentucky Glass Works began operations with twenty-one glassblowers who had likely migrated from Pittsburgh, Wheeling, or southeastern Ohio, where they had learned the trade. The company reorganized and refinanced later that year, and by 1855 it was known as the Louisville Glass Works. A local pharmacist purchased an interest in the works soon thereafter, intending to supply apothecary wares and bottles to druggists. The company operated until 1873 and produced a range of glass products, including "Green and Black Glassware for Druggists, Grocers, Confectioners and Families." The works also made telegraph wire insulators, eagle flasks, and beer and wine bottles, although the company's advertising made no explicit mention of supplying glass containers to the spirits industry.[98] The Falls City Glass Company operated in Portland, on the Ohio River just west of Louisville, from 1884 to 1889. Distiller and whiskey dealer Thomas H. Sherley & Company operated three distilleries and leased the Kentucky Glass Works and the Southern Glass Works (1878–1889) for a short time to make handblown bottles. These and other Louisville glass firms made cylindrical whiskey bottles, glassware, fruit jars, and medicinal bottles.[99]

Kentucky's five glass plants employed some 522 workers in 1880, of whom 147, or 28 percent, were children under the age of sixteen.[100] By 1900, only two Kentucky glassworks remained, but both were idle; neither firm reported glass manufacturing information to the US Census that year.[101] The glass industry had always closely followed the fuel supply, and Louisville glass plants could have obtained Pennsylvania or Kentucky coal delivered by river barge. White quartz glass sand was available in significant quantities from Ohio River deposits in eastern Jefferson County and Oldham County and from Big Clifty sandstone formations in Hardin and Grayson Counties. The Kentucky Glass Works used sand from these sources, as did glass plants in New Albany, Indiana, and West Virginia.[102] Although Kentucky's distilling industry presented a rapidly expanding market for bottles, the nodes for Middle West and Ohio Valley glass bottle production continued to be northern Ohio and Alton, Illinois.

Well drillers discovered natural gas near Mt. Sterling, Kentucky, in the early 1890s. A front-page editorial in the *Mt. Sterling Advocate* in February 1893 observed that the gas field would allow residents to follow the example of Gas City, Indiana, which in 1892 had added 4,000 industrial jobs at fifteen newly constructed glassworks and other factories.[103] But Mt. Sterling did not become a glassworks center. By 1900, eighty-one glass plants in fourteen states manufactured flasks and liquor bottles. Kentucky did not manufacture bottles in 1900, although the bordering states of West Virginia, Ohio, Indiana, and Illinois were all significant producers.[104]

Kentucky did host a glassworks, albeit briefly, beginning in 1906, when Frankfort city promoters convinced William F. Modes, operator of the Modes Glass Works in Cicero, Illinois, to move to Kentucky. Modes was among the many experienced eastern glassmakers who made their way west to Indiana's gas-field cities in the late nineteenth century, and he managed the Modes-Turner Glass Company in Terre Haute, Indiana, before moving to Cicero. In Frankfort, Modes planned to manufacture flint glass bottles and other glassware at a factory sited in the Thorn Hill neighborhood, about one mile northeast of the old state capitol.[105]

William Modes formed the Modes Glass Works Company in Frankfort in July 1906 with capital stock of $50,000. Company officers included president R. D. Armstrong, who was also vice president of the Frankfort and Versailles Traction Company; vice president George B. Harper, railroad superintendent and chief engineer on the Frankfort & Cincinnati Railroad (also known as the Whiskey Route Railroad); and secretary-treasurer John W. R. Williams, a local farmer. Area "capitalists" formed the board of directors, which included Modes.

The Modes Glass Works was the product of relocation. Nineteenth-century industrial processes that relied largely on craft skills acquired through long apprenticeships and experience were not readily transferred to other locations unless the craftspeople themselves migrated. Allowing for the instruction manuals and pattern books used in some trades, the costs of building and profitably operating an industrial business were too high for investors to risk hard-won capital on the proposals of inexperienced amateurs.[106] Instead, local investors, be they bankers, businesspeople, or farmers, recruited glassmakers from eastern glass centers in West Virginia, Pennsylvania, and New England to move their works west to take advantage of cheaper energy or larger markets. In other cases, glassworkers moved as individuals or groups seeking higher wages. Thus, Middle West and Ohio Valley glass plants were not started by novices who hoped to learn the technical side of the trade while building the required structures, buying and installing equipment, training a neophyte labor force, and arranging transportation connections to ensure the delivery of raw materials and shipment of finished products.[107]

In January 1907 the workers at Modes Glass Works started to make flint glass in trial runs to test the new furnaces.[108] Two weeks later, the factory was fully functioning and employed "a large number of skilled glass blowers, a [large] delegation having arrived and more will be put to work as soon as possible." Modes Glass Works was scheduled to operate twenty-four hours per day.[109] Some glassblowers may have moved to Frankfort seeking higher wages, and others might have been out of work. Glassblowing jobs were often terminated by business failure—a frequent occurrence—or by newly introduced machines. Those new factories and existing works that continued to use traditional glassblowing techniques offered employment.

Glass manufacturing in Kentucky and elsewhere was seasonal. In June the Modes Glass Works shut down because the workers could not tolerate the furnace building's heat during the summer months. The works remained closed until September.[110] By the time operations resumed, the Modes Glass Works had recruited workers in sufficient number to outstrip the local rental housing market. In October 1907 the company announced plans to erect a forty-room hotel at Thorn Hill to accommodate its employees.[111]

By early December 1907, after operating for less than a year, Mr. Modes placed operations on hold, although workers maintained the fires under the glass tanks. The manager assured employees and the community at large that the shutdown was temporary, but later that month, the fires were extinguished and the plant closed. Production had exceeded demand by a considerable measure, and only large orders sufficient to exhaust the finished stock on hand would permit the plant to reopen.[112] Editorial comment in the *Frankfort Roundabout* attempted to reassure residents that the economy would not falter because of the loss of jobs at the glassworks; the paper noted that Frankfort's distilleries and saw and planing mills continued to operate, perhaps implying that unemployed glassblowers might find work with those industries.[113] The glassworks did resume production, and by 1909, the plant employed 350 people.[114]

Kentucky distillers were never directly served by a large-scale home-state glass industry dedicated to manufacturing whiskey bottles. And nineteenth-century distillers did not build and operate their own glassworks, although many did have dedicated cooperages. Glass manufacturing was administratively unwieldy—problems related to fuel supply, labor force skill level, glass chemistry, and mechanization required experience and expertise—and as distillers made the transition from barrel to bottle sales, they elected to purchase their glass from established industry suppliers.

Paradoxically, overproduction and other problems encountered by whiskey distillers, both in Kentucky and nationally, led to business failures and plant closings in the 1880s and 1890s. The Whiskey Trust aggregated some operations, halted production, and shuttered their works. Temperance activists were increasingly successful in passing local and state prohibition laws, and the market for whiskey bottles and flasks began to wane just as machine production quickened. In 1904, prior to the introduction of the Owens automatic glassblowing machine, the glass industry produced some 2,157 thousand gross of liquor bottles and flasks. By 1919, that number stood at 993 thousand gross, as orders were sharply curtailed on the eve of national Prohibition.[115]

Agricultural Implements

New patents in complementary industries such as grain mills and lumber and planing mills increased production and product quality in those industries, but technical advancement was uneven, and innovation in some industries lagged. The development of mechanical corn picking and husking machinery, for example, was less robust than that of other farm equipment. Farmers harvested whole ears of corn, which they husked by hand before storing the ears in cribs until they were fed to livestock or sold. Hogs could thrive on whole ear corn, but farmers fed shelled or cracked corn to cattle, horses, mules, and poultry. Distillers preferred to buy shelled corn by the sack, so to maintain their on-farm feed rations and supply the distillers' market, farmers had to shell corn by hand or buy shelling machines they could afford.

Shelling large volumes of corn by hand was strenuous, tedious, and incredibly time-consuming. The demand for mechanical corn shellers stimulated their development, and the number of inventions roughly paralleled the technical advancements in corn planters and cultivators. Early corn shellers were simple devices consisting of rotating toothed "picking wheels," turned by a hand crank or belt, that shelled one ear at a time. Most such machines were capable of shelling less than fifty bushels per day.

John Whitford's corn sheller was representative of the machines available in the 1840s.[116] To operate the Whitford sheller, one fed an ear of corn through an opening in the lid. The ear dropped into a chamber containing a metal cone fixed with spikes in spiral rows. As the operator turned the crank, the cone rotated against the ear, scraping the kernels off the cob. A current of fanned air removed dust and chaff as the shelled corn fell onto an angled exit chute at the bottom. The cob was ejected out the side. Such machines were adequate to shell enough corn to grind into meal and feed a few head of livestock or a

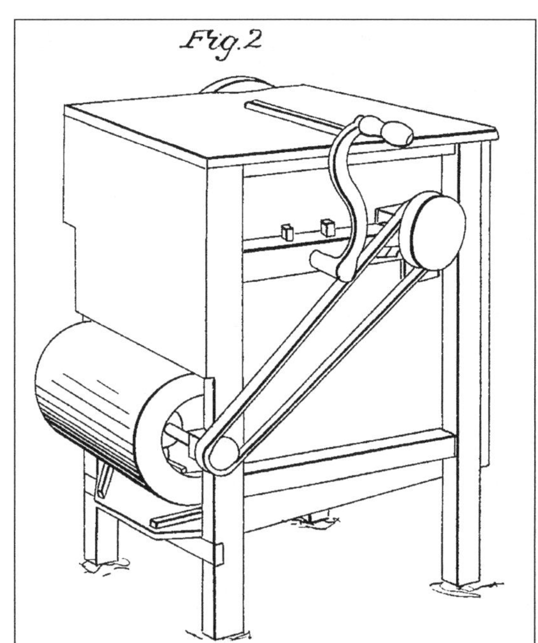

John A. Whitford's corn sheller (patent no. 1,946, granted January 23, 1841). (J. A. Whitford, "Corn Sheller," US Patent and Trademark Office, http://www.uspto.gov)

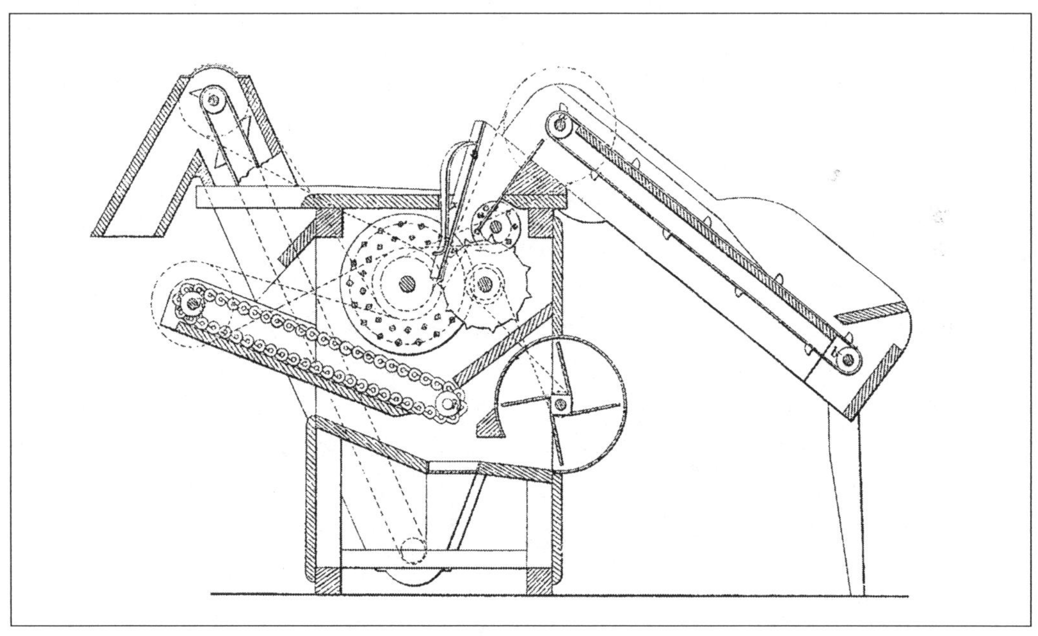

Augustus Adams's corn sheller (patent no. 32,971, granted August 6, 1861). (A. Adams, "Corn Sheller," US Patent and Trademark Office, http://www.uspto.gov)

brood of laying hens, but they could not supply an industrial distillery mashing hundreds of bushels of grain per day.

The A. Adams & Sons Company manufactured corn shellers in Sandwich, Illinois, southwest of Chicago, from the 1850s to the 1880s. Inventor Augustus Adams's first corn shellers could handle four ears at a time and, if operated by two expert laborers, could shell about 800 bushels per day. Adams made a series of improvements to the basic picking-wheel sheller, and in 1861 he patented a self-feeding sheller that allowed the operator to shovel cobs into a continuous-feed chute rather than hand-feed the cobs individually. Cobs entered the machine on a conveyor, where they were shelled by conical, spurred shelling wheels. Seeds dropped to a bottom hopper and were elevated to two exit chutes for bagging. The shelled cobs were elevated out the back by a conveyor crafted from wire rods.[117] With additional improvements by Adams's sons, this type of sheller attained a capacity of 2,500 bushels per day.[118]

The mechanization of farm equipment increased grain production and reduced market prices; this was a boon for all grain consumers, especially distillers. By 1870, distillers were building corncribs at their distilleries. Adding storage capacity to their works allowed distillers to purchase corn when prices were low, and having an entire season's corn requirements on hand before distilling began guaranteed continuous operation, provided no other shortages materialized.[119] Without agricultural mechanization, industrialization of the distilling process during the mid-nineteenth century could not have proceeded. Likewise, mechanization in other areas, be it power generation, lumber processing, mining, or glassmaking, also enabled and enhanced whiskey production. Mechanization across the spectrum of American production was, on the one hand, the result of inspired inventors; on the other hand, it was stimulated by increases in market demand for finished goods that had previously been handmade. Mechanization had the effect of devaluing traditional skills, and it resulted in the discarding of specialized hand tools used by generations of craftsmen—a self-cannibalizing process whereby new equipment replaced the old in pursuit of higher, more efficient production. Out-of-date tools were retired or resold. The retreat of handcraft skills accompanied the mechanization of many trades that supported nineteenth-century distilling, although bourbon distillers would long thereafter contend that their product was "handmade."

8

Signatures of Risk

$150,000 Damage to the Cooperage Plant.
Destructive Fire Visited Manufacturing District.
225 Men Out of Work.
Damage Fully Covered by Insurance—Will Rebuild.
—*Paducah Sun*, 1905

Innovation and Risk

For nineteenth-century craft and commercial distillers, risk was plentiful but refuge was scarce and, usually, expensive. Few aspects of human economic activity are certain, and for distillers, the introduction of innovations and inventions was accompanied by the introduction of some level of risk. In a general sense, risk implies vulnerability—environmental, mechanical, or financial—that has the potential to be costly or lead to loss.[1] Insurers, if they are willing, can provide some measure of financial protection from loss.[2] Individuals and organizations attempt to identify and mitigate risk so they can attain an acceptable level of certainty in their operations, and activities can proceed. Risk mitigation may involve simple adjustments in procedures and behavior, or it may require solutions that leave imprints or signatures on the landscape. Distillers may choose to build fire-resistant works and warehouses with brick and iron, forgoing traditional wooden structures and thereby avoiding both the risk of loss by fire and high insurance premiums. Railroads may build sidings into distilleries to draw boxcars off the main line to avoid traffic delays and accidents. Such risk mitigation measures can be subtle or radical.

New methods and machines require testing and training to improve reliability and elevate performance—that is, to reduce the fluidity of uncertainty and risk. Operational proficiency, acquired through knowledge and practice, can be attained only if one can produce desirable and repeatable results at an acceptable level of consistency. When distillers adopted steam power in the nineteenth century, they were confronted with a complex new technology that was fraught with potential hazard if applied by untrained amateurs. Prior to industrialization, tradespeople working primarily with hand tools learned their crafts through long apprenticeships with accomplished masters. A practiced craft enjoyed a certain predictability. Industrial inventions and innovations reduced complex tasks to semiautomated mechanical processes that increased production but also amplified flux and uncertainty.

Adopting new technology required knowledge of how such machinery performed in different environmental conditions and operational circumstances—hot or cold, wet or dry, fast or slow, tight or loose, expensive or cheap. Manufacturers that adopted steam power were confronted with machines based on applied physics and chemistry and fueled by fire. Steam boilers—like pressure cookers—could explode if they built up more internal pressure than their metal vessels were designed to withstand. Bearings failed if they were not properly lubricated or were improperly aligned. Fluid temperatures other than 32°F and 212°F were important for the optimal performance of chemical processes. Recognizing and addressing risk became important to business success; until cause and effect were clearly understood, implementing new technology represented a risky exercise that fostered the growth of mutual and corporate insurance companies.

Further complications arose from the new business linkages that industrialization invited. Manufacturers could obtain raw materials from much greater distances via steamboats and railroads, which meant that they shared in the uncertainties and contingencies experienced by those distant producers. As distillers increased the capacity of their works, they sent agents into neighboring counties to find reliable supplies of quality grain; eventually, they extended their reach to distant states. Thus, crop failures in those markets became as critical to sustaining production as a prolonged drought at home. When coopers exhausted the white oak stands in their vicinity, they sought staves and barrelhead stock from increasingly distant suppliers. But forest fires could terminate a timber harvest, and a train derailment could delay a lumber shipment. Increased production encouraged distillers to market their products beyond traditional local and regional markets, only to find themselves competing in national markets with other distillers and other products while confronting unpredictable demand and pricing.

Distillers reacted to uncertainty and risk in different ways. A strong desire for economic stability, or sheer survival, might motivate some to take radical action, while others denied or ignored the presence of risk. After making a large investment in new steam-powered equipment, a distiller might not anticipate or suspect that the equipment could fail catastrophically if operated incorrectly. But new technology did not come with a guarantee of perfection.

Environment and Risk

The physical environment visited both direct and indirect influences and risks on distillers. Heavy rain caused steep-gradient streams to flood, destroying creek-side stills, dams, and millraces and carrying away livestock.[3] Floods on trunk rivers inundated floodplain industries—distilleries, sawmills, boatyards, and other works. Heavy rain in February 1883 caused flooding on the Kentucky and Ohio Rivers. At Frankfort, the Kentucky River inundated the Hermitage Distillery and its aging warehouses, even though the works stood on a bedrock terrace some fifty feet above the river's low-water mark. The Cedar Run and E. H. Taylor Jr. Distilleries on the northwest side of Frankfort also flooded.[4] At Louisville, the Ohio River flooded the aging warehouses at two distilleries owned by the Mattingly & Sons Company in Portland and Shippingport, causing significant damage to stored grain and whiskey. A flood the previous year had drowned the cattle confined to feeding pens, prompting the distiller to move the livestock before the floodwaters rose again.[5] A thunderstorm in May 1886 brought strong winds that tore off a portion of a warehouse roof at Frankfort's E. H. Taylor Jr. Distillery; lighting strikes injured two employees and set fire to a whiskey tub in the cistern room.[6] The following year, even though the Taylor Company's New Market Distillery in Mt. Sterling had installed copper lightning rods, lightning struck not the rods but the chimney, causing it to collapse and damage the structures below.[7] Cold temperatures in January 1903 killed twenty-eight feeder cattle in pens operated by the Lillard brothers near the Cedar Brook Distillery, east of Lawrenceburg in Anderson County.[8]

Difficult weather—be it drought, flood, hail, or wind—also caused indirect problems for farmers and distillers. Poor weather conditions could interrupt spirits production and delay transportation schedules. In December 1851 a barge hauling 8,000 bushels of coal sank in the Ohio River near Milton, Kentucky. The coal was intended for the distillery at Snydersville, likely the James O. Snyder Distillery at Hunters Bottom, east of Milton.[9] Farmers hauling grain to a distillery or teamsters moving filled whiskey barrels to a warehouse, depot, or wholesaler could be stymied by impassable roads.[10] Heavy rain in Anderson County in December 1902 caused landslides and heavy erosional damage to rural turnpikes. A newspaper editorial urged timely repairs because "there is no tax so burdensome on the farmer as that which is caused by bad roads."[11] Each problem, whatever its cause, created a situation that affected short-term production and marketing. If these problems compounded, long-term disruption could ensue. Distillers' raw material expenses increased when droughts or hail reduced the grain supply. But high grain prices could also reduce the overall demand for grain and grain products, resulting in oversupply. Alternatively, a distiller might be forced to use inferior grain or suspend distillery operations for a season. Unless one could anticipate this coalescence of contingencies and develop a response plan, the net effect was a reduction in distillery efficiency, production, and profits.[12]

As long as overland transportation remained undeveloped, sources of raw materials and labor were severely curtailed. But this situation was not necessarily improved by better transportation. New connections to a railroad or hard-surface road could enhance whiskey production, but this might exacerbate production surpluses. Distilleries would then be forced to curtail production, resulting in fewer barrels used, thereby affecting cooperages and lumber mills, as well as the potteries making stoneware jugs.

Particularly daunting was the perception that the potential for risk was pervasive, lurking in almost every aspect of the business endeavor: environmental variability, national economic stability, technology reliability, accident and liability, health and mortality. Even malicious behavior by individuals who might wish one ill had to be considered.

Accidents

The preindustrial Kentucky distillery was often a small, unpretentious operation. Housed in a simple shed-like building, the farm distillery contained a small still and a few mash tubs and fermenting barrels. Water powered some operations, and others were worked by hand. When distillers adopted steam power, they often installed the new machinery in an existing still building, making as few structural modifications as possible. Steam engines installed and operated in an old building by those "utterly ignorant of statics or dynamics, or the laws of equilibrium and impulse," caused a "trembling under the revolutions of its heavy equipment."[13] The awkward combination of complex mechanical equipment, hot liquids, high-pressure steam engines, old buildings, and unlettered employees invited distillery accidents. Product was lost when mechanics installed equipment incorrectly, allowing whiskey to overflow holding tanks.[14] On February 19, 1857, William Snyder's distillery at Petersburg blew up, badly scalding employee William Olds, who died the following day.[15] Nine months later, on November 30, Snyder's distillery boiler exploded, scalding Billy Cannon. The following day, three men arrived by steamboat from Cincinnati to repair the boiler.[16] In Harrison County, George Moore fell into a tub of hot slop at the Mill Creek Distillery and was "very badly scalded." In September 1869 James McDaniel fell into the flake stand at the William Appleton Distillery—formerly owned by William Snyder—and was scalded; he died six hours later.[17] An employee at the George G. White Distillery on Stoner Creek in Paris fell into an open slop tub; his misstep proved fatal.[18] A twelve-year-old boy driving a slop wagon at a large distillery in Uniontown, Kentucky, was covered in hot slop when the wagon overturned. He died several hours later.[19] Individuals toppled into open grain and meal bins or caught their limbs or clothing on unshielded rotating shafts, gears, and belts.[20] In 1881 workers at the Appleton Distillery in Petersburg forgot to close a valve, and some forty gallons of whiskey spilled out onto the distillery floor before it could be shut off.[21]

High accident rates prompted insurers to seek legal relief to avoid payouts if an accident was caused by negligence or incompetence. The US Treasury Department, responding to an 1838 resolution in the House of Representatives, conducted a study of steam engine accidents. The report concluded that "severe penalties should be imposed on owners and officers for neglect" should a steam engine boiler burst, and it recommended denial of "the recovery of any insurance by them on the boat and her stores, if injured."[22] Although this recommendation addressed steamboat accidents, it could be applicable to any steam engine, given the high number of explosions and other incidents.

Pestilence as Biological Risk

Animal diseases afflicting domesticated livestock had beleaguered farmers for centuries, and understanding the cause, prevention, and treatment of such diseases remained elusive well into the nineteenth century. Industrialization promoted innovations in transportation that allowed the expeditious movement of livestock over long distances. Livestock farms increased in size in response to a growing demand for beef, pork, and dairy products in urban markets, further abetting the movement of stock and the spread of disease. When confronted with livestock maladies, farmers, milk producers, meatpackers, stockyard owners, and politicians from livestock-producing states often denied the medical evidence and lobbied vociferously against the state and federal governments' attempts to quarantine diseased animals and restrict their interstate transportation.[23] Disease deniers, as they were known, cloaked concerns about their own potential economic losses in a disingenuous concern for state's rights.[24]

Domesticated livestock are susceptible to a wide variety of illnesses, but three diseases stand out because they were highly contagious, had high mortality rates, and were exceptionally difficult to control. Each malady was responsible for the deaths of millions of livestock in the eighteenth and nineteenth centuries. The virulent rinderpest affected cattle primarily in Africa, Europe, Canada, and Australia. Pleuropneumonia infected cattle in Europe and North America, where the disease became a widespread concern from the 1840s through 1890s. Hog cholera killed thousands of animals in the Middle West and South beginning in the 1830s.[25] Pleuropneumonia and hog cholera posed the greatest risk to distillers.

Cattle

Animal disease events in America worsened during the nineteenth century and were largely caused by the importation of infected stock from Europe. Contagious bovine pleuropneumonia (CBPP) affected primarily cattle and may have been introduced in Brooklyn, New York, by infected cattle imported from England in 1843. The disease spread to herds along the Atlantic Seaboard from Connecticut to North Carolina; eventually, infected cattle were transported into the Ohio Valley and the Middle West, where the disease appeared in milk cows in 1866.[26]

Jersey cattle were a popular milk-producing breed during the nineteenth century. Veterinarians identified pleuropneumonia in Jersey cows at a farm in northern Illinois in 1884, and the disease quickly spread to the dairy and feeder cattle at four large Chicago distilleries.[27] From September through October 1886, dozens of cattle died at the Chicago, Phoenix, and Shufeldt Distilleries. The livestock owners attributed the deaths to wrongdoing by the distilleries, including feeding the animals chemical-laced, high-temperature slop. Health inspectors finally convinced the owners that the cattle had pleuropneumonia.

Veterinarians reasoned that a quarantine was required to stop the movement of all cattle and isolate the disease. They quarantined the infected Illinois cattle and slaughtered the entire 3,000-cow herd where the initial infection had been detected.[28] The directed slaughter of other infected and exposed Chicago distillery cattle began in late November, and by mid-December, some 2,270 cattle at the three distilleries had been killed.[29] The Chicago eradication program was part of the federal Bureau of Animal Industry's comprehensive effort to quarantine and slaughter infected cattle. By 1887, the disease had been contained, and in 1889 the secretary of agriculture reported that the disease had disappeared from the country.[30]

Unfortunately, the disease had already spread to Kentucky. It is thought that pleuropneumonia was introduced into Harrison County, Kentucky, when brokers H. D. Frisbie and J. K. Lake purchased cattle from northern Illinois and shipped them to their Cynthiana-area farm in 1884.[31] The disease had spread throughout central Harrison County by 1885, much to the consternation of area cattle breeders and the Kentucky Shorthorn Breeders' Association. After an unwarranted delay, in March 1886 the Kentucky legislature finally authorized the State Board of Health to quarantine and slaughter infected cattle. Farmers began to kill diseased animals in mid-March, and by the end of the month, all exposed stock had been slaughtered.[32]

Dr. W. H. Wray of the Kentucky State Board of Health directed the eradication. Inspectors found thirty-three infected herds, only two of which were associated with area distilleries; four cattle belonging to distiller T. J. Megibben had been infected or exposed to the disease and were destroyed, as were two animals belonging to distiller T. H. Robertson.[33] Unless a formal veterinarian's report was made, there was no way to determine how many cattle were infected and destroyed. Livestock owners were loath to inform the public that their herds were infected; even the suspicion of infection could effectively terminate sales of meat, milk, and breeding stock. Rather, livestock owners often attempted to obfuscate an inspector's findings or, as was apparently the case at the Frisbie and Lake farm in Cynthiana, the owners quietly killed their own stock, or perhaps an "accident" occurred that had the same result.[34] There was no accurate count of the number of cattle culled to halt the spread of the disease. It is probable that the active interdiction program implemented by veterinarians at the Bureau of Animal Industry greatly reduced the incidence of widespread infection. As a result, it is likely that many more Kentucky distillery cattle died in fires and floods than from pleuropneumonia.

Hogs

The highly contagious, often fatal, viral infection known as swine fever or hog cholera was a plague feared by nineteenth-century farmers. Distilleries were particularly susceptible because they usually confined their hogs in large pens at comparatively high densities. Livestock brokers and distillers might feed local stock, but they also received hogs shipped from distant sources, increasing the risk of introducing infected stock into their operations.

From the 1850s through the 1890s, Ohio Valley hog producers and distillers lost entire herds to periodic cholera epidemics. William Snyder's distillery in Petersburg, Kentucky, lost more than 2,500 hogs to cholera in the summer and fall of 1856. That same year, the distillery at Aurora, Indiana, across the Ohio River from Petersburg, lost between 6,000 and 7,000 head. Upstream, at New Richmond, Ohio, the outsized Beck and Fosdick Distillery mashed 3,000 bushels of grain per day in 1856 and fed the slop to 11,000 hogs. The distillery's entire herd died of cholera that year. The distillery itself failed the following year, likely because of the colossal loss, and it was sold to another operator.[35]

In 1857 a report in the *Louisville Weekly Journal* noted the obvious: the fearful malady had disrupted business at large distilleries in the Middle West and South for more than a year. The newspaper pledged a generous $20 toward a fund to support research into the cause of hog cholera. But advances in animal disease science were slow to come, as illustrated by a Maryland chemist who recommended treating stricken hogs with barilla salt and soda ash, a remedy regarded by more thoughtful scientists as ineffective quackery.[36] Herd mortality continued, seemingly unabated. In January 1860 Harrison County distiller T. J. Megibben lost 300 hogs, and James A. Miller lost more than 600 hogs at his distillery in Paris (later known as the George G. White Distillery). By December, Kentucky distillers and farmers had lost stock worth several million dollars, and the state government offered a $1,000 reward for a remedy.[37] During the Civil War, the Union army required Bluegrass distillers to surrender a portion of their mashing grain to government quartermasters. Megibben sold 1,500 barrels of corn to the quartermaster in Paris, and when he was asked to provide more, Megibben requested permission to keep enough grain to feed his hogs, which could provide meat for the military—if he could avoid the cholera plague.[38]

In 1873 a medical charlatan circulated throughout the northern Kentucky countryside in a one-horse wagon selling patent medicine that he professed would cure hogs, cattle, and chickens of cholera.[39] Claims of effective cholera treatment were infrequent, but some unusual remedies were publicized. In 1882 farm newspapers enthusiastically reported a sure cure discovered by Illinois distiller E. E. Lock. One could eradicate the disease, Lock said, by feeding cholera-infected hogs corn that had been scorched or burnt. The supposed cure was republished several times over the next two decades, but apparently Lock's success was never replicated by other hog feeders.[40] In 1889 cholera revisited the Miller (George G. White) Distillery in Paris, where some 800 penned hogs were fed on slop. Almost 200 died initially, and the carcasses were burned as the animals succumbed. The remedies administered in an attempt to save the remaining 600 animals were ineffective.[41] Lincoln County distiller W. H. Traylor lost hogs to cholera in 1894, and the following year distillers in nearby Crab Orchard lost entire herds to the disease.[42]

In 1884 Daniel Salmon of the US Department of Agriculture's Veterinary Division and epidemiologist Theobald Smith initiated hog disease research on behalf of the Bureau of Animal Industry. The following year, Salmon and Smith isolated a bacterium that appeared to cause hog cholera. But subsequent research in the early 1900s found that a virus was the primary source of cholera; the other organisms were secondary invaders. Within a few years, researchers had developed a vaccine to prevent the disease, although an effective eradication campaign would not be initiated for another five decades.[43]

The feeding of livestock involved risks that could terminate a distillery business if they were not addressed. Some feeders refused to recognize the authority of the emerging science of medicine and persisted in applying folk remedies and nostrums or denying the possibility of disease. Others suspected collusion on the part of health officials to drive one person out of business while advantaging others. Animal health science would eventually prevail, but shortsightedness, confusion, denial, and wishful thinking cost distillers millions of dollars in livestock losses and may have driven some from the business.

Distillery Fires—Accident and Arson

Given the potential for fire, nineteenth-century distilling was a high-risk business. Several other industries shared similar vulnerabilities, including mills producing flour, cotton, wool, and flax; cordage works; woodwork manufacturers; and grain elevators.[44] From 1875 to 1896, 365 distilleries burned in whole or in part across the country.[45] Detailed informa-

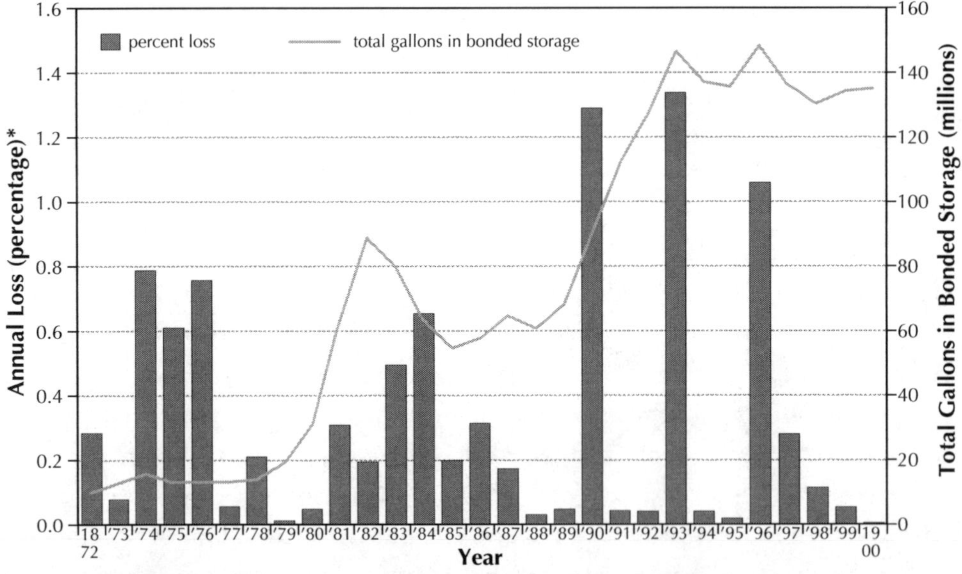

Distilled spirits lost by fire in the United States, 1872–1900. (Compiled from Report of the Commissioner of Internal Revenue, 1901, Treasury Department Document 2247)

tion on distillery fires is thin. Individual fire events were not systematically reported or recorded.

In the early 1870s a strong desire by Congress to account for lost tax revenue prompted the commissioner of internal revenue to start to collect and publish figures on distilled spirits lost to fire. During the quarter century from 1872 to 1900, some 7.5 million gallons of stored spirits, primarily whiskey, burned.[46] The amount lost seemed to track, in part, the total gallons of whiskey in bonded storage, but the trend was idiosyncratic rather than linear. The volume of spirits in bonded storage was consistent through the 1870s, at roughly 15 million gallons per year, yet fire losses surged from 1874 through 1876 to nearly 0.8 percent each year. The smallest annual loss of taxable whiskey during the period occurred in 1879, when about 2,500 gallons burned. In the early 1880s the amount of spirits in storage began to increase rapidly to 80 million gallons. Fire losses lagged behind storage totals by about four years, with losses peaking at more than 400,000 gallons in 1884. The amount of spirits in storage dropped to about 60 million gallons through the late 1880s. This was followed by a sharp rise to more than 140 million gallons beginning in 1889 and persisting through the 1890s. Dramatic fire losses occurred in the 1890s, but at three-year intervals—1890, 1893, and 1896—when fire consumed more than 1 percent in storage, or more than 1 million gallons, each year. The greatest loss occurred in 1893, when nearly 2 million gallons burned.

These peaks in the trimodal fire pattern were likely related to three different causes: environmental hazards, new industrial equipment, and arson. Natural hazards, especially lightning, were a threat during any period, but especially before distilleries installed lightning rods and grounding systems. As rapid industrialization in the 1870s brought new steam-powered equipment into distilleries, accident rates increased. Coal-fired boiler design and construction may have been inadequate for their application, or the boilers may have been operated at unsafe pressures to increase production. Inexperienced operators and mechanics may have overlooked equipment maintenance needs. Machinery-related accidents likely began to taper off in the 1880s as equipment quality improved and proficiency in operation increased, but temperance-related arson fires did extensive damage in the 1890s. Even if arson fires did not increase in frequency, they resulted in greater damage at each event. As industrial distilleries increased their mashing capacity during the 1880s and 1890s, they also built large aging warehouses that contained 30,000 barrels or more. A fire in one such warehouse could consume more than 1 million gallons of whiskey.

The extent of damage caused by a single fire event was related to the inherent fire hazards in a distillery, which were not difficult to discern. Distilleries were combustible structures: wood-framed and wood-sided buildings with thick planked floors. Whiskey, the principal product, was typically composed of 50 percent highly combustible ethyl alcohol. Raw material by-products—grain dust and white oak shavings and sawdust, for example—were readily combustible, as was the provender in distillery livestock barns. Ignition could originate in malfunctioning equipment—dry bearings, seized pulleys, slipping belts. Open wood or coal fires to heat copper pot stills and high-pressure steam boilers were primary sources of ignition. That same fire that might send cinders and brands up the smokestack for airborne distribution across the distillery property. Burning whiskey and wood produced high-temperature gases that put nearby structures at risk, especially in windy weather.

Distillers began to install slop dryers in the 1880s and 1890s. The first versions were coal- or wood-burning direct-heat dryers that could be effective but also posed a fire risk if operated improperly. In May 1898 the new slop dryer at the George G. White Distillery in Paris caught fire, destroying the entire drying plant. The four-month-old facility, which had cost $6,000, was a total loss.[47] Even the candles and lanterns that illuminated the pre-electricity works were implicated in distillery fires. Once ignited, distillery structures tended to propagate fire rather than contain it, given their inventory of combustible materials, ignition sources, and lack of fire-suppression architecture, equipment, and personnel.[48]

Many fires were not self-inflicted by carelessness or technical failure but were purposely set by arsonists—incendiaries, in the idiom of the day. An incendiary fire consumed the large grain warehouse and mill owned by the Brandeis & Crawford Company at First and Main Streets in Louisville in August 1869. The company was one of the largest grain merchants in the South and supplied corn and rye to distillers across Kentucky. The loss was covered by insurance, and the distilling season would not begin for three months, but the company was pressed to reconfigure its storage and milling facilities.[49] Of the 120 distillery fires reported nationwide in 1896, 67, or more than 55 percent, were attributed to incendiarism.[50] Fire destroyed private property, but it also deprived the federal treasury of tax moneys, prompting the Secret Service and the Bureau of Internal Revenue to assign agents to investigate distillery fires. When fire destroyed the R. M. Clark & Company Distillery near Dixon in Webster County, the federal investigator concluded that the fire had been set by an incendiary. "There is a strong temperance feeling in that section," the agent said, "and it is generally believed that the distillery was set on fire by some fanatic. The stillhouse was almost entirely destroyed and the warehouse badly damaged."[51] Some arson fires were public events intended to threaten and intimidate business owners. During the Black Patch Tobacco Wars in western Kentucky (1904–1909), farmers organized an embargo on tobacco sales to the American Tobacco Company. In February 1908 a militant group of "night riders" in Crittenden County burned a distillery and tobacco warehouse owned by Henry Bennett in an effort to pressure him to join the boycott.[52]

Sometimes, a witness or evidence of an arsonist's presence allowed distillers to identify the offender. In November 1904 Joe Chambers, a distiller on Slate Creek east of Mt. Sterling in Montgomery County, saw his distillery burn—for the second time. Chambers contacted V. G. Mulliken of Wilmore, Kentucky, who bred bloodhounds and charged $50 per day plus expenses to track thieves and jail escapees. Mulliken arrived at the Mt. Sterling train station the following morning with two of his dogs. The hounds followed the trail for two miles to a blacksmith's shop and residence, where Chambers and Mulliken confronted the smith. The outcome of the encounter was not reported.[53]

Fireproofing

Traditional distilleries and warehouses used wood post-and-beam construction with wood sidewalls, three- or four-inch plank floors, and shingle roofs. Frederick Stitzel's storage rack system allowed distillers to build multistory warehouses with no interior walls; each floor was a wooden thicket of vertical posts and horizontal beams. By the 1840s, warehouse builders in large urban areas gradually adopted cast-iron columns and girders to enhance fire resistance.[54] Increasingly mindful of risk, distillers built their works with exterior brick walls, covered their plank-and-shingle warehouse roofs with sheet iron, and secured windows with iron shutters. Distillers would learn, though, that the use of seemingly non-combustible building materials did not render a building fireproof.

Iron window shutter at the Elkhorn Distillery's brick aging warehouse.

In 1890 W. S. Hume & Company operated a distillery on Silver Creek in Madison County. Workers covered the 37,500-square-foot warehouse roof with tinned metal sheeting, a strategy intended to reduce the risk of fire caused by airborne embers from the steam engine smokestack. The cost of materials and installation was $2,250 (or more than $60,000 in current value).[55] A fire in the Hume Distillery meal room twelve years later consumed the entire distillery structure. The nearby L & N Railroad depot was damaged, a freight car was burned, and neighboring houses were threatened. Although the whiskey aging warehouse was also set afire, people from the nearby village managed to save the structure and its contents.[56] The company declared bankruptcy three years later, in 1905.

Large urban fires around the country prompted both cities and distilleries to buy and install fire-suppression equipment during the nineteenth century. In May 1884 a residential fire started in an African American neighborhood in Frankfort, adjacent to the Hermitage Distillery. The city's fire engine, drawing water from storage cisterns, ran out of water, and a church and several homes burned before the distillery's steam pump system was activated and workers directed water from the Kentucky River onto the flames.[57] Six years later, the distillery established a small fire company of its own and kept 800 feet of fire hose on the premises to fight distillery fires until the city firefighters arrived.[58]

The telephone and telegraph allowed prompt communication between distillers and fire companies. In April 1884 the J. Swigert Taylor Distillery on Glenns Creek in Woodford County completed installation of a telephone line that connected the works with Frankfort, significantly enhancing operations. The line also permitted ready contact with the city's fire department.[59] Later that summer, the W. A. Gaines Company installed a telegraph line between its office in Frankfort and the Old Crow Distillery on Glenns Creek. The line was connected to an electric distillery clock that was activated by a night watchman every half hour. The telegraph, in turn, registered that signal on the clock in the Frankfort office, permitting management to ascertain that the night watchman was on duty and performing appropriately.[60] Small-town fire companies found that the telegraph allowed them to collaborate with fire companies in nearby towns or at well-equipped distilleries. When a fire started in the Midway commercial district in 1891, city officials sent a telegram to the fire chief in Frankfort, thirteen miles away. The Frankfort fire company loaded a steam-powered fire engine, a hose reel, and six firefighters on a railroad flatcar and proceeded to Midway. The fire destroyed five buildings before the Frankfort company arrived, but by using water from the L & N Railroad's water tank, firefighters were able to save the remainder of the commercial district.

The high cost of fire insurance prompted the invention of new fire-suppression equipment, and by the 1870s, reliable building sprinkler systems were available.[61] In 1892 the W. A. Gaines Company of Frankfort installed an automatic sprinkler system in the Hermitage and Old Crow Distilleries. Three large water pipes ran along the ceiling of each floor, with sprinkler heads every ten feet.[62] Whether arson, accidents, or inept operation caused distillery fires, as the number of catastrophic blazes mounted, insurance companies responded by raising rates. By the turn of the twentieth century, Kentucky distillers were paying corporate insurance companies roughly half a million dollars in premiums annually. In 1902 the Kentucky Distillers' Association met in Louisville to consider organizing its own mutual fire insurance company in the hope that they could reduce what they considered "exorbitant rates" charged by insurers on distilleries, warehouses, and whiskey in storage.[63]

Tax Relief

Large fire losses often forced distillers to close their businesses. But the risk related to distillery fires also extended to others, and this chain of contingency included government tax collectors, corporate insurers, banks, and wholesale liquor dealers. If a burned distillery ceased production, the federal government was deprived of tax moneys. But if the taxes had already been paid and these tax-paid or "free" spirits were lost to fire, distillers often appealed to the federal government for tax refunds. Such requests were numerous. The Congressional Committee of Claims received a petition from distiller and future president Andrew Jackson on February 12, 1803, requesting a refund of taxes paid on lost whiskey:

> Report made, considered and prayer of the petition rejected. The Honorable Senate, and House of Representatives in Company Assembled. The remonstrance of Andrew Jackson of the State of Tennessee, show that on the first day of December, Seventeen hundred and ninety nine, your remonstrant obtained license to work two stills, for the space of one year from the said first day of December. One still capacity one hundred & twenty seven gallons, the other seventy gallons, that on the night of the first Monday of June Eighteen Hundred, the still House of your remonstrant was consumed with fire, with upwards of three hundred gallons of whiskey and the said stills rendered entirely unfit for use, and of no value, and were never made use of after in the distilling of your remonstrant, your remonstrant paid up the tax due the first Monday in June, which was about six months, and was of opinion that the duties would cease to exist, at the period of time the stills were rendered unfit for service.[64]

Jackson's petition demonstrates the appeals process, and although his petition received a sympathetic hearing, it was not approved. Thereafter, however, distillers regularly petitioned Congress for tax refunds following distillery fires, and they were often successful.[65]

Insurance

Depending on the site of origin, the presence of combustible materials, weather conditions, preparedness, and other factors, fires could be comparatively minor and readily extinguished, or they could be major conflagrations that not only destroyed the distillery but also threatened nearby people and property. Many distillers bought insurance to cover both product and property losses, be they limited or total. In 1907 a fire in the livestock feeding barns at Old Crow Distillery in Woodford County destroyed the buildings and the stored feed. Insurance covered the $1,000 loss.[66] Sometime around 1880 the Allen-Bradley Distilling Company, owned by principals in New York and Frankfort, built the Elk Run Distillery near Beargrass Creek on Louisville's east side, in what became a cluster of five distilleries. By 1890, Elk Run was one of the largest sour-mash bourbon distilleries in the world, but one early July morning, the works burned. Witnesses saw a flash of blue light that was followed by a massive explosion; a dozen alcohol tanks ignited and erupted in "severe cannonading." Burning whiskey spread rapidly, and within a few minutes, the entire complex was ablaze and quickly burned to the ground. No employees were hurt, but three boys who had been walking nearby died from "terrible burns." Some 60,000 gallons of whiskey were lost, and preliminary estimates placed the value of the loss at $100,000 to $150,000. The distillers carried $70,000 in insurance.[67]

The Allen-Bradley fire confirmed that distilleries, whatever their size or age, were susceptible to total destruction by fire. The volatile character of burning whiskey also placed adjoining property at risk. In August, one month after Elk Run burned, fire destroyed the expansive Kentucky Distilling Company nearby. That fire spread to the Kentucky Tannery and the Great Western pork slaughterhouse, destroying the structures and all the cattle and hog pens. More than 1 million gallons of whiskey burned. The financial loss, including $981,000 in unpaid taxes to the government, was estimated at $2 million, about $700,000 of which was insured.[68] Several large distillery fires occurred during the fall of 1908, resulting in the loss of some 3.6 million gallons of whiskey. The loss to distillers was $1.8 million (at $25 per barrel); the loss in federal tax revenue was twice that amount, totaling $3 million to $4 million.[69]

The relationship between a fire's magnitude and a distillery's placement relative to slope and weather conditions is aptly illustrated by two events. Late in the distilling season in 1893, four bonded warehouses belonging to the Glenmore Distillery at Owensboro, in Daviess County, caught fire. They were filled with more than 19,000 barrels of whiskey when cinders from the distillery's smokestack some 200 feet away ignited the fire. Fanned by high winds, the fire consumed the warehouses, which collapsed, smashing the barrels and setting their contents ablaze. Nearly 1 million gallons of burning whiskey flowed across the distillery property into the Ohio River, where the flames were only partially quenched. The distillery itself was saved, as were 500 cattle penned in adjacent feeding barns. The city's waterworks was just downstream from the distillery, and the water supply was strongly tainted with whiskey for some time thereafter.[70] The E. J. Curley Distillery at Camp Nelson in Jessamine County burned in July 1894. The stable and a coal shed with 3,000 bushels of coal also burned. Although the aging warehouses stood 300 feet away, they caught fire several times. Firefighters managed to save the warehouses and the wooden bridge that spanned the Kentucky River at the site.[71]

As distillers increased the scale of their works and added more aging warehouses, it became increasingly apparent that, depending on wind speed and direction, the intense heat and flames produced by burning whiskey and well-seasoned wood could ignite other buildings located considerable distances away. The Old Crow Distillery in Woodford County, for example, was built on the narrow Glenns Creek floodplain. The distillery buildings were compactly arranged, and the storage warehouses, though made of brick and having metal roofs, were only a few feet apart. Rising insurance premiums in the 1870s motivated distillers to increase the space between buildings, as site dimensions permitted, and to use more fire-resistant materials in construction.[72] Railroad access, which was a critical asset to industrial-scale distilling, became a liability in the eyes of insurers. They deemed buildings sited near railroad tracks to be poor risks, given the proclivity of steam engines without smokestack spark arrestors to belch flying sparks and cinders.[73]

The rapid adoption of new industrial equipment and processes by American manufacturers during the second half of the nineteenth century presented insurance companies with the problem of setting exposure schedules and insurance rates. Accurate estimates of risk employed the laws of probability and a large enough sample of incidents to judge physical risk. Lacking that information early on, insurers often based insurance rates on guesswork.[74] They found that distilleries belonged to a group of businesses exhibiting a "high level of uniqueness." That is, the range of possible causes of damaging events was quite broad, making it difficult to judge the number and importance of relevant contingencies on which an insurer should base a policy that charged fair premiums but also yielded a profit for the company.[75] And, as distillery works grew in size, so did the concentration of value subject to loss in a single fire event.[76] Nevertheless, insurers developed detailed fire insurance schedules for distillers that assessed risk levels and assigned premium rates accordingly.[77] Given the experience gained in large urban fires such as those in San Francisco (1851), Montreal (1852), Chicago (1871 and 1874), and Boston (1872), insurers advised nineteenth-century manufacturers to give careful consideration to building placement and to maintaining safe distances between buildings filled with combustible contents. Some insurers reduced premium rates if structures were separated by at least 100 feet. Yet, as demonstrated by the E. J. Curley Distillery fire, warehouse fires could readily ignite other buildings more than 175 feet away.[78] By recording the dates of fires, insurers found that the potential for large, sweeping fires was seasonal. Wooden buildings were at greatest risk of burning in late summer and early autumn, when they were thoroughly dry.[79]

Factory fires demonstrated that new industrial equipment installed in traditional wood-frame buildings increased the hazard unless manufacturers made extensive modifications to their structures and processes. Insurers increased premium rates for industrial structures with frame walls, shingle roofs, boiler rooms without iron doors, iron chimney stacks that extended through wooden roofs, and a long list of other potential hazards. Fire events conclusively demonstrated that distilleries with integrated flour mills and cooper shops were at greater risk than works where these functions were housed in separate buildings.[80] Cooper shops that adopted steam-powered equipment, drying kilns, and heaters were especially prone to fire, and they were assigned higher rates than traditional craft shops where such work was accomplished by hand. Industrial buildings that lacked night watchmen also received higher assessments.[81] The high rate of incendiary fires during the latter third of the nineteenth century made it difficult to estimate the danger of fire and assign appropriate premium rates using only objective actuarial information about a distillery's structural materials. Nevertheless, intense competition among corporate insurance companies pressured underwriters to issue high-premium fire insurance policies on high-risk buildings. Critics alleged that while insurance companies attributed more than half of all fires to arson, they also persisted in selling high-coverage policies that invited self-incendiarism on the part of business owners experiencing financial difficulties.[82]

Kentucky Distillery Fire Insurance Rates, 1884

Distillery	County	Fire Insurance Rate per $100 Value	Year of Fire
Elk Run	Jefferson	$0.85	1890
Glenmore	Daviess	$0.85	1893
E. J. Curley	Jessamine	$0.85	1894
G. G. White	Bourbon	$0.85	1898
W. S. Hume	Nelson	$0.85 and $1.00	1902
Old Crow	Woodford	$0.85	1907
Greenbaum	Woodford	$1.25	1908
Tom Moore	Nelson	$1.00	1908

Sources: Compiled from William Mida, *Location of Distilleries with Insurance Rates on Whiskies in Bonded Warehouses in Kentucky,* 1884, and newspaper accounts.

Insurers deemed whiskey stored in bonded distillery warehouses a high risk and assessed insurance premium rates accordingly. Rates on Kentucky's bonded warehouses in the 1880s were typically $1.25 to $1.50 per $100 of valuation, with a range from 85 cents for warehouses located at least sixty feet from a distillery to $5.50 for a frame warehouse with a shingle roof standing less than fifty feet from a distillery building. Distillery buildings were assessed at $3.25 if brick and $5.50 if frame, with additional charges for shingle roofs or other potentially hazardous features. The small distillery of S. G. Foster at Willisburg, in Washington County, was likely built of wood; it was one of several distilleries that received the state's highest assessment of $7.[83] Apparently, whiskey warehouses that were insured at the lowest rate burned with the same frequency as those subjected to higher rates. In 1904, without explanation, the Kentucky and Tennessee Board of Underwriters in Memphis raised insurance rates for all bonded whiskey warehouses in Kentucky, much to distillers' dismay.[84]

In the interest of ensuring the objective assessment of hazard and assignment of premiums, insurance company executive George T. Hope began to compile a large-scale map of New York City in 1850, as an aid to calculating fire risk on individual properties. English engineer William Perris drew the maps according to exacting technical requirements, showing structure placement, building materials, and other pertinent information. After hand-coloring the maps, the company published them at scales of 1 inch to 50 or 100 feet.[85] Following Hope's example, surveyor D. A. Sanborn incorporated the Sanborn Map and Publishing Company in 1876, and over the next quarter century, his surveyors compiled fire insurance maps for all of America's major urban centers. In 1894, apparently in response to requests by insurance underwriters, the Sanborn-Perris Map Company published *Sanborn's Surveys of the Whiskey Warehouses of Kentucky and Tennessee* as an atlas of individual distillery maps.[86] Sanborn updated the whiskey atlas in 1910, and the maps were used extensively by insurance underwriters to assess risk and fix premium rates.[87]

Because insurance companies regarded manufacturing works to be especially hazardous properties, Sanborn surveyors recorded them in precise detail. Distillery maps recorded the exact building size and position relative to topographic features such as streams and steep slopes. Color codes indicated building materials. Special symbols recorded window and door placements. Lettering specified building uses, such as office, corn house, corn mill, coal elevator, cooper shop, and livestock shed. Rooms within buildings with dedicated purposes were also noted: engine room, yeast room, fermenting room, flake stand, cistern room, and bottling room. Symbols and names identified interior equipment inventories, such as stills, vats, tanks, and steam engine boilers; all were properly positioned on the maps. Barrel warehouses included notations on capacity and type of stacking. Special symbols identified pipelines, smokestacks, roads, railroads, and bridges. Text notes described the products produced, the times of operation, the types of lighting used (lard lamps, candles, kerosene or coal oil lanterns, or electricity), and whether the still was heated by an open fire.

Sanborn maps did not include topographic contour lines, but surveyors did indicate hill slopes for some sites. Warehouse fires quickly became catastrophic events when burning structures collapsed, smashing barrels and igniting whiskey that flowed downhill to adjacent structures and property. Attentive to this probability, surveyors recorded likely flow direction with dashed lines and lettering. For instance, the George G. White Distillery in Paris stood on a narrow site positioned on a low-angle slope near Stoner Creek. The aging warehouses were closely spaced, and the surveyor recorded with dashed lines two natural drainage ways between the buildings.

In the early 1880s Thomas Edwards built a new distillery near the Southern Railroad tracks in Midway, in Woodford County. S. J. Greenbaum purchased the distillery, and after he died in 1897, Greenbaum's family continued to operate the distillery under his name.[88] The distillery fed cattle in downslope pens situated near a creek called Lee's Branch. In November 1887, after hearing complaints, the Woodford Circuit Court fined the Greenbaum Distillery $1,500 for "suffering still slops and offal from [the] distillery and cattle pens to run into a branch near by."[89] To partially address the slop dumping problem, the distillery installed a new slop dryer. But in January 1904 the dryer and boiler room caught fire, causing $25,000 in damage.

Four years later, in August 1908, an arsonist set fire to the distillery's six bonded warehouses, which contained 47,500 barrels of whiskey. As the blazing warehouses collapsed, burning whiskey poured from the smashed barrels and ran downhill into Lee's Branch, incinerating the homes of several African American residents. When the liquid fire reached the creek, it ignited the Southern Railroad trestle and two county bridges. For a time, the fire threatened the entire northern section of town. Lexington residents, fourteen miles away, reported seeing the fire on the northwest horizon. Onlookers arrived from across the county to watch the blaze. Even though the Midway fire company possessed only one fire engine, it was able to save the distillery works and offices, the superintendent's home, the bottling house, the Southern Railroad depot, and the Cogar Company grain elevator, all adjacent to and uphill from the burning warehouses.[90] When the burning whiskey in Lee's Branch reached South Elkhorn Creek, it set the creek afire for several miles, burning several creek-side houses. Burning whiskey stained the creek water a dark brown and killed thousands of fish and waterbirds.[91] The downstream collateral damage, likely uninsured, included a "water famine" for residents along the creek, who were unable to fill their cisterns because whiskey contaminated the water for an extended period.[92] The warehouse and whiskey loss was estimated at nearly $1 million, for which the Greenbaum family had insurance coverage.

As shown on a Sanborn map, the Greenbaum Distillery's six warehouses were positioned on a slope above Lee's Branch. Warehouses B, C, and D were housed in a single frame building with iron siding. Warehouses E and F, also iron sided, stood eight and ten feet away, respectively. Warehouse A, built of brick, stood about 200 feet away, separated from the main warehouse cluster by railroad tracks. The Sanborn surveyor described the situation with two notes: "Ware Ho's B, C, D, E, and F are subject to common Whiskey Fire," and "burning whiskey from Ware Ho 'A' will doubtless flow S. on R. R. Tracks and not reach other Ware Ho's."[93] As the Sanborn surveyor prophetically noted, the closely spaced warehouses were consumed, and the burning whiskey did indeed surge downhill into Lee's Branch during the 1908 conflagration.

Depending on the size of the property and the scale of construction, a distillery might be densely spaced or spread across several acres. The greatest fire hazard, of course, was at older wood-frame distilleries built on confined sites. When building new works, distillers altered the site configuration, where possible, isolating buildings and positioning stills and warehouses on slopes with natural or dug drainage channels to direct burning whiskey away from other structures. Sanborn map surveyors depicted these accommodations. George T. Stagg's site in Frankfort included two distilleries: O.F.C. (Old Fire Copper) and Carlisle. Former owner E. H. Taylor Jr. had built the Carlisle Distillery on the east bank of the Kentucky River in about 1880.[94] Over the next decade, workers installed a network of multipurpose surface and underground drains that the Sanborn surveyor portrayed on the map as labeled dashed lines. The drains could have carried away rainwater, drained the site after a river flood, or channeled burning whiskey away from the works and directly into the river or a nearby tributary.[95]

By the second half of the nineteenth century, stored whiskey had become a medium of financial exchange and security, and the risk of warehouse fires affected not only distillery owners and insurers but also bankers, wholesalers, and others who had issued credit to distillers based on whiskey collateral. Once it was stored in warehouses, wholesalers and other dealers—many from beyond Kentucky's borders—bought barreled whiskey using bank loans secured by warehouse receipts issued by distillers. The receipts entitled the holder to a specified amount of whiskey and could be held or traded; therefore, the whiskey in any particular warehouse might have multiple owners besides the distiller. In his testimony before Congress, Louisville commission merchant and distiller Thomas E. Sherley declared, "The banks regard the receipt of a Kentucky Distiller as the best collateral they can get."[96] Although Kentucky whiskey warehouse receipts enjoyed an envious reputation, everything was in jeopardy when a fire started, unless the receipt holders themselves had bought insurance against the risk.

Yet another dimension of nineteenth-century industrialization, the development of trans-Atlantic steam-powered shipping, allowed distillers and wholesale dealers to consider another strategy to reduce the costs of insuring against risk. To raise money for operations and to pay federal taxes, distillers often sold barreled and stored whiskey to dealers or used it as collateral for bank loans. Distillers who could not otherwise raise money might ship whiskey to European ports such as Bremen and Hanover, Germany, where it would be held in warehouses or sold in European markets. Louisville wholesale whiskey dealer W. H. Thomas was one of many who loaned money to distillers and was therefore interested in ensuring the security of the stored whiskey he held as collateral.[97] Thomas found that insurance on whiskey stored in European warehouses cost 25 cents per $100 in value, whereas insurance on the same whiskey in American warehouses cost 75 cents to $1. European insurance companies, wishing to secure their investments and reduce their risk, had instituted very strict policing regulations on stored whiskey, which resulted in lower rates.[98]

When considered in the aggregate, several types of risk presented nineteenth-century distillers with varying levels of uncertainty. Traditional distillers operated within the context of long-standing folkways that stood them in good stead when dealing with pragmatic day-to-day problems such as procuring and moving raw materials. But in dealing with complex issues such as livestock diseases, traditional beliefs contributed to delays in the search for solutions. Science and engineering produced the equipment that animated the industrial distillery, but parallel professionalization of the labor force lagged, increasing the risk of loss by accident or fire. Some solutions arose externally, through the efforts of government agencies seeking to maintain the flow of tax revenue and contain massive financial losses or the efforts of insurers seeking to achieve precision in assessing and insuring risk. These efforts to effectively address risk often left obvious imprints or signatures on the distilling landscape—signatures that insurance underwriters could see and map as distillers adopted new building materials, equipment, and structure and site configurations.

These landscape signatures of risk can be thought of as simple and binary. Livestock were either present or absent in distillery feeding sheds. A distillery warehouse was either standing and in use or a bed of ashes. But risk signatures were also imbued with human behaviors, making them complex and multidimensional. People invested capital and emotion in the places that provided their livelihoods. Their efforts to make a living involved others as laborers, customers, and financiers, all of whom were linked with the distiller in an archipelago of risk. A nineteenth-century whiskey warehouse was not simply an objective structure; it had authors who planned and built it, workers who filled and emptied it, watchmen who monitored it, bankers who loaned money against it, and insurers who gauged its vulnerabilities. A risk signature, then, might have been as obvious as iron shutters and lightning rods or as subtle as a quarantine sign or dashed lines drawn on an insurance map.

9

By-products

In a distillery of raw grain, the swine reared and fatted are not a small consideration—some distillers prefer horned or black cattle—in either case it is a very profitable branch of the business.
—Michael Krafft

Mr. Simon Well, the big cattle dealer of Lexington has 115 cattle in the sheds of the Buffalo Springs Distillery. He has purchased the slop from the distillers.
—*Frankfort Roundabout*, 1907

Spent Grains

Distilling's primary by-product is a slurry of spent grains and stillage or water. The vernacular term for this mixture is "slop." Farmers have long used the same term to describe a gruel-like liquid composed of milled grains, water, and perhaps skim milk that they fed to hogs. In reference to post–Civil War distilling in Nelson County, James S. McKenna, son of Fairfield distiller Henry McKenna, said, "When I was a boy people claimed that there were twenty-six distilleries within a mile and a half of Fairfield; everybody had a little still then; it wasn't so much the whiskey they wanted—the slop was so good for 'fattnin' hogs."[1]

A late-nineteenth-century whiskey distillery produced about forty-four gallons of spent grains and water for each bushel of grain mashed.[2] A medium-sized industrial distillery processing 300 bushels of grain per day produced about 13,200 gallons of slop each day of operation, which required some form of use or disposal. The Kentucky distiller's slop disposal problem was summarized by Howard, Barnes & Company, operators of the Montgomery Distillery near Mt. Sterling in Montgomery County. Their 1868 advertisement in the *Kentucky Sentinel* offered this plaint: "Having constantly on hand more Still Slop than we can profitably make use of, we have concluded to dispose of a portion of it to persons having stock of any kind to feed." The ad suggested that farmers with hogs, cows, and horses would find distillery slop to be superior feed, as well as the cheapest and most profitable provender available. Slop could be purchased at the distillery for 25 cents per barrel, cash or monthly credit.[3]

Livestock Feeding

Dumping slop into a stream, perhaps the very stream that supplied the distillery's fresh water, was a common disposal strategy employed by many early farm distillers. The same stream may have supplied potable water to downstream farms and towns. As distillery output increased, especially after the Civil War, dumping slop in streams became more common and drew voluble complaints from neighbors, so distillers were obligated to come up with other methods of disposal. Early on, many distillers adopted livestock feeding arrangements that not only addressed the slop disposal problem but also offered the possibility of a second source of income as a hedge against the variable market demand for their whiskey.[4] But feeding their slop to cattle and hogs in large-scale, onsite feeding operations required expanded management and considerable investment. Livestock feedlots had to be readily accessible. If a site had sufficient level land, distillers built large shelter sheds adjacent to the distilling works. Optimally, they used pipes to pump slop from the distillery into large wooden or metal holding tanks and then directed the fluid via gravity through overhead pipelines to feeding troughs in the stock pens. This pen-and-feed arrangement resembled the twentieth-century confinement feedlots for beef cattle found across the Middle West and the central and southern Great Plains. The scale of operations was directly related to the volume of grain processed each day. The Paris Distilling Company in Bourbon County, for example, produced eighty barrels of whiskey per day in 1904. The slop produced by the distillery fed 800 cattle that were kept in "the latest imported cattle pens."[5]

Distilleries fed livestock until the animals reached market weight or the distilling season ended. Then they shipped the animals to area stockyards. Whereas farmers marketed livestock throughout the year, the distillers' livestock feeding calendar was seasonal and mirrored the distilling cycle, which generally began in late fall and ended in late spring or early summer. Because distillery feeders preferred to purchase stock for fattening at the beginning of the distilling season, livestock sales were generally very active in November.[6] By the 1870s, four different but related types of livestock feeding strategies had developed. First, some distillers sold slop or "liquid feed" directly to neighboring farmers at low cost. Although such sales were financially straightforward, they were geographically awkward. Farmers had to drive their teams and wagons, often over poor roads, to the distillery, fill barrels or tanks with slop, and return to their farms over those same roads. F. E. Robertson of Crittenden County operated a small distillery and offered to deliver "thick distillery slop at 20¢ per bbl."[7] Few other distillers offered to deliver the slop and simply advertised that they had a supply on hand. J. H. Rogers's Limestone Distillery at Maysville invited "parties wishing to slop cattle [to] please call and get a bargain."[8] The Glenmore Distillery at Owensboro advertised its slop in area newspapers with the slogan: "Hog Raising Is the Poor Farmer's Fortune and the Rich Farmer's Protection." Each day during the distilling season, Glenmore sold 1,200 barrels of slop to Daviess County farmers who waited in line with their teams and tank wagons.[9]

Distilleries operating in or adjacent to towns, especially county seats, usually had direct connections to the best roads. Those farmers with livestock feeding operations within a few miles of such towns had ready access to a large volume of slop. Small rural distilleries did not enjoy this advantage. John Atherton operated a distillery on Knob Creek in Larue County, nineteen miles south of Bardstown. In 1888 Atherton testified before the US House of Representatives Committee on Manufacturers about the operational advantages enjoyed by large urban distillers over small rural distillers. The geographic advantage resided with the urban distilleries, Atherton said. "I refer to the sale of the slop. The small distillers in the country have no market for their slop and can only use it for their farm stock; while in the case of proximity to the cities it can be sold. . . . I know of no other difference between large and small distillers."[10] Atherton might have mentioned that small rural distilleries tended to feed hogs and beef cattle for the slaughter and meatpacking trade, whereas large urban distilleries could also feed livestock for butchering and had the additional option of feeding dairy cows and producing milk for the large urban consumer market.

A second type of slop consumption strategy involved contracting with local farmers, cattlemen, and stockyards to ship their cattle and hogs to distilleries, where employees would feed the animals to market weight. The Union Stockyards in Cincinnati, for example, sent cattle to a Maysville distillery to be fattened.[11] Other stock owners worked as brokers, buying livestock as an investment specifically for distillery feeding. Brokers' profits were predicated on buying the animals at a low price, feeding them at a low cost, and selling them at a higher market price. Cattle brokers Furst & Summers fattened 1,000 head of cattle at the Glenarme Distillery in Midway, operated by S. J. Greenbaum in 1891. Although the brokers paid $11,000 for Glenarme's slop, they realized a net profit of $17,000 when the cattle were sold. In Paris, a Mr. Kahn fed 396 head of Tennessee cattle at the G. G. White Distillery. When the animals reached 1,000 pounds each, Kahn sold them to another broker for more than $20,000.[12]

The seasonality of distillery operations often worked against both farm and brokerage slop feeding. If distilleries ended production in the late spring, thousands of market-weight stock would be brought to market at about the same time, depressing prices. Although low-cost slop made brokerage feeding an attractive arrangement, cattlemen often found that stock prices at the end of the distilling season were so low that the sellers owed more money on the stock than the animals were worth. This risk could be mitigated by feeding stock under an arrangement whereby farmers and brokers shared the risk with distillers and divided the proceeds from the sale of the fattened livestock.[13]

As transcontinental railroads expanded west across the Great Plains in the 1870s and 1880s, western cattle entered eastern markets, often through the Chicago stockyards, where they were butchered and sent to fresh meat markets across the country. Kentucky cattlemen complained that western cattle producers had an unfair advantage because their stock grazed the open range "for free," and they claimed that refrigerated meat shippers such as Armour were marketing beef that was "unwholesome and in many cases positively deleterious." Cattlemen and distillers urged the Kentucky legislature to prohibit the sale of such meat; this, they reasoned, would put state livestock feeders on an equal footing with western producers.[14] Kentucky distillery cattle shipped to the Chicago stockyards in the mid-1890s sold for 15 percent less than grain- and grass-fed beef from the Middle West.[15] The Cincinnati livestock market handled primarily Middle Western stock, which were largely prime corn-fed beef cattle that commanded the highest market prices. Absent such cattle, market buyers paid the highest prices for distillery-fed cattle. Slop-fed dairy cattle sold at prices considerably higher than those paid for standard grass-fed milk cows.[16]

Some distilleries adopted a third type of livestock feeding strategy and purchased their own livestock from farmers or brokers and moved the animals to the distillery for feeding.[17] These distilleries marketed their fattened stock through local stockyards or sold them to brokers for shipment to regional or national markets. Improved railroad connections allowed brokers to ship cattle to Louisville, Cincinnati, or Chicago in search of top prices.[18] In 1887 the Hume Distillery Company of Garrard County sold cattle to a Lehman Brothers broker, who shipped the cattle to Jersey City, New Jersey.[19] Livestock broker Charles Byrne sent twenty railcars of Lincoln County distillery cattle to Jersey City in April 1891, at the end of the distilling season, and he had 300 additional railcars positioned near the large distilleries at Tyrone, on the Kentucky River east of Lawrenceburg, awaiting the end of their distilling and feeding season.[20]

Some livestock dealers—not necessarily farmers—engaged in a fourth livestock feeding strategy: they purchased the seasonal slop production from a distillery and then bought livestock and shipped them to the distillery for feeding. In December 1902, at the beginning of the distilling season, S. F. Ruble of Garrard County bought cattle at the Mercer County Court Day in Harrodsburg. The Ruble Brothers Company intended to buy a total of 1,200 cattle to be fed on slop produced at Edward J. Curley's Bluegrass Distillery at Camp Nelson on the Kentucky River.[21] Two years later, in January 1904, Ruble Brothers again bought cattle to be fattened at the Bluegrass Distillery, which, at the time, was mashing 400 bushels of grain per day.[22] Jonas Weil of Harrison County sent his stock to be fed at the Lair Distillery pens; he had purchased all the slop produced by Lair during the 1902–1903 distilling season, as well as the season's slop production at the J. Walsh Distillery and Rectifying Plant on the Ohio River at Covington. Weil subsequently shipped eight carloads of cattle to the Walsh works for feeding.[23] The following year, Weil bought 400 cattle to feed on slop produced by the Old Lewis Hunter Distillery on the South Fork of the Licking River in Harrison County. The Hunter works was mashing more than 200 bushels of grain per day at the time.[24]

**Cattle and Hogs Fed
at Kentucky Distilleries,
1887**

Collection District	Number of Cattle Fed	Number of Hogs Fed
2nd	2,848	591
5th	13,215	2,694
6th	3,695	3,350
7th	2,066	1,770
8th	1,556	2,582
Total	23,380	10,987

Source: Tabulated from Joseph S. Miller, *Report of the Commissioner of Internal Revenue, 1887*, Treasury Department Document 1031, 2d ed. (Washington, DC: GPO, 1887).

Importantly, those rural counties with the largest number of distilleries also became centers of farm livestock production, as farmers took advantage of the nearby distillery feedlots. The Fifth Internal Revenue Collection District, for example, included one of the state's largest distillery concentrations in Nelson, Marion, and Washington Counties. In 1887 district distillers fed nearly 16,000 cattle and hogs to market weight.

Some livestock feeders preferred the efficiencies inherent in feeding hogs rather than cattle. Farmers who arranged to feed their hogs at a distillery found the arrangement most effective if the distance between farm and distillery was minimal. In Anderson County, W. T. Bond's farm was only half a mile from the Bond & Lillard Distillery on Bailey Run, east of Lawrenceburg. Bond sent twenty-one hogs to the distillery to be fed on slop. Several weeks later, the hogs weighed more than 500 pounds each. The animals were so fat that they could hardly walk, and it took two days to drive them back to Bond's farm for slaughter.[25] Some distillers purchased "slop hogs" from brokers. Those who did included the Traylor brothers, who operated the Edgewood Distillery near Stanford, in Lincoln County, which mashed about 100 bushels of grain per day, and D. L. Moore, who fed hogs at his Anderson County works.[26] The English Berkshire pig, imported into Kentucky in 1824–1825, was reportedly the breed favored by British royalty for its flavorful meat. Many Kentucky hog feeders believed that when Berkshire pigs were fed the thick corn slop produced at distilleries, that breed outperformed others in terms of weight gain and quality of pork produced.[27] But some livestock buyers were less than sanguine about the advantages of slop feeding and paid higher prices for corn-fed stock. In the late 1850s and early 1860s the national market price for distillery-fed pork was 10 to 15 percent lower than that for corn-fed stock.[28]

Slop Dumping and Stream Pollution

Those distillers that did not feed livestock or sell their slop to farmers often dumped their slop directly into adjacent creeks or sinkholes, either openly or surreptitiously. And those distillers that did feed livestock produced a secondary by-product: animal waste. Livestock fed in confinement at distillery pens produced large volumes of waste that had to be routinely removed. If the manure-laden runoff from their feedlots flowed into nearby streams, this became another form of pollution. The Kentucky State Board of Health conducted a meeting with distillers in July 1910 to consider the problem of manure accumulation and drainage at distillery livestock pens. The contaminated runoff not only killed fish but also endangered the health of people living along the watercourse.[29]

"Nuisance": the word implies a petty inconvenience, a niggling and tiresome irritation. In legal idiom, though, nuisance is elevated to the level of a minor crime and misdemeanor. Common law gave citizens the right, without official authority, to seek abatement of a public nuisance without first obtaining a formal legal injunction.[30] Under formal statutory law, a public nuisance was defined as "anything that unlawfully causes injury to the rights of the public resulting in legal damage." A nuisance must "produce a tangible and appreciable injury to [a] neighboring property."[31] In an exemplar case, a court found Louisville butcher Frank Seifried guilty of public nuisance because of the unsanitary conditions at his slaughterhouse on the city's east side. In 1881 neighbors complained that the foul and nauseous stench produced by his operation poisoned the air and endangered the health of the families living nearby.[32]

Distilleries discharging liquid slop or animal waste into a waterway were subject to nuisance abatement regulation, as specified by section 1253 of the Kentucky Statutes. The statute prohibited anyone from putting into a stream or other body of water any substance that would sicken or kill fish, render the water unfit for use, or produce a stench. Such action was a misdemeanor offense punishable by a fine of not less than $10 or more than $100 and by imprisonment for thirty days to six months for each offense.[33] Even though distilling was an economically important industry, in nuisance cases, the state courts apparently did not favor private business over the public interest. More broadly, public nuisance cases were prosecuted across the country as courts identified and sanctioned noxious trades and industries. Livery stables and swine yards, slaughterhouses and tanneries, bone-boiling plants and limekilns, blacksmith shops and brick-burning yards, soap factories and candle shops, and many other businesses were subject to the power of public complaint and sanitary regulation.[34]

Slop dumping was a serious matter for those situated downstream, and notices of lawsuits and legal citations were often publicized in newspapers and trade journals. During the 1880s and 1890s the Headley and Peck Distillery operated beside an expansive karst sinkhole in south Lexington, near a small tributary of Wolf Run Creek. The distillery had a mashing capacity of 300 bushels per day, and when it was offered for sale in 1901, the *Wine and Spirit Bulletin* noted that the distillery had been the subject of numerous injunctions for dumping slop into the stream that ran through adjoining farms.[35]

The Paris Distillers Company in Bourbon County did not have direct access to an all-season surface water supply. George G. White's 547-acre Gilt Edge Stock Farm was situated adjacent to the distillery and was bounded, in part, by an extended frontage on Stoner Creek. For several years, the distillery paid White $300 per annum for the privilege of pumping water from Stoner Creek. The distillery also paid White $300 each year for permission to dump cattle pen waste into natural karst basins on his farmland near the distillery. When White offered Gilt Edge Stock Farm for sale in August 1904, the notice declared, "This farm is well suited for the large handling of cattle because of proximity of distillery." The purchaser would also have the option of extending the contract to supply water to the distillery and provide it with a waste disposal site, for an additional income of $600 per year.[36] John T. Hinton bought the farm, and nearly a decade later, the commonwealth of Kentucky brought a lawsuit against the owners of the distillery for "making and maintaining a common public nuisance." The suit stated that Kentucky Distilleries & Warehouse Company and Julius Kessler & Company, operators of the distillery on a small tributary of Stoner Creek north of Paris, allowed the accumulation of distillery waste and manure and filth from several hundred cattle being fed on slop at the distillery. As per its agreement with Hinton, the distillery pumped this material into karst sinkholes on the adjoining Gilt Edge Stock Farm. The sinkholes functioned as surface receptacles, but the material drained through passages in the bedrock into the Stoner Creek tributary. The sinks, tributary, and creek were fouled, and "the waters of said Stoner creek were so corrupted and charged with filth, stenches, and smells as to be unfit for domestic uses."[37]

In May 1901 landowners in the Hinkston Creek valley south of Mt. Sterling, in Montgomery County, voiced complaints about water pollution caused by slop dumped into the creek by the McBrayer Distillery. "We could not help it," a distillery representative said, "for while we slept some despoiler of property cut our dam and let the slop out into the stream." The *Mt. Sterling Advocate* suggested that the distillery should post guards at the dam and that the guilty parties should be subjected to the full penalty of the law.[38] The McBrayer works was likely holding slop in a dammed pond beside the distillery, a common practice intended to avoid the penalties associated with dumping waste directly into a creek. This containment technique was an early precursor to the "lagoons," or surface cesspools, used by twentieth-century hog producers to contain animal waste.[39] The following year, in April 1902, the *Mt. Sterling Advocate* published an editorial stating, "We are informed that distillery slop is being turned into Hinkston Creek at the distillery here, polluting the water so that stock refuse to drink and the smell of decaying matter is becoming past endurance."[40] The McBrayer Distillery site was about one mile south and upstream of Mt. Sterling. During the distilling season, the works mashed more than 600 bushels of grain per day.[41] In November 1902 W. Gay & Company bought the distillery's slop to feed 400 to 500 cattle onsite. The *Advocate* offered this anticipatory comment: "we may be sure that the offensive smell from the distillery will from this time on be a thing of the past."[42] Five years later, W. Gay & Company continued to slop cattle at the McBrayer Distillery and sold 500 head in July 1907 to Simon Weil of Lexington for 5 cents a pound. Although the cattle apparently flourished—they averaged 1,200 pounds when sold and would have returned about $30,000 to Gay & Company—the smell and stream pollution likely continued, although community criticism seems to have waned.[43]

The Peacock Distillery operated on Stoner Creek in Bourbon County, north of Paris near Kiserton. Peacock contracted with cattle feeders Lair and Houston to buy the distillery's slop for the 1904 distilling season, provided that they house and feed the livestock at the distillery's stables and pens. The contract also specified that the cattle feeders would hold the distillers harmless for any prosecution resulting from slop or manure polluting the stream, and they agreed not to use the premises in any manner that created a nuisance. The distillery operated from March 1 through June 10, mashing 300 bushels of grain per day. The slop was pumped through a pipe across Stoner Creek, where it fed 370 cattle and 70 hogs. Lair and Houston directed the surplus slop and manure into a 100- by 200-foot feedlot holding pond. Warm spring weather accelerated the decomposition process. Apparently, the holding pond leaked or overflowed, fouling Stoner Creek and killing fish, snakes, turtles, and crayfish. Buzzards gathered on the creek banks to consume the carcasses. Livestock would not drink water from the creek, and neighbors living miles away complained of the stench. When summoned to appear before the Bourbon County Circuit Court, the distiller contended that the feeding contract signed by Lair and Houston made them responsible for the nuisance, even though the spill took place on distillery property. The court rejected Peacock's argument and fined the distillery $1,500.[44]

Some distillers attempted to address the runoff problem by relocating their feedlots away from creeks. In the early 1880s Old Taylor Distillery No. 51 near Millville on Glenns Creek, south of Frankfort, fed some 2,700 hogs on slop.[45] By the late 1890s, the distillery, operated by E. H. Taylor Jr. & Sons, was feeding upward of 300 cattle owned by Lexington livestock broker and investor Simon Weil. Millville residents raised "great complaint about the filth from the distillery running into the creek. Many fear that fever will follow unless there comes a big rain to wash the creek out."[46] But distillery operations continued. In 1903 Old Taylor was mashing 300 bushels of grain per day, and Weil again contracted to buy slop at 6 cents per bushel for the period from October 21, 1903, to May 31, 1904. The distillery agreed "to furnish the usual feeding lot and pens, and to deliver the slop to said pens to the satisfaction" of Mr. Weil. In return, Weil agreed to make a cash payment on the first of each month for the slop used the previous month.[47] The following year, Weil again contracted to purchase Old Taylor's slop for 6 cents per bushel for the 1904–1905 distilling season, which would "begin as soon as water supply at the distillery is sufficient to start and to continue through June 1905 if desired by Mr. Weil to run that length of time." This flexibility suggests that the distillery was willing to accommodate Weil's concerns that his cattle might reach market weight before distilling operations concluded for the season.[48] In

1906 Weil signed a similar agreement with the Old Taylor Distillery, which was then mashing 500 bushels per day. The distillery again consented to provide "a properly equipped feeding lot, and to furnish slop to said lot in a satisfactory manner." The contract included an important addendum, however. E. H. Taylor Jr. & Sons also "agreed that under like conditions of the laws and their [*sic*] being no interference by the neighborhood in regard to slopping cattle we agree to make the six cents per bushel the price for three years." The agreement, written on company letterhead, suggested that residents in the neighborhood were continuing to voice their displeasure with the outsized industrial operation in their midst.[49] The distilling company acquiesced to public concerns later in 1906, when it built new cattle pens "high above the valley and away from the creek."[50]

Established in 1878, the Kentucky State Board of Health began an extended campaign after the turn of the century to deal with waterborne diseases such as typhoid fever through statewide inspections of stream water contamination and sewage sources. Almost every private residence, farm, health resort, hotel, industry, and public institution routinely discharged sewage into streams, but one of the board's primary concerns was to address the disposal of slop and livestock waste by distilleries. From 1910 to 1912 the board's sanitary engineer investigated distillery disposal practices at works in Bardstown, Eminence, Owensboro, and Midway. The board formally cited the Kentucky Distilleries and Warehouse Company for creating a public nuisance by dumping cattle waste and slop into creeks at the Bond and Lillard and Cedar Brook Distilleries in Anderson County, three distilleries at Athertonville in Larue County, and two distilleries at New Hope in Nelson County. Some distillers heeded the board's admonitions to revise their waste disposal practices. For instance, the S. J. Greenbaum Distillery in Midway agreed to build a drying plant to dispose of its slop after the engineer inspected the works and tested pollution levels in nearby Elkhorn Creek. The board recommended that the state compel other distillers to cooperate and to contribute to a fund to finance experimental work that would provide solutions to distillers' disposal problems.[51] When some distilleries failed to comply, the Board of Health appointed Drs. John G. South and H. S. Keller of Frankfort to visit the state's distilleries and determine how long it would take to halt the draining of cattle pens into creeks and rivers. Based on their reports, the Board of Health called for noncomplying distillers to dispose of their slop in some manner other than feeding it to cattle confined in pens at the distillery.[52]

Consider the historical context of citizens' complaints about distillery stream and air pollution. Nineteenth-century American towns and cities were noisome places to live. People dumped garbage in any convenient space. Outdoor privies stood in residential backyards, polluting the groundwater. Those few residences with indoor plumbing discharged their household waste into the nearest surface drain. People without privies or plumbing dumped their chamber pots into yards, alleys, and streets. Horses were ubiquitous sources of motive power, but they turned city streets into equine latrines. Noxious trades—slaughterhouses, bone boilers, soap makers, tanneries, butcher shops, and the like—fouled the air and waterways.[53] If people living in such circumstances complained about distillery slop dumping and feedlot runoff, the situation must have been dreadful by present-day standards.

Animal Health and Pure Food

Farmers living near distilleries could feed their beef cattle to near market weight and then send them to the distillery feedlot to "finish" on slop. But aside from environmental contamination, many critics found the practice of feeding distillery by-products to livestock, especially dairy cattle, a problematic process. Two primary concerns arose early on: slop's effect on the animals' health, and the quality of milk produced by slop-fed dairy cattle. The Kentucky Agricultural Experiment Station's 1902 annual report cited a study on the effect of feeding hogs dried slop. It concluded that the most effective feeding ration comprised no more than one-fifth distillery grains to four-fifths corn. Hogs fed a ration lower in corn and higher in distillery grains did not thrive or gain weight at an acceptable rate.[54]

The healthful quality of milk produced by slop-fed dairy cows was another matter. Some physicians thought that milk produced by distillery cattle was unwholesome if not injurious to those who consumed it, especially children. The New York Academy of Medicine undertook a study of the issue in 1858, but the results were inconclusive. The question of milk quality and human health was larded with complex interrelationships. Any scientifically valid study would have to control for several factors: overall animal diet, feedlot sanitation, and other environmental variables such as temperature, as well as the exact chemistry of the slop and the milk and the age, gender, and health of the human subjects. A study based on a sterile laboratory experiment would have little value, given that distillery livestock often lived in unhygienic conditions.[55] The scholarly journal *Science* asked readers' opinions about the efficacy of feeding distillery slop to dairy cattle and utilizing the milk they produced for human consumption. The opinions offered by medical practitioners and other professionals were divided, but the consensus seemed to be that if cattle living in sanitary conditions were fed slop in conjunction with other feeds, there was little risk of harm to the animals. But if slop were fed to cattle living under ordinary, unsanitary conditions, the result would be unhealthy stock, possibly putting those who consumed their milk at risk.[56] Some scientists conjectured that cows fed nothing but watery slop tended to suffer from deranged digestive tracts, and the milk they produced was low in fat and unfit for consumption.

The popular view that distilling by-products affected milk quality persuaded the Kentucky legislature to pass a pure food statute in 1899 that prohibited feeding distilled grains and slop to dairy cattle.[57] Under the law, the State College of Kentucky's Agricultural Experiment Station was responsible for conducting research on food purity and enforcing the statute's provisions. In 1904 Robert M. Allen of the Agricultural Experiment Station published an essay in *Kentucky Farmer and Breeder* on the use of distilling grains as livestock feed. He cited a report by the Food Department of the Massachusetts Experiment Station, which had found that distillers' grains did not affect the quality of cows' milk. Rather, when used as a feed for dairy cows and horses, distillers' grains had nutritive value equal to cornmeal or bran. Off flavors in milk were caused by the filthy conditions in which cattle were kept at distilleries. Allen concluded that distilled grains were "largely wasted by Blue Grass distilleries," and he implied that instead of feeding their slop to livestock, distilleries were often "letting the waste into the streams," for which they were fined.[58]

In 1906 Agricultural Experiment Station representatives, the milk commissioner of the Jefferson County Medical Society, and the Health Departments of Louisville and Jefferson County recommended that county attorneys prosecute eighty cattle feeders operating filthy dairies and feeding their cattle distillery slop. The feeders pleaded guilty in Jefferson County court and paid fines of $100 each. The judge suspended their fifty-day jail sentences, pending their promise to cease feeding distillery slop within two months and to ensure thoroughly sanitary conditions in their dairies. The problematic state of Jefferson County dairies prompted an investigation of animal husbandry practices in all counties shipping milk into Louisville.[59] Sanitation at Louisville dairies improved dramatically once law enforcement procedures were in place.

The slop disposal problem and, coincidently, the issue of animal nutrition were also being addressed in another quarter. From the 1880s to the early 1900s, several mechanical innovations allowed distillers such as S. J. Greenbaum of Midway to screen and dry slop. Through a kind of stepwise invention process, manufacturers developed large-scale dryers utilizing both direct heat and indirect steam heat (see chapter 6). These inventions convinced Robert Allen of the Agricultural Experiment Station that most grains could be dried immediately after completion of the distilling process. If properly handled, spent distilled grains that were high in corn content were 28 to 32 percent crude protein and about equal in nutritive value to gluten feeds.[60] Distillers' grains that were high in rye, in contrast, were considerably lower in feed value. Early-twentieth-century research scientists recommended that dried corn-based distillers' grains, though not especially palatable, could be mixed with other concentrates to provide an excellent high-protein feed for dairy cattle, steers, and sheep. They also suggested feeding it to horses and hogs in limited amounts, given its bulkiness.[61]

Prior to the 1850s, few Americans seemed to be aware of the profoundly unsanitary conditions in which they lived. Commonly held beliefs associated disease with bad air or miasmas given off by organic waste, and it was thought that epidemics affected primarily poor people. The application of scientific research methods to the problems of sanitation during the second half of the nineteenth century exposed disease-causing conditions and pathogens, motivating activists to promote public awareness and the passage of public health laws. While some industrial processes contributed to environmental pollution and the risk of disease, family households, be they urban or rural, added to the hazard. Nineteenth-century distillers were largely comparable to the rest of industrial America in their creation of waste materials and their seemingly casual attitude toward environmental degradation. Yet when presented with appropriate technology (spent grain dryers, for example) to address their waste disposal problem, many keenly pursued that solution.

10

Connections

Transportation facilities were an especially powerful instrument in shaping new patterns of settlement and development; steamboats enlivened riverbanks with aspiring ports at every feasible landing, and the railroad was soon acknowledged as the great determinant of local fortunes, founding strings of new centers and blighting all it bypassed as the depot became the new focus of American social and political life.
—D. W. Meinig

Conjunctions—Transportation and Logistics

Landscape elements connect with one another by pathways and waterways, roads and streets, highways and railroads. Kentucky's eighteenth- and nineteenth-century landscape comprised diverse but integrated elements—lands and waters; farms, fields, and fences; towns, commercial buildings, and industrial plants. The landscape was also a theater of culture—a stage on which people established social relations, developed and used technology, created economies, practiced the politics of governance, and cultivated a mentality or a way of thinking about the material world. People established routeways consisting of single tracks or networks and thereby connected landscape elements so that they could be developed and worked in concert. People and their ideas and goods all traveled along these routeways that provided the linkages to maintain cultural vitality and promote development and modernity.[1] In the eighteenth and nineteenth centuries, many Kentuckians invested time, labor, and wealth in routeways: clearing and building them, and devising ways to use them. In retrospect, the connective infrastructure they created was as epic in scale as it was long-lived; their early routeways were foundational, and many remain in use.

Historical routeways connected places to overcome distance, a fundamental obstacle to many forms of interaction. Thought of in the abstract, distance is a component of time and place: one consumes time to overcome distance. Distance, therefore, is a barrier to commercial enterprise, just as it is a hindrance to communication and human contact more generally. To change or increase the scale of a business operation, one must overcome distance by whatever means available, most commonly by applying the technologies of movement—whether pack animals, wagon trains, hard-surface roads, or steam-powered boats and railroads. Improving transport technology does not shorten distance between origin and destination—except perhaps in the case of straightening a crooked road or dredging a new channel across looping river meanders. Rather, actual distance remains the same, but effective distance is reduced as the time expended to traverse it is shortened.

Transport Infrastructure Mode I: By River and by Road

By River

Kentucky and Tennessee pioneers shipped $225,000 worth of produce to New Orleans by way of the Ohio and Mississippi Rivers in 1788.[2] Such general pronouncements disguise the varied human actions required to realize such a notable accomplishment, given the transport technology of the day. Let us consider a seemingly minor event, but one that illustrates the complex and idiosyncratic layering of participants, experience, and environment involved in delivering goods to New Orleans in the early nineteenth century: In 1809 two carpenters, Porter Clay and Harrison Monday, built flatboats on the north bank of the Kentucky River, 15 miles south of Lexington and some 158 miles upstream from the Ohio River. Porter Clay, the younger brother of Henry Clay, had moved from Virginia to central Kentucky in 1792. A talented cabinetmaker, he operated a shop on Mill Street in Lexington during the early 1800s. In 1808 Clay redirected his skills from cabinetry to building and operating a large sawmill on the Red River, a tributary of the Kentucky River some 30 miles east of Lexington. The following year, Clay and Monday started to build flatboats at a small industrial settlement at the mouth of Tate's Creek on the Kentucky River, using lumber supplied by Clay's sawmill. Thomas Hart, the agent for the two men and Clay's brother-in-law, announced in the *Kentucky Gazette:* "The boats built by Monday are known to be a superior quality; they will be sold at the usual price, and where it will be more convenient to the purchaser the payment will be received at Natchez or New Orleans, giving him sufficient time to dispose of his cargo."[3] Though one cannot be certain why Porter Clay chose to defer cabinetmaking to focus on boatbuilding—perhaps making boats was more lucrative, given the volume of goods being shipped—this riverine enterprise proved to be a reasonable adjunct to his sawmill operation.

Moving Kentucky farm products to market in the late eighteenth and early nineteenth centuries, whether grain, tobacco, livestock, salted meat, or whiskey, was difficult and expensive in terms of time, labor, and tractive power. Commodity transport was also seasonal—in part because farmwork and the goods produced were seasonal, and in part because the water levels in Kentucky's navigable creeks and rivers were so variable. Usually the levels were high enough in late winter and early spring to accommodate cargo boats. Everyone engaged in waterborne commodity shipment was subject to the vagaries of the weather and the seasons. For instance, Clay and Monday built boats through the winter to have them ready for sale by March. The annual river navigation cycle in the Ohio River Valley included three primary high-water shipping periods: a short season in November, followed by another short season in February. High water in May and June usually allowed sustained shipping for several weeks. One expected an extended low-water period during the summer and early fall and a shorter low-water winter season.[4] Given the size of the Ohio River watershed—nearly 190,000 square miles—and the valley's whimsical weather, no one could reliably predict when the high-water season would begin nor divine its duration. Only experienced crews piloting shallow-draft vessels could risk navigating the river in summer, given the low water level and the numerous hazards of exposed rocks and sunken snags.

Riverboats towing loaded barges between Cincinnati and Louisville carried all manner of goods, including large volumes of coal, corn, and rye for delivery to riverside distilleries. In turn, the distillers shipped whiskey by river and rail to distant rectifiers and wholesale markets. In midwinter, streams and rivers often froze over, and the solid ice halted riverboat traffic. Once the waterways began to clear, the open water often contained large ice floes moving downstream with the current, their mass sufficient to crush and sink boats and barges tied up at the river's edge and tear wharf boats from their moorings. The Boone County Distillery at Petersburg, Kentucky, operated on an elevated Ohio River terrace about twenty-four miles below Cincinnati. Distillery laborers made frequent whiskey deliveries to the railroad depot across the river at Lawrenceburg, Indiana; they transported the whiskey by boat when the river was open and by sled when the river was ice covered. In January 1877 nighttime temperatures fell to −10°F for several days, and an ice gorge blocked the Ohio River at Petersburg. On January 14 "the Ice was from Shore to Shore & over 100 barges past down in the Ice Some 8 or 10 of them loaded some Barges with the end Stoved in Some with their sides mashed in &c." The floating ice hazard was especially pronounced that winter. The river was blocked by ice for twenty-three days, and moving ice caused "the Greatest Destruction of Boats by the ice from Pittsburg to Louisville that ever has been Known by the Oldest River Men."[5] Barges bearing barreled whiskey often lost part or all of their cargo when smashed by floating ice, and errant barrels were sometimes seen floating with the current.[6] Because insurers were reluctant to offer coverage on high-risk shipments, low water levels and ice floes presented a deterrent to transporting high-value cargo. Some distillers held their whiskey in riverside warehouses or on open docks, awaiting favorable conditions for safer navigation. Commodity shipping by inland rivers, then, was not a stable and predictable endeavor, even for those who were willing to accept the risk.

Kentucky lay near the headwaters of the Mississippi-Ohio River system and therefore had access to a significant number of downriver markets, the most attractive of which was New Orleans. Boatbuilders frequented the Ohio River from Pittsburgh to Paducah, producing flatboats, keelboats, and barges. Farmers and other goods producers in the watershed could build or buy boats and float their cargo downstream with the current. A major boatbuilding center developed at Brownsville, Pennsylvania, at the mouth of Red Stone Creek on the Monongahela River south of Pittsburgh. In the middle Ohio Valley, large boatbuilding centers were established in Cincinnati, Ohio; Newport, Louisville, and Portland, Kentucky; and New Albany and Jeffersonville, Indiana.[7]

Smaller operations such as the Clay and Monday boatyard on the Kentucky River were commonplace. Hart, their Lexington agent, sold their flatboats on short-term credit. Buyers loaded their boats and drifted down to the lower Mississippi ports of Natchez or New Orleans; once they arrived, they sold their cargo and their boats and paid one of Hart's representatives. This generous financial practice acknowledged that local shippers did not have sufficient operating capital to purchase a boat outright. It also suggests that, despite the elementary state of banking and communication at the time, Hart understood the rather sophisticated business procedures practiced on the lower Mississippi, even though that market lay more than 1,400 river-miles away. Even more amazing is that Hart employed this credit policy knowing that if the boats did not survive the hazardous trip with their cargo intact, payment would likely not be forthcoming.

Overland by Road

Transportation costs were a primary concern for farmers, millers, and distillers, who had to move grain, construction materials, and cooperage to their works and then ship the finished product to market. During the colonial period, the cost of transporting common bulk commodities such as wagonloads of wheat or hemp by road from a Maryland or Virginia farm to a coastal port city seventy-five miles away was twice the cost of transporting the same commodities by ship across the Atlantic to a European destination.[8] A farmer or miller who lived twenty miles from a market would pay at least 22 cents to haul a barrel of flour that distance by wagon, versus 2 cents by watercraft. Given freight costs in the Ohio Valley in 1820, a farmer hauling corn by wagon could not make a profit if the trip was longer than twenty miles.[9] Because commodities shippers placed a premium on water transport, nineteenth-century farmer-distillers had to compete with many other interests for riverboat shipping space. In 1819 Kentucky farmers shipped 25,000 hogsheads of tobacco as well as quantities of flour and hemp down the Ohio. An estimated 40,000 to 50,000 tons of produce passed the Falls of the Ohio at Louisville that year, heading for New Orleans wharfs and markets.[10] The cost differential between river and overland transport was an important factor in distilling's transition from a craft to a commercial industry. Although small craft distilleries operated throughout the Bluegrass countryside, distillers built the first large-scale industrial works beside navigable rivers.

Frontier farmers who produced a surplus of corn might shell it and pack it onto a mule or a horse for a trip by path or trail to the nearest mill or village market or to industrial settlements at iron furnaces and salt licks.[11] Alternatively, farmers so inclined could distill the corn into whiskey. Thoroughly dried, shelled corn weighed about 56 pounds per bushel, so four bushels, or 224 pounds, would have been a tolerable burden for a stout pack animal. By distilling those four bushels of corn into whiskey, nineteenth-century farmers could produce as much as three and a half gallons per bushel, or fourteen gallons. A gallon of whiskey—ethyl alcohol is lighter than water—weighs about 7.75 pounds, so fourteen gallons would have weighed about 108 pounds. This whiskey might have been hauled in stoneware jugs or wooden casks. The eight-gallon keg was a common size, although coopers made wooden casks of various volumes. An eight-gallon keg weighed about 16 pounds empty, so two eight-gallon kegs plus sixteen gallons of whiskey would have weighed about 155 pounds—almost 70 pounds less than four bushels of shelled corn. Corn and whiskey prices varied over time and from place to place, but an average price for whiskey was 50 cents per gallon. A bushel of corn was similarly valued. If one added a third 16-pound keg to the pack animal's load, that would bring the cargo volume to twenty-four gallons of whiskey that collectively weighed about 233 pounds, marginally heavier than four bushels of corn but valued at $12 compared with $2. Whether one transported high-value whiskey or low-value grain, moving commodities and merchandise overland by packhorses or wagons cost from 20 to 60 cents per ton-mile, depending on the distance traversed and the weather and road conditions.[12]

For livestock producers, a compelling alternative to selling grain to a miller or distiller was to feed the grain to livestock and drive the stock overland to a regional packinghouse or a distant market. Pioneers fed hogs in central Kentucky by the 1780s, and by 1810, farmers had established a cattle feeding industry in southern Ohio and central Kentucky. During the autumn of 1810, Ohio Valley farmers drove more than 40,000 hogs to Baltimore and Philadelphia.[13] As the Ohio Valley livestock production industry matured, it supplied hogs to packinghouses in regional urban centers such as Cincinnati and Louisville, where processors salted pork and packed it in barrels for shipping. Drovers herded cattle overland to markets until railroads developed specialized livestock cars in the late 1850s.[14]

Though it is seldom acknowledged, the flow of bulky, low-profit farm products and distilled whiskeys out of Kentucky was greatly exceeded by a counterflow of finished consumer goods into the state. Kentucky merchants ordered all manner of imported goods from Philadelphia and Baltimore wholesalers, including French sherry and brandy, Holland gin, Jamaican rum, Portuguese port, Madeira, and other spirits. Philadelphia furniture and wagonloads of Irish lace, Chinese tea, Spice Island nutmeg and clove, Barbados muscovado sugar, and English cloth goods and ceramics passed through Pennsylvania's Endless Mountains and the Allegheny uplands on their way to the Ohio River near Pittsburgh. Loaded onto river craft, the goods moved downstream to Limestone or Maysville, central Kentucky's early entrepôt. After being transferred once more to wagons, the goods moved overland along the Maysville Road, a mud track at the time, sixty-seven miles to Lexington. Alternatively, shipped goods might continue downstream to merchants in Covington or Louisville.

Imported goods were exceedingly expensive, in part because of high transportation costs. The price to move goods from Middle Atlantic ports to Kentucky was about $65 per ton, and the state's merchants were spending upward of $10,000 per day on imported goods during the 1820s.[15] Deeply concerned that this capital drain was hindering the development of a local industrial economy, Kentucky's political leaders formed an association to discourage the purchase of imported goods. At the association's first convention in 1789, members pledged to reduce their purchases of imported goods and declared, "We will not make use of ourselves or suffer to be used in our families, except in the case of sickness, any wine, rum, brandy or other spirituous liquors, which shall not be made within the District of Kentucky."[16] Although the association succeeded in encouraging the local manufacture of basic goods such as cotton and hemp products, the volume of imported finished goods increased. By 1810, Henry Clay and other business and political leaders were compelled to petition the state legislature to protect domestic manufacturing. Contained within these events was the kernel of Clay's American Plan, which advocated tariffs on imported goods to protect American industry, while lowering transportation costs through an expansive program of large-scale, centrally planned improvements in river navigation and road construction.

Transport Infrastructure Mode II: Turnpikes, Steamboats, and Early Railroads

Locational inertia—caused by the difficulty, slowness, and cost of movement—constrained the expansion of Kentucky's early commercial development, especially distilling. Early roads were largely unimproved, unsurveyed tracks made by pack animals, teamsters' wagons, and drovers' herds; they were deeply rutted in dry weather, impassable mires in rainy season, and hard on horses' hooves when frozen in winter. A few roads followed Indian paths or buffalo traces if they led in desirable directions. The term "improved" implied clearing a road of stumps and stones or digging out a steep creek bank to ease the anxiety of fording a stream. Stone-surfaced roads or turnpikes enabled all-weather cross-country travel and thereby broadened the area from which distilleries could draw resources, be it grain, fuel, cooperage, machinery, or hardware. Inventions and innovations led to better and more economical transportation: turnpikes, steam engines, railroads, and the tools and techniques to build and operate them.

Turnpikes

In 1797 the Kentucky legislature enacted a road construction law that largely replicated Virginia's internal improvement act of 1785. It placed responsibility for roads at the local precinct level and established county-level authority over road construction and maintenance through a court-appointed precinct surveyor.[17] Residents who wanted to build or improve a road petitioned the county court for permission to do so, but they were given no technical instructions other than minimum dimensions and an admonition to clear the way of obstacles. The law empowered surveyors to muster local residents into crews to work on precinct roads—bringing along their own tools and draft animals. Road projects of primary interest led to river landings and mills; other priority roads connected farm neighborhoods with one another and to the county seat. In 1817 Kentucky legislators enacted a law authorizing the construction of two trunk-line turnpikes connecting Lexington to the Ohio River—one to Louisville, the other to Maysville. A financial crisis in 1819 cut short road construction, but the legislature reauthorized the road in 1824, and by 1830, laborers had completed the section of the Maysville Road from the Ohio River four miles south to Washington, Kentucky. This was the first stone-covered road built to Scottish engineer John McAdam's specifications in the trans-Appalachian West.[18]

In 1835 the state legislature created the Board of Internal Improvement, whose purpose was to bring science and engineering to road construction. The board hired several trained and experienced engineers, including H. I. Eastin, who became the state's primary road engineer. Thereafter, road construction required formal petitions and a state charter of incorporation. Road builders followed detailed directives, which included clearing a twenty-five- to thirty-foot-wide right-of-way, raising the roadbed above the surrounding land by cutting down hills and filling depressions, and creating lateral side ditches for drainage. The road surface, following McAdam's plan, had to be constructed of carefully sorted gravel and broken stone laid on the track several inches thick, with a compacted convex crown to direct runoff into the side ditches. Anyone wishing to construct a road—usually a group of local residents—had to petition the state for authorization to form a road company and permission to issue stock to provide funds for construction. State law granted road companies the authority to collect tolls to underwrite construction and maintenance and to build tollhouses and gates for toll collection. Tollhouses had to be spaced at least five miles apart and at least one mile from the nearest town.

Trip duration for teamsters driving loaded wagons over unimproved tracks was protracted, and slow travel increased the costs associated with drivers' wages and overnight accommodations for them and their teams. But these costs dropped and travel habits changed when toll roads—or turnpikes, as they became known—opened. Stagecoaches traveling the hard-surface, all-weather toll roads could manage eight miles an hour. The trip from Lexington to Maysville, which had taken four difficult days on the old dirt track, could now be completed in one day and in relative comfort.[19] The travel costs for large farm wagons loaded with forty bushels of grain (about 2,400 pounds), a full load for four horses, dropped dramatically. Distillery wagons loaded with 400-pound whiskey barrels moved easily on stone-covered turnpikes compared with dirt tracks.[20]

Turnpike construction brought the advantage of all-weather movement, but at the cost of having to pay tolls at regular intervals. And the road-building process was fraught with financial and management problems. From 1834 to 1837 the Kentucky legislature approved seventy turnpike construction charters.[21]

A national financial panic began in 1837 and lasted until the mid-1840s, largely trun-

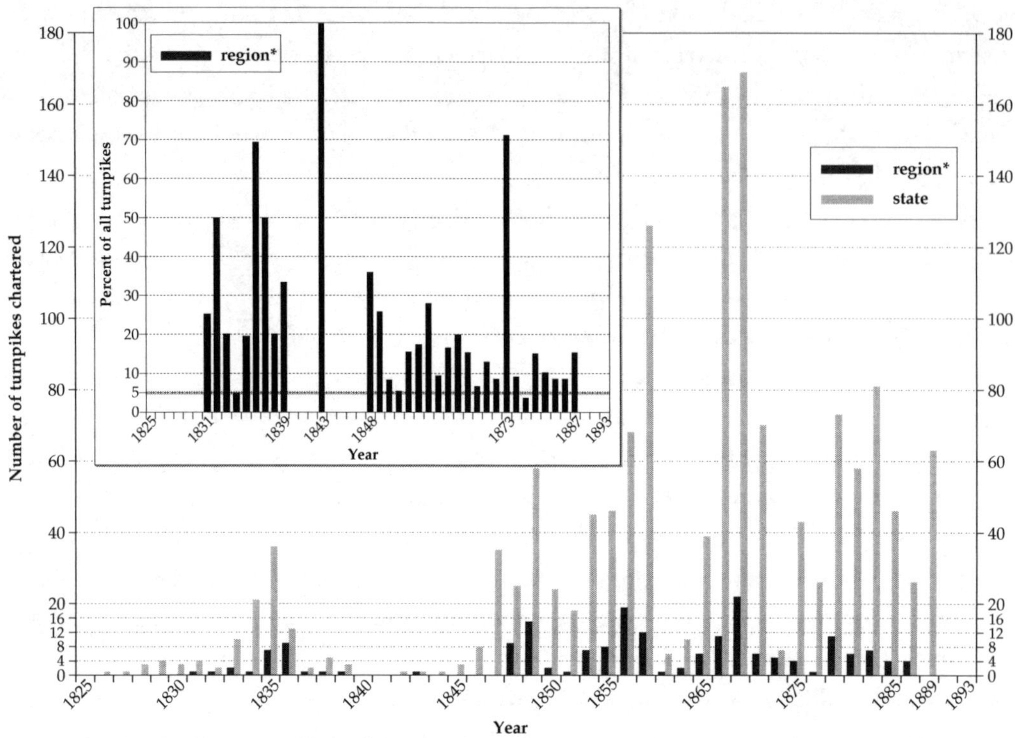

*Consists of six Bluegrass counties: Anderson, Franklin, Marion, Nelson, Washington, and Woodford.

Kentucky turnpikes chartered, 1825–1889.

cating most road construction. Beginning in 1847, though, the number of charters increased rapidly, and more than 120 were granted in 1860, the eve of the Civil War. Turnpike construction resumed in earnest in 1866 and 1867, when Kentucky approved more than 160 charters each year. Thereafter, road-building projects stabilized at forty to eighty per year until construction stopped in 1889. From 1826 until the turnpike era ended in 1898, the state chartered 1,460 turnpikes.

Distillers, along with farmers who sold grain to distilleries, took considerable interest in turnpike construction. During the turnpike era, road commissioners in six leading distilling counties—Anderson, Franklin, Marion, Nelson, Washington, and Woodford—applied for construction charters at a higher rate than did people elsewhere in the state. Only in the mid-1870s did these six counties (constituting 5 percent of Kentucky's 120 counties) submit fewer than 5 percent of the state's total turnpike charter applications. From 1831 to 1843, these six counties accounted for 20 to 70 percent of all state turnpikes chartered in most years. And from 1848 to 1887, these distilling counties constructed from 10 percent to more than 70 percent of all turnpikes in a given year.[22]

Turnpike charters provided corporate standing for turnpike companies, declared the price of stock shares, specified the number of tollgates permitted, and set out procedures for electing board officers and scheduling stockholder meetings. Once a charter application received legislative approval, road construction could commence, provided the commissioners had raised sufficient capital to pay for materials and hire contractors. Charters instructed road surveyors to "run the most practicable route." Board officers had a legal right to obtain rights-of-way through private land, as provided for in the Civil Code of Practice, and to award damages to landowners where appropriate. Access to private land allowed contractors to quarry stone at convenient intervals along the route and to obtain other building materials as necessary. Company officers were allowed to determine a road's width and grade, but charters set a minimum width for the stone or "metal" covering, typically eight to twelve feet. Turnpike boards contracted surveyors, engineers, and workmen to build the road to specifications, fabricate bridges, and erect tollgates. Boards also set the tolls, at rates fixed by state law, and hired and paid toll collectors. Some charters permitted county courts to subscribe road stock up to $500 per mile; otherwise, the companies sold stock to private parties.[23]

Active distillery development was under way in the New Haven–New Hope–Loretto area of southern Nelson and western Marion Counties by the 1850s and continued into the post–Civil War period. Charter applications for three turnpikes there illustrate the relationship between road construction and distillery operations. Investors received a state charter for the New Haven and Howard's Mill Turnpike Road Company in 1871. Although the charter did not describe the road's route, it may have run southeast from New Haven to the Rolling Fork at Athertonville, where it followed present-day Blanton Road to KY 247 and KY 84, the Howardstown Road, to Howard's Mill. The road likely passed the George Elliott Distillery, situated on a small tributary of the Rolling Fork. The charter required the sale of $40,000 of capital stock at $50 per share, and it permitted the election of officers, a land survey, and the start of construction when the commissioners had subscribed $3,000. The roadway was not to exceed fifty feet in width, and the stone-covered roadbed could not be less than eight feet wide. The eight commissioners were Lucratius Blanton (possibly the namesake for Blanton Road), William Boon, Park Cameron, and W. S. Ford; farmers James Boon and Joseph Howard; New Haven merchant Sylvester Johnson; and distiller John Atherton. The Atherton Distillery stood near Knob Creek and Rolling Fork on present-day US 31, south of New Haven in Larue County, and it was served by a spur of the Louisville & Nashville (L & N) Railroad. Atherton also served as a commissioner on the Greensburg, Columbia, and New Haven Turnpike Road, chartered in 1871. The New Haven and Howard's Mill Turnpike linked southern Nelson County to the Green River by way of the Bardstown and Green River Turnpike, which was in operation by 1838, and generally corresponded to sections of present-day US 31E and KY 61.[24]

In 1870 the New Hope and Rolling Fork Turnpike Road Company received a charter authorizing it to issue shares at $50 each and to build a pike that extended from the Bardstown-Greensburg Road, at the Marion-Nelson County line, north five miles to the New

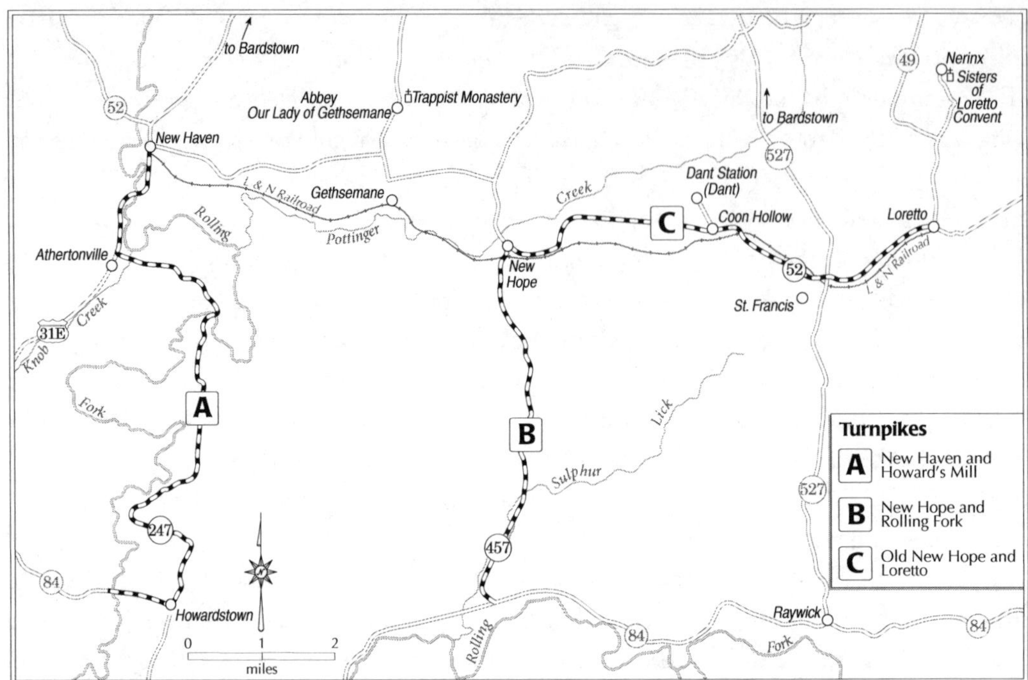

Turnpike routes in the New Haven–New Hope–Loretto area of Nelson, Marion, and Larue Counties, 1870–1884.

Hope Station on the Lebanon branch of the Louisville & Nashville Railroad. Present-day KY 457 generally corresponds to the old pike's alignment. The eight commissioners included three distillers. Edward L. Miles operated the E. L. Miles & Company Distillery at the intersection of the L & N Railroad and the Rolling Fork, New Hope, and Gethsemane Turnpike. In 1867 Miles and his business partner began construction on the New Hope Distillery near the same intersection. Turnpike commissioner Thomas J. Miller was associated with three distilleries at New Hope in 1873. Thomas J. Pottinger operated a distillery on Pottinger Creek near the Gethsemane Station of the L & N Railroad, and in 1884 he served as a commissioner on the New Haven and New Hope Turnpike. The other turnpike commissions were A. J. Beall, John Cravens, farmer Thomas Hutchins, B. H. Miller, and Lafayette Stiles, who had served as postmaster at the Rolling Fork post office.[25]

Chartered in 1884, the Old New Hope and Loretto Turnpike started at the Rolling Fork, New Hope, and Gethsemane Turnpike in New Hope and ran east, roughly paralleling the L & N Railroad, to Dant Station and terminated at Loretto Station. Although railroad surveyors had laid out track routes to connect existing towns, the new rail lines compelled residents to build or upgrade roads leading to the new railroad stations. The Old New Hope Turnpike charter set the stock price at $25 and specified that "the entire capital stock shall not exceed the amount necessary to construct said road and toll-houses."[26] The road likely followed a route very similar to present-day KY 52. The turnpike linked several distilleries: the E. L. Miles and New Hope Distilleries in New Hope, owned by Edward L. Miles; the Kentucky Belle Distillery south of Pottinger Creek, owned by George Hagan; the Coon Hollow Distillery, associated with distiller Thomas J. Miller; J. W. Dant's Distillery at Dant Station; and the Ballard & Lancaster Distillery, operated by Richard Cummins. The turnpike's five commissioners were distiller J. W. Dant, Miles Hagan, Thomas W. Price, G. W. Masterson Sr., and Green B. Masterson. Hagan, Price, and the Mastersons operated farms north and east of New Hope.[27]

These new pikes underwrote a reciprocal relationship between area farmers and distillers. Farmers could readily deliver grain to the distilleries and, if they wished, haul slop in tank wagons to their farms. Moreover, once the turnpikes were completed, the commissioners enjoyed toll-free travel.

Turnpike construction schedules depended in part on the success of stock subscriptions. Once turnpike commissioners had raised sufficient capital and surveyed the route, they hired local contractors to build the road. Many of Kentucky's road laborers or turnpikers were Irish immigrants. One of them was David Durnan, who lived at New Haven in Nelson County in 1870 and operated a turnpike construction business. He and his turnpikers likely built several roads in the New Haven–New Hope–Loretto area. The Durnan family included David's wife Mary, also born in Ireland, and a daughter and son, both born

David Durnan Household, New Haven, Nelson County, Kentucky, 1870

Name	Age	Relationship	Occupation	Place of Birth
David Durnan	45	Father	Road contractor	Ireland
Mary Durnan	30	Mother	Keeps house	Ireland
Mary E. Durnan	10	Daughter	Attending school	Kentucky
Peter D. Durnan	4	Son	At home	Kentucky
James Cox	50	Boarder	Turnpike hand	Ireland
Mike Murphy	50	Boarder	Turnpike hand	Ireland
Jacob McLaughlin	32	Boarder	Turnpike hand	Ireland
John T. McLaughlin	5	Boarder	At home	Kentucky
Edward Smith	32		Turnpike hand	Ireland
Henry Smith	7		At home	New York
John O'Brian	44		Turnpike hand	Ireland
Pat Geohagan	56		Turnpike hand	Ireland
John Dewhar (?)	35		Turnpike hand	Ireland
Kate Dewhar	30		Keeps house	Ireland
Mary Dewhar	10		Attending school	New York
Jenny Dewhar	3		At home	Kentucky
Kate Dewhar	1		At home	Kentucky
David Connell	52		Turnpike hand	Ireland
Mary Connell	48		Attending school	Ireland
James Connell	16		Attending school	Virginia
Mary Connell	14		At home	Virginia

Source: US Census of Population Manuscripts.

in Kentucky. Given that daughter Mary was ten years old in 1870, the family must have moved to Kentucky sometime before 1860. Although the exact date cannot be determined, Mr. and Mrs. Durnan may have come to America as part of the Irish diaspora of 1845–1850, during the Great Famine. The Durnan household comprised at least four families and included seventeen boarders, likely representing members of Durnan's construction crew. Eight of the men were turnpikers, and all the adults were Irish immigrants. In addition to the Kentucky-born Durnan children, three other children were born in Kentucky, two in New York, and two in Virginia.

Many other Irish turnpike contractors lived in central Kentucky. They built the primary roads as well as the shorter neighborhood turnpikes that eventually became secondary county roads. When a turnpike was completed and the tollhouse was built, Irish immigrants were often hired to collect tolls and maintain the road. The early turnpikes linked established towns and connected farms to towns, mills, and distilleries, but they also served as local intermodal connectors to rural railroad depots and enabled the growth of villages at those sites. In this way, a countryside road network developed. Roads promoted general commerce, of course, and they gave distillers access to raw materials and markets for their finished products. Farmers might haul their grain to Irish-operated distilleries by way of Irish-built turnpikes.[28]

Steamboats

Innovations in the application of steam power to water- and land-based modes of transport transformed nineteenth-century America by efficiently overcoming the barrier of distance. Steamboats turned inland rivers and the Great Lakes into transportation corridors and greatly enlarged the hinterlands from which early riverside settlements could draw resources and commodities.[29] Before steam power—from the late eighteenth century to well past the mid-nineteenth century—Kentucky freight moved, in volume, on Ohio and Mississippi River flatboats and keelboats. During sixty days of the winter high-water season in 1810–1811—from November 24 to January 24—197 flatboats and 14 keelboats descended the Falls of the Ohio at Louisville. They carried cargo that was in demand by the cotton and sugar plantations in Mississippi and Louisiana, including 286 slaves and sizable quantities of farm products, including 681,900 pounds of pork, 64,750 pounds of lard, 18,611 barrels of flour, and 2,373 barrels of whiskey.[30] From 1831 to 1847, more than 5,700 boats passed Louisville, although most of the cargo continued downstream.[31]

At Pittsburgh in 1810, Robert Fulton and Robert Livingston directed construction of a new steamboat called the *New Orleans,* which successfully descended the Ohio and Mississippi to Natchez and New Orleans in January 1812.[32] Three years later, Henry Shreve's steamboat *Enterprise* made the first upstream voyage from New Orleans to Louisville. Innovations in steam engine technology and boat design transformed Ohio Valley commerce.[33] Boat makers embraced the Ohio Valley for its timber and metalworking industries, and several towns along the river became boatbuilding centers. Between 1811 and 1835, boatbuilders assembled and launched more than 600 steamboats on the Ohio River and its tributaries.[34]

The Ohio River and its Kentucky tributaries were no longer one-way chutes that emptied the region of resources and processed products; they were two-way avenues along which high-volume shipping operated. The region's economic development accelerated rapidly, with far-reaching consequences. On a national scale, economic and political influence and industrial innovation began to shift away from the East Coast to the middle of the country. Many lake ports and river cities in the trans-Appalachian West became centers of rapid commercial development and modernization. The Ohio Valley attracted entrepreneurs of many persuasions, and skilled mechanics and common laborers, many of them European immigrants, moved to the region.

In 1810, a year before the steamboat *New Orleans* arrived at the Falls of the Ohio, Louisville's population was 1,357. As the city industrialized over the next four decades, the population grew rapidly. By 1850, Louisville had 43,194 residents, an increase of more than 3,000 percent. Upriver, Cincinnati's growth was even more dramatic. The small river port of 2,540 inhabitants in 1810 grew into a commercial and industrial center with a population of 115,435 by 1850. To illustrate the importance of river access to economic development, note that landlocked Lexington's 1810 population was 4,326, three times larger than Louisville's; by 1850, Lexington's population was only 8,159, or one-fifth that of Louisville.

Kentucky distillers shipped barreled whiskey by flatboats down the Ohio and Mississippi Rivers to New Orleans markets in the late 1780s.[35] Small-scale whiskey production was commonplace, but the home consumption of whiskey and its use by farmer-distillers for barter limited shipment volumes until the 1820s. An increase in the number and size of riverboats invited larger shipments of commodities to New Orleans, and flatboat and keelboat transport flourished from the early 1800s to late 1840s. Steamboats were in common use by the 1830s, so the eras of flatboat (drift) navigation and steamboat (powered) transport overlapped from the 1830s through the 1840s. In the four years from 1822 to 1826, whiskey shipments increased seventeenfold—from 1,978 to 33,704 barrels. Distillers maintained annual whiskey shipments of roughly 40,000 barrels from the late 1820s through 1835. Thereafter, as distilling industrialized, river shipments to New Orleans from all upriver sources progressively increased—doubling from 1842 to 1847 and reaching 179,000 barrels in 1857.[36] High-water seasons were a boon to flatboat travel, but during low water, as occurred in 1850 and 1851, boat tonnage waned. Periods of national economic crisis, such as the financial panics of 1819–1824, 1837–1842, and 1857–1862, led to declines in transport consignments. Markets weakened, and distillers' access to capital to fund purchases of grain, cooperage, and insurance or to meet other operating expenses was curtailed.

Early Railroads

The first railroad in the trans-Appalachian West, the modestly engineered Lexington & Ohio (L & O), was envisioned as an overland connection between Lexington and Louisville that would link Lexington businesses to riverine commerce and enhance the cattle-raising industry in central Kentucky.[37] Chartered in 1830, the L & O opened a track between Lexington and Frankfort in 1834. In 1835 it obtained farmland about halfway between those two cities to build a railroad town, Midway. The L & O failed in 1847. Another line, the Louisville & Frankfort Railroad, completed the Frankfort-to-Louisville link in 1851. The line was rarely in good repair, and the rolling stock was undependable. By late-nineteenth-century standards, these early railroads had a limited capacity and were generally troublesome. Innovations in steam equipment and track design nevertheless proceeded apace, and several new rail companies, including the Lexington & Maysville, Covington & Lexington, and Louisville & Nashville, adopted new technologies. State and private investment funded the building of improved lines and provided for maintenance and upgrades. By the late 1850s, a rudimentary rail network was emerging in central Kentucky, although less than 100 miles of track were in operation. Farmers' corn continued to move to market by wagon and road, an expensive proposition.[38] Many farmers engaged in the more profitable practice of feeding their corn to livestock and driving the animals to market.[39] As rail transport improved and trunk and spur lines extended into the rural countryside, small inland distilleries offering quality products could compete with large commercial distilleries located on the navigable rivers at Frankfort, Louisville, and other Ohio River ports. But rail freight charges were high, and the service was sometimes uneven.

At midcentury, dams and locks constructed on the Kentucky River and the Green River created slack-water pools that reduced the hazards of low-water navigation. The state's Board of Internal Improvement set toll rates on river freight to pay for dam and lock construction and maintenance, thereby increasing transport costs. As railroad companies extended their track networks, ready accessibility to rail lines and competitive freight rates forced the board to reconsider river toll rates. In 1851 it reduced the tolls on whiskey being shipped on the Kentucky River to enable steamboats to compete with the railroads. On the Green River, the board reclassified liquor shipped in kegs and barrels and classified certain other commodities, such as sugar, coffee, and hardware, as "pound freight," thereby reducing the tolls by 33 to 80 percent, depending on whether the freight was moving upstream or down.[40]

As early railroad reliability improved and competition among railroad companies increased, freight charges began to drop. The cost of shipping freight via railroad might be as low as 3 cents per ton-mile, or about one-tenth the cost of shipping freight overland by wagon.[41] Completion of the extensive trans-Appalachian and Middle West canal systems in the 1820s to 1840s affected both grain prices and the cost of shipping products by canal and river. By the 1840s, grain transport costs had declined by as much as 90 percent, but grain prices had fallen as well. Small farmers on marginal land found it increasingly unprofitable to grow grain crops other than as feed for their own livestock. The center of wheat production had once been in western New York, but it now moved resolutely west into the Ohio Valley, where laborers were digging new canals across Ohio and Indiana.[42] Regional railroads entered the trans-Appalachian West in the 1850s and 1860s—the Erie, New York Central, Baltimore & Ohio, and Pennsylvania Railroad among them—allowing shippers to send barreled whiskey from the Ohio Valley directly to East Coast ports and avoid the lengthy, risk-filled steamboat trip to New Orleans.[43] Railroads and canals effectively enlarged the supply shed for grain sellers and buyers, greatly extending the distances grain could be shipped economically. The improved transport net also allowed buyers to base their purchases on quality, even if the grains they sought were produced some distance away. Distillers could now acquire in volume the quality grain produced in Kentucky and surrounding states at favorable prices.

Long-distance shipping required weatherproof warehouse storage and handling facilities, especially at canal and rail terminal points. Before the 1850s, grain was usually shipped in sacks, not loose. Shelled corn was shipped in sacks, and ear corn was shipped either loose or in sacks or barrels. Most grain was therefore considered general cargo because it was shipped in a bag or container of some sort, and each container had to be handled individually. With the development of the terminal storage elevator and the railroad boxcar, grain could be handled mechanically in bulk; this dramatically increased the speed of grain handling and significantly reduced the number of laborers required.[44] The first mechanized grain storage elevators adapted Oliver Evans's flour mill conveyor belt technology and appeared in Buffalo, New York, in the early 1840s; thereafter, that elevator design and technology rapidly spread west into the Great Lakes region and the Ohio Valley.

Regional entrepôt cities such as Louisville and Cincinnati became primary railroad origination points, enhancing their manufacturing industries and their agricultural commodity processing businesses and concentrating interregional commercial activity. Because settlement preceded railroad construction in Kentucky, rail lines often connected existing towns, reinforcing their market and service functions for local and subregional hinterlands and bolstering distilleries and other industries that processed locally produced farm products.[45] Although railroads greatly enhanced commerce in those inland towns that lay within railroad corridors, competitive railroad freight rates and punctual delivery schedules became the bane of the steamboat freight industry.

Whiskey shippers eagerly transferred their business from river to rail transport. Small river ports experienced substantial reductions in steamboat freight commissions, especially for general cargo such as barreled whiskey. Whiskey distillers in river port towns increasingly shipped their product "by the cars," and port managers found that their general cargo business suffered from "reduction by railroad." In 1882 Kentucky's operational railroad tracks totaled 1,807 miles. North-south rail connections linked the state to Chica-

Value of Ohio River Commerce at Kentucky's River Towns
1869 and 1886

Place	Primary Exports	Established Dollar Value, 1869	Established Dollar Value, 1886	Remarks*
Springdale Landing†	Whiskey, leather	200,000	45,000	
Maysville	Tobacco, whiskey, lumber, hemp, flour, pork	8,000,000	1,750,000	Reduced by railroad
Rock Springs Landing	Whiskey, tobacco	20,000	40,000	
Petersburg	Whiskey, pork	100,000	75,000	Distillery not in operation
Carrollton	Corn, tobacco, whiskey	2,500,000	250,000	
Westport	Tobacco, flour, whiskey	25,000	6,500	Reduced by railroad
Louisville	Leather, cement, whiskey	15,000,000	53,042,601	
Concordia	Cooper stuff, bark	10,000	15,000	
Grissen's Landing	Whiskey, grain	10,000	12,000	
Henderson	Hominy, tobacco, whiskey	3,000,000	792,000	Reduced by railroad

Source: Compiled from William F. Switzler, *Report on the Internal Commerce of the United States* (Washington, DC: GPO, 1888).

*Remarks are those listed by W. F. Switzler and relate to changes in export dollar values.

†Likely incorrectly recorded as "Springfield" by Switzler.

go on Lake Michigan and New Orleans and Mobile on the Gulf of Mexico, supplementing the east-west rail lines that connected to Atlantic coast port cities.[46] By 1882, Cincinnati wholesalers shipped only about 72,000 barrels of whiskey by boat and more than 395,000 barrels by rail; that trend increased over the next four years, as boat shipments dropped to less than 17,000 barrels in 1886. Louisville riverboat whiskey shipments stood at a meager 4,200 barrels in 1880 and declined further to 1,300 barrels by 1886. Whiskey shipments by rail—Louisville was the company headquarters of the Louisville & Nashville Railroad— stood at 127,000 barrels in 1882. Overproduction and distillery closings reduced whiskey production sharply after 1884, and by 1886, only about 100,000 barrels were shipped from Louisville railroad warehouses.

Whiskey and grain were shipped primarily via rail by the 1880s. Railroad company profits increased, as did employment opportunities in depot towns. And, importantly, the railroad network allowed distillers to greatly extend their reach as they searched for high-quality rye and malting barley—grains that were not readily available in Kentucky.

Many distillers eagerly adopted the railroad to ship their grain, coal, construction materials, livestock, and whiskey. But as laborers completed each new trunk and spur line, distillers also had to consider whether to remain at their traditional sites or move their

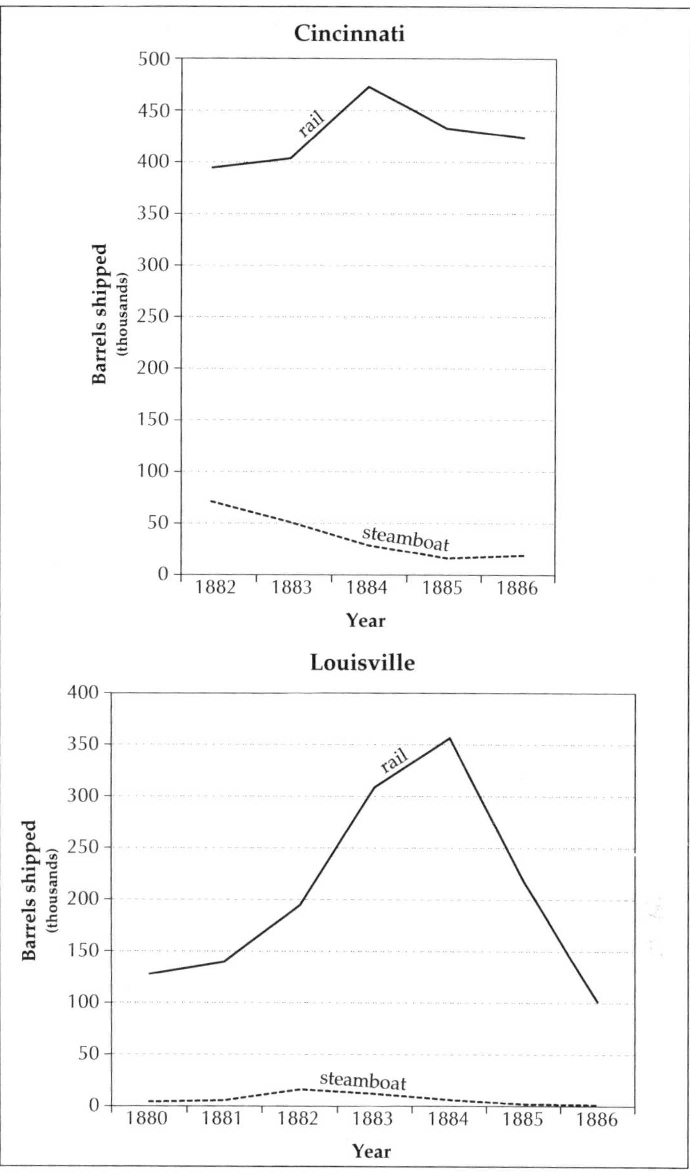

Whiskey shipped from Cincinnati (1882–1886) and Louisville (1880–1886) via water and rail. Note that from the 1860s through the 1880s, Kentucky's whiskey market experienced a series of recessions related to overproduction. Large surpluses in the early 1880s led to sharply reduced production for several years thereafter. Thirty-two distilleries operated in Louisville in 1885, but only three were in production in 1886. (Compiled from US Treasury Department, Bureau of Statistics, *Report on the Internal Commerce of the United States* [Washington, DC, 1888])

works to a trackside location near a station or freight depot. Relocation would permit them to send and receive shipments directly by rail, without a costly trip over a rutted track or expensive turnpike to reach the rail line. Some small-scale distillers decided to stay in their traditional rural locations, but once rail lines reached the small towns that served them, they saw the opportunity to minimize their transportation costs by relocating to rail towns and rail station sites. Traditional shed-type distillery buildings could easily be moved—torn down or salvaged and rebuilt trackside—or new works could be built with materials delivered by rail.

Local farmers who had traditionally supplied grains to distillers remained site bound.

To compete with rail-connected grain producers, though they be fifty miles or more distant, farmers needed upgraded local roads. Whereas a loaded farm wagon might hold 40 bushels, or twenty sacks, of grain, the capacity of a nineteenth-century railroad boxcar was about 325 bushels. And trains moved at many times the speed of mule-drawn wagons. Distilleries and depots employed laborers to load and unload railcars and farm wagons. The empty sacks constituted a counterflow to the grain trade—distillers always returned them to the grain supplier for reuse.

Though widely dispersed in the early decades of the nineteenth century, distillers were coalescing their works into clusters by the 1850s. Three county clusters offer examples of how farm and small-town distilleries became linked to transportation networks: Nelson, Marion, and Washington; Franklin, Anderson, and Woodford; and Bourbon and Harrison. Certainly, railroad officials considering branch line construction weighed the potential for distillery freight income. Railroad surveyors directed track construction to some distilleries and designated station sites near others.

Outer Bluegrass Railroads

In 1851 Louisville & Nashville Railroad executives were considering the main line's southbound route from Louisville to Nashville, and they announced that monetary subscriptions by local communities would determine the exact route. Two routes held promise for profitable operations. The first, an upper or "air-line" route, projected a track through Bardstown near the center of Nelson County before turning south to New Haven and then extending south and west to Larue, Hart, Barren, and Allen Counties. Bardstown's residents had long advocated for this line, but Nelson County officials rejected a proposed $300,000 bond issue that would have met the railroad's subscription expectations and supported construction.[47] Instead, the railroad chose a lower or southern route through Hardin, Grayson, Warren, and Simpson Counties. The L & N's main line to Nashville opened in 1860. Because Bardstown had failed to gain a position on that main line, the remedy seemed to be a branch line.

In 1850 the Kentucky legislature chartered an L & N branch from the main line at Lebanon Junction near Rolling Fork, east thirty-seven miles to the Marion County seat at Lebanon. The Lebanon branch crossed Nelson County well south of Bardstown and passed through Boston, New Haven, Gethsemane, and New Hope. Completed in about 1857, Nelson County's section of the Lebanon branch soon attracted distillers. Edward Miles built a new distillery at New Hope beside the L & N tracks soon after the line opened; finding the location favorable, he enlarged his distillery's production capacity in 1875.[48] J. Bernard Dant began operations at a new distillery near Gethsemane Station on the Lebanon branch in about 1865.[49] And in 1870 Richard Cummins opened a new distillery at the Coon Hollow Station.[50]

The L & N Railroad completed the first section of its Springfield branch to the Early

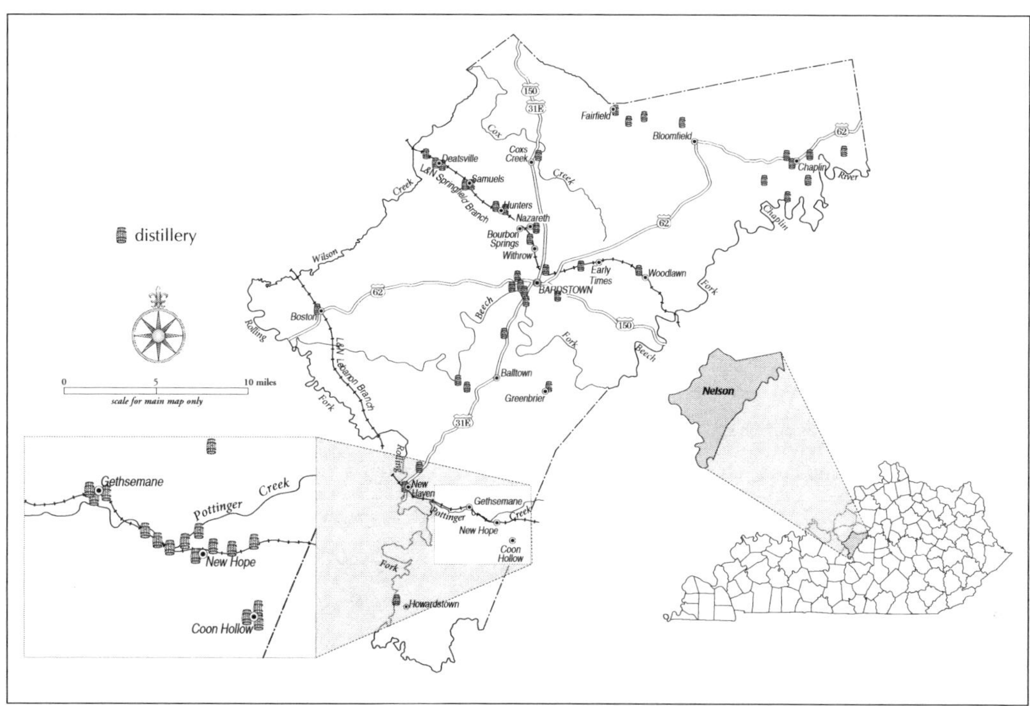

Nelson County distilleries, ca. 1790–1950. Distillery site locations are approximate. Some distilleries were short-lived, while others were in production for decades. (Compiled from *An Atlas of Nelson and Spencer Counties, Kentucky* [Philadelphia: D. J. Lake, 1882])

Times Station just east of Bardstown in 1860. Sometime before 1860, David Beam built a new distillery about three miles east of Bardstown. When laborers completed the L & N's Bardstown-Springfield branch in 1860, the tracks ran directly in front of Beam's distillery; one dedicated siding delivered coal and grain to the distillery, and another shipped barreled whiskey to market.[51] Sam Lancaster built a small distillery on Plum Run Road north of Bardstown in 1850. Lancaster relocated his works to a site on the L & N tracks in 1881.[52]

Upon crossing into Marion County, the Lebanon branch connected the small settlements of Chicago, Loretto, and St. Mary with Lebanon and eventually extended southeast to Crab Orchard and Knoxville, Tennessee. Marion County residents subscribed $200,000 for the Lebanon branch construction.[53] The first L & N train reached Lebanon in 1858, and regularly scheduled trains served the line soon thereafter.[54] J. W. Dant built a distillery on his farm west of Loretto in 1836. The L & N surveyor routed the tracks near the distillery, and the railroad established Dant Station there.[55] Reliable trains running on punctual schedules lent a new urgency to improving road connections to the new railroad stations, and charter applications for turnpikes increased during the 1850s.

The L & N eventually extended the branch rail line from Bardstown east to Springfield, in Washington County. The national financial panic of 1873 and the subsequent

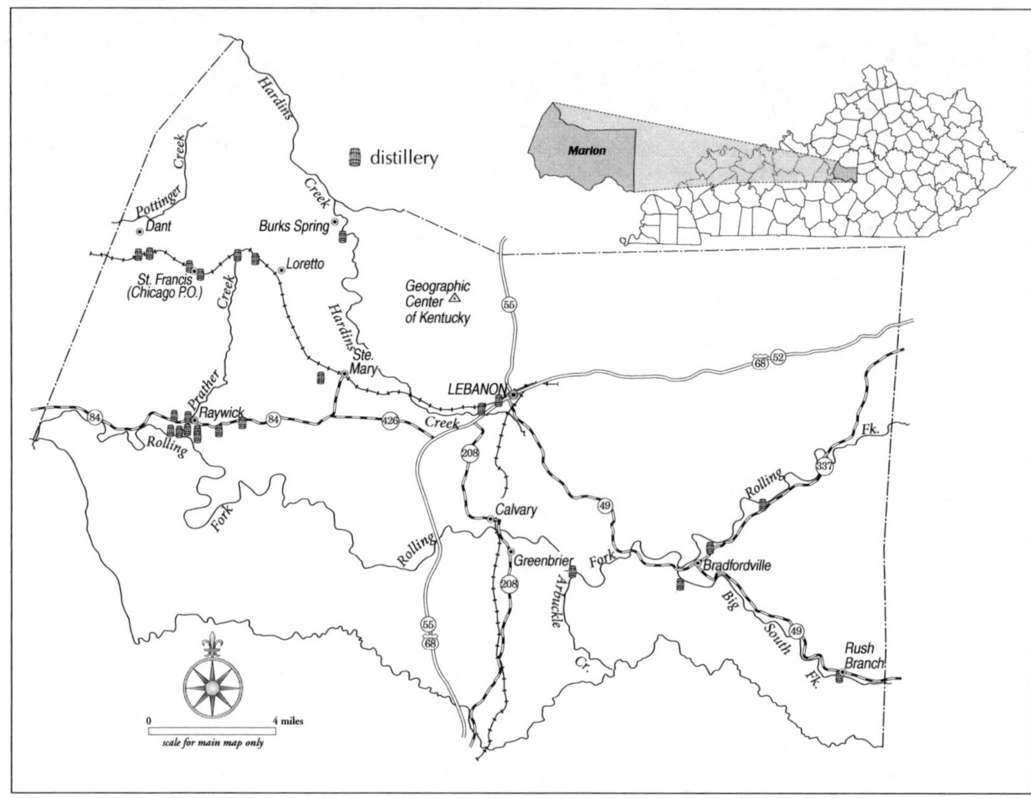

Marion County distilleries, ca. 1790–1950. Distillery site locations are approximate. Some distilleries were short-lived, while others were in production for decades. (Compiled from *Map of Marion and Washington Counties, Kentucky* [Philadelphia: D. G. Beers, 1877]; Chester Zoeller, *Bourbon in Kentucky: A History of Distilleries in Kentucky,* 2d ed. [Louisville, KY: Butler Books, 2010])

depression halted most railroad construction. By the late 1870s, a business recovery was under way, and railroad construction resumed. The extension to Springfield was not completed until 1888, twenty-eight years after the tracks reached Early Times Station. This last track section may have arrived too late to attract distillers. Although nearly a dozen distilleries operated at roadside or creek-side sites in Washington County by the mid-nineteenth century, nearly all had closed by the early 1880s.

Inner Bluegrass Railroads

The Covington & Lexington Railroad completed a track linking the Inner Bluegrass counties to the Ohio River at Covington in 1853. The company later consolidated with other lines to form the Kentucky Central Railroad (KCR). The KCR built county seat stations or depots at Falmouth (Pendleton County), Cynthiana (Harrison County), and Paris (Bourbon County). It also established country depots in Harrison County at Berry, Robinson, and Lair and in Bourbon County at Shawhan and Kiserton. Unlike railroads in the Middle

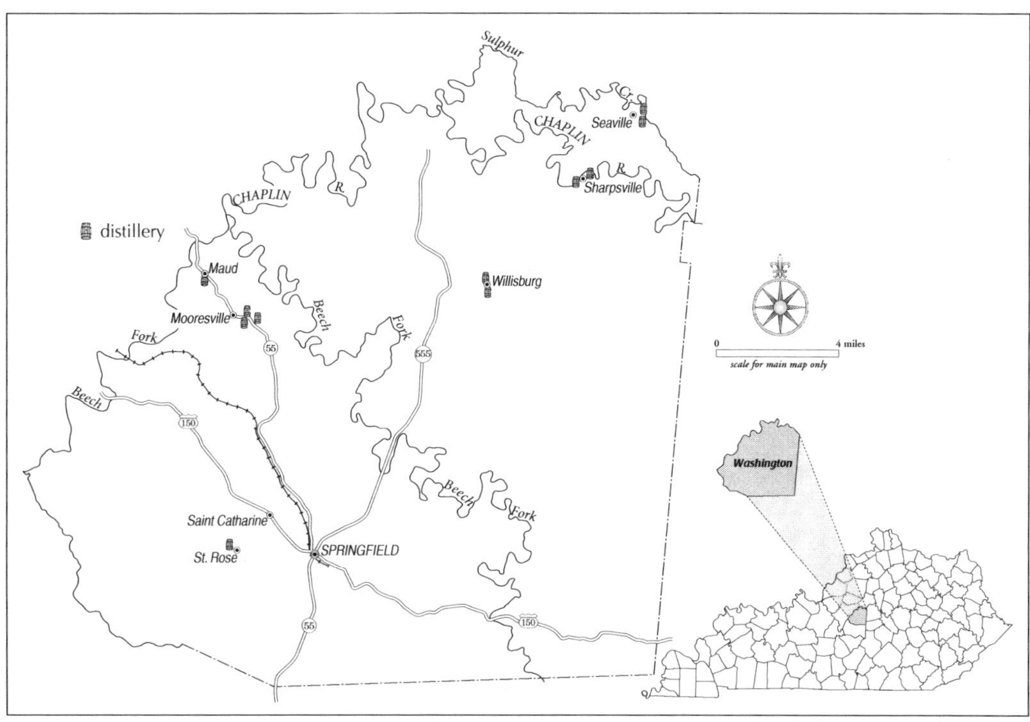

Washington County distilleries, ca. 1790–1950. Distillery site locations are approximate. Some distilleries were short-lived, while others were in production for decades. (Compiled from *Map of Marion and Washington Counties, Kentucky* [Philadelphia: D. G. Beers, 1877]; Chester Zoeller, *Bourbon in Kentucky: A History of Distilleries in Kentucky,* 2d ed. [Louisville, KY: Butler Books, 2010])

West and West, the KCR did not build towns, although commercial businesses and small villages grew up around the country depots.

John January built a new distillery at Berry Station (later renamed Berryville or Berry), north of Cynthiana, the same year the Covington & Lexington line was completed.[56] John Poindexter rebuilt a distillery near the rail line at Poindexter in 1856.[57] South of Cynthiana, Benjamin Brandon built Edgewater Distillery west of the future Lair Station site in 1836 or 1837, some seventeen years before the railroad arrived.[58] A subsequent owner, Thomas Jefferson Megibben, owned the nearby Edgewater Stock Farm, which produced purebred shorthorn cattle and Thoroughbred and Standardbred horses. When the rail line opened, Megibben expanded his distilling and farming operations.[59]

In 1879 three distilleries operated near Lair's Station: T. J. Megibben; Megibben, Bramble & Co.; and G. R. Sharp Distillery. The M. L. Lair & Brother Hotel, the S. T. Reynolds flour mill, and three blacksmiths conducted business near the railroad station.[60]

In Bourbon County, railroads converged at Paris: initially, the Covington & Lexington from Covington and the Maysville & Lexington lines, followed later by the KCR, the L & N, and the Frankfort & Cincinnati Railroad. There were several early rural distilleries in

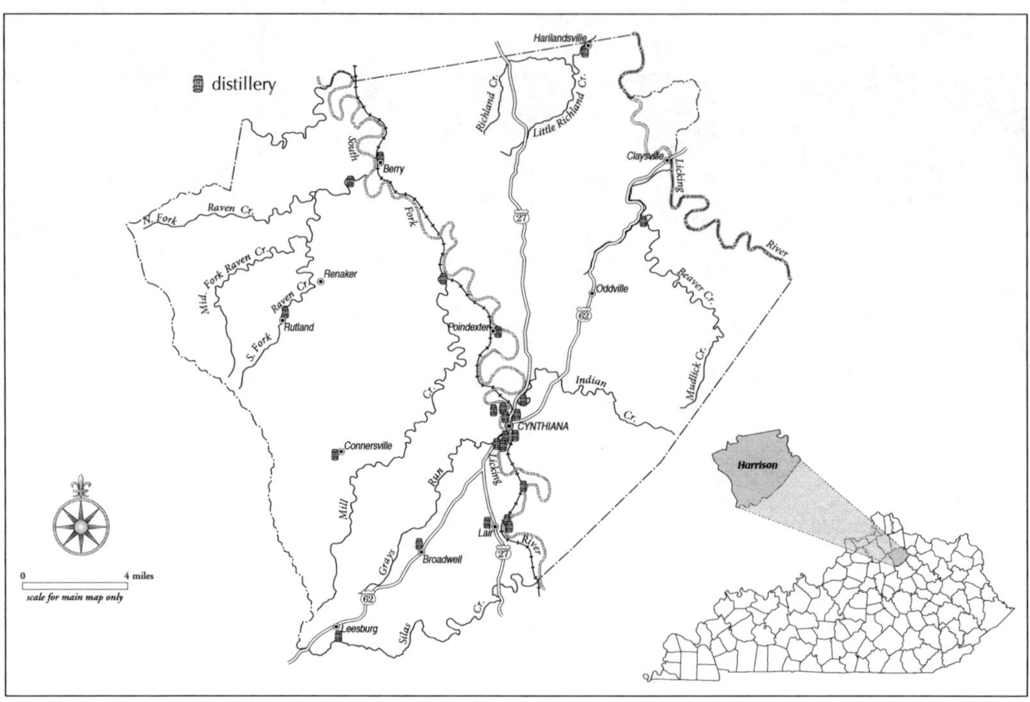

Harrison County distilleries, ca. 1790–1950. Distillery site locations are approximate. Some distilleries were short-lived, while others were in production for decades. (Compiled from *Map of Harrison County, Kentucky* [Philadelphia: D. G. Beers, 1877]; Chester Zoeller, *Bourbon in Kentucky: A History of Distilleries in Kentucky*, 2d ed. [Louisville, KY: Butler Books, 2010])

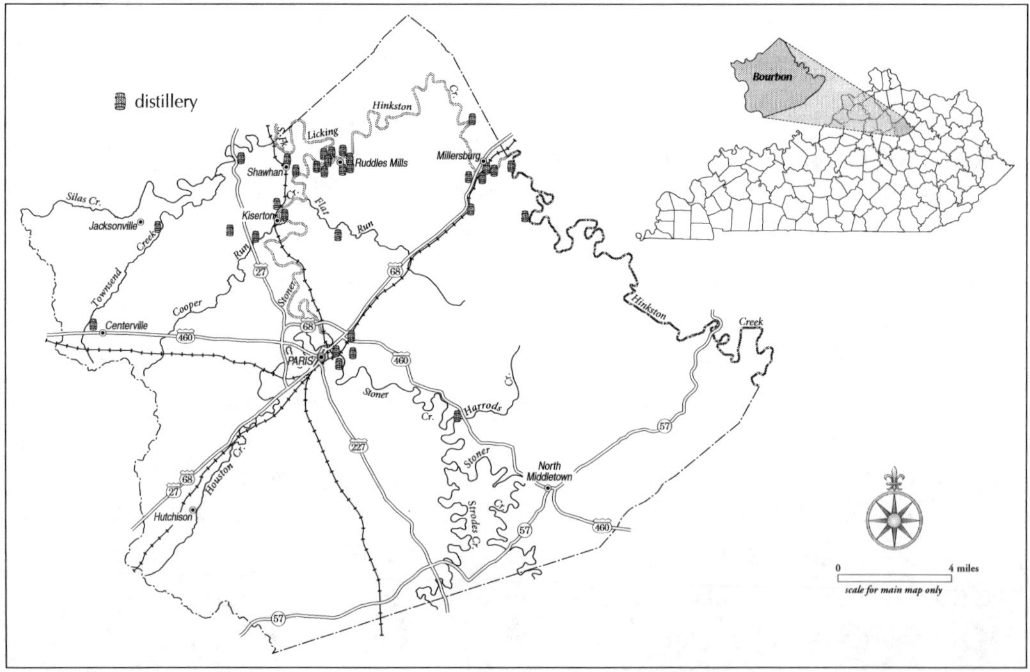

Bourbon County distilleries, ca. 1790–1950. Distillery site locations are approximate. Some distilleries were short-lived, while others were in production for decades. (Compiled from *Atlas of Bourbon, Clark, Fayette and Woodford Counties, Kentucky* [Philadelphia: D. G. Beers, 1877]; Chester Zoeller, *Bourbon in Kentucky: A History of Distilleries in Kentucky*, 2d ed. [Louisville, KY: Butler Books, 2010])

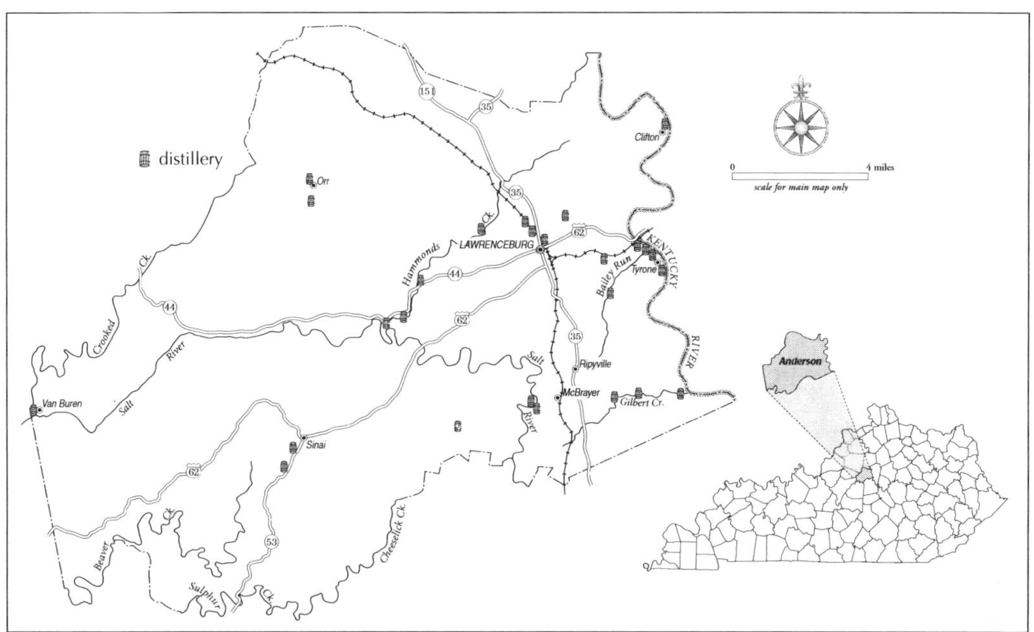

Anderson County distilleries, ca. 1790–1950. Distillery site locations are approximate. Some distilleries were short-lived, while others were in production for decades. (Compiled from Ripy Family Papers, 1888–1963, Special Collections, University of Kentucky; Chester Zoeller, *Bourbon in Kentucky: A History of Distilleries in Kentucky*, 2d ed. [Louisville, KY: Butler Books, 2010])

the county, especially near Ruddles Mills, and distillers built new works as railroad lines were completed across the county. In 1868 the Paris Distillery commenced operations beside the KCR tracks on the north side of Paris. The railroad built a dedicated siding to the distillery building's scales.[61] Distillers also built works at other trackside sites, including Shawhan Station, Kiser's Station (or Kiserton), and Millersburg.

In 1886 the Louisville Southern Railroad started construction on a track from Louisville to Harrodsburg by way of Shelbyville and Lawrenceburg. The line opened for traffic in 1888. A year later, W. B. Saffell began operating a new distillery on a Louisville Southern Railroad siding on the north side of Lawrenceburg, in Anderson County, drawing water from Hammond Creek.[62]

In 1889 the railroad completed a twenty-four-mile branch line to Lexington known as the Lexington-to-Lawrenceburg (LL) line, crossing the 280-foot-deep Kentucky River gorge by way of a 1,600-foot cantilever bridge. At the west end of the bridge, the LL line passed the Old Hickory Springs Distillery, which later became the site of Wild Turkey's distilling works. Three other distilleries operated within a mile: the Thomas B. Ripy Distillery on the expansive bedrock terrace above the Kentucky River at Tyrone, the Waterfill & Frazier Distillery just to the south on Bailey Run, and the Cedar Brook Distillery on Cedar Brook Run. The LL track entered Lexington along the industrialized Town Branch

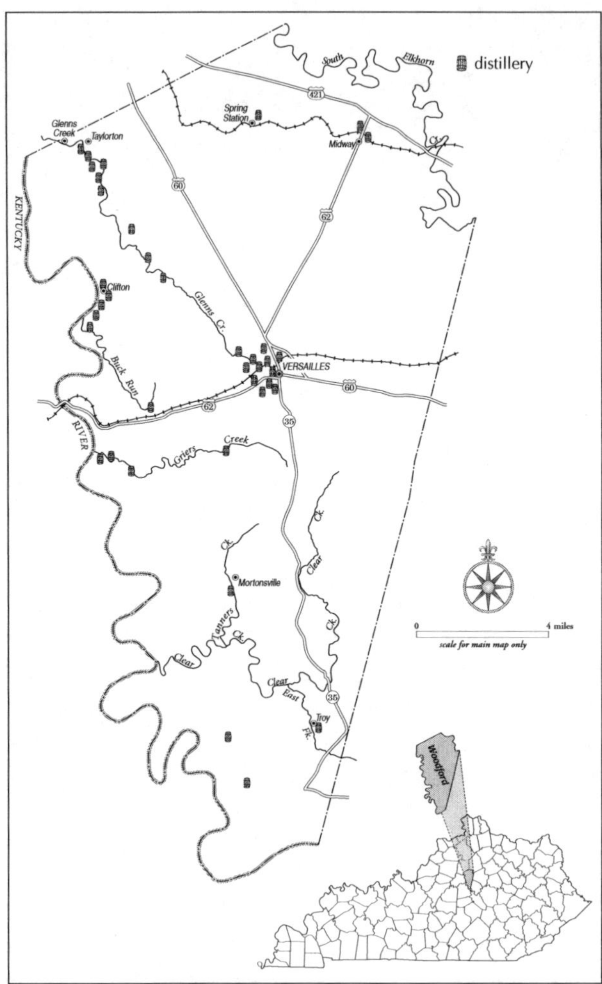

Woodford County distilleries, ca. 1790–1950. Distillery site locations are approximate. Some distilleries were short-lived, while others were in production for decades. (Compiled from *Atlas of Bourbon, Clark, Fayette and Woodford Counties, Kentucky* [Philadelphia: D. G. Beers, 1877]; Chester Zoeller, *Bourbon in Kentucky: A History of Distilleries in Kentucky*, 2d ed. [Louisville, KY: Butler Books, 2010])

valley, passing the James E. Pepper and William Tarr Distilleries operating along Manchester Street.[63]

Woodford County bordered the Kentucky River on the east. Small creeks flowed west toward the river, cutting deep ravines into the limestone upland. Each steep-gradient creek offered potential waterpower sites, and by the late eighteenth century, these were being developed by millers and distillers.

Distilleries operated on Griers Creek, southwest of Versailles, a short distance above a Kentucky River landing where flatboats could be loaded for downriver shipments. A still house operated on Glenns Creek in 1794. In about 1850 J. Swigert Taylor built a distillery on the creek and appointed his nephew, E. H. Taylor Jr., its manager. Oscar Pepper built a distillery on Glenns Creek near Millville in 1860, producing Old Crow whiskey. In the late 1870s W. A. Gaines & Company bought the Old Crow Distillery; soon thereafter, the works was mashing some 500 bushels, or twelve wagonloads, of grain per day.[64] Although Glenns Creek offered waterpower potential, access was awkward and cumbersome. To avoid driv-

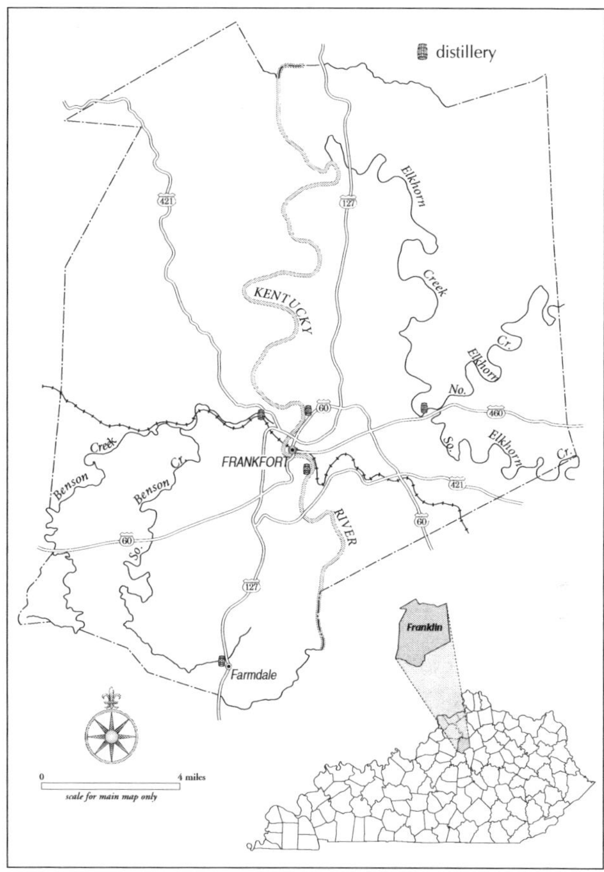

Franklin County distilleries, ca. 1790–1950. Distillery site locations are approximate. Some distilleries were short-lived, while others were in production for decades. (Compiled from B. N. Griffing, *An Atlas of Franklin County, Kentucky* [Philadelphia: D. J. Lake, 1882])

ing wagons along the constricted creek bottom, laborers hewed a narrow road into the base of the limestone bluff. In 1907 Gaines contemplated building a five-mile branch railroad from the Old Crow works on Glenns Creek to Jetts Station on the L & N. "The road will only be about five miles long, and will be of immense advantage to this big distillery as they do an immense amount of shipping, both of grain into the distillery, and of the finished product of their stills from the distillery," reported the *Frankfort Roundabout*.[65] The Kentucky Highlands Railroad Company completed the new rail spur connecting to the L & N line in December 1907.[66] The spur not only provided the high-capacity access required to operate an industrial-scale distillery but also permitted E. H. Taylor Jr. to open the site to visitors. In 1911 laborers completed a rail line extension from Millville southeast eight miles to Versailles, including a spur line to the whiskey warehouses at the Labrot & Graham Distillery.

The Frankfort & Cincinnati Railroad, popularly known as the "Whiskey Route" or the "Bourbon Road," ran from Frankfort east some thirty-five miles to Paris by way of Georgetown, where it crossed the north-south Cincinnati Southern Railroad. Construction on the Frankfort to Georgetown section began in 1888 and was completed in June 1889; in December laborers completed the Georgetown to Paris segment. The branch from

Frankfort to Georgetown served four distilleries. The track passed near two distilleries on the Kentucky River at Leestown in north Frankfort: the O.F.C. (Old Fire Copper) Distillery, later known as Ancient Age–Schenley and Buffalo Trace; and the Carlisle Distillery, later operated as the Kentucky River Distillery. East of Frankfort at the Forks of Elkhorn, the railroad passed the Baker Brothers Distillery, later known as the Frankfort, James B. Beam, and Old Grand-Dad Distillery. Initially, the Frankfort & Cincinnati did not serve distilleries directly; trains stopped at the nearest station, where freight wagons provided the link between the distilleries and the tracks until spurs were built.[67] In 1902 the railroad built a spur line into the Baker Brothers Distillery to complete a direct connection. At Stamping Ground in Scott County, the railroad passed near the Buffalo Springs Distillery; a post-Prohibition rail spur connected the distillery to the main line in 1933. The railroad's eleven rural stations handled agricultural commodities and distillery freight, including carloads of grain, empty barrels (both new and used), full barrels, coal, glass bottles, and livestock headed to and from the distilleries' feeding pens.[68]

Louisville and Jefferson County Railroads

The rolling limestone uplands of the Outer Bluegrass abruptly end in western Jefferson County at the edge of the Knobs and the Ohio River floodplain. Meltwater runoff from Pleistocene glaciers to the north swept down the Ohio River Valley, filling the old channel with more than 100 feet of cobbles, gravel, sand, and clay and creating a nearly level alluvial outwash that became the river's floodplain. The present-day river channel runs along the west side of the outwash. Louisville and Portland are sited at the Falls of the Ohio on the alluvial surface. As Louisville grew during the nineteenth century, developers extended the city's street grid south across the outwash and east onto the brow of the Outer Bluegrass surface. Railroad surveyors were particularly captivated by the Louisville outwash plain. Minimal grade changes allowed them to lay track in long, straight runs, provided the right-of-way was available. City engineers, in contrast, saw the outwash as a potential hazard; because its surface was only a few feet above river level, it was susceptible to flooding, as dismayed residents, storekeepers, and manufacturers found during the great floods of 1884, 1913, and 1937.

At the turn of the twentieth century, four primary railroad lines served Louisville and Jefferson County: the Louisville & Nashville, the Illinois Central, the Southern, and the Pittsburgh, Cincinnati, Chicago & St. Louis, a branch line of the Pennsylvania Railroad. From central Louisville, L & N branches ran southeast, east, and northeast. An L & N line ran north-south along Tenth Street to the Union Terminal on West Broadway and served several large railyards. The Irish neighborhood of Limerick grew up along St. Catherine Street, centering on the St. Louis Bertrand Catholic Church, convent, and parochial school, just east of the L & N's expansive railyard. The Illinois Central ran north-south along Fourteenth Street before turning southwest at the city limits toward Bernheim Lane.

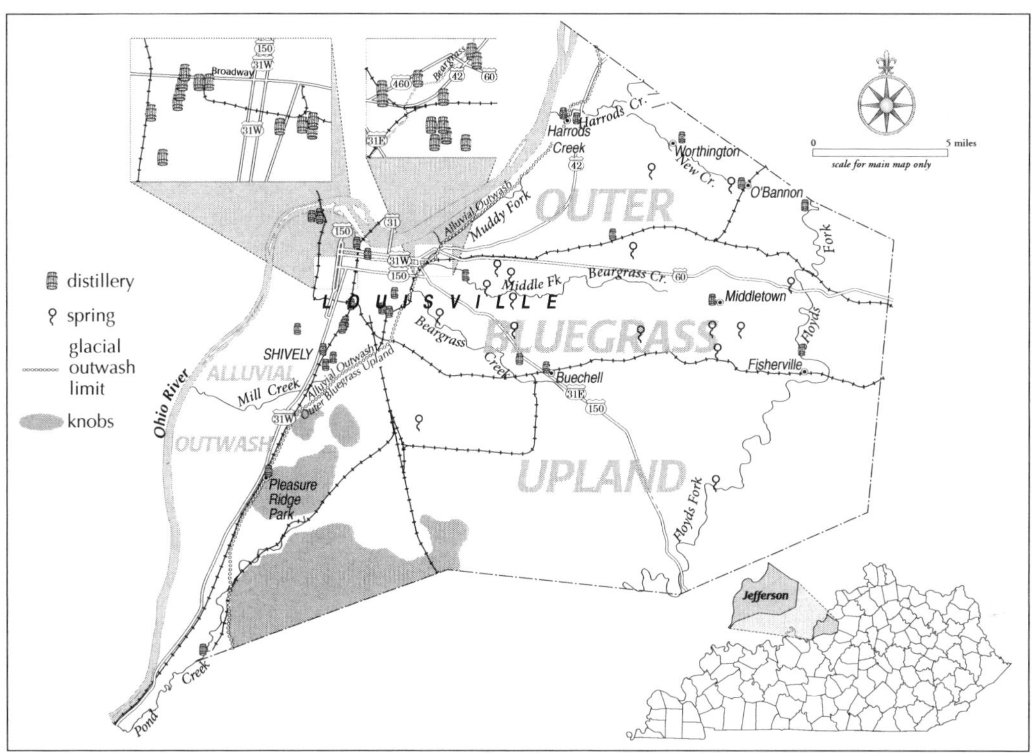

Louisville and Jefferson County distilleries, ca. 1790–1950. Distillery site locations are approximate. Some distilleries were short-lived, while others were in production for decades. (Compiled from *Atlas of Jefferson and Oldham Counties, Kentucky* [Philadelphia: D. G. Beers & J. Lanagan, 1879]; M. I. Rorabaugh, F. F. Schrader, and L. B. Laird, *Water Resources of the Louisville Area, Kentucky and Indiana,* US Geological Survey Circular 276 [Washington, DC: GPO, 1953]; Chester Zoeller, *Bourbon in Kentucky: A History of Distilleries in Kentucky,* 2d ed. [Louisville, KY: Butler Books, 2010])

The Southern line ran southeast from a westside industrial area, and the Pittsburgh, Cincinnati, Chicago & St. Louis crossed the river upstream from the falls and ran south along Fourteenth Street.[69]

Although Jefferson County's earliest distilleries were likely a congeries of backyard craft works, the first generation of commercial distillers arranged their operations in an orderly pattern, largely along Outer Bluegrass streams such as Harrods Creek and Floyds Fork. The landmark springs in the eastern half of the county attracted few large distilleries. Louisville's post–Civil War industrial expansion included a spate of distillery construction. Distillers had built several high-capacity commercial works before the rail lines arrived. Some distilleries that were established after modern railroads entered the city were not enticed by trackside locations. In 1865, for example, the Swearingen & Biggs Company built the Mellwood Distillery on Louisville's east side between Bardstown Road and Frankfort Avenue, where it was not served directly by a railroad. The Mellwood Distillery mashed 1,200 bushels of grain per day, drawn from a 100,000-bushel-capacity grain eleva-tor.[70] During the distilling season, Mellwood required thirty wagonloads of grain per day, if it were running at full capacity; much of this grain was likely delivered shortly after the summer and fall harvest seasons. Moreover, distilleries had to ship out spent grain if they had no livestock to feed, so streets around the distillery were likely congested with horse- and mule-drawn tank wagons.[71]

Louisville distillers also built several rail-oriented, industrial-scale works. In about 1893 John Roach built a distillery in West Louisville at Thirtieth Street and Garland Ave-nue. The distillery mashed 800 bushels of grain per day, drawing from a 7,000-bushel-ca-pacity grain elevator that stood beside a Southern Railroad siding. The works included a cooperage shop, a large cattle feeding yard, and three bonded warehouses.[72] Isaac and Bernard Bernheim built a new distillery on Bernheim Lane in Shively, a southwestern suburb of Louisville, in 1897. Served by an Illinois Central Railroad siding, the distillery mashed 600 bushels per day, aged whiskey in six warehouses, and fed cattle on slop in a large shed.[73]

Transport Mode III: The Multimodal Network

In the six decades from about 1830 to 1890, Kentucky and the nation experienced a trans-portation revolution. Steam-powered boats supplanted flatboats powered by river cur-rents. Construction crews transformed dirt tracks into stone-surfaced toll roads. Iron-horse short-line railroads became steel-rail, long-haul, high-capacity lines running on standard-gauge tracks.[74] Each transport mode—road, river, and rail—was linked to the others, forming an interconnected network of transport routes that, with state and federal support, were engineered to universal standards and maintained to sustain traffic.

From the 1790s through the 1830s, transport costs were a significant percentage of consumer goods' prices. As regional and national transport lines contested for customers, freight rates declined, and the percentage of product cost related to transport charges fell. Early distillers used their own grain or grain produced nearby. As hard-surfaced turnpikes and early railroads improved accessibility and distillers industrialized their works, they bought corn from neighboring counties, rye from distributors in Louisville or Cincinnati, and malted barley from more distant sources. In the late 1860s James Stone, the propri-etor of Elkhorn Distillery in Scott County, sent agents into Shelby and Owen Counties to purchase or contract for corn. He dealt mostly with farmers in the Outer Bluegrass to the northwest, who shipped the corn via railroad some forty miles to an L & N depot near his distillery.[75]

By the 1880s, John Atherton was purchasing corn for his Larue County distilleries from farmers in Iowa, Illinois, Nebraska, and Kansas through dealers in Louisville. Al-though Kentucky farmers produced corn, the western corn was often higher quality, and it cost Kentucky distillers more to purchase and ship Kentucky corn than it did to obtain corn produced hundreds of miles to the west, which was priced at 30 to 33 cents per bushel in the mid-1880s.[76] Atherton and other Kentucky distillers purchased their rye almost exclusively from western farms. Northern plains farmers, especially in the Dakotas, produced the best malting barley. Distillers did not malt their own barley but bought it malted from grain dealers, largely in Louisville and Chicago.[77]

Kentucky distillers often found that good-quality grain was not necessarily local, and they employed the transport net to obtain the best and cheapest grain available. Superior grain quality was often a function of both husbandry practices and transport modes that were vermin free and weatherproof. Fine whiskey distillers found that the taint introduced by musty grain afflicted with mold or smut ruined the commercial value of their product.[78] Large-volume rectifying distilleries often used lower-grade moldy grains that had suffered water damage when shipped on river barges or Great Lakes steamers or had been stored in damp granaries. Rectifiers treated the musty whiskey with permanganate of potash to cover off flavors.[79]

The multimodal transport net that emerged during the last third of the nineteenth century created interaction nodes where road, river, and railroad routes focused and coalesced, thereby stimulating urban growth at these central places. Industries and businesses found that these nodes offered significant commercial advantages to efficient, large-scale manufacturing, including a skilled labor force, access to capital and raw materials, and commensal relationships with supporting industries and business. For distillers, commensal businesses included coopers, coppersmiths, glass manufacturers, and wholesalers. Distillers who operated at large transport hubs such as Louisville benefited from these locational advantages. The expanding transport net greatly increased the size of the supply shed from which distillers could draw raw materials and extended the market range for their finished products, enabling Kentucky whiskey distilling to transform from a vernacular craft with neighborhood suppliers and customers into a national-scale commercial-industrial enterprise.

And what future could farmer-distillers operating at traditional sites outside the transportation network anticipate? John Atherton saw no advantage for small-scale remote distillers whose growth and efficiency were limited by their locations. "The distillers located in the best grain country and with the best railroad facilities would drive out of existence all others," he said.[80]

Making It Work

Who built Thebes of the seven gates?
In the books you will find the names of kings.
Did the kings haul up the lumps of rock?
—Bertolt Brecht

Conjunctions—Labor, Management, and Residence

Nineteenth-century American laborers occupied complex social and cultural settings. A better understanding of these settings can expand and enrich our interpretations of how folklife—living and making a living—worked and how it contributed to the making of distinctive landscapes. A perspective on landscape construction that centers on cash flow and economic balance sheets gives short shrift to the intricate interweaving of cultural attributes and social experiences that influenced how places were established, how they functioned, and how they related to people's lives. The people who fashioned the predominant modes of living in the trans-Appalachian West were pioneer farmers. Some were migrants from Atlantic coast colonies; others were immigrants from Europe. A substantial number were peoples of African lineage. Some of these individuals brought generations-old knowledge and skills; others brought little more than a willingness to work. Some worked their own land and businesses; others hired on to do heavy labor. Some learned a craft by apprenticing with skilled craftspeople; others were enslaved.

The word "work" has numerous meanings. As a verb, it denotes action, the carrying out of a process that results in the making of something. The quality or character of what is made can be described as the product of "workmanship." Similarly, yeast introduced into a mash-filled fermenting vat is said to "work." As a noun, it can describe the product of labor, or something a worker has made. It can also refer to a place of work, such as when a large industrial structure or group of related factory buildings is referred to as the "works." Landscapes are both the product of "work," the verb, and a "work" in progress, the noun. The landscape of work is not a completed entity that, when finished, permits close study of the agents who made it, its functionality, or its meaning. The landscape of work is really a sequence of ongoing events—delineation, construction, occupation, abandonment, decay, destruction, and erasure or, alternatively, rejuvenation, reoccupation, and reuse.[1]

Working, at whatever task, takes place in a fickle and fluctuating environment.[2]

Some tasks involve tedious repetition, such as hoeing weeds from the rows of a cornfield. Other jobs require spontaneity and improvised solutions to unanticipated problems or unpredictable environmental conditions or the use of unfamiliar tools. What is a teamster to do when a sudden thunderstorm washes away a turnpike bridge, thereby stopping the progress of his wagon filled with sacks of shelled corn or barrels of tax-paid whiskey? What mechanical abilities are required to convert a gristmill from water to steam power? Building and operating a simple home distillery might be common knowledge to some and seem like alchemy to others. Building and operating a large commercial distillery in the nineteenth century was a complex business filled with uncertainties, but it was less daunting if one possessed experience and access to the knowledge attained by generations of like-minded forebears. It required workers with a range of abilities: people who could perform heavy physical labor, and others with a basic understanding of applied chemistry and physics, knowledge of business and marketing, and a keen awareness of local and national politics and law. Early Kentucky distillers who had the best chance of commercial success often built on experience gained in the whisky distilleries of Ireland and Scotland or in Virginia, Maryland, and Pennsylvania. Coopers who possessed the knowledge and skill to make tight barrels drew on apprenticeships served in Europe or in colonial distilling centers to the east. Turnpikers who built Kentucky's nineteenth-century roads applied their heritage of stone construction work in Ireland. In an era in which one learned by doing, unskilled day laborers gained an understanding of how distilling worked by absorbing what each task taught them about the overall process, be it hauling or pumping water, grinding grain, or stirring and cleaning mash tubs.

The Labor Force

Nineteenth-century farming and manufacturing were codependent and complementary. Until the early twentieth century, more than 90 percent of Americans lived on farms or in small towns. They produced staple crops and raw materials used in manufacturing: flax, hemp, and cotton fibers for woven goods; grains for milling into flour and distilling into alcohol spirits; and tobacco for smoking and chewing products.

Local Workers

Farmers, millers, and distillers drew on the unskilled labor force in a given locale, depending on the season and the tasks at hand. The day laborers who worked on farms during planting and harvesting might work at mills after the harvest season and at distilleries through the winter and late spring. Occupations complementary to farming, milling, and distilling included woodcutters, sawyers, carpenters, blacksmiths, coppersmiths, teamsters, and stonecutters and stonemasons; each trade's work was in demand by other segments of the producing and consuming population. Skilled craftspeople such as coopers, millwrights, and distillers likely did not seek seasonal work outside of their professions, provided their businesses were stable. And as long as personal mobility depended on walking or riding a horse or a mule to work, the households of such craftspeople often included employees and apprentices.[3] Millwright Carrol Blincou lived in Nelson County in the mid-nineteenth century, and in addition to his wife and three children, his household included three boarders—a millwright and two laborers.

Carrol C. Blincou Household
Nelson County, Kentucky,
1850

Name	Age	Relationship	Occupation	Place of Birth
Carrol C. Blincou	39	Father	Millwright	Kentucky
Emily Blincou	32	Mother	Keeps house	Kentucky
Sarah C. Blincou	13	Daughter		Kentucky
James W. Blincou	9	Son		Kentucky
Thomas J. Blincou	7	Son		Kentucky
Wirod Rhodes	23	Boarder	Millwright	Kentucky
John Rhodes	20	Boarder	Laborer	Kentucky
William Temple	20	Boarder	Laborer	Kentucky

Source: US Census of Population Manuscripts.

Training in farming and milling work was local and vernacular. Millers accepted local men as apprentices. Children learned about farming by doing basic chores and day-to-day work on the farms where they lived or, according to labor-exchange traditions, at neighboring farms. When they were old enough, children might perform more demanding work, such as clearing land of brush and trees, logging and cutting stave wood for sawmills, hauling cargo for teamsters, and helping drovers take livestock to market. Farmers also employed town residents as day laborers for planting, harvesting, hauling, and butchering. Young men who wished to become farmers worked as laborers to acquire the capital to rent land and buy tools, implements, vehicles, draft animals, and livestock.[4] Kentucky's population in 1850 included 114,715 farmers and 27,047 laborers. The "skilled" trades comprised 5,157 carpenters, 3,638 blacksmiths, 2,044 cordwainers (leather shoemakers), and 1,239 saddle and harness makers. Most other trades employed fewer than 500 individuals.[5]

Unlike the Deep South's one-crop farming system—the regionally circumscribed production of either cotton, tobacco, sugar, or rice—the nineteenth-century farm economy in Kentucky's Bluegrass and Pennyroyal regions was diversified. Farmers grew tobacco, hemp, corn, and small grains. They also produced livestock: cattle, hogs, mules, and horses.[6] The annual hemp and tobacco production schedule was punctuated by high labor requirements over ten or eleven months each year. Tobacco seedbeds were prepared in the early spring—first by tilling, then by sterilizing, which involved piling and burning brush atop the beds. When the seedlings were ready, workers transplanted them from the seedbeds into prepared fields. Tasks during the growing season included hoeing and cultivating, removing suckers, topping or breaking off seed heads, and picking worms. During the harvest, workers cut each plant by hand; the plant stalks and leaves were then hung in a barn to dry and cure. All these tasks were repetitive and tedious but did not demand advanced skills.[7] In the late fall, farmhands removed the cured tobacco from the barn and stripped and graded the leaves; this required experience and the ability to judge leaf quality. Then they hauled the processed crop to city auction markets, which operated from late November through early winter.

Given the commercial demand for the fiber, hemp farmers grew the crop in large fields. After the initial soil preparation, they planted hemp by hand-broadcasting seed. The hemp plant's rapid growth quickly shaded the ground, limiting weeds. Harvest was a labor-intensive process of cutting and stacking the stalks, then unstacking and retting the hemp—a decomposition process that loosened the fibers—and regathering the stalks into piles or stacks. Removing the fiber from each stalk involved breaking or crushing; this was followed by cleaning and baling and, finally, delivery to market. Nineteenth-century accounts describe the hemp harvest as exceptionally laborious and dirty, and these tasks were often relegated to enslaved people or, after the Civil War, to African American laborers.[8]

Farmers planted small grains, especially wheat, rye, and oats, in fall or spring. Until the midsummer harvest, grain required little attention unless the crop was stricken by rust or damaged by hail; either misfortune required replanting. The July or August harvest was labor intensive and involved cutting the grain, tying it into sheaves or bundles, stacking the bundles into shocks, and threshing by hand or machine. Small-acreage farmers on productive farmland might hire day laborers only at peak demand times, such as wheat harvest in July or August, tobacco harvest in August and September, and corn harvest in October and November.[9] Large-acreage farmers usually employed one or more year-round hired men and engaged additional workers as needed; they also employed hands in the winter season for tasks such as grain "fanning" or cleaning, corn shucking, land clearing, and woodcutting.

Employment opportunities for rural laborers were therefore varied, depending on the size of farms and the diversity of crops and livestock produced. Some farmers hired part-time workers for four months or less per year, whereas tobacco or hemp farmers might hire workers for six months or longer. Importantly, hemp, grain, and whiskey production were complementary tasks. Hemp farmers produced the fiber that was woven into coarse cloth from which other manufacturers produced sacks. And farmers used hemp sacks to ship their grain to distillers.

Bonded Labor

Sometime between 1774 and 1777, Joseph Bowman of Frederick County, Virginia, at the northern end of the Shenandoah Valley, moved to Kentucky's frontier settlement at Harrodsburg, along with a number of other colonists. Bowman's party included twenty-year-old Jacob Spears from Rockingham County, who served as a captain of the group.[10] At the beginning of the American Revolution, Spears returned to Virginia to join the militia. In 1782, after hostilities ended at Yorktown, he moved back to Kentucky and settled in Bourbon County, north of Paris. Many pioneer settlers from Maryland and Virginia brought enslaved people with them to Kentucky. Evidence that Jacob Spears was a slaveholder is indirect, but in 1790, on the Clay-Kiser Road north of Paris, two black men—either slaves or employees—worked felling trees and cutting the trunks into logs, which they hauled to the Spears farmstead. Jacob Spears hewed and notched the logs to shape, positioned them on a fieldstone foundation near a small flowing spring, and built one of the first distilleries in the county. The output of the Spears farm distillery was small—less than two barrels per day. Several years later, a local stonemason built a stone storage warehouse–malt house along the road near the distillery. Although the log distillery building stood for only a few years, the stone warehouse remains in good condition and is used to store livestock feed.

Mr. Spears and his wife Elizabeth had six children, including Noah and Solomon. As young adults, these two sons shipped the family's whiskey to New Orleans by flatboat. The nearby Stoner Creek and Licking River proved too shallow for flatboating, so the Spears brothers likely hauled their cargo southwest to the Kentucky River for shipment to the Ohio River at Carrollton.[11] Noah became an abolitionist, and he and his wife "bought land in Greene County, Ohio, [near Xenia] before the war and settled their people upon it."[12] The Spears family experience offers a general sketch of how Kentucky's early farm-distillery industry began and how its labor force developed. Jacob Spears (who died in 1825) possessed practical skills, be they woodworking, building construction, farm management, or distilling. Slaves or African American employees constituted at least part of the labor pool and worked split seasons on the farm and at the distillery.[13]

William and John Snyder moved from Albemarle County, on Virginia's Piedmont, to Petersburg, Kentucky, on the Ohio River in Boone County, in 1833. The brothers bought a riverside steam-powered gristmill and later added a distillery. Capitalizing on ready access to steamboat shipping, William Snyder steadily expanded the milling-distilling operation, drawing water from both the Ohio River and a deep well. By 1850, the works employed sixteen hands and produced 7,000 barrels of flour and 164,000 gallons of whiskey. John Snyder's son David worked as a distiller; another son, James, was a cooper. Snyder employed two millers—one from Germany and one from Pennsylvania. The distillery dispatched whiskey to river markets, especially Cincinnati. Corn shipped from St. Louis arrived at the Petersburg wharf by steamboat barge, upward of 2,000 sacks in each delivery.[14] Cincinnati coal merchants delivered coal to the distillery by river barge several times a year. Given the distillery's demand for white oak barrels, it is likely that river shipments included barrel staves in large volume. The works likely employed part-time laborers to load and unload cargo.[15]

In 1860 William Snyder's household included his wife and daughter and five boarders, who may have been extended family members. Snyder's twenty-year-old son, William T., was a clerk, probably at the distillery; he lived next door with his wife and child. By 1860, William Snyder had likely fitted the Petersburg distillery with high-volume column stills, because his annual whiskey production had increased to 1.125 million gallons.[16] His mill had on hand 80,000 bushels of wheat and 300,000 bushels of corn. The works employed sixteen coopers and thirty distillery hands and day laborers.[17]

William Snyder Household, Petersburg, Boone County, Kentucky, 1860

Name	Age	Sex	Occupation	Place of Birth
William Snyder	60	M	Miller and distiller	Virginia
Amanda Snyder	46	F		Kentucky
Viranda E. Snyder	22	F		Kentucky
Martha C. Riddell	37	F		Virginia
James Riddell	14	M		Kentucky
Marcilline Riddell	10	F		Kentucky
Alfred E. Chambers	13	M		Kentucky
Mildred Jenkins	58	F		Virginia

Source: US Census of Population Manuscripts.

Note: In 1869 William Snyder's real estate and personal estate were valued at $42,000 and $7,000, respectively. Son William T. Snyder lived next door. His occupation was listed as "clerk," and his personal estate was valued at $1,200.

Distiller–Slave Owners, Boone County, Kentucky, ca. 1860

Owner	Number of Slaves	Ages of Adult Slaves: Male	Ages of Adult Slaves: Female	Ages of Children
William Snyder	9	47	56	5
			21	3
			18	1
			16	
			15	
William T. Snyder	3		22	4
				4

Sources: US Census of Population, 1860, Kentucky Slave Schedules; Sam K. Cecil, *Bourbon: The Evolution of Kentucky Whiskey* (New York: Turner Publishing, 2010); Chester Zoeller, *Bourbon in Kentucky: A History of Distilleries in Kentucky,* 2d ed. (Louisville, KY: Butler Books, 2010).

Note: The Snyders (father and son) owned the Boone County Distillery, Petersburg, Kentucky.

William Snyder owned nine slaves in 1860: a forty-seven-year-old man, five women aged fifteen to fifty-six, and three children. His son, William T., owned three slaves: a twenty-two-year-old woman and two children. We do not know the slaves' work assignments, but it is unlikely that any of the women were employed at the mill or distillery; they may have worked at domestic tasks.[18] It is also possible that some of the distillery laborers were slaves rented from other owners.

**Distiller–Slave Owners,
Anderson County, Kentucky,
ca. 1860**

Owner	Number of Slaves	Ages of Adult Slaves: Male	Female	Ages of Children
M. S. Bond	4	11	25	2
W. F. Bond	9	28	40	2
		28	17	
		11	16	
		11		
W. H. McBrayer	6	40	42	3
		18		
James Ripy	8	38	30	1
		26	10	
		22		
		13		
		12		

Sources: US Census of Population, 1860, Kentucky Slave Schedules; Sam K. Cecil, *Bourbon: The Evolution of Kentucky Whiskey* (New York: Turner Publishing, 2010); Chester Zoeller, *Bourbon in Kentucky: A History of Distilleries in Kentucky,* 2d ed. (Louisville, KY: Butler Books, 2010).

Note: M. S. Bond co-owned the Old Joe Distillery on Gilbert's Creek, south of Lawrenceburg; W. F. Bond co-owned the Bond & Lillard Distillery on Cedar Brook, east of Lawrenceburg; W. H. McBrayer owned the Cedar Brook Distillery, east of Lawrenceburg; and James Ripy co-owned the Walker, Martin & Company Distillery at Tyrone on the Kentucky River, east of Lawrenceburg.

The "Souvenir Supplement" published in 1906 by the *Anderson News* summarized the county's business history. "Distilling was a more laborious business in those days than it is today," read one essay. It depended on "manual labor, instead of steam and modern machinery; grain [was] ground by waterwheel, fed by the old race, often augmented by the primitive tread-wheel, and carried in sacks from granary to distillery; [it was] mashed by strong muscle and old-fashioned mash sticks, consuming time as well as hard labor."[19]

The *News* made no mention of enslaved laborers, although at least four Anderson County distillery operators were slave owners in 1860.

John Bond, perhaps following his family's tradition in England, built a distillery near the mouth of Bailey Run (also known as Baileys Run), about four miles east of Lawrenceburg, Kentucky, in 1810. Bond died in 1842, but his works operated for more than half a century. One of his sons, Medley S. Bond, owned four slaves and acquired the Old Joe Distillery on Gilberts Creek, southeast of Lawrenceburg, in about 1857. William F. Bond (Medley's half brother) operated the John Bond Distillery on Bailey Run in 1849, and in 1869 he formed a partnership with his brother-in-law Christopher C. Lillard, whereupon they changed the distillery's name to Bond & Lillard. William Bond owned nine slaves— four men, three women, and two children—who may have been distillery laborers and household servants.

Born in 1809 in Tyrone, Ulster, Northern Ireland, James Ripy migrated to the United States and settled in Lawrenceburg in 1839, working as a merchant and a wholesale liquor dealer. Ripy and his wife Artimesa had four children: two daughters, Ema and Beebe, and two sons, James P. and Thomas B. James Ripy moved his family to a farm north of Lawrenceburg in about 1855; a census enumerator listed him as a farmer in 1850 and 1860. In 1860 the recorded value of Ripy's real estate holdings was more than $28,000; his personal estate was valued at $12,000. By 1870, the value of his personal estate had increased to $30,000, and he was one of the wealthiest farmers in the county. Ripy owned eight slaves in 1860: five men aged twelve to thirty-eight, two women aged ten and thirty, and one child. The male slaves may have worked on Ripy's farm. In the 1860s James Ripy was a business partner with S. P. Martin and Monroe Walker, who operated a distillery at Tyrone on the Kentucky River east of Lawrenceburg. In 1869 the partners sold the distillery to William H. McBrayer and Thomas B. Ripy, son of James.

William McBrayer, a descendant of Scots-Irish immigrants, was a Lawrenceburg dry-goods merchant and livestock broker. He established a small distillery on Cedar Brook Run, east of Lawrenceburg, in about 1847; he served as a county judge from 1851 to 1856 and as a member of the state senate from 1857 to 1861. McBrayer owned six slaves: two men, a woman, and three children. In 1870 William McBrayer was a merchant and a farmer, with real estate valued at $20,000 and a personal estate valued at $4,000. Newton Brown was McBrayer's distiller for many years. At the time, distillers were paid $300 to $400 per year. The 1870 census recorded the value of Brown's real and personal estate as nil.[20] In 1876 McBrayer's Cedar Brook whiskey reportedly won a medal at the Philadelphia Centennial International Exhibition.[21] For his efforts, Brown received a gold watch from McBrayer.[22]

In 1860, the year before the Civil War began, five prominent Nelson County farmer-distillers were slaveholders. Raymond B. Hayden owned ten slaves in 1860 and operated the Hayden Distillery south of Bardstown. He was one of the wealthiest farmers in the

Distiller–Slave Owners,
Nelson County, Kentucky,
ca. 1860

Owner	Number of Slaves	Ages of Adult Slaves: Male	Female	Ages of Children
Raymond B. Hayden	10	35	35	5
			35	
			12	
			11	
James Mahoney	3	26	35	1
T. J. Pottinger	12	60	33	3
		30	20	
		25	17	
		16	14	
		13		
Taylor W. Samuels	5	38	19	2
		22		
William Sutherland	15	55	25	5
		45	23	
		35	20	
		25	13	
		25		
		21		

Sources: US Census of Population, 1860, Kentucky Slave Schedules; Sam K. Cecil, *Bourbon: The Evolution of Kentucky Whiskey* (New York: Turner Publishing, 2010); Chester Zoeller, *Bourbon in Kentucky: A History of Distilleries in Kentucky*, 2d ed. (Louisville, KY: Butler Books, 2010).

Note: Hayden owned the R. B. Hayden Distillery near Greenbrier Station, south of Bardstown, founder of the Old Grand-Dad brand in about 1840. Mahoney operated the Fred Brey Distillery at Howardstown, south of New Haven. Pottinger operated the T. J. Pottinger & Company Distillery on the Nashville branch of the L & N Railroad at Gethsemane. Samuels established the T. W. Samuels Distillery near Deatsville in about 1844. Sutherland operated the H. Sutherland Distillery on Sutherland Road, southwest of Bardstown, possibly near Beech Fork.

county, with real estate worth $30,000 and a personal estate valued at $5,250, as estimated by the 1870 census. Thomas Jefferson Pottinger farmed and operated a small distillery near Pottinger Creek west of New Hope in southern Nelson County. In 1850 Pottinger owned fifteen slaves; a decade later, he had reduced his bonded workers to twelve. Pottinger expanded his farming operation by acquiring land in the 1850s.[23] Taylor W. Samuels and William Sutherland were both slaveholding farmers, and they both operated Nelson County distilleries. Sutherland owned fifteen slaves, including six men of working age and four women. A census marshal recorded the value of his landholdings in 1860 at $15,500 and his personal estate at $16,790.

Nelson County farmer-distiller James Mahoney of Howardstown owned three slaves in 1860: a man aged twenty-six, a woman aged thirty-five, and an infant child. Mahoney operated the small Walnut Hollow Distillery near the Rolling Fork, south of New Haven, from the 1830s through the late 1880s. Mahoney's distiller was Richard Bowling, an African American who started working at the distillery in about 1845. The Kentucky Slave Census Schedules listed slave owners by name; slaves were anonymous and listed only by sex and age. Bowling could have been free or enslaved. One of the state's largest wholesale liquor dealers, Sylvester Johnson, conducted his business in nearby New Haven and knew Richard Bowling well. In Johnson's opinion, Bowling "knew more about making straight old-fashioned whiskey than any distiller in Kentucky."[24] James Mahoney sold whiskey to C. Henry Finck & Company of Louisville, wholesale dealers specializing in handmade sour-mash Kentucky whiskeys.[25] Mr. Finck was one of eleven American exhibitors who sent whiskey to the Vienna International Exhibition in 1873, where his whiskey won a premium. It is plausible that the whiskey had been made by Richard Bowling, and the award was therefore an acknowledgment of Bowling's distilling expertise.[26]

Diaries such as the one kept by John W. Jones in Bourbon County during the pre– and post–Civil War period outline the various tasks performed by slaves and freedmen working as day laborers. Farmer-distillers likely assigned their male slaves all manner of seasonal tasks on the farm or at the distillery, as well as off-season maintenance and improvement work. Female slaves were often assigned household work: gardening, cooking, washing, cleaning, and care of the children and elderly. But beyond the straightforward accomplishment of cyclical work assignments, enslaved people provided their owners with an unearned increment of capital gain, whereby the slaveholder's land and personal property increased in value without any labor or expenditure by the owner. A farmer who bought a slave obtained that person's ability to gain personal wealth and accrued that potential gain to himself. The slave owner not only profited from the sale of harvested crops or distilled spirits but also benefited from enhanced land and personal property values. Some owners cleared land and otherwise improved farms and built and operated distilleries with enslaved labor. Others won prizes at exhibitions with slave-produced whiskey— awards that enhanced the distiller's prestige and may have increased sales and profits. All such gains in capital were part of slavery's unearned increment.[27] It is unclear whether the availability of enslaved workers allowed some distillers to avoid or delay adopting new steam-powered equipment and other labor-saving technology.

Slaves for Hire

Kentucky's seasonal cycles were predicated on latitudinal position and distance from the ameliorating effects of large bodies of water—the Gulf of Mexico and the Atlantic Ocean. The extended summer growing season was complemented by the possibility of mild winters, which, outside of short periods of cold and snow, permitted farmers to engage labor to build and repair fences, dig wells, clear land of trees and brush, saw wood for lumber and fuel, and perform related tasks. The pattern of antebellum slavery in Kentucky closely mirrored soil fertility, with the Bluegrass and Pennyroyal regions counting the largest number of enslaved people. Although many farmers and townspeople were slave owners, some farmers, millers, and distillers preferred to hire slaves rather than own them.[28] The availability of slaves for hire also allowed employers to dismiss their white laborers, who may have been unwilling to work extended hours or in difficult conditions.[29] Slave hire could be annual or seasonal and task specific. Enslaved people performed hire work according to formal agreements between owners and those requesting laborers, and contracts often spanned the calendar from January 1 to December 25.[30]

Farmer John Cleaveland owned some 1,100 acres of land near Keene in Woodford County, which he farmed with twenty-seven slaves. When he died in 1853, Cleaveland's five-year-old daughter inherited his estate. The child's guardian prepared an inventory of slaves that listed their values and hire-per-annum rates. That inventory included:

Jack, 45 years of age, worth $500.00. Hired for $120.00.
Ben, 40 years of age, worth $750.00. Hired for $125.00.
Lewis, 37 years of age, worth $650.00. Hired for $120.00.
Simon, 22 years of age, worth $650.00. Hired for $120.00.

Slave rental rates, as represented by this inventory, were approximately 15 to 20 percent of their estimated value, or about $10 per month.[31] Teenage boys could be hired for about half the rate of an adult. Hiring a slave therefore cost about one-third to one-half the cost of hiring a white employee to perform the same tasks. Importantly, rental rates for healthy adult male slaves, such as those owned by John Cleaveland, were comparatively high, and only those with large, financially robust businesses and farms could afford them.

In Bourbon County, farmer John Jones worked his diversified farm with several slaves prior to the Civil War; from 1868 through 1871, Jones continued his farming operations by hiring four or five former slaves. He paid some a yearly wage of $140 to $150, and he paid others $50 plus clothing. During peak labor demand times, Jones hired workers by the month, paying them $15 to $20. He also hired day laborers, often "Irish" or "Dutch," for specific jobs such as rock and rail fence building or well digging; the pay, regardless of the task, was usually $1 per day.[32] Prior to 1865, a distiller wishing to hire a white day laborer for a six-month distilling season, assuming ten-hour days and six-day weeks, would pay wages amounting to about $156. Hiring a middle-aged slave for the same season cost $90.

Industrialization and Labor Practices

Traditional agriculture, milling, and distilling depended on a readily available pool of cheap labor in markets with minimal competition for workers. In central Kentucky, the principal factors in regulating labor availability and cost were the character of crop and livestock production and seasonality.[33] Early-nineteenth-century agriculture was labor intensive. Farmers sowed grain by hand and harvested with handheld scythes and cradles; they thrashed with hand flails. They used axes to clear trees from farmland and mauls and wedges to split the trunks into fence rails and firewood.

Low wages in towns and cities such as Louisville, Lexington, and Cincinnati were a weak attraction for rural laborers to relocate to urban places, a situation that may have been exacerbated by the prevalence of child labor. Farm families often had three or more children; some families had eight or more. When they became old enough, boys entered the flexible local labor force, working on the family farm as tasks required and working for neighborhood farmers, gristmills, sawmills, timber cutters, and distilleries as jobs became available. Girls might become servants for other farm families; those who remained at home tended gardens and orchards, cared for laying hens and milk cows, and performed household chores that included washing clothes and preparing meals. Whether working on a neighboring farm or at a nearby mill or distillery, rural laborers might board with their employer or, if convenient, walk the distance between home and the work site.

As farm, mill, and distillery mechanization and industrialization proceeded after the Civil War, operational capacities increased, requiring additional labor. Three large distilling works in Bourbon County provide examples. In 1882 William Davie employed thirty hands at $2 a day to mash 500 bushels of grain at his Millersburg distillery. George G. White employed thirty-five hands to mash 400 bushels each day, while also feeding 500 cattle and 800 hogs at his Paris distillery. And H. C. Bowen employed twenty hands at a daily wage of $2 to mash 444 bushels at his Ruddles Mills distillery. Bowen's large cooper shop produced thirty-five to forty barrels per day.[34] In addition to the comparatively stable labor force required for routine day-to-day distilling work, distillers hired other hands for special projects, such as filling large whiskey orders by retrieving, gauging, and moving barrels to railroad loading docks or other shipping facilities. Calculating rough employee–to–mashing capacity ratios would be meaningless unless task allocations could be specified. Distilleries with new steam-powered equipment, for example, should have been much more efficient than water-powered works. And distilleries with associated cooperage shops had a higher proportion of laborers engaged in "indirect production" relative to the volume of grain processed than a distillery not so equipped.

As distillery works industrialized, their employment profiles changed, as illustrated by the distilling-related occupations in Anderson and Nelson Counties from 1850 to the beginning of Prohibition in 1920. The Kentucky River forms Anderson County's eastern boundary, and nineteenth-century distillers established large-scale distilleries on elevated river terraces and smaller works along tributary creeks. The largest riverfront settlement began at a steamboat landing about two miles east of Lawrenceburg, and the site grew into a large distilling works that was served, in part, by the village of Tyrone and its attached neighborhood, Peanickle. A high proportion of Peanickle's population was African American, and the neighborhood originated as a post–Civil War settlement similar to many other villages in the Bluegrass region that were established to provide residential accommodations for freedmen.[35]

Distilling–Related Employment,
Anderson County, Kentucky

Occupation	1850	1860	1870	1880	1900	1910	1920	1920	
Distillers	0	0	0	4	0	3	P	0	(Tyrone City*)
	0	4	14	12	15	12		1	(Lawrenceburg and County)
							R		
Distillery employees	0	0	0	21	1	48	O	0	(Tyrone City)
						14		0	(Tyrone City–Peanickle Village†)
	0	0	1	23	3	81	H	38	(Lawrenceburg and County)
Coopers	5	0	0	5	21	4	I	0	(Tyrone City)
	0	5	0	22	5	4		1	(Lawrenceburg and County)
							B		
US government gaugers	0	0	0	0	1	0		0	(Tyrone City)
	0	0	0	0	2	17	I	7	(Lawrenceburg and County)
Bottling house employees	0	0	0	0	0	14	T	0	(Tyrone City)
	0	0	0	0	0	3		0	(Tyrone City-Peanickle Village)
	0	0	0	0	0	12	I	7	(Lawrenceburg and County)
	0	0	0	0	0	0	O		(Tyrone City)
Whiskey dealers	0	0	0	0	2	0		0	(Lawrenceburg and County)
							N		
Total county population	6,260	7,404	5,449	9,361	10,051	10,146		9,982	

Source: US Census of Population Manuscripts.

Note: Prior to 1910 Anderson County census enumerators recorded few laborers by their occupation, such as "mash hand." They were listed simply as day laborers. The occupations of cooper, distiller, manager, and related professions were recorded. The 1880 enumerators recorded a broad group of professions, including carpenter (6 in Tyrone, 10 in Lawrenceburg and the remainder of the county), sawyer (1 in Tyrone), and engineer (1 in Tyrone, 1 in Lawrenceburg and the county).

*Tyrone City was an incorporated industrial distilling settlement sited on a terrace above the Kentucky River about two miles east of Lawrenceburg. Tyrone's population in 1910 was 596, 17 of whom were African American. By 1920, Tyrone City was no longer tabulated separately but was included with Tyrone Precinct.

†Peanickle Village was a small subdivision of Tyrone in 1910, with 17 households and 73 residents, 54 of whom were African American.

Although distilleries operated in Anderson County by 1810, the 1850 census was the first to identify individuals by name, occupation, and place of birth. Five coopers worked in Tyrone in 1850, but census marshals identified no distillery employees there before 1860. Moreover, following official Census Bureau instructions, enumerators recorded workers according to their primary occupations; consequently, individuals working at seasonal jobs in mills, at distilleries, and on farms may have been identified only as "laborers." It is therefore plausible that distillery employees were undercounted. Similarly, a distillery owner who also engaged in another business might have been tabulated as a merchant, a farmer, or a distiller. Employees who worked as distillers were usually so identified, although it might not be possible to discern the difference between a distillery employee and a distillery owner without examining the corresponding value of landholdings and personal estates. Despite such incongruities, nineteenth-century censuses can provide insights into industry growth and change.

Fourteen distillers resided in Lawrenceburg and various rural locales in Anderson County in 1870. The count increased to sixteen in 1880, with four of them living in Tyrone, reflecting the rapid development of large distilling works there. Distillery employees increased at a rate roughly parallel to the number of distillers. By 1910, Tyrone and Peanickle Village had 62 of the county's 143 distillery workers. The number of bottling house employees, primarily women, was nil in 1900, probably reflecting the late arrival of machines capable of manufacturing glass bottles. In 1910 twenty-nine people worked in distillery bottling shops, seventeen of them in Tyrone and Peanickle Village. Although Internal Revenue gaugers were stationed at the state's distilleries by the late 1860s, few were tabulated until the census listed seventeen in 1910.

In 1900, seventeen distillers and whiskey dealers resided in Anderson County; that number fell to fifteen in 1910. National Prohibition went into effect in January 1920, and by the time the 1920 census was conducted later in the year, only one distiller remained to be counted, possibly someone who had received permission to continue distilling products for the medicinal spirits market.[36] Prohibition, of course, forced most distillery employees to seek alternative employment. Some closed distilleries were left to rust and rot; others were dismantled, and their wood, copper, and equipment were salvaged for use elsewhere. But large warehouses filled with barreled whiskey remained, requiring security to prevent theft or report fire and maintenance to fix roof leaks and broken windows and make general repairs.

Nelson County's nineteenth-century census marshals reported no distillers in 1850, likely because farmer-distillers were identified by their primary vocation, which was farming; twenty-six were present in 1900. Distillery employees increased from 7 in 1870 to 174 in 1910, 44 of whom were African American. Marshals did not record bottlers and labelers there until 1910, the same decade in which those occupations materialized in Anderson County. The number of bottlers and labelers decreased during Prohibition, but eleven bottlers remained in 1920, a higher proportion of 1910 levels than any other employment category. Some bottlers were presumably employed to fill flasks with medicinal spirits for sale to pharmacies. Except for 1860 and 1900, coopers enjoyed full employment in Nelson County until Prohibition, after which enumerators counted only one. Millers, whose product was in demand by both the public and distillers, almost doubled in number between 1850 and 1910. Given that many distillers installed their own grain storage and milling equipment, especially after the 1870s, the increase in millers was likely attributable to the

**Distilling-Related Employment,
Nelson County, Kentucky**

Occupation	1850	1860	1870	1880	1900	1910	1920	1920
Distillers	0	1	6	15	26	24	P	3
Distillery employees	0	0	7	43	22	174*	R	22†
							O	
Coopers	28	7	17	30	11	33		1
							H	
Coppersmiths	0	1	1	0	0	0		0
							I	
US government gaugers	0	0	2	16	12	19		6
							B	
Bottling house employees	0	0	0	0	0	23		11††
							I	
Labelers	0	0	0	0	0	15		0
							T	
Whiskey dealers	0	0	0	2	0	3		0
							I	
Millers	8	11	13	6	12	15		9
							O	
Millwrights	8	3	2	0	2	0		0
							N	
Tollgate keepers, toll road laborers	6	2	9	9	0	5		0
Cattle and livestock buyers	0	0	0	0	4	9		7
Total county population	14,789	15,799	14,804	16,609	16,537	16,830		16,137

Source: US Census of Population Manuscripts.
*1910 distillery employees consisted of 130 whites and 44 African Americans.
†1920 distillery employees consisted of 1 African American and 13 whites and 8 women.
††1920 bottling house employees consisted of 11 women.

overall increase in county population. Millwrights, in contrast, declined in number, and by 1910, none remained. By that time, mill construction was no longer the purview of people with combined expertise in structural timber framing, mill-wheel construction, and the carpentry skills necessary to build the chutes, elevators, and sifting boxes required. Finally, industrial distilleries greatly increased their slop production, necessitating an expansion of their livestock feeding operations. Distillers often contracted with livestock brokers or buyers to furnish feeder stock at the beginning of the distilling season and then sold the fattened animals to slaughterhouses and fresh meat markets when the season ended in the late spring. Nine livestock buyers or brokers worked in Nelson County in 1910.

People with distilling-related occupations often clustered in single domestic households and in neighborhoods. In 1880 the nine-member household of George Elliot, a farmer-miller-distiller in southern Nelson County, included three sons or stepsons who worked on the farm and three boarders: John Willitt, an African American who worked at the distillery; cooper Joseph Lang, an immigrant from Bavaria; and woodcutter William James. Neighborhood clusters of distillery employees were no doubt related primarily to proximity and the convenience of working close to home. In addition, household clusters may have been the product of friendship and trust, as owners hired people recommended by valued employees.

African Americans after the Civil War

By 1870, a significant proportion of central Kentucky's distillery labor force was African American, as suggested by the residents of Peanickle and Tyrone in Anderson County. John W. Butcher, an African American, was the distiller at the J. W. James Distillery in Crab Orchard in Lincoln County; he was regarded as "an expert in the business" until his death in 1905.[37] Black distillery workers tended to cluster within domestic households, as did whites. John Henry Beam's Early Times Distillery operated some four miles east of Bardstown on the Springfield Branch of the Louisville & Nashville (L & N) Railroad. One of his neighbors, Austin Kaufman, a black man, was likely a laborer at Beam's distillery. Kaufman's nine-member household included a son whose occupation was listed as "day laborer" and a nephew who worked as a distillery laborer. Kaufman's next-door neighbor Becky Kinslow, who was also black, worked as a cook and took in four black boarders—two distillery laborers and two day laborers. John Beam's residence was three dwellings away, and his household included two white boarders who worked as distillery laborers; one was a nephew, Arch C. Beam.[38] An Internal Revenue warehouse storekeeper and a gauger lived nearby.

**Austin Kaufman Household,
Nelson County, Kentucky,
1900**

Name	Age	Relationship	Occupation
Austin Kaufman	40	Head	Distillery laborer
Ada Kaufman	35	Wife	
Edward Kaufman	17	Son	Day laborer
Henry Kaufman	16	Son	At school
Beatrice Kaufman	8	Daughter	At school
William Kaufman	6	Son	
Alice Kaufman	19	Sister	Wash and iron
Arthur Kaufman	20	Nephew	Distillery laborer
Bula Kaufman	14	Niece	At school

Source: US Census of Population Manuscript.
Note: All household members were African American.

Blacks were also hired, albeit rarely, to fill administrative positions in the Bureau of Internal Revenue. The following notice was posted in the *Frankfort Roundabout* in February 1884: "Rev. J. W. Asberry, colored, has been appointed a U.S. Store-keeper-guager [*sic*] in this, the seventh, district, and assigned to duty at the Cedar Run Distillery of Messrs. J. & J. M. Saffell, in this county. He relieved Mr. J. H. Bailey on Saturday, who has gone to

Hot Springs, Arkansas, for his health."[39] Richmond's newspaper, the *Climax,* carried the following announcement in 1889: "Colored Man Appointed. George W. Haynes, an intelligent colored man, has been notified by the Internal Revenue Department at Washington that he has been appointed storekeeper and gauger in the 8th District of Kentucky. He has given bond in the sum of $20,000 and [is] assigned [to] Collector Burnam [at the] Warwick distillery, Silver Creek station, this county, and went on duty as night watch Monday night. He is one of the few colored men that have been recognized by the Republican administration."[40] The Bureau of Internal Revenue set a gauger's salary according to the capacity of the distillery at which he was stationed, with a maximum amount of $5 per day. If Haynes had excellent references and credit records, his bond may have cost him upward of $400 a year, not an insignificant portion of his salary.

Women

Prior to 1900, few women worked in the distilling industry. Office jobs such as bookkeeper or secretary were commonly filled by men. As distilleries added automated whiskey bottling lines and women trained in stenography and bookkeeping, their employment at distilleries increased, but almost always in jobs that were not traditionally filled by men.

**Distillery Employment Categories,
Nelson County, Kentucky,
1910**

Category	Total Employees	Women	Men
Bookkeeper	1	0	1
Bottler	23	22	1
Cattle slopper	1	0	1
Cooper	33	0	33
Distiller	24	0	24
Engineer	1	0	1
Fireman	8	0	8
Foreman	1	0	1
Labeler	15	15	0
Laborer	23	0	23
Miller	15	0	15
Stave wetter	1	0	1
Storekeeper-gauger	19	0	19
Stenographer	1	1	0
Warehouseman-manager	2	0	2
Warehouse superintendent	1	0	1
Watchman	0	0	0
Teamster	4	0	4
Yeastman	1	0	1
Total employees	174	38	136

Source: US Census of Population Manuscripts.
Note: The 1910 census required enumerators to list specific occupations and types of workplaces such as sawmill, law office, or retail store. Nevertheless, numbers cannot be considered absolute, given the variations in coding between enumerators.

In 1910 Nelson County distilleries employed thirty-eight women: twenty-two bottlers, fifteen labelers who worked in the bottle shop, and one stenographer. One man worked as a bottler—the only job filled by both females and males. The only bookkeeper was male.[41] By late spring or early summer 1920, most Nelson County distilleries had closed, except for those making medicinal spirits. Twenty-two women worked in distilleries in the summer of 1920, primarily as bookkeepers and bottlers. The Bureau of Internal Revenue employed twenty-two warehouse managers, five of whom were women.

**Distillery Employment Categories,
Nelson County, Kentucky,
1920**

Category	Total Employees	Women	Men
Bookkeeper	6	6	0
Bottler	11	11	0
Cooper	1	0	1
Distiller	3	0	3
Laborer	5	0	5
Storekeeper-gauger	6	0	6
Warehouse superintendent	3	0	3
Warehouseman-manager	22	5	17
Watchman	8	0	8
Total employees	65	22	43

Source: US Census of Population Manuscripts.

Note: The 1920 census required enumerators to list specific occupations and types of workplaces such as sawmill, law office, or retail store. Nevertheless, numbers cannot be considered absolute, given the variations in coding between enumerators.

Women distillery workers tended to cluster in domestic households, perhaps to a greater degree than men. In 1910 Sidney O'Bryan owned and operated a barbershop in New Hope in southern Nelson County near several distilleries. Mr. O'Bryan and his wife Mary had five daughters. The three oldest, aged thirteen, fifteen, and seventeen, worked as labelers in a distillery bottling house. A seventeen-year-old boarder, Ella Bartley, also worked as a labeler.[42]

German-born Joseph Werner was a distillery warehouse foreman in Nelson County in 1910. Mr. Werner and his wife Carrie had seven children. Four daughters, aged eighteen to twenty-three, worked as labelers at a distillery, likely the same one that employed their father. The Werner family may have engaged in some type of farming in addition to their distillery occupations, as suggested by two sons who worked as farm laborers.

In Nelson County, Richard and Teresa Gatewood's household included eight children. Mr. Gatewood was a distillery watchman; two daughters worked in the bottling house, likely filling medicinal whiskey bottles; and two sons worked as distillery laborers, probably moving barrels to the bottling house, maintaining the warehouses, and

Women bottler-labelers leaving work at the Yellowstone Distillery. (Oscar Getz Museum Collection, Bardstown, KY)

performing related work. The Gatewood household exemplifies the domestic side of distillery work. Often several household members were distillery employees or worked for the industry indirectly as laborers on nearby farms or as teamsters. Distillery-employee households tended to cluster in town or open-country neighborhoods, often within walking distance of the distillery. This residential pattern mirrored that of other rural non-

Joseph H. Werner Household, Nelson County, Kentucky, 1910

Name	Age	Relationship	Occupation
Joseph H. Werner	47	Head	Foreman—whiskey warehouse
Carrie S. Werner	37	Wife	
John L. Werner	25	Son	Farm laborer—home farm
Minnie Werner	23	Daughter	Labeler—whiskey house
Nidia Werner	21	Daughter	Labeler—whiskey house
Rick M. Werner	20	Daughter	Labeler—whiskey house
Fannie Werner	18	Daughter	Labeler—whiskey house
George Werner	16	Son	Farm laborer—home farm
Robert Werner	14	Son	None

Source: US Census of Population Manuscripts.
Note: Joseph Werner was born in Germany.

farm households, which tended to assemble at points of potential employment: important crossroads, railroad depots, blacksmith shops, drovers' and teamsters' yards, and roadside inns and stagecoach stops.[43] Large distilleries attracted numerous such households, and a few of these clusters grew into small villages or towns.

Labor and Residence: "Distillery Towns"

Athertonville—Larue County

In February 1907 newspaper correspondent Fred McLean drove his horse and buggy south from New Haven, in Nelson County, to the village of Athertonville, two miles away. He later described the place as "a nice little distillery town, with some very picturesque scenery surrounding it."[44] The distillery settlement McLean saw wreathed in midwinter snow was not new; it had been some four decades in the making. Athertonville was an unplanned, handmade community lacking any semblance of corporate uniformity. Distillers and residents had pieced the town together, parcel by parcel, in the shadow of one of Kentucky's largest distillery complexes.

John McDougall Atherton built and operated his first distillery on Knob Creek in northeastern Larue County in 1867. The works stood beside the Bardstown and Green River Turnpike in open farm country. New Haven was the nearest town, and Athertonville, the "distillery town" McLean later described, did not exist. In 1907 McLean lived two counties away and was visiting friends who had moved to Athertonville to work for the Bureau of Internal Revenue as distillery gaugers and storekeepers. Although the bureau tried to assign employees to distilleries near their places of permanent residence, they sometimes had to find lodging at or near their assigned workplace. Boarding with nearby farmers was one choice; boarding with a town resident or at the village hotel was another.

Knob Creek is one of dozens of small streams that slice into the western face of the Muldraugh Hill escarpment, the limestone bluff that marks the divide between the Outer Bluegrass to the east and the Pennyroyal upland to the west. Fronting the escarpment, a narrow belt of conical hills, Pennyroyal remnants, forms a topographically distinct area called the Knobs. Neither Bluegrass nor Pennyroyal, the steep-sided Knobs stand apart, separated by alluvial floodplains created by the region's rivers. From its head atop the escarpment, aptly named Knob Creek flows north, dropping some 450 feet over five miles and passing Thomas Lincoln's family farm on the way to its junction with the Rolling Fork, the regional trunk stream.[45] One mile south of its junction with the Rolling Fork, near the foot of the Muldraugh Hill escarpment, the Knob Creek floodplain is about 1,000 feet wide, with an elevated alluvial terrace on the west side above the creek. The Bardstown and Green River Turnpike followed the terrace. Another tributary, westward-flowing Pottinger Creek, joins the Rolling Fork slightly more than a mile above Knob Creek. Each stream has created its own broad, nearly level floodplain between the Knobs. At their junction, south of New Haven, the streams' composite floodplain is more than 7,000 feet wide.

Though it was forested when Anglo pioneers arrived in the 1780s and 1790s, the bottom-land was fertile and productive when cleared of trees. By the mid-nineteenth century, the Rolling Fork–Pottinger Creek–Knob Creek lowland was producing grain surpluses that supported the development of John Atherton's distillery.

Peter Lee Atherton of Fauquier County, Virginia, obtained a land grant on the Knob Creek drainage and moved to the area in about 1791. The fertile lands produced grains and corn that Atherton processed at a small farm distillery he operated on the west bank of Knob Creek from 1800 to 1830. Peter Atherton's son, John M. Atherton, was born there in 1841 and later owned much of the land in his father's original grant. John Atherton attended St. Joseph College in nearby Bardstown and Georgetown College in Scott County.

In 1865 Alexander Mayfield, a relative of John Atherton's, leased the family farm for three years, paying the rent by mortgaging his livestock, haystacks, oats, wagon, 1,000 pounds of pork, and household furnishings. In August 1867 Atherton formed a partner-ship with Miles Hagan of Nelson County and Marshall Key of Larue County to build a distillery on land they had bought from W. S. Ford and Joseph Dawson. Atherton donated firewood for the distillery, designating that it be cut from the 180 acres of land he owned along Knob Creek; from the same property, he also donated "as much as will be needed for a stick slop yard (say 15 or 20 acres)." The partnership agreement stipulated that the distill-ery could be operated by "any one or two of the owners," provided they paid a reasonable rent and took proper care of the machinery.[46] Designated Registered Distillery No. 87 by the Bureau of Internal Revenue, the works stood on the terrace beside the turnpike and began making sweet-mash whiskey, each day processing about 100 bushels of grain, which yielded about seven barrels of spirits.

In September 1868 Atherton's business partner Miles Hagan sold him a distillery for $5,878, one of his first distilling-related property acquisitions. The price included a distill-ing works, a five-acre distillery lot, and one-third interest in a storage warehouse.[47] The following year, Atherton paid W. S. Ford and Joseph Dawson $150 for a two-acre lot with turnpike frontage adjacent to the Knob Creek distillery. The property may have been used for distillery expansion or residential development.[48] Over the next twenty years, Ather-ton acquired more land near his Knob Creek distilleries in large and small parcels. Some tracts he likely subdivided into small residential lots; others he may have used to expand distillery operations. The larger tracts he may have rented to neighboring farmers.

Alexander Mayfield operated Atherton's distillery in 1874, and in July 1878 Mayfield paid the Atherton Company $37.50 for a small lot near the distillery, likely for residential construction.[49] The following year, Atherton built a new brick distillery with brick ware-houses across the turnpike from his 1867 works. Mayfield operated that distillery under his own name. In June 1879, to facilitate access to the expanding distillery complex, the Atherton Company leased for ninety-nine years a strip of land two miles long and forty feet wide as a right-of-way for a spur rail line connecting the distillery to the L & N Railroad's Knoxville branch line in New Haven. The Atherton Company built and operated the spur.[50] Between 1880 and 1882 Atherton added two more brick distilleries to his distilling complex: Clifton Distillery No. 337 and Winsor Distillery No. 378.[51] Steam powered the distilleries and heated the warehouses in the winter. Automatic sprinklers provided fire suppression, and a gravity-flow pipe delivered water from a hillside reservoir.[52]

In concert with railroad and distillery construction, John Atherton and his partners reorganized their business. In 1881 the J. M. Atherton Company incorporated, with John Atherton as president; son Peter Lee Atherton, vice president; Frank Miller, secretary and treasurer; and William Miller, superintendent.[53] In August of that year, Alexander Mayfield became a partner in the J. M. Atherton Company, and Atherton and Mayfield subsequently formed a partnership to create A. Mayfield & Company. The Atherton Company then leased Distillery No. 229 to Mayfield & Company, "with all the land thereto attached and all buildings and other improvements, sheds, pens, vats, tubs, stills, mills, pipes, worms, pumps, machinery and apparatuses of every kind" to distill whiskey from July 1, 1881, to July 1, 1886.[54] Also in August 1881, the Atherton Company leased two other distilleries. Sylvester O'Bryan and William Miller, partners in S. O'Bryan & Company, leased Distillery No. 378 from the Atherton Company for five years.[55] Miller and O'Bryan, operating under the name William Miller & Company, also leased Distillery No. 377 for five years.[56] Finally, John Atherton himself leased Distillery No. 87 from the J. M. Atherton Company. Atherton had moved his residence to Louisville in the 1870s, but the Atherton Company did not move its headquarters to Louisville until about 1893.

Considering its agrarian surroundings, the distillery complex that John Atherton and his business partners assembled must have seemed an industrial marvel. Masons built the entire works, except for the cattle sheds, with 11 million bricks. The distilling machinery was the newest available. Two batteries of boilers fed 60,000 feet of steam pipe installed in the aging warehouses. Atherton believed that maintaining uniform interior warehouse temperatures year-round accelerated aging and resulted in a uniform, high-quality product.[57] For some fifteen years, from 1879 to 1895, Atherton's four distilleries operated five to seven months a year, drawing water from wells and Knob Creek.[58] Collectively, the distilleries processed roughly 2,200 bushels of grain per day; the aging warehouses stored more than 150,000 barrels.[59] Assuming that grain was delivered to the Atherton distilleries in large farm wagons with a forty-bushel capacity, each day's operation would require fifty-five wagonloads. If each distillery had grain storage space, grain could be delivered almost any time of the year. For those lacking storage, the period between grain harvest—July to October for wheat and corn—and distillery startup in November likely would have been the most active hauling time. Wagons drawn by four-horse teams were awkward to maneuver, and traffic of this magnitude would have congested the distillery grounds. The grain requirements of high-capacity distilling would have consumed locally produced grains quickly, obligating the distillers to deal with grain brokers at increasing distances from the works. A railroad connection would alleviate many of these concerns, and once it was completed, the two-mile Athertonville rail spur was active. Trains regularly delivered grain, coal, and other supplies and carried away barreled whiskey.

During the distilling season, the Athertonville works produced slop in large quantities. An elaborate system of tubs and delivery pipes carried the liquid slop some 900 feet to three large feeding sheds built near Knob Creek, north of and downstream from the distillery buildings. As whiskey production increased during the 1880s and 1890s, Kentucky distillers expanded their livestock feeding operations, which focused on fattening feeder cattle. But the regional availability of livestock was contingent on benign weather and overall national market demand. In 1898 the Atherton distilleries in Larue County, and distilleries in adjacent Nelson County, experienced a cattle shortage that was exacerbated by a severe drought. During the 1890s area distilleries fed slop to about 6,000 cattle annually, but during the 1898 distilling season, they fattened only 2,000 head.[60]

Managing the movement of cattle to and from distilleries had initially been part of the distilling business, but as the scale of distilling operations increased, the larger works began to patronize livestock brokers who purchased animals on the market, shipped them to distilleries, and paid the distillers feeding fees. The Ford & Burch livestock brokerage company of Athertonville maintained this type of business arrangement with the town's distillers. In 1904, for example, Ford & Burch sold more than 300 head of slop-fattened Athertonville cattle to the Bourbon Stock Yards in Louisville. The total sale exceeded $20,000, or more than $500,000 in current monetary value.[61] Four years later, the Elizabethtown cattle brokerage firm of Watkins, Carithers & Company shipped cattle to the Athertonville slop sheds.[62] At the end of the distilling season in May 1909, the company sold 1,600 slop-fed cattle to a Chicago packing firm. The total sale exceeded $100,000, or roughly $2.6 million in current monetary value. It was the largest livestock sale of the year and one of the largest ever made in Kentucky to that date.[63]

As an industrial distillery complex, Athertonville was a hub of business activity. And even though it was situated in a rural area, it required an appropriate urban infrastructure to function: residential, educational, social, retail, fabrication, and transportation facilities. Employee productivity was significantly enhanced if the workers lived near the distilleries. Housing workers at rural industrial sites was a common problem for nineteenth- and early-twentieth-century mining and manufacturing companies. Industrial firms often addressed the issue by building what were referred to generically as "company towns." Nineteenth-century textile mills in New England and the Carolina Piedmont were tethered to stream-side waterpower sites that were often remote from established urban centers, ports, and roadways. To provide housing for their workers, these companies often built mills, housing, service buildings, and streets and roads as centrally planned and company-managed projects. In the early twentieth century, coal and steel companies bought mineral rights and opened coal mines in remote mountain valleys of central and northern

Appalachia. Their potential labor force lived on small valley farms at inconvenient distances from the mining operations. The county seat might be the only town in the county, and roads were often vague tracks along a creek bottom. To house their workers, the mining companies built and operated towns—hundreds of them.[64]

John Atherton situated his distilleries on family farmland, ten miles from Hodgenville, the county seat, and two miles from New Haven and the L & N Railroad depot. Athertonville, the town that developed to serve the distilleries, was unplanned. Rather, it was the product of the ad hoc conversion of farmland into residential lots that were dissimilar in size, configuration, and orientation. In 1880 Athertonville's 118 residents included two engineers and a baker, bookkeeper, brick mason, fireman, foreman, merchant, and miller. Initially, a stave maker and nine coopers made barrels; by the mid-1880s, the works employed more than twenty coopers and consumed some 600,000 staves annually.[65] Some laborers lived in Athertonville, and others lived on neighboring farms; they may have worked both on the farms and at the distilleries during the appropriate seasons. Federal Internal Revenue gaugers and warehouse managers lived in the hotel or boarded with residents. Several households had two or more distillery employees. Taylor Whitehead, Atherton's distiller, and his wife Alice boarded twelve men in their home, including six coopers, a stave maker, a steam engine fireman, and a government warehouseman and storekeeper. Brick mason J. W. Eblin operated a brick kiln in Athertonville, and his household included nine African American boarders who worked as brick molders, setters, and laborers.[66]

By 1890, Athertonville, reportedly the largest employment center in Larue County, had a three-story clapboard hotel; it was later used as a boardinghouse, a schoolhouse, a general store, and a railroad depot. Although John Atherton, through his company partnerships, owned the distilleries and other buildings, he apparently did not own the entire town, nor did he manage Athertonville as a company town. The distilleries employed 150 to 200 people during the fall-to-spring distilling season.[67] Laborers who worked there only during the distilling season may have lived on nearby farms and walked or rode horses to work. But other employees in skilled positions who worked at the distilleries year-round likely preferred to live nearby. For instance, federal revenue agents monitored the bonded warehouses, recorded tax payments, and approved the withdrawal of whiskey from bond for shipment to customers.[68] Engineers and mechanics repaired and updated equipment. Bookkeepers (primarily men) and stenographers (primarily women) worked year-round at record keeping and correspondence.[69]

The town's residents, about evenly divided between Catholic and Protestant, established Catholic and Methodist Episcopal churches. John Atherton built the public school—said to be the largest in the county—and contributed funds toward its operation.[70] At the turn of the century, Athertonville businesses included the hotel, a general store, blacksmith shops, a saloon, a livery stable, and a post office. Counted among the residents were John Lougherty, the thirty-year-old distillery superintendent; an engineer; and twenty-one distillery laborers. Also living in Athertonville were a dressmaker, shoemaker, harness maker, millwright, teacher, slop feeder, and seven railroad employees. Given Athertonville's spur connection to the L & N line two miles north, some distillery employees may have lived in New Haven and commuted to the distillery each day.[71]

From 1867 to 1892, John Atherton and his business partners were active real estate brokers, buying and selling properties of various sizes in and near Athertonville, facilitating land and home ownership. Atherton bought farmland parcels on Knob Creek and the Rolling Fork floodplain ranging from 20 acres to more than 100 acres.[72] Most of his land purchases and sales were smaller and were adjacent to the turnpike and his distilleries. The Atherton Company did not buy a large block of land on which to plat lots and lay out streets. Rather, as it bought and sold various properties, the town of Athertonville materialized. People moving into town bought lots, and people moving away sold their lots; Atherton entertained both markets.[73] A quarter-acre building lot cost $42 in 1882, and property owners built their own houses.[74] In January 1882 one of Atherton's business partners, Sylvester O'Bryan, sold a vacant lot along the turnpike and near the distilleries to Elizabeth Greenwell and her husband.[75] The Greenwells built a house on the lot, and fourteen months later, in March 1883, they sold the house and the lot to Thomas R. Ford for $750.[76] Athertonville's real estate market was active and multifaceted. Distillers and residents alike bought and sold properties.[77] A few of the original houses remain from nineteenth-century Athertonville. The structures are nonstandard—a mix of idiosyncratic single-story rectangular cottages and two-story L-plan homes. This assortment of housing types was likely found in small towns across Kentucky, unlike the Appalachian coalfield towns, where company-built houses were based on a very limited number of standardized plans.

The J. M. Atherton Company continued to engage in the distilling business in Athertonville until 1899, when the Kentucky Distillers and Warehouse Company bought the works and operated the distillery until Prohibition.[78] The distillery was rebuilt in the mid-1930s. Production stopped in 1987, and the buildings are now occupied by a cooperage company. Though it was not a company town per se, Fred McLean's 1907 reference to Athertonville as a "distillery town" was a fair and descriptive representation.

New Hope—Nelson County

The Pottinger Creek valley, extending for some seven miles east of New Haven in southeastern Nelson County, had harbored farm distilleries since the 1790s. In 1857 construction crews built the Lebanon branch of the Louisville & Nashville Railroad—later renamed the Knoxville branch—which followed the creek toward Lebanon, the county seat of Marion County.[79] Just north of the height-of-land divide between Pottinger Creek and Sulphur

Lick Creek, five miles east of New Haven, the L & N built a station and a railroad siding that became known as New Hope Depot.

Henry Miles, a Scots-Irish Catholic from Maryland, operated a distillery in the valley in 1796.[80] His son, Edward L. Miles, also operated a small distillery and was an active real estate investor. Edward Miles bought 172 acres in the area in 1849, paying $15 per acre.[81] In 1853 he bought a 12-acre parcel on Pottinger Creek beside the Springfield Road for about $45 per acre.[82] In making that purchase, Miles apparently anticipated the alignment of the L & N Railroad branch that construction crews completed four years later. He continued to acquire land in the vicinity of his distillery and the depot, and in June 1865 his mother, Ann Miles, bought 1,140 acres in the area between Pottinger Creek and the headwaters of Sulphur Lick Creek that likely included the developing settlement of New Hope.[83]

Together with several other land developers, Miles was a member of B. H. Miller and Company, which in April 1869 bought a seven-and-a-half-acre parcel on the north side of the railroad track from Ann Miles; the price paid was $500, or about $67 per acre.[84] Five months later, in September, Edward Miles and partner James Pike bought the parcel from

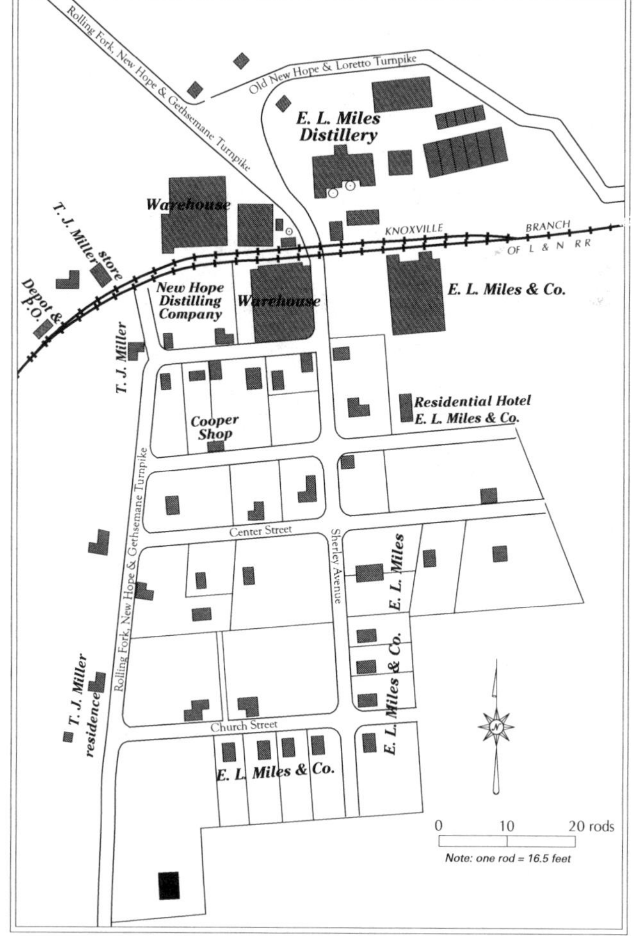

E. L. Miles and New Hope Distilleries, Nelson County. (Based on *An Atlas of Nelson and Spencer Counties, Kentucky* [Philadelphia: D. J. Lake, 1882]; USGS Topographic Quadrangle, New Haven, KY, 1953)

B. H. Miller and Company at a considerable premium. The land likely adjoined Miles's large new distillery, and the purchase would allow him to expand his warehousing, livestock feeding sheds, and other support facilities. In 1870 E. L. Miles was one of eight commissioners who formed the New Hope and Rolling Fork Turnpike Road Company, which built a five-mile-long turnpike that connected New Hope to the Bardstown-Greensburg Road at the Marion-Nelson County line.

Sometime in the late 1860s or 1870s Miles, in partnership with whiskey producer and wholesaler Thomas H. Sherley of Louisville, built a new distillery beside the railroad tracks just west of the existing works. The partners operated it as the New Hope Distillery, with a mashing capacity of 200 bushels per day.[85] On Saturday, May 19, 1883, a fire destroyed both the Miles and the New Hope Distilleries. The conflagration also consumed two granaries and nine cattle pens. Two warehouses containing whiskey worth more than $500,000 caught fire but were saved. The works were insured for $30,000, but the loss was valued at $40,000.[86] Miles and his business partners rebuilt their distilleries in tandem and adjacent to each other on the east side of the Rolling Fork, New Hope & Gethsemane Turnpike and north of the railroad tracks. The E. L. Miles Company Distillery operated as Internal Revenue Registered Distillery No. 146, mashing 300 bushels of grain per day to make sweet-mash bourbon and rye. The New Hope Distillery, Registered Distillery No. 271, stood immediately to the east, mashing the same amount of grain but producing sour-mash bourbon and rye. After the fire, the distilleries apparently stopped cattle feeding and converted to slop drying, which was then abandoned in favor of storing slop in tanks for sale to area farmers.[87]

New Hope's distilleries embraced the railroad tracks. A 6,000-bushel grain storage elevator stood on the north side of a dedicated siding, and the distillers positioned storage warehouses on both sides of the tracks. The distilleries were connected by umbilical-like pipes elevated above the track to carry spirits to barrel and bottling facilities and warehouses on the south side. The railroad depot attracted service businesses related to moving freight, telegraph operations, and a large lumber company. The E. L. Miles and New Hope Distilleries offered seasonal employment, and their size was sufficient to attract permanent residents; they included merchants and grocers who operated three different stores, a physician, a cooper, a blacksmith and wagon maker, and a schoolteacher.[88] By the 1880s, New Hope resembled a small town sited at the junction of three turnpikes and the railroad tracks. The E. L. Miles Company owned a residential hotel and houses on nine lots, including Edward Miles's residence. From 1873 to 1886 Miles sold ten lots, ranging in size from one-quarter acre to two acres, for residential construction.[89] Some purchase agreements required the buyer to donate a fifteen-foot strip of land along one boundary line that, when paired with a similar easement on the adjoining lot, accommodated the construction of a thirty-foot-wide street. Sherley Avenue became the town's primary north-south street.[90]

Curiously, Miles insisted on adding to each deed a restrictive covenant that prohibited liquor sales. Edward Hagan, for example, had to agree not to "keep or permit to be kept on the above described premises, or any part thereof, intoxicating liquors of any kind whatsoever for the purpose of selling the same at retail or in less quantities than by the barrel."[91] Miles provided no rationale for this provision, but it is possible that he operated a "quart store" at one or both of his distilleries, where he sold whiskey to local customers, and wished to neutralize competition.

Although New Hope's origins and morphology bore some resemblance to a company town, the site was not purpose-built; it was an accretive product of real estate development around the railroad depot. New Hope's growth was steered by the expanding operations of two large distilleries. E. L. Miles and other real estate investors sought to profit from developing residential lots, but they played no overt role in creating an overall town plan (other than subdividing lots) or providing centralized services. By requiring residents to donate land for a through street, Miles was operating not as a benevolent town owner but as a calculating real estate developer. Several houses dating from the pre-Prohibition era still stand along New Hope's streets, although some sidewalks end in the center of vacant lots. The brick town bank still stands in relict condition along Sherley Avenue. The residential hotel is gone, and the old bed alignment is all that remains of the L & N Railroad tracks. On the north side of the track bed, at the entrance to town, a carved stone statue of St. Benedict holding a Bible stands atop a stone retaining wall at a leveled site that may have been a railroad depot at one time. The distilleries and warehouses are gone; no foundations or other visible remains suggest that they were ever there.

Tyrone—Anderson County

Three miles east of Lawrenceburg, in Anderson County, the Kentucky River flows north, flanked by near-vertical limestone palisades. Several small streams have cut steep valleys into the palisades' west face, including Bailey Run and its northern branch, Cedar Brook Run. Before flowing into the Kentucky River, Bailey Run flows across a limestone terrace that lies some thirty feet above the Kentucky River. The bedrock terrace is 1,000 feet wide and about 6,500 feet long, an exceedingly rare bit of nearly level topography. In addition to being adjacent to central Kentucky's primary navigable river, this terrace was bisected by the shortest-distance rail line between Lawrenceburg and Versailles, the county seat of neighboring Woodford County. Rafts and steamboats frequented the north-flowing river, and the mouth of Bailey Run provided access to the river and a suitable place for a boat landing. W. H. Dawson and Joel Boggess built a wharf house at the landing in 1850, and the people working the riverboats referred to the small cluster of houses near the landing as Steamville. As distillers established works near the landing, Steamville became the village of Tyrone, named after the Ripy family's home county in Ulster, Northern Ireland.[92]

The valley cut by Bailey Run and Cedar Brook Run offered waterpower potential, and the river terrace below provided transport access not found elsewhere in this section of the Kentucky palisades. Millers and distillers built works in the area early on. Samuel P. Martin established the first distillery there in the 1860s. In July 1869 Martin and his business partners Monroe Walker and James Ripy sold the distillery, the whiskey aging warehouses, and a house and lot to William H. McBrayer and Thomas B. Ripy, James Ripy's son, for $13,000.[93] The three-acre property was situated just upstream from the mouth of Bailey Run, and the distillery operated under the name of Walker, Martin and Company.

Turnpikers completed the Lawrenceburg and Versailles Turnpike in 1857. The stone-covered macadam road descended the bluff to the Kentucky River near the mouth of Bailey Run and then ran the length of the Tyrone terrace to a ferry. Thomas Ripy's Clover Bottom Distillery and warehouses stood beside the pike. Distillery slop flowed through troughs to elevated slop tubs at two cattle sheds on the terrace, north of and downstream from the distillery. The roofed sheds covered almost two and a half acres and housed 2,500 cattle when filled to capacity.[94] Livestock broker Charles Byrne shipped to market some 300 boxcars of cattle that had been slop-fed at Tyrone during the 1891 distilling season. In late June 1892 the distillery sent 76 carloads of cattle, some 1,500 head, to eastern markets over the Louisville Southern Railroad. At the time, the consignment was thought to be the largest single shipment of cattle ever made from Kentucky. Ripy's distillery also fed hogs.[95]

Freight companies operated wagon lines from Frankfort to Lawrenceburg with connections to Tyrone, but when the water level in the river was high enough to permit steamboat traffic, packet boats made daily trips hauling goods between Tyrone and Frankfort, bypassing the turnpike and Lawrenceburg. In the 1880s Thomas Ripy's packet boat *Little Dora* made regular freight runs between his distillery at Tyrone and Frankfort.[96] The comparatively low cost of water transport was especially important when moving low-value bulk commodities such as coal, which the J. B. Ripy Distillery at Tyrone bought in three-barge lots.[97] High water in late winter and early spring brought rafted timber from southeastern Kentucky's forests downriver to sawmills at High Bridge and Frankfort. In April 1895 two 150-foot-long log rafts outfitted with large plank boxes landed at Tyrone. Loggers had filled the boxes with eastern Kentucky coal, which was delivered to the Ripy Distillery. Lumber mills in south-central Kentucky supplied the Tyrone distillery with staves for its cooperage—100,000 or more staves in a shipment; they arrived by water or by railroad, once the Louisville Southern line opened.[98] During the summer, when the river level was high enough, Frankfort steamboat companies offered passengers Saturday evening excursions up the Kentucky River to Tyrone for 50 cents a round-trip.[99]

From 1870 to the eve of Prohibition, development of Tyrone's residential lots and retail and service businesses was similar to that in Athertonville in Larue County and New Hope in Nelson County. Tyrone was not formally planned or platted as a town; rather, the site accrued streets, residential lots, shops, and churches through numerous real estate transactions. Several individuals bought and sold property in Tyrone. Thomas Ripy, for example, acquired several small farms and residential lots on the river terrace between 1870 and 1885. He bought a half-acre lot above the mouth of Bailey Run in November 1870 for $225; the following year he bought another small lot nearby for $312. Both lots fronted the terrace turnpike. Ferry operator Thomas Shryock bought a small lot near the ferry landing for $85. In 1872 Thomas Ripy paid J. H. McBrayer $1,600 for twenty-nine acres adjoining the river. Farm-sized properties were much cheaper per acre than residential lots, even though the farmland was likely to be converted to distillery building space or residential lots.[100] Ripy actively bought and sold house lots in Tyrone immediately across the turnpike from the distillery works.[101]

In October 1877 Ripy purchased the Cliff Springs Distillery at Tyrone from J. Tom Vandyke. The sale included the distillery works, water access and use privileges, and an adjacent forty-one-acre parcel of land with houses that were likely a large segment of the village of Tyrone.[102] Two distillery company partners bought small lots near the river in Tyrone in 1880; John Dowling bought two acres for $63, and Edward Murphy paid $200 for a house and a lot.[103] In 1881, the same year Tyrone incorporated, Thomas Ripy and his partners William Waterfill and Edward and John Dowling built the Clover Bottom Distillery at Tyrone and incorporated their company as Waterfill Dowling & Company.[104] Incorporation papers stated that the company "shall be distilling hand made sour-mash whiskey and conducting and operating in the same place a cooper shop and saw mill."[105] Ripy later bought out his partners and expanded production in both his distilleries, which were mashing more than 1,500 bushels per day and employing about 145 people in the 1890s.[106]

Thomas Ripy's primary vocation was distilling, but he was also an active real estate broker. He bought and sold property in Tyrone as well as in Lawrenceburg, Gilberts Creek, Fox Creek, and other locations across the county.[107] Ripy may have viewed real estate and corporate partnerships as diversifying investments that lessened the risks inherent in depending on distilling as one's sole economic activity. By 1900, Tyrone's population exceeded 500, and 150 children were enrolled in the school built by S. P. Martin two decades earlier. By 1910, Tyrone's population was 596. Benevolent residents built several town buildings. Distillers T. B. Ripy and S. P. Martin built a Methodist church in 1880, and store owners built a Christian church in 1890.[108] The town also included an Odd Fellows Hall, a large general store, several retail shops, a barbershop (operated by an African American barber), machine and blacksmith shops, a medical doctor, a courthouse, and a jail. The Tyrone Dramatic Club presented plays and musical performances.[109] Ironically, given the town's dependence on distilleries for employment, Tyrone was "dry," allowing no liquor sales, until 1906, when residents voted to go "wet."[110]

In 1888 J. P. Ripy bought a distillery atop the bluff above Tyrone; after rebuilding, he operated it as the Old Hickory Spring Distillery.[111] That same year, the Louisville Southern Railroad Company, whose primary rail line ran from Louisville to Lawrenceburg and Harrodsburg, began work on a new branch to connect Lawrenceburg to Versailles and Lexington. The line required a new railroad bridge across the Kentucky River palisades near Tyrone. Completed in August 1889, the trussed cantilever deck bridge had a 550-foot central span, the longest of any bridge in the nation at the time, and stood more than 280 feet above the water. Named the Young Bridge after railroad company president William Bennett Young, the span and its new rail line bolstered the bluff-top distillery operations. The J. P. Ripy Distillery now stood beside the tracks.[112] Following the repeal of Prohibition, the works was rebuilt, and its descendant continues to operate in the twenty-first century as the Wild Turkey Distillery.

The riverside Tyrone terrace distilleries operated by Thomas Ripy continued to expand production in the late nineteenth century. He sold his works, but not his 40,000 barrels of aging whiskey, to the Kentucky Distilleries and Warehouse Company in 1899. That company succeeded in buying Ripy's whiskey the following year for 50 cents per gallon, or more than $1 million.[113] That organization continued production until Prohibition, when it closed and dismantled the distilleries.[114]

Residents might have described Tyrone as a "distillery town" akin to Athertonville and New Hope. But the reference would have been one of identity and association, not a formal category of company-planned and company-owned places such as those established in eastern mining and milling areas. In 1900 Tyrone was the state's leading whiskey manufacturing center outside of Louisville. Closing and dismantling the distilleries truncated the town's livelihood. Residents and businesses moved away. Fire and teardown removed many of the abandoned buildings. In the twenty-first century, a few houses, trailers, and church buildings remain, but the town is now quiescent; its concrete sidewalks cross open grassy spaces, and driveways end at vacant lots.

Millville—Woodford County

Glenns Creek rises near Versailles in Woodford County and flows northwest, incising a 300-foot-deep valley to the point where it joins the Kentucky River about two miles south of Frankfort. The stream gradient was sufficiently steep to attract nineteenth-century water-powered industries, and between 1811 and 1830, Isaac Miles, Roderick Perry, and Randolph Darnell built gristmills and sawmills along the creek. Small distilleries also operated among the mills, and these industries attracted laborers who erected houses, churches, and stores, creating a mile-long string town named Millville.[115] This lower section of Glenns Creek was known to some as Crow Valley.

By 1861, in addition to distilleries, Millville included Johnson's Grist Mill, the Gorbut & Whittington Sawmill, Darnell's Mills, a woolen factory, a blacksmith shop, and several residences.[116] Small farms along the valley's wider sections produced crops and livestock.

The valley's most successful distilleries included Oscar Pepper (1812), John Shields (1844), Old Taylor (1850s), and Old Crow (1860). Each works was sold, purchased, upgraded, and expanded several times during the nineteenth century.

Many Millville residents worked for distillers during the production season.[117] The distilleries operated much like other works across Kentucky—milling and distilling grain, feeding the spent grain to livestock, and warehousing spirits for aging. Given the steep topography and narrow valleys near Frankfort, the logistics of delivering grain and coal and shipping aged whiskey to customers proved a substantial problem for Glenns Creek distilleries. In 1882 turnpikers completed a road from the Versailles Turnpike and L & N Railroad station at Jett to Millville.[118] When surveyors plotted a pathway for the Frankfort-to-Versailles interurban electric rail line in 1906, investors gave some consideration to routing the line through Millville. Instead, the surveyors recommended a direct upland route.[119] The interurban traction line became a favored shipper for Frankfort distilleries, but the valley distilleries at Millville remained tethered to their turnpike. In 1908 the Kentucky Highland Railroad completed a line from Frankfort to the Old Crow Distillery on Glenns Creek, with service extended to Old Taylor and Millville the following year. The line served the distilleries and permitted employees living in Frankfort to commute to work via the railroad.[120]

Unlike those in Athertonville, New Hope, and Tyrone, Glenns Creek distillers did not take an active role in Millville's development, other than constructing a road and a rail line to improve accessibility. The town's first settlers included gristmill and sawmill operators as well as distillers and farmers. After the Civil War, when valley distilleries expanded into large-capacity industrial works, the number of seasonal jobs available to Millville residents increased, and Frankfort residents were attracted to employment there as well. Millville was not a company town, yet, as its name suggests, the valley's mills and distilleries provided seasonal employment and, decades after initial settlement, transport infrastructure.

Across Kentucky, some towns and villages developed organically in proximity to mills and distilleries, such as at Ruddles Mills in Bourbon County. Others grew up near distilleries with associated railroad stations, including Lair in Harrison County and Deatsville and Gethsemane in Nelson County, where teamster and railroad transport jobs supplemented the employment offered by distilleries. In Frankfort and Louisville, where distilleries and related industries operated on the edges of the urban area, real estate brokers and developers built housing for workers, creating neighborhoods that resembled subdivisions. In north Frankfort, several employees of the Taylor and Carlisle Distilleries lived in Thorn Hill, an edge-of-town housing development within walking distance of the distilling works.[121] In Louisville, distillery laborers clustered in the Valentine Schneikert subdivision of Irish Hill, in Butchertown neighborhoods near the Beargrass Creek distilleries, and in Parkland and Portland on the west side, close to the distilleries that operated near the main rail lines and the river landing.[122]

Management and Residence: The Houses that Bourbon Built

A residence, whether built according to vernacular folk tradition or architect-designed in high style, can be thought of as serving two purposes: function, through design, construction, and use; and communication, a purposeful or incidental private-public representation to which people assign meanings and attach memories as elements of their social identity.[123] Houses are functional if they are useful and utilitarian; in addition to providing shelter from the elements, they safeguard residents and enhance the processes of domestic life. Houses communicate by presenting an image to inhabitants, neighbors, and passersby. A house's exterior appearance communicates its inhabitants' values through building materials, paint color, and construction details. Nineteenth-century American landscape architect Andrew Jackson Downing, author of the first treatise on landscape design and advocate for public appreciation of the beauty and sublime qualities of nature, believed that buildings, architecture, and landscape convey meaning to residents and observers. He wrote, "Much of the character of every man may be read in his house. If he has moulded its leading features from the foundation, it will give a clue to a large part of his character. If he has only taken it from other hands, it will, in its internal details and use, show, at a glance, something of the daily thoughts and the life of the family that inhabits it."[124]

Victor Hugo, John Ruskin, and G. R. Lewis, among others, believed that formal nineteenth-century architecture embodied meanings and communicated them with precision through codes incorporated into the form of a building.[125] We can readily accept that a building's appearance might imply social standing and economic status; it can suggest qualities of an owner's character or exemplify an architect's creativity. But the people who interacted with buildings did so based on their varied and fragmented experiences and preconceptions; thus their perception and understanding of structures and their contextual landscapes also varied.[126] A home's meaning, therefore, is fluid and qualified by time, place, and social and psychological associations. A nineteenth-century vernacular house may have been well maintained or ramshackle. Passersby who saw only the exterior may have ignored the building or made assumptions about the character of its occupants. An observer concerned with "progress" may have interpreted the building as stagnant and regressive. The residents, in contrast, may have thought that concern about whether the design and condition of their house represented progress or anything else was abstruse. Yet they self-consciously personalized the interior in a way that reflected their concerns and standards and also served as the fount of memories related to living there. Houses that departed from ordinary or vernacular forms toward high style may have suggested to observers that the owner-builder had in mind some purpose other than basic function, perhaps as an overt communication that the owner had achieved some measure of taste, wealth, and social standing. The historical and biographical dimensions of interpreting landscape artifacts from the past and present, such as houses and industrial buildings, can be important and enriching if this activity provides insight into ourselves, others, and the world around us. But landscape antiquities stand as remnants of the past, unaccompanied by their original context, and present-day interpretations of historical landscape artifacts can be precarious unless the people who made and used those artifacts left some record of how they understood and interacted with them.

Nineteenth-century industrialization led to fundamental changes in the way Americans obtained and processed raw materials, made goods, and marketed and sold products. As industry supplanted craftwork, the new factories attracted job seekers, and cities grew up around industrial centers. Factory owners, merchants, and professional people accumulated wealth at unprecedented levels, prompting a growing association among social standing, city morphology, and residential representation.[127] In preindustrial cities, merchants and manufacturers lived in or near the center of economic activity, but as transportation improved, especially with the development of steam railroads and electric trolleys, work site and place of residence could be separated, sometimes by a significant distance. Cincinnati's nineteenth-century industries, for example, clustered in Over-the-Rhine and the Mill Creek Valley. Factory owners lived near their businesses amidst the dirt, smoke, noise, and congestion. Christian Moerlein's financially successful brewery operated on Elm Street in Over-the-Rhine, and his son Jacob built a roomy French Second Empire–style house across the street from the brewery works, in the middle of the industrial cacophony. After the Cincinnati & Clifton Incline Railway opened the surrounding bluffs to development in 1876, Christian Moerlein built a large brick house atop the bluffs, joining other businesspeople who had erected elegant showplace homes away from the city's rough-and-tumble industrial districts.

Prior to the Civil War, most Kentucky distillers lived in ordinary farmhouses and town homes. When necessary, they enlarged their homes by attaching rooms rather than razing the structures and building new ones.[128] As distillers formalized and industrialized their businesses after the war, they also engaged middle- and upper-class sensibilities, which included cultivating an appreciation for style and quality. Valuing quality, whatever its provenance, was akin to what Russell Lynes called "taste." And, he observed, "millions of Americans discovered 'taste' at the Centennial Exposition in Philadelphia in 1876."[129] Interestingly, the Philadelphia Exposition was the same venue where Anderson County's William H. McBrayer reportedly won an award for his fine whiskey. Some distillers who were also farmers demonstrated an appreciation for quality through their ownership of pedigreed livestock and fine equipment. Imported shorthorns, color-matched mules, race-winning Thoroughbreds, and weight-gaining Berkshires were desirable additions to a farm's menagerie. Farmers organized livestock shows and fairs to judge and reward the best animals. They also valued quality equipment, including Concord coaches manufactured by Lewis Downing in New Hampshire, carriages made by George C. Miller's Cincinnati works, and wagons produced in the South Bend, Indiana, Studebaker factory. As distillers, they demonstrated their appreciation of taste by outfitting their works with the newest technology and touting their products as being "of the best quality."

The types of distillers' houses discussed below are just small selection, yet they serve as exemplars of the house construction process. They also suggest possible relationships between home owners' presumed work ethic or industriousness and business success.

Single Structures

Rural Sites

During the late eighteenth and early nineteenth centuries, farmer-distillers built ordinary houses on their farm property, often made of clapboard or brick and conforming to the unadorned Federal style of the day. Two-story houses were commonly two rooms wide and one room deep and served by a central hall—a configuration that is often termed an I-house—with an ell addition on the back. Larger homes followed a similar form but were two rooms deep or square in plan. Some successful Kentucky distillers built large houses that were not exceptional in scale or design and were similar to those built by their contemporaries whose financial wherewithal was based on farming, business, or a professional vocation in law or medicine.

The early houses built in Kentucky by wealthy Virginia land-grant holders were often part of large estates. Farmer-distiller Jacob Spears of Bourbon County, whose migration to Kentucky in the 1780s was outlined earlier in this chapter, built a house that serves as an example. Sometime around 1810, Spears engaged expert stonemason (and Kentucky governor from 1828 to 1833) Thomas "Old Stonehammer" Metcalfe to build a two-story, five-bay Federal-style limestone house adjacent to Clay-Kiser Road north of Paris.[130] The symmetrical main building still stands on a low rise above the site where Spears built his distillery; it has a central hallway and two interior gable-end chimneys. The square-cut ashlar masonry represents exceptional craftwork. A one-and-a-half-story stone kitchen ell extends from the back wall, and a small, single-story addition was attached to the right wall. Sometime during the nineteenth century, carpenters added a two-story Greek Revival portico and eave brackets to the front façade.[131] Jacob Spears died in 1825, and his property passed to other owners, but his stone house was—and continues to be—a Bourbon County showplace.

At Clermont, in eastern Bullitt County, the T. Jeremiah Beam house stands on a low, knobby hill on the site of the old Murphy-Barber Distillery on Big Level Road, near the Bardstown branch of the L & N Railroad. The house overlooks the Long Lick Creek valley and the modern and expansive Jim Beam Distillery. Built in 1911 as a boardinghouse for coopers, gaugers, and visiting salespeople, the building became a farmhouse and Jeremi-

Jacob Spears house, Bourbon County.

ah Beam's residence when the Murphy-Barber operation closed in 1917. The country residence is a T-plan, two-story frame house sheathed in clapboard. The primary south-facing façade has three bays on the ground floor and two bays on the second floor. An unadorned front door opens into the structure's right-side wall. A two-story cross-gable section is attached to the rear. A large wrap-around porch extends across the front of the house and along each side.[132] The Beam house is an ordinary frame structure, similar to those found on farms and in small towns across the region, and it is a reminder that distillers lived in common dwellings as well as outsized mansions.

Urban Sites

In 1883 distiller, farmer, horseman, entrepreneur, and politician Thomas Jefferson Megibben built his home, called "Monticello," in Cynthiana, the seat of Harrison County. The ninety-acre hilltop property came to be known as Monticello Heights. Megibben worked at, owned, or had a financial interest in six Bluegrass distilleries from about 1849 until his death in 1890. Two of his distilleries, Excelsior and Edgewater, operated about four miles south of Cynthiana near Lair's Station. The Samuel Hannaford & Sons architectural firm of Cincinnati, one of the nation's leading design firms at the time, designed Megibben's home. The $300,000 structure was likely built by the L. P. Hazen Company of New York and Cincinnati.

The large structure included a hipped roof and ell, multiple Victorian gables, three extended bays across the front façade, and a wrap-around porch built of turned posts and spindles and a wrought-iron balustrade. A Palladian arched door opened to a second-floor balcony. Curved Baroque volutes, extended brick pilasters, and Grecian urns decorated the tallest gable. The roof featured an eyebrow vent and tall, corbeled chimneys. The exterior brick was imported from England, as were the woods used in the lavish and elaborate interior. An artist finished the ceilings in frescoes featuring still-life paintings, trompe l'oeil cartouches, and stenciled panels. An entry in an 1882 county history lauded Megibben's business accomplishments, noting, "Kentucky does not present a more striking exemplification of the old maxim, 'Industry brings its own reward,' than in the life of this gentleman who, by his own efforts, became the most prominent farmer, distiller, thoroughbred stock-raiser, etc., in Harrison County."[133] Megibben's reputation as a successful businessman resided not only in his persona but also in his house. Although the home earned entry on the National Register of Historic Places in 1974, it has since been razed. All that remains is the carriage house, which has been refurbished and is now used as a church.[134]

Other urban distillers' houses include the Berry Hill Mansion at 700 Louisville Road in Frankfort, built in about 1900 for George Franklin Berry, a distillery executive for W. A. Gaines Company; the Henry D. Pogue II house (also known as the Michael Ryan house), built in about 1845 at 716 West Second Street in Maysville, overlooking the Ohio River; the John Atherton house on Cherokee Parkway in Louisville's Cherokee Triangle, an early trolley-car suburb; and the Arthur P. Stitzel House at 9707 Shelbyville Road in Louisville.

Multiple Structures: Distillers' Row

At least three Kentucky towns—Owensboro, Bardstown, and Lawrenceburg—had residential streets that boasted clusters of large distillers' houses. Distilling in Owensboro and Daviess County began in the early 1800s. By the 1880s, eighteen distilleries operated there—some in Owensboro proper, and others along the Ohio River up- and downstream from the city. Of French heritage, the Monarch family moved from Maryland to Owensboro in the 1830s and was prominent among Daviess County distillers. Thomas J. Monarch Sr. was born in Owensboro in about 1836. He and his wife Eliza Jane had a daughter, Miranda, and sons William, Daniel, Henry, Thomas, Richard, Martin, and Sylvester. Several Monarch sons became distillers. Peter E. Payne, a family member by marriage, was a distilling company partner. Payne and Monarch brothers Richard, Martin, and Sylvester built large Queen Anne mansions on Highland Avenue in the 1880s and 1890s. The houses built by Richard and Martin Monarch were razed, but the homes of Sylvester Monarch (now known as the Le Vega Clements house) and Peter Payne (the Monarch-Payne house) remain.

Sylvester Monarch owned the Eagle Distillery and a large stockyard. He purchased open land atop the bluff overlooking the Ohio River, where builders erected his brick-and-stone Queen Anne–style home between 1894 and 1897. The site prompted Monarch to name the house "Highlands." Its narrow front façade has an inset three-story hexagonal tower with a double-capped slate roof. A central, gabled two-story section extends the depth of the house. The front door is topped by a Greek Revival pedimented porch and flanked by a one-story turret. Ionic-capped columns support the porch and turret. The long left-side wall has a two-story inset round-sided tower. A large two-story bay window extends from the right-side wall. The window openings are supported by segmented brick arches and covered by stone hoods. The house's central block has a hipped roof covered in patterned slate. The interior is finished in elaborate woodwork, with stone mantels and stained-glass windows.[135]

After Prohibition, the house was sold to attorney Le Vega Clements, who founded the Kentucky Buggy Company and served as mayor of Owensboro. The large lot's unbuilt acreage was subsequently subdivided into small residential lots. The house overlooks the present-day Glenmore Distillery site.

The Peter Payne house stands a short distance west of Sylvester Monarch's home. Also in the Queen Anne style—but with a much simpler façade—this two-story brick home has a projecting front gable. The windows have segmented arches—linked by a stone belt course—and a third-story attic window in the gable has a Roman arch. A one-story porch with a turned-spindle balustrade wraps around the left half of the front façade. The interior features thirteen-foot ceilings, inside shutters, parquet and quarter-sawn oak floors, elaborately carved and turned cherry woodwork, carved stone and encaustic tile mantels, beveled-glass doors, and stained-glass windows. Each house on Owensboro's "Distillers' Row" represents period style, artisanal craftwork, exotic building materials not obtainable locally, and significant financial wherewithal derived from the distilling business.

Anderson County was one of Kentucky's leading whiskey producers during the second half of the nineteenth century. Several distillers built large homes in Lawrenceburg, the county seat, William B. Saffell, John Dowling, and Thomas B. Ripy among them. Saffell owned 500 acres of farmland at the edge of town, where today a street and a school are named for him. He may have initiated real estate development on Lawrenceburg's South Main Street when he sold land to distillers and others who wished to build houses fronting the thoroughfare. In 1880 Saffell was superintendent of the Cedar Brook Distillery near Tyrone and served as vice president of the Lawrenceburg National Bank. In 1889 he built the W. B. Saffell Distillery along the Southern Railroad tracks north of Lawrenceburg, near Alton.[136] Saffell's two-story brick Queen Anne–style house on South Main Street was constructed circa 1900. The 7,000-square-foot structure had a circular turret at the right front corner. Capable carpenters and other artisans finished the interior with cherry and walnut woodwork, including full-length window shutters and hand-carved fireplace mantels. Gas lighting illuminated crystal chandeliers. A wrap-around porch covered the paneled front door, which was inset with a large beveled-glass panel and surrounded by circular stained-glass sidelights.

John Dowling migrated from Ireland during the Great Famine of the 1840s. In the early 1870s he opened a cooper shop in Lawrenceburg and, with partner Thomas B. Ripy, operated a distillery east of Lawrenceburg at Tyrone. In 1886 Dowling completed construction of a two-and-a-half-story brick Italianate-style home at 321 South Main Street. The front door is set with beveled glass and opens into the base of a prominent three-story pavilion that is centered on and extends out from the front façade. Brick quoins accentuate the pavilion's front corners; a metal mansard roof caps the entire assembly. Dentils and large paired brackets hang below the roof cornice, and paneled and corbeled chimneys extend well above the roofline. The interior includes marbleized fireplace mantels, doorways with stained-glass sidelights, and paneled wainscoting and doorjambs. A carriage house stands at the rear.[137]

John Dowling house, Lawrenceburg.

Across the street from the John Dowling house, Thomas B. Ripy purchased a ten-acre residential lot from farmer-distiller William Saffell for $7,500 in August 1885. The *Anderson News* announced that Ripy intended to build a residence on the lot and opined, "there will be a building there that would be an ornament to any city," and "you may be sure that [the house] will be done in a style proportionate with the cost of the grounds" and "will be something worth going to see."[138] In September of that year, Ripy bought an adjoining ninety-acre tract from distiller W. H. McBrayer for $125 per acre. A large force of workmen began laying the foundation during the spring of 1886, and in April the *Anderson News* triumphantly pronounced, "Lawrenceburg will soon have more palatial residences within its borders than any town of its size in the state."[139] Artisans completed Ripy's outsized Queen Anne–style home in 1888—it had thirty rooms and approximately 11,000 square feet of interior space.[140]

Thomas Ripy house, Lawrenceburg.

The façade of the Ripy house is finished in brown brick, with a limestone water table and stone belt courses above and below the windows on each floor. A projecting gabled porch supported with square, fluted wooden pillars covers the front door. The porch connects to a veranda that wraps around the right (north) wall. The polychrome slate roof has a small street-facing gabled dormer and two tall brick-and-stone gables. At the left (south) corner, a three-story octagonal tower extends above the roofline. The tower's mansard roof has five small gabled dormers, capped by a small enclosed "widow's watch." The interior stairs, doors, and wainscoting are mahogany and walnut. Several doorways have arched transoms with stained glass. The fireplaces are finished with marbleized cast-iron mantels stenciled in gold leaf. Elaborate moldings crown the fourteen-foot ceilings, and chandeliers illuminate the largest rooms. In 2018 the Ripy house was undergoing restoration.

The Country Estate

Louisville became a major industrial distilling center after the Civil War. Distillers, whiskey rectifiers, tanners, wholesalers, and others engaged in related businesses built large, expensive residences in St. James Court, Old Louisville, and other exclusive neighborhoods. The general improvement of streets and roads, together with the development of a trolley system and interurban railroads, allowed wealthy businesspeople to build summer homes and country estates on Ohio River bottomland and the bluffs northeast of the city. Farmers in the area had cleared land and grown hemp and other crops from the 1790s through the 1830s. A small milling and distilling industrial center developed at Ashbourne, utilizing waterpower provided by the steep-gradient Harrods Creek. In the 1890s three businessmen, including George Garvin Brown, founder of Brown-Forman Distillers, developed a communal summer colony named "Nitta Yuma" (high ground) that became a large family estate on Harrods Creek. Developers subsequently purchased adjacent farms on three miles of hills and bluffs along the river on River Road and Wolf Pen Branch Road. The land was subdivided into large lots of nine acres or more for development as country estates, which included large residences hidden away on heavily wooded grounds accessed by winding drives. These estates also had worker and guest cottages, carriage houses or garages with chauffeurs' apartments, formal gardens and greenhouses, gazebos, tennis courts, and swimming pools. From 1900 to the 1930s, architects designed and built large residences in Colonial, Georgian, Tudor Revival, Craftsman, and eclectic styles. William E. Chess, president of Chess and Wymond Cooperage Company, began work on his house, "Winkworth," in 1906. Louis Wymond, another member of Chess and Wymond, completed his house 1912. In 1911 Owsley Brown, son of George Garvin Brown, built a 20,000-square-foot house called "Avish" on a twenty-four-acre lot. Landscape architects Frederick Law Olmsted Sr. and Bryant Fleming designed the grounds for these properties and several others. Even though the Louisville country estates along the Ohio River were removed from the main thoroughfares and largely out of public view, the larger community knew of their presence, and their seclusion transmitted the cachet of exclusivity.[141]

Residence and Representation

The social context within which distillers operated is important to an understanding of their values and motivations. An active farmer-miller-distiller was usually appreciated as a community asset—though not by temperance movement activists. Distillers offered employment; consumed locally produced grains; purchased tools, equipment, and cooperage from local sources; and supplied their product to local customers, storekeepers, and saloons. Residents often identified positively with distillers who built large, high-style homes, much as Kentuckians might identify a Kentucky Derby winner as "our horse," even though the animal belonged to someone they had never met and had been bred on a farm they never visited.

Andrew Jackson Downing's advocacy for a considered and informed appreciation of the physical environment and the organic association of landscape structures and personal character preceded by two decades America's post–Civil War economic boom. This development would be largely unsympathetic to his concerns but would not alter the relationship he saw between structures and character. From the 1870s to the 1900s, the nation transitioned from an agrarian to an urban-oriented society through advances in technology and the development of industries that built new machines and transport systems. Business and financial institutions provided strategy and working capital. Working-class wages increased, albeit modestly. Businesspeople amassed wealth; some of them accumulated great fortunes. Geographically favored urban centers experienced rapid economic growth. Small towns and cities sought to participate by fostering development of their own industrial bases or attracting businesses from elsewhere through self-conscious and enthusiastic boosterism.

Community promoters influenced and directed public perception by arguing that business success was the product of intelligence, energy, hard work, and other estimable personal qualities. Building a large, expensive house provided tangible evidence that one possessed those qualities. Though members of the public might not be privy to one's annual income, bank balance, or investment portfolio, those measures of success could be inferred by assessing the house one built. Iconic houses also suggested that their owners possessed admirable attributes other than wealth. A perceptive public conferred on these owners positive character traits such as refinement, gentility, sophistication, solidity, and an appreciation of beauty, style, and fashion. Surely the visual evidence of financial success could not be attributed to restlessness, instability, and sloth. Distillers' mansions could be seen as monuments that prompted the observer to make associations—whether accurate or far-fetched—among a building's appearance, the owner's financial wherewithal, and his or her political power and social influence.

Newspapers and professional organizations of the post–Civil War period developed a language of self-promotion that often touted commercial accomplishments. In Lawrenceburg, the *Anderson News* lauded effective businesspeople and identified the characteristics that led to their success and reflected positively on the community. Consider these descriptions from a special supplement published in 1906: "The officers [of the Lawrenceburg Commercial Club] are active and earnest in their boosting of the old town." They are "the best and most prosperous among us." "Lawrenceburg, though small in population, is yet making metropolitan strides in industries, and possesses the necessary brains and energy required to bring a town such benefits." The town's businesspeople "have worked untiringly and indefatigably for the success of these enterprises and the results have been a handsome compliment to their perseverance and industry." Local businesspeople are "true and loyal citizens," and every opportunity is "eagerly grasped by our hustling and enterprising population." Distiller W. H. McBrayer was described as "a fine person" with a "genial and social disposition, and possessed of cultured intelligence, he ranked among the solid men of influence in Kentucky." Others possessed "boundless zeal," "energetic business vim," "pluck, hustle, energy and determination," and an "indomitable will and strong body and constitution." They were "linked with the spirit of aggressive and progressive business," demonstrating "honesty, integrity, and faithfulness"; they were "whole-souled and energetic—splendid types of the businessmen of the New South." And all the better if they "live[d] in a picturesque home." The paper thus lauded in florid language the builders of large houses and posited a causative linkage among character, business success, wealth, and home. The T. B. Ripy house, for example, was a "handsome home, the noblest monument he could have left to his splendid family of children." "He started as a poor boy, and by his industry and thrift has scored."[142]

The *Anderson News* editor's prideful pronouncements about Lawrenceburg's Ripy house suggest that while the building was the product of individual achievement, it reflected positively on the larger community, and both the structure and its owner were worthy of admiration. Lawrenceburg's Distillers' Row houses were not to be envied; they should be considered a source of community identity and pride—a kind of collective achievement that allowed residents to participate vicariously in the business success and accomplishments of the town's distillers.

Writing in 1854, art and architecture critic Clarence C. Cook posited that the large Manhattan mansions built by New York City entrepreneurs were unjustly disparaged. "Luxury is a vice," he noted, "only when it is extravagance in an individual: the private vices of ostentation and extravagance become public benefits to trade and industry." Further, "viewed in a public light, every external indication of prosperity tends to add attractions to a city, and to promote its increase and influence in more important objects." Cook believed that such mansions beautified a city's streets, and to walk among them was akin to visiting an art gallery.[143]

It is unlikely that most laypeople in the nineteenth century could identify the intricacies of Queen Anne– or Italianate-style architecture. Descriptions that employed formal architectural terminology—Romanesque arches, Gothic windows, Second Empire mansard roofs—likely meant little to them, although they might surmise that a new house was either "in style" or "old-fashioned." Extrapolating measures of status and wealth from a house or other structure was most readily accomplished by reference to surrogate values such as total cost; dimensions, such as square footage and the number of floors and rooms; and the nature of the building materials—the stone was imported from . . . , the stained-glass windows were crafted by . . . , or the copper bathtubs were fashioned by. . . . Such associations placed the Kentucky public in a position to interpret Distillers' Row houses as boosters wanted them to—as monuments representative of the personal qualities that yielded financial success, as akin to Cook's art gallery—or to reject them as icons of privilege and excess.[144]

External Control and Landscape

On the forms you sent to us demanding taxes
We painted the most astonishing pictures for you.
Bellowing in chorus our horatory verses
The people, as always, paid the taxes you claimed were due.
—Bertolt Brecht

Conjunctions—Government Regulation, Taxes, and Surveillance

A long-accepted political convention is that federal, state, and county governments will impose laws on the production, sale, and consumption of alcohol. Some regulations might require distillery licensing and oversight of operations; others might impose strict audits of grain processed and whiskey produced and levy taxes on alcohol stored in aging warehouses. In 1791 Secretary of the Treasury Alexander Hamilton proposed, and the newly formed federal Congress approved, an excise tax on distilled spirits to retire debts incurred during the Revolution. Distillers objected, especially those operating across the western frontier from Pennsylvania south to the Carolinas, where farmers distilled surplus grains and whiskey often served as a medium of exchange.[1] Over the next two centuries, there followed a series of federal laws that regulated or, in the case of Prohibition, severely curtailed the production of distilled spirits. Nineteenth-century Kentucky statutes imposed liquor controls primarily through a valuation tax and retail licensing fees and fines. As directed by the *Kentucky Revised Statutes* of 1894, distillers were required to pay a $25 fee to obtain a license to sell their own product in "quart houses" on or adjacent to their premises—though it could not be consumed there. Those who failed to comply were subject to misdemeanor charges and a fine of $50 to $1,000.[2] In 1900 the Kentucky State Board of Valuation and Assessment fixed the state tax on whiskey stored in distillery warehouses at $7 per barrel. The following year the tax rate was raised to $10, but after protests by the Kentucky Distillers' Association, the per-barrel assessment was reduced to $8.[3] Some Kentucky counties apparently levied their own taxes on whiskey stored in aging warehouses. In the 1880s and 1890s W. A. Gaines & Company regularly moved as much as 2,000 barrels of "free" or tax-paid whiskey from the Hermitage Distillery warehouses in Frankfort, in Franklin County, to the Old Crow Distillery on Glenns Creek, a few miles south of Frankfort and over the county line in Woodford County. Gaines & Company owned both distilleries, and apparently the lower Woodford County tax rate made it cost-effective to move the whiskey back and forth rather than storing it in Frankfort, where the tax rate was higher.[4] This chapter focuses on federal statutes enacted to regulate the production and warehousing of distilled spirits and the landscape modifications made by Kentucky distillers to comply.

The increase in distilled spirits production during the nineteenth century was nurtured by wholesale and retail commercial interests. Yet, at the same time, public objections to the spirits industry became increasingly strident, and Congress sought to control the industry's output to ensure a constant flow of tax payments to bolster the national treasury. The public's concerns focused primarily on three different but related issues: taxation, temperance and prohibition, and product purity. Those distressed by such matters developed sufficient political leverage to prompt the state and national governments to impose various types of official controls on the spirits industry. Legal codification implied a normative operational standard with little room for local variation or interpretation; yet the often captious legal code was created by legislators whose political agendas might have been pro-business or pro-temperance. When viewed from a national perspective, distilling, rooted in folk tradition, was increasingly subject to intervention by state and national governments, in large part to extract taxes, but also to control product quality.

Law and Landscape

Americans are immersed in an environment of laws that govern social and professional behavior. When considering the geographic context of places and events, law can be a foundational yet often unseen element. Nevertheless, these unseen or invisible legal constraints can direct people to construct landscape elements that conform to certain configurations and allow only selective performative functions. Consider some obvious examples of the law's influence on the landscape. Traffic laws intended to ensure our safe navigation along city streets and country roads are often abstracted as signs with symbolic colors and a few words—Stop, One Way, No Passing. Property development covenants and zoning laws proscribe certain land uses but invite others; they prohibit some architectural features while requiring others. Although the visible configuration of landscape elements—for instance, the arrangement of distillery buildings—may seem directed by vernacular tradition, topographic constraints, available technology, or clever engineering, the arrangement may have been prescribed by invisible lawful edicts. Statutes can have a determinative influence on property-making and how landscape elements are built. The observer should remember that landscapes contain more information than one might gather from simply accepting the visual material evidence.[5]

Two brief examples illustrate the imprint of federal survey and agricultural commodity control laws on the American landscape. To procure revenue for the national government and stabilize land sales and claims, the Continental Congress passed the Land

Ordinance of 1785, which created the Public Land Survey System, more commonly referred to as the Township and Range Survey. The survey created an expandable grid, oriented to the cardinal directions, composed of thirty-six-square-mile townships. This system delineated a simple and replicable method of surveying land parcels, and surveyors applied it to the unsettled lands west of the Appalachians. The cadastral survey's imprint on the landscape is ubiquitous and is evident today to anyone who drives the highways of the Middle West and the Great Plains, where the farms are served by a mile-by-mile grid of roads and rural property boundaries; the entire warp-and-weft ensemble is oriented north-south and east-west. A second example of federal regulation's imprint on the landscape was a consequence of legislation passed in the 1930s during Franklin D. Roosevelt's New Deal administration. The Agricultural Adjustment Act sought to control farm crop and livestock production and thereby stabilize commodity prices. The law placed strict production and marketing controls on tobacco and other crops. Through an allotment system, the act assigned tobacco acreage to farm properties, not to individual farmers. Allotment acreage could be decreased but not increased, and new allotments could not be created. The only way a new prospective grower could raise and sell tobacco was to rent an allotment or purchase a farm with an assigned allotment. Government restrictions effectively stopped the areal expansion of tobacco production and froze production districts in place. Rural tobacco production landscapes were thereby institutionalized and continued to exhibit a high density of tobacco fields and curing barns, while auction warehouses tended to cluster in towns that were central to high-production districts. By discouraging the expansion of tobacco production away from traditional production centers, tobacco districts developed abrupt boundaries or edges.[6] In considering how people use land and construct landscape, law and governmental regulation cannot be ignored or acquitted.

Regulating Distillers: Federal Interdiction

Taxation

In July 1862 Congress passed, and President Abraham Lincoln signed into law, the Internal Revenue Act, in large part to raise money to fund the Union's Civil War effort. Twice before, in seeking pragmatic ways to fund war debts, the federal government had levied excise taxes on ardent spirits: the first was a 7-cents-per-gallon tax at the end of the Revolutionary War, and the second was a 20-cents-per-gallon levy during the War of 1812. Congress discontinued both taxes soon after the need for compensatory funds had passed.[7] The 1862 act levied a tax of 20 cents per gallon on spirits; Congress increased the tax to $1.50 in 1864 and to $2 the following year—about $33 in present-day currency. After the Civil War, in 1868, federal legislation reduced the tax to 50 cents. The tax was attached to the distilled spirit when it was produced, and the distiller was required to pay the tax before removing the barreled product from the distillery's storage warehouse. Thereafter, the tax was raised to 90 cents per gallon in 1875, and it was raised again to $1.10 to underwrite expenses incurred in the Spanish-American War (1898).

In 1862 Congress established the Office of the Commissioner of Internal Revenue within the Treasury Department. The commissioner divided the country into tax collection districts and directed and monitored tax collection on spirits. Commissioner-appointed tax assessors and collectors managed operations in each district. In 1866 federal spirits taxes were collected from nine districts in Kentucky. To reduce operating costs, the commissioner consolidated collection districts in 1883. Kentucky's Ninth District was combined with the Seventh and Eighth Districts, and the First, Third, and Fourth Districts were merged into the Second District.

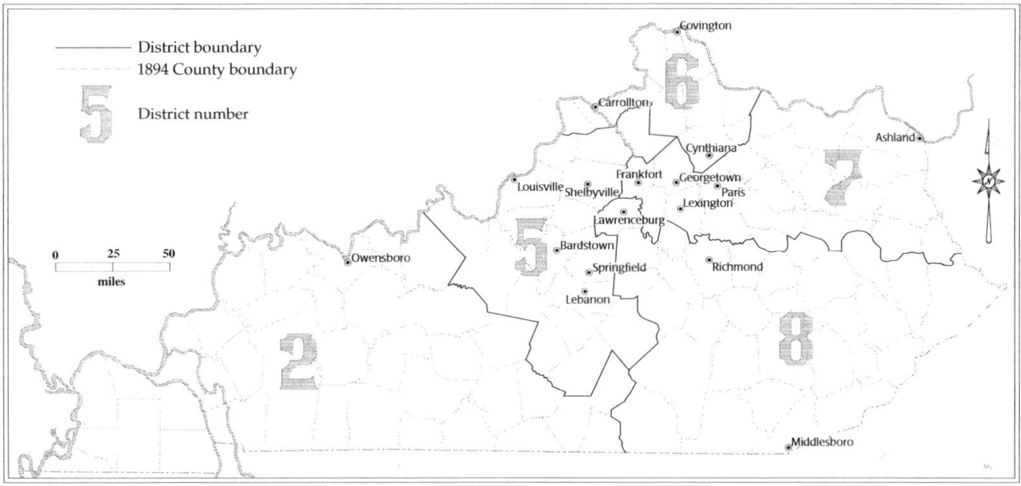

Kentucky's registered distillery tax collection districts, 1894. Federal law changed to consolidate districts, which explains why there are no districts numbered 1, 3, and 4. (Compiled from *Sanborn's Surveys of the Whiskey Warehouses of Kentucky and Tennessee* [New York: Sanborn-Perris Map Company, 1894])

The tax collection program was so effective that by the 1870s, approximately half the nation's revenue was derived from the manufacture and sale of spirits.[8] These tax regulations also represented a significant intrusion on distillery operations and financial management. Responsibility for regulating distilling operations and extracting taxes was assigned to internal revenue operatives; they were often political appointees, and their titles included collector, storekeeper, and gauger. These revenue agents controlled access to distillery property and structures, directed the distilling process, and monitored the modes of transport for raw materials and finished products. Even equipment manufacturers that built the copper stills and related machinery were identified and taxed. During the nineteenth century, distilling became one of America's most regulated industries.

The commissioner of internal revenue issued detailed specifications for monitoring the distilling process, including equipment procurement, grain distilling, distillery configuration and operation, and product preparation for sale. The Internal Revenue Act largely disregarded certain traditions, such as aging newly distilled spirits in oak barrels to improve their flavor. And although the act placed government agents in operational control of distilleries, Congress made no provisions for operational safety—personal injuries and distillery fires were commonplace—and did not address environmental degradation, especially the pollution of waterways by spills and runoff from the livestock feedlots found at most large distilleries.

Using approved instruments, federal agents determined the proof of the spirits in each cask or barrel by measuring the product's specific gravity. The agent then assessed taxes on each container accordingly and held distillers liable for the taxes imposed. Tax assessments included an automatic first lien on the spirits, distillery, equipment, tools, and tract of land on which the distillery and related buildings stood until the taxes were paid.[9] Although the revenue law was not designed to shape the distilling landscape per se, it did require federal agents to monitor and control the flow of grains and spirits throughout the distilling process. Compliance with federal directives therefore influenced how distillery buildings and warehouses were constructed, organized, and accessed. The law further required distillers to engage professional surveyors to draw detailed maps of distillery property and buildings to aid the surveillance and enforcement process.

The 1868 tax law imposed additional levies based on the number of forty-two-proof-gallon barrels a distiller produced.[10] Distillers making fewer than 100 barrels each year paid a $400 tax, equivalent to about $6,500 in present-day currency; those producing more than 100 barrels paid an additional $4 for each barrel in excess of 100.[11] In addition to taxes on manufactured spirits, federal law levied a $2-a-day tax on distilleries with the capacity to mash and ferment twenty bushels of grain per day; another $2 a day was assessed for every additional twenty-bushel processing capacity.[12]

The internal revenue commissioner often prescribed the installation of flow meters to measure spirits production. The law obligated the distiller to purchase and install these meters, including the cost of plumbing supplies and labor. The commissioner was empowered to order changes in the placement and configuration of distillery components such as stills, mash tubs, fermenting vats, whiskey cisterns, fittings, valves, and pipes.[13] Commissioner-appointed assessors surveyed each registered distillery to ensure that all spirits produced were accounted for.

Manufacturers that built stills, boilers, or other distilling-related vessels had to notify the district tax collector of their equipment sales and identify the purchaser, location, capacity, and time when the equipment was moved from the factory to the distilling site.[14] The tax code also required manufacturers to pay a special $50 tax on each still and a $20 tax on each condensing tube or worm.[15]

Federal tax law mandated that distillery aging warehouses be bonded, sealed, and secured by the appropriate government agents. The 1868 tax law required distillers storing whiskey to "give bond." That is, upon depositing whiskey in a warehouse for aging, the distiller would execute a bond to the United States, which secured the tax due at the time specified by the government. Although government storekeepers and gaugers controlled warehouse access, distillers were required to furnish warehouse locks at their own expense, and the federal storekeepers held the keys. Distillers and distillery employees were not permitted to enter a warehouse except in the company of a federal agent. Nor could distillers place their own locks on a warehouse.[16]

During the Fortieth Congress (March 1867 to March 1869), congressional committees conducted lengthy hearings concerning the Internal Revenue Act's directives, a process that was repeated with subsequent liquor control legislation in succeeding decades.[17] Much of the testimony and debate represented three distinct perspectives that were only marginally related to political party affiliation, although "dry" voters seemed particularly devoted to the Republican Party.[18] A reform group represented by Congressmen Charles H. Van Wyck (R-NY), John V. L. Pruyn (D-NY), and William D. Kelley (R-PA) argued that a reduction in the whiskey tax would prove to be an anticrime measure.[19] Many in Congress believed that the Whiskey Ring—a group of politicians, government employees, distillers, and distributors—was siphoning from the national treasury millions of dollars in unpaid liquor taxes each year. Van Wyck and others thought that reduced whiskey taxes would deprive the ring of its illicit income:

> The ring has stalked through the antechambers of this Hall, through the corridors of the Senate, and controlled the avenues to the Executive Mansion. While it directed the action it defied the power of the Government. I then believed that the only way to destroy the ring and check the immense frauds and demoralization permeating that branch of the service was to reduce the [whiskey] tax to fifty cents per gallon; to take away its wealth by depriving it of the source of its ill-gotten gains, and that the tax be collected at the worm of the still, thereby dispensing with the great means of fraud—bonded warehouses.[20]

Tax reduction as a method to reduce fraud and increase collections proved to be a discerning strategy. With the spirits tax set at $2 per gallon in 1868, collections totaled $18,655,000. The following year, with the tax set at 50 cents, $45 million was collected; in 1870 the tax yielded $55 million.[21]

A progressive pro-business group that included several representatives from distilling states, such as James B. Beck (D-KY) of Lexington and Elbon C. Ingersoll (R-IL) of Peoria, supported lower taxes and relaxed regulatory policies that would allow distillers to make and market traditional whiskey products using established methods while realizing a sustaining profit. Ingersoll argued against a provision that would require a distiller to pay tax at the distillery as soon as the whiskey was distilled or, in the case of whiskey that had already been distilled, within 100 days of its placement in a bonded warehouse.[22] Ingersoll stated:

> I do not believe that any distiller in the West can raise the necessary amount of capital with which to run a fifteen hundred barrel house which will produce six thousand proof gallons per day; at fifty cents per proof gallon tax, he will have to pay $3,000 per day. New York is a great market. He ships to New York, and he can not expect to receive returns from the sale of the whisky sooner than thirty days, if you grant that he finds the price in the market satisfactory. He will have to run his house thirty days before he gets any return. It will be seen that he has paid $90,000 tax in the thirty days. The actual cost of the production for that time is not less than $45,000 more, which makes $135,000. Now, it often happens that a distiller does not find the market price satisfactory, and is compelled to hold the whisky for a year, and in this event he would require over one million five hundred thousand dollars capital to [do] business [for] only one year. At the present time no distiller can draw on his commission merchant for advances on whisky shipped.[23]

"Why should we kill the goose that lays the golden egg?" reasoned Congressman Beck. "This interest, from which the Government expects to raise its largest revenue, ought to be fostered and encouraged, instead of being crushed out by hostile legislation."[24] Others noted that most of the whiskey in bonded warehouses was owned not by distillers but by small rectifiers, alcohol dealers, and druggists. Commission merchants held bonded whiskey as collateral on loans. If these businesses were forced to pay taxes on their holdings before they could be sold, the businesses would likely fail, and the government treasury would realize the expense of seizure and auction sales for little tax revenue. "You will break down these parties," said Ohio representative Benjamin Eggleston, "and outrage the feeling of the community where this tax is to be paid."[25]

A third, largely pro-tax group with little sympathy for the practicalities of operating a successful business supported strident policies that would ostensibly provide significant tax revenue to the government without regard for product quality, market conditions, or the distillers' financial survival. In debating the proposal that all distilled spirits in bonded warehouses should, upon passage of the act, be taken out of bond and the tax paid within 100 days, Representative James Mullins of Tennessee argued, "Everyone of us is interested in the revenue that is to accrue from this tax, and I believe that it ought to be paid promptly so the Government shall get the benefit of it in order to as rapidly as possible extinguish our great national debt."[26] Mullins did not stand for reelection in 1869, and President U. S. Grant appointed him internal revenue collector of the Fourth District of south-central Tennessee, the district Mullins had represented in Congress.[27] The pro-tax stance of this group was likely strengthened by sympathies for the temperance movement, whose ultimate goal was to curtail, if not stop, the production and consumption of alcoholic beverages. If this view prevailed, it would certainly ruin distillers financially, but paradoxically, it would also eliminate the industry as a source of revenue.[28] In the end, Congress approved a pro-distiller amendment to extend the whiskey tax collection time to six months after storage in a bonded warehouse. The bill was passed by the House of Representatives on June 26, 1868, and was forwarded to the Senate the following day.[29]

As the Senate considered various sections of the bill, Senator Charles Buckalew (D-PA) argued that the six-month interval between placement of whiskey in a bonded warehouse and tax payment was much too short and should be left indefinite. "Manufacturers in Pennsylvania and Kentucky of what are called fine whiskies, men of capital and character also, who desire to hold in store for a number of years the very fine article which they manufacture . . . , should be permitted to do so." If the time limit was intended to discourage fraud, he argued, fraud was more likely among "the hasty operators in the cities and at other points who desire to manufacture in haste, and dispose of the spirits in haste also, and realize returns; adventurers in the trade; men without solid capital; men who intend to conduct their business upon the absence of honest principles."[30] Senator John Sherman (R-OH) clarified that the bill did not relate to spirits being placed in distillery warehouses but applied to stock on hand that was already held in bonded warehouses. "We are about to change from one system to another," he said. "By this bill we abolish what is called the bonded warehouse system; but we have now in bond twenty-five million gallons of whisky. Hereafter there will be no bonded warehouses in which whisky can be stored. It must be stored hereafter in a distillery warehouse; it can not be taken from that until the tax is paid."[31] Sherman was the chair of the Committee on Finance and an important voice in the government's post–Civil War efforts to revise the internal revenue system while reinvigorating the economy. The Revenue Commission, established in 1865, recommended reducing taxes on the distilling industry while diversifying the tax base by taxing other commodities such as gas, illuminating oils, tobacco, and banking institutions, measures that Sherman supported.[32]

In March 1878, with the support of the commissioner of internal revenue, Congress passed Joint Resolution No. 16, increasing the bond period for whiskey in distillery warehouses from one to three years.[33] The extension was prompted in large part by the national financial crisis that began in 1873 and led to a long-term business depression. Whiskey sales had dramatically slowed, and upward of 5 million gallons of bonded whiskey would have been seized by the government for nonpayment of duties and sold at a significant loss. In April 1880 the Forty-Sixth Congress considered H.R. 4812, which would further amend the internal revenue laws related to distilled spirits by including an allowance for barrel leakage and evaporation, often termed "outage," during storage and in transit. The provision became known as the Carlisle Allowance.[34] By taxing the volume of spirits sold rather than the volume of spirits distilled, the allowance placed whiskey in the same category as beer, tobacco, medicines, and all other products subject to internal revenue taxes on the quantity sold.

While the immediate impetus behind the 1878 bond-period extension was to save the distilling industry from financial collapse and protect the treasury's most important source of revenue, the extension also permitted distillers to store their products for longer periods, which not only improved the whiskey but also potentially reduced market volatility. Likewise, an unintended but important effect of the Carlisle Allowance was to reduce the "spirits aging penalty," thereby encouraging distillers to age their whiskey for longer periods. When considered together, these two developments had important potential effects: they would assuage nervous creditors, reduce production costs, and, ominously, increase production. Heartened by this favorable legislation, some distillers reopened distilleries they had shuttered, and those who had curbed production resumed their output. Some distillers enlarged their works and constructed additional aging warehouses with increased capacity. Collectively, these changes enhanced the industrialization of the industry and added to the coffers of the federal treasury.[35]

Nationally, spirits production in 1880 increased by 18,462,649 gallons over the previous year. Bourbon accounted for 6,827,067 gallons of the increase. Rye whiskey producers increased their output by an additional 2,340,943 gallons. Together, bourbon and rye whiskeys accounted for 49.7 percent of the total increase in spirits production. Tax revenue paid by distillers increased by $8,615,224. The commissioner of internal revenue explained that the dramatic rise in spirits production was partly the result of an improving economy, but he also credited the legislation that had extended the bond period, established a leakage allowance, and eliminated interest payments on unpaid taxes while whiskey aged in bonded distillery warehouses.[36]

In 1880 warehouses across the nation stored 31,363,869 gallons of distilled spirits, 87 percent of which, or 27,311,138 gallons, was bourbon and rye whiskey. Two years earlier, in 1878, the total stored spirits had amounted to less than half this amount, or 14,088,773 gallons.[37] The dramatic increase in whiskey production in the early 1880s led to an oversupply and a market breakdown. Kentucky distillers produced in excess of 6 million gallons in 1883; by 1888, production had increased to 18 million gallons. The accumulation of unsold whiskey was so great that many Kentucky distillers agreed to a "lock-up," or a cessation of distilling, from July 1, 1887, to July 1, 1888. Most small distillers, though, did not support the limitation agreement, and some of the largest urban distillers refused to abide by the restrictions as well, Mellwood Distillery and J. B. Wathen & Brother Distillery among them.[38] Distiller John Atherton estimated that some 150 small Kentucky distilleries operated in conjunction with their owners' farms, and on an average day, they each consumed sixty bushels of grain or less and produced no more than three barrels of whiskey. Atherton also estimated that in 1888 there were some 33 million gallons of "Kentucky whiskey" in storage warehouses on which taxes had not been paid.[39] In addition, distillers had shipped some 4 million gallons to foreign warehouses, especially in Bermuda; Liverpool, England; and Bremen and Hamburg, Germany. Distillers exported this whiskey because they wanted to age it longer than the three-year bond period then in effect, and they could avoid paying taxes on exported whiskey unless it was shipped back to the United States, at which time taxes would be payable.[40]

In 1894 a Progressive Era Congress again debated the national alcohol taxation policy. The House of Representatives recommended extending the bond period from three to eight years, increasing the tax from 90 cents to $1 per gallon, and extending the allowance for barrel leakage and evaporation from three to six years. The secretary of the treasury and the commissioner of internal revenue favored the proposal and recommended its passage.[41] The Senate amended the per-gallon tax to $1.10 and changed the outage allowance to four years. The bill became law in 1894.[42] According to some sources, between 1868 and 1913, taxes on distilled spirits accounted for nearly 90 percent of all US internal revenue taxes collected. Other sources place this value at closer to 50 percent.[43] In 1888, for example, the total US tax on distilled spirits was $69,306,166.41. The following year, 1889, the taxes paid on spirits increased to $74,312,206.33.[44]

Surveillance

Federal internal revenue law was predicated on the government's desire to extract all possible tax moneys from the distilling industry, and on the presumption that all distillers would attempt to subvert tax collection by some type of duplicitous behavior. Distillers wishing to conceal their secrets of whiskey production from competitors found that federal law enforcers could now expose their proprietary ingredients lists and procedures to a curious public. To ensure compliance and maximize tax revenue, the law required the commissioner to assign federal internal revenue representatives, known as storekeepers and gaugers, to all producing distilleries to monitor and direct distillery operations.[45] Spirits in barrels of ten gallons or more could be removed from the distillery premises only between sunrise and sunset and only in the presence of an internal revenue representative.[46] The storekeeper controlled the movement of all materials into and out of the aging warehouses and had primary responsibility for keeping detailed daily account books listing the amount, type, and quantity of all materials brought onto the distillery property, including fuel; from whom it had been purchased; and the time of delivery. Storekeeper accounts also listed employees and the timing and output of the mashing, fermenting, and distilling process.[47] Distillers were required to provide office space for the shopkeepers in the distillery building proper or in a separate building that allowed oversight of the grounds. Even now, perhaps as a testimony to the days of unrelenting surveillance, a 1930s storekeeper's office, sided in sheet iron, stands at the Willett Distillery south of Bardstown, although it has been repositioned.

Storekeeper's office at the Willett Distillery, Nelson County.

Some legislators thought the level of federal oversight was oppressive and that the high personnel costs required to enforce the law would eat up much of the anticipated revenue before it reached the treasury. Pennsylvania representative George F. Miller stated, "One great drawback to the revenue [collection] is the numerous useless officers employed, which absorbs a large amount of the revenue, and this abuse ought to be corrected at once." Miller's argument did not convince the law's supporters.[48]

Distillers paid taxes on the volume of alcohol as measured by 100 proof gallons, or 50 percent alcohol, not the total volume of alcohol-bearing liquid a distillery produced, which might have been higher or lower than 50 percent. Gaugers were responsible for measuring the proof of each barrel of spirits before assessing the tax. By taking these actions prescribed by law, the gauger transformed anonymous bulk liquids into individual salient entities by marking each barrel with the distiller's name, date of warehouse placement, proof and quantity, and serial number, a process that gave the previously nameless a unique identity.[49] When distillers paid the taxes due, gaugers carved or branded each barrel head with the date of payment, proof gallons, place and person to whom the barrel was to be delivered, and number of the tax-paid federal stamp. The person emptying the barrel was responsible for canceling the stamp.[50] Creative distillers did not ignore the barrel-branding process or the transformation of spirits from anonymity to individuality. Rather, they carried it forward into the twentieth and twenty-first centuries by identifying certain barrels with superior flavor characteristics and decanting the contents into numbered bottles as "small-batch" and "single-barrel" bourbons for sale at premium prices.[51]

Distillery Registration and Compliance

Legally, distilling could take place only at a registered distillery and only within a designated distillery building, not in any other buildings on the property. Only an authorized distiller could operate the distillery and carry out distilling operations. Registration was an information-intensive exercise. A distiller had to disclose in writing the type and size of the still, the number and type of boilers, the number and size of mash tubs and fermenting vats, and the number and size of whiskey-receiving cisterns. Registration also required a description of the property on which the distillery was situated, including the number and size of auxiliary buildings and the types of materials used in their construction. Distillers were obligated to disclose the number of hours each mash tub was used in fermentation and the estimated quantity of spirits the distillery could produce in a twenty-four-hour period. Fines for noncompliance ranged from $500 to $5,000—about $8,100 to $82,000 in present value—and violators were subject to imprisonment for six months to two years. Distillery laborers were often paid $1 per day or less, and owner-operators paid their distillers $300 to $400 per year in the 1860s and 1870s, which meant that the *minimum* noncompliance fine was more than one year's income.[52] Some violations required seizure of the distillery facilities by government agents. Seized properties could not be reclaimed by the distiller until an official government judgment had been rendered. Regulations provided an exception for those distilleries feeding more than fifty head of livestock; those distillers could reclaim their works upon posting an appropriate bond.[53]

Enforcement of federal distilling regulations resided with the commissioner of internal revenue. District supervisors appointed by the commissioner were responsible for tax assessments and collections and for preventing, detecting, and punishing fraud related to distilling. Supervisors were obliged to reside in the district to which they were assigned and to establish an office at a place determined by the commissioner. Supervisors' salaries were not to exceed $5,000 per year, a considerable stipend at the time. Commissioner Andrew Johnson nominated tax assessors and collectors to each of Kentucky's collection districts. For example, Johnson nominated David S. Goodloe of Lexington as Seventh District assessor and Willard Davis as collector; William M. Spicer of Lebanon became the Fourth District assessor, and Currans C. Smith was designated collector. District boundaries were later adjusted, and the Fourth District became part of the Fifth District, whose assessor was from Louisville.[54]

Surveys, Drawings, and Maps

The internal revenue commissioner wanted accurate estimates of each distillery's potential production capacity as a base against which to measure actual production. Distilleries producing below capacity would be suspected of surreptitiously siphoning whiskey from the production flow and diverting it to hidden cisterns or holding tanks. Regulations required distillers to hire surveyors to determine each distillery's capacity. Surveyors did so by measuring the total number of cubic gallons in each fermenting tub, on the basis of 232 cubic inches per gallon. Every forty-five gallons of mash, wort, or wash capacity was estimated as one bushel of grain, at fifty-six pounds per bushel. Any surveyor who willfully reported an inaccurate distilling capacity—a greater capacity might defraud the owner, while a lesser capacity might defraud the government—would be charged with a misdemeanor and subject to a fine of $1,000 to $3,000, followed by dismissal from office. Any distiller who knowingly operated under an incorrect survey would forfeit his license and be subject to a $2,500 fine and imprisonment for five years.[55]

To assist storekeepers in detecting all possible illegal whiskey take-out points, the 1868 internal revenue regulations required each distiller to provide a detailed drawing and precise description of the distillery executed by an accomplished surveyor and drafted in ink and water color. This drawing of the distillery building and equipment had to include the location of all workings that contributed to the distilling process, including all pipes, valves, joints, and the components to which the pipes connected.[56] In 1877 a revised law updated this requirement by specifying that the drawings had to be "on good paper or tracing-cloth, fifteen by twenty inches in size and where the distillery occupies more than one floor or story, each floor or story should be represented on a separate sheet and none but competent draughtsmen should be employed to make it."[57] Each distillery drawing had to illustrate, to scale, the exact location of each pipe and its connections to the still and other vessels. The drawings had to present piping and equipment cross sections, as well as the position and cubic capacity of every still, mash tub, fermenting vat, and receiving cistern. Prescribed color codes identified each pipe, indicating the type of material it carried: red pipes carried mash or beer; blue pipes moved low wines to the still or doubler; and water and whiskey pipes were white and black, respectively.[58] The law further required distillers to paint all distillery pipes using the same color codes. Any pipe not so painted warranted a fine of $1,000.

Another measure imposed "enforcement lines or boundaries" to facilitate distillery property security. Government surveyors charted each registered distillery according to three levels of site security access and control, which they demarcated by color-coded lines on maps. The first security level included the perimeter lines of the premises, which designated those parts of the distillery property to which federal tax law applied. Second- and third-level lines referenced areas and individual buildings with restricted access. The maps were apparently intended to be used by distillery employees, who were prohibited from crossing certain lines unless accompanied by a storekeeper, gauger, or other federal representative or with their permission. Apparently, these lines were not staked or fenced on the distillery grounds, so employees would have to familiarize themselves with the security boundaries by referring to the maps.[59]

Buildings and Premises

The internal revenue law of 1868 prohibited the use of any distillery building for purposes other than distilling. Relatedly, distillery equipment could be used to grind wheat flour and cornmeal only if those products would be used for distillation on the premises.[60] To collect and store distilled whiskey before barreling, distillers had to fabricate two or more receiving cisterns that were not connected to one another, each of which was capable of holding a single day's production. The 1877 law extended this requirement by mandating that cisterns be housed in a separate and secure cistern room or building. The cistern room's walls and ceilings had to be plastered or cased with matched boards. If constructed of wood, the exterior of the cistern room or building had to be covered in clapboard or weather boarding. Cistern room floors could be concrete, brick or stone laid in cement, or doubled tongue-and-groove flooring planks, with the upper planks laid at right angles to the base planks. Windows had to have wood or iron shutters that could be barred on the inside. Thick doors had to be securely mounted with proper lock hasps, and all hinges had to be affixed so that they could not be removed from the outside.[61] Because the gauger required access to all cisterns, connecting valves and pipes, faucets, and stopcocks to facilitate inspection and locking, cisterns had to be positioned according to exacting specifications: no less than eighteen inches off the floor or less than thirty-six inches between the top and the floor above, and with sufficient lateral space to allow movement between vessels. Whiskey stored in receiving cisterns for three days could, under the supervision of the gauger and storekeeper, be drawn off into barrels and moved directly to a storage warehouse. Mash tubs, fermenting vats, doubler tanks, and condenser or worm tanks had to be spaced to provide agents full access. Furnaces and boilers had to be fitted with doors that could be locked.[62]

The storekeeper weighed all grain processed by the still, and the distillery had to be constructed in such a way that it would be impossible to add ground grain to the mash tubs without the storekeeper's knowledge. Regulations required that ground grain or meal be kept in a dedicated and secure meal room, and the hoppers through which the meal was conveyed to the mash tubs had to be secured with locks.[63] If meal hoppers were not practicable, the law recommended that a second story be constructed to include a secure room that permitted ground grain to flow by gravity into the first-floor mash tubs. In single-story distilleries, the meal room could be constructed in a space adjoining the mash tubs. Distilleries could not commence production until a tax collector thoroughly inspected the works for compliance with all regulations.

Storage warehouses could be built only on distillery premises and could be used to store only spirits manufactured at that distillery. Warehouses had to stand alone, unconnected to other buildings, and they could have no doors, windows, or other openings that permitted direct access to the distillery or other buildings. Warehouses meeting these requirements could be designated bonded warehouses by the commissioner and placed under the district collector's custody. Because the internal revenue storekeeper controlled warehouse access, distillers could not enter their own warehouses except in the presence of the storekeeper.[64]

The law also prohibited distillers from conducting any other business on distillery premises. Revenue officers could enter and inspect the premises at any time and, in certain cases, levy fines on a distiller for improper procedures. Egregious violations resulted in forfeiture of the distillery premises to the government. Each phase of production was subject to isolation and lockdown control; the storekeeper's locks were federally owned and budgeted. From the federal government's perspective, the law's primary objective was to create a standardized and secure distilling space that was resistant to unsupervised operation.

Identity and Access

Regulations required distillers to identify their works as a "Registered Distillery" with painted or gilded signs with lettering no less than three inches high. Anyone working in a distillery or delivering grain or other raw materials to a distillery that was not so identified would forfeit all carts, drays, wagons, horses, and other draft animals used to carry those materials and be subject to a fine and imprisonment. All tubs, vats, and cisterns carried identifying numbers. Distillers were forbidden to erect a fence higher than five feet around the distillery property that might impede access by federal assessors or revenue officers, and the distiller had to provide revenue officers duplicate keys for all fence gates and door locks to allow them access at any time. Any federal officer denied access to a distillery by distillery personnel or who encountered locked gates or doors was authorized to gain entrance by force, breaking doors, windows, or walls if necessary. Distillers had to provide all ladders, lights, tools, or scaffolding required by federal officers wishing to inspect any part of the distillery property.[65]

Record-keeping Requirements

Upon filing with the commissioner of internal revenue a notice of intent to commence production, the distiller posted a bond of $5,000 or double the amount of tax due on spirits produced over a fifteen-day period, whichever was larger. To ensure that taxes would be paid and the bond was guaranteed, the distiller could not mortgage the distillery, accept a lien, or otherwise encumber the property.[66] The 1868 tax law required the distiller to keep a daily inventory of all grains, fuels, and other materials used in the distilling process. Record book entries had to include quantities in pounds or bushels, costs, names of suppliers, dates of purchase, and type of delivery conveyances used. Distillery equipment repairs were listed, along with the names of those who performed the work and the cost. Distillery records listed all employees by name, job, and place of residence. The law also required the distiller and storekeeper to record the quantity of grain used each day, the time of day yeast was introduced into the fermenting process, and the quantity of mash in each tub. Required inventory entries included information on fermentation rates in each tub or vat, as gauged by "the number of dry inches." Specifically, records had to show the following:

> the number of inches between the top of each tub and the surface of the mash therein at the time of yeasting, the gravity and temperature of the beer at the time of yeasting, and on every day thereafter its quantity, gravity and temperature at the hour of twelve meridian; also the time when any fermenting tub is emptied of ripe mash or beer, the number of gallons of spirits distilled, the number of gallons placed in warehouse, and the proof thereof, and the number of gallons sold or removed, with the proof thereof, and the name, place of business and residence of the person to whom sold.[67]

The 1877 tax law refined the regulation of fermentation rates to differentiate between sweet-mash and sour-mash distilling. Sweet mash used yeast as a fermenting agent, and the fermentation process was usually completed in forty-eight hours. Sour-mash distilling used spent mash or slop as the primary fermenting agent; this was a slower process, with an expected fermenting period of seventy-two hours.[68]

Inventory record books were to be kept at the distillery and were subject to inspection by any revenue officer. The penalty for omissions, false entries, or loss of record books was a fine of up to $5,000, imprisonment for up to two years, and forfeiture of the distillery and all distilling equipment and property to the government.[69]

A distiller who was suspending production because of a shortage of grain or fuel or terminating production at the end of the distilling season had to notify the federal assessor, who then locked or sealed the doors to the still, boilers, and furnaces. The distiller could not commence distilling until notifying the assessor, who would then remove the locks and seals.[70] Should a distiller wish to reduce production capacity, the assessor would cover and seal the appropriate number of fermenting vats to prevent their surreptitious use.[71]

Enforcement

Although federal internal revenue regulations were seemingly objective, enforcement was, at least in part, subjective and susceptible to interpretation and various levels of strictness. Distillers found that the storekeepers and gaugers assigned by federal assessors and controllers were not equally capable of carrying out their responsibilities or equally motivated to do so. Some storekeepers and gaugers were strident disciplinarians; others were friendly and cooperative. Some federal enforcement officers were open-minded about alcohol production and use; others were pro-temperance. Distilleries' capacities and operating schedules differed, so storekeepers and gaugers might have been assigned to oversee large, complex works or small, low-volume distilleries. The largest distilleries tended to be located in towns or cities, where storekeepers and gaugers could acquire living quarters in boardinghouses, hotels, or private homes. Rural distilleries offered no such amenities, leaving federal officers to accept whatever accommodations they could find. Some obtained lodging near their assigned distilleries; others had to deal with the inconvenience of living five miles or more from their work sites and had to absorb the expense of buying a horse and perhaps a carriage to facilitate travel. Because the distilling season extended through the winter months, roads blocked by snow or covered with ice could prevent officers from reaching their assigned distilleries on a given day. Such situations were more than inconvenient because tax law prohibited distillers from operating if their storekeepers and gaugers were absent.[72]

Whether to survive financially or to resist regulations on principle, some distillers attempted to undermine the law and avoid its enforcement. Some of their subversion techniques were simple, others elaborate. A distiller might wait, for example, until the federal storekeeper took a lunch break and then load extra ground grain into the mash tubs so that additional whiskey could be made. A few days later, when the fermentation and distillation process had been completed, the distiller would wait for the same opportunity to fill extra barrels that had been secreted in the distillery. To curb corruption, supervisors, storekeepers, and gaugers were subject to substantial fines and dismissal if they were found to be collaborating with distillers in fraudulent activity. Yet storekeepers and gaugers often took lodging in a local boardinghouse house or hotel, or in some cases, they lived as boarders in the distiller's home. Such social propinquity certainly invited fraternization between distillers and their federal overseers. To avoid, to the extent possible, any temptation to falsify records or ignore violations to a distiller's benefit, federal supervisors periodically reassigned storekeepers and gaugers to different distilleries.

Federal law initially imposed taxes on distilled spirits at the point of production, not necessarily after aging, when the product was ready for sale. This taxing method placed small-scale producers at a disadvantage because it required them to pay taxes on a product that would not be sold for four years or more; this policy largely eliminated farm stills. Large industrial-scale enterprises could adjust to the taxing schedule, provided they had access to capital through bridge loans, the sale of warehouse receipts, or other financial arrangements. Distillers could sell unaged whiskey to private investors, provided the buyer paid the taxes due and stored the whiskey in a closely monitored bonded warehouse, such as the Louisville Public Warehouse, built to serve that purpose.

The Effect of Governmental Regulation on Landscape

The relationship between landscape and law can be bold and discernible or subtle and concealed, but it is always complex. Federal law became an exercise in property making by refashioning or manipulating the landscape distillers had created. Some regulations shaped building interiors; others required compliance in the arrangement and placement of structures. Some building materials were proscribed, while others were prescribed. Invisible survey lines demarcated restricted spaces; visible color codes identified pipelines by their contents and vats and tanks by their capacity. What had been an idiosyncratic landscape couched in vernacular traditions and individual subscription to inventions and innovations was corseted into conforming configurations.

As repositories of identity, thought, and meaning, landscape and law are not discrete or independent; rather, they are closely interrelated and, to a degree, collaborative. Constructed landscapes are unique, given their diverse environmental contexts and their builders' eclectic preferences. Where law was introduced, it was often cast as a medium by which differently constructed landscapes were normalized, or perceived in the abstract and treated as though they conformed to common standards. Only then could law apply equally to each landscape unit, despite their obvious differences one from another. Laws that relate to place and landscape, therefore, had to be sufficiently elastic to have broad application, never mind the particularities of the discrete landscape units to which they were applied. No two nineteenth-century bourbon distilleries were alike beyond formalizing the natural processes that created potable alcohol. Each had a unique environmental context, history, level of technology, and construction. Although the product they produced shared a generic name, the objective manipulation of chemical reactions and the subjective aroma and taste of each distillery's product were different. Indeed, distillers often deliberately differentiated their bourbons by varying yeast varieties, adjusting grain and malt mashing formulas, or positioning barrels in aging warehouses so that whiskeys aged at different rates.

Internal revenue law was long argued in Congress. Legislators considered, contested, and amended policy at length, and in the end, they drafted cumbrous regulations that were intended to maximize tax revenue while simultaneously holding fraud in abeyance. The straitjacketing law that emerged, phrased in absolutist terms, increased governmental assets, but the monetary cost and managerial inconvenience to the industry can only be characterized as punitive. Revenue law granted the government dominion over America's distilling industry. Although distilleries were owned by individuals or companies, they were effectively controlled by federal agents who introduced and enforced standardized operational measures. Internal revenue collectors and distillery storekeepers became enforcers who exercised full control over the distilling process, an exercise of power epitomized by the government-owned locks that denied distillers access to the buildings and equipment that permitted spirits production. Unless distillers complied with internal revenue regulations, they could not commence operations. Thus, the landscape they created transformed invisible laws into visible structures; the abstract and cerebral became tangible.

From the vantage point of Washington's congressional representatives who approved the aspatial regulations and reallocated the tax revenue they produced, writing the law may have seemed a straightforward technical act, an appropriate application of federal power.[73] The federal officers who enforced the law likely saw it as complex and thorough. The distillery owners who were compelled to comply with the law saw it as intrusive, to the point of requiring them to conform to a schedule not their own and imposing other managerial and operational difficulties that discouraged investment and triggered bankruptcies of the very businesses from which the government sought to extract budget-balancing capital.

13

Temperance Troubles

The temperance reformers were wealthy men, and they possessed enormous power.
But they preferred to translate power into authority, and to reform lesser men by persuasion rather than by force.
—Paul E. Johnson

Conjunctions—Temperance versus the Distilling Industry

Distilling ardent spirits in early-nineteenth-century America allowed farmers to use their surplus grains. Whiskey was widely produced and easily preserved; it gained value when barreled and stored and was commonly interchangeable for other goods and services. During the four decades from 1770 to 1810, farm-distilled liquor served as currency in some rural areas. Spirits production was often viewed as a positive contribution to the community, and many believed that consumption was beneficial to heath and a prophylaxis. Alcohol was not generally consumed to achieve intoxication. Improvements in transportation and rapid advances in industrial technology in the nineteenth century allowed distillers to participate in the market economy and concentrate production in those regions best suited for grain production: the Ohio Valley in particular and the Middle West more generally. By the mid-nineteenth century, industrial distilling in those regions began to supplant the farmer-distiller, and whiskey production increased. Although the making of whiskey was progressively refined, it was also increasingly encumbered by taxes and production controls.

While many believed that making and consuming liquor were cultural traditions and an economic good, others were concerned that money spent on spirits would be better applied to purchasing food, clothing, and shelter or investing in land, business, and industrial expansion.[1] Drinking to excess could be socially disruptive and physically debilitating, and many began to see distilleries and related businesses as symbolizing loose morals and decadence and taverns as places where such character flaws were proclaimed. Certain philosopher-physicians and scientists such as Daniel Drake concluded that the "intemperate consumption of alcohol was . . . the moral equivalent of the open sewer, not just contributing to corruption, but sufficient alone to excite inflammation and disease."[2] Public opposition to whiskey production and consumption increased by the 1830s, rousing local and regional groups to organize; moderates among them advocated for social and legal curbs on drinking, whereas radicals wished to prohibit making and using—or abusing—liquor entirely. These collective efforts became known as the temperance movement.[3]

A Temperance Tempest Arising

The American Temperance Society, established in 1826, organized the desultory efforts of sobriety supporters to encourage individual abstinence and control, if not eliminate, spirits production.[4] There followed the gradual development of a countrywide campaign for national prohibition by the Washington Temperance Society, the Women's Christian Temperance Union (WCTU), the Independent Order of Good Templars, the Father Mathew Catholic Total Abstinence Society, the Catholic Total Abstinence Union of America, and, in 1893, the Anti-Saloon League. These and other state and local organizations advocated total abstinence from alcoholic beverages and aggressively pursued state and federal legislation to prohibit the manufacture, transportation, sale, and consumption of ardent spirits. In the early 1820s, 222 temperance societies were active in the northern states, claiming more than 30,000 pledged members. Temperance leaders' recruitment efforts were so successful during that decade that by 1830, the number of societies had increased to 1,000, and they boasted more than 100,000 members. They also measured their success by the 400 or so merchants who had pledged to discontinue the sale of spirituous beverages and the fifty or more distilleries that had closed.

Although popular support for temperance reform lagged in the South, county-level temperance societies began to organize in Kentucky by the late 1820s.[5] By 1833, the state's ninety temperance auxiliaries claimed that their collective efforts had resulted in the closure of forty-six distilleries.[6] The true cause of such closures cannot be ascertained. Certainly, some distilleries stopped production for a number of reasons, including the passing of a generation of knowledgeable distillers; destruction of facilities through fire, flood, or other misfortune; bankruptcy brought on by poor management or crop failure; and a host of other causes. The temperance movement gained considerable moral and behavioral traction in the state when activists formed the Kentucky Legislative Temperance Society in 1834, with Governor John Breathitt (D) as president and Lieutenant Governor James Morehead (R) as one of five vice presidents.[7]

By 1838, America's temperance societies numbered more than 8,000, with over 1.5 million members.[8] In 1900 the national temperance movement included twenty-three state societies. At the urging of New England temperance activists, the Massachusetts legislature passed a local liquor sales option law in 1840, and within five years, more than 100 towns had passed local prohibition ordinances. Maine instituted statewide prohibition in 1851, followed by New Hampshire in 1855.[9] In the Middle West, Iowa passed prohibition laws in 1882 and 1884. Considered in the aggregate, though, the temperance movement did not progress inexorably toward universal prohibition. Public sentiment vacillated; improvised political and legal strategies shifted. From 1846 to 1855, thirteen states adopted some form of prohibition, including Illinois and New York. The strictures lasted one year in New York and two years in Illinois, and by 1863, only six of the thirteen states retained their prohibitory statutes.[10]

As regional and transcontinental railroad companies extended their lines into the lightly settled Great Plains and western mountains, land speculators, farmers, shopkeepers, and saloon proprietors followed the tracks. Trains supplied this new regional market with goods, comestibles, and, of course, alcoholic beverages. Whiskey followed the tracks west from one rail town to the next, and local and regional temperance activism emerged soon thereafter. Kansas passed a state prohibitory law in 1884. Local option laws created "dry" cities or counties, but these strictures were often ineffective because consumers could readily travel to adjoining jurisdictions that remained "wet." The resilience of local temperance statutes was also mitigated by lost tax revenue on alcohol sales, and many dry counties eventually reversed themselves and voted to permit alcohol sales again.

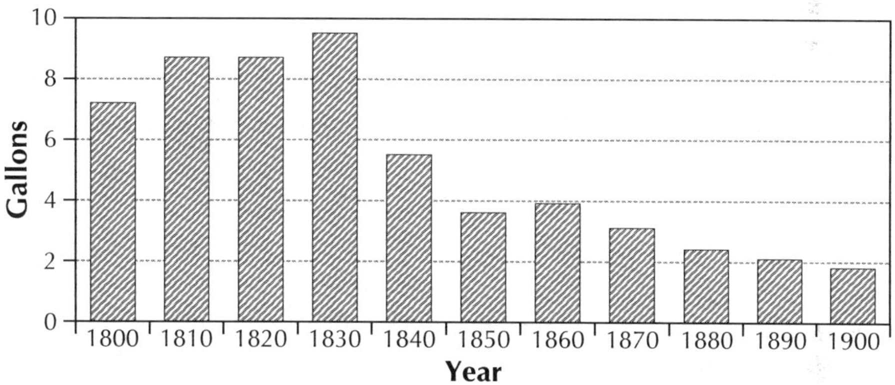

Ardent spirits consumption in America, 1800–1900. Ardent spirits include aged and rectified whiskeys, rum, gin, brandy, etc. The graph depicts gallons per capita among those aged fifteen years and older. (Compiled from W. J. Rorabaugh, *The Alcoholic Republic: An American Tradition* [New York: Oxford University Press, 1979], 233)

It is impossible to state with certainty the regional and national impact of temperance efforts, but the personal use of distilled spirits declined steadily after 1830; by 1900, per capita consumption had dropped to less than 20 percent of the 1830 level. The collective resistance to the manufacture and consumption of alcohol likely affected each distiller, wholesaler, and tavern owner differently, but the tenacity of temperance activists, however biased the invidious hyperbole of their literature and lectures, cast doubt and uncertainty on business decision making.

Temperance campaigners themselves were conflicted when the objects of their scorn were local businesses operated by friends and neighbors. Newspapers editorialized for and against distilling and liquor consumption. Many Kentucky communities experienced a kind of collective cognitive dissonance among temperance leanings, church teachings, and chthonic friends whose clothing seemed to reek of brimstone and whose participation in the distilling economy was deemed loathsome. Yet nineteenth-century distillers and their employees were often active church members; other parishioners grew and sold grains to distilleries, made whiskey barrels, and built and sold distilling equipment. Conservative religious groups were less ambivalent about allowing those with connections to the distilling industry to remain "within the fold." The pious people of a Cynthiana church expelled a member for retailing whiskey, and other congregants adopted the deportment of a hostile jury that claimed "the broad ground that all who have any connection with the trade are equally obnoxious."[11] Perhaps as a belated rejoinder, on Wednesday, October 12, 1898, Rutherford E. Douglas, pastor of the Nicholasville Presbyterian Church in Jessamine County, and Reverend Ziegler, a Presbyterian minister at nearby Spears, in Fayette County, held a meeting in the mash room at the E. J. Curley Distillery on the Kentucky River. While the conclave's purpose was not divulged, it was, apparently, a small revival meeting, similar to those held at the distillery in previous years. Passersby reported singing and preaching emanating from amongst the fermenting vats.[12]

Temperance work was aimed at achieving individual and collective behavioral change through personal persuasion or the application of comprehensive legal constraints. Newspapers and other print media increased in number and areal coverage after the Civil War. Temperance activists amplified and transmitted their messages via newspaper editorials and in pictorial form with cartoons such as "The Drinker's Progress," widely circulated broadsides, and even stereographic pictures.[13] But activists also worked toward their objective through interpersonal social experiences, especially at organized meetings that served to inform, convince, motivate, and reinforce solidarity for the cause. Movement leaders invited motivational speakers to these meetings. Some were locally and regionally prominent in the movement and were considered brilliant, high-voltage orators, such as Frankfort attorney Sidney French and "silver-tongued" George W. Bain. Others were nationally celebrated, including Iowa attorney Judith Ellen Foster and former Kansas governor John P. St. John, who spoke at a Franklin County temperance meeting in February 1885.[14] Perhaps the most effective nineteenth-century temperance advocate was Irish priest Father Theobald Mathew, who was credited with persuading more than 2 million of his compatriots to take an abstinence pledge. He visited the United States beginning in 1849 and reportedly administered the pledge to more than 251,000.[15] Father Mathew visited Kentucky briefly in 1851 and called on Henry Clay at his farm in Lexington, but he did not lecture publicly or recruit pledges, likely because of physical infirmities.[16] Nevertheless, activists named at least one Kentucky temperance hall after the efficacious Irish priest.

Kentucky's temperance themes initially emphasized the physiologic, social, and economic "evils of drink," but activists gradually recognized that the most effective strategy would be to gain sufficient political leverage to pass local option laws. And the passionate pursuit of that goal was best fueled by group meetings, where personal interaction could motivate participants and whip up fervor, rather than through passive print appeals or interpersonal campaigning. Temperance groups held meetings in churches, city halls, and other public and privately owned buildings.[17] In Louisville, seven chapters of the Good Templars held midweek meetings in different parts of the city, including Templar's Hall at Hancock and Jefferson Streets; though closed and in poor condition, the three-story brick building was still standing in 2018.[18] The Ancient Order of United Workmen (AOUW) provided its members with social and financial support by offering insurance and death benefits, and some chapters were active in temperance work. The Maysville AOUW chapter met at the Father Mathew Temperance Hall at Second and Market Streets, in the heart of the business district.[19] The Grand Division of the Sons of Temperance in Maysville held twice-weekly meetings at its rented space in the prestigious Cox Building at East Third and Market Streets. In 1906 Cloverport residents formed a temperance organization at a mass meeting held at Oelze's Hall. Attendees elected the Reverend B. M. Currie president, and the town's pastors were made members of the executive board, which also included officers of the WCTU, the Baptist Young People's Union, and the Epworth League, a Methodist young adult association.[20] Some rural communities raised funds to build dedicated temperance halls. Members of the Thespian Club of the Independent Order of Good Templars of Hustonville, in Lincoln County, raised money to construct a new temperance hall by selling tickets to dramatic plays and musical performances.[21] Some privately owned restaurants and billiard parlors declared their support for temperance principles by providing meals and recreational activities at liquor-free venues.[22]

Greenville's temperance hall dated from the early 1850s. The Muhlenberg County structure was a two-story frame building on South Main Street, with four rooms on the first floor and a meeting hall on the second level. Edward Weir built the hall as a "rendezvous" for the Sons of Temperance, but when the order failed to sustain itself, the building was occupied by doctors, lawyers, and other businesses, and the upstairs hall was opened to the public for meetings, entertainment, receptions, and other events. The structure was later disassembled and moved to a local farm, where it was converted into a tobacco and stock barn. The First Presbyterian Church of Greenville was built on the vacant Main Street lot and dedicated in 1885. When the old frame temperance building was finally demolished in 1913, its passing merited nostalgic commentary in the local newspaper. "There were many rousing meetings held in the building," the editor noted, and "its passing will be a matter of regret to many of our older citizens."[23] Such testimonials serve as reminders that although temperance meetings had political agendas, they were also social occasions, and landscape elements as seemingly prosaic as a many-purposed wood-frame building can serve as a symbolic place and a basis for communal or collective history, as well as an anchor for memory and meaning for those who participated.[24]

By the 1880s, especially during the summer months, temperance hall venues were supplemented by weekend and extended meetings held at rural camps. The High Bridge Camp in Jessamine County held Temperance Day meetings in August. In Lewis County, trains from Lexington and Maysville brought people to Park's Hill Camp Ground, at reduced ticket rates, where they were inspired by prominent temperance orators. People attended August Temperance Day meetings at nearby Ruggles Camp, a Methodist Church facility established in 1873; some guests camped in tents, and others booked lodging in cottages or at the camp's hotel. An onsite sulfur water well provided medicinal drafts for those so inclined. A camp at Ashland held ten-day temperance meetings during the month of July in the 1880s, as did the Sulphur Springs spa at Beaver Dam in Ohio County. At Dawson Springs in nearby Hopkins County, temperance advocates gathered at an arboreal campsite in an expansive grove west of town.[25] In fervor and moral suasion, these outdoor abstinence assemblies may have been akin to Kentucky's religious revival gatherings held at the Red River Meeting House, Cane Ridge, and other locales in the early 1800s, or perhaps they resembled a temperance Chautauqua.

Women's Christian Temperance Union camp meeting, ca. 1880. (Bullock Photographic Collection, Transylvania University Library)

Some radical Kentucky temperance activists aggressively pursued their goal by saloon smashing, in the manner of Garrard County native Carrie Nation.[26] Others took more drastic action. William L. Berry operated a distillery west of Owensboro, and in 1871 a mob with suspected temperance proclivities set his works afire. Berry subsequently appealed to the US House of Representatives' Committee on Claims for indemnification for the destruction of his distillery.[27] But for temperance advocates who wished to advance their cause in a morally acceptable manner, acts such as smashing saloons and incinerating distilleries were counterproductive. Saloon keepers had an allegiant clientele, after all, and fires threatened life and property; such activities portrayed the instigators as lawless if not unstable.

Advocates recognized that for temperance to succeed as a movement, they would have to do more than simply attend meetings and socialize with like-minded colleagues; such actions would not lead to behavioral change on a broad scale. Legal leverage was required, be it at the local, regional, or national level. Kentucky's temperance organizations, led by the Grand Lodge of Good Templars, initiated a movement to establish local options to prohibit the sale of alcohol. In 1874 the group presented a petition with 147,000 signatures to Governor Preston Leslie, requesting the passage of local option legislation that would permit counties to compel referenda on the issue. The governor forwarded the request to the General Assembly, which passed a law later that year. Initiating a local option vote was straightforward. Residents of any county, precinct, or town could compel a referendum election to establish a local option by gathering a mere twenty names on a petition. A vote for approval would stop the retail sale of intoxicating liquor in the petitioning jurisdiction. Led primarily by the WCTU, numerous referenda followed, especially in the Bluegrass counties, where much of the state's whiskey was distilled; the first local prohibition law passed in 1877.[28]

Temperance supporters had strongly encouraged the state to permit local option voting, but practical application of the law led to jurisdictional questions related to defining the term "local." Precincts, towns, cities, and counties were all types of localities that could vote as discrete units. The courts, in ruling on the matter of jurisdiction, stated that if a precinct voted for prohibition (dry), an adjoining precinct could not contravene that decision. If a precinct voted against prohibition (wet), the precinct's status was unchanged; it remained as it had been before the vote was taken. But if a town, city, or county that included that (wet) precinct voted for prohibition (dry), the entire political unit became dry. Conversely, if a precinct had previously voted dry and its encompassing town, city, or county voted wet, the precinct could retain its dry status. Small political divisions with temperance empathies were therefore motivated to vote themselves dry, in contradistinction to a larger jurisdiction's interest in maintaining the wet status of its aggregated units.[29]

Temperance activism also influenced the passage of state tax legislation. Prior to

1882, Kentucky did not levy a tax on whiskey as property.[30] Beginning in 1850, state regulations required businesses selling liquor to pay point-of-consumption license fees. Coffeehouses, boardinghouses, and restaurants selling liquor paid a $10 license fee in 1850; this increased to $25 the following year and to $50 in 1874. A tavern owner paid an operator's license fee of $10 in 1874, but if the tavern sold liquor, the fee was $50; the latter fee increased to $100 in 1886 and to $110 in 1902. Retail merchants selling liquor paid a $5 license fee in 1850, $15 in 1851, $25 in 1874, $100 in 1886, $75 in 1892, and $100 in 1902. Druggists who wished to sell liquor in 1892 paid a license fee of $50; a decade later, the fee increased 50 percent to $75. The different fees paid by retailers and druggists suggest that legislators considered spirits to have medicinal value. Distillers who sold their own products at the distillery were first charged a $25 license fee in 1892; by 1902, legislators had increased that payment to $100.[31] In 1882 the state legislature levied a special tax on distilled liquor at the regular property tax rate of about 47 cents per $100 evaluation.[32] Rather than assess taxes based on the amount of whiskey on hand on the assessment date, the new regulations required distillers to report all deliveries of newly distilled whiskey to aging warehouses on June 1 and October 1 of each year. Distillers and whiskey owners paid their taxes to the state auditor. Whether licensing fees and taxes imposed on area businesses inspired temperance in the business owners or the consuming public is unclear, but it is unlikely. These fees and taxes did underwrite the formation of an administrative infrastructure to collect payments and enforce compliance, and the collected money became a source of income for thinly sustained jurisdictional budgets.

Meanwhile, the ardent spirits industry was not without a counterbalancing voice in the matter of temperance. Larue County distiller John Atherton was a founding member and president of the National Protective Association (NPA), an organization formed in 1886 by wholesale liquor dealers and distillers in Chicago for the purpose of opposing by lawful means the enactment of prohibitory legislation. The organization was not involved in distillery operations or management; rather, members contributed to a general fund to support legal and publicity efforts.[33] The NPA leadership focused on the public discussion of sumptuary legislation, presuming that such debate would be more effective than lobbying state and local officials. Members argued, in public forums and in literature distributed throughout the country, that prohibitory legislation would put the distilling business in the hands of less reputable and less responsible people and would bring "mean whiskey and doctored whiskey" into those places that had passed such laws. "Half of Kentucky, I am sorry to say, is under prohibition by local option," Atherton argued. "They vote 'dry' today and 'wet' tomorrow, and change about; and yet I never struck a Kentuckian anywhere that wanted a drink and could not get it."[34] In Kentucky, the wellspring of America's bourbon industry, the NPA's success was limited at best. Worse, at least from the distiller's perspective, temperance as an idea had metastasized into an acrimonious drive toward full prohibition, one county at a time.

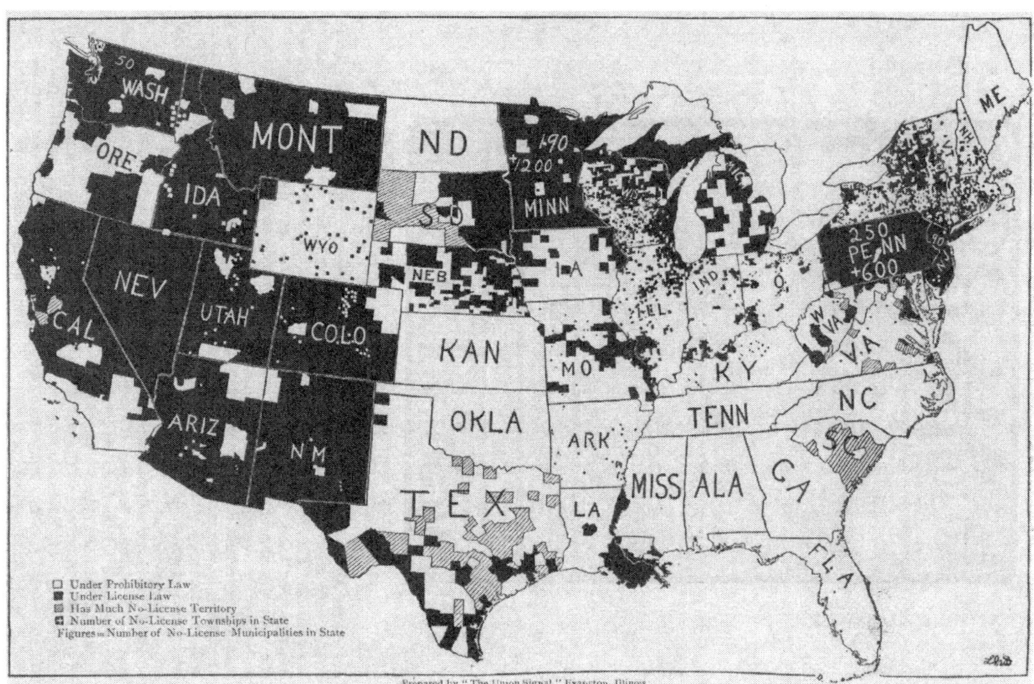

Prohibition map of the United States, 1910. (P. J. Mode Collection of Persuasive Cartography, Division of Rare & Manuscript Collections, Cornell University Library)

By 1908, 92 of Kentucky's 119 counties were dry; the ratio rose to 106 of 120 counties by 1915.[35] In 1910 the WCTU's newspaper, the *Union Signal,* published a map depicting the nation's wet and dry political jurisdictions. The cartographer depicted in white those state, county, and local jurisdictions where prohibition laws were in effect; wet districts and those operating under licensing laws appeared in black.[36] Several states, including Kentucky, Illinois, Massachusetts, New Hampshire, New York, Ohio, and Wisconsin, were largely dry in 1910, but there were many isolated "license law" jurisdictions that remained wet. In Kentucky, only twenty-three places, principally towns and cities, could legally issue licenses for liquor sales. Though poorly delineated, the map appears to portray three or four wet counties: Campbell, Kenton, Jefferson, and possibly Meade. Several towns in the Inner Bluegrass and western Pennyroyal regions remained wet. Temperance had become, by this time, the preeminent issue in Kentucky politics.[37]

The state legislature had long seemed ambivalent about temperance, but it did pass a law in 1916 that forced saloons to close on Sundays. Three years later, it passed Amendment VII to the Kentucky Constitution, enacting state-level prohibition. Church groups—especially the Methodist Episcopal Church, South; the General Association of Kentucky Baptists; and the Cumberland Presbyterian Church—staunchly supported the amendment.[38] In 1919 four states voted on constitutional amendments to prohibit liquor sales. Kentucky and Texas adopted prohibition, Michigan voters defeated a proposal to repeal their prohibition amendment, and a prohibition amendment was defeated in Ohio. Kentucky's legislation closely paralleled that approved at the national level at about the same time. The state's amendment read: "After June 30, 1920, the manufacture, sale or transportation of spirituous, vinous, malt or other intoxicating liquors, except for sacramental, medicinal, scientific or mechanical purposes, in the Commonwealth of Kentucky, is hereby prohibited. All sections or parts thereof of the Constitution, insofar as they may be inconsistent with this section are hereby repealed and nullified. The General Assembly shall enforce this section by appropriate legislation."[39]

As more states responded to temperance advocates by establishing statewide prohibition laws, people wishing to circumvent those laws, especially in border counties and larger cities, employed the railroads to import spirits from adjoining wet political jurisdictions. An Iowa decree prohibiting railroads from shipping liquor into the state was challenged in court in 1888. When *Bowman v. Chicago & Northwestern Railway* came before the US Supreme Court, the justices ruled the Iowa statute unconstitutional in large part because, under federal law, liquor shipments were protected by statutes governing interstate commerce.[40] Therefore, states could not prohibit the shipment of liquor across borders into their jurisdictions because such prohibition restricted interstate commerce. To provide a point in time when state regulation of liquor distribution could begin, the court adopted the sixty-year-old "original package" doctrine. In 1827 the Supreme Court had ruled in *Brown v. Maryland* that alcoholic goods imported into a state, "while remaining the property of the importer, in his warehouse, in the original form or package in which it was imported," could not be taxed or prohibited without running afoul of interstate commerce statutes. State regulations applied if the original packaging—whether barrel, keg, or box—was opened and the contents sold by wholesalers or retailers.[41] Accordingly, Kentucky distillers could ship liquor into dry states provided it remained in the original shipping containers and was sold in those same containers. Some Kentucky distillers, including Henry McKenna, had extensive lists of mail-order customers, and they were particularly adept at shipping goods into dry counties or into prohibition states such as Kansas.

Marketing Medicinals

Kentucky's nineteenth-century distillers did not originate the idea of selling whiskey as medicine or claiming that it had curative or preventive qualities as a way to circumvent temperance-inspired prohibition laws. Colonists from Europe brought with them the habit of drinking cider, fruit brandies, wine, beer, and whiskey; spirits consumption was traditional, embedded in agrarian values. Whiskey and other spirits were taken with meals and were a long-standing part of both European and American diets. Spirits consumption was also a prerequisite in many types of social interaction. And whether in the Old World or the New, eighteenth-century people held the deep-seated conviction that alcoholic beverages had medicinal merit.[42] Either intentionally or incidentally, people who drank spirits and beer avoided many of the common waterborne maladies of the period. During the nineteenth century, physicians prescribed spirits for their patients, and druggists routinely used whiskey to create medicinal compounds and tinctures.[43] Spirits taken in moderation, especially "pure" whiskey, were thought to have a palliative effect in patients suffering from dyspepsia, dropsy, rheumatism, gout, palpitations, hysteria, and many other afflictions. Temperance propagandists, citing the writings of Benjamin Rush and other abstemious physicians and scientists, argued that spirits caused these same maladies.[44] Tradition and habit seemed the more powerful argument. By 1830, Americans aged fifteen years and older were consuming nearly ten gallons of spirits per person every year.

Temperance physician Charles Jewett chided the many doctors who, he claimed, were in the habit of making casual "sidewalk prescriptions" to acquaintances who seemed unwell. "Oh! Take a little whiskey occasionally—that is what you need; and if you get 'a little by the head' now and then, as the sailors say, it won't hurt you."[45] Many Americans willingly accepted their physicians' recommendations, whether these were rendered during office visits or social occasions, confident that spirits taken in moderation would benefit certain medical conditions. In 1868 Cyrus H. McCormick, America's most successful reaper manufacturer, wrote a letter to his Kentucky cousin J. B. McCormick of Woodford County in which he enthusiastically declared, "I am now using the best whiskey, I think that I have found—Ky. made—at $10 a single gallon here. . . . You may send me some of the best for purity & health."[46]

But alcohol, sometimes in the form of whiskey, was also used extensively by the patent medicine industry. In 1886 the *Frankfort Roundabout* carried an announcement of a WCTU speaker's visit; on the same page there was an advertisement for Electric Bitters, sold by druggist Joseph LeCompte. The ad promoted the product's supposed medicinal properties, claiming that it was "the best and only certain cure known" for biliousness, jaundice, constipation, and weak kidneys. This "Great Family Remedy," manufactured by H. E. Buckleu & Company of Chicago, contained 18 percent alcohol (36 proof), which qualified the product as brandy if it also contained fruit extracts.[47] Even indulgent temperance advocates remained convinced that patent medicines containing alcohol were effective in treating disease. Charles D. Barker, publisher of the *Temperance Advocate* in Atlanta, Georgia, posted a short note in the *Frankfort Roundabout* about his "typhoid malaria" affliction. "Two weeks ago," he testified, "I purchased a bottle of Swift's Specific which has proven a sure cure for this dreadful malady." He continued: "Would to God that all the afflicted people residing in the malarous counties of Georgia, Florida, and Alabama could read this and try the S.S.S [Swift's Sure Specific] instead of dosing themselves with quinine and mineral remedies."[48] Barker's endorsement appeared immediately below the "Frankfort W.C.T.U. Notes," which carried an announcement of the group's next meeting. Swift's Sure

Specific was manufactured in Atlanta and was advertised as a cure for syphilis, blood poison, and rheumatism, among other ailments. While the elixir's other ingredients were of questionable curative value, the concoction contained 15 percent alcohol.[49]

The National Prohibition Act became federal law in 1920; thereafter, spirits could legally be made and sold only under the pretext that they possessed medicinal qualities. Patients who claimed to have the appropriate disorders could obtain from approved physicians prescriptions for a pint of fermented liquor to "afford relief from some known ailment" when "necessary."[50] The legislation thereby acknowledged traditional medical practice but obligated the Treasury Department to require that doctors use US Internal Revenue Form 1403, Prescription Blank—National Prohibition Act, to write their prescriptions. And according to federal regulations, whiskey prescriptions could be filled only by authorized pharmacists. The cynical observer might have concluded that Prohibition had the salutary effect of transforming participating pharmacists' shops into saloons.

Given distillers' struggles with overproduction, federal regulation, and predatory actions by the Whiskey Trust in the nineteenth century's last quarter, many thought the industry would be vulnerable to decisive temperance actions. The cumulative negative influence on the distilling industry had a debilitating impact on Kentucky's economy and a liquidating effect on the landscape. As distilleries foundered, employees and supporting craftspeople lost jobs, farmers lost markets, and uncertainty plagued commerce leaders, whether their businesses were directly related to distilling or not. All this was reflected in the landscape in ways that were obvious to residents and visitors alike. In 1907 a *Breckenridge News* editorial predicted that the entire state would be dry within a year. "The distilleries that have made the state famous have not been running full time this year, and many of the once valuable plants have been abandoned and are rotting down. The evidences of the passing of the great Kentucky industry is to be seen on every water course in the state."[51] In 1920, when the Eighteenth Amendment to the US Constitution formalized Prohibition, the distilling industry's deconstruction seemed complete.

II

Knowing from the Inside: A Tale of Two Distilleries

14

Making and Selling Whiskey at the Henry McKenna Distillery

I challenge the world to produce a Purer Whisky than mine.
—H. McKenna

To better understand a human cultural undertaking like Kentucky's distilling industry, it is useful to get to know it from the inside, to learn how people lived, worked, built, and transformed their environments and thereby created knowledge and tradition. In historical situations, it is best if we can hear people's voices and understand their choices through first-person narration rather than acceding to third-person interpretations of general information. In the case of distilling, the objective is not simply to document the facts of when and where it took place but to inquire about the people who created and fostered the industry, as well as those who contributed to its operations. Distillery records allow us to make detailed observations and comparisons of production and management techniques. But personal records also contain information that, if treated with mindful attention, may provide insight into what these people believed and knew, their reasoning and thinking, and their intentions, concerns, creativity, and decision making. Such records may also reflect how these people wanted to make and shape the world in which they lived. We aspire to understand both distillers and their distilleries within their social and cultural contexts.[1]

Two nineteenth-century Kentucky distilleries—Nelson County's Henry McKenna Distillery in the Outer Bluegrass and Scott County's Elkhorn Distillery in the Inner Bluegrass—preserved for our consideration detailed chronicles of their day-to-day operations through correspondence and account ledgers. Though both distilleries produced bourbon, Kentucky's signature whiskey, they had dissimilar origins and operated quite differently. The Elkhorn Distillery failed soon after it began operation, and its brand name has long since vanished. It is the subject of chapters 16 and 17. The Henry McKenna Distillery continued to operate up to Prohibition and then resumed production after Congress repealed the Eighteenth Amendment in 1933; its brand continues to be produced today. The Henry McKenna Distillery is the subject of this chapter and the one that follows.[2]

Origins of the McKenna Mill and Distillery

Draperstown, O'Derry—known today as County Derry—in Northern Ireland was Henry McKenna's birthplace. He was born in 1819 and migrated to Philadelphia in 1837 or 1838.[3] McKenna later moved to Kentucky and worked in Lexington for a time as a laborer and a turnpike builder. In 1850 he obtained a contract to build a new eleven-mile-long turnpike from Bloomfield through Fairfield to High Grove, where the road would connect with the Bardstown-Louisville Pike. The following year, he moved his family to Fairfield to commence work on the pike.[4] Upon completing the new stone-covered pike, McKenna and carpenter Richard Constantine bought a lot in Fairfield from John and Elizabeth Renshaw in 1854. They paid $72 for the property, which was just over an acre in size and positioned near the top of a low hill at the head of the East Fork of Cox Creek and beside "the big road." The two men built a modestly sized flour mill that produced, in addition to flour, a large quantity of "middlings," a by-product for which there was no market. About a year later, the two men added a small distillery and began making wheat whiskey, using the middlings as a distilling grain. In 1857 McKenna converted from making wheat whiskey to making a sour-mash whiskey from corn.[5]

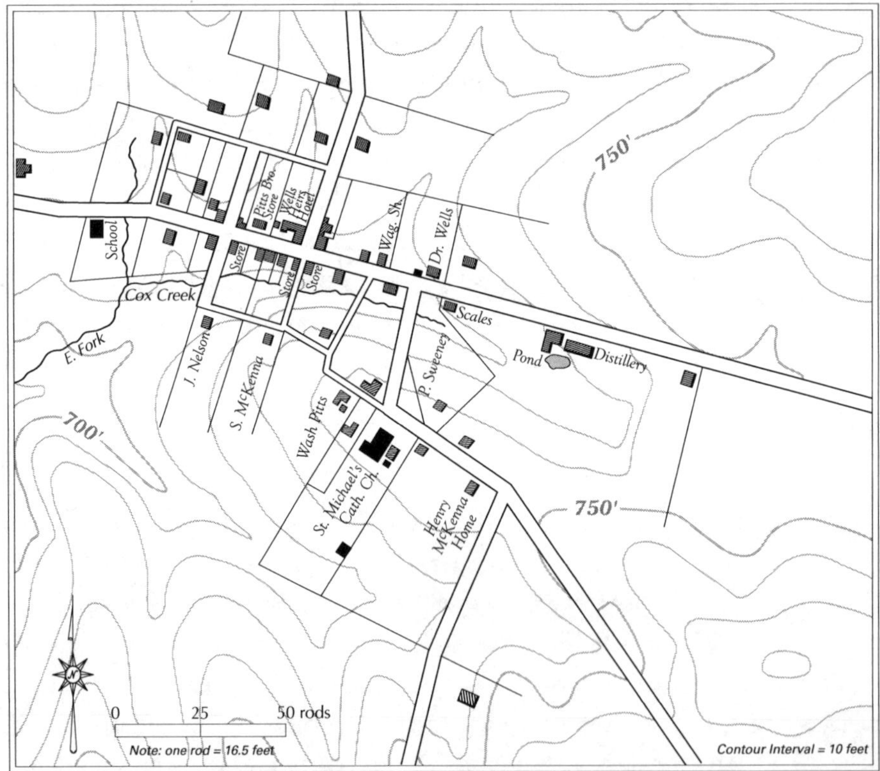

Site of the Henry McKenna Distillery, Fairfield, Nelson County. (After D. J. Lake, *An Atlas of Nelson and Spencer Counties, Kentucky* [Philadelphia, 1882])

The East Fork was a small creek that flowed intermittently—only a meager trickle in August and September. Its fall and flow volume would have been sufficient to turn a small-diameter undershot waterwheel until a wood-fueled steam engine could be installed to power the mill and distillery. McKenna did not covert the engine to burn coal until several years later. Initially, the simple oak still house produced only about twenty gallons, or roughly half a barrel, per day, so the water required could have been hand-carried in pails. In addition, McKenna may have directed creek water into the distillery by means of a pipe or small wooden trough for use in mashing and fermenting operations. He dammed the creek to create a small holding pond, from which he could draw water for cooling purposes and for his penned livestock. The records make no mention of "limestone springwater"; nor do they indicate that the creek was the primary reason for choosing that site for the mill and distillery. In later years, McKenna and his distillers used distilled water to reduce the whiskey to the desired proof for sale.[6]

McKenna's mill stood beside the four-mile-long Bloomfield-Fairfield Turnpike, which the Kentucky General Assembly had chartered in 1836.[7] The pike connected with the Bardstown-Louisville Turnpike at the crossroads settlement of Cox Creek to the southwest. Construction of McKenna's mill and distillery preceded the extension of railroad branch lines into Nelson County, and the site was never directly connected to the railroad. After 1860, when the L & N Railroad line reached Bardstown, McKenna's teamsters hauled goods on the Fairfield–Samuels Depot Turnpike ten miles south to the T. W. Samuels depot. They also moved goods on the Bloomfield-Fairfield Turnpike to the L & N depot in Bloomfield after that branch line opened in 1880.

Henry McKenna's primary income, initially, was from milling farmers' grains into flour, meal, and bran. He operated the mill as needed and was especially busy after grain harvest in July and August and corn harvest in October and November. The distillery operated during the winter and early spring months. McKenna marketed his whiskey as Old McKenna Sour Mash Whiskey, a brand he originated in 1855.

Ten people constituted the Henry and Elizabeth McKenna household in 1860: the McKennas, their five Kentucky-born children, and three Irish boarders. Boarder Patrick Sweeney started work as a laborer and became McKenna's distiller. In 1870 the US census marshal identified Henry McKenna as a distiller and miller. Son Daniel worked on the family farm, and son John worked at the mill. Another son, Stafford, had been born since the 1860 census. Four African Americans boarded with the McKenna family in 1870: Rhoda Camus, a household servant, and her son Willie; and farmworkers N. Bodine and Sandy Bodine. In 1874 Henry McKenna's family built him an eleven-room brick house near St. Michael's Church on a low hill overlooking the mill and distillery works. A two-story brick dormitory-type structure behind the house may have housed the farmworkers and their families.

The household's composition had changed again by 1880. The oldest sons worked at

**Henry McKenna Household,
Nelson County, Kentucky—Fairfield District,
1860–1880**

1860

Name	Age	Relationship	Occupation	Place of Birth
Henry McKenna	41	Father	Miller	Ireland
Elizabeth McKenna	39	Mother		Ireland
Daniel McKenna	11	Son		Kentucky
Henry McKenna	9	Son		Kentucky
Mary McKenna	7	Daughter		Kentucky
James S. McKenna	4	Son		Kentucky
Peter McKenna	2	Son		Kentucky
Patrick Sweeney	23	Boarder	Laborer	Ireland
Frank Goodwin	23	Boarder	Laborer	Ireland
Margaret Dumfrey	21	Boarder	Servant	Ireland

Source: US Census of Population Manuscripts.

1870

Name	Age	Relationship	Occupation	Place of Birth
Henry McKenna	51	Father	Distiller & miller	Ireland
Elizabeth McKenna	49	Mother	Keeps house	Ireland
Daniel McKenna	21	Son	Farmer	Kentucky
John McKenna	20	Son	Works at mill	Kentucky
Mary McKenna	17	Daughter	At school	Kentucky
James S. McKenna	14	Son	At school	Kentucky
Peter McKenna	12	Son	At school	Kentucky
Stafford McKenna	8	Son		Kentucky
Rhoda Camus*	19	Boarder	Servant	Kentucky
Willie Camus*	9/12	Son		Kentucky
N. Bodine*	35	Boarder	Farm worker	Kentucky
Sandy Bodine*	28	Boarder	Farm worker	Kentucky

Source: US Census of Population Manuscripts.
*African American.

1880

Name	Age	Relationship	Occupation	Place of Birth
Henry McKenna	61	Father	Distiller	Ireland
Elizabeth McKenna	59	Wife	Housekeeping	Ireland
Daniel McKenna	31	Son	Works in mill	Kentucky
Mary McKenna	25	Daughter	Housekeeping	Kentucky
James S. McKenna	23	Son	Works in mill	Kentucky
Stafford McKenna	17	Son	Student	Kentucky
James McKenna	58	Brother	Shoemaker	Ireland
John Burke	16	Servant	Works on farm	Kentucky
Adaline Cosby*	30	Servant	Cook	Kentucky
Sandy Bodine*	39	Servant	Works on farm	Kentucky
William Bryant*	34	Servant	Works in distillery	Kentucky
Milton Downs*	44	Servant	Fireman	Kentucky

Source: US Census of Population Manuscripts.
*African American.

the mill. Henry McKenna's brother James, a shoemaker, had joined the family in Fairfield. Again, four African Americans lived with the McKenna family: Adaline Cosby worked as a household cook, and three men worked at the mill and distillery.

Henry and Elizabeth McKenna's oldest son, Daniel, was active in the business as a young man, and according to their granddaughter Marcella McKenna, he was managing the business by 1881. When Daniel McKenna died in 1918, his brothers James and Stafford managed the distillery until Prohibition and then resumed doing so after repeal.[8] In 1933 James and Stafford incorporated the business as H. McKenna Incorporated, with assets that included a cash bank balance of $90,000; fifteen acres of land, including a water supply, valued at $8,122; buildings and structures valued at $49,091; and distilling equipment and machinery valued at $17,756.[9]

McKenna made no mention of the Civil War in his ledgers and letters that appraised the conflict's impact on his businesses or the wider community. But Marcella McKenna recorded the period in a 1966 presentation: "Certainly we are aware of the guerrilla scourge that swept over our county in force beginning its height in the autumn of '63 and reaching a diabolical crisis through '64," she wrote. "Our old H. McKenna distillery, in the original mill section, bears still the bullet holes with of course, the bullets still deeply embedded in many of [the] heavy old oak upright supports, put there by a ruffian band of guerrillas, giving evidence of the harassing 24 hours they gave the town of Fairfield and particularly the distillery."[10]

Henry McKenna Distillery, Fairfield, Nelson, County, 1886. (Marcella McKenna Papers, Oscar Getz Museum, Bardstown, KY)

After the Civil War, Henry McKenna gradually expanded and improved the distillery. He built a new still house in 1883 and increased his mashing capacity to about three barrels per day. The mill and granary stood beside the Bloomfield-Fairfield Turnpike, with the still house attached at the back. A brick chimney marked the location of the engine room. A cooper shop stood to the east, flanked by piles of staves stacked beside the building. The livestock feeding shed had a monitor roof and stood to the west, downstream from the distillery. The works eventually mashed 100 bushels per day before Prohibition suspended distilling.[11] Federal regulators designated the works as Kentucky Registered Distillery No. 111.

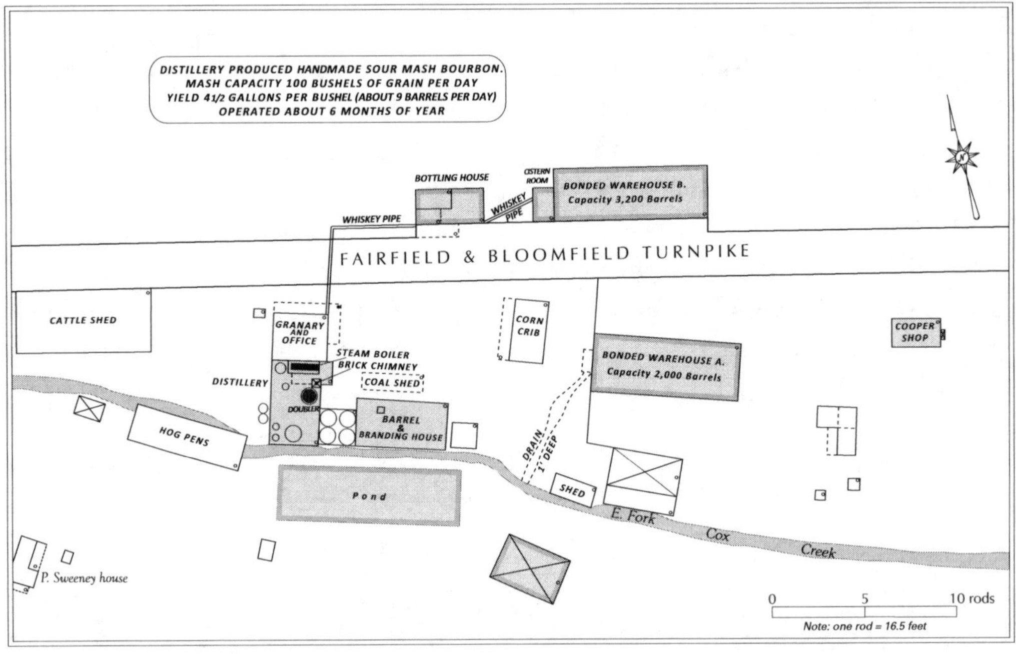

Henry McKenna Registered Distillery No. 111, Fairfield, Nelson County, ca. 1910.

By 1910, the distillery works included a still house, a mill and granary, two bonded warehouses, a bottling house, cooper shop, woodworking shop, ice house (location indeterminate), corncrib, coal bin, dwellings, and cattle shed and hog pen. An overhead pipe carried distilled whiskey over the turnpike to the bottling house. Relict distillery buildings still stand on both sides of the highway in Fairfield.[12]

The Granger

Soon after he established his mill, Henry McKenna began acquiring farmland. In 1858 he bought seed potatoes and fifteen bushels of rye from James Brewer, a twenty-nine-year-old farmer who lived near Bardstown. To this he added a disk and a sow, for which he paid $8 and \$10, respectively. As the farm developed, it became an adjunct enterprise to his mill and distillery.[13] Notably, Henry McKenna did not establish his distillery as a subsidiary business to a farm, as had been commonplace a generation earlier. Rather, he used his mill and distillery income to subsidize land acquisitions. In 1863 McKenna paid neighboring farmers \$150 for hogs. A year later, his breeding program was sufficiently successful that he sold \$1,220 worth of hogs.[14] He fenced in a feedlot adjacent to the mill and distillery, where he fattened stock on slop; he owned some of the hogs and cattle in the lot, and he also fed stock belonging to townspeople and laborers.

Within fifteen years of starting his milling operation, Henry McKenna had acquired 360 acres of farmland on the south side of Fairfield near his mill—140 acres of improved land and 220 acres of woodland (in 1870 the average farm in Kentucky was 158 acres). The farm's cash value was \$15,440. His livestock included 7 horses, 7 mules, 4 milk cows, 24 feeder cattle, and 225 hogs. He produced 700 bushels of winter wheat, 1,500 bushels of corn, 150 bushels of oats, 10 bushels of Irish potatoes, 15 tons of hay, and 250 pounds of butter. McKenna placed a value of \$2,660 on his farm products. His distillery used 50 bushels of grain per day during the distilling season. Discounting the corn he may have fed to his livestock, his entire annual corn production would have provided roughly one month's supply to his distillery.

McKenna likely had regular employees, such as a farm manager, as well as laborers who divided their time among the farm, the mill, and the distillery. He also hired part-time laborers during spring planting season and the summer and fall harvest. Other laborers tended livestock and built and maintained fences. Livestock butchering was a winter task that required several hands. Patrick Sweeney, McKenna's primary distiller, worked on the farm when the distillery was not in operation and helped with tasks such as butchering.[15] Jesse Lewis, an African American farm laborer, worked during the harvest season shocking grain. In 1871 McKenna paid Lewis 10 cents each for 117 shocks.[16] Benjamin Wiggington worked for more than two weeks harvesting wheat and rye, for which he received about \$16. Twenty-five-year-old farm laborer Henry Hughes lived five miles north of Fairfield, on Camp Branch Creek in Spencer County. He worked for McKenna during harvest season and received \$1 a day for two and a half days of threshing. He also dug post holes for \$1 a day and earned \$1.50 for one day of killing and butchering hogs, probably working at that task with Sweeney.[17] McKenna fenced his farm fields with split rails, the most common fencing material before the invention of barbed and woven wire.[18] Unless one used rot-resistant wood, such as chestnut, fence rails deteriorated and required periodic replacement. During the fall of 1871 Henry Hughes split enough logs to make 725 fence rails—each rail was typically ten to twelve feet long. McKenna paid Hughes a penny per rail, or \$7.25, for his work. The following spring, E. F. Tucker bought two loads of old fence rails from McKenna at \$2 per load, likely the rails he had replaced the previous fall. Tucker might have reused the rails in his own fences or cut them up into firewood.[19] But, good fencing did not always confine livestock. In September 1874 some of McKenna's mules strayed and got into a neighbor's oats and cornfields; he credited that person's account $30 in recompense.[20]

McKenna also offered area residents farm-related services. His farm, being adjacent to Fairfield, offered convenient pasturage for nine months of the year. Several Fairfield residents paid him to pasture their cattle or horses, for which McKenna charged $1.25 to $1.30 per month for each animal.[21] He also bought and sold farm machinery and rented it to other farmers. He may have held machinery as collateral on loans as well. In 1865 Syms Hutchings of Chaplin, in eastern Nelson County, rented a corn sheller from McKenna. Hutchings returned the sheller after two seasons and paid McKenna a $7 rental charge. The transaction suggests that McKenna had purchased a new sheller and could spare the old one.[22]

By the eve of the Civil War, a cadre of Irish-born residents had settled in northern Nelson County, perhaps drawn by the Catholic community already living there. Catholic migrants from Maryland, Clement Gardiner and Nicholas Miles, had established the settlement in 1795. Some immigrants worked for McKenna as laborers, tilling his farm and cutting his wood. Several carried accounts at McKenna's mill-distillery store. McKenna also provided assistance to some who wished to work their own farms. He likely helped Mr. A. Cleary establish a small farm near Fairfield in 1867. Cleary raised chickens, and he fed hogs on slop at McKenna's distillery. He bought staples from McKenna, as well as chicken feed and a hand plow. In May he hired McKenna's employees to plow, harrow, and plant a cornfield. They also hauled lumber in February and fence posts in August, and they plowed and sowed Cleary's winter wheat field in October. McKenna charged Cleary $22 for the farmwork and an additional $14 for pasturing a cow during the year. At the end of the year, McKenna bought Cleary's slop-fed hogs for $56, balancing his account.[23]

Within three decades of his arrival in Kentucky, road builder and miller-distiller Henry McKenna had also become, in traditional Irish idiom, a "strong farmer." The expressive term was an honorific reserved for independent farmers who had farmed lands owned by Irish and British landlords. As rent-paying tenants on expansive acreages, they managed to accumulate capital and property through skillful land and livestock management, and many became landowners.[24] By about 1900, the McKenna farm encompassed nearly 1,000 acres.[25]

The McKenna Business

In its most elementary form, Henry McKenna's business was an industrial market that bought and processed grain and a supply center that sold wheat flour, cornmeal, and bourbon and rye whiskey. McKenna milled farmers' corn, wheat, rye, and corn into flour, meal, and other products. He was also a merchant miller, in that he purchased grain that he milled and then sold the products to his customers. He expanded his sales by opening a small general store that offered a wide variety of items, including foodstuffs such as sugar, honey, molasses, salt, pork, and beef. McKenna sold his own whiskey as well as gin, rum, wine, and champagne; one could buy wagon bows, harness lines or reins, and plows and plow handles, as well as wheat, clover, and timothy seed and sweet potato plants. McKenna also maintained a wood yard, where he purchased and sold lumber and shingles. He bought cordwood to fire his steam engine, and he sold cordwood to customers for home heating and meat curing.

Henry McKenna kept financial ledger books for his mill, distillery, and cooperage from the 1850s through the late 1880s. The ledgers listed more than 350 different customer accounts for laborers and farmers. His patrons also included retail and wholesale outlets in Fairfield, Bloomfield, Mount Washington, and the surrounding area, as well as in Bardstown and Louisville.[26] In 1858, when the entries in Ledger 1 began, McKenna charged his customers 20 cents per bushel to mill their wheat. He sold wheat flour for about 2 cents per pound—a household of ten people might consume thirty pounds of flour per week. Cornmeal cost a penny per pound or 50 cents per bushel. He sold bran, the hard outer layer of wheat seeds that contains some starch, protein, and fiber, for about 50 cents per bushel. He priced shorts, a milling by-product that contains particles of bran and wheat germ, at 25 cents, the same price he charged for chicken feed. Customers who purchased by the barrel were charged for the barrel, but they received a 50-cent credit for barrels that were returned.[27]

Some customers made purchases several times a week; others shopped once a month or at irregular intervals. Many individuals bought staples and whiskey on credit. They paid their accounts in cash every six to twelve months, traded their labor for the goods they had purchased, or sold commodities such as cordwood or beef to McKenna, which he used or resold and for which he credited their accounts. In December 1860 McKenna's unpaid customer accounts totaled more than $1,000, which he carried, by and large, interest free.[28] Two years later, the December account summary included McKenna's personal debt of more than $2,800, moneys he owed to fifteen different individuals, including $900 to son Stafford and $300 to son James. His cash and stock on hand at the beginning of 1863 included more than $1,000 from whiskey sales and fourteen barrels of "old" whiskey, which he sold to a Louisville client for $600. He retained thirty barrels of new whiskey, which he valued at $540, and the wheat that remained in his granary would be worth $480 when made into flour. At the start of 1863, McKenna's net balance stood at just over $200.[29] Mail-order sales extended the distribution of McKenna whiskeys to a nationwide market after the Civil War, yet northern Nelson County remained the primary source of raw materials processed by the mill and distillery. The town of Fairfield and the surrounding farms provided the basic market for the grain-based staples and whiskeys the McKennas offered for sale.

Henry McKenna developed a strong economic and social relationship with the townspeople of Fairfield and the residents of the larger rural community in northern Nelson and southern Spencer Counties. Customer accounts reflect the personal arrangements he made, which varied according to vocation and living and working patterns. Farmers' accounts were different from those of day laborers, and artisans' accounts differed from those maintained by retail stores and saloons. The purchase patterns established by people in each group reflected adjustments required by their economic circumstances and by the impact of the vagaries of weather and other environmental conditions. But even within vocational groups, individual accounts were idiosyncratic, unique to people's interests and capabilities. Some customers deviated from their regular purchasing patterns by requesting that McKenna's employees plow their gardens, butcher their stock, shell their corn, or deliver goods to their businesses or residences. Some of these special requests suggest changes in the lives and economies of individual customers: the demise of a draft animal, a broken wagon wheel, or the purchase of a new piece of equipment; the birth of a child, death of a spouse, or change of residence. Events external to the community could also influence purchase patterns, such as the completion of a railroad branch and the construction of a depot. Even events half a world away, such as Ireland's Great Famine, had an effect: the massive out-migration that catastrophe prompted and the political movements that sprang from it, such as the Fenian Brotherhood's protests against British occupation and oppression.

A customer's account at Henry McKenna's store was more than a simple balance sheet of debits and credits; it was a kind of social barometer that reflected stability or change, lives lived in comparative financial comfort or at the margin of hardship and distress. McKenna's ledgers also reflect a dense social network. Many town residents maintained accounts at the McKenna works, and his farm accounts covered an expansive area of Nelson and surrounding counties. McKenna's ledgers were detailed and carefully tabulated. He sought to settle each account at some point during the year, yet he recognized the vagaries of customer need and the wherewithal to pay and was willing to accommodate the individual customer, whether a laborer, a farmer, or a wholesaler. When considered together, the accounts were grouped into general types according to ability to pay, skill and experience, the requirements of mill and distillery operations, the idiosyncrasies of seasonality, and the luck of largess or misfortune. An account with Henry McKenna functioned as cash payment or short-term credit transactions for some customers and as a lending institution or charity for others. Accounts were also open, in the sense that individuals could charge their purchases to the accounts of others—with permission, one presumes.[30] People also made payments on behalf of others, either to repay a debt or as a matter of friendship.

McKenna's mill and distillery became something of an organizational fulcrum for people's lives. The mill purchased the grain they produced and converted it into edible staples. It also fostered employment; McKenna not only hired hands to work at the mill and the distillery but also provided a market for those who labored to produce items that were useful to his works and to the larger community. It was, perhaps, the community's center of formal economic exchange. Although McKenna's production and accounting system sustained elements of an age-old barter economy, it also offered the opportunity to adjust the pattern of economic and social life to one based on monetary exchange.

While area farmers and townspeople likely perceived the McKenna operation as one of the most important institutions in their part of the county, patrons and employees probably understood the mill and distilling operations differently—the place held different meanings for different individuals and groups. A wholesaler's source of profit was a function of the price McKenna charged him and, in turn, the price he could demand—the primary concerns being product quality, transport costs, and a customer's ability to pay. For a day laborer, a McKenna credit account might have meant the difference between sufficiency and deprivation.[31]

Evolution of the McKenna Distillery

Henry McKenna's 1855 distillery was a simple oak timber–framed structure likely covered in clapboards. The original two-stage still was the product of folk tradition and somewhat simplistic by late-nineteenth-century standards, but it was representative of a process developed by distillers who could not afford a full complement of metal stills, pumps, and piping. McKenna's first-stage beer still was made of three nine-foot-long logs of tulip poplar (*Liriodendron tulipifera*), hollowed out with a "devil's axe" or long-handled scoop. The squared-off logs were then stacked atop one another, end to end, and fitted with an interior coiled pipe to convey indirect heat to the liquid contents; the assembly was referred to as a "three chamber still."[32] The top log chamber had a rounded top end, likely so shaped to better direct rising vapors into an extracting funnel and condensing coil. A "heater" brought the beer to the boiling point, whereupon the distiller drew off the liquid from the bottom chamber into large buckets and carried them across the distillery floor to the sixty-gallon mash tub. A laborer added middlings or cornmeal to the mash tub and stirred the mixture with a "mash stick" or rake until it thickened. This "held-over" mash was broken up by adding buckets of water, grain, and yeast and was left to ferment for three days. When fermentation was complete, a laborer used a dipping bucket to pour the mash into a wooden trough that carried it to the top of the beer still, where the heater brought it to a boil. The steam moved upward into the extracting funnel and copper coil or worm, where it condensed into singlings or low wine. The product then entered a wood-fired copper pot, the second-stage still, which brought the mixture to evaporative temperature. The following day the distiller ran the product through the wooden beer and copper stills again to produce the finished whiskey.[33] By the mid-1860s, McKenna's still was capable of producing one to three barrels per day.

McKenna updated his three-chamber log still in about 1878. The new works was an upright three-chamber still made of heavy poplar staves; the liquid moved via pipes and steam-powered pumps, replacing wooden buckets and troughs. A new brick distillery was built in 1883, and by 1912, the works had evolved into a modern plant with seventy mash tubs, a copper column still, a traditional pot-still doubler, and external flake stands or condensers. The still's daily grain-mashing capacity reached 200 bushels.

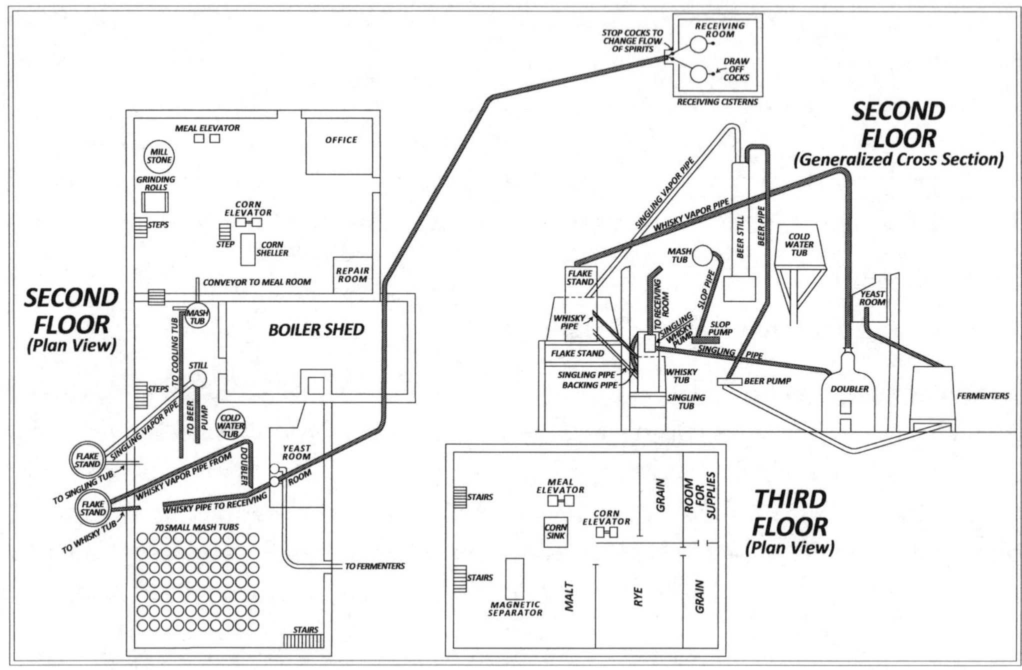

Supplemental plan for H. McKenna Registered Distillery No. 111 by J. P. Curtin, draftsman, 1912. Note that the various rooms were necessitated by internal revenue regulations requiring separate rooms for different functions. (Based on the original document in Marcella McKenna Papers, Oscar Getz Museum, Bardstown, KY)

Internal revenue regulations passed by Congress in 1868 and 1877 had a dramatic effect on mill and distillery operations. Under the law, milling and distilling had to be separate functions, and distilleries could not conduct ancillary businesses such as retail stores selling mill products. McKenna's business had begun as a flour and gristmill, and he continued to conduct business transactions there after adding the distillery. The mill and distillery buildings were adjacent but apparently did not violate the 1877 law, which specified: "Where there are grist-mills adjoining a distillery they must be so entirely separated by solid walls and otherwise as to be in fact independent premises, although both premises may be driven by the same power. They must be so arranged that the description [plan drawing] of the distillery premises does not include the mill."[34] The plan for McKenna's new brick distillery built in 1883 would have required internal revenue approval. By 1912, the two-story mill was separate from the distillery, and the distillery's third floor was divided into separate rooms for different grains. Yeast preparation took place in a dedicated room set off from the mashing floor, and one entered the engine and boiler room through a single doorway. The whiskey-receiving cisterns were in a separate building connected to the flake stand by a long pipe. The official distillery plan, drafted by J. P. Curtin, gave no indication how fuel was delivered to the engine.

Operations

Henry McKenna distilled both rye and wheated whiskeys and corn-based bourbon whiskey; the conversion from wheat to corn allowed him to take advantage of the price differential between the two grains. A sale of ten pounds of hops to farmer Samuel Lancaster in April 1858 suggests that McKenna not only bought and sold hops but also may have used them in distilling, especially in the preparation of three types of yeast: jug yeast, day yeast, and night yeast.[35] The recipe for jug yeast required the following:

> 2 buckets hops, 2 bushels cracked malt, 20 gallons of water, and salt. Commence at 9 AM. Scald hops in 20 gallons of water. Let them come to a thorough boil and boil for 10 minutes. Then strain and with this juice mash the malt at about 150 [degrees]. Let this mash stand 3 hours. Then, by night, cool down to 80 [degrees]. Then add 1 gall[on] of jug yeast and about a double-handful of salt. Then cool to 74 [degrees] and wrap Can with several sacks; then in the morning put Can in a cool place for about 2 days, until fermentation ceases. Note. If there should not be enough of the Malt-Extract to fill jug pour from 3 to 5 gall[on]s of Hops-Juice, at intervals of one-half hour, on the mashed Malt.

The night yeast recipe also called for a "bucket of hops" to make hops water. Day yeast required rye but no hops.[36]

The distillery initially produced slop, its primary by-product, in small amounts. James Marshall bought staples from McKenna as well as slop, for which he paid one-half cent per pound. In 1858 Marshall bought meal, flour, and 60 to 140 pounds of slop each month, which he may have fed to livestock, although the small volume suggests that the slop was a supplement and not his primary feed source.[37] Sometime in the late 1850s McKenna began feeding distillery slop to cattle and hogs onsite.[38]

By the early 1880s, the McKenna family was conducting a wholesale and retail liquor business at the distillery and at a business office at 245 Fourth Street in Louisville. H. McKenna was the company's official name. Henry's son Daniel served as distiller, and he and his brother Stafford managed the business. Sales correspondence originated at both the Louisville office and the Fairfield distillery. In 1882 the D. J. Lake atlas of Nelson and Spencer Counties carried a business advertisement that read: "H. McKenna, Distiller of

Pure Old Line Sour Mash Whiskey. Wholesale and Retail Liquor Dealer. Old whiskey for medical use."[39] An announcement in *Bonfort's Wine and Spirit Circular* in October 1883 proclaimed, "this establishment limits its stock to the actual wants of fine drinkers, making a specialty of old sour-mash whiskies."[40] Four months later, the H. McKenna Company announced receipt of an order from a Dr. Griffiths for two gallons of its best bourbon for presentation to Albert Edward, the Prince of Wales.[41] By 1889, the company reported that sales were far ahead of expectations and, as a consequence, it had "run short of all ages of the 'McKenna' whisky." Daniel McKenna was distilling 1,000 barrels a year, although the reported demand was for twice that amount.[42]

Equipment

Henry McKenna and his sons periodically updated the mill and distillery buildings and equipment. In 1892 Daniel McKenna contacted the Dean Brothers Steam Pump Works in Indianapolis, Indiana, concerning the purchase of a new distillery pump and boiler. Dean Brothers, one of the largest pump manufacturers in the country, sent a catalog and promised to sell anything listed at 40 percent off.[43] Inventor Nelson P. Bowsher patented a new type of feed mill in 1894, which McKenna apparently purchased. The new steam-powered design ground grain and corn and could also be used to crush corncobs. In 1897, when the worn parts required replacement, McKenna contacted Bowsher's company in South Bend, Indiana, manufacturer of Bowsher's Combination Feed Mills and the Globe Sweep Feed Grinder, to order a new set of "knives." The company responded:

> Dear Sir:
>
> Your order calls for only a set of knives; but we are inclined to think you want both knives and grinders. We have several different dress of grinder patterns and before making shipment would like to have a little more information in regard to the work that you now intend to do. If possible, we would like you to mail us a small sample of the material to be operated upon. Let us know if it is dry or damp and also about how fine a reduction you will need to make; then we will provide you with the dress which we think will be best suited to your work. We are anxious that the mill should give you the best service possible. We will ship the goods promptly on receipt of your reply.
>
> Yours truly, the N. P. Bowsher Co.[44]

While McKenna relied on local carpenters, brick masons, blacksmiths, and other tradespeople for basic building construction and other artisanal work, he also kept abreast of inventions and improvements available from national manufacturers, and he employed new technologies to make his works efficient and reliable. The Bowsher Company's letter also suggests that mill equipment manufacturers had progressed well beyond mere

"brute-force" milling technologies and were capable of nuanced engineering that accounted for differences in grain moisture content and other customer requirements.

Cordwood

Henry McKenna bought and sold locally produced wood. He used cordwood in large
quantities to fire his still boiler and to fuel his steam engine. He also likely heated his home
with wood. He sold small quantities of wood to some customers, likely for cooking and
heating and possibly for smoking meats.[45] McKenna bought wood from itinerant laborers
who cut trees from their employers' land and from farmers who cleared their own land
during the winter months. Edward Fitchoner had a small account with McKenna for flour
and bran. He also sold a dozen cords of wood—one cord of ash and eleven cords of "sugar"
(maple)—to McKenna during the summer and early fall of 1858. Fitchoner received $1.75
per cord.[46]

In September 1859 Jack Wiggington delivered seventy-one cords of wood in fifty-sev-
en wagonloads, for which McKenna credited his account $124.25. If laid in a linear stack,
the seventy-one cords would have been 4 feet wide, 4 feet high, and 568 feet long. Add to
that the fifty-two cords McKenna bought from Spencer County farmer Benjamin Miller, and the stack's length would have approached 1,000 feet.[47] To store and air-dry this
much wood, McKenna probably had a dedicated wood yard near his mill's engine room.
Although Wiggington purchased no staples from McKenna, he used his account to designate payments to other account holders to whom he owed money.[48] The following year,
Wiggington delivered seventy loads of wood to McKenna in August, September, and October, for which he received $1.75 per cord—tabulated this time as loads, but paid in total
cords. Wiggington drew cash from his account and again designated payments to other
account holders, including $19 to cover a bill from Fairfield physician Dr. B. B. Wootten. In
December 1860 Wiggington owed McKenna $175. He apparently continued to haul wood
until his credits matched his debt, because his account was balanced at the end of the
year.[49] Again, in the fall of 1861, Wiggington delivered wood, but this time only twenty
cords in sixteen loads, for $35. Still, that amount was sufficient to balance his debits for
staples and meat and to pay another bill from Dr. Wootten.[50]

As some wood suppliers disappeared from McKenna's account ledgers, others entered to fill the demand. Farmer B. Wiggington, perhaps a relative of Jack Wiggington, cut
and delivered forty-six cords of wood from January through May 1870 and earned $2.50 per
cord. The following year, he sold 1,400 bushels of corn to McKenna, much of it hauled to
the mill and distillery from March through May.[51] John Offutt, a farmer living near Camp
Branch in southern Spencer County, occasionally worked as a teamster for McKenna and
chopped wood for the distillery in 1870. He bought a horse harness from McKenna and
delivered forty-two cords in November of that year, earning $3.50 per cord, or $147; that
was enough to cover his $126 whiskey bill.[52] Laborer Jack Steel cut ninety-eight cords of

wood and sold it to McKenna for 75 cents per cord. The steep price differential between Offutt's wood and Steel's may have been based on wood quality and the cost of hauling, which McKenna likely had to provide. If Steel was cutting wood on the dissected hills near Jacks Creek, Simpson Creek, and Camp Creek in Spencer and Nelson Counties or in the shale hills to the east, he may have been cutting stands of black oak and beech, species that favored the sour soils in those areas but were not desirable as heating fuel.[53]

Farmer Edwin Bridell was one of Henry McKenna's largest wood suppliers. Bridell delivered one to four cords of wood per day from early November 1871 through May 1872, for a seasonal total of 160 cords.[54] Farmer Archibald Pitt lived on the Fairfield–Samuels Depot Turnpike southwest of Fairfield. In 1875 his account with McKenna reflected only wood credits. By early October, "his team" had delivered thirty cords of wood for $90. From October 25 to November 2, McKenna's teamster picked up three cords of wood per day from Pitt, paying $1.50 per cord, or $124.50, which McKenna remitted at the end of the year.[55] Both Bridell and Pitt were likely clearing their farms of trees to increase their tillable acreage.

Logistics

Henry McKenna's mill and distillery initially served as a local and regional customer market and expanded to serve a national market as the nation's nineteenth-century railroad network developed. Fairfield residents could walk to McKenna's store and return home with their staples or have them delivered by wagon. Early on, he served the larger regional market, primarily Bardstown and Louisville, by wagon. McKenna's ledgers often recorded hauling payments made to laborers or teamsters, and in some cases they listed destinations. In addition to serving Fairfield and the surrounding farms, McKenna's retail and wholesale customers came from Spencer County to the north, Chaplin to the east, Bardstown to the south, and Louisville to the northwest.

McKenna made at least one whiskey shipment via the steamboat *Fannie Fraser* in December 1883–January 1884 to Ransdell & Birchett at Monterey, Kentucky, a town on the Kentucky River between Frankfort and Owenton. The record does not indicate from which river or landing the shipment originated. The small intermittent stream that flowed past McKenna's distillery, the East Fork of Cox Creek, flowed west for several miles before joining the Rolling Fork, the area's trunk stream. The imperfectly navigable Rolling Fork was shallow, filled with snags, and obstructed by overhanging trees that hindered boat passage. Though it was too crooked and encumbered for steamboat navigation, small flatboats descending the stream in high water may have carried local cargo to the Ohio River.[56] It is more likely that McKenna shipped his whiskey to Monterey by wagon east to Bloomfield, then by train to the Kentucky River at Frankfort, and finally by steamboat. Each stop and transfer would involve unloading and reloading, which would have added considerable expense to the overall transport bill. Whatever route this shipment traveled,

McKenna received a confirmation letter from Ransdell & Birchett stating that the whiskey had arrived four days after leaving the distillery.[57]

Distillery-fed livestock walked to market under the care of drovers. Laborer Farney Simpson worked for Henry McKenna for two months in 1860. His tasks included droving McKenna's hogs to Louisville. Simpson returned to Fairfield via stagecoach, and McKenna credited his account $3 for two coach fares.[58] All other raw materials processed by McKenna's mill and distillery, as well as the products they produced, had to be hand-carried or hauled by pack animal or some mechanical contrivance.

If nineteenth-century millers and distillers wished to make the transition from waterpower and take full advantage of the mechanical efficiencies inherent in steam power, they had to make parallel improvements in allied technologies. Because the highest waterpower potential was available only at fixed locations, often in steep-gradient stream valleys away from productive farmland, water-powered mills required a link to the grain supplier and another link to the market for the mill's products.[59] Steam power offered mechanical efficiencies over waterpower and removed concerns about drought, seasonally low water levels, and territorial and riparian water rights. Steam engines could be installed almost anywhere well or surface water was available, but they also required a reliable link to a fuel source as well as a market. Links mean transportation, of course, and industrial milling and distilling could thrive only if they had access to industrial transportation—steam trains and steamboats. But Henry McKenna's enterprise lacked direct railroad access; consequently, his transport links were in large part animal powered. In the 1850s personal travel was by foot, horseback, carriage, or stagecoach. Commercial travel required a cadre of well-equipped teamsters if steam trains were incompetent or beyond reach.

Turnpikes

One could travel about Nelson and the surrounding counties on old country tracks—vernacular roads—or on turnpikes. Track roads were usually poorly maintained by local residents, and most were impassable by coaches and wagons in wet weather. Well-built turnpikes were all-weather roads, comparatively smooth, and resistant to heavy loads if well maintained. One had to pay a toll to use turnpikes, but stagecoach lines and freight haulers preferred them to the alternative. Turnpike companies built several roads in Nelson and surrounding counties, including the Bloomfield-Fairfield Turnpike that ran past McKenna's distillery. Five pikes connected Bardstown, the county seat, with outlying towns, including the Bardstown-Louisville Turnpike and pikes to New Haven, Boston, and Bloomfield. The Fairfield–Samuels Depot Turnpike connected the McKenna Distillery to the L & N Railroad northwest of Bardstown. Irishman Frank Sheehan was the tollgate keeper on one of the turnpikes near Fairfield. In addition to collecting tolls, the gatekeepers were traditionally hired to maintain the sections of pike close to their tollhouses. Sheehan had small whiskey account with McKenna in 1873.[60]

HORSES

McKenna's wagon teams were likely made up of mules and light horses. His ledgers contain few details, except to mention that he paid George Royalty $8 for a jack to service one of his mares.[61] Lightweight cart and carriage horses lacked the heft required to pull heavily loaded wagons over indifferent roads. Commercial hauling, be it on city streets or country roads, required heavy horses, appropriate harnesses, and stout wagons. America's Conestoga and Morgan horses were too light for industrial work, and their breeding produced inconsistent results. European draft breeds, such as the Belgian, Percheron, Clydesdale, and Shire horses, began to appear in the Middle West after the Civil War, and their blood stock was present in Kentucky by the 1880s.[62] In 1889 the editor of the *Maysville Evening Bulletin* noted that a farmer in nearby Bracken County had purchased a pedigreed Percheron stallion in Illinois. "Farmers will now have a splendid opportunity to improve the size of their horses," he enthused, as "the horses of the country have been growing too small."[63] Newspapers and livestock journals urged city draymen and farmers to consider breeding their best stock to "heavy horses," noting that "the heavy draft horse will stay with us till electricity or some other motive power can drag monster trucks and express and beer wagons through city streets."[64] Many farmers and teamsters considered the Morgan the best horse for light work, and some deemed a cross of three-fourths Percheron and one-fourth Morgan ideal for most draft work. Such animals weighed up to 1,400 pounds, which, though considerably less than the weight of a full-blooded Percheron, was sufficient to haul a wagon filled with heavy whiskey barrels.[65]

TEAMSTERS

Just about everything arriving at Henry McKenna's mill and distillery—grain, cordwood, lumber, hardware, paint, oak staves, and iron hoops—moved by wagon, as did shipments of flour, meal, and whiskey to near and distant customers. McKenna likely owned at least one delivery wagon, a horse or mule team, and the appropriate harnesses. In the 1860s he bought a new harness from James Whealon's shop in Bardstown and returned there when the harness required repair.[66] McKenna employed laborers to load and drive his wagon to make deliveries, but he hired teamsters or farmers who owned their own wagons to do most of his hauling on an "as available" basis.

In the early 1860s farmer Raphael Hagan had an active account for purchases of flour, sugar, salt, and lime, which McKenna's teamster delivered at a charge of $1 per load. McKenna also advanced cash and paid drafts against Hagan's account to people Hagan owed money to. At the end of one year, Hagan's debt balance stood at almost $100. To balance the account, McKenna drew $25 against the account of another person who owed Hagan money. He also bought a hog from Hagan for $16 (a high price at the time) and forty-eight bushels of wheat, bringing Hagan's total credits to about $93. Hagan settled his McKenna account with a small cash payment.[67]

McKenna installed a large wagon scale at his mill, and he often paid teamsters according to the weight of the freight they hauled. He also paid by the number of barrels or other bulky items in a load, and he may have adjusted his charges based on the distance traveled. In 1858 McKenna sold a set of wagon bows to John Cassel, one of his haulers.[68] Bentwood bows formed an arched framework above the wagon box to support a canvas cover. That transaction suggests that McKenna's freight moved in covered wagons, a simple mode of securing perishable cargo from the elements. He may have required this type of setup of all his teamsters.

Most teamsters who maintained accounts with McKenna paid them off, at least in part, by hauling. Teamster Francis Bowels bought modest amounts of flour, meal, sugar, whiskey, bran, and chicken feed on credit. He hauled freight for McKenna throughout 1858 and 1859, averaging a load every two weeks. His freight weights ranged from as small as 50 pounds to as large as three-quarters of a ton, although most loads were between 100 and 250 pounds—likely local deliveries to people who had ordered flour by the barrel or meal by the bushel.[69] McKenna paid haulers 4 cents per pound, so Bowels earned $8 for delivering a 200-pound load. McKenna did not identify most of the freight that Bowels carried, although two loads contained "iron" and plow handles. Most hauls credited to Bowels's account were recorded simply by weight and payment. Conveniently, at the end of the year, Bowels's credits and debits balanced out.[70]

For teamsters who made longer trips and used their own wagons and draft animals, McKenna paid them by the day. Edward Bennit received $3.50 per day, or $10.50, for "3 days of team hauling" to Samuels Depot in 1871, the mill's nearest railroad connection at the time.[71] Laborer Stephen Smith lived in Bloomfield in the early 1860s and did general hauling for McKenna throughout the year.[72] Smith made frequent trips to Louisville, hauling barreled flour to the city and returning to the Fairfield mill with the empty barrels. In 1863 he made several trips to Louisville to haul back empty whiskey barrels, a dozen barrels per load. For each trip, McKenna credited his account $1.80. Smith continued to do general hauling through the summer, and in September he drove his team to Louisville and returned with 440 pounds of paint and oil that McKenna used to paint his mill.[73] William Osburen also made trips to Louisville in 1863, returning with empty whiskey barrels, and he received $4.50 for a day spent hauling sawmill lumber to the distillery, possibly stave stock for the cooper shop.[74] Teamsters making long-distance hauls likely patronized turnpikes, incurring toll fees and expenses for overnight stays at roadside taverns. S. B. Finley of Bloomfield hauled freight to Louisville for McKenna in 1863, using McKenna's team and wagon. McKenna credited his account $5.15 for the cost of room, board, and stable facilities for the team at the Thirteen-Mile House on the Bardstown-Louisville Pike.[75]

In the late 1850s Alfred Bodine was one of the most prosperous farmers in northern Nelson County. The combined value of his real and personal estate stood at $57,000.[76] He owned sixteen enslaved people: seven males aged four to forty-three, and nine females aged one to fifty-eight. Bodine's account with Henry McKenna in 1861 included credits for hauling freight to and from Louisville. A trip in March hauling 3,600 pounds of unspecified freight earned a credit of $9. In September he was credited $2.50 for hauling plank from Louisville to the distillery.[77] Given the enslaved labor at his disposal, it is unlikely that Bodine was doing this loading and hauling himself.

Fairfield area farmer Alfred Thomas bought flour, bran, and whiskey from McKenna during the 1870s; in return, he sold timothy seed to McKenna. He also provided stud services, breeding one of McKenna's mares to his stallion and a sow to his boar for stud fees of $8 and $3, respectively. Thomas also earned credits toward his account by hauling. In 1872 he received $15 for "three days of team hauling whiskey to Samuels Depot."[78] From 1873 through 1879 Thomas made repeated deliveries of whiskey, nails, bottles, sugar, molasses, coffee, oats, potatoes, and salt. For some hauls, McKenna paid a daily rate of $5. Other trips were likely local deliveries, for which Thomas was paid according to weight and distance.[79] In the fall of 1880 another teamster used Thomas's team to haul a variety goods for McKenna, including 2,500 pounds of hoop iron, presumably from a Louisville supplier. McKenna also obtained 265 bushels of malt divided in three loads—one of the few mentions of malt in the ledgers—in November and December, likely in preparation for the winter distilling season. And in early January 1881 McKenna credited his teamster with $5.20 for "hauling [an] engine back and forward," presumably the steam engine used at the mill and distillery.[80] Several other farmers and laborers hauled freight for McKenna during the 1870s and 1880s, each earning sufficient credits to balance their accounts for purchases of staples and whiskey.[81]

The Railroad Connection

The Kentucky legislature chartered the Cumberland & Ohio Railroad Company in 1869. The line was to run from the Ohio River at Covington-Newport south to Lebanon and Greensburg by way of Bloomfield. The north section was never built. The Cumberland & Ohio obtained a right-of-way and partially constructed a rail bed, which the company then leased to the Louisville, Cincinnati & Lexington Railroad in 1879; the company completed the twenty-seven-mile branch line from Shelbyville to Bloomfield the following year.[82] The Louisville & Nashville Railroad acquired the branch line from Shelbyville to Bloomfield, and by 1888, the US Post Office Department had designated the track mail route number 20,026.[83] The Bloomfield depot was about four miles east of the McKenna Distillery at Fairfield and sixty miles from Louisville. Freight and passenger trains arrived and departed each day, expediting mail delivery and whiskey shipments. Importantly, the comparatively close railroad connection allowed McKenna to consider switching his fuel source from wood to coal. He contacted P. C. Kelleher, general manager of the Main Jellico Mountain Coal Company in Whitley County, Kentucky, to request a price quote for lump coal. The cost to McKenna, for the "Best of Bituminous Coals Kensee," was $1.65 per ton FOB.[84]

Employees' and Customers' Accounts

Operating Henry McKenna's mill, distillery, and farm required employees who possessed a range of skills and experience. McKenna hired some full-time workers who lived nearby in Fairfield. He hired other people as day laborers; some lived nearby, and others came from area farms. Almost all these people were engaged in seasonal tasks. Milling activity increased in association with the grain and corn harvests. Preparation for the distilling season began in October, and the distillery operated during the cold months of winter and early spring. McKenna's farm required the usual assortment of tasks that changed with the season. Adapting to this vocational variety required that McKenna recognize ability and apply it to those activities where it would be most productive. Even though he was a serious businessman who methodically entered even minor transactions in his account ledgers, McKenna was generous and charitable in extending credit and providing work for those who needed to pay off their balances. A detailed review of a sample of his accounts demonstrates that making flour at his mill and whiskey at his distillery involved much more than processing grain and selling and transporting the finished product. McKenna created a kind of benevolent social and economic institution that became the hub of community employment, finance, and social interaction.

The Distiller

When Henry McKenna began distilling in 1855, he was "aided by an old Negro slave," who had likely been hired from a nearby slaveholder; no records verify that McKenna himself was a slave owner.[85] By 1860, nineteen-year-old Patrick Sweeney had become McKenna's distiller. Sweeney was born in County Derry, Northern Ireland, in 1841, making him a countryman of McKenna's. Sweeney likely came to America as a child in 1844, on the eve of the Great Famine. He lived in Philadelphia for a time and eventually, as a young adult, made his way to Fairfield, where he found work with McKenna and boarded in his house. Whether instructing his sons or others employed at his distillery, McKenna's operational credo was: "Never in making whisky sacrifice quality for quantity." He likely passed this work ethic along to young Patrick Sweeney.[86]

Sweeney's pay schedule was not specified in his employer's ledgers, but his earnings totaled $450 per year. It is unlikely that he was paid an annual salary. If he worked six-day weeks and received time off for holidays and personal business, a year of roughly 300 workdays at $1.50 per day would total $450, a wage substantially higher than the $14 per month McKenna usually paid part-time laborers. This arrangement would explain the payments McKenna entered as credits to Sweeney's account for various nondistilling jobs he performed. The distilling schedule was seasonal and likely irregular, allowing Sweeney to work at tasks related to McKenna's farm and mill.[87]

Sweeney maintained a charge account at McKenna's store, where he regularly purchased flour, meal, beef, pork, salt, tobacco, whiskey, and cordwood and drew cash from that account against his salary. McKenna also paid Sweeney's account at Simpson's Dry Goods Store in Fairfield, other bills that Sweeney incurred in Louisville, and charges for medical care. On January 30, 1865, for example, McKenna paid Dr. H. Wells $17 to cover Sweeney's medical bills and entered the amount in Sweeney's payment-due account.[88] In the 1860s and 1870s Sweeney sent periodic donations of $1 to $5 to the Irish Fenian Brotherhood, an organization founded in the United States in 1858 by John O' Mahony and others seeking an end to British rule of the Irish Republic. He drew cash for these donations from his account with McKenna.[89]

In 1863 Patrick Sweeney married Ellen Dunphy, originally from Ireland but more recently a resident of Spencer County, north of Fairfield. He purchased a small house and a five-acre property on the hillside between McKenna's house and the distillery's hog pen and cattle shed along the East Fork of Cox Creek. The Sweeneys' son, John, was born later that year. Patrick Sweeney planted and tended a large garden, and on March 28, 1865, one of McKenna's farmhands plowed Sweeney's garden, for which McKenna charged him $3. Sweeney then bought three to four bushels of potatoes at the McKenna store for $2.25, likely to use for seed.[90] Sweeney also kept a few hogs; McKenna charged his account $2 for feeding the hogs, possibly on slop in the distillery's hog pen. In early winter 1865, butchering season, a McKenna employee butchered Sweeney's hogs, for which McKenna charged his account $1.75. Sweeney bought a calf for $15 from George Drake (who earned $14 a month working as a laborer for McKenna), which McKenna charged to Sweeney's account.[91]

For more than two decades, Patrick Sweeney worked as Henry McKenna's distiller and performed a variety of other tasks for which he earned credits to his account. In addition to the usual staples, his expenditures often included purchases on behalf of other people or for personal items that McKenna did not stock. From August through December 1868, his purchases included a pair of pants and a coat ($19), 700 pounds of pork at 10.5 cents per pound ($73.50), fifty-six pounds of beef at 8 cents per pound ($4.48), and a cord of wood ($4). The wood may have been for home heating and cooking. Sweeney also paid $10 for pasturage, likely for his cow. In November Sweeney purchased $3 worth of flour for "orphans."[92] The pasturage charge continued into 1869, but Sweeney's cow produced two calves, which he sold to McKenna for $20, an amount sufficient to pay his annual pasturage bill.[93]

In January 1870 Sweeney sent a $1.20 donation to the Fenian Brotherhood; his charity also included purchases of cornmeal and several cash withdrawals for local individuals.[94] Sweeney's purchases during the fall months included a barrel of salt ($5.75), perhaps for curing the 95 pounds of beef and 400 pounds of "hog meat" he purchased at about the same time. Ellen Sweeney also made cash withdrawals on her husband's account, some for $1 and one for $22.[95] For the year, Sweeney earned $393 for nine months and twenty-four days of work, which, not coincidentally, was exactly the total of his purchases and cash withdrawals, leaving him with a zero balance at the end of the year.[96]

In early January 1871 Sweeney milled grain for McKenna, earning a cash credit of $21.50. On January 19 Sweeney started work at the distillery; his pay was now set at $65 per month. He worked at the mill and distillery until June 24, "leaving out some days lost by him," and earned $325 for the five months. Sweeney's whiskey account was very active during the year; from January through July he purchased whiskey in pints, quarts, and gallons every few days.[97] Sweeney's nine-year-old son, John, may have been ill; he died the following year.[98]

The records for 1872 and 1873 are incomplete, but McKenna hired Sweeney in early February 1874 for four months (less three days) at $65 per month, or $252.50.[99] McKenna did not specify which tasks earned this pay rate, but given previous years' activities, Sweeney was likely working at the mill and the distillery. Sweeney bought shingles and paid carpenter Richard Constantine to shingle his house in the spring of 1874.[100] On June 30 Sweeney borrowed $227.25 from McKenna to purchase a threshing machine. In late December McKenna paid Sweeney $38.85 for threshing and $26 for "one month's work at different times such as gathering corn."[101] In December 1875 Sweeney earned $195 for three months of corn shelling, and he received an additional $27 for working extra days. His distillery work was modest, as the season was just beginning; he mashed 162 bushels of corn and 75 bushels of wheat, for which he was paid $15.89.[102]

In 1877 Sweeney's account contained regular entries for purchases of flour, meal, and whiskey and for cash withdrawals. On May 26 he received credit for two months and three days of "stilling" at $65 per month, or $137.50 total. On August 25 Sweeney withdrew $12.25 in cash, which he gave to Father O'Calahan, the priest at St. Michael's Church, located across the street from Sweeney's house on the east side of Fairfield.[103] In November McKenna provided a rare detailed summary of Sweeney's employment during the second half of the year. Beginning with a credit of $152.35 carried over from the first six months of the year, Sweeney worked thirty-nine days at 75 cents per day ($29.25), thirty-seven and a half days "before stilling" ($27.50), six and a half days harrowing at $1.50 per day ($9.75), two and a half days mashing ($2.50), and forty-six days at 75 cents ($34.50). Harrowing was a high-paying chore, perhaps because Sweeney provided the mule and the harrow to complete the job.[104] The following year, 1878, Sweeney worked at the distillery for five and a half months, earning $330; McKenna billed him $12 for pasturage.[105]

In December 1882, at the beginning of the distilling season, Sweeney bought six hogs from farmer Thomas Grigsby for $67.80. He may have intended to feed the hogs distillery slop and supplement that ration with corn because soon thereafter, he started buying corn every two weeks in single-bushel lots at 40 cents per bushel. He also bought two cords of wood in March, perhaps for home heating and cooking or for smoking meat.[106] Although McKenna built a new still house in 1883, there are no ledger entries that catalog the expenditures for labor and materials. Sweeney worked 121 days in 1884, ending on November 29, for $120, or about $1 a day. He began stilling on December 1 and earned $65 for that month, as he had the previous year.[107] His 1884 account included charges for flour, wood, whiskey, and corn and cash withdrawals totaling $385.35, and he carried a credit balance of $817.83 forward to the beginning of January 1885.[108]

Sweeney's 1885 account listed the usual charges for flour, meal, wood, whiskey, and cash, but he also purchased potatoes from McKenna and fish from Louisville. His wife Ellen made four cash withdrawals in March, April, August, and October totaling more than $91. For his part, Sweeney earned $415 for six months and ten days of stilling at the standard rate of $65 per month, and a dollar a day when he worked seventy-eight days during the summer and fall months.[109] By January 1886, Patrick Sweeney's account balance totaled $920.63. His weekly expenditures for meal, corn, cash, and whiskey continued, but he also bought more wood—several cords in January and May. In May, perhaps reflecting home improvements, he bought a door for $3.50. Importantly, he made his first coal purchase—a significant amount—in September, and he ordered a new stove and some pans for which he paid $18 plus 70 cents for freight. Most noteworthy, however, the McKenna Distillery was in extended production that year. In addition to working for twenty-four days during the late summer and early fall for $24, Sweeney worked in the distillery for eleven months, earning $406. In August he subscribed to the *Catholic Advocate* ($4.50), and his account balance at year's end stood at $1,490.[110]

The last two years of Sweeney's account ledger, 1887 and 1888, continued the familiar pattern of cash withdrawals and purchases of meal, flour, and whiskey. In 1887 Sweeney worked twenty-four days and earned $24, but he also worked at the mill and distillery for six months and four days, for which he was paid $400.[111] He undertook a fencing project in November, for which he bought plank ($5) and posts ($2.50) from McKenna; he paid J. A. Allen $6.40 for wire and staples. Sweeney's account balance at end of 1887 was $438.91, from which McKenna deducted purchases totaling $401.50, giving Sweeney a credit of $37.41 to carry forward into 1888.[112]

Henry McKenna's account ledgers do not provide an intimate view of Patrick Sweeney's personal life, but they portray him as a valued worker who performed several different types of tasks and engaged in other activities that supported his family, such as gardening and raising livestock. The ledgers also demonstrate that McKenna's account system was sensitive to the variability of the community's cash, barter, and credit economy.

Ellen Sweeney was fifty-six when she died in 1897; Patrick Sweeney died in 1904 at age sixty-three. His successor as distiller, Allen "Coleman" Bixler, began working for the McKennas the following season and remained with them until Prohibition in 1920. After ratification of the Twenty-First Amendment in 1933 repealed Prohibition, James and Stafford McKenna set about reconditioning the distillery for operation; they also rehired

Bixler. Sometime thereafter, the H. E. Pogue Distillery at Maysville engaged Bixler as its primary distiller.[113]

The Coopers

A skilled cooper who could build slack barrels for flour and tight barrels for whiskey was essential to the operation of a nineteenth-century mill and distillery. Two coopers worked for Henry McKenna: Abner Blanford and Henson Hopewell. Hopewell's sons William Henry and Samuel also became coopers. The coopers paid for their own tools, bought stave lumber and hoop iron, and rented McKenna's cooperage shop. The prices McKenna paid for barrels were variable, but an average was 45 cents for a flour barrel and $1.85 for a whiskey barrel.

Farmer and cooper Abner Blanford sold cordwood to Henry McKenna in 1858 and 1859, and each month he purchased flour by the barrel and meal by the bushel. He also bought whiskey by the gallon, every two weeks, and purchased garden produce from McKenna that may have been grown by a local farmer. For his household, Blanford purchased salt in barrels for $2.45 and sugar in barrels for $21.20, as well as coffee and candles. He bought chicken feed and kept livestock, feeding them, at least in part, distillery slop, for which McKenna charged him 70 cents for 200 pounds. In September 1859 McKenna paid Blanford $13.25 for putting fifty-three hoops on his mash tubs and $495 for 1,100 new flour barrels. In March 1861 Blanford sold McKenna 198 flour barrels at 45 cents each, or $89.10; two whiskey barrels earned him only $1.35 each.[114]

Blanford made and sold sixty-eight flour barrels in 1862. In 1866—the year his account ended—he bought flour, meal, and plow handles. Importantly, he also paid $75 to buy McKenna's cooper shop.[115] Abner Blanford's account did not reappear until 1877, when his purchase priorities were different. He no longer bought staples at McKenna's store; nor did he sell barrels or supply cordwood. He purchased only whiskey, for 55 cents per quart. From August 8, 1877, to September 24, 1878, he bought thirty-five quarts, some of it, perhaps, for friends or for resale.[116]

Cooper and carpenter Henson Hopewell's first account entry appears in McKenna's 1863 ledger. "Hense," as he was known, was forty-six years old; his wife Susan was forty-three. They had two daughters—nine-year-old Lizzie and four-year-old Pamelia—and two sons—ten-year-old William Henry and seven-year-old Samuel. A third son, Joseph, was born in 1865. Hense Hopewell's purchases were similar to those of McKenna's other customers; for his household he bought flour, meal, and bran in barrel lots, as well as potatoes, sugar, coffee, vinegar, and lard. His only whiskey purchase in 1863 was a $30 keg, which apparently had a long shelf life. Hopewell made slack flour and tight whiskey barrels as piecework, which he sold to McKenna. In April, eighteen flour barrels earned Hopewell 45 cents each. In May, he sold forty-two flour barrels and three whiskey barrels for $1.25 each. From June to mid-August, Hopewell made 126 slack barrels for which McKenna paid $90.45.[117] This pattern of store purchases and barrel sales continued through the 1860s.

In 1870 Hopewell and his wife lived in Fairfield, near Henry McKenna's home. His sons William Henry and Samuel were also coopers and lived in the family home.[118] Hopewell's account at McKenna's store included frequent purchases of staples in volume; he flour and bran in lots of fifty to seventy pounds and meal by the bushel. About every two weeks, Hopewell purchased wheat shorts in amounts of 70 to 170 pounds.[119] Shorts were commonly used as livestock and poultry feed. On February 6 he bought 215 pounds of pork for $18.77.[120]

From 1871 to 1874, Henson Hopewell's account deviated little from his established routine. He periodically bought flour, bran, meal, and shorts in quantity, together with butter, bacon, beef, salt, and coffee. He made cash withdrawals, some of which he directed to other individuals, and he bought shoes from a local store on his McKenna account. He bought whiskey each week in quarts and half gallons. In 1873 he sold 252 tight barrels to McKenna at $1.75 each, together with 21 mash tubs at $1.50 each. He also paid McKenna for hauling, perhaps for delivering stave wood and hoop iron to the cooper shop. McKenna paid Hopewell $26 for repairs on a house, building sixteen panels of fence at 25 cents per panel, and making a gate for 75 cents.[121] By December 1874, Hopewell's debits had reached $941.80, and his credits for cooperage and labor totaled $681.60, leaving him owing $260.20 to start 1875.[122]

In 1875 the Hopewell account appeared to shift to Henson's son William Henry, known as Henry. He bought staples, including eighteen pounds of "middling meat" at 15 cents per pound and nine and a half pounds of shoulder at 10 cents. Business expenses included hoop iron, a hammer, and payments for hauling wood and staves. Henry also purchased corn by the bushel, suggesting that he was feeding one or more hogs. In June he sold McKenna 27 mash tubs for $1.50 each and 197 whiskey barrels at $1.75 each. That transaction balanced his account. In October he paid McKenna $1 for hauling a load of timber.[123]

Through 1876 Henry Hopewell patronized McKenna's store every few days. He also shopped at the Pitt & Simpson General Store in Fairfield and charged the groceries and clothing he bought there to his McKenna account. Every three months, Hopewell paid McKenna $60 house rent, $20 rent for the cooper shop, and $18 for pasturing a cow and a horse.[124] The year's accrued debit reached $359 on June 1. In the meantime, Henry Hopewell made and sold 225 whiskey barrels to McKenna at the established price of $1.75 each, or $393.75. He earned another $21 for "burning the heads," or applying the distillery brand marks, on 210 barrels and making barrel repairs, for a total income of $434.75. McKenna also agreed to pay Hopewell $37 for some of the wood he had used, for a total midyear income of $471.75.[125] During the second half of the year, the familiar staple purchases continued, although Hopewell's whiskey procurements increased. From August through December he charged to his account four to eight quarts of whiskey each month. One has to assume that Hopewell was purchasing some of this whiskey for friends or that they bought the whiskey themselves and charged it to Hopewell's account. After paying for pasturage, cordwood, and other expenses in December 1876, Hopewell carried a debt of more than $42 into the new year.[126]

During the spring of 1877, Henry Hopewell paid to haul staves in March and April, and in May he purchased 2,000 staves from a Mr. Seyers for $40. He also worked on McKenna's corncrib for eighteen days, earning $18. On June 1, 1877, Hopewell's credit account stood at $674.75, but his debts totaled $704. For the remainder of the year, the frequency of charges to his account diminished; Hopewell made only nineteen transactions from June through December, compared with fifty-eight the previous year.[127]

From 1878 through 1882 Henry Hopewell continued to build and sell barrels. He charged purchases to his account with McKenna, including several at Pitt & Simpson in Fairfield and at a store in Louisville. He also used his account to pay debts owed to others or to advance them small loans—his sister Lizzie and brother Sam were such benefactors. In March 1878, Hopewell withdrew $5.50 in cash, which he gave to John Elder; in June he allowed Elder to charge two cords of wood to his McKenna account. From July through September Hopewell charged four gallons of whiskey for a Mr. Mason, probably a relative. He bought a cooper's knife for $1 and charged hoop iron, staves, house rent, and pasturage for a cow and a horse.[128] In September 1880 Hopewell sold McKenna 501 slack barrels at 75 cents apiece, or $375.75. He also received a $33.40 credit for "charing" (likely charring) old barrels and making other barrel repairs. This is the first suggestion that Henry McKenna and Patrick Sweeney may have been reusing barrels to store whiskey. McKenna had paid teamsters to haul used barrels from Louisville to the Fairfield distillery. It seems likely, given this account entry, that Hopewell was recharring those barrels for reuse. Hopewell's credit account balance at the end of September 1880 was $411.27, but his charges amounted to $470.62. His debt to McKenna stood at $59.35.[129]

Late fall was the traditional season for butchering beef cattle and hogs. Air temperatures below 45°F permitted the animal carcasses to cool rapidly, reducing the risk of souring the meat through bacterial action. In late December 1881 Henry Hopewell charged to his McKenna account 798 pounds of "hog meat" for $59.85, or 7.5 cents per pound. Just nine days later, on January 2, he bought an additional 423 pounds of hog meat for $31.72. The following day he bought fifty pounds of salt, and on January 17 he bought one and a half cords of wood at $3.75.[130] Hopewell's meat, salt, and wood purchases were likely associated with his intention to produce dry-cured and smoked meat for his extended family and friends. In addition to salt, Hopewell may have had other common meat-curing ingredients on hand—sugar, saltpeter, pepper, and spices—or he may have bought them at another store (although those purchases often appeared in his McKenna ledger).

Hopewell periodically purchased cooper's tools through McKenna. In January 1882 he bought a spoke shave and a tress hoop for $5. A spoke shave was a small double-handled plane used to shape staves. The tress hoop was a requisite tool for forming a tight barrel. The heavy iron tress hoop was the same inside diameter as the outside diameter of a whiskey barrel and triangular in cross section. Coopers used the tress hoop as a form or jig into which they placed standing staves in a wide-narrow sequence to "raise" the barrel's sides or body. When all the staves were in place and fitted properly, the cooper removed the tress hoop and replaced it with permanent galvanized or painted iron hoops. Hopewell's barrel work earned him $987.05, and his charges for 1882 totaled $862. Likely for the first time, Henry Hopewell ended the year with a positive carryover balance of $125.[131]

The following year, 1883, Hopewell's earnings plus his credit carryover totaled $1,653.90, whereas his purchases and cash withdrawals totaled $1,597.90, leaving him with a positive balance of $56 to begin 1884. Hopewell sold McKenna 128 slack barrels for $1 each and received $45.50 for other work. His debit account for the remainder of the year was primarily cash withdrawals. He purchased six quarts of whiskey during the year and bought corn by the bushel every two weeks. In November he sold McKenna an additional 418 slack barrels and received $9.10 for additional work, for a credit of $427.10. His expenditures added up to $258.25, leaving him with a credit balance of $168.85, which McKenna paid in cash to end the year with a balanced account.[132] Hopewell did not pay house or shop rent to McKenna after 1882. In 1900 William Henry Hopewell was single and lived with his uncle, William Mason, in southern Spencer County on Camp Branch, about four miles north of Fairfield and the McKenna Distillery. He was still working as a cooper.[133]

Laborers—Long and Short Term

Some Kentucky distillers owned slaves; others may have hired slaves owned by neighbors. The state's antebellum slave population was disproportionately distributed. The highest densities of enslaved people in 1860 were found in Kentucky's prime agricultural regions: the Inner and Outer Bluegrass, the alluvial plains along the Ohio River west of Owensboro, and the Pennyroyal's limestone lands west of Bowling Green. The number of enslaved people in these areas often exceeded 30 percent of the total population. In the Inner Bluegrass, for example, 40.3 percent of the Scott County population was enslaved in 1860. In Nelson County in the Outer Bluegrass, the proportion stood at 35.7 percent; in Henderson County on the Ohio River, 40.7 percent; and in Christian County in the Pennyroyal, 46.1 percent.[134]

When US census marshals tabulated Nelson County's enslaved people in 1860, Henry McKenna was not a slave owner, but he did participate in the common practice of hiring slaves. An entry in McKenna's personal ledger on January 17, 1859, lists a "Negro hire to Aug, $15.00." The pay rate of $15 per month was about $7 per month—or roughly 30 percent—lower than the rate for white laborers.[135] McKenna hired Bill, who belonged to a Mrs. Blinco, on January 1, 1859, to work for four months. After Bill missed some days,

McKenna paid Mrs. Blinco for two and a half months' "work of boy, $37.50."[136] In August 1860 McKenna's account for J. Barkly (or Barkley) included this entry: "I am to furnish Louisa 2 skirts, 2 Sunday dresses, 2 calicos, 2 pair of shooes, 2 pair of stockings, 1 blanket." Louisa may have been a housekeeper or a servant in McKenna's household. Slave owners also traded their bondsmen's labor for goods. In December 1862 customer M. Izack bought six barrels of flour from McKenna for $35. In return, McKenna hired Izack's "boy" to card wool for two and one-third months at $15 per month, or $35, the exact value of the flour purchase.[137]

On Monday, May 10, 1858, a "black boy" working for Thomas Miller began hauling seven and a half cords of wood from Miller's Spencer County farm to McKenna's distillery wood yard. The farm was a few miles north of Fairfield, and the task required two days to complete. From June to September, Miller cut and delivered an additional seventy-eight cords of wood by wagon team, for which McKenna credited his account $128.47. Miller was a prosperous farmer, and he may have hired laborers to cut trees and clear his fields. In turn, Miller purchased flour from the McKenna mill, in five lots of about 100 pounds each, from April to December.[138]

Stephen Wickam, a thirty-three-year-old Irish farm laborer, boarded in Thomas Miller's home. A few days before Christmas 1862, Wickam began to work for McKenna at a wage of $200 per year. McKenna also gave Wickam the rent-free use of a dwelling to house his wife and six children, as well as an allotment of six barrels of flour and twenty-five bushels of bran.[139] During the year, Wickam drew on his account with McKenna for mutton, meal, sugar, tobacco, wood, and whiskey. In May he requested a $2 cash advance that McKenna recorded as "relief for Ireland." Wickam's remuneration, therefore, was partly in kind and partly in cash, a common arrangement between McKenna and the laborers who worked for him.

Laborers who worked for other employers also maintained accounts with the McKenna mill and distillery and often "worked off" their debts by hiring themselves out to McKenna sometime during the year. No matter who employed them, few of these laborers were able to carry positive account balances forward into succeeding years. Although it appears that McKenna provided make-work to help them balance their accounts, they often subsisted from one accounting period to the next, dependent on their credit accounts for staples, and using cash and credits to pay debts owed to others.

The Irish immigrant labor force is a compelling backstory that is often explicitly addressed in McKenna's ledgers. Distiller Patrick Sweeney, himself an immigrant, made contributions to the Fenian Brotherhood, suggesting that he had not abandoned his heritage. McKenna may have hired immigrant Irish laborers to work in his mill and distillery to help individuals in need of social and economic support. Patrick Dumphrey, originally from County Kilkenny, started working for Henry McKenna on April 16, 1858, at the rate of $22 per month. An Irish border in McKenna's home, Margaret Dumphrey, may have been a relative. Dumphrey worked for nine months, for which McKenna paid him $198. Beginning January 15, 1859, McKenna increased his pay to $25 per month. Dumphrey worked an additional four and a half months and was credited with $112.50. He also collected $1 for working in the distillery on a Sunday. Dumphrey's pay for this period totaled $323; although the ledger does not specify exactly, this probably covered fifteen months of work. Dumphrey also rented a house from McKenna for $3 per month and purchased flour, bran, shorts, and slop. McKenna charged him $15 for "8 months slop[p]ing hogs and bran and shorts," suggesting that Dumphrey owned hogs and was feeding them in the distillery hog pen. When McKenna closed the account, Dumphrey had a positive balance of about $80.[140]

Irish immigrant Frank Goodwin lived as a boarder in the McKenna household in 1860 and may have been a relative of Mrs. McKenna. Peter Goodwin, another relative perhaps, had been hired in October 1858 to work in the distillery at a wage of $400 per year. McKenna's ledgers did not specify Goodwin's job, but his wage was roughly equivalent to that paid to distillers. Although Goodwin charged a small amount of flour to his account, he bought no other items from McKenna's store. Rather, he made cash withdrawals and designated cash credits to other accounts to cover debts, including quarterly payments of $25 to a P. Smith, suggesting that Goodwin was boarding rather than maintaining his own household. In the aggregate, his debt totaled $377 at year's end, and McKenna marked the account "settled in full by his work to this date."[141]

Years later, in 1881, Henry McKenna's son James hired Patrick Goodwin, possibly a relative of Peter, at a rate of $200 per year; this was increased to $240 in 1883. Thereafter, Goodwin's account entries are inconsistent and lack specifics; a general entry for 1884 simply stated: "To cash at different times $304.70." At the time, James McKenna was managing the distillery sales office in Louisville and visited Fairfield only periodically, which may explain the record's recession into informality. Yet, by August 1884, Goodwin's credits stood at $670, and his debts were only $319. James McKenna paid Goodwin $350 cash to balance the account.[142] Assuming that Goodwin was solely employed by the McKenna works, he was apparently frugal and able to sustain himself on less than $320 over three years.

Two days after Christmas 1864, S. Hiskem began to work for Henry McKenna, receiving remuneration of $35 per month and six barrels of flour—a wage roughly double that received by other laborers. Hiskem worked through the distilling season and quit in June 1865; he received $175 for five months' work. Later in the year he returned to work for twenty-five days, for a total annual income of about $212. Meanwhile, Hiskem maintained a charge account with McKenna for flour, which he bought in 100- to 200-pound lots, as well as meal, bran, sugar, and coffee. In February he bought 312 pounds of pork at 13 cents per pound. Hiskem's purchases and cash withdrawals totaled more than $217 for the year, including the $1 he sent to the Irish Fenian Society in September. At the end of the year, he owed McKenna about $5.[143]

George Drake began working for Henry McKenna on January 3, 1866, the approximate beginning of the distilling season. Drake's pay was $14 per month, and by August he had earned $101.90. McKenna also arranged three additional weeks of work for Drake and sold him a mule worth $10.50. Drake's purchases from McKenna for January to August included flour, meal, hominy, bacon, "meat," chickens, and whiskey by the quart. Drake also took several cash advances, and by August, he had accrued a total debt of $97.40. When McKenna closed his books that month, he paid Drake $4 cash to balance his accounts. Drake then worked from August 17, 1866, until Christmas Day, earning about $59. His charges for that period, for flour, whiskey, and cash advances, totaled about $60. McKenna ignored the difference and considered the account settled in full.[144]

Michel Henry also began to work for McKenna in January 1866, although his pay rate was $20 per month, or $6 more than Drake's. McKenna's ledger does not specify the types of work assigned to the two men, but both the mill and the distillery were likely in production. Henry received an immediate advance of "cash to Send to Ireland, $85.00"; he also sent $1.50 to the Fenian Society. McKenna gave him $1 to have his shoes repaired. Henry did not draw cash from his account until September 18, when he received $2.50. McKenna advanced him another $11 on October 8, which Henry sent to Ireland. The ledger does not identify where Henry was living or how he sustained himself—perhaps he was boarded with someone who lived close to the distillery in exchange for work. In late October McKenna advanced Henry $3.75 to purchase five yards of flannel and 50 cents to pay for "making drawers." The last draft was for shoes and tobacco on October 27, by which time Henry's debt totaled $224. If Henry worked ten months from January to the end of October, his income would have been about $200. The account ended with no record of how Henry settled his remaining debt.[145]

Day laborers, in sizable numbers, worked to support Henry McKenna's farm, mill, and distillery operations. He routinely paid short-term workers a daily rate or for piece-work rather than a monthly or annual wage. Most short-term jobs supported McKenna's businesses either directly or indirectly. Thomas Conley had a small whiskey account with McKenna in 1863. He brought to the mill thirty-two pounds of wool to be carded, for which McKenna charged 6.25 cents per pound. Carding was the process of disentangling raw wool and aligning the fibers in preparation for spinning the strands into thread. Conley paid about $2 for the carding, but he also asked McKenna for a $5 advance, which he sent to "the poor of Ireland."[146] Farmer Samuel Wilkerson brought 285 pounds of sheared wool to be carded in 1863. He paid McKenna 7 cents per pound, and later that year he settled his account balance with a cash payment of $15.[147] McKenna hired fourteen different people to card wool for him in 1862 and 1863, paying them 7 to 10 cents per pound for their work. Most of these laborers carded 20 to 50 pounds of wool, but Izack Stone processed 283 pounds, earning more than $19.[148] Carding work supported McKenna's businesses indirectly by providing wages that workers could use to pay their account balances in kind or in cash or to produce income for other purposes.

James Heady's 1870 account with McKenna included charges for flour and whiskey—one pint for the entire year—for a total debit of about $36. Heady performed work for McKenna that directly supported the mill and distillery, such as cutting wood, for which he received $11, and one day of "harvesting" at $2—an ample rate of pay. Heady split fence rails for $1.50 and completed other tasks to earn a total credit balance of more than $56 for the year.[149] Many of McKenna's accounts for short-term laborers were layered with several transactions and, rather than being settled periodically, were carried over from one year to the next.

Mathew Wickham, his wife Johanna, and their nine children lived in Fairfield near Patrick Sweeney and Henry McKenna. Wickham had migrated from Wexford, Ireland, to New York in 1851. He made his way to Indiana and eventually settled in Nelson County, where, by 1870, he had established a farm. In addition to working his own farm, Wickham hired himself out part time to Henry McKenna. Wickham made monthly purchases of flour and meal in 100- or 200-pound quantities. His distillery account listed whiskey purchases of one quart to half a gallon four or five times a month—quarts cost 55 cents, and a gallon was $2.60. In December 1870 he sold twenty-four bushels of potatoes to McKenna at 65 cents per bushel. He also earned credit toward his account by working for three days on McKenna's icehouse; for that he made $6—a higher pay rate than most part-time laborers received. The following year, in November 1871, Wickham paid $18 cash on his account, but he was unable to pay the balance at year's end.[150] The whiskey account continued through 1873 and 1874, and by January 1, 1875, it totaled more than $77. In December McKenna hired Wickham to repair his milk house, paying him $11.25 for seven and a half days' work. By January 1876, Wickham's unpaid account balance exceeded $88. With the exception of a quart of whiskey that Wickham allowed George Shields to charge to his account, all the purchases were apparently his own.[151]

Community Account Holders

African Americans

In 1880 census marshals listed two African American men, Samuel Knight and Dick Skinner, who worked at McKenna's distillery. Milton Downs, another African American, was a fireman—the person who managed the distillery steam engine. Unlike the Hopewells, who worked as coopers, none of these men kept charge accounts with McKenna.

Many account holders were not employees of McKenna but members of the rural community in northern Nelson and southern Spencer Counties. Several, including some farmers, were employers themselves who hired African Americans full or part time or engaged them in other social and economic relationships. Farm laborer John Offutt worked varied jobs, including teamster and woodcutter, in 1870. In November he sold forty-two cords of wood to McKenna at $3.50 per cord and bought a harness, "wagon lines," and whiskey on credit. On occasion, Offutt sent a "black man," perhaps his own employee,

Distillery–Related Employment
Fairfield Precinct,
Nelson County, Kentucky,
1880

Name	Age	Race	Occupation
Henry McKenna	61	W	Distiller
Samuel Knight	35	AA	Works at distillery
Dick Skinner	30	AA	Works at distillery
Milton Downs	44	AA	Fireman
George Hobbs	19	W	Cooper
Henson Hopewell	72	W	Cooper
L. Hopewell	47	W	Cooper
William H. Hopewell	31	W	Cooper
Payne Cox	45	W	Cattle dealer

Source: US Census of Population Manuscripts.
Note: Most of these individuals lived in the village of Fairfield, the location of the McKenna Distillery. The distillery was steampowered, and the fireman may have been responsible for engine operation.
AA, African American; W, white.

to the distillery to retrieve whiskey he had purchased.[152] The McKenna ledgers recorded numerous such transactions.

Sam Brewer maintained accounts with Henry McKenna for nearly two decades. In 1871 he sold pork and onions to McKenna and bought flour "per black man."[153] Woodcutter Ed Bridell bought small amounts of whiskey on credit; a purchase in April 1876 was entered in the ledger as "whisky per order by black boy," and in June he purchased whiskey on his account for "George" and "Henry."[154] Those black men known to McKenna were usually listed in the ledger by their first names; others were simply identified as "black man," or the ledger listed the transaction as "whisky for colored man charge $1.20."[155]

In 1878 McKenna charged farmer Charles I. Wathen's account for regular purchases of whiskey for "Jess" and "Saul" from January 7 through November 15—about seven quarts for each man. Jess and Saul always bought their whiskey on the same days. In May, Wathen charged to his account two bushels of meal for Saul, and in September, he charged two bushels of meal for Jess.[156] "Saul" remains anonymous; his name appears in no other account ledger. "Jess" was Jesse Lewis, an African American laborer who lived with his wife Maria and three children in farmer Daniel Ludwick's household near Fairfield. Lewis also worked for Wathen, who lived about two and a half miles south of Fairfield. During the late fall months of 1870, Lewis purchased flour, meal, and whiskey from McKenna for $13.70. To balance his account, he worked in one of McKenna's grain fields and made 117 shocks of grain at 10 cents per shock, or $11.70. Thus, he still owed McKenna $2 at the end of the year.[157]

FARMERS

Henry McKenna's milling and distilling businesses depended on area farmers to supply grains. Prior to the Civil War, the distillery's output was small, a barrel or less per day in some seasons; more grain was processed into flour and other milled products than was consumed at the distillery. McKenna also bought cordwood, meat, wool, and honey from local farmers.[158] Farmers, in turn, depended on the credit extended by McKenna to cover their purchases of staples, spirits, and seed for planting. In addition, McKenna may have rented machinery and acted as a pawnbroker, buying or holding farm machinery as collateral on large debts.

Most farmers sold only small amounts of wheat and corn to the McKenna mill and distillery. In July 1858 McKenna bought 152 bushels of rye from Bardstown area farmer Samuel Lancaster at 55 cents per bushel. If Lancaster had planted rye as a winter cover crop the previous fall, this sale could have been from a freshly harvested crop.[159] Samuel Bell sold McKenna eighteen cords of wood at $1.75 per cord in September 1859, and in November he delivered thirty-three bushels of wheat, for which he received a credit of $33. For wheat, McKenna generally paid 85 cents to $1 per bushel; the high-end price Bell received may have been based on availability or quality, or it might have been related to a personal relationship between the two men.[160] Farmer and slaveholder Asher Bodine—he owned twelve enslaved people in 1860—sold wheat to McKenna. After the August harvest in 1858, Bodine brought 308 bushels of wheat to the mill and distillery. He agreed to accept a 10 cents per bushel less than the prevailing wheat price in Louisville and received just over $260.[161] James Murry produced wheat and corn on his farm near Fairfield. He sold small amounts of grain to McKenna in 1858: 120 bushels of wheat and some 50 bushels of corn for about $84. When McKenna began distilling in the mid-1850s, he used wheat as his primary grain; by 1858, he had switched to corn and used smaller amounts of wheat and rye as secondary grains. Wheat's high market price may have been the rationale for the change, rather than a concern for the whiskey's flavor or customer demand.[162] The ratio of wheat to corn in Murry's 1858 transaction was likely related to the mill's flour output, although it may have reflected the differential in overall grain consumption as McKenna made the transition to corn as his primary distilling grain.[163]

William Grigsby farmed in Washington County, southeast of Chaplin. In 1859 he paid Henry McKenna $3 to grind ninety-three bushels of corn; he also sold McKenna about twenty-six bushels of wheat at 85 cents per bushel and twenty-three bushels of corn at 50 cents. In January, during the distilling season, Grigsby began delivering wheat and corn by the wagonload, for a total of 75 bushels of wheat and 203 bushels of corn. In March 1859 James Marshall, a regular customer whose account included frequent purchases of shorts, brought clover seed to the mill for cleaning, for which McKenna charged 25 cents.[164]

In 1858 farmer James Brewer started an account at the McKenna works. His charges included whiskey in quarts and seven bushels of salt for $2.10 plus $1 for delivery. By December, his debt to McKenna totaled more than $41. In February, to balance his account, McKenna accepted from Brewer cabbages worth 30 cents, potatoes valued at 87 cents, and a sow worth $10. In September Brewer sold fifteen bushels of rye and also "mortgaged" his disk to McKenna for $8.10. This brought Brewer's credits to more than $42, which allowed him to settle his account. Selling or mortgaging an important farm implement such as a disk was a risky move. Had Brewer bought a new disk and could therefore afford to sell his old one? The ledger offers no explanation, but selling a disk without having a replacement would have been like making flour from next year's wheat seed: no implement, no field preparation; no seed, nothing to plant. Nearly thirty years later, in 1886, Brewer still had an account with the McKennas, and all his charges were for whiskey bought in quarts and half gallons. At the end of December of that year, his debt totaled $56.20; his only credit was a sale of honey for $1.95.[165]

Syms Hutchings of Chaplin, in eastern Nelson County, rented a corn sheller from McKenna in 1861; he used it for two seasons and returned it in 1863, paying $7. McKenna bought some of his distilling corn unshelled and inspected the individual ears for smut or diseased kernels, which he removed by hand before shelling the entire ear. Renting his corn sheller for two years suggests that McKenna had purchased a new one and made the old one available for his customers' use.[166]

By the 1870s, some grain and corn farmers were able to enlarge their production, and their sales to the McKenna mill and distillery increased accordingly. Thomas Bryant bought clover seed from Henry McKenna in 1858 and paid almost $23 for the seed, plus $1.20 for delivery. He also bought whiskey in amounts varying from one gallon to one barrel. In 1870 he sold sixty-seven cords of wood to McKenna and bought $18 worth of clover seed. In June 1872 McKenna credited Bryant's account for nineteen cords of wood plus 406 bushels of corn at 50 cents per bushel, or $203. The following year, Bryant sold some fifty-six bushels of wheat to McKenna at $1 per bushel; Bryant also sold him ten cords of wood chopped "by his man," for which he earned a credit of $10. With the price of wheat double that of corn, McKenna's transition to making corn-based whiskey must have seemed like a sound financial decision.[167]

William Hughes had a small charge account with Henry McKenna for flour and whiskey. For instance, he bought one quart of whiskey on March 30, 1872, "by order of Dr. Wells." The doctor's office was across the turnpike and a short distance west of the McKenna Distillery. McKenna also consolidated the notes for loans Hughes had made to three other farmers, totaling $139.58; this brought Hughes's total debt to $360. In May and June 1871 Hughes sold McKenna 600 bushels of corn at 60 cents per bushel in 50-bushel lots over a three-week period. The corn sale produced exactly enough credit, $360, to pay off Hughes's account in full.[168] A standard-size late-nineteenth-century horse-drawn farm wagon would have measured approximately ten feet long, twenty-six inches high, and thirty-eight inches wide, or about seventy-two cubic feet. Its full capacity would have been about fifty-seven bushels of loose shell corn or twenty-nine bushels of ear corn. If Hughes was hauling loose shell corn, he likely would not have filled the wagon to the top, to reduce the chance of spillage. If the shell corn was in sacks, he could have stacked the filled sacks above the top of the wagon box. In any case, delivering 600 bushels of corn would have required at least twelve wagonloads of fifty bushels each.

Rufus Jones operated a farm in the Waterford District of western Spencer County, about three miles from McKenna's mill and distillery. In addition to carrying Jones's small debt account for whiskey, McKenna loaned him $327 in April 1873. To pay off the four-month interest-bearing note, Jones sold 726 bushels of corn to McKenna for an average price of 45 cents per bushel. Jones delivered the corn in lots of 50 to 100 bushels during the spring distilling season. Five years later, in 1878, Jones delivered to the McKenna Distillery 436 bushels of corn at 35 cents per bushel. McKenna credited him $152.80, which, combined with a note of credit, balanced his account.[169] Farmer Edwin Bridell was one of McKenna's largest corn and cordwood suppliers. From mid-March to mid-April 1872 he sold McKenna 1,230 bushels of corn at 45 cents per bushel, for a total sale of more than $550.[170]

The McKenna Distillery ledger-book entries ended in 1886, but a letter from 1896 illustrates that the system of direct grain sales between farmer and distiller extended until near the turn of the century. J. C. Drake of Whitfield, Kentucky, a crossroads community about ten miles north of Fairfield, wrote a letter to McKenna Distillery manager Daniel McKenna in May 1896. It read:

> Dear Sir:
>
> I have between four and five hundred bu[shels] of good corn for sale on my family farm near Ben Millers. Would be glad to arrange for the sale of it at the most convenient point, and time that you can handle it at fair price, and without delivery, also including delivery. Hoping to have your statement at an early date I remain yours very respt J. C. Drake.[171]

By the 1870s, many industrial distillers were buying their grain from large brokerages in Louisville and Cincinnati. These brokerages aggregated the grain that farmers sold to local grain merchants and grain elevators and resold it to distillers for rail or steamboat delivery. The original McKenna works was not a high-capacity distillery and likely gleaned much of its grain from local farmers. The McKennas completed work on a new distillery in 1883 and thereafter likely began to obtain grain from greater distances. The rail depot nearest the McKenna Distillery in the 1890s was four miles east at Bloomfield, so it is possible that Daniel McKenna was purchasing grain from both local sources and regional brokers by 1900.

Artisans

The flexibility of Henry McKenna's accounting system is exemplified by the accounts he kept for the community's artisans—stonecutters, carpenters, blacksmiths, painters, and sawmill operators. Stonecutter Paul Besinger lived in Mount Washington, a village on the Bardstown Turnpike, fourteen miles northwest of Fairfield in Bullitt County. In addition to general stonework, Besinger made gravestones. He also bought whiskey from McKenna by the barrel. In June 1858 he bought a forty-one-gallon barrel for which he paid $20.50; he bought four more barrels in August and September. Buying whiskey in this volume suggests that Besinger was a spirits retailer; he might have been in the saloon business himself, or perhaps he was supplying whiskey to other saloon keepers. He also bought flour in 100-pound lots. During the summer of 1858 McKenna bought two small gravestones from Besinger for $20 each. The stones were "for Dick" and his wife Julia. Dick may have been a slave or a freedman. One of Fairfield's physicians, Dr. Wooten, contributed $10, and "Thomas" gave McKenna a note for $20 toward the purchase; McKenna paid the $10 balance.[172]

Carpenter Thomas Brewer maintained an account with McKenna from 1858 to 1863. He bought about fifty pounds of flour each month and smaller amounts of meal and bran. His whiskey consumption was modest. In May 1858 he hauled 500 feet of wood flooring for McKenna at $3.50 and presumably installed a new floor, perhaps in the still house. Brewer made a bed for McKenna, as well as screens and gates; he also repaired hogsheads. He sold a watch to McKenna for $18 and settled his account in October 1863. McKenna noted, "This account is settled by work done on still house and garden and settles all work done by Brewer for me."[173] Some of Brewer's carpentry work may have been painted by Samuel Mills. In September 1863 McKenna paid Mills $4.50 to make paint and $25 to give the mill three coats. Mills's account included no purchases of food staples or whiskey, but he charged $3 for a pair of shoes and $1 for a brush. McKenna also carried Mills's doctor bill for $1 and gave him two cash advances of $1 each. Mills likely finished painting the mill in early November, when McKenna settled his account. Mills may have been itinerant, for he does not reappear in the McKenna ledgers.[174]

Henry McKenna credited carpenter and handyman Ennis Doran more than $100 for "work for my boys houses" in 1870. Over the next decade, Doran maintained an account with McKenna for flour, bran, meal, and whiskey. In December 1880 Doran received an advance of $1 "cash for Parnell defense." Parnell does not appear elsewhere in McKenna's ledgers, but he may have been a friend or relative of Doran's who was charged with some legal infraction. In January 1881 McKenna charged Doran's account for "one seasons pasturage for cow $12.00" and credited him "by his account against me $40.90." The credit was likely for carpentry work, and it gave Doran a $15 credit balance for the year.[175] In 1878 and 1879 carpenter Charles Conlin had a small account with McKenna for whiskey that he bought for other people, including one gallon for "C. Duncan colored." He drew $2.50 cash from his account to pay for medicine in Bardstown in 1879. McKenna credited Conlin for seventeen and one-quarter days' work on his house for $17.25, plus nine and one-quarter days for $9.25. The expected number of hours worked per day was not recorded.

George Royalty farmed and operated a sawmill south of Bloomfield in Washington County. In 1876 and 1877 he maintained an extended account for whiskey. He preferred "old" whiskey, which he bought in pints, quarts, and gallons. Although McKenna began selling spirits in bottles in 1871–1872, the Royalty account is the first in which he included a container cost: glass bottles cost 10 cents, and earthenware jugs 15 cents.[176] In January 1876, for example, Royalty charged three quarts of whiskey at 65 cents per quart, and he paid 10 cents each for the quart bottles. In late March he charged seven pints and bottles at $2.55, and in June he bought a quart of "old" whiskey for $1.70, with 10 cents added for the bottle. In October 1876 Royalty purchased a horse from McKenna for $110. He paid his account in full the following month by selling McKenna 11,000 feet of oak and poplar plank at $14 per thousand, or $154, as well as 200 feet of clapboards at $18. In April of the following year, Royalty delivered 1,630 feet of lumber and fencing plank for McKenna's farm, for which he was credited $32.60. The transaction suggests that McKenna was converting some of his old split-rail fences to post-and-plank fences. Royalty charged a hand saw, a drawing knife, and a box of copper rivets to his account in 1877, and he bought more than $32 worth of clothing and meat from the Pitt & Simpson store in Fairfield, which McKenna paid and carried on his account. McKenna's teamster delivered some of Royalty's shingles to a customer for $4.[177] In April Royalty delivered to McKenna 2,500 feet of lumber for a barn, followed in June by an additional 1,800 feet. In an account summary in late July, McKenna noted that Royalty "supposes the amount [of] staves furnished last year to be 12 thousand . . . at $25 per thousand," for which he credited Royalty's account $300. The reference to staves suggests that Royalty was one of the primary suppliers of barrel staves to McKenna's cooper shop. Royalty's credits for lumber and staves through October 1877 amounted to $396.50. His debt for whiskey, cash withdrawals, and store purchases exceeded his credits by more than $200.[178]

Fairfield blacksmiths John Nelson and his son William J. Nelson made buggies and offered "blacksmithing, horse shoeing, smith work, and repairing done in first-class style. Also handle the chilled Iron Plows, which we keep on hand and for sale."[179] William Nelson kept a modest account with Henry McKenna. From 1880 to 1882 he purchased little more than one to two bushels of meal every five to eight weeks. On January 1, 1881, McKenna entered a note on Nelson's ledger: "Bill as against me for work for 1880 $73.15" and "2½ cords of wood by Jack Weaver $6.25." Weaver was apparently employed by Nelson part time and cut wood on his behalf, which he then sold to McKenna for credit toward his account. At the beginning of 1881, Nelson's credits exceeded his debts by about $75. In March Nelson sold twelve pounds of hoop iron to McKenna for $1.60, presumably for the cooper's shop. In 1882 Nelson's purchase pattern changed; from January through June he bought more than thirty bushels of corn in two-bushel lots every seven to ten days. McKenna's price for a bushel of shelled corn was 85 cents through the winter months but increased to $1 in April. Nelson may have been feeding a hog on his residential lot in Fairfield, as did several other people. If that was the case, the reference to "corn" may have meant ground corn rather than shelled corn. And those customers who fed chickens purchased McKenna's milled chicken feed, not whole-kernel corn. In July McKenna paid Nelson $36.15 for his "work at shop," the exact amount of Nelson's debt. By 1884, Nelson had stopped purchasing corn but continued to buy meal and pay for "hauling." In early March Nelson's debt totaled $32.70. McKenna again balanced the account by noting, "settled in work at shop $32.70."[180]

PHYSICIANS

Five doctors practiced medicine in or near Fairfield from 1860 to 1890. Neither Drs. B. J. Moore nor George W. Hedges, both of Fairfield, patronized McKenna's mill and distillery. Drs. B. B. Wootten of Fairfield, Harrison Wells of Fairfield, and Francis S. Reid (also spelled Read) of Bloomfield all kept active accounts. Dr. Wootten made thirty-eight purchases of whiskey in pints, quarts, and gallons in 1870; he likely consumed most of it himself, but some of the whiskey was designated for other individuals identified in the ledger and may have been medicinal prescriptions.[181]

Dr. Harrison Wells lived with his wife Frances and their children in Fairfield, a short distance west of the distillery and across the turnpike from the distillery's cattle and hog feedlots. Dr. Wells's 1866 account included meal and a barrel of flour every six to eight weeks. He bought only one bottle of whiskey in the entire year. In the 1870s Dr. Wells pastured an animal, perhaps a carriage horse, on McKenna's farm during the spring, summer, and fall, paying $1.50 per month. Dr. Wells's patients included distiller Patrick Sweeney. Henry McKenna paid Sweeney's $17 medical bill in January 1866 and entered the amount in Sweeney's debit account. In 1870 McKenna's household included Rhoda Camus, Willie Camus, Nace Bodine, Sandy Bodine, and William Bryant, all African Americans. Dr. Wells submitted two medical bills for Lundy Bodine during the year, one for $17 and the other for $17.50. McKenna paid both bills. It is not clear whether Lundy was related to the Bodines in McKenna's household. Sandy Bodine, William Bryant, and Henry McKenna all required Dr. Wells's medical care in 1878. McKenna paid the bill, which totaled more than $41.[182] Dr. Wells also dispensed prescriptions for medicinal whiskey. On January 9, 1879, Mrs. Daniel Langsford bought from McKenna "1 qt. of whisky per Dr. Wells, .55 cents." Two days later, she bought another quart, "per Dr. Wells, .55 cents."[183] James Crumes received one-half gallon of whiskey that McKenna charged to Dr. Wells's account. Perhaps the doctor had prescribed the whiskey and paid for it; more likely, he gave it to Crumes for work performed.[184] Dr. Wells also provided a medical endorsement for McKenna's whiskey in an 1877 advertisement; a decade later, he again endorsed the McKenna product as medicinal whiskey in an advertisement that appeared in the *Southwestern Medical Gazette.* The endorsement was something of an irony, given that the physician's one whiskey purchase had lasted him an entire year.

Dr. Francis Reid practiced medicine in Bloomfield, some four miles east of Fairfield, from the 1860s to the 1880s. He maintained an account with Henry McKenna for flour, meal, and whiskey. Dr. Reid preferred "old" whiskey, which sold for $1 per quart; he rarely bought the cheaper two-year-old whiskey for 55 cents. Most whiskey sales carried on his account were for personal consumption, although he did buy two gallons for a Mrs. Hume, likely for medicinal use, in August 1881. McKenna apparently patronized both Drs. Wells and Reid. In April 1881 McKenna credited Dr. Reid's account for a $19 medical bill he had incurred.[185] Like Dr. Wells, Dr. Reid endorsed McKenna's medicinal whiskey in the *Southwestern Medical Gazette* in 1877. It is not clear whether McKenna patronized the two doctors for their medical specializations or because he wished to solicit their endorsements.

Merchants

Retail merchants in Fairfield, Bardstown, Mount Washington (Bullitt County), and Louisville patronized McKenna's mill and distillery, primarily for flour and other milled grain products. Washington (Wash) Pitt sold general merchandise at the Pitt Brothers Store on the Bloomfield-Fairfield Turnpike, Fairfield's main street. He maintained an account with McKenna for two decades, from the early 1860s to the 1880s. He purchased small quantities of flour, meal, and whiskey, consistent with personal or family consumption rather than retail sales.[186] F. H. Bowlds was also a Fairfield shopkeeper. Like Pitt's, his accounts reflected purchases of family-sized amounts of flour and meal. Bowlds also pastured cows and other livestock on McKenna's farm for ten months in 1879, for which he paid $15. McKenna apparently made purchases at Bowlds's store because his year-end account summary in January 1879 read: "account against me at shop for 1877 and 1878 $114.05."[187] Presumably, the Fairfield merchants did not sell McKenna's mill products because residents could walk a short distance to the mill and purchase these staples directly.

The distance to the McKenna mill and the community market size were factors in sales: merchants whose businesses were farther from the mill and who served larger markets were more likely to buy his flour and meal in quantity for resale than were those in Fairfield or nearby towns. The Finley Company of Bloomfield purchased flour and meal in 90- to 100-pound lots once or twice a month in the 1860s.[188] Collins & Company of Bardstown ordered twelve to sixteen barrels of flour every two weeks in 1860, paying $6.50 per barrel.[189] During the 1860s, Mattingley & Brother of Bardstown bought flour in even larger lots, eleven to twenty-five barrels at a time, which it sold on commission. With each order, the merchants returned their empty barrels for refilling.[190] In 1868 McKenna loaned the Mattingleys $300 cash, which they repaid in full later that year.[191] A few small retail stores in Louisville also carried McKenna's mill products. Mickel O'Connel and Henry Phelps bought flour in quantity and sold it on commission in the late 1850s and early 1860s. The largest purchase by Louisville grocers was eighty-four barrels ordered by Stye & Reeling in the fall of 1859.[192]

Saloon Keepers

Five Louisville saloon keepers submitted an unusual group order for nine barrels of McKenna's whiskey in 1860. The five orders, dated November 28, were sold on credit, with cash payment due 120 days later on March 28. McKenna's delivery instructions to his teamster listed the saloons' locations, which were clustered near the Ohio River. Mickel Dorson bought two barrels of whiskey, as did Golden of Louisville. McKenna described the Golden saloon as being "opposite [the] artesian well" near the corner of Monroe and Tenth Streets, not far from the Ohio River. The saloon was a prominent landmark near the head of the Louisville and Portland Canal; it was one of the busiest sections of the riverfront, where boats tied up to await their turn to navigate the canal's lock. Nugens Daghertey's saloon was "below Dorsens [Dorson's] on Levy [levee]," and Phil McGure "lives on Levy below 7th," a short distance east of Golden's saloon. A Mr. Curen purchased a single barrel; he lived "below Daghertey's," also at the levee.[193]

Wholesalers

Louisville wholesalers Griffith & Rilly, John Queen, and John Martin bought McKenna's flour in quantity to sell on commission. A barrel of flour traditionally contained 196 pounds, and Griffith & Rilly bought 312 barrels at $4.50 each from January to August 1858. McKenna's teamsters delivered the flour in ten- to thirty-barrel lots, making from one to three trips each month.[194] John Martin ordered flour in quantities of ten to forty barrels in 1859. His orders increased to fifty-eight barrels in January 1860 and ninety-six barrels in February. From August through November he bought 462 barrels. He also bought "old" whiskey in twenty-barrel lots. John Queen's flour purchases were smaller and occurred only in 1862.[195]

Louisville foodstuffs wholesalers had access to several flour mills in Jefferson and surrounding counties in the 1850s and 1860s. By 1880, more than forty mills operated within Louisville's immediate hinterland—including the McKenna mill—twelve of them in Louisville itself.[196] That Henry McKenna was able to market his products in Louisville, more than thirty-eight miles from his Fairfield mill, suggests that he actively pursued sales there and that his prices were competitive, despite his comparatively high hauling costs.

Boardinghouses, Hotels, and Schools

For much of the nineteenth century, women had few occupational opportunities they could pursue without social ostracism. Managing boardinghouses and hotel and turnpike taverns was an accepted occupation, perhaps because, in principle, these establishments operated somewhat like private family homes. Certainly, boardinghouses were often women's private homes that they opened to paying boarders. During the turnpike era, women often operated their husbands' roadside taverns and became the sole proprietor if he died.[197] Most Kentucky turnpike towns, even those of modest size, had a tavern or a hotel.

Fairfield's hotel, known as the McKenna Hotel by 1917, stood in the center of town facing the Fairfield- Bloomfield Turnpike.[198] Boardinghouses could have been operating almost anywhere in town. Taverns and hotels traditionally served patrons according to the American plan; that is, meals were furnished as part of the daily charge and were served in large dining rooms.[199] Feeding guests three meals a day would have required access to food staples in quantity.

Mrs. Elizabeth Renshaw was the widow of shoemaker John B. Renshaw, who died in 1855. Recall that McKenna had acquired land for his mill from the Renshaws the previous year.[200] Mrs. Renshaw may have operated a boardinghouse in Fairfield in 1858, and she was one of the few women who made regular purchases at McKenna's mill store. Beginning in January and continuing throughout the year, she bought staples in larger amounts than a single-family household would likely consume. She purchased some 400 pounds of flour at 2.5 cents per pound, cornmeal for 10 cents per pound, thirty-three and a half bushels of bran at 8 cents per bushel, more than 300 pounds of distillery slop at half a cent per pound, and chicken feed. She probably kept chickens to provide eggs for the boarding-house table, and she may have fed the bran and slop to a hog or a dairy cow.[201] At the end of the year, her account was paid and balanced. Elizabeth Renshaw's name does not reappear in McKenna's ledger. She remarried in 1860 and may have given up her business or moved from Fairfield.

Frances Elizabeth Simpson was the wife of Fairfield grocer Henry Simpson. Either he died or they were divorced sometime before 1860. Mrs. Simpson lived in Fairfield in a commodious house near Henry McKenna; a US census marshal valued her real estate at $3,000 in 1860. Beginning in 1861 she maintained an extensive account at the McKenna mill for meal by the bushel, flour by the barrel, and salt by the pound; the scale of her purchases suggests that she was operating a boarding facility in her home or elsewhere in town. From January 1 to June 30, 1861, she bought 502 pounds, or roughly 10 bushels, of cornmeal.[202] She also bought twenty-one cords of wood between August and October, perhaps for cooking and baking as well as for heating her home during the winter months. She made similar purchases through 1863 and 1864. From January to September 1864 she bought about 360 pounds of meal and some 720 pounds of flour, as well as whiskey in dram-sized amounts.[203] She also took advantage of McKenna's moneylending policy by drawing cash to make payments on other accounts or to give to her son James. Her account, like that of Mrs. Renshaw, balanced at the end of the year.

The Sisters of Charity of Nazareth founded the Nazareth Academy, a Catholic educational institution near Bardstown, in 1814. In the mid-1860s the academy maintained a large account with Henry McKenna and bought flour by the barrel. From July through October 1864, for example, a teamster delivered twenty to twenty-two barrels each month at a per-barrel price ranging from $7.50 to $11. The academy's account also included charitable contributions. Four private individuals paid for a $350 order of thirty-five barrels of flour in October, and in 1866 McKenna sent fourteen barrels to the academy, which he charged to his personal account. Every two weeks, the academy returned empty slack barrels to McKenna, fourteen to thirty at a time, for a credit of 25 cents each. In addition to the donations made to its account, the academy submitted periodic cash payments, and at the end of the year a final payment of nearly $200 settled the $1,400 bill in full.[204]

The Community's "Banker"

Reviewing the extensive accounts maintained by employees and residents with Henry McKenna over two decades requires patience, but persistence is rewarded. In the ledgers one finds a rare window into the day-to-day operation of one of the Fairfield community's essential institutions. Before McKenna established his mill in 1855, the village may have had a general store, but the local economy likely depended on a traditional system of barter between producers of different goods, as well as itinerant peddlers who sold small items and specialized services. It would not be an exaggeration to say that the economy of Fairfield and the rural community in northern Nelson County more generally would have included subsistence farming and gardening as well as elementary commercial exchange activity with merchants at the county seat and larger turnpike towns. McKenna's mill and distillery became an extension of a system of trading surplus foodstuffs such as cured meats, dried fruits, and pickled vegetables for linsey-woolsey thread, handmade cloth and clothing, furniture, woven baskets, horseshoes and iron tools, and folk medicines and whiskey.[205]

Farmers cleared trees from their land to create tillable fields and sold the wood to neighbors for heating, cooking, curing meat, and steam engine fuel. The mill and distillery were the community's first industrial works, and by offering employment and a market for locally produced grains, they provided a base on which the community developed a consumer culture and economy. In lieu of periodic village markets, such as those that served county seats, rural people transacted business at the mill's store. As Fairfield grew, more stores opened and likely worked out their own credit systems for carrying customer accounts. McKenna's accounts, underwritten by profits from milling, distilling, farming, and property rental, offered customers the ability to withdraw cash to pay bills or make purchases elsewhere. Employees made withdrawals against anticipated wages, while community customers made withdrawals based on the good-faith understanding that they would pay their balances in cash or in kind and be charged little if any interest.

The mill store initially handled a narrow range of bulk products produced by the mill—flour, bran, and meal—or by its customers—cordwood, wool, potatoes, and shell corn. Village stores sold only small amounts of the mill's flour but offered a wide assortment of other goods. Until stone-covered turnpikes replaced vernacular wagon tracks, bulky commodities were awkward and expensive to transport. Mill service areas were necessarily local. Water-powered mills were largely confined to creek-side sites within reach of a grain-producing farming population. Siting a steam-powered mill was more flexible. Favored locations were those that permitted straightforward procurement of raw materials, access to equipment, a labor force seeking employment, and a convenient mode of transport for product distribution. McKenna sited his mill beside a turnpike that was linked to other improved roads, providing ready, though not trouble-free, access for suppliers and customers alike. The mill's supplier and customer base was limited more by the time required to travel than by the weight of the freight being moved or the inconvenience of navigating a steep-gradient wagon track to a water-powered mill. McKenna readily extended his market to Louisville by way of turnpike connections that allowed him to sell his products in a large market and obtain other goods to sell to customers at his mill store.

One might describe Henry McKenna's commercial ethics as commendable or admirable, but such platitudes fail to articulate the community-building power of his business practices. It is not likely that McKenna intended to alter or restructure the northern Nelson County economy with his system of credit accounts. These accounts greatly advantaged his distiller, coopers, and other patrons. Perhaps to encourage them to remain in the neighborhood, he underwrote their debt, often allowing them to carry negative balances from one year to the next. He was a kind of community banker, yet he did not charge interest on unpaid accounts. He was also a broker, in that he purchased locally produced products such as cordwood, shell corn, and wool and either used or processed these items himself or sold them to others. A small bank operated in Fairfield for a time, but the institution had no recorded interactions with Henry McKenna. A bank also operated in nearby Bloomfield. The primary banking institution serving the northern half of Nelson County was managed by Wilson and Muir in Bardstown. In Fairfield, McKenna provided important financial services that enabled the local economy. Henry McKenna made flour and whiskey, but he and his family also engaged in making community.

Recognizing His Irish Roots

Henry McKenna became a central figure in the larger rural and small-town community of northern Nelson County and a principal representative of Irish immigrants and those of Irish heritage.[206] McKenna hired Irish laborers. He extended them credit and lent them cash. He loaned money for remittance payments to laborers' families and coordinated contributions to the "Irish poor" and for the "relief of Ireland." He was likely at least a tacit supporter of the Fenian Brotherhood, and Patrick Sweeney and others in his employ contributed to the Fenian movement through cash advances on their accounts. In 1881 McKenna and his son Stafford traveled to Draperstown, Henry's hometown in Ulster, for an extended stay. They returned to Kentucky in 1883, the year the new distillery was completed, and Stafford joined the Ballinascreen-Draperstown branch of the Irish National League. The league advocated for Irish nationalist interests and home rule.[207] Although McKenna and his family bore a strong allegiance to their Irish heritage, their beneficence toward their compatriots was not simply ethnic loyalty; it was often an imperative to ensure their countrymen's sustenance.

McKenna's Irish harp emblem.

McKenna's whiskey brand and letterhead featured an Irish harp—"our trademark, the Irish harp of McKenna Whisky," in the words of granddaughter Marcella McKenna.[208] The Irish or Celtic harp has ancient roots as the symbol of survival through regeneration; the instrument has been a definitive subject in Irish lore and life for at least a millennium.[209] The harp was traditionally associated with the mythical Celtic god Dadga, regarded as one of the founders of Irish culture. The metal-stringed instrument became an icon, a metaphor, and a musical tradition of increasing significance in national identity and resistance against British rule, and it became the official emblem of Ireland for both Irish residents and expatriate immigrants in America.[210]

McKenna's pride in his heritage is evident in St. Michael's Cemetery, west of Fairfield. The graves of Henry and his wife Elizabeth are marked by a tall granite obelisk on which stonecutters carved the McKennas' names and the words "Draperstown, Ireland," their birthplace. In 1906 the *Kentucky Irish American* newspaper carried a story summarizing the sixty-year history of McKenna whiskey. "Henry McKenna was an Irish gentleman of the old school," the story read. "He was proud of his birth and ancestry, proud of America and nobody had any greater pride in Kentucky."[211]

15

The McKenna Family Distillery

Marketing, Medicine, and Temperance

Dear Sir

I herewith send you 2.60 two dollars & sixty cents Please send me one gallon of your 3 year old whisky send it by express to Fairfax, Minn as I must have Ky whisky. They do not have nothing and only cheap stuff here. I am slowly improving. Hope when those few line come to hand will all enjoying good health.

I remain yours Truly

Thos F. Brown

Local and National Markets

Henry McKenna's Fairfield mill and distillery served as a rural market for staples, spirits, and a wide selection of other goods. As a financial institution, McKenna's works practiced business and credit policies that enabled rather than constrained the community's economic and social organization. Local customers purchased primarily mill products, although many patrons purchased both staples and whiskey. In 1880 the McKenna family opened an office on Market Street in Louisville; two years later, they moved to a larger space at 245 Fourth Avenue (now Street).[1] From their urban location, the McKennas engaged another market, one of regional and national extent, whose clientele was interested in purchasing only spirits, not staples. This whiskey market was expansive; it began at the fringe of northern Nelson County's rural community and extended east to the Atlantic Seaboard, south to the Gulf Coast, and west across the Great Plains and Rocky Mountains to the Pacific Northwest. Advertising and personal endorsements informed wholesalers, retailers, physicians, druggists, and private citizens of the estimable qualities of McKenna whiskey. The US Postal Service, railroads, and express companies relayed communications and delivered orders for products. The McKenna family's success in developing this dispersed and far-flung national market seems extraordinary, given the distillery's modest capacity and the era's broader political and economic circumstances. State and national temperance activism gained strength during the nineteenth century. The Kentucky legislature passed a local-option bill in 1874, and states in New England, the Middle West, and the Central Plains instituted prohibition ordinances by the 1880s. Paradoxically, the distilling industry's Whiskey Trust inflicted controls on itself; through its associated organizations, it bought and shuttered a number of Kentucky distilleries during the 1890s. The McKenna family not only negotiated such obstacles but also developed a reputation for producing excellent whiskey and conscientious customer service.

Mail-Order Sales

In May 1881 Henry McKenna signed over power of attorney to his oldest son Daniel, who had become the active manager of milling and distilling operations.[2] He was later joined by younger brothers James and Stafford. Their Louisville office on Fourth Avenue, the primary center for their mail-order and shipping activities, was convenient to railroad freight warehouses and a short distance from Louisville's Whiskey Row on West Main Street. Family friend Henry Bosquet operated the Old Blue House, a retail liquor store and saloon, one block north of the McKenna office; in a back room he served a buffet and McKenna whiskey. As a McKenna sales representative, Bosquet placed advertisements in the *Kentucky Irish American* declaring that he accepted mail-order requests for "Old McKenna Whisky" and shipped whiskey "neatly cased in new unmarked boxes to all towns of the United States."[3] After Henry McKenna's death in 1893, the McKenna brothers expanded their company's advertising and marketing efforts and continued to build their mail-order business, while also attending to the traditional Nelson County customer base their father had served since the 1850s.

A Diverse Customer Base

During the fifteen-year period from 1892 to 1907, the McKenna Distillery received mail orders from forty states and territories and the District of Columbia. The ten states with the most orders were Kentucky (53), Mississippi (39), Illinois (28), Ohio (24), Kansas (19), Texas (18), Georgia (16), New York (14), Indiana (12), and Missouri (12). The distillery received six or more orders from Alabama, Colorado, Iowa, Louisiana, Michigan, Minnesota, Pennsylvania, Tennessee, Virginia, and Wisconsin. States missing from the customer list were Arizona, Delaware, Idaho, Maine, Nevada, New Hampshire, Rhode Island, and Vermont.[4] The McKennas' mail-order customers included individuals, businesses, and organizations, among them bankers, clergy, doctors, druggists, and state and federal government officials; hotels, restaurants, saloons, and military posts. Most of the whiskey ordered was for personal use, but patrons also bought whiskey for friends and as promotional gifts for customers.

The McKennas received mailed inquiries from local, regional, and national markets. For example, a druggist in Los Angeles requested a catalog and price list for "Old Fashioned Sour Mash, etc."[5] Mt. Sterling, Kentucky, resident W. J. Queen requested that the McKennas send a gallon of whisky to his sister, Mrs. Elizabeth Smith of Bardstown, who was unwell.[6] Similarly, Everitt Osburn of Nelson County ordered three gallons of whiskey to be sent to his father, Isaac Osburn, in Madisonville, Kentucky, who was ill. This request was unusual for two reasons. First, Everitt Osburn lived in Deatsville, and his address was the T. W. Samuels Distillery. Even if Osburn was not an employee of the Samuels Distillery, he easily could have ordered the whiskey from T. W. Samuels, or he could have bought it at the distillery and sent it across the state to Madisonville himself. Second, there were distilleries in Owensboro, only some fifty miles from Madisonville, whereas whiskey sent from the McKennas' Louisville office had to travel about 160 miles to reach Isaac Osburn, incurring roughly three times the freight cost. The choice of whiskey purveyors suggests that Osburn believed McKenna whiskey to be a superior medicinal product. Or perhaps Osburn was a friend of one of the McKennas, and that relationship was more important than freight rates.[7]

The McKenna correspondence includes several other examples of friendship directing whiskey orders. Mrs. S. Molly O'Casey wrote to her "Friend Dan" McKenna, requesting that a gallon of good whiskey and four bottles of sweet grape wine be sent to The Fair general store in Madison, South Dakota, for her husband Tom, who had been "quite sick for a week and he cannot drink the whisky we get here." Mrs. O'Casey concluded her letter with the friendly phrase, "Hope you are swell." One cannot assume that distance and perceived remoteness meant that personal friends and acquaintances were beyond reach. Madison was a railroad town near Sioux Falls, and letters and express deliveries between South Dakota and Kentucky required only a few days' travel time in the 1890s.[8]

The Reverend Adolph Wesseling of St. Peter's Church in Council Bluffs, Iowa, ordered three gallons of five-year-old whiskey "in jug packed in box by express" in 1890, even though Iowa had passed temperance laws in 1882 and 1884.[9] And the Reverend W. F. Boden of St. Paul's Church in Athens, Ohio, requested four and three-quarters gallons, "well boxed."[10] It is doubtful that the religious sacraments in these two parishes were celebrated with whiskey, so the clergymen likely ordered the spirits for their personal use or to serve on social occasions.

C. H. Clason was a real estate officer at the Plankinton Bank in Milwaukee and a regular mail-order patron. One order included his check for $20 and a request that Mr. McKenna ship five gallons of "good whisky" to be divided among himself and three friends, one of whom was P. D. Lamoreux of Waupun, Wisconsin, the state prison warden.[11] F. S. Clank of the New York Infant Asylum sent an acknowledgment that his whiskey order had arrived by express in time for Christmas and noted, "My medical friends that have sampled it say that it is all that I claim for it which is a great deal."[12] V. H. Robinson of Summit, Mississippi, aggregated orders from Dr. W. W. Moore and other friends and purchased a half barrel of five-year-old whiskey, a strategy he repeated in succeeding years.[13] B. M. Burdett and his wife operated a general store in Killarney, Florida, in 1902. They sold "Staple and Fancy Groceries, Cigars, Tobacco, Hardware, Queensware, Dry Goods, Notions," and Kentucky whiskey. Killarney was a small depot town in central Florida's Lake District west of

Orlando; it was sited at the intersection of two rail lines—the Tavares & Gulf Railroad and the Sanford & St. Petersburg Railroad. In 1892 Mr. Burdett aggregated orders from eight of his customers and sent Henry McKenna a request for fifteen gallons of whiskey. Burdett also requested a $3 commission and sent instructions to "ship the same route as the last and prepay the freight." He included an appeal to "send at once as the parties are anxious to have it for Christmas."[14] The following year, Mrs. Burdett sent another aggregated order for ten gallons of whiskey, again requesting a commission and shipment by express. Notably, Mrs. Burdett was aware of Henry McKenna's death and requested that the person who received the order "deliver the letter and money to Mr McKenns's [sic] successor in business and have him send it."[15]

The McKennas' Louisville office location allowed them to enlarge their sales options and operate as a general retail liquor business. Their expanded inventory included Holland gin, Jamaica rum, Hennessey brandy, champagne, wine, and liquors. These products proved popular with several mail-order customers, including Alfred Mullin of the Kentucky Lumber Company in Burnside.[16] A California whiskey customer, John Moran, was also the "Manufacturer of Pure California Wines" in Los Angeles, and he filled wine orders from H. McKenna in 1892. In an effort to reduce freight rates, Moran urged the McKennas to combine orders with two or three other parties and ship by the carload, which would save about 10 cents per gallon. "Try and see if you could not take 20 or 25 barrels," he said.[17]

A Preference for Age and Taste

An affluent segment of the McKennas' mail-order customers demonstrated their preference for aged whiskey by stipulating specific distilling seasons, number of years in the barrel, or a certain price point; they often added a qualifying phrase such as, "send me best you can at $2.00 per gallon." Quinn, Woodland & Company of Fall River, Massachusetts, imported dry goods and millinery. Mr. Quinn of that company ordered twelve-year-old whiskey from the McKennas and shared it with some of his friends, one of whom, salesman Alexis Menarge, later requested the same for himself.[18]

Topeka, Kansas, attorney Eugene Hagan sent the McKennas a check for $22 to cover the cost of "a case of your 16 year old whisky." He requested shipment by express to the Topeka Club. Presumably, Hagan invited some of his friends to sample the whiskey because six weeks later, on October 15, 1890, he wrote to the McKennas and enclosed another $22 check for a case of "your whiskey brand of 1874 (16 year old)," which he wanted to be sent by Wells Fargo Express to Robert Dunlap of Topeka.[19]

From 1873 to 1875 the USS *George S. Blake,* a navy oceanographic research and survey ship, was deployed by the US Coast and Geodetic Survey to take part in the first deep-sea floor mapping of the Gulf of Mexico. In March 1891 the *Blake* was at the navy yard in Washington, DC. William C. P. Muir, son of Dr. James Muir of Bardstown, Kentucky, served aboard the *Blake* as a ship's officer.[20] William Muir sent a letter on behalf of C. E. Vreeland, the ship's captain, to the H. McKenna office in Louisville. The letter explained: "The captain of this ship wishes to make a friend a present of your very fine old whisky. He wishes you to send him one gallon of your best ten year old by express," addressed to the navy yard.[21] The letter did not explain how Captain Vreeland had learned of McKenna whiskey, but it is plausible that Muir, who had come of age in Nelson County, was his source. These letters and others make it clear that customers who could afford aged whiskey were keen to order it, often specifying the preferred maturity and acknowledging that it was intended as a gift or to serve some social purpose.

Retailers, Wholesalers, and Agents

The McKenna Distillery offices in Louisville and Fairfield sold whiskey in case and barrel lots to wholesalers and saloons, and they entertained queries from prospective agents who wished to represent the McKenna brand. Orders from saloon keepers and wholesale houses often specified a preference for "medium whiskey," two to four years old and of sufficient quality to satisfy their customers while keeping prices low enough to ensure continuing sales. Orders varied in size from a few cases to a few barrels, but several wholesale customers bought whiskey in much larger quantities and often did so with warehouse receipts. These receipts allowed the distillery to raise money, possibly to pay federal taxes to release whiskey from bond. Wholesalers could then take delivery when their existing supplies were drawn down or when the whiskey had aged to the maturity they preferred. Wholesale liquor dealer Boekhoff & Mack of Omaha, Nebraska, held a McKenna warehouse receipt for twenty-five barrels, which it claimed in July 1889.[22] Wholesale dealers in Kansas City, Missouri, and Eau Claire, Wisconsin, made similar claims on warehouse receipts they held for five and ten barrels, respectively.[23] On rare occasions, competing distillers ordered whiskey from the McKenna Distillery. S. O'Bryan & Sons, "Distillers, and Wholesale Dealers in Hand-Made, Sour-Mash, Open-Kettle-Distilled Whiskey" at Hodgenville, Kentucky, ordered two-year-old McKenna whiskey by the barrel, at $1.80 per gallon.[24] O'Bryan's distillery was quite small; it mashed only six bushels of grain per day, and its warehouse held only ninety barrels. Because this order was placed a year after Mr. O'Bryan was dropped from the tax records, the whiskey might have been purchased to fill remaining customer orders or for use by family and friends.[25]

Established wholesalers and retailers often purchased printed stationery with letterheads that included the liquor brands they carried, a form of advertising that implied the listed products had merit. The Elliott-Loeffler Company of La Crosse, Wisconsin, was a wholesaler of imported wines and liquors. The company letterhead carried this generic statement: "We carry the leading Brands of Kentucky Whiskies, Goulet Champagnes, Imported Spanish Ports and Sherries." The company paid about $650 for eight barrels of McKenna whiskey in 1895, implying that it was considered a "leading brand."[26]

A. C. Paul operated the Olympia Saloon in Wharton, Texas, a county seat southwest of Houston served by the Southern Pacific Railroad. His saloon letterhead included a list of products sold, including "McKenna's Old-Fashioned Hand Made Sour Mash and Rye Distilled in 1891" and "Atherton Rye." In 1900 one of his orders requested a case of sour mash and a case of rye, both distilled in 1893. He also inquired about prices for barrels of 1893 and 1895 whiskeys, which McKenna quoted at $2.40 and $2.10 per gallon, respectively. In addition to being a saloon keeper, Paul was a liquor merchant. His wholesale business letterhead made no mention of McKenna whiskey but declared that his firm carried "Famous Echo Spring and Rose Valley Whiskies," produced by the "S. Grabfelder & Company Distillery 401" in Clermont, Kentucky. Even so, Paul's wholesale business bought McKenna whiskey, and a letter with payment instructions stated emphatically, "We want the whiskey."[27] Because Grabfelder & Company operated the Echo Spring Distillery until Prohibition, it is likely that Paul's interest in McKenna whiskey was based on quality, not on the prospect of losing a supplier to the Whiskey Trust.

Miners found deposits of gold, silver, and lead in some low hills about twenty-five miles south of Santa Fe, New Mexico Territory, in the late 1870s.[28] The site was a railroad work camp called Cerrillos. The mineral assays were of sufficient quantity to warrant construction of a smelter in 1879. News of the strike and the promise of a smelter led to a rapid influx of miners and other workers. Within a year, Cerrillos had become a town. A hotel, opera house, general stores, and twenty-six saloons—some in tents, others in adobe and clapboard buildings—served residents and visitors.

N. E. Crenshaw operated one of the general stores in Cerillos, and his letterhead declared that he was an agent for "Morey's Common Sense Gasoline Lamps" and "Strauss Bros." jeans, needful items in a frontier mining town. Remarkably, Crenshaw was also a special territorial salesman for McKenna's sour-mash whiskey, as boldly stated on his letterhead. In 1902 Crenshaw submitted an order to the McKenna office in Fairfield for a barrel of 95 proof whiskey at $2.50 per gallon, to be shipped to an Albuquerque liquor dealer and paid from an account at the First National Bank of Albuquerque.[29] The order made no mention of how a territorial storekeeper 1,300 miles from Fairfield, Kentucky, had become an agent for McKenna whiskey. It is possible that Crenshaw moved his business from one mountain mining town to another and, in the process, had encountered McKenna whiskey or an advertisement for it. Or perhaps he accepted the recommendation of someone familiar with the product. Nothing in the records suggests that the McKenna brothers sent salespeople on trips to recruit agents or wholesalers, a common strategy practiced by other distillers.

Liquor wholesalers and retailers often sent unsolicited requests to the McKenna offices seeking to become sales agents. Some queries came from unserved areas in the East and South; others came from remote locales on the western frontier. Some inquiries were

AGENT FOR STRAUSS BROS.,
AMERICA'S LEADING TAILORS CHICAGO.

AGENT FOR A. G. MOREY'S
COMMON SENSE GASOLINE LAMPS.

❧❧ N. E. CRENSHAW, ❧❧

SPECIAL TERRITORIAL SALESMAN FOR
McKENNA'S CELEBRATED OLD FASHION-
ED HAND-MADE ————————

SOUR MASH WHISKEY,

NELSON COUNTY, KENTUCKY.

Cerrillos, New Mexico, _________________ 190___

Letterhead of N. E. Crenshaw, salesman for the McKenna Distillery. (Marcella McKenna Papers, Oscar Getz Museum, Bardstown, KY)

prompted not by straightforward business references but by social referrals—wholesalers' and retailers' friends, relatives, or fellow townspeople who were also friends of the McKennas. Other requests were open solicitations that might have been sent to any distiller. Mr. W. L. Caldwell of Gibbs & Hancock Seed and Commission Merchants in Lynchburg, Virginia, sent an unsolicited inquiry to the McKennas' Louisville office and offered, "If you have any vacant territory in which you would like to place a salesman, I can work if you say to and should you feel disposed to entertain this I will see you in person."[30] Mr. Merit Street, a banker and exchange broker in Bluff Spring, Alabama, wrote to the McKennas asking for "prices on corn whiskey (also rye & bourbon). I have some customers (dealers) whom I buy for, not less than 5 to 6 bbls at a shipment and if terms & prices satisfactory would be pleased to handle your goods. PS Name present freight rate to Oxford and Alexander City."[31]

Similar queries arrived from several wholesalers and retailers in the South, including businesses in Tadmore and Corsicana, Texas, and Natchez, Mississippi. Considered together, these inquiries suggest that the McKennas were successful in distributing their product information to a broad group of wholesalers, including those in major urban centers such as St. Louis, but also those conducting business at crossroads hamlets where there was little more than a general store. Consider a query, handwritten on a printed letterhead, sent by S. Web Sandifer of Sandifer & Walker Dry Goods and Groceries in Walker's Bridge, Mississippi: "Gents. Will you please quote us prices and send samples of your whiskey, say from too [sic] to four dollars for gal. We have bin [sic] purchasing of late from the Hayner distilling co., Springfield Ohio we only buy in small quantities say from 4 to 10 gal at a time Please quote prices deliver at Magnolia Miss."[32] Walker's Bridge was somewhere near Magnolia in southwestern Mississippi, and it may have been a plantation service center, although the site never developed and it eventually disappeared. Notably, the Sandifer & Walker store had been ordering from the Hayner Distillery in Springfield,

Ohio. In 1890 Hayner claimed to be the largest whiskey mail-order outlet in the United States. We do not know why Sandifer decided to contact McKenna. It is possible that four- to ten-gallon orders were too small for Hayner to entertain. Or perhaps Hayner's prices were too high or its product did not sell well in Mississippi.

Another unsolicited inquiry concerning prospective whiskey sales came from Mr. S. B. Hopkins of the Dallas Distilling Company, Wholesale Liquor and Cigar Dealers, in Dallas, Texas. "I am desirous," he wrote, "of finding a small distillery that has the crop of two or three years of some good brand of sour mash whiskey that I can handle in Texas. I have been in the whiskey business twenty three years, ten in the state of Texas and I can dispose of from 1200 to 2000 barrels per year. If you think favorable of a proposition of this kind, please let me hear from you."[33] Hopkins may have sent similar letters to other distillers, but this inquiry was curiously contradictory. He wished to do business with "a small distillery," yet he wanted to purchase up to 2,000 barrels annually. The McKenna Distillery's maximum yearly output at the time was about 1,000 barrels.

Referral orders received at the McKenna offices, like the open inquiries, came from diverse places. A letter from J. McCormick's Livery, Feed and Sale Stables and Liberty House Lunch Room in Liberty, Virginia, read: "Your whisky has been recommended to me. Let me know your prices & terms & co, maybe I can handle some for you."[34] John E. Penn, proprietor of The Saloon on St. Charles Street in New Orleans, volunteered that his friends in Summit, Mississippi, "have spoken very highly to me about your 6 years old Straight goods as I am about to reopen my place about the first of September please send me your prices."[35] R. S. Campbell of Whitesboro, Texas, a Red River Valley railroad town north of Dallas, was a personal friend of James McKenna. Campbell worked with Bland Bennett, a dealer in staple and fancy groceries and farm and ranch supplies. Reflecting on past business dealings, Campbell wrote to James McKenna: "You have often shown me accommodation & I will endeavor to return same. A. F. Hornback here can use 50 Bbls at time of your whisky, & will *make a yearly contract* for the *same.* This is simply a friendly pointer & you may profit of it if you choose."[36]

Inquiries and referrals also arrived from distant frontier sites. W. T. Martin sent a request for whiskey prices from Fort Fetterman, Wyoming Territory. The fort, one of the US Army's remotest Great Plains installations, was on the North Platte River north of Douglas in northeastern Wyoming. In 1886 isolation turned into accessibility when the Chicago, Burlington & Quincy Railroad and the Wyoming Central Railroad completed lines to Douglas, with an intent to pass near Fort Fetterman. Martin's letter presented his situation, and one can sense his excitement at the prospect. "As I am about to start in the wholesale liquor business," he wrote, "I would like to have a few samples of your medium whiskey. Having been referred to you by an old townsman of yours J. B. Russell by name. If your liquor proves satisfactory I think I can give you an order for several Barrels this summer as there is two Rail Roads coming through here soon and I expect a large trade."[37] Farther west, W. T. Robb, another prospective wholesaler, informed the McKennas, "I am visiting the trade of the north west from the Rockies to the [Pacific] coast and am desirous of making arrangements with a producer in your line to represent them in the territory I visit. If you desire to extend your trade in above named section I should be pleased to hear from you and to enter into particulars."[38] The expansion of whiskey sales along railroad routes is covered in more detail below, but whether the McKenna Distillery represented its business by advertising, by friendly referrals, or by the serendipity of personal conversations, its mail-order business extended far afield from Fairfield, Kentucky.

Unsolicited whiskey orders may have been stimulated in part by general disorder in the national economy and in the distilling industry. Consider that two post–Civil War financial panics stymied the national economy; the first began in 1873 and continued for five years, and the second persisted from 1893 to 1897. Distillery overproduction inundated the whiskey markets, the Whiskey Trust bought and closed distilleries, and individual distillers were often forced into bankruptcy by problematic business practices or onerous federal taxes and regulations. Few distilleries could maintain continuous production over several decades and thereby be considered reliable long-term business partners for wholesale and retail whiskey dealers. The McKenna Distillery was one of the few resilient businesses; operations continued uninterrupted from 1855 until 1918, when Daniel McKenna's death, World War I, and the specter of Prohibition prompted its closing. The distillery reopened in 1934, a few months after Congress passed the Twenty-First Amendment repealing Prohibition.

Hotels and Restaurants

The McKenna Distillery provided whiskey to small and large hospitality businesses in Kentucky and over an expansive national territory. The Central Hotel in Maysville and the St. Asaph Hotel in Stanford, Kentucky, ordered whiskey from the McKennas.[39] Fred Wooten, a former resident of Bloomfield, Kentucky, worked at the Charlotte Harbor Hotel in Charlotte Harbor, Florida, and ordered whiskey in five-gallon lots.[40] The McKennas sent whiskey samples to the Calumet Club Restaurant & Café for Ladies & Gentlemen in St. Louis and to W. F. Shaner, proprietor of the Shaner House in Ottawa, Kansas; the Shaner House, "a New Hotel, elegantly furnished, and second to none in the state," paid top price for five gallons of "good rye whiskey."[41] The Denver Hotel and the Windsor Hotel in Denver, Colorado, purchased McKenna whiskey.[42] In February 1888 the management of the Windsor Hotel requested three barrels of the 1881 distillation and a price list for whiskeys produced from 1881 to 1886, both tax paid and in bond. The order included a compliment: "We are very much pleased with your whiskey, and we thought if we could make a satisfactory trade we would lay in a stock for future use when our present supply is exhausted."

The McKennas' response included a listing, by year produced, of the whiskeys on hand, the prices, and the amount remaining in inventory:[43]

    1881 @ $3.00   5 Barrels on hand
    1882 @ $2.64   4 [Barrels on hand]
    1883 @ $2.35   11 [Barrels on hand]
    1884 @ $2.15   8 [Barrels on hand]
    1885 @ .75 In bond, original gauge 75 Barrels
    1886 @ .65 [In bond, original gauge] 200 Barrels

The whiskey inventory for the specified six-year period totaled only 303 barrels, although the distillery had both older and younger whiskeys on hand. The 187 percent price increase for whiskeys distilled in 1884 compared with those made the following year reflects the taxes paid and an additional year of maturity. Thereafter, the price increases were more reasonable: 9 percent from 1884 to 1883, 12 percent from 1883 to 1882, and more than 13 percent from 1882 to 1881. These incremental increases reflect, presumably, maturity, demand, and stock on hand.

In 1907 the Fairfield distillery office received an order from the US Naval Academy Officers' Mess in Annapolis, Maryland. The Navy Club treasurer, H. Y. Yates, wrote to the McKennas: "Your seven year old Bourbon whisky having been recommended to us by Commander W. C. P. Muir, U.S. Navy, this Club is desirous of trying it. Will you kindly inform us at what price you can furnish it to us in ten gallon lots (in wood), and also whether it could be had any cheaper if purchased by the octave or half barrel."[44] Recall that sixteen years earlier, Commander Muir had recommended McKenna whiskey to his ship's captain. Daniel McKenna replied that 1901 bourbon would cost $3.35 per gallon for a half barrel and $3 per gallon, or $73, for a barrel.

Advertising

Henry McKenna started to advertise his distilling business early on, likely by the 1860s. He adopted two ancient Irish symbols, the Irish or Celtic harp and the Maiden of Eire, as his brand's logo and carved or burned the image into the tops of his whiskey barrels.[45] His engraved business stationery featured a harp on the left and, on the right, a cartouche featuring a cornstalk set against a sunrise backdrop and overprinted with the words: *H. MCKENNA DISTILLER OF NELSON CO. OLD FASHIONED HAND MADE SOUR MASH WHISKY.*

Employees and customers regarded Daniel McKenna as greatly gifted; he was an insightful and energetic person with an excellent business sense. By the early 1880s, he was likely playing an important role in developing the distillery's printed circulars and the advertisements that appeared in medical periodicals and other professional journals.[46]

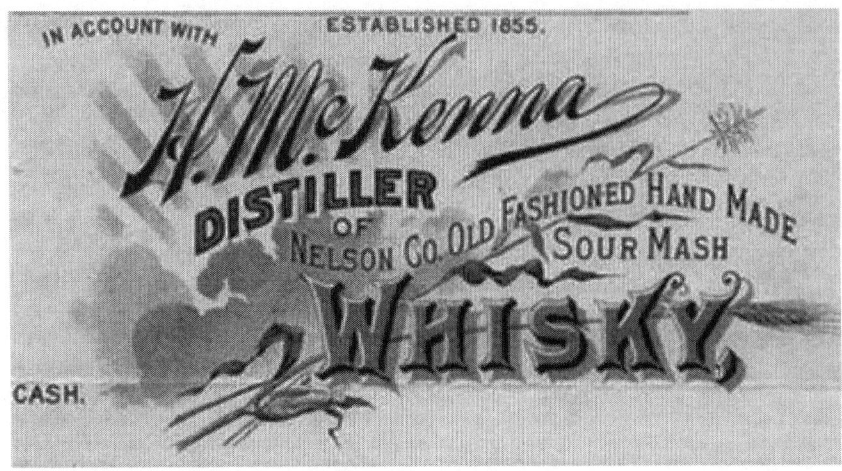

Letterhead of H. McKenna, Distiller. (Marcella McKenna Papers, Oscar Getz Museum, Bardstown, KY)

The Demonstration

The strategy of promoting wines and spirits through tastings and catered dinners was developed long before the late twentieth century. When Dr. Dudley Reynolds, representing the Cahalan Brothers drugstore in Wyandotte, Michigan, ordered a case of McKenna whiskey in August 1888, he added a praiseworthy comment that provides insight into spirits promotions in the nineteenth century. "Please send one case," he wrote, "like that which inspired us at Crab Orchard by express as soon as possible."[47] Crab Orchard Springs was one of nineteenth-century Kentucky's most fashionable "pleasure resorts"; it boasted mineral springs and a hydrotherapy spa. The Lincoln County resort was 115 miles southeast of Louisville, near the geologic splice between the Outer Bluegrass and the Knobs. The site was an important frontier settlement on the Wilderness Road and was served by stagecoach lines and, beginning in 1866, by the Knoxville branch of the L & N Railroad. In the 1870s the Crab Orchard Springs Company, headquartered in Louisville, built a 250-room hotel, a restaurant, and a spa facility near the springs at a cost of more than $125,000. In addition to taking advantage of the medicinal mineral pools, Crab Orchard management offered various organized social events, including formal dinners, concerts, and horse racing at the Spring Hill racecourse.[48] Board cost $17.50 per week during the peak visitor season in July and August. Professional organizations, including physicians and druggists, held group retreats at the springs, and representatives of the McKenna Distillery likely attended, providing samples and touting the medicinal qualities of their "pure" whiskey.[49]

The Personal Endorsement

The McKennas received numerous letters that included unsolicited testimonials about the quality of McKenna whiskey and of Kentucky whiskeys in general. Some offered comparisons to other products and requested that the McKennas arrange for sales agents in their areas to ensure local availability. Most such endorsements came from businesspeople, doctors, druggists, attorneys, and politicians, including members of the federal judiciary and Congress. Flattering assessments also arrived from stockmen, whether they were located in Kentucky or in the high plains and mountains of the western territories.

Some personal recommendations were anonymous. For example, the postmaster at Ellettsville, Indiana, a railroad town near Bloomington, wrote, "I learn from some party that you have some fine whisky," and "I would like to get a few gallons if it is fine."[50] Others named their sources. W. J. Alvey bought and shipped livestock in Morganfield, Kentucky, and operated a saloon on the town's courthouse square. Alvey credited "Mr. L. B. Campbell who has lately moved here." He "tells me you make the best whiskey and I want that kind to run my business. Will you please send me a sample of your 4 year goods. If satisfactory I can use a good lot of your goods."[51]

J. M. Archuleta, a cattle and sheep farmer in Amargo, New Mexico Territory, also operated a general store near the tracks of the Denver & Rio Grande Western Railroad, where he sold a variety of dry goods, groceries, and spirits. Basque sheep and cattle farmers were common in the mountain West. Amargo was just a tiny settlement centered on Archuleta's store; the place no longer exists. The site is located just east of Dulce, which was a Denver & Rio Grande Railroad town and is the "capital" of the Jicarilla Apache Nation. Dulce has a display of the old narrow-gauge tracks and a few of the original railcars. "Mr. J. P. Hogan has recommended your house to me for good whiskies," Archuleta wrote in a letter to the McKenna office, and "I would like to have your price list on your goods. Also, three sample bottles of your best whiskies."[52]

Some primary endorsements offered the possibility of secondary testimonials, which was the intent of an order from Everett Osbourn of Samuels Depot, ten miles south of the Fairfield distillery. Osbourn, who was a friend of James McKenna, wrote: "Please send by Southern Express to Hon Polk Laffoon, Madisonville, Ky two (2) gallons of whisky." At the time, Laffoon was a member of the US House of Representatives. "Send him something fine," Osbourn continued, "and as he like the critter pretty well—I think he will want more of your whisky and he will be a good man to advertise your whisky in that place."[53] On occasion, regular McKenna mail-order customers went beyond personal recommendations and sent whiskey directly to their friends, perhaps to allay skepticism or convey a special gift. Sherman Fry of New York City requested that a half gallon of whiskey be shipped to his friend C. C. Herrick in Troy, New York. Another special request came from Felix Brannigan, an attorney with the US Department of Justice. He placed an order for five-year-old McKenna whiskey to be shipped to an English friend by way of a British ocean liner, the SS *City of Berlin.* Brannigan's request conveyed a tone of urgency; the *City of Berlin,* rated the fastest liner on the Atlantic, was scheduled to sail for Liverpool the following week.[54]

Some mail orders were accompanied by testimonials based on personal experience with the McKenna product. W. A. Sullivan lived in Clinton, Illinois, a railroad junction town south of Bloomington, in the 1880s. He sent an order to the McKenna office requesting the best whiskey available at $3 per gallon. He included a revealing plaint about Middle West whiskey: "I was Raised in Shelby Co Ky and cant use the whiskey made in this country nor any that is sold here either."[55] Wholesale lumber dealer E. L. Edwards of Dayton, Ohio, offered a similar rationale for ordering McKenna whiskey: "When I moved here from Lexington, Ky, I was unable to get the same whiskey that I have obtained at Jim McAllister's at Lexington and got some of the people here to start buying it so I could have what I had been use to and from this I think there has been quite a little trade started on it in Dayton, and they are using quite a little of it."[56] At the time, many whiskeys made in the Middle West were rectified. Their blends contained little or no aged whiskey but harbored all manner of ingredients contrived to create an acceptable, low-cost product.

A. T. Gerard of the Adams Express Company's Western Department Office in Cincinnati wrote to Green Duncan, express agent at Bloomfield, Kentucky, asking Duncan to act as an intermediary with the McKenna Distillery so that he could obtain whiskey at a "cut rate." Gerard affirmed his tasting experience in dramatic fashion: "Route Agent E. L. Ware was here a few days ago," he explained, "and brought with him a portion of a bottle of McKenna's whiskey, stamped 1889, which he gave me when leaving. I took it home," he continued, "and we have pronounced it the finest article of its kind in existence. I have been getting Green River whiskey made by McCullough, Owensboro, Ky., and our agent at that point, Mr. J. H. Small, was able to get a special price for me through his acquaintance with Mr. McCullough. I thought possibly you might have a pull with Mr. McKenna and could get a cut rate. If so, I would prefer to use the Fairfield brand to the Owensboro article." Duncan imprudently forwarded the letter directly to the McKennas, who sent the price list without acknowledging being privy to Gerard's cheeky inquiry.[57]

All the letters in the McKenna record that address product quality were complimentary. No patrons complained about off flavors or inferior taste, and no one requested a refund.

Print Media and Medicinal Whiskey

The *Maysville Evening Bulletin* published a feature on the value of newspaper advertising in 1890 in which the editor quoted Seth W. Fowle, a Boston purveyor of patent medicines and cures. "It is an established fact that however useful or valuable a medicine may be," Fowle had written, "the sale of it can only be kept up by constant advertising."[58] The McKennas advertised their whiskeys to the general public in a limited way by placing a small, column-wide ad in the *Kentucky Irish American*, a weekly newspaper published in Louisville. The ad ran from 1903 to 1919 and read: "Be Sure to Call for McKenna Whisky. It is Always Pure."[59] Saloon and hotel purveyors also advertised in local newspapers that they stocked McKenna whiskeys.[60]

Physicians and druggists had long believed that alcohol spirits, taken in judicious amounts, possessed medicinal benefits. The McKennas addressed this potential market with promotional circulars and ads placed in professional journals. For instance, the "Thirty Years Record," a full-page ad, included physicians' endorsements and referenced the distillery's three formative decades from 1855 to 1885. The advertisement appeared in medical and professional journals and as single-sheet circulars. McKenna solicited endorsements from ten area physicians, including Dr. Harrison Wells and Dr. Francis S. Reid, both of whom had accounts at the McKenna mill and distillery. He also obtained an endorsement from Dr. Dudley S. Reynolds, an ophthalmic surgeon and one of the state's most prominent physicians. Reynolds was a founder of the Louisville Academy of Medicine and published the *Medical Herald,* a monthly literary and scientific journal, beginning in 1879. The *Herald* became one of the most influential medical publications in the country.[61] The McKennas ran full-page advertisements in the *Herald;* each ad carried a quote from Dr. Reynolds that read, "H. McKenna's whisky is the purest and best I have ever seen." Physicians around the country responded to the advertisements. Dr. G. E. Matthews of Halifax County, North Carolina, wrote to request a sample. "I notice your adv in *Med*

Advertisement for H. McKenna that appeared in a medical journal. (Marcella McKenna Papers, Oscar Getz Museum, Bardstown, KY)

Herald," he explained, and "I have a number of patients in whose care *pure* whisky such as you advertised is needed. I have several consumption patients who can find no alcoholic product that fairly agrees with them. I think yours the product in such cases."[62]

When the proprietor of the Rees House Hotel in Winchester, Kentucky, entertained a physicians' meeting in June 1886, the McKennas provided complementary boxed whiskey samples for the attendees. Mr. D. M. Raymond represented the distillery and reported, "The MDs all took it eagerly and walked off with it. There is between two and three hundred doctors here and there will be very little difficulty in disposing of what I have—each one recd a copy of *S Progress* with their sample & I called their attention to your ad. I trust good will come of it."[63] Raymond was likely referencing the *Southern Progress,* a newspaper published in Bowling Green, Kentucky. The McKennas apparently placed one of their medicinal whiskey ads in the paper and obtained copies for meeting attendees.

Nineteenth-century medical practitioners and developers built resorts and hydrotherapy spas at the most accessible mineral and thermal springs in the Blue Ridge Mountains and in the ridge and valley country of southwestern Virginia and eastern West Virginia. For those people with sufficient time and adequate means, these resorts offered treatments and remedies for a wide variety of afflictions. William Cotton of Normandy Station, Kentucky, lived north of Bloomfield, only a few miles from the McKenna Distillery. Cotton's brother James was planning a trip to a West Virginia mineral spring to regain his health. Fearing, perhaps, that quality medicinal whiskey would not be available there, Cotton requested with some urgency that the McKennas "pack & seal in a box so [James] can place in his trunk 2 gallons of your \$3.00 whisky."[64]

Physicians attended to the guests at some resorts, but many proprietors simply advised their clients to bathe and drink the bitter waters as suited them. In addition to extended medicinal treatments, the large resorts, including Bullitt Hot Springs, Little Warm Springs, and Warren White Sulphur Springs, offered exercise facilities, social programs, and well-stocked lounges.[65] In 1896 C. W. Cullen, owner of Warren Springs near Front Royal, Virginia, saw an advertisement for H. McKenna's medicinal whiskey in the *Army & Naval Magazine,* in which he also advertised. Cullen ordered a case of McKenna's ten-year-old bourbon to stock his bar. How better to facilitate the healing process than by offering medicinal whiskeys to enhance the spa's therapeutic waters?[66]

The McKennas' whiskey advertisements appeared in different sizes and designs to accommodate different publishing formats and inform a variety of audiences, but they all included the same basic product message: identity ("old fashioned hand-made sour mash"), location (Nelson County), quality ("strictly pure"), experience ("Thirty Years Record"), disclaimer ("no patent yeast or lime"), and endorsement ("the purest and best I have ever seen"). The half-page ad combined the standard message with an engraved image of an avuncular Henry McKenna formally attired in coat, shirt, and tie. The ad emphasized "family use," using the phrase twice: "I put up Whisky from five to twelve

Half-page advertisement for H. McKenna, 1887. (Marcella McKenna Papers, Oscar Getz Museum, Bardstown, KY)

years old in cases for family use," and "To any person sending $9.00, I will send a case of my Whisky put up especially for physicians and family use." Customers' orders often replicated the ad's language or requested clarification.[67] Dr. J. A. Smith of Norway, Iowa, a Chicago & Northwestern Railroad town west of Cedar Rapids, asked, "How much whiskey [is] in your 9.00 case[,] how old is it, & what kind?" He explained, "I stand in need of some good *old whiskey* & should like to know how much it would cost to get the 'kg' (keg) delivered at this place." As though to confirm where he had seen the advertisement, Dr. Smith wrote in the left margin, "Weekly Med Review," referring to the *Weekly Medical Review*, a journal published in St. Louis.[68] South Illinois physician R. F. Vaughan requested a similar explanation: "If in the event I should send you $9.00 how much of your *pure old hand made whisky* would I get? Your advertisement as it stands in the *Medical Gazette* is very indefinite as to quantity."[69]

Direct Advertising to Medical Professionals

As the temperance movement gained support in the 1870s and 1880s, many states and localities voted to go "dry," passing laws that prohibited whiskey shipments and sales other than for medicinal purposes. Henry McKenna accommodated the changing wholesale and retail spirits market by developing an active mail-order business for medicinal whiskey to supplement his income from local and regional sales. As an ethical distiller, McKenna could attest to his whiskey's "purity" because he controlled all aspects of the grain procurement, milling, distilling, and aging process. Arguing that the product also had medicinal value was another matter.

H. McKENNA,
Distiller of and Wholesale and Retail Dealer in

NELSON COUNTY PURE OLD LINE SOUR MASH WHISKY,

245 FOURTH AVENUE,
LOUISVILLE, KENTUCKY.

Distillery and Warehouse, Fairfield, Nelson County, Ky.

CIRCULAR TO DRUGGISTS AND THE MEDICAL PROFESSION

The universal adulteration of imported spirits and the rectifying process of cheaper grades of American distillation render the utmost caution necessary in securing a PURE stimulant for medicinal use. Having manufactured only the very best grade of Whisky for over thirty years, and using no patent yeast or other deleterious drugs to produce quantity at the expense of quality, I can unhesitatingly offer it to the community at large as STRICTLY PURE Hand-made Sour Mash Whisky, manufactured in the old-fashioned method of a past generation when adulteration was comparatively unknown. The reputation of the

"McKENNA WHISKY"

Has led the medical fraternity of Kentucky to give it an unqualified endorsement as *strictly pure, fine* grade Whisky. A few of these testimonials are offered for your consideration.

TESTIMONIALS.

We, the undersigned, have been using H. McKenna's Whisky for medical purposes in our practice for years, and we find it pure as represented by him, and we can recommend it to any person needing Whisky for medical use:

H. WELLS, M.D., Fairfield, Ky.
J. L. POPE, M.D., Chaplin, Ky.
F. S. REID, M.D., Bloomfield, Ky.
Drs. SHADBURN & STRAUS, Waterford, Ky.
Prof. M. F. COOMES, Louisville, Ky.
N. Cox, M.D., Cox's Creek, Ky.

L. N. HUME, M.D., Taylorsville, Ky.
Prof. J. M. MATHEWS, Louisville, Ky.
DAVID CUMMINS, M.D.
HUGH D. RODMAN, M.D. New Haven, Ky.
Drs. S. M. & N. H. Hobbs, Mt. Washington, Ky.

Mr. McKenna's Whisky is the purest and best I have ever seen.

DUDLEY S. REYNOLDS, M.D.

And in this connection, showing the importance of using only A1 grades of whisky for the sick, an extract is given of Prof. J. M. Mathews' recent lecture on this subject:

[Abstract from a lecture delivered by Professor J. M. Mathews, M.D.)

"It often becomes necessary for the physician to employ alcoholic stimulants in his practice. The very best of these, in my opinion, is *pure Whisky*. It should contain from 48 to 56 per cent. of absolute alcohol, so that its sp. gr. Should not be greater than 0.922 at 60 degrees F., nor less then 0.904. In low fevers, in prostration and insensibility, in syncope and asphyxia, etc., etc., it becomes necessary to administer stimulants in order to sustain and save life. The use of it, however, sometimes proves to be the abuse of it, and you should employ only that in your practice which you know to be *absolutely pure*. The indiscriminate use of the vile adulterations that are thrown upon the market produces much disease; certainly they should not be used by the physician combatting disease."

I have in stock a full supply of

McKenna Pure Sour Mash Whisky

Of the crops of 1874, 1875, 1876, 1877, 1878, and 1879, *in glass*, and 1880, 1881, 1882 and 1883 *in wood*. Handling no adulterated goods, I can recommend all Whisky shipped from my establishment as *strictly pure*. Prices low and QUALITY GUARANTEED.

Send me your orders or write for quotations.

Yours truly,
H. McKENNA.

Circular distributed to druggists and medical professionals. (Marcella McKenna Papers, Oscar Getz Museum, Bardstown, KY)

Purveyors of patent medicines typically issued questionable claims about their products' medicinal qualities, so McKenna deemed it necessary to obtain endorsements from medical professionals. His full-page "CIRCULAR TO DRUGGISTS AND THE MEDICAL PROFESSION" opened with a repudiation of adulterated spirits and the process of rectifying cheap, unaged whiskey with all manner of additives to enhance sales. Because his traditional distilling process introduced no chemicals or additives, his product was a pure "stimulant for medicinal use." Other McKenna advertisements included physicians' endorsements, but this circular featured an extended section of testimonials by local physicians, declaring that they had used the product in their practices and that they recommended it for people "needing Whisky for medical use." The circular then juxtaposed a statement by Dr. Dudley Reynolds attesting to the purity of McKenna whiskey with an extended quote from Dr. J. M. Mathews, outlining the qualities and applications of medicinal whiskey. Dr. Mathews did not mention McKenna whiskey as an example of a pure medicinal whiskey, but he did warn against the "indiscriminate use of the vile adulterations that are thrown upon the market" and caused rather than treated disease. The McKennas sent the circular in direct mailings to doctors and druggists and placed the ads in public venues, including aboard passenger trains. The advertisements prompted queries from physicians and pharmacists, which often led to an exchange of correspondence. Doctors, in turn, recommended the whiskey to their patients, who sent mail orders to the distillery's offices.

Physicians often responded positively to the McKennas' direct mailings. Dr. W. A. Statan of Peytona, West Virginia, wrote, "Your price list received would say you may ship me one case of 7 year old Bourbon whisky at $9.00 Should it give satisfaction will favor you with a larger order in the near future." That Dr. Statan ordered premium seven-year-old whiskey invites speculation as to whether the spirits were for his patients or himself.[70] Dr. R. D. Robinson of Calburn, Indiana, did not obfuscate. "I want a pure whiskey to use in my practice," he said, "an article about six years old and one on which I can rely. I want to order in lots or cases from one to five gallons from time to time."[71] Dr. H. V. Donovan of Lovington, Illinois, referenced a recommendation from his brother, a physician in a neighboring town, who said that McKenna whiskey was "always the best." Following his brother's lead, Dr. Donovan ordered a gallon of "whiskey that I can make my own Tinctures with perfect satisfaction."[72]

Physicians often recommended McKenna whiskey to druggists, and the bourbon became known as a reputable medicinal product. Druggist and watchmaker Hiram Burgess of Goodland, Indiana, wrote an urgent note to the McKennas, stating, "Dr. Cheatham of our place had a correspondence with you about last Aug. regarding a Sour Mash Whisky. You sent him a sample of spring of 1888. Which we think you quoted at 2.25 by Bbl. Please send a barrel at once by most direct route and over as few [rail]roads as you can. We are strangers but would refer you to Brodstreet & Dunn. The grip (influenza) makes a draft on the whisky."[73] John L. Cardin of New Haven, Kentucky, an apothecary and a dealer in "drugs, paints, oils, brushes, wall paper, window glass, lamps and everything pertaining to the drug trade," ordered a barrel of "5, 6, or 7 year old whiskey for medicinal purposes." Cardin's order is noteworthy because, at the time, more than a dozen distilleries were operating within a few miles of New Haven, yet he chose to order from the McKenna Distillery twenty-six miles to the north.[74]

Patients also received recommendations from their doctors to obtain McKenna whiskey on their own initiative. T. H. Sisk of Dalton, Kentucky, wrote that his family physician, S. M. Lieper, had received an advertisement from the McKennas concerning their "good medical whisky." On Dr. Lieper's recommendation, Sisk ordered a full barrel.[75] Druggists who had satisfied customers ordered more whiskey. D. T. Davenport of the Davenport Drug Company in Americus, Georgia, reported that the "1891 McK Rye seems to be very good & we will take ½ barrel of the same goods.[76] Perhaps the most extraordinary order came from W. H. Furr, who operated a general store in Flatwoods, Kentucky. Furr was an agent for Dr. Brown's family remedies, liniments, and Oani-Aleta, a "woman's wine." "Dr. Burnett of Lancaster refers me to your good pure whiskey for medical purposes," he wrote. "I want it for no other purpose except medicinal."[77] Despite being a purveyor of cure-all patent medicines and other nostrums, Furr chose not to ingest his own products but to accept his doctor's advice and take McKenna's medicinal whiskey instead.

By 1890, many locales regulated or prohibited mail-order whiskey shipments. Distillers continued to ship to those places that were not yet locally restricted, following federal regulatory guidelines. When Congress passed legislation restricting medicinal whiskey consignments to less than five gallons, distillers complied by dividing large orders into multiple shipments. In September 1917 the government stopped distilled spirits production and imposed a war tax on warehoused stock. Shipments continued, but only for medicinal goods; the inventory of distilled whiskey remained. The McKennas received an order from E. H. Walker of Orlando, Florida, on October 30, 1917, for one quart of twelve-year-old whiskey. Their response revealed how regulation had affected the McKennas' operations: "We beg to say that you do not state that you want the goods for Medicinal Purposes Only. The law forbids us to accept orders that do not state they are for Medicinal Purposes Only and when we make shipment [the] package has to be thus marked." The letter also noted that, due to higher taxes, the price had increased by $2.80 per gallon, and it concluded by outlining a further restriction on advertising: "In certain dry districts and states we are not allowed to send advertisement of our class of goods through the mail, so fearing that your town may come under such, we won't mail price-list, and are also tearing our letterhead off in order to be on the safe side."[78] Though constrained by restrictions, the McKennas continued to ship their remaining whiskey stock, provided the orders contained the "medicinal clause."[79]

Whiskey Rides the Rivers and the Rails

From the first decades of Kentucky's commercial whiskey production, oak barrels and earthenware jugs filled with spirits followed the state's navigable rivers and the Ohio and Mississippi River systems. By the 1850s, high-capacity railroad trunk and branch lines connected river ports to regional hinterlands. Inland consumer corridors developed along rail lines, and depots became nodal shipping points for turnpike wagon traffic. Few railroad companies in the long-settled Northeast and South built railroad towns, as did the Illinois Central and other Middle West and western transcontinental railroads. Rather, most rail-oriented towns in the East and South grew up organically around railroad depots, especially those that were served by one or more cross-country turnpikes. Henry McKenna's nearest railroad connection lay ten miles to the south at Samuels Depot until the L & N's branch line from Shelbyville to Bloomfield opened in 1880; thereafter, the distillery's proximity to the Bloomfield depot made it the primary focus of McKenna's whiskey shipments. Volume whiskey shipments to wholesalers and large retailers went primarily to larger urban centers that were well served by rail and river transport. Most mail orders, though, were shipped to individual customers, and as those requests increased, the problem of shipping access and freight costs shifted away from the distillery and toward the out-of-the-way customer. In the early 1880s, most mail orders came from rail-side customers. Riverton, Iowa, for example, was a small railroad town south of Council Bluffs on a Burlington Railroad spur. General store owner G. D. Batcheldor sent an order to the McKennas for a five-gallon keg of "good whiskey." He gave explicit shipping instructions: send "by Express to Martin Gleason, Agent American Express Co. Riverton Fremont County Iowa."[80]

Long-distance railroad lines became extended customer corridors, whether the tracks ran into the South or across the Middle West and Great Plains. Shipments to depot towns were straightforward, and one can follow the extension of railroads across Kansas, Texas, and Mississippi by the orders the McKennas received from ever-increasing distances. John Marbach was a bat guano dealer in Bracken, Texas, northeast of San Antonio. He also conducted business at Sabinal, some sixty miles west of San Antonio. Both towns were on the Southern Pacific Railroad. Marbach ordered four and a half gallons of whiskey to be delivered to his Bracken address and ten gallons conveyed to his Sabinal address. Marbach's intent was not clear, but delivery would have been comparatively straightforward, given that the Southern Pacific served both places.[81] Those patrons with no direct railroad connections had to make other arrangements. Several customers in central Mississippi lived off the primary lines served by the Illinois Central and Yazoo & Mississippi Valley Railroads, so they had their whiskey shipped to depot agents or friends who lived close to the track.[82] H. D. Daniel, who operated a farm store in Altamonte, Florida, requested that his five-gallon order be shipped via the Grange Belt Railroad to Palin Springs, his closest depot.[83]

When rail customers realized that not all freight charges were equal, they began to request least-cost shipments. John Eddins, a US Land Office agent in WaKeeney, Kansas, ordered two gallons of whiskey with instructions to "prepay express charges, for by paying it there [Kentucky] the expense will be much lighter, as the Union Pacific R.R. from Kans City stick the tarrif on the express and the other being $3.00."[84] Unless one paid close attention, freight rates could equal or exceed the cost of the whiskey. Hardware and stove dealer N. S. Bridge of Assumption, Illinois, ordered five gallons of whiskey sent to the town's freight agent. He offered this clarification of his location: "I am on the Ill Cent RR," he wrote, but "if they have no connections with your city send by Big 4 RR [Cleveland, Cincinnati, Chicago & St. Louis Railway] or C H & D [Cincinnati, Hamilton & Dayton Railroad] any way to get it here the best & cheapest."[85]

Independent short-line railroad companies built lines to serve comparatively small areas bypassed by the regional railroads. When Dr. O. B. Bush of Mayhaw, Georgia, ordered a case of seven-year-old rye whiskey in 1895, he gave instructions to ship on the "A.M.R.R. to Donalsville," a town ten miles away but the closest railroad connection. The 160-mile-long Alabama Midland Railroad connected Bainbridge, Georgia, to Montgomery, Alabama, by way of Donalsville.[86]

The McKennas received only one order that required an extended river shipment. The Cox brothers operated a general merchandise store in Blue Point, Mississippi, a remote plantation area north of Tunica. In 1902 the brothers requested information about the cost of shipping whiskey to their store. Blue Point was on the west side of the Mississippi River, about forty miles southwest of Memphis on the Council Lake river meander. A Corps of Engineers project to straighten the channel had cut off the meander in about 1874.[87] The McKennas sent a freight estimate for whiskey shipped from Bloomfield to Blue Point in wooden barrels versus glass bottles. Whiskey shipped in barrels cost 27 cents per hundred pounds (a full barrel weighed roughly 400 pounds), conveyed by river to Memphis; an additional 3 cents per hundred pounds for wagon drayage from Memphis to Blue Point; plus $1 per barrel for bulk handling. Thus, the total freight cost for whiskey shipped in wooden barrels would be only about 55 cents per hundred pounds, whereas a glass bottle shipment cost 75 cents per hundred pounds to Memphis, plus more than $4 per hundred pounds for drayage to Blue Point. Although river shipment for both glass and wood was cheaper than rail shipping, the high cost of hauling glass by wagon from Memphis to Blue Point made for some very expensive spirits.[88]

Security and Pilferage

The cost of shipping whiskey, whether in barrels or bottles, was not simply a matter of freight charges. Securing whiskey shipments from theft and pilferage was difficult, whatever the stage of the trip. Theft of spirits en route was not complicated and could be accomplished in several ways. The wooden boxes used to ship small sealed kegs, jugs, and bottles could easily be pried open and the contents appropriated. Full barrels were more formidable to handle and required strength, balance, ramps, and waiting wagons—unless one simply extracted the bung and siphoned off the whiskey. The distillers tried packing small jugs and kegs inside full-sized barrels to deter theft; this succeeded, provided the contents could be not be identified and distinguished from other tight barrels containing salt pork, pickles, molasses, and other comestibles. Salt Lake City druggists Druehl & Franken requested this packing technique "to keep it from being tampered with on the road."[89] W. T. Seymour of the National Yeast Company at Seneca Falls, New York, requested that his four-case order be shipped in a single box. Each of his previous orders had been shipped by rail, and he lost one or two bottles to thieves each time the cases were shipped separately.[90]

While the theft of entire bottles or jugs out of a larger shipment was readily evident, other methods were more subtle. When W. S. Langley of Jackson, Mississippi, reordered five gallons of whiskey in January 1888, he made a simple request: "Will you be kind enough to seal the bung hold with sealing wax as I fear it is tampered with sometimes after leaving your hands as the last three kegs have had bagging wrapped around the stopper in the bung hold and the keg does not seem full." He then explained, "I am obliged to have good whisky as I have an invalid in my family who cannot do without it."[91] J. M. Clayton lived in Canmer, Kentucky, nine miles from the railroad depot at Rowletts. Clayton sent $5 to the McKennas for two gallons of their best whiskey and requested, "Please box & seal jug as some goods have been opened at Rowletts."[92]

Detecting the amount of liquid stolen from a sealed barrel was not always straightforward, as most barrels leaked, and all barrels lost alcohol through evaporation, albeit in small amounts and at a relatively slow rate. When Danville, Kentucky, druggists Davis & Forman received a barrel of McKenna whiskey, they found that they had been billed for thirty-eight gallons, but only thirty-two gallons remained when they took delivery. Either "there was some mistake or the package was robbed on the train," they wrote. They requested an explanation, noting that the six-gallon loss would "eat up our profits."[93] When one of Stafford McKenna's friends received the "tonic" he had ordered, he found that it was one quart short. He wrote to McKenna with an undertone of resignation, "I guess some railroader thought it too much for one man."[94]

Some whiskey shipments were not simply pilfered; on occasion, entire orders went missing. Charles Price of the Palace Hotel in San Francisco ordered two gallons of McKenna whiskey. When the shipment failed to arrive after a three-month wait, Price asked the Santa Fe Railroad to trace the package. The whiskey could not be found, and the railroad reimbursed Price $8, which he passed along to the McKennas to pay his bill. "The whisky never reached me," he wrote, "but it was no fault of yours, it having been stolen by some warehouse or railroad thief."[95]

Testing Temperance: Surveillance, Control, and Clandestine Marketing

Prohibition arrived in small-town America well before a constitutional amendment made it the law nationwide in 1920, complicating matters for distillers who wished to expand their business through mail-order sales. The nineteenth-century temperance movement gathered support through local activism and local-option votes in small towns, which eventually led to state-level sanctions against the manufacture and sale of distilled spirits. State prohibition laws were first passed in Iowa, Maine, and Kansas, and support for the movement's objectives spread across the country after the Civil War. State and local regulations truncated retail sales by druggists, restaurants, and saloons, but shipping whiskey into dry states proved more difficult to stifle.

The Kansas legislature adopted statewide prohibition of retail liquor sales in 1884. The liquor trade continued, but almost entirely by mail. An editorial in the *Kansas City Journal* in 1887 spoke to temperance activists' frustration with a legal policy that failed to cover all contingencies: "For a time after the adoption of Prohibition in Kansas liquor-dealers in Kansas City did a large business with drug stores, but since they have been stopped from retailing liquor the trade has dwindled to almost nothing. Still some business is done in Kansas but it consists entirely of private orders by mail."[96] Four years after Kansas passed prohibition legislation, George H. Gibson of the Kanopolis State Bank ordered some "stogies" and whiskey from James McKenna. Kanopolis was a small railroad town southeast of Ellsworth. Gibson wrote, "I tasted some excellent whiskey which Mr. Curry said came from your place. . . . Mr. Curry told me to use his name in any dealing with you."[97] Such personal endorsements helped guide the McKennas' mail-order business directly to those potential customers who were willing to ignore the state's proscriptions, and a steady flow of McKenna whiskey entered Kansas in disguised packaging by way of the railroads. A. L. Lyman of Goddard, Kansas, was a regular customer. In April 1889 he requested a $12 keg of whiskey shipped in a box. "Do not bill it as whiskey for owing to the laws of this state it is hard to get whiskey from the express office," he warned. Ten months later, Lyman ordered another five gallons of whiskey sent by freight (not express) and noted, "Do not bill it 'Whiskey' as the Kansas laws on R.R. and whiskey are strict." A postscript added, "The last I got gave perfect satisfaction."[98]

Unsanctioned Temperance: Social Controls

Formal legal strictures were not the only methods employed by temperance activists to prevent whiskey shipments from entering their communities. Self-appointed informers publicly exposed those who consumed spirits in private, leading to their ostracism and thereby imposing informal or unsanctioned social controls over liquor distribution. When C. H. Fenell of Ponoma Nurseries in Humboldt, Tennessee, ordered three gallons of McKenna bourbon, he included instructions to ship by New York Express and warned,

"Do not brand as anyone can tell what is in the box."[99] Tennessee liquor dealers also resorted to surreptitious shipments to continue their business.[100] Dr. C. J. Donovan of Arthur, Illinois, was more explicit in expressing his concerns about possible discovery and the court of public opinion: "Please send me a gallon of your 3 year old whisky at once. And pack in plain box and put no bills on so that any body will know what it is as this is a very strong Temperance Town." He added for emphasis: "Now don't put any labels or any thing else on the box so the people will know what it is as they are very strict here in regard to Dr. getting spirits of any kind."[101] The Reverend W. J. Doummy of Elkhart, Illinois, ordered a small keg of seven-year-old whiskey from the McKenna Distillery with instructions to ship prepaid by express. He added: "Be sure to bill goods as 'Alter Wine.' Too I will say that this is local option territory."[102]

Sanctioned Temperance: Legal Controls

Officials in dry towns shut down saloons and forced restaurants and hotels to stop serving alcohol. When the residents of Owenton, Kentucky, voted to go dry, some tried to bypass the restrictions by purchasing whiskey through mail order. But whiskey shipments to individuals could be confiscated, leaving shippers and customers with no recourse. Willie Hicks saw a McKenna advertisement for sour-mash whiskey and requested a price list. "This is a Prohibition Town," he cautioned, so "send in a plain envelope with no heading address."[103] Even the discovery of a price list could raise suspicions.

With his order, S. W. Offett of Madisonville, Kentucky, sent along positive reinforcement of the McKennas' advertising efforts. "The circular within came to hand on last train. We are under the heels of Prohibition in our county and are forced to send elsewhere for our supplies." Offett and other Madisonville residents had been supplied by the A. S. Winstead Distillery in Henderson, Kentucky, but Winstead's "Silk Velvet" brand was not well regarded and, in Offett's opinion, "cant say [it was] very satisfactory." Offett requested a sample bottle and a price list and promised, "Doubtless I can get many orders for you."[104]

Trenton, Tennessee, wine and liquor dealer W. J. Bennett received regular whiskey shipments from the McKennas until 1903. In April of that year he wrote, "This place is now a *dry town* & the people that has been using the Old Kentucky that I bought of you think that there is no other goods like it." Despite the local ordinance, Bennett submitted an order for ten gallons of his favorite "98 goods" and cautioned the McKennas to "say nothing about the package."[105]

Banker LeRoy Davidson of Charlotte, North Carolina, was anxious to join the mail-order queue. "This is a dry town, and consequently we have to look to other places for our supplies. Awaiting the favor of your advices." Shortly thereafter Davidson received a price list and a handwritten note from the McKennas, who assured him, "We can ship you ten or twenty gallons, or by the single gallon. You will notice that we charge extra for jugs, kegs, and boxing. In the event you should desire a shipment to reach you *under cover,* we bag."[106] The exact meaning of "we bag" is not clear, but it likely referred to either anonymous generic packaging or placing the order in a canvas railroad mailbag that carried only an identification number or the destination address.

South Carolina created a particularly aggressive prohibition system designed to give the state control of "legal" liquor sales while furthering the temperance movement. Its guiding principle was to develop and operate a municipal distribution network that removed the profit from liquor sales while generating revenue for the state. In 1893 Governor "Pitchfork" Ben Tillman created a state board of control to administer and regulate spirits sales. He banned alcohol production within the state and directed the control board to order alcohol from distillers outside the state; there is no indication in the McKenna records that they had any dealings with South Carolina officials. When whiskey orders arrived, the spirits were repackaged and shipped to county and city dispensaries operated by county liquor control boards. To purchase spirits, a patron had to submit a formal request to the board, and those deemed to have intemperate habits or flawed social reputations were refused. State agents charged with enforcement became known by the citizenry as "Tillman Spies."[107]

By November 1893, the McKennas began to receive letters of inquiry concerning private mail-order shipments to South Carolina. D. D. McColl, president of the Bank of Marlboro in Bennettsville, requested a freight shipment of "one case of your 7 year old whiskey for which I will send you ck on receipt of same—you must so ship and mark as to prevent contents being known, otherwise it will be seized & or confiscated by the Til[l] man Spy's. It might be best to ship to your own order & send me bill [of] lading. If you can deliver this to me & send me a first class article can give you more orders." Three weeks later, McColl wrote again to advise the McKennas that the whiskey had arrived and he had taken delivery. He included some cautionary news. "It was well that it came by freight," he noted, "as the Til[l]man Spy here raided the Express Office yesterday and found, seized & confiscated eight packages of xmas whisky—You distillers should do all possible to break down this South Carolina Monopoly—and the best way to do this is to under sell the state and furnish a better article—Every whisky man in the whole country should furnish whisky to people in this state at cost til this odious law is killed."[108] McColl's suggestion that altruistic distillers could stop South Carolina's blockade on liquor shipments may have been impractical, but it hints at the breadth of objection to the state's policy among well-to-do residents.

Selling Whiskey in Indian Country and Indian Territory

Congress passed the Trade and Intercourse Act in 1834, which, in a general sense, prohibited any person without a federal license from trading with Indians living in what was then termed "Indian country."[109] The act included prohibitions against introducing spirituous liquors into Indian country. Enforcement resided with the superintendent of Indian affairs, Indian agents, or military commanders. Penalties included a $300 fine and forfeiture of all "goods, boats, packages, and peltries" possessed by the guilty party. The act also prohibited the use of a distillery to manufacture ardent spirits in Indian country; the penalty for doing so was $1,000, and authorities were empowered to destroy the still.[110]

The Pacific coast of northwestern Washington, the narrow littoral between the ocean and the Olympic Mountains, was the traditional home of at least seven Indian tribes when white fur trappers and settlers first arrived. In 1855 the Washington territorial governor directed that one of the tribes, the Quinaults, be settled on a small reservation near the mouth of the Quinault River. The federal government then established a US Indian Service Agency to manage reservation affairs. Thirty-four years later, in 1889, Indian agent Edwin Eells reported to the commissioner of Indian affairs on reservation life and well-being. He noted that although some Quinault people worked in sawmills and some at fishing, others were "living hand to mouth." One of Eells's primary concerns was "the tendency of the Indians to drink" whenever presented with the opportunity. He reported that a local police force and a court of "Indian offenses" enforced prohibitions and that drinking was "measurably restrained on the reservations." But he cautioned, "it is doubtful to what extent our authority extends over American citizens even if on a reservation," although he reiterated that "it is very much kept in check here."[111]

Agent Eells may have believed that the Indians' use of ardent spirits on the reservation was in check, but he did not know that the white residents were seeking supplies for themselves. On September 9, 1889, R. J. Huston of the Quinault Agency sent an order to the McKenna Distillery office in Louisville for two gallons of seven-year-old whiskey, three pints of fourteen-year-old whiskey, and one gallon of sherry wine. "I am very particular as to how I would like this packed," he wrote, "as it has to go a long distance and over several roads. Whiskey is not allowed on any of the agencies, so do not let your *name* or the *contents* of the box go on the outside. Would prefer it to be shipped in an old dry goods box. This is very important *so do not neglect it.* I enclosed a check for the amount. I am buying this with several of my friends, and if it proves satisfactory I think that it will be the means of you getting several orders."[112] Huston made it clear that he wanted the spirits for his own use and that of his friends—not for sale to the Quinault people—but the records do not confirm that the McKennas responded to this order.

The McKennas also received orders from residents of the Indian Territory, the area of multiple reservations on the southern Great Plains that later became eastern Oklahoma. J. H. Wright, cashier at the First State Bank of Sulphur, sent a letter to the McKennas in 1904. The town of Sulphur was about fifty miles south of Oklahoma City, and in the late nineteenth century it was part of the Chickasaw Nation. Wright sent a check for $18 and requested delivery of two dozen quarts of whiskey. His shipping instructions were explicit and typical of the era: "Ship by St. Louis or Memphis care Wells Fargo & Co, express me at Sulphur, I. T. [Indian Territory]. This is Prohibition Country and will request that you pack well in plain package with no marks on it or letters other than the name and address." Wright also mentioned a whiskey dealer in Glasgow, Kentucky, so it is possible that he knew the McKennas personally.[113] Another order from the Chickasaw Nation came from J. W. Kirsch, who operated a shoe store in Ardmore, about seventy miles south of Oklahoma City and twenty miles south of Sulphur. He introduced himself to Dan McKenna as "a son of L. Kirsch at Bloomfield and have bought several gallon from you," indicating that he may have been a family acquaintance. Kirsch's order was accompanied by a common plaint: "We can get no good liquor out here [and] I wish you would send me a list of prices especially on your best grade as I want to get my boss a gallon to show him what good whisky is and I no you have it if any body had."[114]

To fill orders sent from Indian reservations and territories, the McKennas had to reconcile two layers of regulatory prohibition: the federal Trade and Intercourse Act's proscriptions, and the territorial or local prohibitions on the sale of spirits. Records do not verify whether the McKennas responded to orders from Indian reservations and territories, although their replies to numerous other orders suggest that they complied with changing federal internal revenue regulations concerning medicinal whiskey and limits on the size of shipping containers.[115]

The McKennas' advertising program was basic and straightforward, and it yielded orders. Although mail-order sales may have seemed a reliable if not lucrative source of income, the business was layered with risk and uncertainty. Temperance activists intimidated potential customers, and governmental regulations, be they local, state, or federal, impeded shipments and threatened the loss of goods. Freight rates could be punitive, while at the same time, whiskey shipped by rail was subject to rough handling and broken containers, as well as pilferage and theft. Nevertheless, anecdotal evidence suggests that the orders the McKennas received exceeded their production capacity by two to one. By emphasizing medicinal whiskey, they could take advantage of regulatory exceptions. As long as physicians prescribed "pure" spirits to their patients, the quality McKenna product would be competitive with cheaply made rectified whiskey and patent medicines.

Payment for goods received was not always assured. Some customers prepaid by check; others promised to pay upon delivery or qualified such statements by adding, "if it satisfies." A few patrons paid months after receiving their shipments because they were experiencing financial difficulties. J. W. Bradley of Ponchatoula, Louisiana, wrote a letter of apology to the McKennas for his delinquent account. "The amount I owe you have been dry for *very very* long time," he acknowledged, "and I have been trying to settle it, but looks as though I can now do it so will you kindly draw on me through the bank of Hammond a 20 or 30 day sight draft and by doing this I will not forget or put off paying you any longer for I must say one always get the worth of his money when buying your goods."[116]

Anticipating Prohibition: Closing the McKenna Distillery

The McKenna family distillery played a central role in transforming the local economy in northern Nelson County from self-sufficiency and barter to a financial exchange system based on producers, processors, and consumers. The distillery was the only industrial business in Fairfield, and it served a broad customer constituency. As prohibition advocates gained public and political support for their cause, many distillers initially reduced and then stopped production entirely. By 1917, in anticipation of national prohibition legislation, brothers Daniel, James, and Stafford McKenna began to curtail production. They could have simply waited until they were forced to shut the distillery and absorb the loss, but this made little sense, given the investment in equipment, buildings, cooperage, and barreled whiskey stored in their government-controlled warehouses. Instead, the McKennas began to dismantle their facilities and offered for sale their equipment and cooperage supplies. The Paragon Cooperage Company of Fort Wayne, Indiana, had been one of the McKennas' primary suppliers of barrel staves and iron hoop materials, and the company agreed to accept unused stock and resell it on the commercial market. Barrel size and construction had changed by this time; the McKennas' eight-hoop barrels were no longer in common use, having been replaced by six-hoop construction. And their stave inventory was of thicker stock than the new standards. Paragon was nevertheless able to salvage some value for the McKennas.[117]

The distillery closed before Prohibition, but it was not dismantled. When Congress passed the Twenty-First Amendment in December 1933, repealing Prohibition, the McKenna Distillery was one of the state's few works with sufficient structural integrity to resume production. Distilling began again in 1934.

Henry McKenna's Legacy

The news of Henry McKenna's death in 1893 spread rapidly through his Nelson County community; it also reached his patrons and commercial acquaintances, wherever they resided.[118] The McKenna family continued his long-standing businesses, which had witnessed dramatic changes in technology and transportation, survived the Civil War and the transition from enslavement to freedom for African Americans, and confronted the rise of temperance as a rebuttal of his distilled products. He integrated a farm, a mill, and a distillery into an effective business that included a general store and a community "bank" that served more than 350 local customers. Tradition was important to Henry McKenna's style of business, but he was not averse to change. Although he did not adopt mass-production methods and technology, he did develop an extensive mail-order customer base. He and his sons were pioneers in advertising, distributing circulars to the public and running ads in medical and other professional journals. Federal tax policies stifled other distillers, but McKenna paid his whiskey taxes (perhaps with income from his mill and farm) and held his tax-free whiskey in warehouses for extended periods—some aged in barrels for as long as sixteen years. This not only increased the prices he was able to set but also enhanced his reputation as a producer of high-quality spirits. His neighbors and patrons deemed him personable and kindly. Certainly the Irish immigrants he employed and extended credit to benefited from his paternity. Sons Daniel, James, and Stafford carried on Henry McKenna's business, following their father's conscientious practices and emulating his gracious business disposition.[119]

16

Building James Stone's Elkhorn Distillery

We have about 1,000 barrels whiskey made but have sold more yet.
—James M. Stone

Post–Civil War Distilling in Kentucky

By the Civil War's close, Kentucky's distillers were industrializing at an accelerating pace. In 1864, 166 distilleries operated in Kentucky, and steam engines powered 95 of them. Many farmer-distillers were consuming more grain than they produced. Railroad companies laid new track and sidings, which facilitated grain deliveries and whiskey shipments. Congress passed an internal revenue law in 1868 that required distillers to radically refit their stills and warehouses. Federal collectors extracted taxes on spirits before they were sold. The expenses associated with industrializing and increasing the scale of operations, together with the new federal tax levies, greatly enhanced distillers' demand for operating funds, and access to lending banks and financiers became a primary concern.

South Elkhorn Creek is centrally positioned in the Inner Bluegrass region. The watershed extends from Fayette County northwest into Scott, Woodford, and Franklin Counties. The low-gradient stream flows along broad meanders that loop across horizontal layers of gray Grier limestone, where fossilized brachiopods are visible during the extended low-water periods in late summer and fall. Beginning in the late eighteenth century, millers dammed South Elkhorn Creek at several points and directed water along raceways to waterwheels that powered small creek-side gristmills and flour mills. Annual precipitation of some forty inches fed the creek and a few small tributaries. Much of the precipitation entered underground karst channels, which fed a few perennial springs, including the high-volume Royal Spring in Georgetown; even so, many springs dried up in the summer. The Inner Bluegrass soils in the South Elkhorn Creek watershed were some of the most productive and highly valued in the state and were the basis for the area's livestock and crop farming economy.

James Stone built the Elkhorn Distillery adjacent to South Elkhorn Creek in the mid-1860s. Moore's Mill operated just a quarter mile upstream from Stone's distillery site. The Elkhorn Distillery operated only a few years before it was closed and eventually torn down—except for a brick storage warehouse that still stands. The distillery's legacy has been preserved, though, by an extensive collection of more than 1,500 letters contained in three letter and account books.[1]

**Slaves Owned by James M. Stone,
Scott County, Kentucky,
ca. 1860**

Owner	Number of Slaves	Ages of Adult Slaves:		Ages of Children
		Male	Female	
James M. Stone	23	60	40	6
		35	22	
		22	16	
		19	15	
		16	14	
		16	14	
		14	14	
		14	12	
			11	

Source: US Census of Population, 1860, Kentucky Slave Schedules.
Note: Stone was partnered with J. H. Shropshire, Elkhorn Distillery, Paynes Depot, KY, ca. 1867–1872.

An Abridged Biography of James Stone

On the eve of the Civil War, James Matthew Stone was a forty-year-old Scott County farmer. He owned twenty-three enslaved people, including eight males aged fourteen to sixty, all of whom were probably assigned farmwork and related tasks. The nine enslaved females, aged eleven to forty, likely worked in the household, garden, and fields as needed. Stone may have freed his slaves sometime after the Emancipation Proclamation or at the end of the Civil War. His first wife, Nancy, died (or perhaps they were divorced) sometime between 1860 and 1870, when the household included his second wife, Sally, and five children. Five African Americans boarded with the Stone family in 1870, including domestic servant Evaline Johnson, her three children, and farm laborer George Washington. The distillery's white night watchman, D. G. Christopher, also boarded with the Stones.

Stone farmed 137 acres of improved land in southwestern Scott County along Elkhorn Creek's South Fork. His property also included 110 acres of wooded land, which was probably open pastureland punctuated by mature oaks, ashes, hickories, and hackberries and grazed by his sheep and a small herd of cattle.[2] (In 1870 the average farm size in Kentucky was 158 acres.) The cash value of Stone's farm was $24,900. He specialized in livestock production and in 1870 reported owning thirteen horses, fourteen mules, seven milk cows, two oxen, seven other cattle, thirty sheep, and thirty-nine hogs, with a total combined value of $3,675. He valued the animals he slaughtered at $2,000. Stone produced 750 bushels of corn and raised no other grains. He marketed 200 pounds of wool and produced 250 pounds of butter; he also produced and sold bluegrass seed.[3] The value of Scott

**James Stone Household,
Scott County, Kentucky,
1870**

Name	Age	Relationship	Occupation	Place of Birth
James Stone	53	Father	Farmer	Kentucky
Sally Stone	39	Wife	Keeping house	Kentucky
Charley Stone	22	Son	Farmer	Kentucky
Bettie Stone	20	Daughter	At home	Kentucky
Emma Stone	16	Daughter	At home	Kentucky
Cary E. Stone	13	Daughter	At school	Kentucky
James Stone	10	Son	At school	Kentucky
Evaline Johnson*	42	Boarder	Domestic servant	Kentucky
Dudly Johnson*	11	Boarder	House boy	Kentucky
Debby Johnson*	9	Boarder	Not recorded	Kentucky
Hanny Johnson*	7	Boarder	Not recorded	Kentucky
George Washington*	29	Boarder	Farm laborer	Kentucky
D. G. Christopher	55	Boarder	Night watchman	Kentucky

Source: US Census of Population Manuscripts.
Note: The value of James Stone's real estate was $54,900, and the value of his personal estate was $5,075.
*African American.

County farmland averaged $65 per acre in 1860.[4] James Stone's farm was worth about $100 per acre in 1869, and the combined value of his real estate and livestock placed him among the county's five most prosperous farmers.

Stone may have operated a small distillery on South Elkhorn Creek before 1865, but by 1867, he had become the leading distiller in Scott County, producing 30 barrels per day and some 200 barrels per week.[5] The Elkhorn Distillery chronicle does not begin until August 1868; although distilling was under way during the winter and spring of 1867–1868, there is no record of those operations. James Stone was a farmer-distiller, but he could also be considered a merchant-distiller, in that he purchased all the grains processed by the distillery, primarily through agents and grain brokers in Henry, Owen, Scott, Shelby, and Woodford Counties and in Cincinnati and Louisville.

Congress passed new internal revenue regulations in the summer of 1868 that forced all distillers to modify their existing works for security and accounting purposes. To comply with the new regulations, Stone made extensive alterations to the Elkhorn Distillery and built a new brick storage warehouse during the late summer and fall of 1868, suggesting that the first distillery had limited production and storage capacity.

Once reliable steam power enabled high-capacity distilling and the volume of whiskey production increased, so did the need for financing. The Elkhorn Distillery's monetary requirements likely would have outstripped the reserves of a small-town bank, necessitating the establishment of new financial relationships with larger institutions in distant cities. The large financial investment required to maintain production and the extensive record keeping and management required to comply with the new regulations may have moved James Stone to enter into a partnership with James H. Shropshire to operate the distillery in 1868. The *Louisville Courier-Journal* reported that Shropshire, "an ex-Bourbon and an ex-Lexington merchant, has formed a partnership with Matt Stone in running the large distillery near Payne's Depot. Mr. Montgomery, of Ruddell's Mills (Bourbon County), who is an attaché of the distillery, says Shropshire is idolized by the hands."[6] Shropshire lived and maintained an office in Lexington, at least twelve miles from the distillery. In addition to providing financing for distillery operations, Shropshire maintained the business records and conducted an extensive correspondence with banks and insurers in Midway, Paris, Maysville, Lexington, Louisville, Cincinnati, and New York; with wholesalers and suppliers in Louisville and Cincinnati; and with whiskey dealers in Boston, New York, and other cities.[7] Correspondence was usually signed "Stone & Shropshire" or "S & S."

James Stone's oldest son, Charley, worked on the farm and as a manager at the distillery. Even though Charley aspired to be a livestock agent, his father placed him in charge of the warehouse, including barrel storage, whiskey gauging, and removal for shipment. In late October 1869 Shropshire wrote to Charley: "I have orders from your father to gauge 317 barrels whiskey tomorrow and need to request that you will please have hands enough to handle it promptly so that we can get through if possible tomorrow. Please don't fail to hire the hands."[8] Filled whiskey barrels weighed 400 pounds or more and were likely stacked in rows in the warehouse. Moving and gauging 317 barrels in one day would have been a formidable undertaking, even for experienced laborers.

James Stone declared bankruptcy in 1873 and reluctantly closed the Elkhorn Distillery. The works was demolished (except for the warehouse) in 1877.[9] Three years after the distillery closed, James Shropshire was an auctioneer and had an office at 30 East High Street in Lexington. Distillery records suggest that, by that time, he had also established a whiskey brokerage business with a wholesaler in Galveston, Texas.[10]

The Elkhorn Distillery: 1868–1869

The Situation

James Stone built the Elkhorn Distillery on his farm in southwestern Scott County near the Scott-Woodford County boundary. The town of Midway was about two miles southwest; Georgetown, the county seat, was six miles northeast. Patronizing the Louisville, Cincinnati & Lexington Railroad depot in Midway required that teamsters move heavy freight wagons across South Elkhorn Creek and navigate a narrow, twisting mill road. Paynes Depot, two and a half miles southeast of the distillery, became the favored railroad connection. Although the connecting thoroughfare did not require a steep-banked creek crossing, the road was of marginal quality.

The Site

Positioned on the north side of a long meander loop, the distillery stood on a low rise between the edge of the South Elkhorn Creek floodplain and the toe of a low-angle hill that sloped down to the creek bank. The site's modest elevation change did not impede the movement of heavily loaded wagons, but the slope's pitch allowed Stone to position his warehouse so that the first floor was at ground level on the west end and about six feet belowground on the east end. This allowed Stone to construct a large door and a low stone ramp on the east side that gave access to the second floor. The cooler inside temperature of the first floor's in-ground section proved to be a disadvantage for storing whiskey because, as both Stone and Shropshire concluded, whiskey aged slower in that part of the warehouse. The distillery was only a few yards from South Elkhorn Creek, and although an intermittent spring issued from the limestone bedrock on a hillside about 200 yards to the southeast, the distillery's water source was an onsite artesian well that flowed in wet weather but required pumping if conditions turned dry.[11]

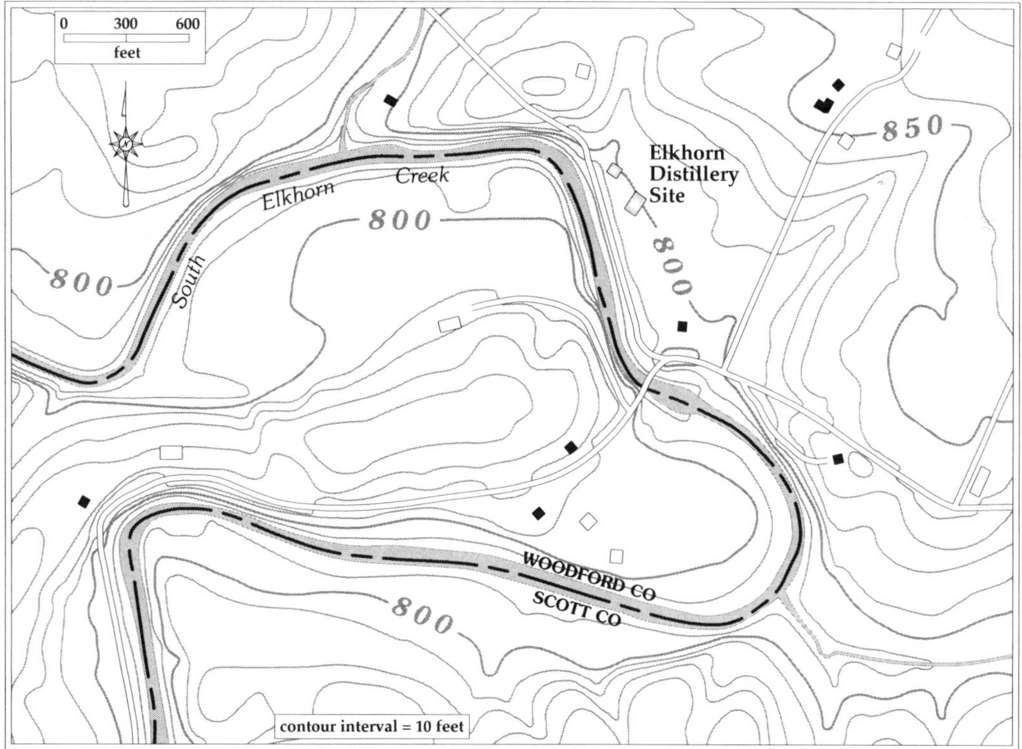

Midway, Kentucky, USGS 7.5' topographic map, 1952.

The Structures

During the fall of 1868, the Elkhorn Distillery was a congested construction site. Laborers enlarged and refitted the distillery, masons erected the new brick warehouse, and carpenters completed various wooden structures. All this activity was under way at the same time because Stone wanted to be ready for the 1868–1869 distilling season, which began in early December. The distillery would be fully industrial and in compliance with the new federal internal revenue regulations.

At the beginning of the 1868–1869 distilling season, the Elkhorn Distillery site included several new or enlarged structures: a large "fireproof" storage warehouse, grain storage bins, corncribs, a coal bin, hog feeding pens and sheds, a cooperage, an office, and slop storage tanks and delivery pipes. The remnants of the distillery's stone foundation convey the building's basic dimensions, and the record books allow one to not only enumerate the other structures but also visualize their configurations.

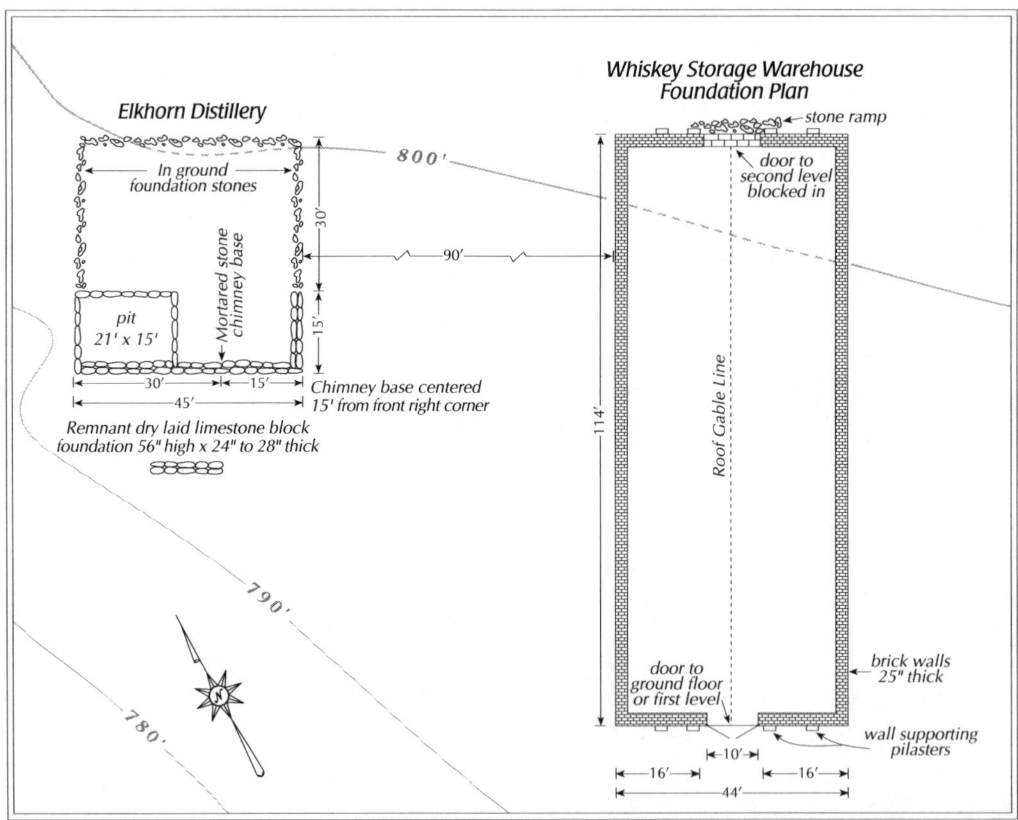

Elkhorn Distillery site plan, ca. 1868.

The Still

The structure that likely served as the distillery was a square wood-framed building that measured approximately forty-five by forty-five feet. The boiler room housed a coal-fired steam engine that heated two boilers that were forty-two inches in diameter and twenty-eight feet long; the boilers were positioned ten inches apart on a tile floor. The copper still was a new column-type, continuous-flow, high-volume still. James Shropshire claimed that it made excellent whiskey. "We know that our whiskey is equal in quality to any steam copper whiskey made in Kentucky," he said, "and as the old fashioned small copper distilleries are not remaining, or at least very few of them, we claim that it is as good as any."[12] But reconfiguring the distillery with new equipment was not a straightforward proposition. The column still required larger pumps and a longer condensing coil or "worm pipe," which Stone & Shropshire ordered from the W. & G. W. Robson Company of Cincinnati. The Robson Company operated a distillery across the Ohio River in Newport, Kentucky, and supplied equipment and parts to distillers.[13] In late October, Stone sent the Robson Company his old worm pipe, along with a drawing of the modifications he wanted; the new worm pipe was to have four additional coils. Later that week, Stone & Shropshire ordered from the Robson Company a brass pump with brass valves but no leather bushings. "This pump is for slop and will soon eat any leather which it may have about it. Please ship it to us as soon as possible. Please send us the articles (the copper worm and another pump) we ordered a day or two ago as soon as you can." Two weeks later, the new worm pipe had not yet arrived, so an impatient James Stone directed his office clerk to send a letter of inquiry. "We are now ready and waiting for the worm we ordered from you," the letter advised.[14] The worm pipe and pumps arrived sometime after November 12.

Internal revenue regulations required that newly distilled whiskey be pumped into a holding cistern made of wood or concrete. Whiskey remained in the cistern awaiting transfer to barrels. The cistern gave the internal revenue gauger a place to measure production and hold the whiskey under the security of a government lock. The cistern also allowed any fine particles in the whiskey to settle out. Workers constructed the Elkhorn cistern on-site, and Stone & Shropshire ordered a barrel of cement from Perkins, Wiard & Company in Louisville "suitable to plaster cistern with."[15] Internal revenue regulations also required that Stone & Shropshire provide a plat or site drawing of the distillery premises. The drawing was to record the position of all buildings and equipment to scale and identify each machine, tank, valve, and pipe by name; the drawing also had to include the location of all government locks. S & S likely commissioned Cincinnati artist and engraver William R. McComas to make the drawing. In late October 1868 McComas submitted a draft of the drawing to S & S for review, and it seems that he had mislabeled the cistern. "The tub spoken of is not really a Rectifier or a Rectifying Cistern," Stone wrote in reply. "It is simply a Clarifying Cistern having nothing but charcoal & blankets in it. The purpose for which it is used is to catch all the sediment which might be in it. Could you not change the name to Clarifying Cistern, and state the use for which it was made, so that it will pass."[16] Stone's plaint about the cistern's misidentification provides two important insights. First, the phrase "so that it will pass" indicated Stone's concern that the internal revenue collector would not approve a drawing with this type of error. Second, Stone wanted it to be clear that the cistern was not used to rectify or mix whiskey with other ingredients; it was a holding tank that also functioned as a settling vessel to filter out impurities.

Grain Storage and Handling

Stone & Shropshire's new column still increased the distillery's production capacity, which in turn required additional grain processing and storage space. To that end, Stone built, at considerable expense, a new grain mill, corncribs, and grain bins. The mill used traditional belt-driven milling stones to crack grain. The new equipment also included a corn sheller, which Stone had ordered from the Bradford & Sharp Company in Cincinnati. Once distilling began, the mill was put to heavy use. By late December, one of the millstones had "become loosened in its frame," and Stone wrote to Bradford & Sharp, requesting advice: would the company be willing to "send me asap a man to fix it," or would it make more sense "for me to ship it to you?"[17]

The corn sheller was a type patented by William Readings in 1852, and although the machine was new, the design was outdated. The Readings sheller was effective but labor intensive. After first shoveling ear corn out of a crib and into a wagon, laborers had to shovel the corn from the wagon into a hopper attached to the top of the sheller's metal case. A rotating, belt-driven cylinder measuring forty-eight by sixteen inches and mounted with metal teeth separated the kernels from the cobs. The shelled corn was then moved to the mill, and laborers shoveled the cobs into a storage bin or crib for future use. First put into use when the distillery began operations in December, the corn sheller's cylinder had worn out by early February and required replacement.[18]

The Warehouse

James Stone built his first whiskey storage warehouse not at the Elkhorn Distillery site on South Elkhorn Creek but near the Louisville, Cincinnati & Lexington Railroad at Paynes Depot. The bonded brick warehouse was large enough to store the 828 barrels he had produced during his short production run in 1868—the distillery operated from March through June, producing about 50 barrels per week. After the refit in 1868, the distillery's expanded capacity required a much larger storage structure. Stone wanted to replace the limited-capacity depot warehouse with a large brick warehouse at the distillery that would be convenient, secure, and fire resistant; he envisioned barrels being moved from the distillery to the warehouse along a rail barrel run instead of by wagon.[19] When a prospective customer indicated that he wished to store his whiskey purchases at the Elkhorn Distillery rather than take immediate delivery, Stone replied enthusiastically, telling the customer that the whiskey would be secure and the insurance rates low because "my warehouse will be strictly fire proof." It will have "iron doors, windows and window shutters. And [a] metal roof. No wood about it will be exposed. If there ever was a fire proof building mine is one of them."[20] Stone also hired a night watchman to oversee warehouse security, boarding the man in his own home.

Elkhorn Distillery warehouse.

Warehouse construction began in the fall of 1868. The solid brick perimeter walls were twenty-five inches thick, and the end walls were reinforced with four full-height exterior pilasters. Builders cross-bolted the load-bearing exterior walls end to end and side to side, a common practice in the construction of large commercial and industrial buildings at the time. The tie bolts were secured on the exterior with large star-shaped nuts. Tinned sheet iron covered the roof, and an ironworks fabricated the purlins, doors, and window shutters.[21] Stone contracted with brick mason William Haskins. By mid-October, shipments of sand, lime, and cement arrived weekly at Paynes Depot—all of which had to be off-loaded from railcars onto wagons and hauled the two and a half miles to the distillery.[22] Stone submitted initial orders for more than 60,000 bricks to the H. I. Todd Company of Frankfort. Carloads of brick arrived at Paynes Depot, but as Haskins discovered, much of it was soft and low quality. Although Stone admonished the Todd Company for supplying an inferior product, Haskins was left with the task of working the substandard bricks into the building's upper story, where the walls carried less of a load. Stone coordinated the construction work crew, teamsters and wagons, and delivery of materials. Late sand or brick shipments stopped construction, and because the laborers were not paid if they did not work, some of them quit. Teamsters quit for the same reason. Stone therefore pushed the Todd Company to expedite brick production and ship on a regular schedule.[23] At the end of October, Stone ordered 58,000 additional bricks and contracted with a roofer to install and paint the tin roof. He boarded the roofer in his home while the work was under way. When the roof was finished, Stone installed lightning rods and grounding cables.[24]

Inventor Miles Greenwood established the Eagle Iron Works in Cincinnati in the 1830s, and the firm had become one of the Middle West's largest foundries by the end of the Civil War. The Eagle Iron Works produced hardware, stoves, wagon boxes, plows, and custom metalwork. When James Stone ordered bolts and iron window shutters for his warehouse, Greenwood's company fabricated them to fit the building's dimensions.[25] To facilitate moving and stacking full whiskey barrels in the cavernous warehouse, Stone ordered from Greenwood a "hoisting machine"; given its description, this consisted of a metal block and tackle operated with a set of hooks and clamps and sixty-five feet of two-inch rope. When the hoist arrived, Stone complained that it was too expensive, too small, and difficult to operate. It will "wear out the patience of the best mannered men on earth that have 200 or 300 barrels of whiskey per week to house and contain," he said. Stone ordered a larger hoist.[26]

Once the warehouse was complete and distilling began in December 1868, Stone ordered from the J. G. Dudley lumber mill in Frankfort 870 poplar two-by-fours, in lengths from twelve to sixteen feet, suitable for "ricking" whiskey barrels. Until Frederick Stitzel patented his whiskey barrel rack system in 1879, most distillers laid barrels on their sides and used two-by-fours and other lumber (called dunnage) as floor blocks to wedge the barrels into place and as spacers laid laterally across the rows of barrels to allow them to be stacked on top of one another. As barrels filled the first-floor space, laborers began to fill the upper floors, and Stone placed additional orders for "scantling" wood from the Dudley mill to serve as dunnage.[27]

The Office

During the fall of 1868, teamsters, laborers, masons, carpenters, and other craftspeople animated the Elkhorn Distillery site. After the new year, James Stone directed the construction of a new office building. From a Paris, Kentucky, lumberyard he ordered 5,500 feet of seasoned one-inch hemlock sheeting with a square edge.[28] Twelve-paned window sashes, eighteen-foot flooring boards, and 100 feet of half-inch poplar tongue-and-groove ceiling planks arrived at the construction site by way of Paynes Depot.[29] As the weather warmed in the spring, a painter arrived to paint the warehouse roof (two heavy coats) and the new office (two coats of "some good color"). The painter furnished the paint, putty, brushes, and other necessities and received $75 in wages and free board in the Stone household.[30]

The Labor Force

When the distillery construction projects were completed, James Stone hired laborers, several of them African American, to work in the distillery—moving grain, working the mash tubs, filling and storing barrels, and performing whatever other tasks were required. Some laborers worked the entire distilling season; others were hired week to week for specific jobs such as cleaning vats, gauging, and shipping the full barrels.[31] Office clerks helped keep the books and wrote most of the distillery's correspondence with suppliers, customers, banks, insurers, and internal revenue officers.[32] Three distillery jobs required specialized knowledge and skills: the distiller, the miller, and the cooper. The records do not mention the distiller by name; the few references to the distiller are related to his judgments about the quality of grain or malt arriving at the distillery.[33] The miller was also anonymous. Sometime in early February 1869 the miller quit, and Stone sent notices to Samuel Clay Jr. and other business acquaintances in Lexington, Paris, Cynthiana, and Louisville announcing that he wished to hire a "good & reliable miller." He stated, "We have two mills and receive about 500 bushels to be ground per day." In addition to "a head miller," he wanted "a good assistant miller who had been around mills enough to watch over them while running. Pay $60 per month plus board." He added this qualification: "We don't need a first class flour miller only want a man that can keep the rollers flawlessly sharpened & grind for the distillery." On March 4, 1869, Stone hired a new miller whose name was not recorded, but he may have been James Howitt.[34]

COOPERS

German-born cooper Adam Mikkle, also known as Michaels, operated the Elkhorn Distillery's dedicated cooperage shop. The cooperage produced tight whiskey barrels and tubs for mash, beer, and fermenting. The business arrangement between Michaels and Stone & Shropshire is unclear; Michaels probably rented the cooperage and produced barrels there, which he then sold to the distillery or other customers. He also likely ordered barrel staves, head wood, hoops, rivets, and bungs through Stone & Shropshire.

Michaels and his family lived near James Stone's house; another cooper, Henry Smith from Bavaria, boarded with them. If the Elkhorn Distillery used 180 to 200 tight barrels each week, and if each cooper produced four to five barrels per day and worked a six-day week, the Elkhorn cooperage must have employed five or six coopers in addition to Michaels and Smith; alternatively, Michaels and Smith worked in the cooperage year-round, stockpiling barrels for the distilling season.

In late October 1868 Stone sent his first order for barrel hoops and rivets to the W. B.

Belknap & Sons Hardware and Manufacturing Company in Louisville. Back in 1840, Belknap started an iron nail manufacturing business in Louisville as an agent for Pittsburgh iron manufacturers George K. and John H. Shoenberger and the Juniata Rolling Mill. In 1849 Belknap purchased a rolling mill in Louisville and began producing a variety of iron products.[35] Stone's order offered the prospect of significant business. "Will you please send us your prices for best Pittsburg Hoop Iron, suitable for mash, beer and fermenting tubs and for barrels, and what is your deduction for cash. We will want from $150.00 to $200.00 worth per week and if your prices and stock are suitable or satisfactory we will favor you with an order. An immediate answer requested." From early November through the end of the distilling season in June 1869, Belknap Hardware supplied the Elkhorn Distillery with "bundles of hoop iron in one and one-half and one and three-fourths inch widths" and with rivets, twenty-four pounds to a box.[36] The distillery's large mash tubs required larger hoop stock. A November 1868 order requested 935 feet of three-inch and 935 feet of two-and-a-half-inch hoop iron for beer tubs.[37] The wood for the tubs was supplied by a Lexington lumberyard—200 feet of one-and-three-quarter-inch poplar in twelve-foot lengths and seventy staves of the same thickness and length. Stone preferred to seal his whiskey barrels with the "best quality walnut bungs," which he obtained from Josiah Kirby at Cincinnati.[38]

Although Elkhorn did not operate during the 1869–1870 distilling season, Michaels's cooperage continued to make and store barrels, some of which he sold to other distilleries. In late January 1870 Stone & Shropshire received a query from the Carter & McDonald Distillery, which operated at a gristmill on Combs Ferry Pike at Boone Creek, near the Fayette-Clark County border. Carter & McDonald wanted to know if Adam Michaels produced whiskey barrels for sale. James Shropshire replied, "I have seen our cooper & he says he will furnish you what barrels you need at $2.50 each delivered at L C & L Railroad and give you 90 days time to pay for them by paying him 10% per annum interest. He says he can load 120 barrels in a car. These are the best barrels I think I have seen [and] he can deliver them any time."[39] Over the following months, Michaels continued to operate the cooperage, ordering hoop iron and rivets from W. B. Belknap & Company. He paid his account, through James Stone, punctually each month.[40]

Michaels produced high-quality barrels, but it was impossible to make every barrel leakproof, especially when they were hauled to the depot in unsprung wagons over two and a half miles of poor road. Leaking bungs could be fixed by proper trimming and resetting.[41] Should stave joints leak, the cooper could reset or replace the defective staves. Stone & Shropshire may have attempted to reduce leaks by applying silicate of soda to the inside of tubs and barrels, an increasingly common practice after 1840.[42]

TEAMSTERS

Managing the Elkhorn Distillery's freight—moving whiskey barrels to the depot and transferring sand, cement, brick, coal, grain, and other supplies to the distillery—required tactical proficiency, an appreciation of the seasonality of labor availability, and an ability to adjust to the vagaries of weather.[43] Teamsters moved freight in wooden wagons pulled by harnessed horses or mules, a seemingly simple process that required heavy labor and a cadre of supporting craftspeople such as wagon and harness makers and blacksmiths to shoe draft animals and repair broken wheel rims. The wagons were presumably lightly built farm wagons, not the heavy-duty Conestoga-type wagons used by cross-country freighters.[44] With few exceptions, teamsters and laborers were anonymous; they were identified only in a general way as people who moved, loaded, shoveled, and hauled. At least three different types of laborers moved freight to and from Paynes Depot. The railroad's freight agent hired day laborers and teamsters to unload freight and haul it to customers; for this service, customers had to pay a conveyance fee in addition to the basic freight shipment charges. The agent at Paynes Depot charged $2.60 to haul one wagonload of grain from the depot to the distillery—a common rate. A second group included private haulers who often operated businesses near the depot. J. H. Beauchamp ran a general store at Paynes Depot and also hauled freight. And finally, farms and businesses such as the Elkhorn Distillery maintained their own teams and wagons and hired teamsters and laborers as permanent, seasonal, or day-to-day workers.[45]

**T. J. Mefford's
Account with James M. Stone,
Elkhorn Distillery**

Year	Month	Day	Transaction	Amount
1866	October	22	To tobacco of Beauchamp	$0.25
	November	17	Coffee, $1.00; sugar, $1.00; tobacco, $0.50 of Beauchamp	$2.50
		17	Soda, $0.50; indigo, $0.25; molasses & plug, $1.40	$2.15
		28	Cash paid to Wm. D. Lang	$4.50
		30	Cash paid on account	$1.00
	December	8	Tobacco of Beauchamp	$1.00
		10	Order on Beauchamp for merchandise	$2.00
		23	Order on Beauchamp for merchandise	$5.00
		27	301 pounds of pork on account @ 9 cents	$27.09
1869	January	31	Cash on account	$20.00
	March	8	Merchandise bought of W. T. Arnold & Company	$4.40
		27	Cash (check on deposit Bank of Midway)	$15.00
	June	5	Tobacco and Sunday Times since April 13, 1869	$2.50
		5	Renting house since Oct 21, 1868: 7 months & 10 days @ $2.00 per month	$14.67
				$102.06
			Credit	
			For 162 days services as teamster @ $1.00 per day	$162.00
	June	5	Balance due T. J. Mefford	$59.94

Source: Elkhorn Distillery Records, 1868–1872, 1997MS419, Book 1, p. 494, Special Collections, University of Kentucky.

James Stone employed teamster T. J. Mefford seasonally from 1866 through 1869 and paid him $1 a day to haul freight and carry freight bills and payment checks between the depot agent and the distillery office. Stone also paid charge accounts that Mefford maintained at J. H. Beauchamp's store at Paynes Depot and at another store in nearby Midway. Though we have only a partial record of Mefford's pay and expenses, the account demonstrates that he drew cash from his account with Stone and bought a variety of commodities from stores at the depot and in Midway. He read the *Sunday Times,* bought pork in quantity, and rented a house from James Stone at $2 per month.

Internal Revenue Controls

The internal revenue law passed in 1868 placed the distilling industry under government control, and all distilling activity at the Elkhorn works had to be approved by the internal revenue collector for the Seventh District, R. M. Kelly. District assessors, storekeepers, and gaugers monitored day-to-day production and record keeping. After spending the summer and fall of 1868 rebuilding the distillery to comply with the new federal requirements, James Stone and James Shropshire submitted papers to the internal revenue office in Washington, DC, in November, documenting the distillery's structures, equipment, and capacities and requesting permission to start operations at the beginning of the distilling season in early December.[46] They also sent a notice to Kelly that they wished to commence milling and mashing on December 1 and requested that he send the federal storekeeper, William Hamilton, to the works so that distilling would not be delayed.[47]

In January 1869 Stone ordered from Benjamin Gratz, the Seventh District's assessor stationed at the Lexington internal revenue office, the official record books required for government reports, including a distiller's book of purchases, a distiller's yeasting book, and a storekeeper's book. Shropshire began submitting the required trimonthly production reports listing the amounts of grain and yeast used and the amount of whiskey produced each day.[48] Stone and Shropshire were required to comply with federal regulations and submit reports and pay fees on time, but they found it difficult to obtain barrel tax stamps and tax papers from the Lexington internal revenue office in a timely manner. Shropshire commented pointedly: "It is like drawing eye teeth to get any stamps or papers of any kind from our collector office on time."[49] Nevertheless, when Lexington wholesale-retail grocers Hutchinson, McChesney & Company ordered ten barrels of Elkhorn whiskey in January 1869, Shropshire followed the rules and sent Kelly a check for the internal revenue tax and barrel stamps.[50] Whether these first awkward encounters with internal revenue requirements gave Stone and Shropshire a sense of foreboding, we do not know. Their discomfiture would begin to emerge soon enough.

Making Whiskey

Elkhorn's distiller made bourbon according to a mash bill of three-fourths corn and one-fourth rye, and a rye whiskey with two-thirds rye and one-third corn. The proportion of malting barley used was not factored into these ratios. Customers who placed large orders could get an all-rye product if they wished. Responding to a query from William T. Walters & Company of Baltimore, James Shropshire sent samples of Elkhorn's bourbon and rye whiskeys and proposed a contract to furnish both, but in a rather unorthodox way. Acknowledging that the distiller preferred to use "new" corn, Shropshire conceived of a way to extend the distilling season. "We cannot commence on new corn usually before 15 November," he advised, "so that we could have about the two best months in the season [September and October], to make rye whiskey."[51] Rye was usually harvested in July or August, and Shropshire proposed starting the distilling season early, in September, using just-harvested rye while waiting for the corn to be harvested, dried, shucked, and shelled. But financial distress truncated any attempt to develop an early-season rye whiskey.

Despite minor mechanical difficulties, the refitted distillery performed well. By mid-March 1869, James Stone advised wholesalers, "We are running regularly & smoothly making 200 barrels per week."[52] To ensure the quality of Elkhorn's whiskey, Stone sought the best grains and admonished his agents and brokers to supply only superior corn, rye, and malt. His whiskey barrels were high quality and made with the "best hoop iron," and he promoted his new brick storage warehouse as the best in the state. Stone was pleased with the distillery's ability to blend traditional production techniques and new equipment to make quality whiskey, yet he was not satisfied. He sought the advice of business colleagues as to how he might configure the production process to produce a more "highminded" whiskey.[53] This quest for quality exacted a price, and one might question whether the costs associated with Stone's pursuit of excellence compromised the business and abetted its eventual financial failure.

The distillery's exemplary warehouse, admired for its fire-resistant qualities, did not consistently produce high-proof whiskey—an inadequacy that caused some concern. In September 1869 Shropshire noticed that when gauged for proof, the barreled whiskey on the first floor of the warehouse, produced in the spring of 1868 and "two summers" old, was consistently lower in proof than that stored on the upper floor. The back half of the first floor was below grade and somewhat cooler, and it had a narrower temperature range throughout the year than the upper floor.[54] Stone & Shropshire preferred to barrel all their whiskey at 5 percent over proof, which comported with consumer preference. Although they did not understand the reason for the difference in proof, they came to believe that the variance was merely an artifact of improper gauging with an inaccurate instrument. Nevertheless, some customers refused to buy whiskey that had been stored on the warehouse's lower floor.[55]

Connections and Logistics

The Elkhorn Distillery, although situated beside South Elkhorn Creek, did not ship or receive freight by way of the creek's shallow, winding channel. All goods moving to and from the distillery moved by overland road or by rail, and all railroad shipments had to be transshipped by road. Located two and a half miles from the Louisville, Cincinnati & Lexington Railroad tracks at Paynes Depot, the distillery was dependent on railroad equipment and employees and had to adapt to the railroad's schedules. Stone & Shropshire shipped all grain, building materials, and coal "on the cars," and their whiskey entered the American market by way of that same depot. Turnpike roads were "built," in the sense that laborers reduced their grades and covered their surfaces in compacted broken stone, according to state engineering standards. Primary turnpikes connected county seats. Secondary pikes, usually funded and built by local landowners, connected to other pikes and to river landings and railroad depots. Local roads were not turnpikes; they were usually unimproved, narrow, single-lane tracks that were poorly maintained and often impassable in bad weather. Teamsters hauled goods between the depot and the Elkhorn Distillery by way of Moores Mill Road and the Lexington-Leestown Turnpike, or Paynes Depot Road.

Suppliers shipped wheat, rye, and shelled corn to the depot or distillery in sacks; hops were shipped in bales. Bulk or ear corn was shipped in open-topped gondola railcars, while sacked grain moved in boxcars. Ear corn weighed seventy pounds per bushel, shell corn and rye fifty-six pounds, and wheat sixty pounds. The largest eight-wheeled wooden boxcars used by the Louisville, Cincinnati & Lexington Railroad in the 1850s and 1860s were about twenty-eight feet long, eight and a half feet wide, and up to seven feet high, with a capacity of nine to twelve tons.[56] Each car could carry about 350 bushels of rye or shell corn, which would have weighed roughly ten tons. The critical value in railroad shipping, though, was not railcar capacity but the length of time and the number of laborers required to load and unload the cars while abiding by railroad schedules. Freight nomenclature categorized loose materials such as grain, sand, and coal as bulk cargo. Goods shipped in boxes, barrels, or other containers were termed general cargo. Until the invention of steam-powered elevators and gravity-flow grain and coal bins, bulk cargo had to be shoveled by hand on to and off of each railcar or water vessel, making it expensive to move. Transporting general cargo required fewer workers and less time, and the reduced cost created an incentive to convert bulk cargo to general cargo whenever possible. This was accomplished by shipping grain and coal in sacks. The sacks themselves were also a commodity; shippers and distillers owned the sacks they used, and protocol required that the sacks be returned to their owners for reuse.

Synchronizing Schedules

The Elkhorn Distillery operated according to a daily schedule that sequenced milling, mashing, fermenting, distilling, and other procedures. Each day the sequence was repeated, although adjustments could be made to accommodate mechanical problems, late grain shipments, or new whiskey orders. Shipping grain and whiskey via the railroad required coordinating the distilling schedule with the railroad's rigid timetable. Stone & Shropshire discovered that the railroad offered little accommodation to any contingencies the distillery might experience. Distillery operations, the hiring of laborers and teamsters, and railroad schedules had to mesh—a complicated process made more difficult by the long haul between distillery and depot, bad weather, and intransigent depot agents.

Stone & Shropshire's preparations to start operations at the Elkhorn Distillery in the fall of 1868 included filling their cribs and bins with grain purchased for them by three agents who shipped from nearby railroad depots at Midway, Spring Station, and Yarnellton. On Saturday, November 14, 1868, James Stone wrote a letter to agent R. H. Staub in Midway, instructing him "not [to] ship us corn so fast. We have several carloads to unload by eight oclock to night. If it is not it will be thrown out on the ground. Please never let anyone load any cars with grain for us *on Saturday.* The R.R. will not give us time to empty them if you do. And will you please give the loading your personal attention. Every one piles on the corn as long as it will stick and the result is that it is strewn all along the R.R. from Midway to Paynes."[57] To clarify matters, Stone wrote a second letter to Staub: "We have made an arrangement with Mr Tarlton at Yarnellton by which he will ship each Monday, Wednesday, and Friday morning the grain he has for us. We wish you to ship on Monday, Wednesday and Friday evenings. By this arrangement we will be kept constantly hauling and not get [crowded] at the depot with loaded cars. . . . On no account whatever allow your cars to be loaded on Saturday, for they will be compelled to remain over Sunday and put us to extra expense."[58] Then on Wednesday, November 25, Stone wrote to Timothy Hughes at Spring Station: "Please do not ship any more until the R.R. will furnish you the cars with sufficient time to load them," he instructed, "and upon no account throw the corn on the ground. We are overrun with freight now and will be for several days to come. As soon as we can get it cleared from the R.R. we will let you know."[59] Each depot—Paynes, Midway, Spring Station, and Yarnellton—had a dedicated railroad siding where depot agents parked railcars that were being loaded and unloaded so they would not impede through trains. The railroad company's primary reason for wanting shippers to load and unload their cars quickly was apparently not related to stationary cars blocking the main track; rather, they wanted to assure a supply of empty cars for use by other shippers.

Most railroad shipments of grain, coal, and other commodities purchased by Stone & Shropshire arrived at Paynes Depot intact, and there are few records of lost shipments or missing railcars.[60] Shipments of whiskey in barrels or sample boxes were usually delivered without incident, but there were several cases of pilferage. When a box of corn and rye whiskey samples sent to a St. Louis wholesaler was "lost," Stone shipped a replacement box and enclosed a note that explained what had likely happened to the first shipment: "These samples are packed more securely and I think will stand the vigilance of "Bummers" of the Express Cos and R.R. Employees."[61] Similarly, when a barrel of whiskey shipped to a New York wholesaler arrived nearly empty, Stone sent a replacement "packed in a cask to prevent its being sampled or tampered with on the road."[62]

THE GATEKEEPER

Charles B. Lewis, the Louisville, Cincinnati & Lexington Railroad agent in charge of Paynes Depot in 1868, lived five miles away in Georgetown. He kept accounts on all incoming and outgoing freight, directed loading and unloading, and collected customer payments. Lewis also operated the depot's hauling business, which employed teamsters and provided wagons and teams of draft horses and mules for hire to area shippers who wished to move freight to and from the depot. When railroad laborers unloaded freight, the standard charge was about $1 for three-quarters of a car, or $1.30 to $2 for an entire car, depending on whether the cargo was loose (bulk) or shipped in sacks or other containers (general). The Elkhorn records offer few details about Lewis's role in the hauling enterprise, but he presumably operated a stable that provided hay and water for horses and mules. In addition, the depot site necessarily included a loading-unloading ramp or dock and a mix of railroad and privately owned freight storage buildings, including cribs and bins; a blacksmith shop for shoeing draft animals and repairing wagons; and housing for employees. Stone & Shropshire paid their freight bills by checks sent to Lewis at the depot. Lewis was apparently somewhat casual in his accounting methods, because the depot and distillery shipping records sometimes disagreed, compelling Shropshire to request a detailed listing of wagonloads hauled each day.[63]

In January 1868 Lewis ordered from W. B. Belknap & Company in Louisville "one set of patent wagon boxes 4 × 12," priced at $9.75. When Lewis failed to pay for them, the company contacted Stone & Shropshire and asked them to intercede and obtain payment from Lewis. Such wagon boxes were commonly used to haul freight, and they may have been intended to expand the railroad's freight hauling capacity or to replace damaged equipment. Lewis received the wagon boxes, but they were apparently not the type he wanted. Only after S & S requested that he return them to Belknap & Company did he comply.[64]

Perhaps Lewis was displeased with Stone & Shropshire's role in the Belknap dispute, or perhaps he was just enforcing railroad company policy, but in any case, in early February 1869 he began to refuse the checks S & S sent to pay their freight bills and demanded payments in cash instead. Stone & Shropshire's checks were drawn on the Deposit Bank in Midway, less than three miles from the depot, but Lewis claimed he could not cash them and declared he would not deliver freight to the Elkhorn Distillery until S & S paid the freight bill in "ready money." Stone did not keep cash at the distillery for security reasons, and he and Shropshire paid workers' wages and all other creditors by check. Shropshire wrote to H. A. Dudley, a Louisville, Cincinnati & Lexington Railroad official, to explain the problem and request that the company clarify its policy on accepting checks.[65] Meanwhile, Stone made his case to the railroad company's senior freight agent in Louisville and concluded his appeal by noting, "We are the heaviest freight incoming at this depot and your agent seems determined to give us (we think) much trouble and as far as we know without any just course for so doing. Cant you give us some relief. It seems to us you might get an agent that lives nearer than 6 or 8 miles from depot."[66] Stone & Shropshire continued to complain to railroad officials about Lewis's errors and inconsistencies in freight billings. By late March 1869, Charles Lewis was no longer the freight agent at Paynes Depot, having been replaced by James W. Craig.[67]

Elkhorn Distillery Grain Freight Charges, Louisville, Cincinnati & Lexington Railroad, February and March 1869

Date	Number of Sacks*	Grain	Total Pounds	Rate per 100 lb.	Total Freight Cost
February 23	147	Rye	18,062	7 cents	$12.64
March 6	158	Rye	18,358	7 cents	$12.86
March 7	125	Malt	10,703	7 cents	$7.49
March 8	146	Rye	18,429	7 cents	$12.90
March 13	343	Rye	32,931	7 cents	$23.05
March 18	129	Malt	11,070	7 cents	$7.70
March 20	96	Rye	[11,690		
March ?	140	Rye	+18,902]		
			=30,592	7 cents	$21.40
March 23	135	Rye	18,150	7 cents	$12.70
		Due S & S			$110.74

Source: Elkhorn Distillery Records, 1868–1872, 1997MS419, Book 1, p. 264, Special Collections, University of Kentucky.

*The standard weight for rye is 56 pounds per bushel; for barley (malt), 48 pounds per bushel. Although sacks varied somewhat in weight, a sack of rye would have contained just over 2 bushels, and a sack of malt just under 2 bushels.

Grain suppliers sent carloads of rye and malt to the Elkhorn Distillery during the spring 1869 distilling season. The shipments were irregular; a car might arrive at Paynes Depot each day, while other deliveries arrived once a week. Laborers weighed each grain sack as it was unloaded. A sack of rye weighed about 120 pounds, a sack of malt roughly 85 pounds. At 7 cents per 100 pounds, the freight charge for a carload of rye usually exceeded $12. Shipping malt cost roughly 40 percent less. The Elkhorn Distillery's monthly railroad freight charges for some ten carloads of rye and malt exceeded $110.

Grain Agents and Brokers

In 1869 the laborers on James Stone's Scott County farm harvested some 750 bushels of corn and several hundred pounds of bluegrass seed. If Stone produced any rye, wheat, or barley, it was of insufficient quantity to report to the census marshal in 1870. A late-nine-teenth-century farm distillery of medium industrial capacity could mash 100 to 300 bush-els of grain per day, and for bourbon producers, most of this was corn. The refitted Elk-horn Distillery would have consumed Stone's entire corn crop in less than a week. The Stone & Shropshire partnership was, accordingly, wholly dependent on grains produced by other farmers and purchased and delivered by agents and brokers. Distilleries on navi-gable rivers or with railroad connections could obtain grain from much larger supply areas than those distilleries served only by roads and wagons.

Stone hired grain buyers or agents to canvass grain-producing communities for farm-ers who were willing to sell their crops, harvested or unharvested. He customarily paid agents a commission of 1.5 cents per bushel to purchase grain, although he also paid up to 5 cents per bushel for purchase and shipping; the fees varied according to the season and grain availability. Agents usually purchased rye as harvested grain; they purchased corn as picked ear or shelled corn, but they also bought entire fields of standing corn before it was harvested to make sure the crop was not acquired by another distiller or miller. Stone & Shropshire also purchased grain from specialized brokers in Lexington, Louisville, and Cincinnati. Brokers bought and sold grain in quantity, often in anticipation of general re-sale rather than to fill a specific order. Brokers might acquire grains locally, but large-vol-ume purchases and specialty products such as malted barley and hops were often produced elsewhere—Canadian farmers grew much of the barley purchased by the Elkhorn Distill-ery. Grain arriving at Paynes Depot had to be stored or hauled to the distillery—preferably in dry weather. The distiller risked large losses if grain shipments arrived at the depot in the rain, unless the loaded wagons could be securely covered with canvas tarps.

Stone & Shropshire initially obtained their primary distilling grains, rye and corn, from two areas, each served by different agents. Inner Bluegrass agents shipped grain from Louisville, Cincinnati & Lexington Railroad depots near the Elkhorn Distillery: R. K. Staub at Midway (about two miles), Charles S. Tarlton at Yarnellton (three miles), and Timothy Hughes at Spring Station (seven miles). B. W. House at Stamping Ground shipped grain by road twelve miles to the Elkhorn Distillery. About forty miles northwest of Paynes Depot, a second grain supply area centered on Henry and Shelby Counties in the Outer Bluegrass, one of the region's most productive farming locales. From Shelbyville north to New Castle, the cornucopian land rises and falls in long whaleback swells. The earth there is nourished from below by moldering Ordovician limestones; corn and grains flourish.[68] Pioneer farmers cleared and cultivated this land in the late eighteenth century. By the last third of the nineteenth century, the area's production was sufficient to provide a ready supply of corn and rye to local millers and distillers.

Stone & Shropshire purchased corn and rye from Outer Bluegrass farmers, primar-ily through four purchasing agents: A. B. & W. L. Crabb (later J. M. Crabb & Brother) of Eminence, in Henry County; Jervins & Howland of Pleasureville, in Henry County;

Wilcox, Sam & Brother of Christiansburg, in Shelby County; and W. S. Myles in Shelby County. The Louisville, Cincinnati & Lexington Railroad and its progenitors laid track from Louisville to Frankfort through southern Henry County. North of Shelbyville, at the height-of-land between the Kentucky River and Salt River watersheds, the railroad built a depot, appropriately named Eminence. The town of Eminence grew up around the depot. The railroad also established depots a few miles away near Pleasureville in Henry County and at Christiansburg in Shelby County.[69] Each agent in this area had ready access to a railroad depot with direct connections to the Elkhorn Distillery by way of Paynes Depot.[70]

James Stone began commissioning agents to buy grain in early September 1868—before storage bins had been built and before the distillery had been completed. Among his first contacts were the Crabbs of Eminence, grain agents who became reliable sources of rye and corn for the Elkhorn Distillery. Both men were farmers and slaveholders before the Civil War. W. L. Crabb partnered with dry-goods merchant David M. Fible in 1872 to build and operate the Fible & Crabb Highlands Distillery, just southwest of Eminence on the headwaters of Fox Run.[71] Stone wrote to the Crabbs:

> You will now commence [buying grain] so soon as you wish to do so, and buy all you can get. I will pay you the usual price, five cents per bushel for buying and shipping. You will please keep a strict account of your transactions and render me a bill for each lot purchased, as I am compelled by law to give the names of all the persons of whom I purchase grain, with the price. Send me rail road receipts [when] you have it delivered. The law is very strict and I do not wish to have any controversy with the government. Buy me a lot of white rye, and ship it as soon as possible, if there is any in your section. I wish to seed part of my crop with it. Is there any better than the black rye? P.S. I do not wish to purchase any old corn at the price [I] can buy it for less here.[72]

When the corn harvest was under way in October and November, Stone reduced the price he was willing to pay: $2.50 to $3 per barrel, "delivered in our cribs being the highest price we are paying."[73]

Stone wished to exercise some control over his agents' districts to avoid overlap and competitive bidding. He instructed one of his Shelby County buyers, W. L. Myles, not to purchase grain in Henry County because "I have given an agency to the Messers Crabb at Eminence to buy rye for me. You must not go in that section, as you will clash with them and put up the price of rye when there is no necessity for it. You must keep the price down and not advance it in this way."[74] Although he instructed Myles to buy as much rye as he could and ship it expeditiously, Stone also scolded him for offering $1.30 per sack when other agents were buying it at $1.20. "Why are you doing so. You are letting other parties impose on you," he wrote.[75] It is likely that Myles was buying rye at a higher price because overall distillery demand was reducing the supply. Six weeks after Stone issued his first rye purchase request, the price had increased to $1.40 per sack delivered.

As Stone's agents shipped carloads of rye and corn to the distillery, he struggled to keep his accounts current, while his carpenters endeavored to build cribs and bins faster than the grain arrived. He was often late paying the agents because record keeping "proved to be a long tedious job the rye coming from so many different parties."[76] When one agent shipped corn that arrived coincident with another's delivery, Stone cautioned them: "Please do not ship us any more corn for two or 3 days until we can provide a place to stow it away. We have nine car loads to unload & haul to day & no pens to put it into built. We are building pens as fast as we can tho and will be ready to receive your shipment about day after tomorrow."[77] The lack of storage space limited the rate at which corn could be delivered, but once it was in a bin or a crib, high-moisture corn was subject to "heating," rapid degradation, and mold. Shropshire voiced this concern about the corn in storage at the distillery, but Stone offered the opinion that while it was not possible to measure the temperature of their stored corn, it was dry enough that "it will not heat any of any consequence in the crib."[78]

Elkhorn Distillery Account with John G. Brooks,
November 1868–January 1869

Year	Month	Day	Transaction	Daily Charges	Summary Balance
			Debit		
1868	November	12	To cash paid for hauling corn		$100.00
		21	To account for hauling corn	$156.50	
		21	To account for corn in field	$75.00	$231.50
					$331.50
			Credit		
			Paid shucking and delivery		
			210 bushels of corn from field @ 75 cents	$157.50	
			198 bushels of corn from field @ 50 cents	$99.00	$256.50
			Due Stone & Shropshire paid on corn in field Nov 21		$75.00
			Debit		
	December	14	To cash paid on corn in field		$504.00
1869	January	1	To note given Brooks		$756.00
			Credit		$1,335.00
			Buy 63 acres of corn in field @ $20.00	$1,260.00	$1,260.00
			Balance due Stone & Shropshire		$75.00
			Debit		
1868	December	23	To cash for 11 loads of corn from Bruchaus(?)		$199.92
					$274.92
			Credit		
	December	7 to 21	Buy 11 loads of corn from Bruchaus(?)		
			23,325 pounds, 333.15/70 bushels @ 60 cents	$199.92	$199.92
			Balance due Stone & Shropshire being amount overpaid		$75.00

Source: Elkhorn Distillery Records, 1868–1872, 1997MS419, Book 1, p. 381, Special Collections, University of Kentucky.

Scott County farmer John G. Brooks lived near White Sulphur, about seven miles from Paynes Depot. He likely sold his own corn to Stone & Shropshire and purchased harvested corn from other farmers, which he then hauled to the distillery using his own wagons and teams. Brooks also bought unpicked standing corn by the acre. In January 1869 he bought sixty-three acres of field corn at $20 per acre and was apparently paid 50 to 75 cents per bushel for picking, shucking, and hauling the corn to the distillery.[79]

Commissioning individual agents to purchase and ship grain became increasingly cumbersome. Stone struggled to locate grain at affordable prices and then coordinate shipments from different sources with railroad schedules and hauling crews in an effort to make the overall endeavor manageable. Stone also required quality barley malt and hops, but neither was available locally. To obtain grain in the volume the distillery needed while also reducing the number of agents he had to deal with, Stone increasingly turned to large grain brokers such as the Louisville firms Brandeis & Crawford, Sampson & Milton, and Schickendantz & Sewell; B. W. Wasson & Company in Cincinnati; and Woolfolk, Craig & Gay and Wolf & Yellman in Lexington.[80] In mid-December Stone & Shropshire contracted with Brandeis & Crawford to deliver 6,000 bushels of prime rye to arrive in four weekly shipments of 1,500 bushels each. Bad winter weather hampered hauling for the first two weeks of January, and the shipping schedule had to be deferred until the roads improved. The distillery's losses were compounded by a distressing shortage of storage bins at the depot, and the weather caused a significant loss when one shipment of corn could not be properly stored.[81]

Stone obtained barley malt from Schickendantz & Sewell of Louisville.[82] A representative order requested thirty bags of the "best fall" malt and ninety-five bags of the "best Canada Spring Malt." Although Stone requested the best, what he received sometimes failed to meet his standards. An April 1869 order was accompanied by this grievance: "Our distiller says the last spring malt you sent out was a very inferior quality."[83] In May Stone requested rye malt, which his distiller wished to mix in a ten-to-one ratio with prime fall barley malt to reduce costs, but both the distiller and the miller pronounced the rye malt inferior and unacceptable. Stone abandoned the rye malt and reordered seventy bags of prime Canadian spring barley malt.[84]

Nineteenth-century distillers often used hops (*Humulus lapulus*) not to complement flavor but to make yeast and to regulate vinous fermentation, hoping to avoid producing vinegar. Stone's distiller used hops, and although the records offer few specifics on the distilling process other than the mash bill, Stone and the distiller may have been familiar with period books such as Michael Krafft's *The American Distiller or, the Theory and Practice of Distilling,* published in 1804, and Harrison Hall's 1818 *The Distiller.* Both authors advocated the use of hops.[85] Twentieth-century distilling practices may have included using hops extract, which was believed to inhibit the growth of bacteria that could compromise the yeast's effectiveness.[86] Stone ordered hops from Lexington brokers Woolfolk, Craig & Gay and Wolf & Yellman, as well as W. Morgan & Company of Cincinnati.[87]

In the spring of 1869, as the end of the distilling season approached, Stone had to rebalance orders to reflect the daily use rate of rye, corn, malt, and hops. Because he sought to avoid distilling with "old corn" from a previous year's harvest, he wanted to use up all his grain and thereby avoid any carryover. Then he could begin the next distilling season with freshly harvested rye and corn. His rye orders in March for April and May distilling reached 5,000 bushels per month, and he requested competitive contract prices from two different brokers. By late May, Stone had reduced his orders to 1,000 bushels as he tried to balance his rye, corn, and malt supplies. Elkhorn's distiller mashed about 125 bushels of rye per day, so, given his supply of corn and malt, Stone calculated that he could close out the distilling season sometime after June 10.[88]

A Confusion of Sacks

At full operational capacity, the Elkhorn Distillery's grain mill, mash tubs, fermenting vats, and continuous-flow column still processed a considerable volume of grain each day.[89] Grain—rye, shelled corn, and malted barley—usually arrived at Paynes Depot in sacks that laborers transshipped by wagon to the distillery. Yeast and cement were also shipped in sacks, and suppliers could dispatch commodities to the distillery only insofar as they had empty sacks on hand. Grain delivered to the distillery could not be stored in the sacks in which it arrived unless those sacks belonged to Stone & Shropshire. If the distillery delayed emptying and returning sacks, future shipments would necessarily be deferred. And to maintain an adequate sack count, grain shippers had to replace burst and lost sacks punctually.[90]

Kentucky farmers produced hemp, the state's primary fiber crop, as early as 1775, and some of the state's first industries produced hemp cloth. Hemp sacks and cotton or "white" sacks were available in central Kentucky for farm and commercial use by the early nineteenth century. The John Hunt and John Brand Company of Lexington made hemp bagging material in 1803, perhaps the first produced in America. By 1810, Fayette County had six cotton mills, five fulling mills, five bagging plants, and thirteen ropewalks.[91] By the eve of the Civil War, the center of hemp bag manufacturing in Kentucky had shifted to Louisville. Operating the Elkhorn Distillery required Stone & Shropshire to keep a supply of sacks on hand. In November 1868 they paid N. H. Prior & Company of Louisville $164.50 for new sacks.[92] Kentucky hemp gunnysacks were priced according to size and were fairly expensive. Two-bushel grain sacks were 48 to 50 cents apiece, two-and-a-half-bushel sacks were 56 to 58 cents, and three-bushel sacks cost 65 to 66 cents.[93] If Stone & Shropshire used two-bushel sacks, Prior & Company filled their order with 329 sacks.

During the second half of the nineteenth century, the demand for grain sacks in the Middle West and Upland South was extraordinary. A single river steamboat towing a barge might carry 10,000 sacks of grain or more, depending on the boat's dimensions. In the 1860s grain shipped by railroad boxcars was usually shipped in sacks, upward of 150 per car.[94] Kentucky distillers began receiving grain shipments in the fall to fill their granaries in preparation for the winter's distilling season, and the demand for grain continued until distilling ended in the spring. Central Kentucky grain agents and brokers stored grain in sacks until they found a buyer.[95] For nine or more months a year, grain sacks were in demand across the region, and accounting for sacks was a substantial enterprise for brokers and distillers. Most firms marked or printed their company names on their sacks to ensure they would be returned for reuse.[96]

At the beginning of the 1868–1869 distilling season, Stone & Shropshire ordered grain, malt, and yeast and sent sacks to the purveyors to enable the shipments. The distillers sent 300 sacks to W. S. Myles in Shelbyville, which allowed him to purchase and ship 700 bushels of rye. They also promised to send Myles 200 additional sacks and return those sacks that belonged to him. Stone & Shropshire matched each grain shipment received from their largest grain suppliers with a return shipment of the seller's sacks. Rather than simply shipping the empty sacks in loose piles, laborers packed the sacks into large shipping bags for ease of handling and to prevent loss or pilferage.[97] On January 22, 1869, for example, Stone & Shropshire returned to Brandeis & Crawford of Louisville "5 bags empty sac[k]s say 134 large white (cotton) and 78 hemp ones." To Skinner & Company in Cincinnati they sent "three bags empty sacks, say 147 in all. There are 3 missing which we cannot find."[98] During the week of February 22, S & S returned eleven bags containing 572 sacks to Brandeis & Crawford but cautioned, "Please recount them and notify us of any mistake."[99]

In the winter and spring of 1869, the Elkhorn Distillery received some 5,000 bushels of grain per month, and the task of sorting and accounting for bags became increasingly difficult and confusing. Suppliers sent reminders when sacks were not promptly returned, and by mid-March, the trade in sacks had become sufficiently muddled that brokers threatened to delay or stop grain shipments. The situation was aggravated, in part, by the suppliers' failure to properly brand or mark their sacks, thereby permitting laborers to identify and sort the sacks by owner. Those sacks that were well marked were readily accounted for and returned. Stone & Shropshire described the difficulty to Brandeis & Crawford: "We think it very possible that some of Messers Sampson & Minton's [sic] sacks are mixed with yours for we are short about 7 or 8 on his and would like to have you examine all the sacks not in [your] mark and if any belong to S & M they will call for return. There are about 2000 sacks here and it is almost impossible to get them sorted correctly. Please mark all your s[ac]ks very plainly."[100] The distillers sent a similar message to Sampson & Milton: "We ship you by Rail Road this day 6 bags empty sacks, say 294 in all and we think the [balance] of 6 bags you will find by calling at Messers Brandies [sic] and

Crawford to whom we have just written on this subject asking them to examine their sacks and hold subject to your inspection. Unless sacks are plainly marked in shipping marks it is impossible to keep them separate here."[101]

Sack returns were also delayed because Stone & Shropshire preferred to empty the sacks directly into the distillery's mill hopper, as the grain entered the milling and distilling process, instead of into a grain storage bin. Unloading grain into a bulk storage bin would have forced the distillery's laborers to either rebag the grain or shovel it into a wagon to haul it to the mill, significantly increasing the labor required to move the grain once it had reached the distillery.[102]

Although the Elkhorn Distillery's office clerk, C. M. Gillespie, sorted and counted rye sacks and returned empties to at least one supplier, the hauling crews and distillery laborers, many of them African Americans, were largely responsible for this task. As the distilling season advanced and the problem of returning the proper number of sacks could not be resolved, Stone & Shropshire shifted responsibility for inaccurate counts and delays to their African American laborers. When returning 700 empty rye sacks to Brandeis & Crawford in May 1869, Stone & Shropshire asked that they carefully recount them, "as they were put up by darkies and counted by them. We hope you will find the count correct." A June letter to Schickendantz & Sewell that accompanied a return shipment of sacks explained that the number of sacks "should be 270 by darkies count." Another note sent to Brandeis & Crawford stated that S & S was sending "20 packages of sacks counted by darkies at 995." Similarly, the distillers wrote to Wolf & Yellman, "We will ship you today 123 empty sacks to balance our bag account with you. These were put up by darkies but one presumes their count is correct."[103] Only when the distilling season ended in June and grain shipments stopped did Stone & Shropshire resolve the confusion of sacks.

Operating the Distillery

Local businesses supplied the Elkhorn Distillery with basic construction materials such as paint (Taylor & Brother of Midway), lumber (E. R. Spotswood of Lexington), and metal parts (Barnes, Ballard & Wood and J. A. Gray of Lexington).[104] Cincinnati firms supplied most of the specialized distilling equipment, although Stone & Shropshire also acquired equipment and parts from Louisville dealers.[105]

Ordering Parts and Equipment

Communication between Stone & Shropshire and their equipment suppliers was often awkward and imprecise. Orders described parts in very general terms, implying that S & S did not have access to a parts catalog or parts numbers and were obligated to itemize their requirements in terms of purpose, material, and function. When James Stone placed orders with J & E Greenwald of Cincinnati for pipe, pumps, and valves, his specifications were often so vague that the company shipped the incorrect parts, which prompted returns and reshipments.[106] Midway through the 1868–1869 distilling season, equipment failures began to increase.[107] In a March letter, Stone requested that J & E Greenwald send "an eccentric rod and shaft made as light as possible as it is surely to brake a light screw from an upright shaft. It will require a three inch mortise. The shaft will require to be 2 ft 6 inches long from the center of the upright shaft."[108] The Greenwald Company apparently advised Stone that he would receive better service if he consulted the company's parts catalog. In late April he ordered a cogwheel from Greenwald, sent by express. The wheel should have "say 72 cogs & keys, 2 inch pitch & 7 inch face," he wrote. His concluding request suggests that he finally had an epiphany: "Send catalogue of your gearing."[109]

This brief exchange illustrates another facet of the adoption and application of industrial processes, including standardized machines and parts, in the craft distillery business. Machinery manufacturers had long recognized that the proliferation of standardized machines and parts would become unmanageable if they were not systematically organized and accessible through standardized parts catalogs. Yet some distillers continued to function according to traditional craft procedures—describe a need or a function to a blacksmith, carpenter, or other fabricator and receive a one-off machine or part. If equipment suppliers found these requests bothersome, they likely also recognized that such requests provided insights into the types of mechanical problems distillers encountered—problems that fabricators might address with special parts or new inventions, thus better serving their customers.

While heavy use wore out and disabled some distillery equipment, parts also broke for other reasons. Some of the corn and rye shipped to the Elkhorn Distillery contained nuts, stones, nails, and other foreign material that abraded or broke close-clearance milling equipment. To address the problem, but not knowing who might manufacture equipment that removed such objects, Stone sent a request to Louisville grain brokers Brandeis & Crawford for "two corn riddles and one rye riddle 19 × 48 inches." He explained, "We want the riddles or screens to be worked by eccentric power from mill spindle at point where corn & rye falls from elevator into hopper. If you have anybody in your city who manufactures such things they can readily make them the right size." An eccentric or shaft offset converted rotating movement to linear reciprocating motion that would cause a screen to shake, facilitating the separation of grain and contaminants.[110]

As the weather warmed in the spring of 1869, Stone & Shropshire ordered a fan from the J. E. Greenwald Company of Cincinnati. Stone's letter described what he wanted in only general terms, relying on the supplier to provide something that might work: "Express ship us just as soon as possible a distilling fan calculated to cool two (2) mashes at the same time and of such size as you think best. We think it ought to run between 800 or 900 pr min [rpm] as our engine runs 65 per minute and there is a 4 ft [illegible] with a 9 inch face for main shaft. You will also please furnish counter shaft & hangers for the same. [Fabricate] a[n] idle pulley [and] fan shaft as we do not know the size of pulley [and] fan shaft you will [send] make your own calculation as to size of pulley."[111] Such imprecise descriptions must have tried the patience of these parts suppliers, which had to be flexible and intuitive in responding to customer communications. The distillers then had to employ mechanics with sufficient technical ability to fit ill-matched parts and fabricate others to assure that the works operated properly.

Stone & Shropshire purchased most of their distilling corn as bulk or ear corn, which required shelling. Their Readings corn sheller had originally come from J & E Greenwald, but when a critical part wore out, Stone ordered a replacement from Radford & Sharp in Cincinnati.[112] To identify the part he needed, he simply described it as a cylinder for a corn sheller patented by "Wm Readings July 13, 1852."[113] This sheller design was dated, and new and improved units were available, suggesting that Stone did not wish to spend the money on a new sheller, or perhaps he did not want to interrupt the distilling operation midseason to install and properly fit a new unit to his existing mill.

The Elkhorn Distillery also required petroleum products to operate: machine oil for lubrication, and coal oil or kerosene for illumination. But again, Stone & Shropshire struggled with imprecise language when ordering these products and trying to specify the quality they wanted. Stone ordered a barrel of the "best machine oil" from the M & S Hodkinson Company in Louisville in late October 1868. "Please do not send the kind you did last season," he wrote. "It was no account." Not knowing the exact specifications of the oil he wanted, Stone requested the same kind of oil the company had sent to Horace Turner and Samuel M. Clay Jr., who operated the Turner, Clay & Company Distillery on Town Branch in Lexington. That oil, Stone found, had been "strained," a term he used in subsequent orders.[114]

Stone used the same strategy to order coal oil. Even in daylight, the Elkhorn Distillery's buildings were dimly lit, and spaces beyond the light coming in through small, unshuttered windows were deeply shadowed. Whether reading steam engine pressure gauges or gauging barreled whiskey, workers required artificial light. Having had experience with poor-quality coal oil and malfunctioning lamps, Stone implored the Hodkinson Company to send via railroad a ten-gallon can of "the best" or "warranted to be the best" coal oil. The "best" turned out to be expensive; Stone paid $39.40 for a ten-gallon can delivered to Paynes Depot.[115] Pouring coal oil from ten-gallon containers into the small oilcans Stone & Shropshire used to fill their lamps was awkward. To reduce spills and the fire risk, the Hodkinson Company recommended that the distillery purchase patented five-gallon oilcans designed to control spillage, although the results were apparently not satisfactory. Elkhorn's laborers also used glass lanterns fitted with candles. Stone ordered "star candles" by the box and six-inch "hotel candles" in 100-pound lots.[116]

FUEL: COAL AND CORNCOBS

When nineteenth-century farmer-distillers converted their distilleries from waterwheel power and an open-flame heat source to steam-powered machine distilling, they often fueled their engines with wood cleared from farmland to create tillable fields. Distillers with access to a navigable river or railroad track could burn coal, delivered, perhaps, from Kentucky's eastern or western coalfields. Stone & Shropshire fired the Elkhorn Distillery steam engine with coal supplemented with corncobs, and they entered into monthly delivery contracts with coal brokers. Initially, in November 1868, S & S engaged Davenport & Taylor of Midway to supply coal. R. H. Davenport was a commercial broker and, before the Civil War, a slave owner. The distillery consumed four railcars of coal each week, and Stone requested that Davenport & Taylor deliver two cars of coal every Tuesday and Friday, a schedule that permitted Elkhorn's haulers to move grain and coal on alternate days. Coal, like ear corn, was likely delivered on gondola cars that were also used for bulk materials such as lumber, stone, sand, pig iron, and other products.[117] Early gondola cars were small, open topped, and flat floored, with sixteen-inch wooden sides and a capacity of about sixty-five to seventy bushels. Such cars did not have bottom unloading chutes, so crews had to shovel the coal from the cars into waiting wagons.[118]

In early December Stone & Shropshire paid $288 for the sixteen cars of coal received in November. The coal they received was low quality, and Stone was not pleased. "The coal we received yesterday was miserable stuff," he complained, "hardly worth cost of hauling from depot."[119] In addition to him selling inferior coal, Stone discovered that Davenport & Taylor was shorting the loads and charging for more than was delivered. The coal shipment in December was short, as measured at the Paynes Depot scale, and Stone responded accordingly:

> We wish to call your attention to the fact that the coal we are receiving from you is falling short to a considerable amt and to request you to attend to it so that it will not occur again. The last two cars rec'd were unloaded from them into our wagons and weighted and fell short one ten bushels, and the other twenty. You are aware that taking it from the cars into the wagons causes no waste worth mentioning. I can therefore speak only of the last two cars with certainty. All the other cars were unloaded upon the ground and if they fell short any of course I cannot say. Will you please give this your earliest attention.[120]

Alerted to Davenport & Taylor's dubious shipping practices, Stone began to weigh every carload.

By the first week of February, Stone had switched coal dealers and was receiving shipments from Thomas James & Company in Louisville. Although this coal was good quality, the shipment was, again, short in weight.[121] Davenport & Taylor sold coal for 12 cents per bushel, or 21 cents delivered at Paynes Depot, a price that Thomas James could not match. Despite the low-quality coal and short weights, Stone continued to query Davenport & Taylor about possible coal contracts for February and March 1869, likely because of the company's low price.[122] When Davenport failed to respond, Stone bought more coal from Thomas James and extended contracts to James W. Craig, the railroad's freight agent at Paynes Depot, who was apparently operating a coal brokerage and hauling business as well.[123] Mr. Craig and Thomas James became the distillery's primary coal suppliers through the end of the distilling season in June, by which time the price per bushel delivered at Paynes Depot had increased from 21 to 24 cents.

When ordering distilling corn, James Stone preferred to buy bulk or ear corn, rather than shelled corn. He wrote to his agent J. M. Crabb in Eminence, "We prefer buying corn in ear too as the cobs are good fuel."[124] Many distillers built slat-sided corncribs to store and dry ear corn.[125] When dried to the appropriate moisture content, the ears were shelled, the corn was distilled, and the cobs were stored for later use.[126] Stone likely did not know the actual heat value in dry corncobs, but he knew that they lit readily, burned with a low flame but intense heat, and assisted in igniting coal. Bituminous coal produces roughly 12,000 BTUs per pound. Corncobs, by comparison, produce between 6,700 and 7,400 BTUs per pound, or about 56 to 62 percent of an equal volume of coal. Stone preferred to mix coal and cobs and could even operate the distillery's steam engine on cobs alone for short periods if coal deliveries lagged. It is also likely that cobs helped compensate for the poor-quality coal the brokers shipped to him.

Slopping Hogs

Many early-nineteenth-century distilleries dumped their spent mash or distillery slop into nearby creeks or sinkholes before recognizing its value as livestock feed. Stone & Shropshire built large hog feeding pens at the Elkhorn Distillery. They purchased hogs and contracted to feed hogs belonging to local farmers. Hogs coming from nearby farms could be driven to the distillery on foot. Hogs were also shipped to the distillery by railroad car.[127] During the last two weeks of November 1868, Elkhorn Distillery laborers were readying the feeding pens and taking delivery of hogs. James Stone notified one of his grain agents that a payment would be late because "we will have to use the money on hand this week in payment for hogs to be delivered." From another creditor awaiting payment, he requested patience: "We must ask you for a few days indulgence," he wrote, "as we have a lot of hogs to secure this week [but] you shall not have to wait long."[128]

Stone & Shropshire contracted to purchase 100 feeder hogs from Fayette County farmer Montgomery H. Parker. When the first shipment of forty-three hogs arrived on December 3, just after distilling operations began, laborers weighed them and found them to be underweight. Stone did not send the hogs back but was unwilling to pay the full price. He paid Parker $300 for the hogs but insisted that the remainder meet the weight called for in the contract, or they would not be accepted.[129]

**Elkhorn Distillery Account with W. W. Hedges,
1868–1869 Distilling Season**

Year	Month	Day	Transaction	Monthly Charges	Summary Amount
1868	December	15	Carryover hogs in pens & expenses to date		$2,489.60
			Credit		
	December	15	Pay cash on above	$1,339.44	
		31	Pay cash on note to J. T. Nutters(?)	$576.70	
1869	March	27	Pay cash on note to J. M. Garst	$230.70	$2,086.84
			Balance due on original contract which is equal amount of our note to P. Mckay(?)		$402.76
			Debit		
1869	January	10	To still slop to this date per contract	$158	
	February	10	To 1 month still slop to date	$500	
	March	10	To 1 month still slop to date	$500	
	April	10	To 1 month still slop to date	$500	
	May	10	To 1 month still slop to date	$500	
	June	1-18?	To 2/3 month still slop to date	$333.33	$2,491.33
			Credit		$2,894.09
1869	February	6	Pay cash for slop to 10 January	$158	
		10	Pay 1/2 of slop to this date assessed by J. M. Stone	$250	
		10	Pay cash for 1/2 still slop to date	$250	
	March	22	Pay cash for 1/2 still slop to 10 March	$250	
		22	Pay 1/2 slop to 10 March assessed by J. M. Stone	$250	
	April	22	Pay cash for 1/2 still slop to 10 April	$250	
		22	Pay 1/2 slop to 10 April assessed by J. M. Stone	$250	
	May	10	Pay 1/2 slop to this date assessed by J. M. Stone	$250	
	June	1	Pay 1/2 slop to this date assessed by J. M. Stone	$166.67	$2,074.67
			Balance due Stone & Shropshire		$819.42

Source: Elkhorn Distillery Records, 1868–1872, 1997MS419, Book 1, p. 371, Special Collections, University of Kentucky.

Scott County farmer William W. Hedges contracted with Stone & Shropshire to feed his hog herd at the distillery during the 1868–1869 season. Hedges paid $500 per month for slop, water, and shelter. At the end of the distilling season in June, Hedges owed Stone & Shropshire a balance of more than $800. Shropshire was eager to close the account and wrote to Hedges: "As the writer is anxious to get the books closed here as soon as possible," he explained, "we hope you will come down next week & let us settle up & besides as we are closing & paying off hands we need the money very much."[130] Because whiskey sales were slow, the money Hedges paid Stone & Shropshire to slop his hogs was one of the distillery's only sources of income in the spring of 1869.

For nearly eighteen months, James Stone and James Shropshire worked at the many tasks required to ready the Elkhorn Distillery and make whiskey during the 1868–1869 distilling season. They rebuilt the works, including extensive mechanical and security modifications, to comply with federal internal revenue law. They contracted with agents and brokers to supply grain and hired a distiller, millers, and laborers. Complicated logistics slowed shipments of building materials, grain, and barreled whiskey; all this was made more difficult by an intractable railroad freight agent. Despite these problems, through clever innovation and thoughtful business decisions, Stone & Shropshire brought the distillery into production and made whiskey. The task was fraught with obstacles and lacking in romance. What awaited would prove much more daunting. Stone & Shropshire had to solve the conundrum of maintaining operations and fiscal solvency while complying with federal tax law. At the same time, they had to vend their whiskey in distant markets that seemed only marginally interested in their product.

17

The Elkhorn Distillery's Demise

You had better sell enough whiskey to bring at least $20,000 cash before you come home if possible.
—James H. Shropshire

For James Shropshire, Christmas Day 1869 was not a festive holiday. He had spent the previous two weeks in his Lexington office writing letters, entreating creditors to accept partial payments on the distillery's debts and imploring prospective whiskey buyers to ignore the flat spirits market and commit to large purchases. In less than twelve months, the optimism surrounding his investment in the Elkhorn Distillery with business partner James Stone had given way to a morose resignation that the operation was likely doomed to insolvency. Shropshire's financial commitment to making the distillery a success had cost him his life savings, and he had no idea how he was going to pay for his children's education, given that their future was tied to thousands of barrels of newly distilled whiskey. "I do not have money enough to go to market," Shropshire wrote in an anguished note to Stone on December 10. "I cannot stand this pressure much longer."[1]

Although the bourbon and rye whiskeys that Stone & Shropshire produced were high quality—made by an experienced distiller from select ingredients—selling them at all, never mind making a profit, proved exceptionally difficult. The Louisville financial markets priced newly distilled raw whiskey at 95 cents per gallon tax paid on Christmas Eve 1869.[2] Shropshire, taking into account commissions, storage, freight, labor, internal revenue fees, and leakage costs, thought the distillery could make a small profit at 75 cents. The asking price in August had been 85 cents, and by December, Stone & Shropshire were considering offers of 60 to 70 cents from gimlet-eyed wholesalers who declined to pay cash but requested, instead, extended periods of credit.

The Monetary Context for Distillery Operations

Both James Stone and James Shropshire brought personal or family funds into their business partnership, but the scale of their operation soon outstripped their resources, and they required additional sources of financing. Stone & Shropshire's distilling and marketing activities were set within a welter of legal and financial requirements—compliance with internal revenue law and taxation practices, access to accommodative banking, and mitigating risk by insuring operations and barreled whiskey stock.

The Vexation of Internal Revenue Regulations

After the passage of taxing regulations in 1868, the commissioner of internal revenue became the gatekeeper to Kentucky's distilling businesses. Ironically, though intended to capture tax moneys for the US Treasury, the internal revenue regulations served as a significant constraint on the distilling industry's economic performance. The 1868 regulations required distillers to pay a special tax of $4 per barrel and a per-gallon tax of 50 cents within one year of the day the whiskey was distilled. Paying the tax thereby removed the whiskey from bond, making it "free," or without tax liability. Any whiskey not removed from bond at the appropriate time was subject to forfeit to the government. The spirits would then be sold and the proceeds applied to the taxes due. Once the Elkhorn Distillery warehouse was inspected and approved, it was officially designated Internal Revenue Bonded Warehouse No. 5, Class A 1.

Barreled whiskey had potential value that the distiller could not obtain until it was sold, but it could not be sold until the commissioner's representatives had collected the taxes due. The regulations were therefore biased in favor of those distillers who could draw financial resources from other vocations or had access to sympathetic bankers from whom they might obtain low-interest loans to pay the taxes. Enforced taxation also distorted local and regional financial resources. In April 1869 Stone wrote to Louisville grain merchants Brandeis & Crawford about "the present condition of money." His tone reflected his aggravation: the "demand for money to pay whiskey tax in our district has produced a lightness in money."[3] Distillers obtained a measure of relief in 1878 when Congress extended the period that whiskey could be kept in bond without paying tax from one year to three. By that time, though, the Elkhorn Distillery had been closed and the works dismantled. Only the iron-shuttered brick warehouse remained standing.

For the distiller, federal control and monitoring of the distilling process began with construction of the works, and only when the taxes had been paid did some internal revenue regulations lapse. Before distilling began, Stone had to apprise the district tax assessor of his intentions. At the end of the season the following spring, Stone notified the assessor of his intent to conclude production. He then submitted his accounts of grain mashed and proof gallons of spirits produced during the year. On June 17, 1869, Stone sent a letter to the assistant assessor informing him of the distillery's production record that season: "In compliance with the law," he wrote, "we hereby give you notice that we cease mashing grain for our distillery this day having mashed 17,744 pounds of grain and will cease running whiskey on Monday the 21st day of this present month."[4] The records do not confirm that the S & S partnership had sold any whiskey, but it was now obliged to pay the requisite taxes according to the production calendar the partners and their labor force had fashioned.

One section of the internal revenue law required that distilled spirits be identified by numbered warehouse tax stamps placed on whiskey barrel heads, attesting to the contents. The stamp identified the distillery, the date, the cask or barrel number, the district, and the names of the collector, storekeeper, and gauger. Some of this information was recorded by personnel at the district tax office; the remainder was entered at the distillery. Revenue office staff was responsible for providing stamps in a timely manner, but when distillers ordered stamps to cover large sales—perhaps for several distilleries at once—the under-staffed office often proved incapable of meeting the demand. Distillers could not remove or ship barrels that were not stamped. In January 1870 Shropshire spent most of one day trying to obtain stamps for 184 barrels and received only a partial order. In late March his order for more than 100 stamps was delayed nearly a week, and even though the tax office apparently agreed to send the stamps to the distillery by express, they did not arrive until early April. The lag in stamp availability reoccurred in late April, when Shropshire wrote to Stone at the distillery: "The clerks at the office promised me 500 stamps to day but as usual did not give me all (only 295)."[5]

In addition to warehouse tax stamps, whiskey barrels had to be identified by searing a brand into each barrel head with a hot iron. The brand carried the distillery's name, place of production, and tax-paid status. Barrel branding irons were custom made and repaired as needed. On November 9, 1868, Stone ordered one set of branding irons for the distillery warehouse from Mercer & Company in Louisville. "We wish such as you furnish other distillers and are required by the Internal Revenue Law with class and number of warehouse and a set of figures from 1 to 0 inclusive. Send also the months from November to April inclusive." Stone went on to say that internal revenue collector R. M. Kelly had advised him that a set of branding irons would cost about $15. Removing whiskey from bond required that distillery identification information, the date of production, and a serial number be burned into each barrel.[6] Heating branding irons to the appropriate temperature also required a forge or furnace of some type, thereby putting fire in close proximity to the stored whiskey barrels. In April 1869 Stone sent four of his branding irons back to Mercer & Company, requesting their immediate repair and return. "They were just in the furnace and neglected," he wrote.[7]

Federal law obligated the commissioner of internal revenue to station federal gaugers at distilleries. Gaugers enforced government controls over the processing of grain, the still and all the distillery's components and parts, and warehouse security. Gaugers also measured and recorded each barrel's volume and alcohol proof. The 1868 revenue law specified that fees for government gauging were prescribed by the commissioner and paid to the district collector by the distillery owner. A portion of those fees was returned to the gauger as salary, which could not exceed $3,000 per year; this was a very generous pay rate, considering that the day laborers who performed the heavy work of operating the distillery received roughly $1 per day. Internal revenue directives also required distilleries to record on official forms all distilling, storekeeper, and gauger activities. Stone & Shropshire submitted these federal forms trimonthly to the Frankfort offices of Benjamin Gratz, the Seventh District assessor, or to assistant assessor R. K. Woodson. The reporting periods covered days 1 to 10, 11 to 20, and 21 to 30. Timely reports were predicated on accurate record keeping by the distillers and punctual reporting by federal gaugers and storekeepers.

Understandably, federal internal revenue jobs such as gauger were aggressively pursued. The Louisville collector's office was particularly set upon by applicants. Maysville's *Evening Bulletin* summarized the quandary: "In the Louisville district collector Ben Johnson has been so overrun with office seekers wanting positions as Storekeeper, Gauger, etc., that he has been compelled in self-defense, to tack up on the board in big letters a notice stating emphatically that the duties of his office must receive part of his time, and applicants will be seen only between 10 and 12 o'clock."[8] Area newspapers published gauger assignments, and the personals columns included prideful announcements whenever a local resident became a well-paid "revenue man."[9]

To measure the proof and volume of each barrel, the gauger required clear access. In a barrel-filled warehouse, this required a considerable labor force to move the barrels around. Each sale required that the laborers find, move, and arrange specific numbered barrels for measurement. Gaugers sometimes made mistakes, caused by either antiquated instruments or careless measurement and recording. In December 1869 the government gauger repeatedly recorded incorrect gauges for barrels being withdrawn from Elkhorn's bonded warehouse for tax payment and delivery to a customer. Shropshire was prompted to write a letter of explanation to creditors January & Wood of Maysville, notifying them that inaccurate gauging had delayed a shipment to a Boston customer by six to ten days. As a result, Stone & Shropshire would be late in making their debt payments, subjecting them to a penalty and interest charges.[10] Liquor wholesalers were aware of the potential for gauger errors and often demanded that their payment be based on their own gauging measurements once they had taken delivery.[11]

The alcohol content and volume of a barrel of whiskey changed during storage, and that change was unique to each barrel, given uneven rates of evaporation and leakage. In one year of warehouse storage, one barrel might lose one gallon of volume, while another might lose three gallons or more. Stone & Shropshire did not favor shipping "light" barrels to customers, but revenue regulations did not specify how they might "top up" those barrels that had lost significant volume. In October 1869 Shropshire sent a letter to the internal revenue commissioner in Washington, explaining the problem. "In withdrawing whiskey of last two seasons make from bond we find the shortage from 1 to 3 gallons & occasionally more and we wish to be advised whether we have the right with out infringing the law to fill barrels one from another before shipping same (after tax has been paid) and if so what notice if any have we to give of such a change and to whom." He added a postscript: "We find a barrel occasionally half or more out & unfit for shipment."[12] A special revenue commissioner provided an answer in January 1870 during a personal visit to the distillery. He informed Shropshire that the distillery could top off light barrels by filling them to the gauger's mark, provided the additional whiskey came from a tax-paid barrel.[13]

Federal storekeepers were also stationed at distilleries to monitor record keeping and tax filings. By law, storekeepers were bound to be at the distillery every day of the week, including Sunday, if they drew pay. The law also specified that a distillery could not operate if the storekeeper was absent, a situation that prompted Stone & Shropshire to complain to the district assessor that their storekeeper was absent at inopportune times. When the situation did not improve, Shropshire wrote directly to the commissioner of internal revenue in Washington in January 1870. "We do not like to trouble you with our small complaints," he began, "but we have been repeatedly charged for store keepers services during the past year for some days when the store keeper was not at the distillery at all on those days and we write to know if you cannot have the amount refunded us by the collector here. We know this charge is wrong and hope you will see us righted in this matter."[14]

Later that spring, James Shropshire, acting on behalf of the Elkhorn Distillery and others, consulted the Louisville law firm of Harlan, Newman & Bustow about filing a lawsuit challenging the constitutionality of Congress's joint resolution that required distilleries to pay storekeepers for Sundays when they did not work. The matter was apparently resolved when the commissioner agreed to remove the Sunday storekeeper salary charges, although the law firm's fee amounted to half of what the distillers later received.[15]

The Financial Imbalance

Stone & Shropshire maintained the Elkhorn Distillery's accounts with the Deposit Bank in Midway—their primary bank—as well as with the Northern Bank of Kentucky, City National Bank, and First National Bank in Lexington; Farmers Bank in Georgetown; Deposit Bank in Paris; and Citizens Bank, National Bank, and Second National Bank in Louisville. They also had accounts or transacted business with banks in Cincinnati and the Bank of America in New York City. One rationale for multiple accounts was to avoid exceeding maximum credit limits with a single bank. In addition, given that some regions were more economically stable than others, federal tax collectors and some banks would not honor notes or checks drawn on small local banks, preferring to do business with banks in Lexington and other cities. New York banks usually accommodated Stone & Shropshire's transactions, but some banks reportedly levied surcharges of up to 25 percent for processing warehouse receipts and other whiskey-related notes.

Stone & Shropshire also arranged their purchases and sales in a way that allowed them to obtain formal and informal credit from their suppliers and wholesalers. The Louisville grain brokerage firm of Brandeis & Crawford was one of the largest grain merchants in the South after the Civil War. W. W. Crawford was a commission merchant; his business partner, Adolph Brandeis, was an immigrant from Prague, Bohemia.[16] The company maintained a large grain warehouse and mill at First Street and Main in Louisville and owned the Ohio River steamboat *Fanny Brandeis*. The steamboat made regular round-trips to Evansville, where the crew on-loaded sacks of Indiana wheat, rye, and corn for delivery to the Brandeis & Crawford warehouse in Louisville or for shipment further upriver to warehouses in Cincinnati.[17] In January 1869 Brandeis & Crawford sent a letter to the Elkhorn Distillery requesting a payment of $2,000 (roughly $34,000 in present-day value) to retire Stone & Shropshire's outstanding balance for grain shipments. James Stone responded, "We are compelled to pay $4 or $5,000 Internal Revenue tax before the end of the month and at present we are not in full knowledge of where it is to come from. Mr. Wilson expects to go to New York next week with hopes of selling some whiskey."[18] Rather than ignoring the request for payment, Stone responded with an honest explanation of the distillery's financial difficulties. His tactic was to suggest that although payment would not be immediately forthcoming, if Brandeis & Crawford were patient, they would be paid eventually. In effect, Stone was requesting an informal loan at no interest and was using the grain broker as a source of credit to continue operations. Brandeis & Crawford, for their part, sought to set aside a contractual provision that the freight charges on rye shipments to Paynes Depot be paid by the shipper, not by Stone & Shropshire. Despite the distillery's debt, Stone admonished the grain broker to honor the contract. "This is not a huge matter," Stone wrote, "but in present condition of money matters we would like to have the use of even that small amount."[19] Stone & Shropshire paid other suppliers with short-term bank loans. In April, near the end of the 1869 distilling season, Stone settled an $857 bill with Louisville malt brokers Schickendantz & Sewell by obtaining a ninety-day bank note from the National Bank of Louisville.[20]

Stone & Shropshire conducted substantial business with Louisville whiskey brokers Newcomb, Buchanan & Company. H. Victor Newcomb, the son of Horatio D. Newcomb, president of the Louisville & Nashville Railroad, was born and raised in Louisville. He became a commission merchant and a partner in Warren, Newcomb & Company in New York City. Upon returning to Louisville, he served as a member of the L & N Railroad's board of directors, and he partnered with whiskey broker George C. Buchanan to form Newcomb, Buchanan & Company. In 1872 the company owned the Graystone Distillery and the Anderson, Nelson, and Buchanan Distilleries on Beargrass Creek in Louisville, the largest distillery complex in Kentucky at the time.[21] On Christmas Eve 1869 George Buchanan sent a note to James Shropshire offering to buy 400 barrels of Elkhorn whiskey. Shropshire accepted the order but, given the distillery's lack of cash flow, specified that until the barrels could be gauged the following week, he would not know the amount of government tax due and therefore how much money Buchanan would have to advance to conclude the purchase.[22] By the first week of January 1870, the distillery gauger and laborers had gauged 200 barrels, which Shropshire promised to ship, provided Buchanan paid the $4,446 tax required to withdraw the whiskey from bond (approximately $80,000 in present-day value). This initial payment did not include any money for the whiskey itself.

Shropshire implored Buchanan to "send us a check on Cincinnati or one of our city banks as our [tax] collector will not take exchange or checks on our neighboring cities."[23]

Distillery operations continued, despite these financial complexities, and Stone & Shropshire continued to order grain, replacement parts, and cooperage supplies. When Adam Michaels, the distillery's cooper, ran short of staves or hoop iron, Stone placed orders with Belknap & Company in Louisville, noting, "We have about caught up with our cooper and any delay for want of material would seriously discommode us." He accompanied this order with a check for $500 toward the distillery's account and apologized for his inability to pay more. "But as our credibility is dull and hard we beg indulgence from our friends," he wrote, adding a notation that Belknap could charge interest on the unpaid debt.[24]

Freight costs concerned Stone, and he requested that his suppliers always ship freight to the distillery prepaid. The Louisville, Cincinnati & Lexington Railroad maintained a strict freight charge policy: all costs had to be paid punctually, or shipments would cease. Interruptions in the supply of grain, coal, and other materials would stop distilling; halting whiskey shipments to fill customer orders would stop cash flow. Both threats were of sufficient gravity that Stone & Shropshire made sure to pay their weekly freight charges promptly, usually for amounts less than $100, to the Paynes Depot freight agent. The Louisville, Cincinnati & Lexington was the only railroad serving Paynes Depot, which made its rigid payment policy especially intimidating. When the distillery's finances were sufficiently dire, Shropshire withdrew funds from his personal bank account to pay the freight bill.[25]

The Insurance Imperative

Stone & Shropshire did not wish to operate without insurance coverage; indeed, the frequency of distillery and warehouse fires demonstrated that forgoing insurance was foolhardy (see chapter 8). S & S routinely purchased two types of policies. One insured the contents of whiskey barrels in storage at the Elkhorn warehouse. The second insured the taxes the distillers paid on the whiskey taken out of bond and placed with commission merchants in St. Louis and in Troy, New York.[26]

Thomas C. Timberlake of the Kentucky Farmer's Mutual Insurance Company in Louisville served as Stone & Shropshire's primary insurance agent. As barreled whiskey accumulated in the Elkhorn warehouse, Shropshire ordered additional insurance. In late February he wrote to Timberlake: "We have in ware house over 2000 barrels and are placing there over 200 weekly & if we have no bad luck will fill it up before 1st June say 5000 barrels in all. Let me know at once what your companies can take at the rate of $26 per barrel and at what rate for 6 mon & 12 mon."[27] As the barrels in storage accrued, Shropshire informed Timberlake and requested new or supplemental policy coverage; he was able to obtain $5,000 policies for twelve months at 1.25 percent. The policy premium was $62.50, a comparatively low rate that was based on the fire-resistant qualities of the distillery's brick warehouse.[28] Insurance coverage of $26 per barrel, based on an average of forty gallons per barrel, would have been less than half the combined value of the whiskey at wholesale, the cost of the barrel, and the federal taxes. It is unclear whether the insurance companies were unwilling to insure for a higher value or whether Stone & Shropshire wanted to avoid high premium charges and were, in effect, insuring half the value of their whiskey themselves.

In mid-February 1869 Stone & Shropshire made one of their largest sales. During a trip to seek eastern wholesale customers, Shropshire sold 550 barrels of Elkhorn bourbon. The buyer, Isaac McConihi & Company, operated an expansive mercantile trading and manufacturing firm in Troy, New York. Troy stood across the Hudson River from Albany near the east end of the Erie Canal; the city also had connections to the New York Central Railroad. During the nineteenth century, Troy was a major manufacturing center— resident Henry Burden, for example, was one of America's largest manufacturers of iron horseshoes and railroad spikes. The city was also a transshipment point for goods being conveyed west by canal and rail to the Great Lakes states and the Ohio Valley, east to New England, and south to New York City. McConihi likely intended to distribute the Elkhorn whiskey to retailers in New York State and New England. The purchase of 23,660 gallons cost McConihi 85 cents per gallon and earned the distillery more than $20,000.[29] McConihi asked Stone & Shropshire to insure the whiskey until it was delivered to New York. Because S & S had already reached their maximum insurance coverage limit, McConihi agreed to apply for insurance with eastern companies and requested a plat of the Elkhorn Distillery showing building placement and distances between structures. McConihi knew that insurance companies would request this information before they would agree that the Elkhorn Distillery was an acceptable risk and issue a policy. James Shropshire assured Mr. McConihi that if their insurance companies would not cover stored whiskey at the rate of $26 per barrel, he knew of several companies that would, including Aetna, North America & Franklin, Enterprise of Cincinnati, Imperial, Home of New York, and Liverpool & London.[30]

Isaac McConihi was also concerned that his whiskey be insured while in transit from Kentucky to New York, so he issued sight drafts to Stone & Shropshire for $1,730 and $4,970 (a sight draft was a type of bill of exchange that gave the seller title to transported goods until the buyer received and paid for them). S & S sent the drafts to the Deposit Bank at Midway with instructions to collect them promptly, "as we need money every day." Because sight drafts obligated S & S to cover any damages or loss during shipment, the distillers had to consider insuring the cargo themselves; hence, McConihi had effectively shifted that responsibility from buyer to seller. Undeterred, Shropshire notified McConihi in early March that the distillery was going to cash the sight drafts and added an appeal: "Can't you take 500 bls more of it. Our market here is still dull."[31]

Stone & Shropshire also attempted to use insurance as a sop to placate creditors. In

April 1869 James Stone received from Thomas Timberlake three new insurance policies—one for $5,000 and two for $2,500—each policy underwritten by a different agency. Stone then assigned the policies to Brandeis & Crawford, making the Louisville grain brokerage firm the beneficiary, in lieu of paying off the distillery's outstanding account balance. As an explanation, Stone could offer only this now-familiar justification: "The demand for money to pay whiskey tax in our district has produced a lightness in money." Brandeis & Crawford were not pleased with the substitution of insurance policies for cash payments, as suggested by a subsequent exchange of letters.[32]

Selling Elkhorn Whiskey: The Marketing and Sales Conundrum

The success or failure of the Elkhorn Distillery would pivot on product sales. But selling whiskey was not simply a matter of convincing customers to buy it; spirits sales in the late 1860s and 1870s were linked in complex ways to federal tax policy, the vigor of regional and national economies, and the cost of transportation, insurance, and borrowed money. From the winter distilling season of 1869 through midsummer 1872, James Stone and James Shropshire corresponded with more than a dozen creditors and over a hundred wholesale customers, some of whom also became creditors. A few wholesale customers were small-scale local dealers in Midway and Lexington, but most were in Louisville and other large urban markets, especially in the middle western and northeastern states. Major liquor wholesale houses sampled Elkhorn whiskey, and several handled the product, some in large volume.

While Stone managed the distillery, Shropshire corresponded with potential customers, describing the qualities and prices of Elkhorn whiskey but also seeking information about their sales success, the general strength of their markets, and strategies for pricing relative to market demand. He often prodded wholesale customers to pursue sales more aggressively, and he was not averse to suggesting sales tactics he thought might succeed in their markets. When sales failed to materialize and income began to trail expenses, he reminded customers that distillers were lobbying Congress to amend the revenue laws and extend the time that whiskey could be kept in bond. He predicted that if the policy could be changed, distillers could withhold whiskey from the market and thereby force prices and profits higher, while at the same time improving the product by additional aging.

In January 1869 Shropshire made a sales trip to visit New York City wholesalers, accompanied by whiskey samples.[33] He also sent barrel samples to prospective wholesalers. Initial purchases were usually small amounts—three to five barrels. Stone & Shropshire sent three barrels to John F. Magale, a successful wholesale liquor dealer who operated a large brick Victorian commission house on Strand Street in Galveston, Texas. Two barrels were newly distilled; one was a year old. The Elkhorn Distillery clerk, C. M. Gillespie, wrote an accompanying letter promoting the aged whiskey. The year-old whiskey will "show you how the whiskey improves [with age]," he wrote. "You will find it a very fair article of whiskey, the old will bring 3 or 4 dollars."[34] In later correspondence, Shropshire outlined Elkhorn's policy of offering the best prices to the best dealers, while also noting that whiskey held in a Kentucky warehouse over one summer improved noticeably in quality, adding 25 to 50 cents to the value of each gallon. If sales were sufficient, Elkhorn would agree to supply only one wholesale house in a city with its product. While not demeaning his Kentucky competitors, Shropshire asserted that Elkhorn's "steam copper" whiskey was as good as any made in the state; he also claimed that it was as good as whiskey made by the few "old fashioned small copper distilleries" that remained in operation.[35] He invited Magale to consider the offer and asked him to estimate the total annual sales in the Texas Gulf Coast market, suggesting that if Magale could sell all the whiskey Elkhorn could make, Stone & Shropshire would guarantee a supply for the next two years.

In February 1869 Shropshire received an order for samples from William T. Walters & Company in Baltimore. Founded by William Walters in the 1850s and operated with his brother Edwin, Walters & Company became one of the largest and most successful liquor wholesalers in the nation. Shropshire sent two samples of Elkhorn's June 1868 rye whiskey (made according to a mash bill of two-thirds rye and one-third corn) and presented a sales proposal:

> In view of the fact that our Ky distilleries are making corn whiskey we would like to open a new market if we can do so on the merit of the whiskey. We can make it any proportion it may be desired to meet needs of buyers or all rye if preferable. If you find they will suit your market we would like to make a contract to furnish you with 1 or 2 thousand barrels. If we could make such a contract it would be with the understanding that you got all we make out of our regular proportion of three-fourths corn and one-fourth rye. The whiskey we send you has its color intensity from the barrels.

In explaining the advantage of holding the whiskey for a year or more, Shropshire added: "One summer on Kentucky whiskey adds from 25 to 50 cents per gallon to value." Shropshire also offered to designate William T. Walters & Company Elkhorn's sole distributor in Baltimore.[36]

Despite James Shropshire's financial concerns, James Stone continued to run the distillery at full production capacity during the late spring 1869 distilling season. In early May Stone added to the operation's debt balance by ordering enough carloads of rye from Brandeis & Crawford to last until the season ended in June.[37] Shropshire sent Adolph Brandeis barrel samples of whiskey, perhaps hoping that he would find buyers in Louisville. Although Brandeis & Crawford were grain merchants, this may have been the beginning of an effort to get them to accept whiskey in lieu of cash for grain purchases and hold it as collateral or attempt to sell it on commission. On May 20 Stone ordered another 1,000 bushels of rye from Brandeis & Crawford, and later that week, S & S shipped the grain brokers five barrels of whiskey.[38] The whiskey order may have been based on the quality of the earlier samples sent by Shropshire, but more likely it was a sober recognition by Brandeis & Crawford that accepting whiskey as an in-kind payment was probably the only value they would realize from the Elkhorn account for some time. A week after Shropshire saw the five-barrel order loaded onto a railcar at Paynes Depot, he sent an anxious note to Brandeis, asking if he had found any buyers.[39] Another note followed a few days later: "What kind of offers have you had for our whiskey would you not wish to buy the 500 barrels of whiskey."[40] The distiller and the grain merchant were now bound together in a disagreeable creditor-debtor arrangement that could be settled only by a considerable infusion of cash. And this was unlikely to come from even a sizable whiskey sale, given a market choked by surplus spirits produced by competitors who were willing to sell at prices that had fallen to less than the cost of production.

Nevertheless, Shropshire persisted. He sent a letter to Brandeis on June 11, stating:

> We find we are about to be short of room to store our whiskey and write to ask [you] to pay tax and take out one to two hundred barrels of whiskey to be shipped down to you & held until sold. If you can do so it will be a great accommodation to us and it will enable us to close out all the [grain] on hand which we are very anxious to do. We will expect to pay for insurance & interest. Our situation is such that we must know by tomorrow for if you or some other firm cannot do this for us we must stop mashing by Monday. If you are willing to do this advise us by telegraph to Lexington tomorrow by 1:00 pm and send us terms by mail.[41]

Brandeis & Crawford acquiesced and purchased an unstated amount of Elkhorn whiskey, but they were not pleased with what they received. On June 30 Shropshire again wrote to Brandeis: "We have heard that your purchase of whiskey did not come up to your expectations & was only partially received. We know that in the general market our whiskey is worth at any age at least 10 cents per gallon more than *many other* brands. We have never offered a bad barrel of whiskey and have not *as many others have* slighted our whiskey on small grain as hard as the season has been."[42] His assertiveness stoked, perhaps, by impending financial collapse, Shropshire reiterated his request that Brandeis & Crawford order 500 additional barrels.

In June 1869 the Elkhorn warehouse was full, and the whiskey inventory stood at 5,000 barrels. At more than forty gallons per barrel and 90 cents per gallon, the whiskey's potential worth was greater than $180,000 (roughly $3.24 million in present-day value). But orders continued to trickle in, and Shropshire had to be content to sell two or three barrels at a time, each barrel adding about $35 to the assets side of his accounting ledger.

Each week in June Shropshire wrote to creditors, requesting permission to convert Elkhorn's debt balance to 4 percent interest-bearing notes. Hardware suppliers W. B. Belknap & Company of Louisville and J & E Greenwald of Cincinnati received nearly identical letters pleading an inability to pay balances of only $400 to $500. Shropshire managed to raise a paltry $115 by selling 105 sacks of surplus rye to a Frankfort grain dealer for $1.10 per bushel.[43]

By early July, Stone & Shropshire were rejecting queries from brokers and wholesalers offering to buy whiskey at prices below the cost of production. They accepted other offers and sent samples when requested. Most buyers paid partly in cash, with the remainder due over two or three months, plus the cost of insuring the whiskey until the balance was paid. Other accounts proved hard to collect. Lexington general commission merchants Woolfolk, Craig & Gay agreed to take 150 barrels of Elkhorn whiskey. The terms of this sale were unstated, but if S & S received 90 cents per gallon cash, the yield from this sale would have been about $5,800. On July 9 Shropshire sent a note to Charles V. Higgins at the Deposit Bank in Paris, with a check for $2,000 to be credited to the S & S account.[44] Higgins was a Bourbon County farmer who operated a tanning and currier business in North Middletown in the 1830s. He had extensive landholdings throughout the county and in Paris, the county seat. He became a director of the Deposit Bank and later served as its president; Higgins was also one of Stone & Shropshire's largest creditors.[45]

In early October 1869 James Stone traveled to Louisville, St. Louis, and Chicago seeking customers; his luggage included Elkhorn whiskey samples.[46] He obtained an agreement with Louisville distiller and wholesaler S. Grabfelder to handle Elkhorn whiskey.[47] In St. Louis, the wholesale and commission merchant warehouses were clustered near the Mississippi River on Second Street. There, Stone visited wine and liquor wholesalers J. G. Wells & Company and J. B. Ghio & Company, as well as St. Louis Mutual Life Insurance Company president D. A. January. Apparently, the samples proved satisfactory and Stone was persuasive, because Mr. Wells and Mr. Ghio expressed interest in buying whiskey to supply the mid–Mississippi Valley market, and Mr. January ordered ten barrels. Upon hearing the news, Shropshire wrote to Robert E. Cary at the Exchange Bank in St. Louis, asking for financial references for Wells, Ghio, and January. Upon receiving a positive response from Cary, Stone & Shropshire shipped whiskey on sight drafts to January. On October 14 Shropshire sent a formal business proposal to Wells. After admitting that S & S did not have sufficient money to pay the revenue taxes required to withdraw whiskey from bond, he assured Wells, "We are anxious to have our whiskey well & properly introduced into your market and believe that it would make for itself a good trade. Would it suit you to permit us to draw at sight [draft] for amount to cover tax on 100 barrels whiskey." If so, he continued, "we can give you satisfactory references and assurance that the money will be properly appropriated."[48] J. B. Ghio & Company became an Elkhorn customer, but J. G. Wells & Company became an active consigner of Stone & Shropshire whiskey and willingly agreed to hold Elkhorn warehouse receipts. The Wells salespeople were instrumental in selling S & S whiskey, advancing money for taxes, and sustaining the business through March 1870.[49]

In mid-October James Shropshire made a fruitless trip to Louisville, seeking whiskey buyers and relief from creditors. Yet he did receive an order for twenty-five barrels from his relatives in New Orleans, Shropshire & Brother Company, on a sight draft, provided that S & S could pay the revenue taxes that would allow the whiskey to be removed from bond. But the distillery was already $4,000 in arrears on tax payments and liable for a 5 percent penalty, with no prospects for income. To aggravate matters, its new St. Louis customer, J. G. Wells, wanted to buy 100 barrels but would not accept a sight draft to cover the tax, forcing S & S to raise the money elsewhere to consummate the sale.[50] Shropshire was able to convince the Northern Bank of Lexington to make a loan of $2,500 for sixty days, which would pay the taxes on the 100-barrel order.

Letters lamenting slow markets and tendering small orders crossed Shropshire's desk during the last two weeks of October. A wholesaler in Philadelphia had made no sales, but a Baltimore company proposed to pay the revenue taxes if it could get Elkhorn whiskey. Chicago rectifier and liquor merchant Fridolin Madlener ordered thirty barrels at 85 cents per gallon, for a sale of slightly more than $1,000. Each order, however small, required S & S to arrange for warehouse labor, federal gauging, and wagon hauling, as well to endure the stress of negotiating with the Paynes Depot freight agent.[51]

On October 21 Shropshire sent a letter to Galveston, Texas, rectifier S. H. Gillman, an old and "esteemed friend":

> I committed myself with one of my old customers and friends in the distilling business last fall (1868), and we made over 5,000 barrels whiskey under the name of Stone & Shropshire and are holders of the larger portion of it at present. We make a good article of Bourbon whiskey and I think our oldest will compare favorably with any in the state of storage and I will send you by express samples of January 1869 & Dec 1868. In starting the [rectifying] house you spoke of however you will need some older whiskey and I can always get it for you here although at present there is but little to be had over 2 years old and all such is held very high. I will do any and everything I can at this [time] to assist you in getting a good house under way. Our whiskey 1 year old kept in your church through 1 summer would be good enough to supply your best trade. We could at present sell you 5 or 10 barrels at $1.50 per proof gallon delivered on the railroad so that it would cost you $1.60 to .65 delivered. Two year old whiskey cannot be had for less than [$]2 to 3.00 here. When samples reach you let me hear what [you] think of it and what I can do for you. As matter of course I am most interested in selling our whiskey but will fill your order in any [way you require].[52]

Shropshire appended a box of samples of aged Kentucky bourbons for Gillman's consideration, including "Stevens & Brunman, copper 2 yrs $3.00; Pepper, copper 4 yrs $4.00; Lillard, copper 3 yrs $4.50; and Ashland, copper 3 yrs $3.00." Shropshire lauded these whiskeys as the best available and offered to broker them on Gillman's behalf and ship them to Galveston. Whatever Gillman thought of the samples, we do not know; his name does not appear again in the Elkhorn records.[53] We are not privy to the social contacts that connected Scott County, Kentucky, and Galveston, Texas, in 1869. S & S customer John F. Magale and Shropshire's friend S. H. Gillman conducted business there, and Shropshire later developed a business relationship with Galveston merchants George Ball and John Hutchings as the Elkhorn partnership foundered.

By late October, James Shropshire's sales tactics were reaching the threshold of urgency. He pressed Shropshire & Brother in New Orleans to make its best effort to sell Elkhorn whiskey at any price that would net S & S at least $1.50 per gallon, but to sell its twenty-five-barrel order as promptly as possible. He then added the expectation that Shropshire & Brother would be ready for an additional 100 barrels by December 1. A federal tax payment on 500 barrels was due in January, and Shropshire desperately wanted to pay the tax out of sales instead of begging cash-strapped banks for expensive loans. He offered to grant his relatives at the New Orleans firm exclusive agency for Elkhorn goods in that city, which, he assured them, "ought to pay you pretty well."[54]

Shropshire also began sending unsolicited letters of inquiry to prospective buyers. A letter to New York merchant William H. Little on October 27 presented a straightforward question: "How would it suit you some time in the next days to make advances to amt of taxes on Bourbon whiskey to be shipped to your market. We have nothing to come out of bond for 60 or 75 days but if it could be sold free before that time expires we would like to [sell] part of the crop in hand. Our whiskey is maturing as well as we could wish and in time will compare favorably with any of the herd made in the state."[55] The response from William H. Little & Company was prompt and apparently positive, although the tone of Shropshire's reply on November 6 hinted that all was not well. Shropshire explained that although Elkhorn had 4,000 to 5,000 barrels in bond, it probably would not want more than 2,000 to go to New York. He asked for an advance payment to cover the $22-per-barrel tax and part of the whiskey's value. He added, hopefully, that Elkhorn would likely have all its January whiskey out of bond by December 1 and would be under no financial pressure before February. The asking price on a 500-barrel lot would be $1.45 to $1.50 delivered in New York. He ended on a disquieting note, revealing that it was unlikely that Elkhorn would be making any whiskey in the upcoming season.[56]

The letters exchanged between Stone & Shropshire and William Little suggested more than a straightforward query and sales pitch. Liquor dealers were entering the whiskey market and offering low prices because they recognized that internal revenue regulations were forcing distillers to sell, regardless of the price. Shropshire stretched the truth when he implied that Elkhorn would be under no financial pressure to sell until February, although the revelation that there would be no production in 1870 was factual. On November 29 Shropshire shared his suspicions with Stone, who was on a sales trip to New York. "Without sales I do not see how we can weather it much longer as some of my debts are pressing as well [as] claim[s] due next month which must be made some way. I think from late indications that whiskey dealers & compounders [rectifiers] have made up their minds to use their influence against any [congressional] extension in bond so as to force distillers to the wall."[57]

How could a Kentucky distiller respond to this national-scale political leverage and economic pressure? Shropshire instructed Stone to visit Little and ascertain under what terms he would advance funds to pay both the revenue taxes and Elkhorn's asking price. He advised Stone to say that Elkhorn had no immediate need to move whiskey, which continued the tactical bluff contained in his own letter to Little, but that he would consider a fair offer. He also pushed Stone to sell 1,000 barrels for the best price he could get. If Little proved intractable, Shropshire advised his business partner not to leave New York until he had found a reliable dealer who would pay the tax and sell Elkhorn's whiskey on commission. Neither J. G. Wells in St. Louis nor Shropshire & Brother in New Orleans had reported any sales.[58]

One month later, on December 28, Shropshire sent another letter to Little in New York, replying to Little's offer to pay an advance against taxes but also a proposed per-barrel price that was so low that Shropshire was obliged to reject it. Employing a full measure of obfuscation, Shropshire wrote: "We have our matters so assigned that we will not be troubled for tax money again until towards the 1st of March by which time we trust there will be a change in the date." He was referring to the hope that congressional legislation would extend the bond period on barreled whiskey. Prices would therefore increase on unsold whiskey, in part because maturing whiskey increased in value—even though evaporation and leakage decreased its volume—and in part because only a few distilleries were operating, and distillers were selling down their stock to pay their debts.[59]

For those whiskey orders that Stone & Shropshire did receive, the shipping process was fraught with frustrating difficulties. Wagoners were otherwise occupied, and they were delayed in getting to the Elkhorn warehouse. An order for more than 300 barrels from J. G. Wells in St. Louis required James Stone's son Charley to hire a team of laborers to unstack, sort, and assist in gauging barrels in the tightly filled warehouse. Hauling the lot two and a half miles to the depot required two days. Then the freight agent at the depot would not issue a bill of lading that would send the shipment directly to the destination without first going to Louisville, where it would have to be unloaded and reloaded onto another train. Despite the size of the sale, Shropshire remained pessimistic, as federal storekeeper and gauger salaries had to be paid, and payments on large debts would soon be due. On the positive side, a convention of distillers was scheduled to meet in Lexington on November

3, where news of market conditions and prospects for legislative action on extending the bond period might be forthcoming.[60]

Although Stone's extended trip to St. Louis in late October 1869 resulted in sales to J. G. Wells, J. B. Ghio, and D. A. January, the prices were lower than Shropshire had been anticipating. All three were selling to the same general market, but Stone had accepted January's offer of $1 per gallon, while Wells and Ghio paid more. Shropshire voiced his concern that such a low price would not only debase their whiskey in that market but also distress the dealers who did not benefit from paying the lowest price. Further, it left S & S with little margin for profit. Commission, consignment, and brokerage fees were about 17 cents per gallon; storage and other expenses were another 3 cents. If the whiskey sold at $1.40 per gallon, they would net only about $1.20. From that amount, there were further deductions for leakage and evaporation, plus the 5-cent warehouse fees; this reduced the net to about 65 cents, with interest on loans not yet included. Shropshire resigned himself to the price, though, reasoning that even if Elkhorn lost money on this initial sale, it would provide entree into that regional market. But Shropshire was also facing a $5,000 debt payment to a creditor on November 11, and he urged Stone to go back to Wells and ask him to advance the money on an additional 150 barrels of tax-paid whiskey in storage.

As the fall season advanced into early November, James Stone and several other distillers began to seriously consider running their distilleries in the spring 1870 season. Shropshire strongly advised against it. Elkhorn would first need to sell its remaining stock of whiskey to recoup distilling expenses and pay down its debt. Coal prices had increased substantially. The expense of hauling barreled whiskey and grain to and from the depot was sizable; this, combined with the high cost of railroad shipping, was a disadvantage not suffered by distillers with more favorable transport connections. Given the depression in whiskey prices that would surely follow if a large number of distilleries started production, the only possible source of profit, in Shropshire's opinion, would be to fill the stock pens to full capacity with enough hogs to consume all the slop the distillery produced.[61]

While attempting to salvage the Elkhorn Distillery, James Shropshire's personal assets had largely been consumed by the Stone & Shropshire partnership. He asked local friends for loans and sent inquiries to distant acquaintances seeking to borrow cash. A Texas friend was one of the few who responded, and although he had problems of his own, he said he had three cribs of corn that he might be able to use as collateral for a loan. Shropshire's response reveals his desperation, yet he closes on a hopeful note:

> You stated that there was some probability that you could get something out
> of 3 cribs of corn try and send me some money as I am especially in need of it
> for day to day family expenses. I am laid up with over 4,000 barrels whiskey in
> bond and the dalliances of the market and near approach of time to release it
> from bond is harassing me as I hardly ever was in my life before and any aid you

can give me now will be more than appreciated. Send me any amount you can on the corn or find any source your obligations will permit you. I still have an old fogy notion that honesty integrity & perseverance with a moderate amount of good sense will succeed and shall continue to fight it out.[62]

The sizable whiskey orders from St. Louis brokers Wells and Ghio were costly to ship. Wells advised Shropshire to ship by way of Cincinnati. The Bishop Brothers freight agency there was known to offer rates as low as 80 cents per barrel from Cincinnati to St. Louis.[63] Anxious to reduce costs and shorten delivery time, Shropshire contacted E. P. Wilson, the Louisville, Cincinnati & Lexington Railroad agent in Louisville, and asked to be given favorable rates.[64] Elkhorn was receiving orders from St. Louis brokers and, he told Wilson, we "want you to place us on the best footing. The first 99 barrels we shipped over your road to Louisville but the freight from there to St. Louis was out of all reason." The agent at the Midway Depot had assured S & S that Elkhorn whiskey was entitled to a through bill of lading from there to St. Louis. Shropshire's request to the Louisville freight agent was to ship fourth class at 35 cents per 100 pounds, but the best rate Wilson would offer was 57 cents per 100 pounds.[65]

By mid-November, Elkhorn whiskey shipments to St. Louis were increasing modestly. J. B. Ghio & Company bought 317 barrels. Shropshire shipped 150 barrels, or more than 6,000 gallons, to J. G. Wells and urged him to push sales as rapidly as possible so that the firm would have sufficient funds to cover the drafts S & S would be making on its accounts as well as to cover taxes on future consignments. Wells ordered 200 more barrels and informed Shropshire that sales were prospering, although some buyers were offering only $1.35 per gallon when the distillery's cost value was $1.50. Shropshire admonished Wells to hold out for $1.50, but he also recognized that cash flow was necessary to conduct business. Banks were low on funds and reluctant or unable to make sizable loans. "Money matters are very close here at present," Shropshire informed Wells, "but we trust when our farmers get cash for their hogs & trading returns etc our banks will have better balances to operate on."[66]

Other brokers were less sanguine. Shropshire & Brother in New Orleans was having little success in advancing Elkhorn whiskey sales. A dull market in the lower Mississippi region might have been expected, given that more than 30,000 barrels of whiskey arrived in New Orleans by riverboat in 1869, likely saturating the local market (see chapter 5). James Shropshire advised Shropshire & Brother to have its salesmen visit the city's saloons and get them to take a barrel or two of the recently distilled Elkhorn whiskey, even if they had to sell it as low as $1.50, as a way of getting started. The tavern trade did not require fine old whiskeys, and Shropshire observed, "We think if they got to using it once they will want it again and it will not be much trouble to raise the price a little after they once commence using it. If you can find [the] parties that use Chicken Cock Whiskey—made by the G. G. White distillery in Paris—offer inducements in price and assure them you can furnish it as they want it we think likely you could build up a trade for it, for it is better than that brand."[67] Shropshire was not given to offering negative critiques of whiskeys produced by other distillers, but muscular measures doubtless seemed warranted.

In early December 1869 an anxious James Stone made a sales trip to four Atlantic port cities. He visited wholesalers in Boston, New York City, Philadelphia, and Baltimore but made only a few small sales in Boston and New York. He found no buyers in Philadelphia, and Shropshire had warned Stone that he would be subject to a heavy penalty by Baltimore officials if he offered samples to prospective buyers in that city. Shropshire advised him to go to Washington, DC, to lobby for the bond extension legislation that was then under consideration in Congress. Stone was aware that payment of an $11,000 loan from Maysville cotton manufacturers January & Wood was due before Christmas, and a debt to another creditor for more than $6,000 was due New Year's Day. In addition, both Stone and Shropshire had personal debts of some $4,000 that could not be deferred. Stone knew he had to sell at least $20,000 worth of Elkhorn whiskey before he could return to Kentucky.[68]

Several prospective buyers, including a firm in Lexington, were willing to purchase 50 to 100 barrels and make interest-bearing advances to Stone & Shropshire to cover revenue taxes, but at usurious rates that Shropshire referred to as "shylock prices."[69] Some buyers wanted to charge interest rates of 19 percent per annum, plus a commission for selling the whiskey to retailers. Shropshire preferred to sell whiskey at low prices rather than pay high interest rates. He received an offer from a Louisville broker to buy 500 barrels at 75 cents per gallon. That sale would be sufficient, provided it could be consummated, to pay half of the note held by January & Wood.[70]

James Shropshire affably acknowledged each order with a statement to the buyer summarizing the payment draft he had received. He promised that each barrel would be specially selected at the distillery and gauged to obtain the exact proof and price. Shropshire assured each buyer that the distillery would "spare no pains" in sending whiskey in good condition. Buyers often specified that they preferred whiskey at 103 to 105 proof, but the February 1869 whiskey varied from 103 to 106 proof on some days and from 106 to 112 on others. Selecting barrels in the desired proof range required a considerable expenditure of labor and money to move, gauge, and restack barrels, with no assurance of finding what had been ordered. A new gauger during the previous season's run had used an old-fashioned glass hydrometer that produced a range of readings from well below 100 proof to 10 to 12 percent above that level. If Shropshire could not locate the desired proof barrels, he wrote to customers and asked if they would accept higher-proof whiskey, adding that it aged better than lower-proof barrels. To other brokers, Shropshire touted low-proof whiskey. The north-central Missouri market did not require high-proof whiskey, he wrote to Yates & Dillen in Chillicothe, Missouri:

Our January '69 whiskey [produced] 40 or 50 barrels that instead of running 5 percent above proof as it was intended run 4 to 5 percent below that is just as good whiskey as any we have except for the difference in proof. In buying this 5 percent below proof you only pay for 38 gallons or 40 wine gallons and it would probably make more money for you than higher proof. We will deliver this at our RR station for you at $1.50 per proof gallon and get you through bill of lading to St Louis or Chicago at 57 cents per 100 lbs if you want as much as 20 or 25 barrels. If you order send us $22 per barrel to pay tax [and] we will send the whiskey balance to be paid in 60 days.[71]

Shropshire was increasingly aware that whiskey markets had a geography. By working with his warehouse gauging crew, he learned the variability of his own product, and by corresponding with prospective customers, he cultivated tactics to sell it all, whatever its qualities. He was developing the skills to be a broker, rather than relegating himself to balancing the books and urging Stone to sell more whiskey. Shropshire was also resolute in trying to shift expenses to the customer; he paid freight only to his own depot and requested that customers pay the freight to their own warehouses.

Several prospective buyers voiced no interest in the whiskey's proof but inquired about the availability of "old" whiskey. In early December 1869 W. J. Walker & Company of Louisville contacted Stone & Shropshire about aged whiskey. Elkhorn had ninety-nine barrels of one-year-old whiskey (distilled in December 1868) on hand, which Shropshire offered to sell for cash or "good paper at short date. The whiskey seems to be well matured for its age and will be something over proof," he wrote in his response to W. J. Walker, and it "is in good iron bound barrels, out of bond and we will sell it to you for $1.35 in the warehouse here."[72]

There were some dissatisfied customers. A Kansas City buyer purchased twenty-five barrels from S & S wholesaler J. G. Wells in St. Louis but refused to pay; the customer said he had sampled and bought March (1869) whiskey but received January (1869) whiskey, which he claimed was raw and not as good as the March product. This must have seemed odd to Shropshire, given the mere two-month age difference. While the flavor dissimilarities in whiskeys distilled in December 1868 and January and March 1869 might seem trivial or even undetectable—especially given that the product was stored in the same type of white oak barrels and rested in the same stack in the S & S brick warehouse—some customers noticed undesirable differences, or at least they said so in order to extract a lower price from the distiller.[73] One broker complained that the barrels he received were leaking badly, and all the hoops had to be tightened before putting the barrels in storage. Another customer claimed that "four barrels were leaking at the bung," to which Shropshire replied that the barrels were "well bunged and trimmed but hauled 2½ miles which can cause a little dampening about the bung."[74] And disgruntled Boston broker C. B. Barnett complained that the leakage and evaporation from his ten-barrel shipment amounted to twenty gallons. Shropshire explained that the "outs" were measured at the distillery, and the barrels were topped off; each barrel was then tightly bunged and did not leak. Before the barrels were shipped, the bungs were drawn and relined to prevent any possible leakage. Shropshire concluded that the loss must have resulted from pilferage en route. Barnett requested that S & S refund the tax on the twenty-gallon outage and deduct it from his bill. Shropshire denied the request.[75]

Although whiskey orders lagged during the fall of 1869, they increased abruptly as the new year approached. In mid-December 1869 Stone & Shropshire received two 100-barrel orders from Boston—one from broker Ayre & Eaton, and the other from Enoch Payne & Company. Shropshire immediately wrote to creditors January & Wood in Maysville and Charles Voorhies in Chicago with the good news that he would be able to pay some $6,500 on the S & S notes they held. Though substantial, this amount did not cover the distillery's debts with either creditor. Shropshire submitted that it showed good faith, and he asked both firms to extend the debt four months; he offered January & Wood a warehouse receipt for 300 barrels of whiskey as further accommodation. Shropshire continued to believe that the House Committee on Ways and Means would recommend an extension of the bond period, and he asked January & Wood to advise their Louisville banker not to move the S & S debt into the courts.[76]

Shropshire had written to McConihi & Company in Troy, New York, in October, urging officers to buy 500 to 1,000 barrels of whiskey and promising a price that would allow for ample profits. In late December McConihi ordered 550 barrels. Shropshire intended to stage the order's gauging inspections and shipments in lots of about 105 barrels and planned to process the entire order by January 3, when the bond period expired. This large order proved logistically difficult, in part because laborers and wagoners were reluctant to work during the Christmas and New Year holiday season. Even with a full labor cohort and benign weather, the clumsy process of sorting, gauging, and hauling to Paynes Depot limited S & S to a maximum of 500 to 600 barrels per week. As part of the payment agreement, McConihi & Company had sent S & S a check for $5,000 drawn on a New York bank, which Shropshire planned to submit directly to the state revenue collector to pay the January bond taxes. The collector refused to accept the check, saying that banks were overloaded with New York funds. Shropshire had to arrange for one of Elkhorn's local banks to accept the check and provide payment to the collector.[77]

Stone & Shropshire made four revenue tax payments to remove the McConihi whiskey from bond in late December 1869 and early January 1870, totaling more than $12,000. This expense fell to McConihi, together with fees for storage, labor, and federal gauging. For the most part, McConihi had paid these expenses by the end of December.

On the last day of December, Shropshire received in the mail 200-barrel orders from J. G. Wells in St. Louis and Newcomb, Buchanan & Company of Louisville. Both firms

Transactions Between McConihi & Company and Elkhorn Distillery, December 1869–December 1870

Year	Month	Day	Transaction		Amount
			----- Debits -----		
1869	December	7	To cash pd tax on 105 bls whiskey nos 131 to 235		$2,312.50
		20	To " " 105 bls " 236 to 340		$2,336.50
		29	To " " 120 bls " 341 to 460		$2,670.50
1870	January	1	To " " 220 bls " 461 to 680		$4,850.50
		3	To " storage 550 bls whisky		
			from 15 Feb 1869 to 15 Dec 1869:		
			11 months at 10 cents per mo		$550.00
			To " labor in handling 10 cents per bl		$55.00
			To " cartage on 548 bls 2 1/2 miles		
			to R Road at 25 cents per bl		$137.50
			To " gauger for gauging and		
			detail statement on 550 bls at 20 cents		<u>$110.00</u>
					$13,022.00
			----- Credits -----		
	July	21	By cash pd our dft	$200.00	
	December	7	By " pd on It' dft	$2,321.20	
			Less discount chg	$8.70	$2,312.50
		15	By cash pd our sight dft		$2,341.50
		29	" " " "		$2,650.00
		30	" Dft on "NY to" order of Kelly*	$5,000.00	<u>$12,504.00</u>

Balance due Stone & Shropshire $518.00

Source: Elkhorn Distillery Records, 1868–1872, 1997MS419, Book 2, p. 396, Special Collections, University of Kentucky.

*Col. R. M. Kelly was the Internal Revenue Service District 7 tax collector.

expressed an interest in negotiating for more whiskey. All parties seemed to believe that the banks would offer more accommodations after January 1 and Congress would extend the bond period for new whiskey. If these events should come to pass, the financial benefits would accrue to those businesses that held whiskey in storage. Brokers and consigners also knew that many distilleries were not operating and that much less whiskey was being made than the previous season, forcing prices higher.[78]

Elkhorn's two principals adopted a timorous optimism that their operation could be salvaged by increasing sales in the new year. Shropshire began to tell customers that Elkhorn did not wish to commit to selling more whiskey at current prices but anticipated that Congress would pass a bond period extension in the next thirty to sixty days. "If this fails and prices of whiskey continue below cost of production & interest," he wrote to Newcomb and Buchanan, "we will want further assistance."[79] Although Stone & Shropshire were obligated to Newcomb and Buchanan and other brokers for their orders—which had slowed if not stopped Elkhorn's slide toward insolvency—they also hoped to control their sales in a manner that would at least cover their costs. Gauging and warehouse inefficiencies continued to delay shipments, and in early January 1870 Shropshire placed a large order for rope with a Cincinnati supplier, likely to replace the warehouse hoist ropes that laborers were putting to heavy use.[80] The S & S cooperage remained in operation, consuming the stock of staves and barrel heads on hand, and building and storing barrels should Stone decide to restart the distillery.[81]

In late January 1870 Shropshire completed the financial arrangements for the orders from Newcomb, Buchanan & Company for $14,500 and McConihi & Company for $13,022. He also ordered thirty barrels of whiskey sent to Shropshire & Brother in New Orleans, with the directive to "sell it at $1.75 per gallon and to keep it well sampled and before your city trade." "It ought to bring as much as 'Chicken Cock,'" Shropshire reminded them. "It's a better quality but we want you to use proper course to make a trade for our brand."[82]

Stone & Shropshire had placed some sample whiskey with George W. Wicks & Company in Louisville. Mr. Wicks operated a cotton factory on Main Street and was a partner in the Louisville Tobacco Manufacturing Company; he also served on the board of a bank and an insurance company. Wicks recommended Elkhorn whiskey to Louisville merchant and wholesale whiskey dealer J. T. S. Brown, and Brown sent a note to Shropshire inquiring about availability and prices.[83] Shropshire informed Wicks that 200 to 300 barrels of February 1869 whiskey remained in bond, and he could take Brown's order for 100 barrels at $1.45 cash. Brown could also buy 100 barrels of March whiskey at $1.40 cash or a sixty- or ninety-day note. To facilitate removing the whiskey from bond, Shropshire required a cash payment for the amount of the tax. Shropshire shrewdly designated Wicks a commission broker for the sale, for which he would receive a fee of $1 per barrel.[84] The offer was apparently not sufficiently attractive to Brown, for he never replied.

On February 5, 1870, James Shropshire traveled to Washington, DC, where he stayed at the prestigious Willard Hotel. The trip's purpose was to join other distillers in lobbying Congress to pass legislation to extend the bond period. He wrote to J. G. Wells in St. Louis, "We feel pretty confident an extension in bond will be granted us very soon which will at once very naturally stiffen the market."[85] The letter was brave in tone but premature in execution because their lobbying efforts came to naught. In late February the House Ways and Means Committee agreed to offer a bill extending the bond period but adding an interest charge of 24 percent per annum across the bond period. Shropshire telegraphed the internal revenue commissioner and asked him to block the bill, if possible, because the net effect would be worse than leaving the existing one-year bond period in place. Resignedly, he told Stone, "I think we will have to row our own boat without any assistance from congress."[86]

Without helpful congressional legislation, S & S faced tax payments amounting to some $13,000 due on February and March whiskey, as well as $2,000 to $4,000 payable on other debts. Beyond that, taxes would soon be due on 2,300 to 2,400 barrels of May and June whiskey, amounting to some $50,000. The partners were now obliged to consider accepting almost any offer for Elkhorn whiskey, and the future of their business seemed to hinge on a few large transactions with their most reliable customers.[87] Undeterred, and true to his ethic of perseverance, James Shropshire wrote to his brokerage customers to inquire, yet again, if they would buy more whiskey. On February 28 Shropshire telegraphed J. G. Wells in St. Louis and asked him to purchase another 125 barrels of February 1869 whiskey. Wells responded the following day with an order for the remaining February whiskey, 127 barrels, which Shropshire agreed to ship immediately.[88] During the first week of March, as Shropshire readied the Wells order for shipment, he received a query from Wells about the price for an additional 200 to 300 barrels. Shropshire quoted a price of $1.60. The apparent reason for Shropshire's optimism in setting a price 20 to 30 cents higher than he had been willing to take a few weeks earlier was a telegram he had received from a friend in Washington. The House of Representatives, the message said, had passed a bill that would extend the bond period and add a one-half cent per gallon per month interest payment; if the bill passed in the Senate, this would stiffen whiskey prices. Shropshire advised Wells to watch for the bill and be ready to raise prices if it passed. Stone & Shropshire received $5,000 from Wells to cover the 127-barrel purchase, but the entire amount was relegated to paying the tax on March whiskey, and they had no immediate prospects of being able to pay April taxes.[89]

On March 12, 1870, Shropshire sent a confirmation of Wells's order. Shropshire paid the tax on the whiskey and planned to ship it in two lots by way of a Covington freight agent the following week, with a shipping charge of 57 cents per 100 pounds. But the letter also included a caution. "Extension in bond has not so bright a face as it had a few days since," Shropshire wrote, "but we think it will be worked through the senate next week. This is the opinion of our atty. If the 127 barrels has reached you sell it if you can at $1.60 but if you have to take a little less don't let it hang fire too long. Sell at depot if you can. We will send you samples of improved whiskey also another sample of 99 barrels December in a few days."[90] Three days later, Shropshire again wrote to Wells:

We have no further advice from Washington only that our atty expects to test the extension before the senate this week. If it fails it will have us to fight it out on our own. We have offer of money to take out as much whiskey as we choose but prefer as long as we can sell at a small profit to keep it moving therefore we desire that you will press sales of the 127 barrels, also of the 205 barrels as we are now shipping you. Next week we will have something over 200 barrels more to take out and if we hear of sale of the 127 will probably ship a part or all of it to you. Advise us by 22nd either by mail or telegraph what sales you have made. Keep sales up. Will send samples.[91]

Through mid-March Shropshire, who was facing an unremitting sequence of tax deadlines, sent appeals to his most reliable wholesalers, imploring them to buy more whiskey. He wrote to McConihi & Company in New York, stating, "Extension in bond hangs fire in Washington and unless we get it we want to sell 300 to 500 barrels of whiskey by or before May 1. We have 100 barrels made Dec '69 don't you want to buy it."[92] Fridolin Madlener in Chicago, Shropshire & Brother in New Orleans, and other wholesalers received similar letters as Shropshire sought to assess the sales potential in each regional market. He also updated wholesalers on production trends, hoping they would use the information to bolster their sales. In late March 1870 he reminded J. G. Wells that "the falling off in production in this district is very heavy for February. Made in Feb 1869 575,787; Made in Feb 1870 196,668; Deficit for Feb 1870, 319,119 [barrels]."[93] Shropshire advised that the production shortfall would continue for the remainder of the distilling season, implying that the effect on whiskey prices would be positive, provided one could retain ownership. By early April, New York liquor wholesalers estimated that the 1869–1870 whiskey production season produced only 40 percent of that distilled in 1868–1869.[94]

Many Kentucky distillers who were required to take whiskey out of bond anticipated an increase in prices and chose to negotiate with banks and other lenders to borrow money to pay their taxes. Provided they could get loans at good interest rates, distillers reasoned that their most profitable option was to hold their tax-paid whiskey for better prices rather than operate their distilleries, which would only depress prices further. Stone & Shropshire made some sales to wholesalers, thereby garnering funds to pay taxes on bonded whiskey. Nevertheless, the chronically mismanaged Lexington revenue office continued to impede whiskey shipments by failing to provide federal tax stamps in a timely manner. And irksome payment problems continued to disrupt cash flow to the Elkhorn Distillery. Shropshire requested that wholesalers pay by express currency or checks drawn on Cincinnati or New York banks, even though checks drawn on New York institutions cost more to cash in Kentucky. Meanwhile, the Elkhorn Distillery's $13,000 debt to Paris banker Charles V. Higgins moved Shropshire to issue Higgins warehouse receipts for 300 barrels to hold as security.[95]

On March 24 Shropshire wrote to Wells with bad news concerning congressional approval of an extension of the bond period:

> Confidentially it now seems likely that we will fail in getting extension in bond [which] makes us a little more decisive to make sales. We still have 2,000 barrels in bond to come out between this and 1st July. We have already shipped or will ship 588 barrels to you. When sold will go a good ways in paying the tax on the remainder. We make this statement fully so that you will see what we want and try to make sales to meet the emergency. We shall want to take out and keep at home some 300 to 400 barrels April whiskey. We have just had a talk with one of the largest N York whiskey dealers whose opinion is that the moment the

whiskey of last season is out of bond and the pressure to sell to meet tax is over that whiskey will be worth [much more] per gallon. You know this confirms the opinion of the writer.[96]

Wells responded that sales had slowed in part because the St. Louis market was depressed and in part because the B. K. Reynolds Distillery at Covington was selling whiskey in the St. Louis market at commission-free prices. Shropshire, though dismayed, provided an insightful response. "We regret to hear Reynolds whiskey is pressing on your market," he wrote. "The fact is *confidentially* this brand is by many considered one of the lowest grades in quality bearing the name of bourbon yet it will have a showing and impose on your market. Your sales at $1.50 after deducting charges & commissions will not net $1.30 cash so you see that if he gets $1.30 and saves commissions he will do as well as we do and we feel very sure his whiskey is not well made." Wishing to avoid public disparagement or calumny of a competitor, Shropshire concluded: "We make this statement confidentially as we know Reynolds well and do not wish to reprove him or his brand."[97]

As James Shropshire worked on the Elkhorn Distillery's books in early April 1870, he could see that the partnership's financial situation was steadily deteriorating, their menu of options narrowing. He wrote to James Stone:

The matter is getting very oppressive and is worrying me *wonderfully* to know how I am to meet the tax and pay something on the notes falling due in the banks which they now begin to think we ought to do. Can't you get Mr. Higgins to help us some in this emergency by endorsing a bill or so. I know you do not like to do this but it would not do for us to make any *slips* just now in way of satisfying our banks that hold our paper. If we do it will give us incalculable trouble & loss. If you find any money amongst your friends at reasonable rates of interest I think you had better try & incur it.[98]

Just two weeks later, Shropshire began to receive queries from wholesalers in Chicago, Philadelphia, and New Orleans, seeking price information and requesting samples. These potential patrons were motivated not by increasing retail demand in their own markets but by the national shortfall in whiskey production, which by this time had become obvious and would increase their costs if they delayed purchases too long. Samuel Tolman and John King operated a wholesale drug trade in Chicago in the 1870s, and King also made and sold patent medicines. In early April Tolman & King sent an inquiry to Stone & Shropshire, requesting information on product availability, especially older whiskeys. Shropshire sent samples by express and offered to sell ninety-nine barrels of December 1868 whiskey at $1.60 per gallon, and fifty to sixty barrels distilled in May 1869 at $1.50 tax paid. He explained that Elkhorn had no two-year-old whiskey left, as all of it had been purchased eighteen months earlier by a New York wholesaler. Few other Kentucky distillers had any two-year-old whiskey on hand; the stock that remained commanded $2 to $2.50 per gallon for "steam copper" whiskey and $2.50 to $3 for traditional "straight copper" bourbon. In an unusual turn of policy, Shropshire also sent Tolman & King a sample of a locally produced two-year-old whiskey called "Eclipse." The distiller (whose name Shropshire did not reveal) claimed it was the best whiskey in the country and set the price at $2.50 per gallon. Shropshire's willingness to provide a sample of and advocate for this product suggests that he was open to selling whiskey for other distillers, another indication that he might become an independent spirits broker should the Elkhorn Distillery fail.[99]

During the third and fourth weeks of April, more letters arrived at Shropshire's Lexington office expressing interest in buying Elkhorn whiskey or negotiating for barrel storage and warehouse receipts. The William Brice Company of Philadelphia offered to advance money on fifty barrels of whiskey, provided that Stone & Shropshire pay interest, insure the lot, and pay commission and storage fees on the money advanced.[100] New Orleans wine and liquor merchant Theodore Bailly Blanchard Jr. operated a retail store on Decatur Street in the French Quarter. He ordered 150 barrels of Elkhorn whiskey and proposed an extended business relationship with Stone & Shropshire. Upon checking Blanchard's references, Shropshire offered to sell him 150 barrels delivered to Paynes Depot at $1.50 per gallon, free of tax, and payable on a four-month note. Shropshire also agreed to allow Blanchard to advertise his firm as the sole agent for Elkhorn whiskey in New Orleans. The agreement specified that Stone & Shropshire would refer all orders from New Orleans to Blanchard and would maintain future prices at the same level as those paid by customers in other markets. Shropshire also requested that Blanchard demonstrate due diligence in introducing and selling Elkhorn whiskey in the New Orleans market. The agreement would remain in effect until January 1, 1871. Shropshire was likely conflicted by this proposition, given that his relatives operated Shropshire & Brother in New Orleans; yet the firm had been unsuccessful in marketing Elkhorn whiskey in that city's market, despite James Shropshire's sales strategy directives. Theodore Blanchard knew of Shropshire & Brother and requested that he be permitted to buy any of its remaining stock of Elkhorn whiskey at a price he was free to negotiate. If Blanchard agreed to these prices and conditions, James Shropshire would be forced to sever his business relationship with his New Orleans kin.[101]

Whiskey sales further improved during the last week of April, when Shropshire received two 500-barrel orders—one from J. G. Wells of St. Louis, and the second from Newcomb, Buchanan & Company in Louisville. Both firms agreed to an advance payment of $20,000. In fact, J. G. Wells may have requested more than 500 barrels, but Shropshire believed it would be counterproductive to introduce a larger amount into the St. Louis market. That these two orders were unsolicited surely reinforced Shropshire's expectation that reduced production in the 1869–1870 distilling season would motivate wholesalers to buy at favorable prices. The infusion of money allowed Shropshire to pay taxes on the barrels removed from bond and reduce some of the distillery's debt with local banks and lenders.[102] But the combined order of 1,000 barrels posed the usual scheduling problems for the federal gauger and the now customary difficulties in obtaining tax stamps from the Lexington revenue office in a timely manner. Elkhorn's warehouse laborers had to top off those barrels that had lost five gallons or more to leakage and evaporation. The wagoners had to deal with the logistics of loading the wagons and negotiating the track to the railroad depot. And incoming grain and coal shipments complicated outgoing whiskey consignments. Shropshire hoped to expedite the entire process because he could not draw payments from the wholesalers until they had received their shipments, and tax and debt payment deadlines were imminent.[103]

The Stone & Shropshire records from late April 1870 to late November 1871 are missing.[104] During the summers of 1870 and 1871, Stone again traveled to Milwaukee, Chicago, and St. Louis and to large urban centers on the Eastern Seaboard, providing whiskey samples to wholesalers and hoping to make some sales. Whiskey wholesaler and consigner J. G. Wells had been a reliable customer. He now held Elkhorn whiskey in a sheet-iron warehouse in St. Louis, and Shropshire sent prospective buyers samples from those barrels. Queries and orders from Dore, Reynolds & O'Neill in Milwaukee; G. H. Eaton & Wells in St. Louis; G. I. Birch & Company in Albany, New York; Charles Beckman in New York City; J. D. Richards & Sons, John McCormick, and John D. & Moses Williams in Boston; and William Brice & Company in Philadelphia arrived at Shropshire's Lexington office. To establish the creditworthiness of prospective clients, Shropshire consulted financial directories and, absent a listing, contacted their banks for references and recommendations. When a Pittsburgh bank refused to give wholesaler J. S. Finch & Company a positive recommendation, Shropshire responded to the company's query by stating the price and demanding cash and "home gauge" before he would approve a sale. Finch accepted the sale but requested and received a 2 percent discount for cash. Though the sale netted the distillery more than $2,800, the sight draft that Shropshire executed on the Finch account was paid to the order of Paris banker C. V. Higgins, one of Stone & Shropshire's primary creditors. No balances remained in the various S & S bank accounts that could be used to pay business operating expenses.[105]

The Elkhorn Distillery did not operate during the 1871–1872 distilling season. By late 1871, the national wholesale whiskey market was glutted with surplus product; the prices obtained for those goods that could be sold were below the profit margin. Congress had failed to act on extending the tax payment period, and internal revenue requirements remained in place. Distillers were still obligated to remove their whiskey from bond and make full tax payments within one year of its manufacture, regardless of whether it could be sold. Those large-scale distillers without alternative sources of income operated in a perpetual state of financial crisis. During the nineteen months from April 1871 to November 1872, James Shropshire—the Elkhorn Distillery's accountant, bookkeeper, and business manager—continued to direct the sale of whiskey produced in 1869 and 1870. Several thousand barrels of bourbon and rye whiskey remained in Elkhorn's bonded warehouse, which was managed by James Stone but controlled by internal revenue collectors, store-keepers, and gaugers.

In November and December 1871, instead of readying the distillery for the next season's production, Stone & Shropshire aggressively pursued spirits wholesalers in the hope of finding local and regional markets where liquor sales were active and brokers were willing to buy Elkhorn whiskey in multiple-barrel lots. Lexington whiskey broker Samuel Clay Jr., a business partner with Cynthiana distiller Thomas Jefferson Megibben, traveled to New Orleans and to Mobile, Alabama, in December 1871, seeking new spirits markets for Elkhorn and other Bluegrass distillers.[106] Shropshire also sent one of his clerks, L. W. Mix, on a sales trip up the Ohio River, where he found buyers in Gallipolis and Pomeroy, Ohio, and in Parkersburg and Wheeling, West Virginia. Some wholesalers sent their own sales representatives afield to pursue markets outside their local areas. Most such sales were small, five to twenty-five barrels. Shropshire set the price of Elkhorn whiskey at $1.50 to $1.60 per gallon cash, but the partnership carried most sales on sixty- or ninety-day credit notes at 10 percent interest. The distillery's credit sales offered the potential of opening new markets but provided no immediate cash that could be used to pay taxes and gauging fees, labor and freight costs, wholesalers' commissions, or insurance premiums.

Shropshire implored prospective buyers to buy 100 barrels or more, but most sales were smaller lots. Nevertheless, the distillery received enough purchase orders each week during the winter months of 1871–1872 to gradually draw down the distillery's stock. Shropshire continued to advise prospective buyers that Kentucky distillers had sharply reduced total production from previous years and that shortages of old whiskey would eventually force prices higher. He began to ask $1.60, then $1.62, and eventually $1.75 per gallon for 1869 whiskey sold in lots of ten barrels or less, paid partly in cash and partly in credit notes of thirty days or less. By February 1872, wholesalers were aware of impending shortages of Kentucky bourbon, and Shropshire started to receive unsolicited queries from dealers as far away as Providence, Rhode Island.

Consignment wholesalers J. G. Wells & Company in St. Louis and Newcomb, Buchanan & Company in Louisville worked closely with Shropshire to sell the Elkhorn whiskey held in their warehouses. By early December 1871, Shropshire was advising prospective buyers that he was confident that the 300 barrels of 1869 whiskey stock held at the Elkhorn Distillery's own warehouse, plus the 200 to 250 barrels at the consignment houses in St. Louis and Louisville, would be gone in sixty days. When writing to wholesale brokers, Shropshire routinely touted his whiskey as high quality, especially that produced in 1869, which by this time was approaching two years in the barrel. The Elkhorn whiskey stored in the Wells warehouse was aging particularly well, he claimed. "We have about 250 barrels of the April & May, '69 whisky that has now been 2 summers under a metal roof in St. Louis," he wrote to a Chicago wholesaler, which is "no doubt much superior to any of our 1869 you have yet seen." In correspondence with another wholesaler, his enthusiasm for the 1869 whiskey approached hyperbole: "We have now 200 barrels spring '69 in St. Louis that has suffered the depleting effects of 2 summers since it was freed from bond under a metal roof and is nearly as good as ordinary 4 year goods."[107] As sales drew down the 1869 stock, Shropshire was no longer willing to accommodate purchasers who wanted to regauge the barrels at their receiving warehouses, hoping to avoid paying for any losses from leakage or evaporation during shipment. He not only increased his asking price but also insisted that buyers accept the gauging values of alcohol proof and barrel content established by the distillery's federal gauger.

Tensions in the Stone & Shropshire partnership began to materialize in the fall of 1870.[108] The division of responsibility for Elkhorn Distillery operations between James Shropshire and James Stone had become compartmentalized. Although much of the early correspondence had been in Stone's hand, he now left bookkeeping and financial management to Shropshire or an office clerk. The brittle relationship between the two men became increasingly strained, and much of their communication was by letter. Stone's longtime residence was likely his Scott County farm near the Elkhorn Distillery, but because he did not respond promptly to written requests, Shropshire began to send Stone's mail to his son Charles, who lived nearby in Midway. James Stone had borrowed money from the partnership for his personal use, and when the loans went unpaid, Shropshire repeatedly admonished his partner to settle his outstanding balance. According to Shropshire, the debt hindered his ability to cover the distillery's expenses and debt obligations, and the business risked being sued by its banks for nonpayment. In early December 1871 Shropshire alerted wholesale customers that he and Stone were planning on "closing our copartnership" and were therefore offering low prices on existing whiskey stock.[109]

Even though the distillery was inactive and their personal relationship was stressed, Stone made sure that the whiskey Shropshire sold was processed and shipped. He hired laborers and wagoners to load and haul whiskey to the depot. He ordered railcars through the depot's freight agent and ensured that the agent weighed each shipment and arranged for routing and bills of lading. Importantly, Stone directed the sorting and gauging of whiskey barrels, a task Shropshire assumed whenever Stone was away on sales trips. Establishing each barrel's alcohol level and volume by weight allowed S & S to sell whiskey based on these "home" measurements; the procedure also gave customers the advantage of paying only for actual contents rather than paying a by-the-barrel price, whatever its contents. Cargo weights also permitted the distillers a degree of control over freight rates. Shropshire could pay either a per-barrel rate or 35 cents to $1 per 100 pounds, depending on the railroad line, number of transshipments, and distance hauled.

Railroad freight agents at Paynes Depot and at a forwarding depot in Cincinnati occasionally sent whiskey to the wrong destination. In November 1871 Stone & Shropshire shipped a ten-barrel order to a Boston buyer that never arrived. Shropshire issued a tele-graph tracer to the railroad company and discovered that the whiskey had been delivered to the wrong depot in Boston, where it sat unclaimed for nearly six weeks. Such mistakes truncated the business's cash flow because buyers would not pay until they received their orders. Shipping to New England by way of the Erie Railroad was particularly unreliable, and Shropshire gave the Cincinnati freight agent strict instructions to avoid shipping on that line.

On Christmas Day 1871 Shropshire sent a letter to Stone, reminding him that the partnership's $4,000 warehouse receipt held by Mechanics Bank of St. Louis was due on January 29, and unless they could make payment or otherwise protect the loan, the bank would likely sell the whiskey. The following day, Shropshire sent to the Northern Bank of Kentucky in Lexington a $1,500 sight draft on the John McCormick brokerage in Boston for collection and credit toward the distillery's debt.

Unable to raise enough money through product sales to meet internal revenue tax deadlines, Shropshire was forced to arrange bank financing. Recall that in December 1869 Stone & Shropshire had obtained three loans from Paris banker Charles V. Higgins with an aggregate worth of $13,000 (roughly $234,000 in present-day value). The churlish Higgins charged 12 percent interest on each loan—or approximately double the national interest rate on secured loans. He later agreed to reduce the rate to 10 percent but subsequently reneged and demanded the full 12 percent for the entire loan period.[110] Eventually, the dis-tillers paid Higgins more than $3,100 in interest on the two largest loans. As collateral for the loans, Higgins required the distillers to move whiskey of equivalent value to a Lexing-ton warehouse operated by Woolfolk & Craig. Stone & Shropshire then issued warehouse receipts to Higgins, giving him ownership of the whiskey. This meant that the distillers could not sell the whiskey without permission from Higgins, whose goal, one suspects, was to extract maximum interest rather than to close the loan punctually. Payment of monthly warehouse storage fees and insurance fell to the distillers.

Stone & Shropshire's perpetual state of financial duress continued into 1872. On Feb-ruary 1 Shropshire sold the last 141 barrels of Elkhorn whiskey stored at the distillery's warehouse to Cincinnati distiller-wholesaler Mills, Johnson & Company for $1.65 cash. The sale yielded $8,303, which was allocated to paying the debt owed to Higgins; at that point, the partnership's unpaid balance stood at $4,400. Subsequent receipts allowed Stone & Shropshire to pay Higgins in full on February 13. Thereafter, Shropshire's accounting and bookkeeping activities consisted of clearing small accounts and collecting charges for sin-gle sample barrels sent to wholesalers.

Timing, Coincidence, and the Aggregation of Risk

The timing of Stone & Shropshire's distillery construction and production start-up unfor-tunately coincided with the opening of several other new post–Civil War industrial-scale distilleries and a rapid increase in production by those distilleries already in operation. As reported by Louisville wholesale firm Newcomb, Buchanan & Company, Kentucky's whiskey production for the year ending in June 1869 reached 9,853,173 gallons, or nearly three times the state's aggregate production in 1865, the year the war ended. With overproduction enters the risk of declining prices and unmet debt obligations. Although overall whiskey production began to decline thereafter, depressed prices forced distillers to curtail production further or halt operations entirely. In 1870 production dropped to 6,791,923 gallons; it fell again in 1871 to 4,452,369 gallons, followed by a 29 percent rebound to 5,750,000 gallons in 1872.

Production variation or volatility also introduced a whipsaw effect into the pricing and marketing system, another form of financial risk. Consider Newcomb, Buchanan & Company's assessment of the Kentucky bourbon and rye whiskey market in 1872:

> The great depression caused by the excessive over-production of the year ending 30th June, 1869, has passed away, and goods of that season's distillation are rapidly tending to a proper level of prices. Notwithstanding the production of the seasons of 1870 and 1871, followed so enormous a production as that of 1869, stocks of 1870 and 1871 goods are now very much broken and command relatively high prices, and as that portion of the 1870 and 1871 goods carried over the summer will constitute our supplies of two and three year old goods next season, they must from their scarcity necessarily rule dear.

Moreover, it noted:

> The relative scarcity and high prices of 1870 and 1871 goods [are] compelling the trade to substitute the younger goods, and as every week of hot weather ripens, and renders the 1872 goods more available for use, and as every day's consumption increases the scarcity of 1870 and 1871 goods, consumption will be forced on to the younger goods in a constantly increasing ratio, and a healthy active trade at a remunerative scale of prices may therefore be confidently anticipated.[111]

James Shropshire had anticipated the increase in whiskey prices—indeed, he was one of the distillers who initiated it—and he had alerted his customers more than six months before Newcomb, Buchanan & Company issued its prediction in July 1872. By late January, Shropshire was asking $1.65 to $1.75 per gallon cash and shipping small orders of ten to twenty-five barrels to wholesalers from the Atlantic coast to the Great Lakes states.

In late January 1872 Shropshire informed a Boston wholesaler that James Stone was making some improvements to his distillery and planned to run the operation independently during the winter months. Perhaps Stone's decision to restart the Elkhorn

Distillery was buoyed by the increasing number of wholesalers who were inquiring about whiskey prices and availability. The Elkhorn Distillery records include no letters from Stone after 1870, but Shropshire's letter books include several letters to Stone detailing transactions, updating accounts, and directing shipments. Shropshire repeatedly requested meetings with Stone to discuss business, sought his approval for pricing, and, increasingly, made pointed reproaches about James Stone's failure to pay his debt to the partnership. Records do not list the amount of Stone's debt, but it likely exceeded $4,500. Stone largely—though not completely—avoided his partner and visited his Lexington office only when Shropshire was out of town on business, much to Shropshire's chagrin. Stone apparently preferred to leave notes and instructions for Shropshire to enter in the partnership's books, thereby avoiding uncomfortable encounters. For his part, Shropshire complained to Stone that the distillery was some twelve miles away, and even a short visit cost him a day's time, which, given the vigorous sales push he was undertaking single-handedly, he could ill afford. Shropshire sent his last letter to Stone on February 22, 1872. "When you were last here," he wrote dolefully, "I proposed to you that in order to reach a settlement of our copartnership business if we could not agree upon basis of settlement ourselves that we submit the matters to arbitration putting the books & all papers etc. pertaining to the business before them each of us to give obligations & security that we would abide the decision of arbitration."[112]

The End

The details of the dissolution of the Stone & Shropshire partnership are lost, but Shropshire apparently took steps to establish his own spirits wholesaling business in late 1871. By May 1872, Shropshire was working with Ball, Hutchings & Company of Galveston, Texas, where he had set up a special deposit account. He had previously traveled to Galveston and was personally acquainted with George Ball and John H. Hutchings, dry-goods merchants and bankers. Although he remained in Kentucky, Shropshire removed himself from the role of salesman to the wholesale trade. His correspondence suggests that he was creating his own "wholesale house" and selling whiskey to retail customers in the South, primarily saloon keepers and druggists, at Montgomery, Alabama, and Grand Gulf, Aberdeen, Vicksburg, and Natchez, Mississippi.

"We are getting very poor" a dolorous James Shropshire had written to a New York wholesaler in May 1869, "but we have great confidence in our whiskey."[113] James Stone had insisted on purchasing the best corn, rye, and malt available, and his distiller made an acceptable if not superior product. Nevertheless, the Elkhorn Distillery experienced a cascade of financial difficulties after only two seasons of operation, prompted by industry-wide overproduction and inflexible internal revenue taxing practices, but exacerbated by mechanical problems, awkward and expensive logistics, poor roads, and bad weather. Stone & Shropshire fed their slop to a swine herd to earn additional income, but in the midst of their second distilling season, in the spring of 1869, they were already falling behind in their payments to grain brokers and fuel and hardware suppliers. Elkhorn whiskey was sold primarily in barrel lots to regional and national wholesalers. Unlike other distilleries, Elkhorn did not operate a retail quart house, where local customers could fill their personal jugs for cash or on credit, thereby yielding a small weekly cash flow. As debt accumulated, Stone & Shropshire were obligated to borrow operating funds from banks and business contacts by issuing warehouse receipts as collateral, but at the cost of exorbitant interest rates.

High purchase prices for grain, coal, and equipment kept operational costs high. But low sales prices and high taxes aside, the distillery's primary vulnerability was the inconvenient location of its nearest railroad depot; although the distillery was fully modern and mechanized, the works did not have ready access to industrial transportation. When railroads extended their lines across Kentucky after the Civil War, many distillers who aspired to industrial-scale production moved their works to track-side sites or locations that were accessible by a short spur. James Stone chose to build the Elkhorn Distillery at a traditional site—on his farm beside South Elkhorn Creek—even though Paynes Depot was two and a half miles away. Haulage costs subtracted profits from all sales. The lag time introduced by repetitive loading and unloading and hauling on poor roads protracted delivery times and often led to conflicts with railroad freight agents, late shipments, and delayed payments to creditors. The freight wagons in use at the time were comparatively small capacity, and in May and June 1869 Stone & Shropshire paid wagon haulage costs of roughly $2.60 per load, or more than $45 in current monetary value.[114] Operating a distillery at the depot with a dedicated siding would have saved most of those charges.

The Stone & Shropshire partnership dissolved after some four years, and the Elkhorn Distillery business terminated in 1872. James Stone declared bankruptcy in 1873. Through the 1870s Stone sold off small pieces of land he owned on the Big Eagle Creek drainage in northern Scott County, some of it for as little as $26 per acre. Land broker J. W. Taylor may have salvaged the distillery in 1877.[115]

Most distilleries operating in Kentucky during the last third of the nineteenth century had to deal with chronic financial problems, even those whose owners had alternative sources of income. Edmund H. Taylor Jr. was a contemporary of James Stone and James Shropshire. He was a banker and a stockholder in E. H. Taylor, Jr., & Company, which operated the highly regarded Old Fire Copper (O.F.C.) and Carlisle Distilleries at Leestown, near Frankfort in Franklin County, only a few miles from the Elkhorn Distillery. Taylor's company was one of the oldest distilling firms in the state and manufactured several brands of whiskey. His successor company, E. H. Taylor, Jr., & Sons, operated the Old Taylor Distillery on Glenns Creek, south of Frankfort, from about 1882 to the early 1900s. Although Taylor's background was in banking and finance and his distilling operation was seemingly successful, his company encountered cash-flow problems and foundered. The company and its principals, Taylor and his two sons, J. Swigert and Kenner, all made "assignments" in July 1893. An alternative to bankruptcy, a financial assignment was a contract that transferred the company's title and property to a third-party trust, which sold the assets and distributed the proceeds to creditors. The Taylors' creditors included Frankfort banks and several whiskey dealers in Chicago, St. Louis, Milwaukee, and other cities.[116] Kentucky's post–Civil War distillers brought different types of knowledge, experience, and financial backing to their businesses, but many experienced a fate similar to that of Stone & Shropshire and E. H. Taylor Jr. Although Kentucky's late-nineteenth-century distilling industry enjoyed a national reputation for producing exemplary whiskey, it was exceedingly difficult to maintain a successful distillery over the long term.

Nineteenth-century Kentucky distilling, as exemplified by the Henry McKenna and Elkhorn Distilleries, juxtaposed success and longevity with fragility and failure. Several factors contributed to successful operations, including a favorable physical environment, especially proximity to potable water; timely adoption of new technology; ready access to reliable, high-capacity transportation; and the availability of a seasonal labor force. Other factors inveighed against success: social restrictions advocated by temperance interests and blunt-force governmental regulations that obliged some distillers to become entangled in intractable financial arrangements. Of these, finance appears to be the cultural keystone to sustainable spirits production.[117] Remove financial access or modify financial stability, and Kentucky's post–Civil War distillers could not operate at an industrial scale, especially after Congress enacted new federal internal revenue regulations in 1868. Without financing, distillers did not have access to large volumes of quality grains, and they could not purchase quality copper stills, distilling equipment, and steam engines. Nor could they build rail spurs or relocate their works to track-side sites. Small-scale distillers who also operated farms and self-financing flour mills could often generate sufficient capital to underwrite low-production distillery operations. They processed their own grains and those purchased from neighboring farms. Their limited whiskey production and comparatively low operating costs allowed them to pay federal taxes according to the prescribed internal revenue schedule. If the whiskey market was weak, they could subsidize distilling operations with profits from their mills and farms or simply stop making whiskey and still realize a sustaining income.

Water, be it from springs, creeks, rivers, or wells, was important, but one water source could often be substituted for another. The primary concern was not ideal water chemistry but access to sufficient quantities for distilling and cooling or as a convenient means of slop disposal. Steam technology enabled increased production, but high-capacity equipment was expensive and often required building expansion and numerous other operational modifications, such as increasing the grain supply and the works' handling, transport, storage, and milling capacities. In some places, distillers and other investors formed turnpike associations to upgrade rural roads by grading and paving them with broken stone. Distillers seeking to increase their production capacity by adopting new industrial equipment and processes found that access to affordable financing was the key to their success. Paradoxically, financial entities, be they banks, insurance firms, or private lenders, were themselves subject to unstable economic conditions.

III

Remaking Bourbon's Contemporary Distilling Landscape

18

Naming and Branding

The external landscape is the one we see—not only the line and color of the land and its shading at different times of the day, but also its plants and animals in season, its weather, its geology, the record of its climate and evolution. . . . These are all elements of the land, and what makes the landscape comprehensible are the relationships between them. One learns a landscape finally not by knowing the name or identity of everything in it, but by perceiving the relationships in it. . . . The difference between the relationships and the elements is the same as that between written history and a catalog of events.
—Barry Lopez

Associating the Past with the Present

Evidence of the transition from craft to industrial distilling resides in its historical ecology—in the necessary relational conjunctions among the environment, cultural heritage, technology, transportation, raw materials, risk, and social, political, and legal actions. The historical process of making bourbon resides in lore and memory, in written records and museums, and it is signposted on the landscape. Family histories preserve stories; documents in various forms often contain objective information that awaits interpretation and context; structures and land-use patterns place signs and marks on the landscape that provide evidence of dimensions, materials, movements, and relationships at scales ranging from a building foundation to a national railroad network.

Distilling technology—some of it new, some traditional—is on display at each contemporary distillery. At the site of Scott County's Elkhorn Distillery, James Stone's brick storage warehouse still stands beside Elkhorn Creek. Henry McKenna's home and the home of his distiller, Patrick Sweeney, stand on the hill above the remnants of the McKenna Distillery in Fairfield. A catastrophic fire at Heaven Hill Distillery in Bardstown on November 7, 1996, destroyed the distillery works, seven storage warehouses, and 90,000 barrels of aging bourbon. The event is commemorated at the distillery's Bourbon Heritage Center Museum. The temperance movement and Prohibition prompted distillers to focus on producing medicinal spirits, a process recorded in an exhibit at the Buffalo Trace Distillery in Frankfort. The logistics of by-product disposal can be viewed at the Maker's Mark Distillery in Loretto, where slop is pumped from steel holding cisterns into tank trucks for transport to nearby cattle feedlots, and at the Barton Distillery in Bardstown, where a large drying unit converts the liquid remnants of distilling into high-protein livestock feed.

Tank truck takes on a load of Maker's Mark slop.

Hundreds of distilleries operated in Kentucky during the nineteenth century. Most have vanished. Yet their heritage is vigorously pursued and preserved, often in the form of reviving historical brand names and sites, by present-day patrons who find it of interest and value to record and maintain the industry's traditions.

Brand Names and Trademarks

Branding one's product begins by naming it.[1] The phrase "making a name" as applied to a commercial product refers to establishing that product's reputation and desirable qualities, and it relates to named or branded goods, not generic commodities. Kentucky whiskey makers have recognized the value of naming their product since the early decades of the nineteenth century. Long before internal revenue regulations mandated the practice, distillers used branding irons to burn their names, or the county's name, into their whiskey barrels before shipping them to downriver customers. Those products so identified that proved popular with consumers gained a market following. The product itself became intertwined and synonymous with its name and source—one representing the other.[2] Distillers recognized that they could establish a relationship between a product's name and its transactional worth; therefore, the names of successful products acquired their own value. When a nineteenth-century Kentucky distillery was sold or consolidated with another company, the purchaser often received the brand name as part of the sale. Henry Shawhan of Georgetown, Kentucky, offered the Shawhan & Atkins Distillery for sale in 1870. "The use of the Brand of Jno. L. Shawhan, deceased for five years will be sold with the Distillery," his advertisement noted. "This Brand has been before the Public for more than Thirty Years, and is highly popular."[3]

Prior to 1870 the federal government did not award trademarks to product brand names and images. Product brands could be defended by the application of common-law trade rights, which applied only to the geographic area in which the name had been in continuous use. Recognizing that brand names were a form of property that could have national or international markets, Congress passed the Federal Trade Mark Act in 1870, codifying long-standing practices.[4] The law was held unconstitutional in 1879; it was substantially revised in 1881 and further amended in subsequent years. Because nineteenth-century distillers such as Mr. Shawhan could not apply for trademarks to formally protect their brands until after 1870, they had to rely on common law in claiming rights to their brands. Through the principle of continuous use—as implied by the phrase "has been before the Public" in Shawhan's ad—distillers could substantiate their proprietorship. This phrase, and the actions it implied, later became a critical part of formal trademark law. Whether distillers were satisfied with the protections inherent in common-law rights or they wanted to avoid the expense of formally applying for a trademark, their response to the 1870 legislation was casual rather than keen. For many distillers, their trademark applications lagged well behind the initiation of production, perhaps because, unlike patents and copyrights, trademarks are based not on a creative effort but on proof of commercial use in the marketplace.

A trademark is a convenient way to identify a product, but government registration does not grant one the exclusive right to its use. One establishes the right to a trademark by using it, and one retains that right by continuing to use it.[5] Abandoned trademarks may be claimed by others. Those distillers with limited production saw little risk in proceeding without an approved trademark. Others produced generic whiskey that they sold by the barrel, unnamed and unbranded, to saloons, to wholesalers, or to rectifiers who blended it with other anonymous whiskeys. J. B. Dant, for example, began making whiskey in Marion County in 1854. One of his bourbons was produced for and distributed by Taylor & Williams of Louisville under the name "Yellowstone." The brand name was first used in commerce in 1878. In 1931 Taylor & Williams took steps to trademark the name, perhaps in anticipation of the repeal of Prohibition.[6]

When distillers discontinued production and did not renew their trademarks, the brand names "died" and were withdrawn from commerce, or they were purchased by other companies. Some names remained dead for years before being revived by new companies. Jacob Beam and his descendants began distilling whiskey on Hardins Creek in western Washington County by the 1790s; their distillery was located about five miles north of the spring that later became the water source for Burks Distillery, now Maker's Mark. By the early 1850s, grandson David M. Beam was marketing the family whiskey under the brand name "Old Tub," and it was sold widely in the Middle West, Great Plains, and South by the 1880s. When Prohibition terminated most spirits production in 1920, many distillers could no longer maintain their brand names in continuous use, and many names were lost or declared "dead" by the US Patent and Trademark Office. This is apparently what happened to Old Tub, because in 1935 the James B. Beam Company discovered that someone else had claimed the Old Tub brand.[7] The brand was declared abandoned by the Patent and Trademark Office in 1996 and was reclaimed by Jim Beam Brands Company in 2013.[8] Today, Old Tub is produced by the Jim Beam Distillery in Claremont, Kentucky, but it is sold primarily at the visitors' center.

Label for Old Tub Brand Kentucky Straight Bourbon Whiskey. This label was likely copyrighted in 1936. It depicts an African American laborer stirring a barrel of mash and declares that the Old Tub trademark dates to 1882.

The Knob Creek brand was first used in 1898. After several decades of inactivity, it was used commercially in 1992 and registered by Jim Beam Brands Company in 1998.[9] In Louisville, George Garvin Brown developed Old Forrester—the spelling was later changed to Old Forester—as a medicinal whiskey. The name was first used in commerce in 1874, and the Brown-Forman Company was apparently issued a trademark in 1905.[10] The Old Forester mark was periodically renewed, and by 2012, it had been registered to Early Times Distillers Company as the many syllabled "Geo. G. Brown Est'd Old Forester 1870 First Bottled Bourbon, Old Forester Est'd 1870 Kentucky Straight Bourbon Whisky."[11]

Both distillers and consumers benefited from brand names, but in decidedly different ways. For the distiller, the brand transformed his whiskey from a generic product into a spirit with specific qualities that became associated with the producer, the place, and, ideally, suitable honorifics. For the consumer, the branded label identified the producer, who, at least in terms of reputation, could be held accountable for the product's quality. Importantly, brands and their trademarks implied product consistency and came to represent the spirit's character and worth.[12]

James S. McKenna, president of H. McKenna Incorporated, applied for two trademarks in 1905 on behalf of the company. (Recall that McKenna was a son of Henry McKenna, founder of the McKenna mill and distillery in Fairfield, Kentucky, in 1855.) The first application requested a trademark for the brand name "McKenna" as it appeared on labeled bottles and barrels sold by the distillery; the second requested registration of the image of an Irish harp and winged Maiden of Eire that appeared in profile with the name "McKenna" superimposed. The application forms affirmed that the McKenna name had been in exclusive use on the corporation's distillery products since 1868, and the harp and winged-maiden image had been in use by 1881.[13]

When the Twenty-First Amendment repealed Prohibition in 1933, many distillers, including James McKenna, sought to reestablish their trademarks. McKenna's affidavit outlined the distillery's history as it related to the reapplication. It stated that, in the 1850s, "the whiskey put out by that distillery was generally known as 'Old McKenna'; that the United States Government assumed supervision of the distillery business about 1867 or 1868 whereupon the Collector of Internal Revenue notified affiant's father [Henry McKenna] that he would be required to designate his whiskey by some specified name; that at this time affiant's father adopted as a name for his product 'McKenna Old Line Pure Sour Mash.' The whiskey was sold under this name in various states and became widely known as 'Old McKenna.'" In about 1881 Henry McKenna adopted the practice of marketing bottled whiskey that carried a white label printed with the harp and winged-maiden image and the name "H. McKenna." In 1893 the McKennas changed to a gold label but retained the traditional name and image. During Prohibition, McKenna whiskey stocks were removed to the Arthur P. Stitzel Concentrated Bonded Warehouse in Louisville, where the whiskey was bottled and sold to the drug trade. W. L. Weller & Sons then bought the remaining whiskey and bottled it as "Old McKenna." Weller also obtained permission to use the label on "any good, well-made, pure, Kentucky sour mash whiskey for the purpose of keeping this trade name before the public."[14] Through these and other business arrangements, the McKennas were able to maintain their brand in the public marketplace, regardless of how that might have been construed during Prohibition; this tactic placed the brand in a favorable position for reapplication as a registered trademark when Prohibition ended. Twenty-seven months after James McKenna submitted the request for trademark re-registration, he received notice from the US Patent Office that the application had been approved and would remain in force for twenty years.[15] The Henry McKenna trademark passed to Joseph H. Seagram & Sons in 1963. Heaven Hill Distilleries Incorporated of Bardstown, Kentucky, acquired the H. McKenna brand and trademark from Seagram & Sons in 1989. Heaven Hill applied to renew the mark in 1994, 2004, and 2015.[16] Trademarks often included only the brand name—in this case, H. McKenna—and did not claim rights to any generic industry-related words such as "distillery," "whisky," "whiskey," or "bourbon," which were also used by other distillers. Some approved trademarks included icons and design elements such as Henry McKenna's harp and winged maiden, as well as a distinctive bottle shape and embossing.

Brand Infringement

Those distillers who adopted brand names, which frequently included their own names, often aggressively defended their brands against infringement by others. During the post–Civil War period of rapidly expanding industrial distilling, two general types of businesses vied for a share of the consumer market: distillers and rectifiers. Distillers engaged in an industrial process—purchasing and milling grain, distilling, aging the final product (often referred to as "fine whiskey"), paying taxes, and, finally, placing the product with wholesalers and retailers or selling directly to customers. Rectifiers either bought whiskey in bulk (whether freshly distilled or aged) or distilled their own product with the intent of blending it, often unaged, or altering it with additives to enhance taste and palatability. Rectifying was a comparatively low-cost process, and the product could be sold at prices low enough to undercut distillers' sales. A tactic sometimes employed by both distillers and rectifiers was to copy or simulate the brand of a successful distiller in an effort to enhance sales. This sometimes led to intimidation and lawsuits.

The Pepper Brand

Elijah Culpepper of Culpeper County, Virginia, settled in Woodford County, Kentucky, in about 1780, and soon thereafter he established a distillery in Versailles, the county seat. Culpepper changed his name to Pepper at some point, and in 1812 he built a small distillery on Glenns Creek in Woodford County; after 1825, the distillery was operated by his son, Oscar Pepper.[17] In 1833 Oscar Pepper hired Scottish physician and chemist James C. Crow as distiller; by applying scientific methods to the distilling process, Crow produced a highly regarded bourbon whiskey.[18] Oscar Pepper died in 1865, after which his son, James E. Pepper, directed distillery operations. The works was later sold to Leopold Labrot and James Graham, who continued to produce bourbon whiskey under the brand name "Old Oscar Pepper Distillery, Woodford County, Ky."[19]

In 1879 James Pepper purchased a distillery on the Middle Fork of Elkhorn Creek (Town Branch) in northwestern Lexington, near the Louisville, Cincinnati & Lexington Railroad tracks (later the Louisville & Nashville). In 1894 James Pepper lived at Meadowthorpe, a large stock farm in what is now northern Lexington, about half a mile from the distillery. Pepper's steam-powered distillery had twenty-one mash tubs and was processing more than 500 bushels of grain per day. The works included three corncribs, a cooper shop, a bottling plant, and five aging warehouses with a capacity of more than 40,000 barrels.[20]

"Old Pepper Whisky" was one of James Pepper's brand names, and it was the subject of a lawsuit he brought against Labrot and Graham in 1891. They contended that the brand name, including the term "Old," rightfully belonged to the original "Old Oscar Pepper Distillery," which they had purchased and now operated.[21] Pepper's bottle label included a red heraldry shield that proclaimed, "Pepper Distillery Hand Made Sour Mash Established in 1780." The back label stated: "This Whisky is distilled under the same formula and process used by the Grandfather and Father of our Mr. Jas. E. Pepper, more than one hundred years ago." James Pepper was apparently willing to forgo any acknowledgment of James Crow's contribution to refining Oscar Pepper's distilling methods. And it seems a bald exaggeration to state that a high-capacity industrial distillery was operating according to the same "formula and process" developed by ancestors a century earlier. In 1892, in apparent reaction to encroachment on his brand by saloon keepers, rectifiers, and others, Pepper placed a four-page color advertisement, cloaked in a broadside, in George Buchanan's periodical *Fine Whisky Facts*. The first page featured an engraved bird's-eye view of the distillery, warehouses, and railroad above the title "Jas. E. Pepper & Co. Proprietors Lexington, Ky." Under the heading "Old Pepper Whisky," the second page read:

> To the Public
>
> It having come to our notice that inferior Whiskies are being sold, put up in imitation of the package we have introduced and sold for years past as the Distillery Bottling of our celebrated old Pepper Whisky, we deem it our duty to warn Dealers and the Public against these base imitations.
>
> The object of the imitators is obvious: "To force upon the public a spurious article under the name and guise of a Whisky, which, having been made by three generations of the Pepper family on the same formula, for more than a hundred years, and introduced to the world at a great outlay of labor and expense by the present firm of Jas. E. Pepper & Co., has become a staple brand, and is in demand everywhere by an appreciative public."[22]

Pepper's reference to "imitators" may have been directed at Labrot and Graham, whose claim to the brand "Old Oscar Pepper" had been upheld by the US Circuit Court in 1891.[23]

The tract's third page described how Pepper had "patented our Trade Mark" and warned, "besides the suits already instituted against certain imitators and infringers, we shall begin others, if necessary, and press them vigorously to the fullest extent of the law."

The trademark encroachment extended to saloon keepers and other dealers who had emptied Old Pepper Whisky bottles and refilled them with a cheap substitute—perhaps a rectified whiskey. The tract also implied that other distillers were making and selling simulated Old Pepper Whisky without the correct labels, white flint-glass bottles, or case design. The fourth page included a color reproduction of a filled and labeled bottle, with the labels shown in detail on a side panel. While readers might have interpreted Pepper's tract as caution to prospective encroachers, one could also interpret it as an advertisement, for it strongly implied that Old Pepper Whisky was of such high quality that both the product and its identifying labels, glassware, and packaging were subject to extensive counterfeiting.

THE TAYLOR BRAND

Frankfort distiller E. H. Taylor Jr. first used a registered trademark for his "Old Taylor" whiskey in 1867. When the partnership of E. H. Taylor, Jr., & Sons was formed in 1887, the company adopted "Old Taylor" as both the trademark and the name of the distillery on Glenns Creek south of Frankfort.[24] In 1899 the label design for the Old Taylor narrow-neck bottle resembled a Greek cross with a foreshortened crossbar that wrapped around the left and right sides of the bottle. The vertical bar contained a white field with the words "Old Taylor" at the top and E. H. Taylor Jr.'s signature written sideways in red ink in practiced Spencerian script. There was a cameo engraving of Mr. Taylor on the left crossbar and an engraving of his new castle-form distillery on the right crossbar. Although the distillery was still under construction at the time, the label's design evoked the traditional Victorian-era concern for personality, as chronicled by the portrait and the ornamental script, and progress, as represented by the distillery's smokestack-punctuated industrial façade.[25] In 1910 Taylor & Sons announced that thereafter the label's predominant color would be richly chromatic—yellow, not white—to readily distinguish it from all other whiskey bottles. Yet the company declined to modernize the rest of the design. "The other features of the label," the company stated, "such as the portrait of COL. E. H. TAYLOR JR., the picture of the OLD TAYLOR Distillery and the facsimile signature of this corporation, as written by Col. Taylor, and its form and shape remain and will remain unchanged." The label was so registered with the US Patent Office.[26]

Mr. Taylor often sued those who marketed whiskey under his name or copied other elements of his trademark.[27] In 1905 Taylor & Sons won a lawsuit against Marion Taylor of the Wright & Taylor Company in Louisville. Wright & Taylor sold its blended whiskey under the brand "Old Taylor" and represented the product as straight unblended whiskey. The company also represented itself as a distiller, even though it did not own or, apparently, operate a distillery. Taylor & Sons requested damages and lost profits from trademark infringement, for which the Court of Appeals awarded an estimated $300,000.[28] Similarly, W. A. Gaines & Company, owner of the Old Crow Distillery near Frankfort, sued W. L.

Weller & Sons of Louisville and A. H. Heilman & Company of St. Louis for trademark infringement when they sold their blended whiskeys under the name "Old Crow." The Gaines suit against Heilman was successful, resulting in an assessment for damages and lost profits. Nevertheless, the company was plagued by brand infringements until the eve of Prohibition.[29]

KENTUCKY WHISKEY

Bourbon brand names have legal provenance as registered trademarks, but they can also be repositories of company identity. Consumers may associate product qualities with a certain brand and its representative labeling and packaging. Historical brands, given their cache of origin narratives and the impression of continuity they convey, are of considerable perceptual value to contemporary distillers. Brands were often fashioned around the distiller's name, but they also included references to the heritage or tradition embedded in the manufacturing process and, significantly, the place of production. The physical qualities of place—water, limestone, seasonal temperatures—were often deemed as important to product quality as the process by which it was made. Kentucky bourbon whiskey was somewhat akin to Cuban tobacco, French burgundy, and Douro Valley port, in that the physical character of the place of production was thought to impart desirable qualities to the product.[30]

Those distillers whose brands designated Kentucky as the state of manufacture were potentially vulnerable to encroachment by distillers in other states who wanted to brand their products as "Kentucky" whiskey. Kentucky distillers' brand claims were bolstered in April 1898 when the US Circuit Court of Appeals ruled that all classes of merchandise, including spirits, could not be branded as a "Kentucky" product unless it was, in fact, made in Kentucky. Nor could distillers brand a product as a Pennsylvania or Maryland rye whiskey if it was not produced in that state. A geographic place name could not be used as a generic term to enhance a brand.[31]

Venerating the "Old"

Late-nineteenth-century distillers often embraced the latest technological improvements in stills and other equipment; at the same time, they marketed their products as "handmade" and "old-fashioned," implying that the manufacturing process was historically authentic and the product was, necessarily, "genuine." As an adjective, "old" could imply traditional recipes and methods—whiskey distilled in small batches, for example. "Old" could also refer to the length of time a whiskey had aged in a charred white oak barrel—the longer the aging, the better the product. "Old" could imply maturity and quality, as in the phrase "Oldest is best." Or "old" might reference a legacy, perhaps an Old World family tradition or the knowledge accrued by an experienced distiller. Used this way, "old" could also be a term of endearment, veneration, and respected heritage, as in an "old friend."

Bourbon's association with place—a county, for example—may also relate to the tradition practiced at that place: an adherence to proven processes, a preference for certain grains, or the use of a certain type of expensive copper apparatus. While each distiller approached spirits manufacturing somewhat differently, the extended family business associations and the density of neighborhood distilleries provided a venue for the exchange of ideas and encouraged competition (friendly or otherwise) among distillers to produce a favored product. When quality bourbon could be attributed to a particular distiller or technique, market demand and prices often rewarded those using traditional, or "old," methods and equipment. In the 1850s the market price for copper pot–distilled whiskey was 75 to 80 cents per gallon, compared to 65 to 75 cents for steam-distilled whiskey.[32] Accordingly, Edmund H. Taylor Jr. named his Frankfort distillery Old Fire Copper (O.F.C.) in 1869.[33] And in the 1870s T. J. Monarch produced a sour-mash whiskey in his distillery at Grissom's Landing, west of Owensboro in Daviess County, that he marketed under the assonant-sounding brand "Old Fashioned Fire Copper—T. J. Monarch Distiller 2nd Dist. Ky."[34] Both distillers underscored their use of traditional equipment and techniques, hoping that potential customers would attribute positive meaning to the brands as a way of maximizing the monetary value of their products.[35]

More than fifty nineteenth- and twentieth-century whiskey brands used the term "Old" or "Olde" in their names, including Old Crow, Old Fitzgerald, Old Forester, Old Grand-Dad, Old Pogue, Old Tub, and Very Old Barton. Other brand names implied age and therefore dignity and venerability, such as Ancient Age, W. L. Weller Antique, Early Times, and Bulleit Frontier Whiskey. Some brand names recalled pioneer distillers such as Elijah Craig and Evan Williams. Several surnames evoked a valued heritage: Beam, Brown, Clark, Clay, Dant, Dedman, Mattingly, Medley, Samuels, and Shawhan. And certain carefully chosen first names gave a brand a sense of tradition: Eli, Elijah, Ezra, Emanuel, Isaac, and Silas. Distillers' company names might include a family reference to imply tradition and heritage: Best & Sons, Hawkins & Son, Howell & Son, Walker & Son, and Merritt Brothers. Relatedly, brand names employed the heritage of place by referring to notable historic sites such as Knob Creek, Rowan Creek, Noah's Mill, Buffalo Trace, and Buffalo Springs. And place names referencing a "pure" water source were featured in brand names such as Cave Springs, Clear Spring, and Silver Spring. Urban distillers often used brand names that ignored their municipal-industrial settings: Old Boone, Old Charter, Old Log Cabin, Old Times, Old 1889, and Sunny Brook. In Louisville (Jefferson County), two distilleries were ironically named Anderson and Nelson, for two largely rural counties that ranked among the state's largest producers of fine whiskey.[36]

Brands and the Perception of Quality

Cleverly and continuously promoted brands could accrue credibility, thereby allowing distillers to demand higher prices if the market and economy allowed. In the late nineteenth century, for example, John M. Atherton and his business associates owned five distilleries at Athertonville in Larue County, Kentucky. J. M. Atherton & Company, A. Mayfield & Company, William Miller & Company, S. O'Brian & Company, and J. T. Carter had been in business for some time before being acquired by the Atherton Company. By leasing their property back from Atherton, the five distillers were able to continue whiskey production and marketing under their own names and therefore retain their respective brands.[37] In 1888 Mr. Atherton was invited to testify before the House Committee on Manufacturers, which was reviewing internal revenue law and practice. Regarding product identity and the market value of whiskey, Atherton stated that brands were purely trademarks. But he also acknowledged that should retail customers become accustomed to a certain trademark or brand, they would pay more for it than they would for another brand made in precisely the same way in an adjacent distillery, even though the two products had the same intrinsic value.[38] In his opinion, two brands of whiskey made in Kentucky by the same process and that looked and tasted alike should be of similar value. Yet one could be sold for $2 a gallon while the other garnered only $1.40. The reason for the variance, he said, was the perceived difference between the brands. "There are makers that can not tell the difference intrinsically," he affirmed, "nor could I tell the difference."[39] Atherton's experience with taste-alike whiskey was not unique. At the approach of the twentieth century, consumer goods of all types produced by American industry, whether bread, toothpaste, or hand soap, were becoming increasingly alike and more difficult for consumers to differentiate.[40]

Atherton also advocated retaining the federal tax on whiskey, in part because the government identification requirements protected fine Kentucky whiskeys from encroachment by rectifiers and distillers of cheap whiskeys. "The marks put upon the barrels by the Government and kept there by the order of the Government, which no man in the United States can erase or destroy, are evidence all over the United States of the character of the whisky which the man buys. The manufacturers of fine whiskies," he noted, "without this safeguard would be utterly at the mercy of the cheaper grades of goods that are made by compounding spirits with a small quantity of good whisky."[41] "We could put our distillery brands on the goods in Kentucky," he continued, "but could not compel a rectifier from erasing them." By internal revenue edict, a whiskey's date of production was burned or branded into the barrelhead as absolute proof of the whiskey's age. The date provided proof or validation of the brand; therefore, both the date and the brand had commercial value. Atherton concluded, emphatically, that repeal of the internal revenue laws that required such identification would destroy the manufacture of fine whiskey in the United States.[42]

The Transition to Modern Branding and Advertising

Early- to mid-nineteenth-century whiskey marketing was based largely on a distiller's reputation and small advertisements placed in local and regional newspapers. By the 1880s, distillers were aggressively advertising in dedicated industry magazines such as the *Wine and Spirit Bulletin,* medical magazines, and other print outlets. Kentucky's nineteenth-century distillers adopted modern branding, advertising, and marketing strategies idiosyncratically but approximately in parallel with the industrialization of distilling. Perhaps following the innovations of nineteenth-century industrialists and advertising pioneers such as E. I. DuPont, the distilling industry extended its reach into popular magazines, newspapers, and retail store signs, but distillers emphasized their "heritage" rather than industrial innovation.[43]

Until the 1870s, saloons sold whiskey from barrels, retailers sold packaged whiskey in handblown glass flasks, and distillers sold their stock in generic ceramic jugs or wooden casks of various sizes. Distillers who sold whiskey at the distillery door often conducted business on credit, offering local customers individual charge accounts, rather than insisting on cash-only sales. Customers and distillers knew one another personally, and they were financially codependent as lenders and borrowers. As glass bottle manufacturing evolved from manual glassblowing to machine production, whiskey sales moved from branded bulk cooperage to anonymous glass flasks and bottles, and product identity became imperative. Distillers responded by requesting brand-embossed bottles or affixing professionally designed and printed labels to their bottled goods.

A distiller's brand could also be enhanced by social actions. Henry McKenna of Nelson County sold whiskey at his distillery, but he also entertained orders from distant customers. The customers' order letters frequently sought to establish rapport and trust by including personal comments about their preferences, local community opinions on temperance, and the names of friends who had recommended McKenna's product. Mail-order customers often visited the distillery when traveling in Kentucky for business or pleasure, thereby reinforcing their personal relationship with the distiller. Those distillers who shipped branded goods via railroad and steamboat often accompanied the cargo themselves or sent a representative who coordinated the sale with an agent, thereby establishing or maintaining personal associations between seller and customer.

Brands were initially "signatures" that permitted distillers to claim ownership of whiskey in storage or in transit. Over time, they became objects in themselves, taking the form of stamps, labels, embossing, or other marks in response to legal claims, business and marketing strategies, and an expanding consumer base.[44] Distillers also recognized that brands were an effective way to separate their products from the cheap blended spirits produced by rectifiers.[45] Distilled spirits purveyors lagged behind patent medicine manufacturers in developing modern marketing techniques. Unfortunately, the public often equated the questionable qualities and brazen claims of patent medicines with the properties of distilled products.[46] Despite operating in the shadow of patent medicine's advertising sensationalism, for two decades or more, distillers sought to convince the temperance-mined public that their whiskeys were pure and had therapeutic value. They sought endorsements from physicians and pharmacists attesting to their products' salutary effects. Some brands used explicit text on their bottle labels to reinforce these medicinal claims. The Sunny Brook brand (later changed to Old Sunny Brook) was first produced in 1891 by the Rosenfield Brothers & Company Distillery in Louisville. The spirit's 1910 label declared that the bottle contained "Pure Food Whiskey," perhaps a reaction to the 1906 Pure Food and Drug Act, and a small red shield centered at the top of the label included the words "Pure Food." The label identified the name of the distillery and added the words "5th District of Kentucky, Louisville, Jefferson County, Kentucky," to reinforce the product's locational provenance.[47] Temperance activists largely ignored distillers' medicinal claims and were increasingly successful in fueling public ambivalence, if not hostility, toward distilled spirits. Some pro-temperance advertising companies such as N. W. Ayer & Son began to reject liquor accounts in the late 1890s, even though Ayer and other advertising firms maintained direct financial interests in patent medicine producers.[48]

Advertising as Strategy

Many late-nineteenth-century whiskey ads were the product of the distiller's imagination, an engraver's chisel, and a printer's press. They appeared in newspapers and magazines as simple single- or multicolumn information blocks juxtaposed with all manner of other advertisements. During the twentieth century's first decades, many large industrial distillers hired professional advertising firms such as Young & Rubicam to design and manage advertising campaigns. Professional advertisers developed marketing strategies that targeted middle- and upper-middle-income consumers with color pictorial ads featuring trademarked brands, labels, and heraldry accompanied by text emphasizing the product's quality, purity, taste, and age.[49]

Traditionally, most people bought food and drink based solely on taste. If the taste of two different whiskeys was similar or indistinguishable, then why, as John Atherton observed, would an individual select one over the other? Advertising psychologists were beginning to realize that the appreciation of food and drink, the actual value a person received from a product, could be increased by appealing to consumer sentiment and imagination. Sentiment could be invoked if a consumer placed importance on the rarity of an item or the reputation associated with one product but not another. Professional advertisers referred to this as emotionalized facts. Articles could be packaged or presented differently—in an aesthetically pleasing and sensuous manner, or unadorned. Customers who enshrined the aesthetically presented product in sentiment were often willing to pay more for it.[50] Whiskey advertised as a prosaic, generic alcoholic spirit was not likely to be as attractive, or to evoke as much sentiment, as the same product associated with aspirational social standing or reverence for a valued past.

Marketers also began to recognize that the container a product was presented in— the bottle, case, or package—was an integral part of the commodity. In the mid to late nineteenth century, merchants sold food and spirits in bulk. General stores sold sugar, flour, cheese, crackers, potatoes, pickles, and a broad assortment of other foodstuffs out of wooden barrels and crates, just as saloons and spirits retailers sold whiskey out of oak barrels to customers who furnished their own jugs. Packaging spirits in glass flasks and bottles provided the opportunity to enhance the product by using colored glass and affixing the containers with artful labels. Early advertising, then, began to mold consumers' ideas about a product's abstract qualities, its appearance, and its associations rather than focusing solely on the quality of taste.[51] An advertisement portraying a whiskey drinker as discerning, sophisticated, gregarious, and genteel was apt to attract a middle-income patron who aspired to such qualities. And the discriminating consumer was also one who knew and appreciated how a quality product was made—and all the better if it was made according to the traditions developed in a dimly remembered but romanticized past.

With the repeal of Prohibition in 1933, distillers sought to reclaim the public spirits market by reopening abandoned distilleries; they revived old brands and created new ones, and they retained professional advertising firms to promote their products. During the preceding half century, from the 1880s to repeal, the manner of professional advertising for a range of consumer products had changed dramatically—from a focus on product quality to an appeal to the interests of prospective lower- and middle-income customers. The new advertising strategies adopted a new form of visual grammar that no longer reproduced likenesses of finely dressed company officers or imposing manufacturing works to represent the firm to the public; nor did officers sign product labels. These "sops to vanity" were thought to hinder the effectiveness of advertising; they consumed valuable space while gaining little sales value in return. The mode of production was seldom mentioned in text or illustrated by images of smoke-belching factories, as had been common in the 1880s and 1890s, whether the product was farm machinery, hand soap, or pianos. Even a product's place of origin was anonymous. Many companies turned to conceptual "human interest" ads that featured iconic characters such as Betty Crocker and the Michelin Man, who communicated with readers about product qualities. The new ads became more effective with the invention of halftone printing in the 1890s, a technology capable of showing infinite shades of gray or color tones that brought realism to print advertisements.[52]

Promoting Tradition

Many Kentucky whiskey makers seemingly ignored national trends in advertising and continued to produce traditional ads that included an engraved likeness of the company's founder or reference to the distiller's name and heritage. Bottle labels included the distiller's name and sometimes his signature; labels, show cards, and newspaper and magazine ads often featured bird's-eye views of industrial distilleries or engravings of traditional log-cabin still houses or oxen pulling wagons loaded with oak whiskey barrels. And labels frequently specified the location of production not only by state but also by county or county seat. Many distillers continued to believe that the place and circumstances of whiskey production and consumption were necessarily related—that images of the production site and references to production heritage were important to customers. Yet the mode of production, which distillers considered critical to making a superior whiskey, was poorly understood by the public, and its nuances were unappreciated. In an effort to educate the consuming public about bourbon production, E. H. Taylor Jr. commissioned film director J. Law Siple of Lexington, a principal in the Gulf Coast Motion Picture Company of Alabama, to make a movie about whiskey making in central Kentucky and featuring interior and exterior scenes of the Old Taylor Distillery on Glenns Creek in Woodford County. The film was seen by audiences in New York, Boston, and Chicago and was presented at the Phoenix Hotel in Lexington in September 1914.[53]

Kentucky's leading distillers continued to believe that portraying the place of bourbon production, and its implied heritage, was not just an acceptable advertising strategy; it was mandatory. As late as the 1960s, the label of the Old Sunny Brook brand included a bird's-eye view of the Louisville distillery, identified as "The Old Sunny Brook Distillery in 1900," and the text: "Sunny Brook is distilled, as in years past, to the highest quality standards in one of Kentucky's historic distilleries. Today, as always, Sunny Brook ranks as one of Kentucky's great brands and one of its best selling whiskies." The other side of the label declared that Sunny Brook had won a grand prize and a gold medal at the Louisiana Purchase Exposition in St. Louis in 1904. By reminding buyers of a six-decades-old prize, the Sunny Brook Distillery was suggesting that its whiskey was still made the same way and was therefore just as good as the 1904 product.[54]

A more subtle form of symbolic context appeared in a 1935 Four Roses bourbon advertisement. The color ad featured the mythically resonant image of a formally dressed young woman sitting beneath a rose arbor and apparently listening to the entreaties of a uniformed young man. The accompanying text described the setting as sometime before the "War between the States," when a "young gentleman named Paul Jones laid siege to the heart and hand of a lovely Southern belle." She responded by wearing a corsage of four red roses. The text then segued to a statement about the whiskey and encouraged readers to consider making it a part of "your own gracious scheme of living." The whiskey, it claimed, was as "fragrant as Southern flowers, mellow as the moonlight shining on them." While this stereotypical scene might have been replicated in select places in the South, what followed was an appeal to the sensibilities of a narrowly rendered heritage, an awareness that might have been lost during the production hiatus enforced by Prohibition. "Four Roses is made by the same company that has always made it—and made in the same way—by the slow, costly, old-fashioned method."[55] Four Roses also promoted its brand name with bold outdoor advertising on a large lighted sign above Times Square in New York City, a vertical space shared with Chevrolet, Planters Peanuts, and Coca Cola. This strategy was intended to return the brand to public consciousness only two years after the repeal of Prohibition.[56]

In the 1930s and 1940s the Seagram Company retained the New York advertising agency Young & Rubicam to develop a campaign to sell its blended Four Roses whiskey during the warm summer months. The ad, which represented a departure from traditional heritage advertising, featured a photograph of four red roses frozen in a block of ice—a visual image suggesting a cool, refreshing drink.[57] The Four Roses brand and label made no obvious reference to Kentucky's distilling tradition. The roses-in-ice ad employed symbolic context as a way to stimulate consumers' desire to buy the product not only because it had intrinsic worth but also because it possessed some form of ethereal value; heritage had given way to modernity.[58]

The mid-twentieth-century advertising strategy that National Distillers pursued on behalf of its Old Sunny Brook brand seems traditional compared with Seagram's attempt to promote Four Roses. Initially, the ratio of liquor sales to advertising costs appeared to favor the more conservative National Distillers' approach. Between 1951 and 1953, for example, the advertising cost per case of whiskey sold was about 76 cents for Old Sunny Brook, with average annual case sales of 1,247; for Four Roses, the advertising cost was $1.55 per case, and annual average case sales totaled 1,635. Both brands ranked in the top twenty in liquor sales; Old Sunny Brook ranked sixth, and Four Roses ranked ninth. A decade later, in 1960 to 1962, Four Roses' case sales had fallen to 1,192, and it ranked sixteenth among the top twenty brands. Heritage-conscious Old Sunny Brook, unfortunately, had fallen out of the top twenty.[59]

Distillers expected that effective advertising and timely marketing techniques would spur retail sales. But sales did not necessarily increase in proportion to the money spent on promotional programs because "people differentiate among competing suppliers less in terms of the use-value of the commodities than according to their ability to assuage some desire created through effective marketing and advertising."[60] Creating what advertisers term a "contagion of desire" for a consumer product, therefore, was a notional process. Advertisers crafted images that communicated by way of symbols, and potential consumers were invited to engage symbols that represented claims on, or a fondness for, social cachet, fashion, power, pleasure, status, or refined taste.[61] Advertising for distilled spirits was most effective if those symbols expressed an existential heritage.

Marketing through Narrative

An important consideration in contemporary whiskey marketing is to give the product a narrative or legendry that gains the attention of potential consumers and links product qualities to experiences, conditions, or situations they find interesting and to which they can assign value and favorable meaning. If a primary motive for buying bourbon is its taste, then a marketing narrative should explain the process by which a particular taste is achieved and link that process to heritage—historic events, characters, and processes— and place—the picturesque limestone lands of central Kentucky.[62] The most effective narratives are those that portray the product as legitimate and authentic. Those brands presented as authentic, as adhering to their traditional production values and maintaining their historical relationship to the place of production, can claim a unique brand identity.[63] Advertisers and producers found that narratives that portrayed a product as authentic increased sales; consequently, that strategy became a cornerstone of contemporary marketing practice.[64] For the distiller, the most effective branding and labeling strategies are those that successfully associate the product with a trustworthy and time-honored narrative. For the consumer, the creditable narrative reinforces an appreciation of the product.[65]

Consumer Base Differentiation

Distillers may follow traditional manufacturing practices that seem to change little over time, and their labels and advertisements encourage that assumption. Nevertheless, these same distillers are also aware that the consumer market changes, and they experiment and adjust their production techniques so as to continue to produce compelling products. The late-nineteenth-century spirits customer bought whiskey for its use value and may have obtained it from a city saloon, a drugstore, or a bulk whiskey purveyor. The contemporary American consumer base for distilled spirits is diverse and comprises a gradient of interest ranging from the ambivalent, the uninitiated, and novices to accomplished critics and connoisseurs. Uninitiated and novice consumers may be attracted by ads that address tradition and heritage. Connoisseurs may ignore advertising altogether and concentrate instead on the recommendations of colleagues or the studied evaluations published in hard copy or communicated through informal online blogs and web pages.[66]

Marketing narratives trend toward continuity and claims of heritage, but spirits marketing strategies have changed markedly. Nineteenth-century distillers and wholesalers practiced a comparatively straightforward distribution of goods to a generic mass of potential customers. Twentieth-century marketing was a purposeful, knowledge-based effort to influence customers' buying habits through advertising and promotional techniques targeting specific groups.[67] Consequently, promotional advertising has progressively changed its emphasis from basic use value to symbolic value—that is, assigning value to a consumer good through the creation of desire. Some sectors of the spirits industry were more successful at promoting symbolic value than others. The 1950s and 1960s saw tepid or even declining sales of American whiskeys, despite increased advertising expenditures; the four major distilling firms—Seagram, Schenley, National, and Walker—were spending 5 percent or more of their total annual sales revenue on advertising.[68] Yet the consumption of vodka, gin, and Scotch and Canadian whiskeys, the ingredients of increasingly popular mixed drinks, grew, indicating an apparent shift in consumer preferences fostered by advertising.[69]

At the same time, whiskey ads often portrayed well-dressed men consuming spirits in comfortable, even luxurious settings that could be interpreted as "exclusive" and "the good life." Rarely, if ever, did whiskey ads suggest that the prospective customer might be a farmer, a trucker, or a tradesman—images commonly employed in beer advertising. If women appeared in whiskey ads, they were most often portrayed as passive bystanders. Moreover, brand-oriented, class-conscious advertising stimulated the conspicuous consumption of whiskey, especially that purchased to be served to guests; in that situation, brand image and symbolic value weighed most heavily on consumer preferences. If Kentucky whiskey makers wanted to effectively compete with other spirits producers and sell their products to the working-class customers of post–World War II America, advertising firms would have to identify symbolic values with which those potential customers could associate.[70]

In an affluent society, advertising is often crafted to both inform potential customers of a product's qualities and expand and refine their sensibilities. Promoting a potentially profitable product requires a subject with superior intrinsic qualities as well as a compelling brand name, packaging, and price.[71] Contemporary marketing techniques often involve packaging premium bourbons in special containers with unique labeling. By burnishing a brand's reputation, effective advertising can usually increase product demand.[72] The producer responds by increasing production volume, which in turn should increase the company's economies of scale and scope. "Economies of scale" refers to the total increase in production costs, which rise less proportionately than output—as one sells more product, per-unit costs should decline. "Economies of scope" refers to the costs of producing a number of products jointly, which should be less than the total cost of each product if they were produced separately. Since the 1980s, advertising campaigns have cultivated the association of certain brands with high quality, and consumers have been increasingly attracted to premium spirits produced by considered distilling techniques and extended aging periods, but turned out in limited quantities. Some distillers produce two or more whiskeys from the same mash bill. One might be aged four to six years and modestly priced; another might be aged twelve to twenty years and produced in small quantities, commanding a very high price. Owning such a rare "designer brand" can confer a measure of prestige and social status on the owner, or it can generate disdain for having paid an unreasonable price for a product intended for consumption.[73]

Innovations

Brand names, product differentiation, and innovations in packaging all play an important role in a company's financial success. Large multinational spirits manufacturing firms produce several types of whiskeys under a number of different brand names. Heaven Hill, for example, markets seventeen whiskeys, including bourbons with the venerable brand names Elijah Craig, Evan Williams, Henry McKenna, and Old Fitzgerald. The Sazarac-owned Buffalo Trace makes several bourbons, including Buffalo Trace, Elmer T. Lee,

George T. Stagg, Pappy Van Winkle, and W. L. Weller. Distillers may also revive historic bourbons, as Campari-owned Wild Turkey has done with its Whiskey Barron's Collection, which features revivals of the Bond & Lillard and Old Ripy brand names.

Some distillers experiment with blending strategies, such as bottling from a single barrel or from a small number of barrels, a production method termed small batch. Experimentation suggests that various aging techniques can produce distinct differences in taste; that is, products with the same mash bill, produced at about the same time, will have different qualities depending on whether the aging warehouse was built of brick, stone, or wood or sided in tinned sheet iron. And distillers have known for more than a century that seasonal temperature variations can be enhanced by positioning barrels in the warehouse: high (hot in summer), in the middle (moderate year-round), or low (cool). This can produce spirits with dramatically different aroma and taste qualities.

While some contemporary distillers realize financial success through traditional branding and marketing programs, others are innovative and bring new products to market promoted by clever, if not professionally rendered, advertising. Taylor William (Bill) Samuels, a sixth-generation descendant of Scots-Irish whiskey makers, bought the 1889 Burks Spring Distillery near Loretto, Kentucky, in the early 1950s, with the intent of producing a new type of bourbon. He was seeking to make a smoother, fuller-flavored whiskey than his family and other distillers were producing. Heeding the advice of Pappy Van Winkle at the Stitzel-Weller Distillery in Shively, Samuels abandoned the old family whiskey recipe that called for rye as the secondary or "flavor" grain and substituted soft winter wheat.[74] Grain milling and the small-batch distilling process remained traditional. Workers rotated barrels in the aging warehouses to equalize the long-term effects of interior top-to-bottom temperature differentials. The whiskey's production, branding, and promotion were a family undertaking. Samuels's wife suggested the product name, Maker's Mark, and the bottle and label design. She also suggested dipping the bottle neck in hot red wax, a practice often used by the makers of fine distilled cognacs. Although Samuels was reluctant to engage a professional advertising firm to promote Maker's Mark Whisky (note the spelling), word-of-mouth endorsements, positive reviews by trade magazines, and tasting competition awards brought widespread recognition and increasing sales. Samuels's son, Bill Jr., joined the business and, despite a quiescent American whiskey market in the 1980s and 1990s, enhanced sales with a clever homespun advertising campaign. Though disparaged by the professional advertising community, the brand's offbeat and self-deprecating advertisements won awards and accolades. The Maker's Mark ads appearing in print media and on billboards are often credited with creating a national awareness of and demand for premium bourbon whiskeys, and Bill Samuels Jr. is regarded by many in the business as an inspired promoter.[75] In the process, Maker's Mark became one of the industry's iconic brands.[76]

Marketing and the Distillers' Landscape

Kentucky's historic distilleries have long been identified with their locales and regions. Those nineteenth-century industrial distilleries that remain have adopted modern production methods, but they still possess historical integrity, despite the intercession of Prohibition and the vicissitudes of market demand for their products. Distillery structures are largely genuine or authentic; little about these places is contrived. Their initial purpose, to produce distilled spirits, remains their reason for being. When considered corporately and juxtaposed with their complementary industries—farming, cooperage, construction, transportation—and the physical environment, the state's distilleries form a distinctive landscape; its unique identity is not replicated elsewhere in the nation. Though trained architects designed some distillery structures, especially in the post-Prohibition expansion of the 1930s, many remaining structures were built by local craftspeople and construction crews following vernacular traditions. And, importantly, distillers embrace this landscape as their heritage and as meritorious of preservation. Contemporary distillers, as we have seen, view this venerable landscape as a reservoir of images that not only document their heritage but also market their products.[77]

Bourbon branding, labeling, and advertising often employ images—drawings or written descriptions—of historical distilling practices. New buildings may be replicas of old structures. Distillers have built few new large-scale industrial distilleries, preferring to rehabilitate old works. A new $115 million distillery began operating in Shelby County in 2017 to manufacture Bulleit Bourbon Frontier Whiskey. Although the works is modern in every way, including solar power and a new fifty-two-foot copper column still, the doubler was reclaimed from the old Stitzel-Weller Distillery in Louisville, and the product is based on an old family recipe. The overall architectural design is modern, but it features gray metal-sided buildings with complex rooflines that echo nineteenth-century distilleries; projecting stone-faced side gables recall historical aging warehouses. The Bulleit brand is named for a long-established distilling family in Louisville and is owned by the multinational spirits manufacturing company Diageo. Since the nineteenth century, distillery ownership has changed with high frequency. Initially, works passed from one local owner-distiller to another, but in the twentieth and twenty-first centuries, ownership extended to regional, national, and international companies. Even though each change in the scale of ownership offered the potential to remove production to other locales or to divorce production from its traditions, new owners have chosen to maintain and upgrade existing works or, in the case of Bulleit, build new works as complements to the iconic landscape of the industry's birth.

Kentucky's distilling landscape documents nearly two centuries of change, accumulating additions and forfeiting erasures. It is a landscape that juxtaposes the old and the new; it is a palimpsest that embeds, and perhaps reflects or suggests, ongoing evolutionary processes that include the accrual of knowledge through innovation and invention. The distillers' landscape, historical and contemporary, is a fount of heritage that links individual and family identity and local and regional business development.

Recognizing that the consumer base for Kentucky's distilled products is diverse and, importantly, that there exists a plentiful population of potential customers who are ambivalent about high-proof alcohol products, how does the industry address that diversity and expand its market? Can heritage and landscape contribute to an effective marketing strategy? Technology-driven industries increase sales and consumption by continual innovation and reinvention, which permits them to promote their products as "new and improved"—a strategy pursued by manufacturers of a broad range of consumer goods. Rapidly evolving production technologies are the driving force in introducing and selling new products. Whiskey distilling is largely market driven, and sales are expanded by increasing the size and purchasing power of the consumer base. Although distillers are innovative, sales are not generally predicated on a continual flow of new products.[78] Rather, a long-term industry priority has been to stress heritage and maintain a traditional image, a strategy that has proved important to its financial successes.[79] But Kentucky's distilling industry is not one-dimensional. Contemporary distillers maintain a market presence by continuing their long-term emphasis on heritage and tradition—the old—yet at the same time they have broadened their product lines and promotional strategies—the new. Distillers have done this by marketing both traditional and new products and by capitalizing on their products' tactile ambience—the place of origin and its landscape. Effective methods include employing the undeniable appeal of the Bluegrass countryside to encourage themed tourism—that is, opening distilleries to tourists, operating whiskey heritage museums, and pursuing a broad range of tourist- and consumer-centered activities.

Theme Landscapes and Marketing Methods

Business and industry traditionally market their products by direct sales and advertising. The geographic association of product and place allows some types of businesses and industries to engage in destination marketing as well. Amusement parks, for example, necessarily employ a form of destination marketing; potential consumers must travel to the park to experience the attractions. Private businesses that position themselves at the entrance to a "natural attraction," such as a national park, are also engaged in an indirect and parasitic form of destination marketing. The destination is, in this case, the national park, not the attendant businesses and synthetic attractions, yet entrepreneurs have found that they can sell souvenir T-shirts and convince some park-bound passersby to buy tickets for such ersatz attractions as snake farms, petting zoos, and wax museums.

The purpose of all commercial places is to sell goods and services. Some businesses are successful in disguising this exchange of money for goods as another type of relationship between the commercial place and the consumer, portraying it as an "experience" or "adventure." An integrated and holistically organized design theme that creates a venue that is attractive to consumers can be an important advantage to an otherwise traditional business model.[80] By introducing some form of entertainment—not random combinations of artificially constructed "attractions," but themed leisure activities and amusements—businesses can sell their goods or services and at the same time associate a pleasant experience with the business name and the place of production. Theme parks employ this tactic, as do some restaurants, hotels, and residential real estate developments, by linking the desirable qualities of the place to the fulfillment of the customer's desire for the product. Themed commercial places may be of sufficient size to constitute a destination rather than a stopover on the way to somewhere else.[81]

Business interests and political boosters that are interested in increasing income and tax revenue through tourism may promote combinations of several themed attractions to compete with businesses in other jurisdictions. Attractions suitable for promotion may include successful college or professional sports teams; named shopping meccas such as Minnesota's Mall of America and Kansas City's Country Club Plaza; "experience" entertainment zones such as those in Reno and Las Vegas, Nevada; renovated historic districts such as the French Quarter in New Orleans and the Art Deco district of Miami Beach; and specialized rural landscapes such as the Amish farm country of Lancaster County, Pennsylvania, the Amana Colony in Iowa County, Iowa, and the Thoroughbred and Standardbred horse farms near Lexington, Kentucky.[82]

Some businesses can readily convert their places of production into themed spaces with direct appeal to potential consumers. Others may erect auxiliary sites that include a museum dedicated to business founders and the historical development of a product, such as Hormel's Spam Museum in Austin, Minnesota. To promote their products, many American vintners have created themed landscapes based on their vineyards and their attendant wineries. Grape production and winemaking in California date to the Spanish colonial era, and European immigrants widely developed the industry in the nineteenth century. Promotions remained focused on traditional product advertising until Napa Valley winemaker Robert Mondavi and others began promoting California's quality wines in the 1970s and 1980s. Mondavi's innovative approach was to advance sales by educating potential customers in fine wine appreciation through seminars, product tasting and culinary experiences, and musical events at the winery. Other Napa Valley wineries followed suit by opening or enlarging tasting rooms, erecting informational signage along primary highways, and encouraging tourism. The inspiration spread to nearby Sonoma Valley and other wine-producing areas. Some winemakers built architect-designed showplace wineries that became attractions in themselves. Given the density of vineyards and wineries, the greater part of Napa Valley, including its towns and country grocery stores, became a themed landscape, albeit unintentionally and voluntarily accretive. Because a themed landscape can also function as a sales landscape, access and information are basic operational requirements. Similar educational and promotional efforts in the Finger Lakes district of western New York and in northwestern Oregon's Willamette Valley have created other vine-and-wine-themed landscapes.[83]

Creating a themed distillery landscape in Kentucky involves the reconfiguration of an industrial site in a way that makes the process of producing a consumer good—from processing raw materials to bottling and shipping the final product—accessible and understandable. In the 1980s and 1990s Kentucky whiskey distillers began to augment traditional marketing techniques with educational programs that emphasized heritage as a primary attraction and the traditional links between the physical environment and production locales—springs and creeks, fertile limestone soils, grain production, and milling and distilling. A themed landscape does not require high site density (notwithstanding Napa Valley). While many Kentucky distilleries are nodes of activity separated by open farmland, they do not represent an archipelago of sites separated by large expanses of unremarkable countryside. The visitor can, with some attention and supplemental information, identify the subtle supportive elements that made distilling possible in the past and continue to serve the industry today. Open farmland is not anonymous; it may constitute the local grain supply or the consumer of the distillery's spent grain by-products. Springs and creeks supplied distilling water; river ports, old turnpikes, and railroads—now perhaps recalled only on a commemorative plaque—provided transport outlets. The themed distilling landscape, therefore, is not an artificial construct such as a casino built in the shape of a city skyline or an Egyptian pyramid or a theme park composed of multiple non-site-related stations such as a five-eighths-scale Main Street or a miniature European castle. The sole purpose of such places is to entertain visitors with a calculated recreational experience based on scraps of myth, perfunctorily rendered legend, or generative imagination. Such places do not possess heritage; nor are they authentic, other than within their own genres.

Distilleries as Attractions

Tourism makes significant contributions to Kentucky's economy. In the mid-1990s visitors spent more than $3.6 billion in Kentucky annually. The tourism industry accounted for nearly $7 billion of total economic activity and employed nearly 150,000 people.[84] By 2016–2017, tourism contributed more than $15 billion to the state's economy; tourists contributed $9.5 billion of that amount in direct expenditures, and tourism-related businesses employed more than 195,500 people.[85] The state's official *Vacation Guide* identifies more than 500 tourist sites, and two-thirds of those places have historical significance.[86] A few are based on the state's coal mining history, such as the Coal-Railroad Museum in Jenkins and the Kentucky Coal Mining Museum in Benham. The state has a broad-based industrial past, but how does a factory, even one that has been elevated to the status of a state-recognized attraction, lure tourists? Kentucky's distilling businesses have achieved that recognition by building on their Irish, Scots, and Scots-Irish past to create an industry of unique character and heritage, one whose landscape offers a legacy of landmarks that the traveling public finds compelling.

Americans have long responded to the appeal of tourist travel, but they have also sought designated attractions. By the early nineteenth century, the East Coast moneyed elite traveled overland into Virginia's interior to visit Natural Bridge and frequented sites such as Kaaterskill Falls, near the Hudson River Valley in New York, in sufficient numbers to make these places fashionable attractions that developers then equipped with accommodations and improved transport. The Erie Canal provided access to Niagara Falls at the New York–Ontario border by the 1830s, and railroads brought tourists to the trans-Mississippi West after 1865.[87] Writers and social critics such as William Cullen Bryant and Anthony Trollope often wrote admiringly about large-scale American industrial landscapes, noting with approval their size, organization, and engineering.[88] But common work sites in general, and distilleries in particular, were usually overlooked by tourists as uninteresting if not degrading; they were often seen as vernacular or blighted, and they did not conform to conventional ideas of what constituted a proper place to visit. Such places were best viewed from afar, if at all.[89] Yet industrial heritage and archaeology had been traditional interests in European tourism for more than a century, and refurbished visitor sites in Great Britain included distilleries, mills, mines, potteries, and related canal and railroad transport systems, the same corridors along which tourists tended to travel.[90]

Yet, representing industrial heritage to the public is problematic. Machinery may be immobile and impractical to transfer to a centralized museum. Yet industrial artifacts placed in a museum are out of context and hence depreciated by abstraction, which narrows the interest base to those with a "trained eye" or mechanical infatuation, such as engineers, industrial archaeologists, and historians. Machines in context and at work are attractions that can educate the observer about time, place, and process; machines out of context and at rest are curiosities that beg for background, and their interest and educational value fade to negligible. Moreover, if the presentation—whether it occurs in a reconditioned factory or a sanitized museum—gives short shrift to the human dimension of industrial work, the exhibit becomes one-dimensional and lacks social perspective and meaning. Industry-related tourism, therefore, commonly follows one of two forms: heritage tourism or industrial tourism. Heritage tourism focuses on dated structures, antiquated machinery, anachronistic processes, and decontextualized ways of life. Industrial tourism engages contemporary industrial processes. Manufacturers interested in linking tourism to sales may provide plant tours in which the modern factory is the main attraction; the Mercedes Benz auto assembly plant at Vance, Alabama, and the BMW Zentrum Museum and assembly plant at Spartanburg, South Carolina, are examples. These two types of tourism may overlap if the modern factory continues to use traditional equipment and processes.[91] Several Kentucky distilleries, for instance, blend the old and the new in a manner that allows them to retain their heritage and authenticity. Moreover, the distillers' search for the authentic is ongoing and manifests in different ways. Where possible, some distillers prefer to use locally sourced, non–genetically modified grains. And one distillery has acquired nearby farmland to raise heirloom corn in an attempt to replicate the grains available for distilling in the nineteenth century.

Historical industrial sites that are narrowly conceived as marketing devices to promote tourism may not present the full context of factory work. They may simply illustrate a mechanical process while ignoring or obscuring workers' social and spatial relationships to management, the equipment, and the place. It is for the visitor to decide whether a restored historical site harbors authenticity or seems staged.[92] But experienced tourists who seek historical industrial sites may readily recognize that a restored and functioning factory bears some degree of authenticity, whereas a museum dedicated to static displays of the same process may lack authenticity and appear artificial and contrived. Functioning historical sites may be designed to educate visitors through signage and other forms of passive engagement. Casual guests may be satisfied with that experience, but purposeful visitors may prefer active participatory learning and entertainment.[93] Select historical industrial sites demonstrate how places can be reclaimed and operated if they are effectively designed and managed and adequately financed. The Saugus Iron Works National Historic Site in Saugus, Massachusetts, and the Paterson Great Falls National Historical Park on the Passaic River in Paterson, New Jersey, are examples of well-preserved colonial industrial sites operated by the National Park Service, although neither is a fully operational industrial works.

Even if industrial history and geography are portrayed at a working industrial site, at a reconstructed site that serves as a museum, or at a stand-alone museum, what the visitor sees is a portrayal—a purposeful representation that has winnowed history, environment, and tradition for images and narratives that can alter human perception of culture and nature. The working site and the museum are, in effect, institutionalizing heritage, albeit in different ways.[94] Functioning distilleries may retain their historical sites and selected elements of their environmental contexts, but with few exceptions, a comprehensive portrayal of historical processes and events is very difficult, given the cumulative loss of information through time and the multiple layers of vested interpretation to which each site is subject. Nevertheless, distilleries can provide the interested visitor with insights into a range of cultural, social, economic, and political events from the past, keeping in mind that the industry has chosen to portray a carefully considered heritage.

The contemporary bourbon whiskey production system is largely focused on long-standing sites, and product marketing increasingly capitalizes on images that evoke heritage and nostalgia for an imagined past. Though the number of distilleries in production is only a small fraction of those operating in the 1880s, several old distilleries have been refitted, and abandoned works are in various stages of rehabilitation. Others have newly constructed visitors' centers that offer tours and onsite tastings, thereby joining the list of Kentucky's tourist attractions. More than $1.1 billion in distillery-related capital projects were planned for the decade 2011 to 2021.[95] These expenditures are earmarked to update distillery production facilities and promote public access, often in the context of preserving heritage and increasing tourism. The moneys include tax rebates approved by the Kentucky Tourism Development Finance Authority.[96]

Accessing the Distillers' Landscape

Distillery tourism has been a pastime in Kentucky for more than a century. At the turn of the twentieth century, E. H. Taylor Jr. erected a new distillery on Glenns Creek in northern Woodford County. Around 1901, masons began building the Old Taylor Distillery by erecting the still house atop the site of an older distillery. They completed the distilling building in 1907, and the site's sunken gardens, peristyle springhouse, and gazebo were finished at about the same time. The castle-form Tyrone limestone still house was the center of a showplace intended, in part, to promote Old Taylor bourbon through visitor entertainment. In 1908 the Kentucky Highland Railroad began passenger service between Frankfort and the Old Taylor Distillery. During the week, trains carried distillery employees, including "a big force of young ladies to attend to the work of bottling and packing," bound for the Glenns Creek valley; the first train departed Frankfort at 6:00 a.m., and the last train returned from the valley at 5:00 p.m.[97] A distillery railroad depot opened sometime between 1910 and 1912, and Taylor and his family kept an "open house" at the distillery and frequently entertained weekend visitors arriving by rail. When the Elks' State Association held its annual meeting in Frankfort in August 1910, the local hosts arranged for the group to travel to the Old Taylor Distillery aboard a special train for a luncheon and an afternoon of leisure on the distillery grounds.[98] As part of the state's gubernatorial inauguration in December 1915, Taylor extended a general invitation to the public to visit the distillery and "roam over his magnificent Woodford County farm and see his champion herd of Hereford cattle and other prize-winning stock." Several other distilleries in the Frankfort area opened their doors to tourists for the occasion.[99]

To encourage train ridership, the Louisville & Nashville Railroad publicized in the early 1900s what might be considered Kentucky's first "bourbon trail." Advertisements urged people to ride the line from Louisville to Bardstown and beyond to the eastern mountains.[100] Near Bardstown, the train passed the John Ritchie Distillery site. The railroad's Passenger Department identified Mr. Ritchie as "the first distiller of that [sour mash] whisky which has made Kentucky famous." Near Ritchie's farm, passengers were told, they could see "the ruins of this primitive plant, the original sourmash house of the Bluegrass State, or, for that matter, of America."[101] The L & N's route juxtaposed a version of the bourbon origin narrative, and the legendary "first" cedar log distillery, with that era's contemporary industry, as represented by several modern, high-capacity distilleries operating beside the tracks. Those whiskey works were attended by livestock pens and warehouses holding 50,000 to 100,000 barrels "ranging in age from one year to eight."[102] By citing the Ritchie story and identifying his distillery's site, the L & N's Passenger Department played a directorial role in defining a person and a place as authentic and worthy of commemoration in bourbon distilling heritage.[103] L & N passenger trains no longer travel this original bourbon trail, but the concept remains: the R. J. Corman Dinner Train, operating restored engines and passenger cars, conducts periodic Bourbon Excursion train rides that traverse the Nelson and Bullitt County distilling landscape from Bardstown to Chapeze.

To establish a place in distilling history, contemporary distillers could simply assert an authentic connection between bourbon production and place of origin. But to meld image, product, and place in the public's mind, one need only invite consumers to visit a still-functioning 125-year-old distillery for a guided tour of the grounds and the structure's interior spaces. Distillers sell whiskey, but they are also selling place and landscape because the contemporary product's character and the heritage of the site are inextricably linked. A distillery visit can convey an emotional immediacy, an attachment to the product *and* the place, but at the risk of trivializing the more than two centuries of human effort required to reach the present. Distillers compensate by opening museums, displaying vintage photographs and dioramas, and creating other visual and audio presentations to help re-create for visitors an indelible holistic experience and expand their appreciation of the industry's long history.

Touring a distillery works can convey a powerful sense of the past, which becomes strongly associated with the product. The Albert B. Blanton home, "Rock Hill," was completed in 1934 and overlooks the former George T. Stagg Distillery on the Kentucky River in Frankfort—now Buffalo Trace. Mr. Blanton was employed at the distillery from 1897 until 1952, including serving as its president; one of the distillery's brands carries his name. At the Jim Beam Distillery in Clermont, Kentucky, a reconditioned boardinghouse became the home of T. Jeremiah Beam, who directed the construction of a new, post-Prohibition distillery works. A bronze statue of famed master distiller Booker Noe stands in front of the Beam home. Both the Blanton home and the Beam home are on the National Register of Historic Places.

These homes suggest that industry icons were also real people whose contributions to their respective businesses were not only foundational but also exemplary. Whether biographical information about business founders or master distillers takes the form of museum memorabilia or artist-rendered statuary, the knowledge provided tends to ground that element of the distillery's heritage and solidify a link to place and product by visually connecting time, workplace, and personnel. In this way, cross-country tours and distillery visits provide experiences that each visitor can use to understand and interpret the historical development of the distilling tradition and thereby assign personally relevant meaning to the product. The understanding gained by a visitor touring a functioning distillery can fundamentally change his or her consumption preferences. Shopping for a distilled spirit in a retail store after a tour is no longer an ambiguous occasion; rather, it may be suffused with meaning and images of where the spirit was made, and memories to which a visitor can anchor future purchases.[104]

Commercial establishments have long used symbols and motifs to advertise their goods or services: a red and white striped barber's pole, a pawnbroker's three-ball sign, a pharmacist's show globe, a mortar and pestle sign announcing a druggist's shop, or an outsized pair of pince-nez reading glasses hanging over an optometrist's door. Contemporary distilling businesses carry this idea a step further by erecting entire buildings to represent iconic spaces.[105] Examples include Heaven Hill's Bourbon Heritage Center in Bardstown, Jim Beam's $20 million American Stillhouse visitors' center at Clermont, and Woodford Reserve's visitors' center in Woodford County, which resembles a large nineteenth-century home but features a tasting room, lounge, restaurant, and chef in residence.

Jim Beam Distillery's American Stillhouse visitors' center.

Some distilleries cultivate tourism by remodeling existing structures as themed spaces, such as the Buffalo Trace visitors' center, which occupies an old brick aging warehouse—likely Warehouse No. 113B of the old E. H. Taylor Jr. and Carlisle Distilleries. It stands amidst other venerable distillery buildings. The need here is not to construct a new distillery but to present existing structures as historically significant and therefore worthy of assuming the role of motif or theme.[106]

The Bourbon Trail

What better marketing strategy could contemporary distillers employ on behalf of Kentucky's distilled spirits industry than to provide admittance to not just one historic distillery but all of them? Onsite distillery tours have proved very effective in promoting product interest and sales since the early twentieth century. With visitor entree, the distilling landscape is no longer an inaccessible abstraction seen from a distance, if seen at all. Access allows personal immersion beyond the surficial patina, and that experience can alter customer perceptions and preferences.[107] A distillery tour's primary attractant is the experience of interacting with an "authentic site" of production, rather than with some ambiguous generic structure that could house almost any type of commercial activity, whatever its architectural provenance or aesthetic appeal. It follows that if the distilling industry linked several historic distilleries together, the marketing effect could be multiplied. The Kentucky Distillers' Association (KDA) accomplished this by formalizing site visits for tourists through the Kentucky Bourbon Trail. The KDA inaugurated and trademarked the Bourbon Trail in 1999 on behalf of six member distilleries to "give visitors a firsthand look at the art and science of crafting Bourbon, and to educate them about the rich history and proud tradition of our signature spirit."[108] The Bourbon Trail is a tour of KDA member distilleries that is complemented by travel instructions, suggested itineraries, route maps, and websites. Participating Bourbon Trail distilleries offer guided tours, tastings, and souvenir shopping. Visitors are encouraged to obtain a "passport" online or at a participating distillery. After paying an admission fee and completing the distillery tour, the visitor's passport is stamped, certifying the experience. Visitors who collect stamps from each participating distillery receive a commemorative T-shirt. A foldout Kentucky Bourbon Trail map shows the location of each distillery, and a mileage chart lists the distances between stops. The accompanying text reminds visitors of the industry's heritage story: "For more than 200 years, Kentucky's legendary distilleries have crafted the world's finest bourbons, using secret recipes and a time-honored process passed down from generation to generation." Tour instructions recommend that visitors set aside three or four days to complete the tour, to avoid missing any attractions and, of course, to patronize area restaurants, hotels, and gas stations. The Bourbon Trail has proved very popular with tourists. Between 2011 and 2016 the participating distilleries entertained 2.5 million visitors.[109]

The Kentucky legislature significantly enabled distillery tourism by passing laws to permit distilleries to conduct tastings and sell whiskey onsite, thereby enhancing visits. Kentucky Revised Statute 243.0305, allowing souvenir package sales by licensed distillers and the sampling and sale of alcoholic beverages on distillery premises, was passed into law in 1996, with subsequent revisions through 2016. Class A distillers with an annual production of 50,000 barrels or more are required to purchase a distilling license, which costs $3,090 and must be renewed each year. Distilleries wishing to offer visitors sample tastings must pay an annual license fee of $110. Distillers holding a sampling license may not charge for the sample and cannot provide tastings in excess of one and three-fourths ounces per visitor per day.[110]

Any licensed Kentucky distiller located in a wet jurisdiction that permits retail liquor sales may operate a retail sales outlet on the premises and offer souvenirs for sale. Visitors may also purchase packaged distilled products, provided they are of legal drinking age. Sales quantities are limited but generous; each visitor may purchase up to four and a half liters of distilled spirits per day. Experienced guides conduct distillery tours and may also lead tastings. Visitors discover that bourbon tasting is closely akin to tastings of fine wines, ports, cognacs, and liqueurs. One experiences fine bourbon in four dimensions or senses: sight or appearance (color and clarity), aroma (honey and spice), taste or palate (caramel, fruit, sweetness, or oak), and experiential, or what might be termed "tasting the place."

Tour guides encourage visitors to understand the various elements of the distilling process in the context of the operating machinery. A distillery visit is larded with sensory impressions. Initially, one experiences the visible and tactile textures of venerable historical structures set into limestone topography. Moving through a working distillery illustrates, in a general way, how the machinery works and how spirits are made; more specifically, the visitor can feel the texture of the grain, smell the saccharine aromatics of yeast fermenting the mash in outsized cypress or steel vats, and hear the blended humming sounds of the machinery at work. Copper or steel stills are served by a puzzlement of complicated steel and brass plumbing that conducts milled grains and gravid fluids to their proper destinations. One can witness raw whiskey flowing into a glass and brass tail box and wade through the quiet gloom and wood-perfumed air of a warehouse where thousands of full white oak barrels recline.[111] Finally, the entire experience reaches a denouement as people and machines bottle, label, and box the finished product. Observing the work dramatizes the process for the visitor and moves the encounter well along an experiential scale, which can range from judging the place and the procedure as contrived to accepting it all as authentic. Upon completing a distillery tour, the visitor is no longer a spirits novice. And with a directed tasting of the company's products, the novice becomes an appreciative critic.[112]

For the visitor, the experience may extend well beyond the sensations of aroma and flavor; it can gather all these place-associated sensory experiences, the historical ecology of the place, and fold them into meaning and memory. Upon completing a distillery tour and tasting, the visitor may find that the brand is no longer an abstraction, mere words on a bottle label or in an advertisement. For the visitor, the brand becomes attached to the place and is at the center of a place-based experience. The experience, for some guests, can be metaphorical, akin to a visit to a secular shrine such as a sports stadium, where something seemingly miraculous occurs and powerful emotions are experienced, where attentive curators have outfitted the site with appropriate relics, and where evocative images of venerated personages are displayed. Such shrines are revered places and are often attended by a pilgrimage of adoring supplicants who accept the product as part of their identity through self-association with a brand and its place of production.[113]

Motoring the Bourbon Trail is enabled by highway signs directing visitors to the various attractions. Distilleries are considered cultural or historical attractions, making them eligible for special highway signage. The Kentucky Transportation Cabinet administers sign-related regulations. If a distillery's annual visitor attendance exceeds 5,000 people and it complies with other distance and traffic-flow requirements, the business may request that the cabinet install a large brown "limited supplemental guide sign" at an interstate highway or parkway interchange, announcing the distillery's location.[114]

Hallmark Events

In addition to the Bourbon Trail, communities and businesses schedule periodic hallmark events each year in support of Kentucky's distilling industry. Although most such events are oriented toward fostering domestic tourism and are intended to fulfill economic and social objectives, they may also promote the development or maintenance of community or regional identity, a form of civic branding that can relate to product branding.[115] Each September, for example, during National Bourbon Heritage Month, Bardstown—the self-proclaimed "Bourbon Capital of the World"—hosts a six-day Kentucky Bourbon Festival. Guests participate in bourbon tastings, black-tie galas, a balloon glow, bourbon cooking classes, history tours and lectures, outdoor arts and crafts markets, musical events, a barrel-making demonstration, and a bourbon barrel relay. Some events are free, and others require reservations and tickets. Louisville hosts the Kentucky Bourbon Affair each June, a weeklong series of distillery-related events featuring barrel tastings, luncheons, and lectures by master distillers. Most events require reservations and tickets. Hallmark events such as these can be very effective in fostering place identity and connecting places to products, to the mutual enhancement of both.[116]

A commercial product's public identity resides in its name or brand. A name need not be simply an identifier, comparable to a brand on a steer's hide. A name can speak to a product's qualities and imply its worth. Kentucky distillers have long recognized that their brands are ineluctably linked to authenticity and heritage. Beyond identifying a certain whiskey, the product's name may imply a family lineage or ethnic heritage and transform a generic product into something unique. Achieving market success with a nonbranded product in the nineteenth century required extraordinary business acumen. Brand names allowed fine whiskey distillers to differentiate their products from the spirits produced by rectifiers. In the late nineteenth century, Ohio Valley rectifiers' first concern was to produce cheap whiskey they could sell in volume. The names of their manipulated products were intended to disguise their generic origins and signal their low cost. Should a deceitful rectifier misappropriate a name from a fine whiskey distiller in an attempt to pilfer heritage, sales might increase, but at the risk of a lawsuit.

Naming is an adjunct to making. Producing a commercial product must be followed by "making a name" for that product through promotion and marketing. The name, in turn, becomes a fulcrum for claiming quality and authenticity. Generic products sell on the basis of their use value and are priced according to costs, supply, and demand. Although a branded product similarly has a use value, it may also acquire symbolic value, for example, if it is claimed to possess a valued heritage. And because heritage is linked to place, a brand name also links the product to its parent landscape.

19

A Reconstructed Past Lives in the Present

> Kentucky's distilling industry is built on heritage; it pays homage to the past. Even the new distilleries have the look and feel of history.
> —Kevin Didio

History, Heritage, and the Contemporary Bourbon Industry

In 2007 Kentucky senator Jim Bunning introduced Resolution 294 in the US Senate, declaring September as National Bourbon Heritage Month. The resolution read in part: "Resolved that the Senate recognizes bourbon as 'America's Native Spirit' and reinforces its heritage and tradition and its place in the history of the United States."[1] The resolution was subsequently passed by the Senate. The Kentucky distilling industry celebrates National Bourbon Heritage Month each year with hallmark promotional events and releases of limited-edition whiskeys.

But such commemorative declarations beg the question: What is heritage? And if heritage is not based solely on legendry but has material origins, whose heritage deserves approbation? Kentucky's bourbon whiskey distilling industry entered the twentieth century with an embedded craft tradition. The transition from folk practice to industrial spirits production was steered by multiple influences—some visible, some unseen. The work was driven by personal aspirations, accomplished by toiling laborers, enabled by insightful technological innovations, beset by environmental and financial risk, controlled and surveilled by federal regulations, and hectored by temperance reformers.[2] The ecological perspective employed here suggests that at the conjunction of these and other interdependent influences, people assembled the materials that make up the contemporary distilling landscape. Though shorn of many of its historical works, the present-day landscape presents still-visible residuals that are testimony to individual and collective accomplishment—or, in some instances, abject failure. And family oral histories and archival papers record unseen dimensions of the lives lived and the meanings the work and the landscape held for their makers. Bourbon distilling in the twenty-first century is a thriving, multidimensional, international business. Product sales advance in response to extensive marketing campaigns, positive product experiences, and, increasingly, visual and tactile connections to the places of production. Old distillery works possess historical cachet, and distillers refit selected venerable buildings with state-of-the-art equipment or erect new works with architectural references to the past.[3]

If an entire industry can be said to have a soul and a usable past, it is Kentucky's bourbon industry. But, given the breadth of influences that have contributed to the industry's survival and present-day success, which physical and cultural elements most fully represent that revered past? Further, which of those elements not only survive but thrive as anchors for the historical narratives and images distillers have chosen to preserve in memory and perhaps redeploy for marketing and advertising? Importantly, in sorting the lost from the found, we must recognize that narratives about the past are likely to be constructed in a manner that reshapes our perceptions of both the past and the present.[4]

The most commonly credited influences on distilling's industrial progress and longevity are environmental, such as pure limestone springwater; cultural, including the influence of the Old World distilling experience of the Irish, Scottish, and other groups; and innovational, such as the charring of white oak barrels. Accepting these explanations or narratives uncritically can be narrowly deterministic and misleading and unsympathetic toward the nuances of how people understand and experience past places and events. The landscape on which much of the industry's past was impressed—corn-producing farm fields, water-powered mills, steam-powered still houses, multistoried wood-frame and brick warehouses, stone-covered turnpikes, muddy riverbank boat landings, and steel-rail railroad sidings—was an expression of culture in the context of the physical environment. One cannot appreciate the presence of an artifact on the landscape without understanding, as archives and memories permit, the making of that artifact. Understanding bourbon production in Kentucky is not abetted by ignoring its landscape context and its geographic history and thereby separating whiskey from its place, past or present. No one's interests are served, especially not those of the contemporary distilling industry, if the meaning of a historical distilling structure that remains on the landscape is reduced to its use value.[5] The narratives we construct to explain that landscape are not given; they require that we seek out and interpret meaningful patterns in how culture and environment interrelate over the course of time.[6] If a historical narrative is the considered interpretation of past events and places, then how does one determine whether that interpretation has credible authenticity or is a fabrication founded in sentimental imagining?[7] When can such interpretations be accepted as legitimate tradition? Or are fungible interpretations of the past acceptable? And who bears responsibility for creating and upholding an industry's historical narrative: the industry, the public, or the curators of historical documents? These questions are not simple speculations; importantly, they relate to the role of authenticity in the distilling industry's understanding and embrace of its past and its portrayal of that past to the public. From a humanist perspective, words are not imprecise if their use is understood and agreed upon. One might define geographic history as the sum of all events and their ecological context, for example, but we must be wary of suggestions that a narrative is "the history" rather than "a history" or terms that ignore ecological context entirely.[8]

We construct historical narratives to describe our traditions by selecting and interpreting available information. History, then, in the words of historian Raphael Samuel, is "a social form of knowledge; the work, in any given instance, of a thousand hands."[9] We may glean a nuanced understanding of historical narratives and tradition from folklorist Henry Glassie, who observes that tradition is the continuous creation of the future out of the past by individuals in accordance with their interests or preoccupations. Tradition and geographic history are congruent but are not the same. "History is not the past; it is an artful assembly of materials from the past, designed for usefulness in the future."[10] History is the resource out of which people create tradition; tradition, therefore, is not a given. In maintaining cultural continuity, in organizing their lives, people create tradition through purposeful choices and the engagement of wills; they retain and embellish some aspects of the past while allowing others to slip into obscurity. Nor is tradition the singular event that advances extensive change. It is, rather, the product of the willful selection of those events and behaviors that people deem to be important and that contribute to their identity. Traditions change as new technologies, behaviors, and events present themselves.[11] If tradition is important to an economic activity such as distilling, then the retention of favored traditional methods may be deemed essential to proper production and, thereby, preferred over the adoption of more modern and advanced procedures. Yet, if distillers wish to produce quality spirits in quantity and at affordable prices, they may be forced to forgo some traditional methods and implement modern industrial processes. In so doing, they may feel compelled to disguise their modernity by adopting marketing narratives that highlight their traditions and their venerable past.[12]

Searching for Heritage and Authenticity

Historically constructed traditions may transcend generational divides and become heritage, whether by inheritance or adoption. Heritage is often embraced through resonant synonyms—age, ancestral, agrarian, handmade, kinship, legacy, old, origin, venerable, vernacular, revered, and stable—or refuted by antonyms such as ephemeral, new, improved, upgraded, and recent.[13] A distiller's past may be winnowed, sorted, and sanitized for marketing or tourism purposes. Though it is the object of a multifaceted historical narrative, the production of fine bourbon whiskey required the work of sundry folk—the farmers who grew the grains, the laborers who produced whiskey, and the denizens of the riverboats that carried the product to distant markets among them. Yet their lives and labors are often anonymous or presented as normalized contractions. This form of recalled past does not depict a legacy but offers simplistic abstractions from which awkward actualities have been removed or simply dissolved through loss of memory.[14] Some businesses and manufacturers may be ambivalent about their past and may reject its authority by emphasizing progress, innovation, growth, and modernity. Others publicly recognize their longevity and imply heritage in subtle ways, such as fast-food restaurants that post signs extolling the number of hamburgers sold, or banks and financial institutions whose letterheads include phrases such as "Serving This Community since 1948." Distillers have often embraced, even cultivated, their heritage endowment as an ideal and have strived to retain its salient elements. This may be achieved by reducing the historical essentials to symbols as represented by brands, trademarks, advertising, structures, and museums that replicate and house traditional forms.[15]

Time-honored knowledge or practice—a yeast recipe or a mash fermentation process, for example—may be deemed worthy of renewal and be passed on from one generation to the next if it is believed to be original, authentic, and meritorious. But selected knowledge of the past may be authentic, or it may be reinvented or reimagined in complex ways by those in the present who are motivated by some kind of economic, political, or social interest. If a firm's heritage, its tradition, is rooted in questionable circumstances or otherwise found lacking, authenticity may be forgone in favor of an invented tradition. Such tactics suggest that deciding which cultural attributes are invented, or reinvented, is a selective process and is likely accompanied by purposeful limitations on who does the inventing and how those inventions are put into practice.[16] Authenticity, therefore, is an inexact concept. It may be genuine or the subject of choice and emphasis; it may even be fabricated as a socially constructed polemic rather than an objective, concrete attribute.[17] And because authenticity may be constructed according to some accepted custom or desired outcome, it is also subject to change if that standard is adjusted.[18] Whatever its provenance, and whether it is presented as authentic or not, heritage is often employed in product marketing to establish importance, value, and a sense of cultivated antiquity. Heritage may draw consumers, but when presented in tangible form, it has also proved to be a prime attraction for tourists.[19] Businesses and communities have learned that heritage can be invoked to market everything from tweed cloth and tailored clothing to burgundy and ardent spirits.[20]

A Convergence of Heritage and Distilling

Kentucky's contemporary distilling industry operates according to modern and proprietary technical knowledge, but it is also physically embedded in its historical landscape. The contemporary industrial distillery, whatever its size and capacity, is a modern facility with contemporary equipment; yet operators often cultivate an aura of age, even antiquity, by refurbishing and maintaining aging structures—heritage is given material form. The patently composite buildings do not elicit a critique of incongruity; rather, they invite the conclusion that preservation of the past, however limited the volume and eccentric the composition, is meritorious. Importantly, rejuvenation of antique landscape elements may also enable reconstruction or approximation of venerable and valued past practices. A visitor may find Kentucky's distilling heritage present in profound places: finely chiseled date stones, a carefully excavated archaeological site that reveals old segmented window arches incorporated into newer brick walls, a distiller's grand Victorian mansion on South Main Street in Lawrenceburg, or the Romanesque-arched tablet stone marking James C. Crow's grave site in the Versailles Cemetery in Woodford County. But heritage can also be as subtle as family memory, the folktales told about a whiskey and its consumers, or the dynastic tradition of master distillers passing their mash bill formulas from one generation to the next.[21]

Since the 1990s, the contemporary bourbon whiskey economy has grown at an accelerating rate. By 2017–2018, sixty-eight distilleries operated in Kentucky—triple the number recorded in 2009—and they produced about 95 percent of the world's bourbon. The distilling industry's direct base employment stood at about 5,000 jobs, whereas total economic impact included about 20,100 jobs and a gross payroll of some $1 billion.[22] This fiscal performance was stimulated, in part, by industry-wide promotion and innovations in product marketing and advertising. Whiskey-oriented websites and blogs evaluate and rate bourbons in a manner reminiscent of the reviews of fine wines that appear in social and print media. And just as the wine industry has promoted the role of terroir—the soil, water, and climatological qualities that contribute to a wine's distinctive character—distillers have correspondingly referenced the environmental qualities of whiskey making. But they have also cleverly deployed historical landscape images to reinforce the layered links between their products, their place of production, and their past.

Marketers selectively sort images of past production landscapes to create icons that accompany contemporary products and sales strategies. Industrial images that become iconic are based not only on a sense of collective social and cultural history—the "old," the "traditional," or "heritage," for example—but also on the sensory aspects of the landscape, especially the signature structures associated with whiskey production. In this way, historical landscapes not only inform the present but are themselves valued for their age and character. Further, a landscape's physical attributes, be they limestone bedrock or hillside springs, can be objectified and presented as unchanging, even ageless. An appeal to authenticity based on the premise that certain environmental characteristics are perpetual and inviolable can be convincing, but it is also subject to question. Groundwater chemistry, after all, is not stable; it is affected by the volume of precipitation, measured daily or seasonally, and can be altered by changing land use in the surrounding watershed.

Kentucky's contemporary bourbon landscape is the repository of places and structures whose precursors were constructed by nineteenth-century farmer-miller-distillers and related businesspeople: grain and livestock producers, timber cutters and lumbermen, coopers, coppersmiths, pottery jug and glass flask makers, pipefitters and steam engine fabricators, turnpikers and wagoners, railroaders and riverboat pilots, bankers and financiers, and government revenue agents. Even though these people built and regulated the industry's base, they are often represented on the contemporary distilling landscape as only symbols or icons, and the depth and breadth of their participation cannot be readily comprehended without the concerted efforts of archivists and museum curators.

Landscapes are never complete and stable; they are always accumulating, changing, evolving, or retrograding.[23] The distilling landscape's form possesses antiquity, but it is not torpid; it is, rather, emergent. It is the product of processes that create ongoing flows of people and commodities, technology and economy, ideas and cultural information, all set within an ecological interplay of human behaviors and physical environmental conditions. Bourbon production techniques and marketing modernized in the twentieth century, in parallel with other industries. There is, at present, a "rebirth" of the bourbon market that includes the revival of lost practices and extinct brand names and the reclamation of relict distilleries. Architects and engineers plan and build new distilleries that deliberately echo the past in design and production techniques. And marketers actively employ heritage and nostalgia to promote both traditional and newly developed products.

Preservation, Reclamation, and Replication

Manufacturing businesses create images out of the commodities they produce and assets out of those images. Whether structures and landscapes are constructed to be solely functional or contrived to portray an image that evokes a certain emotion or behavior, they all convey meaning to those living with them and to visitors who see them only momentarily. Commercial buildings on the nation's Main Streets have long embodied the essence of image making; they may be designed to inspire confidence in an institution or elicit desire for a specific product or service. Greek Revival porticoes and pillars on nineteenth-century bank façades and courthouses were intended not to provide shade or shelter from inclement weather but to signify that the occupant was aware of current fashion trends and wished to cultivate an image of permanence, strength, wisdom, and trust. Although central-city commercial buildings were functional, their façades and interior designs were often guided by fashion and style and the desire to convey certain messages to both occupants and observers. Such structures were also authentic, in the sense that they functioned as banks, department stores, city halls, or office buildings. Industrial buildings, unlike commercial structures, were often built primarily to house manufacturing activities as functionally and economically as possible. Nevertheless, workers and passersby may have assigned meaning to such structures—as sources of reliable employment, perhaps, or as places of demeaning working conditions and environmental degradation.

Over time, the iconic value of structures and landscapes changes, and as one generation of observers gives way to another, the meanings assigned to those places may also change. Abandoned Main Street commercial buildings are deemed, by some, as mere placeholders, occupying lots that can be cleared, consolidated, and rebuilt with high-rent office or residential buildings. Others view the same structures as worthy of stabilization and rehabilitation. Federal, state, and local preservation programs cultivate heritage appreciation by allotting funding and design expertise to save these buildings, reaffirming the centrality of and economic and social stability engendered by historic Main Streets and other venerable structures.

Saving Sites through Historic Preservation

Many Americans express ambivalence about preserving historic buildings. Preservation seems to fly in the face of the American Dream, which celebrates progress, growth, and a better future. Geographer Peirce Lewis has outlined a logical rationale that might be used by those interested in preserving historic structures and landscapes. The elements of his argument that are most relevant and applicable to industrial sites involve cultural memory, antique texture, environmental diversity, and economic gain.[24]

Individuals may wish to cultivate a personal sense of history, a cultural memory, and this can be abetted by tangible reminders in the form of historic structures. But such reminders need to be truthful to be plausible. Can a reclaimed or refurbished industrial building provide true insight into past manufacturing processes? Management and labor relationships? Health and safety? Environmental impacts?

Structures may merit preservation because they have an antique texture; they possess venerable designs and finishes and worn and weathered qualities. Old buildings can lend environmental diversity to a built environment that has become increasingly homogeneous. Builders assemble subdivisions according to conventional floor plans, corporations stipulate standardized building designs for chain restaurants and shops that populate highway retail strips, and contractors construct anonymous big-box stores out of synthetic materials and prefabricated parts. People come to appreciate historic properties because they are different, both visually and tactilely, from modern landscapes. A census of nineteenth-century industrial buildings might include timber-framed mills, stone and brick warehouses, and, once the materials became available, general-purpose manufacturing buildings sided in sheet metal. A functioning distillery with a nineteenth-century skeleton set in a remote creek valley amidst modern farms lends unexpected diversity and cardinal interest to that landscape.

Many Americans find experiential value in tradition and permanence, but retaining and reusing old structures may be possible only if the economic incentive is significant. Preservationists hold that building rehabilitation conserves resources—in terms of materials, labor, land, and time, it often costs less to preserve an old structure than to clear the site and build anew; rehabilitation and maintenance of historic properties, therefore, can have a pragmatic rationale. Preserved historic buildings and landscapes may attract tourists. Real estate values may stabilize and even increase, leading to higher tax assessments. The prospect of preservation may appeal to speculators, developers, contractors, and realtors because of the potential for economic gain. The risk here is that reclaimed and repurposed sites are subject to multiple, perhaps conflicting, incentives that are detached from the structure's original use. An industrial building repurposed as a flea market or office lofts may retain its exterior shell but yield its internal space to unconventional configurations and uses. Preserving a functioning industrial distillery involves few such detractions, and rehabilitation can maintain a significant degree of authenticity.[25]

If legitimate preservation is motivated primarily by economic gain, concerns might arise as to whether the preservation process conserves both the structure and its authenticity. But how does one confirm authenticity?[26] In one sense, accepting a structure as authentic is subjective and resides in the judgment of the individual observer. Another view is that authenticity has objective qualities that can be assessed by trained professionals, be they architects, archaeologists, or conservators. In 1966 Congress passed the National Historic Preservation Act as part of a domestic program to engage professional preservationists in the identification and protection of historic resources. The act authorized the National Park Service and state preservation offices to determine the historical legitimacy of structures and landscapes through the National Register of Historic Places and the National Historic Landmark program.[27] To qualify for the National Register of Historic Places, structures must be at least fifty years old and associated with significant people and events, or they must be deemed noteworthy examples of architecture, landscape, or engineering history. Properties must also retain structural integrity. To nominate a structure for listing in the National Register, the owner, or the owner's representative, must submit an application to the state historic preservation office for evaluation. In Kentucky, that office is the Kentucky Heritage Council, an agency of the Kentucky Tourism, Arts and Heritage Cabinet.[28] Structures approved for register listing are certified, and any rehabilitation work, if it is completed according to National Park Service standards, is eligible for a 20 percent tax credit.[29] Since 1974, twelve distilleries or distillery districts, five distillers' homes, two distillery-related residential districts, a warehouse, a bottling plant, and an office district have been nominated and placed on the National Register of Historic Places.[30]

Historic properties that are deemed to have exceptional value as illustrations of America's heritage may be designated National Historic Landmarks. The standards for National Historic Landmark designation are more stringent than those for listing on the National Register of Historic Places. The National Register listed more than 90,000 historic properties nationwide in 2017, but only 2,500 had been approved as National Historic Landmarks.[31] To nominate a structure or district as a landmark, one must submit a comprehensive, expertly prepared nomination. The nominee must represent a nationally important event, place, or person in American history. For example, the place may illustrate the achievements of a significant individual, or it may be an exceptional representation of a particular building or engineering method. After a two- to five-year process that includes extensive research and other formalities, the secretary of the interior, upon recommendation by the National Park Service, may approve the designation. In addition to tax credits, landmark properties are eligible for grants to maintain the property's character and integrity.[32]

Kentucky has thirty places designated as National Historic Landmarks, and three are distilleries: Burks Spring (Maker's Mark), George T. Stagg (Buffalo Trace), and Labrot & Graham (Woodford Reserve). The Maker's Mark Distillery was listed on the National Register of Historic Places in December 1974 and designated a National Historic Landmark in December 1980. It was the first distillery in America to be so recognized while the landmark buildings were in active use for distilling.

National Register and National Historic Landmark status is actively pursued in Kentucky; in 2016 the Heritage Council tallied more than 3,400 historic properties comprising more than 42,000 buildings and structures that had been approved for registration since the program's inception, the fourth largest state total in the nation.[33] National Register status provides tax credits that help underwrite rehabilitation and maintenance and confirms that the site or district is "real," thereby corroborating claims of authenticity. Register status also creates a target for citation and publicity in tourist guides and places properties in a different category from destination tourist sites like theme parks, which are intended to attract tourists looking for synthetic recreational experiences. Tourists who want to visit authentic places have their interests confirmed by National Register status, which is backed by scholarly research. The operating distillery, then, is not a "tourist site" or "setting" but a private space that gives visitors insight into an authentic place and production processes that would otherwise be hidden from view. Tourists not only gain access to an authentic historical site but also are rewarded with the knowledge that both the place and the experience are rare. Such singular experiences may, conveniently, also bolster product sales.[34]

Reclamation of Historic Distillery Sites

From the early 1980s to 1997, some 1,370 historic structures in Kentucky underwent rehabilitation, representing a private-sector investment of $432 million. The effort created some 19,000 jobs.[35] Distillers have worked to maintain heritage and authenticity by actively participating in different types of historic structure restoration projects. The term "obsolescence" is not in contemporary distillers' operative vocabulary; rather, they work at rehabilitating relict distilleries, refurbishing distillers' home sites, and reclaiming venerable equipment and building materials.

The Archaeology of Heritage Recovery

Heritage abides in oral form in stories and tales and in tangible and tactile form on the landscape's surface and, often, underfoot. At the Buffalo Trace Distillery in the frontier community of Leestown, northwest of central Frankfort, the present riverside works is flanked by a once decommissioned brick building that shares the site with at least three older distilling structures, all likely built by E. H. Taylor Jr. Taylor's notes and company records, Sanborn insurance maps, and other documents suggest that the contemporary site may still contain evidence of the extraordinary distilling works Taylor built in 1882.

In 1869 Taylor built his first brick distillery next to a perennial spring. He named the works Old Fired Copper, or O.F.C. That works was torn down in 1873 and rebuilt with brick exterior walls rising above foundations crafted from cut blocks of Tyrone limestone laid in a random ashlar pattern; stone quoins accentuated the corners, and tall Romanesque Revival arched windows, sometimes referred to as the American round-arched style, admitted light. This building was struck by lightning in 1882 and burned, leading to immediate reconstruction as a two-story building with a full basement. Apparently following federal internal revenue regulations, the distillery building was divided into discrete rooms dedicated to milling, mashing, fermenting, distilling, and power functions. Much of the lower level was covered by a concrete floor when a subsequent owner, Schenley Distillers, converted the building into storage space sometime after 1942.

At the invitation of the distillery, archaeologist Nicolas Laracuente and a team of like-minded colleagues began research at the old building site in 2012. Their work involved a thorough archaeological excavation of the building's interior, which revealed a "bourbon Pompeii." When a section of the concrete floor was removed, they found remnants of the 1869 and 1873 buildings, as well as a remarkable set of eight rectangular fermenting vats that were apparently an innovation implemented by Taylor. To avoid the buildup of contaminating acids and other elements in wooden vats, Taylor commissioned the construction of eight, 14,600-gallon brick vats faced with portland cement and covered with copper sheeting. By 2018, Buffalo Trace had reclaimed this historical structure and installed metal stairs and catwalks over the old fermenting vats and wall remnants. The building's heritage is documented by information display boards and two video screens that play digital recordings of the building as it was constructed by Taylor in 1882 and the recent archaeological excavation. The distillery plans to offer special tours and open the building for singular social events.[36]

The Restoration of Relict Industrial Distilleries

Rehabilitating a large industrial distillery can be a difficult and expensive undertaking. The Brown-Forman Corporation bought the Labrot & Graham Distillery site in Woodford County in 1993 and expended $7 million to restore the works, part of which dated from 1838.[37] Speaking to the restoration strategy, Bill Creason of Brown-Forman said, "We, as a company, owe our beginnings to bourbon. Because of our heritage we have a very important place in the bourbon industry. It's fitting that Brown-Forman should have a facility like Labrot & Graham, a place where Kentuckians and tourists alike can come to learn about bourbon's rich heritage. The site is the most historically important place in bourbon distilling." The works is presently known by its primary bourbon product, Woodford Reserve.[38]

Labrot & Graham Distillery site.

The relict remains of E. H. Taylor Jr.'s Old Taylor Distillery stand along Glenns Creek, four miles north of the Labrot & Graham site. In May 2014 investors purchased the 113-acre Old Taylor property, which had been abandoned since 1972. They planned to spend $6 million to refurbish the works and put it back into production as Castle & Key Distillery. The *Lexington Herald-Leader* quoted Governor Steve Beshear, who said the investment would "further solidify Kentucky's bourbon legacy and create more jobs but it also signals the renaissance of one of the state's most historic and iconic distilleries."[39] By 2016, when the property was nominated for the National Register of Historic Places, the expansive site retained more than forty architecturally distinctive buildings, most of them constructed of stone or brick. Their condition ranged from relict to good. Reconstruction included reconditioning some old equipment, as well as installing new gear where needed. After the repeal of Prohibition in the early 1930s, National Distillers Products reopened the distillery and installed new high-capacity steel fermenting vats—six 22,000-gallon tanks and fifteen 11,000-gallon units. The steel tanks that replaced Taylor's original wooden vats still stand. Six have been refurbished, cleaned, and relined. The old still has been replaced with a new fifty-foot copper column still manufactured by the Vendome Copper and Brass

Works in Louisville for distilling bourbon, and there is a forty-foot stainless-steel column still for making gin and vodka.

Old Taylor Distillery site.

Warehouse B was built of brick, with stone quoins at the corners. At four stories tall and more than 530 feet long, it is thought to be the largest whiskey aging warehouse of its type in the world. The turreted castle-form still house was constructed of Tyrone limestone, sometimes referred to as Kentucky River marble; with the passage of time, it has weathered to a bright white. The iconic building not only housed the distilling apparatus but also became the symbol of E. H. Taylor's distillery, appearing in publicity photographs and on postcards. A large sunken garden, built in 1906 and central to the works site, has been reclaimed from overgrowth; some hundred-year-old yews (*Taxus* species) now serve as rare specimen plants. Artisans have restored the peristyle springhouse or pergola, which, along with the garden, was a favorite site among early-twentieth-century visitors. When fully operational, the renovated works will be capable of producing about 12,000 barrels of spirits per year.[40]

RECLAMATION AND CRAFT DISTILLERIES

Maysville, known as Limestone Landing in the late eighteenth century, was an early Ohio River shipping point for whiskey distilled in northeastern Kentucky's Mason and Bourbon Counties. In 1869 O. H. P. Thomas operated a whiskey distillery at the river's edge on Maysville's west side. Henry Pogue bought the Thomas works in 1876 and renamed it the Old Pogue Distillery. Pogue produced several different branded whiskeys, including Sour Mash Copper Whisky and Fire Copper Whisky. The site included four large aging and shipping warehouses. Production ended, except for a small amount of medicinal whiskey, with the enforcement of Prohibition in 1920. Production resumed after 1933, but the distillery and warehouses were torn down sometime thereafter. The Pogue family home remained. The bluff-side Greek Revival house had been built in 1845 for hemp merchant Michael Ryan. The three-story, central-hall brick house featured a distinctive stepped-gable roof and, at 4,800 square feet, was among Maysville's largest and finest homes. The Pogue family sold the property in 1955, and subsequent owners allowed the house and grounds to deteriorate. Henry Pogue's descendants repurchased the house and surrounding property and began renovations in 2005. The house is now listed on the National Register of Historic Places. A new double-garage-scale building houses the Old Pogue Distillery, a revival craft operation that began producing a bourbon and two rye malt whiskeys in 2012.[41]

In Lexington, the James E. Pepper Distillery stands on a narrow property pinched between Town Branch Creek and Manchester Street. The Headley & Farra Distillery operated at the site in 1858 but burned in about 1873. James Pepper built a new replacement works on the site in 1880. Whiskey production stopped in 1918, although the warehouses were used during Prohibition to consolidate and store whiskey recovered from closed distilleries elsewhere. Schenley Distillers purchased the site in 1933, after the repeal of Prohibition, and the company completed construction on a new distillery the following year. A few months later, a fire destroyed several buildings, including six new warehouses. The replacement structures were basic buildings of utilitarian design constructed of common modular brick, load-bearing structural tile, and reinforced concrete. Whiskey production resumed in 1935 and continued until 1958. Thereafter the warehouses were used to store bonded bourbon until 1976. All but one of the warehouses were subsequently torn down or burned. The present site consists of the distillery building and its mashing, fermenting, and distilling spaces; an office building; a warehouse for barrel storage and cooperage; a five-story warehouse; a water tower; and related structures. The works was scheduled to reopen as a limited-production craft distillery in 2018 bearing the James E. Pepper brand name.[42]

The James Pepper Distillery is part of a larger twenty-eight-acre site slated to be redeveloped as the Lexington Distillery District. The urban–county government project, first proposed in 2008, would blend private and public capital investment and cost an estimated $190 million, including the rehabilitation of the remnants of a second distillery, the Old Tarr works. The plan includes extensive residential and infrastructure improvements and construction of retail, market, and hotel spaces. The design proposal invoked the language of historical cachet and heritage by stating that the area would "establish the identity of Lexington's former bourbon corridor, thematically creating a dynamic Bluegrass asset that leverages history to drive tourism and conventions, attract the creative class, and serve as a local destination."[43]

RELICT DISTILLERIES AWAITING RECLAMATION

Kentucky's distilling history has been punctuated by cycles of construction, operation, closure, abandonment, and, ultimately, erasure from the landscape. The nineteenth-century market for Kentucky's distilled spirits was volatile; business could be competitive in the extreme, and distillers experienced severe recessions and even a whiskey depression during the 1880s. When production soared and sales plummeted, distilleries failed. A combine of distillers formed the Whiskey Trust in the 1890s and engaged in the serial purchase and closure of distilleries in an attempt to control production capacity. Only six distilleries remained at the repeal of Prohibition in the early 1930s. Abandoned distilleries were often salvaged for their brick, wood, and metal—their sites stripped to bare ground. The D. L. Moore Distillery in Burgin, dating to 1871, closed in 1959. A large warehouse remained standing on the site until it collapsed in 2014. Salvage of the century-old timbers began soon thereafter, leaving few visible clues that an operating distillery ever existed there.

Few of the bourbon industry's bones—old, relict industrial distilleries—remain as units or even as single identifiable structures. Should a nascent distiller wish to reclaim an extant distillery site, resume production, and thereby secure the promotional advantages of heritage and authenticity, the list of sites is short, and works with more than a few remaining structures are scarce. The candidates are outlined below.

Coon Hollow Distillery, west of New Hope in Nelson County. Production stopped at this rail-side works in 1920. Other than an aboveground iron cistern, only circular and rectangular concrete foundations and portions of metal-framed buildings remain. Farther east, along State Road 52 at St. Francis—about one and a half miles west of Loretto, where contemporary Maker's Mark warehouses stand in clusters—the concrete foundation of a distillery warehouse remains on the south side of the highway.

Stitzel-Weller Distillery in Shively in southwestern Louisville. The distillery was built in about 1936, and production ended around 1992. The rail-side works, including a cooperage building, is in good condition; the warehouses are in active use for whiskey storage. The owner, Diageo, also uses the works as a visitors' center for the Bulleit Frontier Whiskey Experience.

Chapeze or Old Charter Distillery near Clermont in Bullitt County. The site includes intact warehouses, a water tower, and a distillery office building. Production stopped in 1951, and the property is currently owned by the nearby Jim Beam Distillery; the company uses the warehouses for barrel storage.

T. J. Samuels Distillery at Deatsville in Nelson County. The post-Prohibition rail-side

Jim Beam truck carrying filled whiskey barrels at the Chapeze Distillery.

works was in production until 1952. The warehouses are still in use, and most of the distillery buildings are intact.

Henry McKenna Distillery at Fairfield in Nelson County. Production stopped in 1974. The post-Prohibition distillery building is merely a shell and the warehouses are gone, but the site includes other buildings, the old bottling house among them. McKenna's large brick home stands on a hill overlooking the distillery site.

Old Lewis Hunter Distillery in Harrison County. Several remnant buildings still stand beside the Licking River on Old Lair Pike. The concrete floor of the old cattle feedlot remains a short distance away. The works closed in 1978.

Atherton Distillery at Athertonville in Larue County. Production stopped in 1984, and the remaining buildings, which are architecturally similar to those at the McKenna Distillery, are being used by a wood pallet manufacturer.

Glenmore Distillery at Owensboro in Daviess County. Although distilling ended in 1993, the Sazerac Company continues to operate the large bottling plant, and the warehouses are used for whiskey storage.

At relict distillery sites, the most common building remnants are warehouses, which may be used for whiskey storage or converted to other storage and industrial purposes. Examples include the small Jacob Spears Distillery warehouse in Bourbon County (hay

T. J. Samuels Distillery.

storage), the Paris Distillery at Paris (industrial and construction equipment storage), the Elkhorn Distillery in Scott County (horse barn), the Midway Distillery in Woodford County (repurposed as an apartment building), the Old Tarr Distillery in Lexington (repurposed as an entertainment venue), and the Commonwealth Distillery on Sandersville Road in Fayette County (formerly used for plant nursery storage).[44]

We see relict distillery structures for what they are: remnants of the past. But unless we make some effort, we will fail to see the full extent of their original social, technological, economic, and behavioral context. Places that retain such remnants were significant for those who lived and worked there, and an informed archival reconstruction of their historical context is required if they are to serve as references to heritage for contemporary occupants and visitors.

Replication of Historic Structures

Architectural replication is a comparatively rare construction strategy. Whether a property owner wishes to add new rooms to a private residence or an art museum's board of directors wants more square footage of display space, the additions often depart substantially from the original design and materials and tend to be readily identifiable. Relatedly, if an old building cannot be restored because of extensive damage, outdated technology, or a dearth of artisanal skill or compatible building materials, perhaps a "faithful" replica can be built and, in some manner, stocked with heritage. All the better if the replica is viewed approvingly by the subjects of its use as a commercial marketing device.

An early use of a replica structure to market bourbon appeared at the 1893 World's Columbian Exposition in Chicago. In 1869 the Old Times Distillery was operating in Louisville at 2725 West Broadway.[45] The idea to commission an exhibit at the Chicago exposition has been attributed to Charles Lemmon (also spelled Lemon), Old Times Distillery president, and Richard Meschendorf, its secretary and treasurer. The exhibit featured a miniature replica of the Old Times Distillery.[46] Like the original, the distillery building was log and included rooftop tanks and an adjacent lean-to shed roof to shelter the fermenting tubs. Exposition organizers lauded the Old Times exhibit, noting, "They reproduced their distillery in miniature on the Exposition grounds near the Cliff Dwellers' exhibit and close to the Anthropological building. Here, since the beginning of the Fair, they have shown to the hundreds of thousands of people who have visited their exhibit, the entire process of distilling their celebrated whisky. Since the Fair opened there has been more interest shown in this exhibit and more people visited it than any other single exhibit at the Exposition."[47] In the minds of Lemmon and Meschendorf, the replica distillery had the desired effect. They said of their exhibit: "Besides giving them the freedom of the distillery, many thousands of visitors have been afforded the opportunity to sample matured Old Times which no doubt will always stick to them as a remembrance of the best thing they saw and tasted at the World's Fair."[48] A querulous critic might have observed that free whiskey was likely to attract approving patrons wherever it was dispensed.

Contemporary replication of historic distillery architectural features is not common, and the few extant examples are often limited in scale and "interpretative" rather than faithful reproductions. The Willett Distillery south of Bardstown has undergone extensive remodeling based on the accretive forms of nineteenth-century distilleries, such as attached rooms of various sizes and ceiling heights and low-angled shed roofs. Clerestory windows and stone-faced exterior walls provide additional historical references.

Interpretative architectural replication is most commonly applied to new visitors' centers. The Wild Turkey Distillery in Anderson County erected a postmodern visitors' center that seems to be inspired by a multistoried storage warehouse, although some interpret the black-painted building as an architectural reference to the burley tobacco barns once found on nearly every farm in the area. The Heaven Hill Distillery visitors' center and museum in Bardstown references the monitor roof and clerestory windows used at the T. W. Samuels Distillery in Deatsville and several others in the area. At the Evan Williams Experience on Whiskey Row in Louisville, designers built a small artisanal distillery that operates in a glass-sided room; they also re-created a Disneyesque "Main Street" of visitor-oriented gift shops.[49]

A most notable distillery replica building is the Jim Beam American Stillhouse visitors' center at Clermont in Bullitt County. The building replicates the distillery's original

Willett Distillery.

Sign at the Jim Beam American Stillhouse.

still house, and a small plaque near the front door states that the building is a reproduction. Inside is a replica within a replica: a miniature diorama made of brass and copper depicting an old distilling works.[50]

Kentucky's distilling industry is notable for the way it engages its past. Preferences align with the retention of old structures, methods, and brands, while also embracing experimentation and modern business practices. The distillers' past lives are foundational to origin narratives in which individuals and their ideas are not only retained but also embellished as memory and historical archives permit. Importantly, distillers recognize places as sites of beginnings and where heritage resides. The distillers' narratives bond with those of associated trades: coopers, coppersmiths, grain farmers, bankers, and a large cadre of other commensal businesses. While appreciative of their heritage, contemporary distillers do not live *in* the past but *with* the past. Recognition of the value of tradition—the realization that heritage is also identity—blossomed in the nineteenth century, as witnessed by legal wrangling over proprietary branding and the concern that unfettered and unlettered competitors would encroach on or even usurp one's heritage. Contemporary distillers welcome innovation in science and mechanics while embracing the matchless textures and sensory expressions of traditional structures and landscapes. They also recognize that, if tutored, the public will come to appreciate these qualities and associate product with place and heritage in a gainful manner. In so doing, distillers not only achieve business success but also validate that maintaining historic places and landscapes has the dual advantage of preserving their heritage and underwriting their product.

Epilogue

Making Bourbon, Making Landscape

A landscape, like a language, is the field of perpetual conflict and compromise
between what is established by authority and what the vernacular insists upon
preferring.
—J. B. Jackson

Understanding, or at least appreciating, the complexity that attends human events in
time and place summons the insights provided by a historical ecology perspective. Kentucky's nineteenth-century bourbon distilling industry, though seemingly modest in scale
and down-to-earth in manner, was interwoven with influences that were often external
to the distillers' operations. The inclusive view invited by an ecological perspective is to
systematically identify the conjunctions between bourbon production and a congregation
of environmental, social, economic, and political factors that impose direct or indirect
contingency on the human and technological aspects of the manufacturing and marketing process. Considered business decisions may produce the desired results in the form
of a superior product that can be profitably marketed. But the best-intended and best-informed decisions can also be mitigated by unanticipated events that spring from factors
and incidents beyond one's control, perhaps beyond one's field of cognition.

Nineteenth-century industrial manufacturing required new technologies and power
sources and heretofore unestablished business relationships. Inventing and adopting machines and related mechanized methods permitted increased output but also entrained
unanticipated contingencies into the whiskey-making process. Consider the relationship
between bourbon production and raw materials supply. Grain production was dependent
on benign weather, and delivering the harvest required reliable and economical transportation. Some distillers built roads to link their works to their markets; others moved their
stills to new railroad lines. Coopers obtained barrel staves from active timber-cutting and
sawmilling operations that were, by necessity, mobile. Once distilling was complete, the
product required aging in appropriate structures built with materials likely obtained from
distant suppliers. These and other factors related to labor supply, government regulation,
and temperance introduced uncertainty and risk into the distilling process.

In everyday life, certainty is uncommon; variance is the rule. Day-to-day distilling operations in the nineteenth century were dynamic and unstable and involved risk in various forms. Production, supply, and demand often increased or decreased for reasons beyond a distiller's control. Even if consistency in spirits production could be established, supporting corollary systems rarely achieved predictability. And risk resided in natural processes as well as human institutions. A distiller might build a turnpike leading to a railroad depot that provided connections to distant grain producers, thereby expediting grain delivery. But if smut attacked the growing corn or rust or the Hessian fly truncated wheat production, the railroad connection lost its importance. As the giver of light and heat, fire was a primary tool in nineteenth-century industrialization, but it was applied unevenly, given people's idiosyncratic habits of selective innovation and adoption or rejection of new equipment and technologies. And how did one safely produce an exceedingly flammable product before electricity was widely available? Distillers heated water and activated pumps with fire-powered steam engines, and workers used candles and lanterns to illuminate dim warehouse interiors whose atmosphere was sometimes suffused with alcohol vapors.

On close examination, one finds that Kentucky's nineteenth-century distilling economy was the product of people of differing experience and background who employed widely divergent techniques, whose rights to production were strongly contested by some, and whose environmental and locational contexts were unique. Understanding how an irrepressible industry grew out of such diversity requires an examination of the historical distilling industry at the level of individual distillers and their workers, be they enslaved or free, who operated the works and contributed to the manufacturing process's overall success. Studying individual behavior assists in deciphering the nuances of decision making and permits one to gauge an individual's priorities, strategies, and adaptability. Such an inquiry requires detailed personal information, and fortunately, the records of Henry McKenna, James Stone, James Shropshire, and others are at hand.

Henry McKenna put an Old World trade into practice when he opened a distillery to supplement his grain milling business in the Nelson County village of Fairfield in the 1850s. The two operations likely become symbiotic: mill-processed grain supplied the mash tubs, and income from the sale of milled flour, meal, and other products funded the distillery. Although his Nelson County mill and distillery were steam powered, McKenna made whiskey according to a vernacular craft tradition; he did not aspire to high-capacity industrial distilling and was thereby spared the expense of dramatically increasing the size of his works and extending his reach for grains, other raw materials, or financing. Moreover, he likely thought of his business operation as related to Old World tradition, family, and community, rather than as a calculated for-profit proposition.

Success for McKenna was measured not by the accumulation of wealth but by the improvement of his neighborhood. He used the income from his mill store and whiskey sales to purchase farmland that he developed for crop production and livestock grazing. Clearing woodland to improve his farms made cordwood available to fuel the steam engines at his mill and distillery and supplied the community with firewood. McKenna practiced a liberal credit policy in terms of accounts due and negative credit balances. These accounts functioned as a community support institution, whether the transactions were based on barter, cash, or mortgage. McKenna had a diverse customer pool, and the accounts illustrate how he helped his customers by buying their grain, which he milled or distilled, and purchasing their wool, wood, vegetables, and other commodities, which he used, processed, or resold. McKenna regularly agreed to mortgage the equipment of those who needed cash. He thereby adapted to local needs and expanded his operation into areas that were related to milling and distilling in only a general way; he turned those requests into opportunities to diversify his business, to the mutual advantage of his neighbors and himself. His diversified income base and his comparatively small spirits production allowed him to cope with emergent federal tax policies and the enforcement and surveillance controls those policies required.

The McKenna Distillery and the town of Fairfield became magnets for immigrant settlers, especially those from Ireland. McKenna gave them jobs and, as documented by census records, even included some of his employees in his family household. McKenna mentored community development as a source of employment, credit, and foodstuffs, and in so doing, his mill and distillery promoted the flow of capital through the local community, adding stability to its economic and social institutions.

The Elkhorn Distillery in Scott County operated as a partnership between James Stone and James Shropshire. Motivated by the punitive federal tax policy implemented in 1868, the two men's business plan called for the construction of a new high-capacity distillery that would produce bourbon and rye whiskey, largely for the wholesale market. To achieve the desired capacity and ensure approval by tax authorities, they built their new distilling works according to internal revenue guidelines. Their expensive new multistory brick warehouse was deemed sufficiently fire resistant to earn reduced insurance rates. Stone and Shropshire likely anticipated that their new Elkhorn works would become a successful and profitable business. Stone's insistence on using the best grains and other raw materials may have reflected his interest in establishing a reputation as a superior distiller and enjoying the prestige associated with that status.

Stone and Shropshire initially used their personal working capital and savings to finance construction of the new works. When those moneys proved insufficient to cover costs, they borrowed from banks and private financiers, and some of the loans carried avaricious interest rates. The business partners conceived of their new, enlarged distillery as a modern industrial facility that offered the potential for sizable profits, but they would need operational reliability and a willing market if they were to recoup their investments.

In constructing the new works, however, Stone and Shropshire also increased the complexity of their logistics, amplified their dependence on the local labor market, and broadened the field of contingencies that could compromise their operation. The distillery's geographic and monetary liabilities proved too great to overcome. Reliance on a remote railroad depot, for example, meant that marginal roads, poor weather, and an officious depot agent heightened operational ambiguity and made the partners' decision-making process susceptible to costly misjudgments. Shifting national government policy, variable railroad freight rates, a lethargic national economy, and unpredictable production by competing distillers all added uncertainty to the Elkhorn distillers' decision making and increased the likelihood that external events could significantly impact their profits and, ultimately, their business's survival.

In the late 1860s and early 1870s sales of Elkhorn Distillery products slowed in response to a national recession in spirits wholesaling and retailing, owing in part to the increasing effectiveness of the temperance movement. Members of Congress debated whether to change the 1868 law to relieve the tax burden on whiskey makers. Wholesalers anticipated both sides of the tax relief issue and whipsawed distillers with low or high price offers, depending on the rumors emanating from Washington. These actions exacerbated uncertainty and risk rather than tempering it, forcing Stone and Shropshire to resort to extraordinary marketing measures in an anxious attempt to sustain their business.

A historical ecology perspective sets one in pursuit of nuance in behaviors and events, detail that would be lost if one naïvely accepted aggregated information and generalization as representing a kind of normal. To sustain operations, rural distillers who lacked experience had to learn the process and the equipment and then assemble and train a local labor force drawn from bondsmen, freedmen, immigrants, neighborhood farm youths, and family. To ensure a flow of raw materials, distillers extended contracts to farmers to obtain grain. The coopers they patronized engaged lumber mills to acquire barrel staves. Distillers wrote descriptive letters to hardware suppliers and equipment manufacturers in search of proper parts and improved machinery. They were beset by government agents whose regulatory actions could enable or impede their operations. Nineteenth-century distilling was far removed from the idealized twentieth-century images of amber whiskey emerging from white oak barrels to grace the bottom of a cut-glass tumbler.

The business was subject to influences from many quarters: agreements, both formal and informal; negotiations from strength and from weakness; appeals to avaricious interests for considerate cooperation; conflicts and protestations; compliance with business contracts and legal constraints; the vagaries of weather and nature's other complications, whether livestock diseases or grain parasites. Some factors enhanced distilling operations; others exacted unforeseen penalties. An ecological perspective invites examination of all those factors that contributed to or subtracted from those operations, in the hope of revealing where contingencies arose, how distillers responded, and how unanticipated events affected the day-to-day lives of *all* those engaged in the distilling process. Our concern is to understand how the business worked, which requires that we identify and examine, where sources permit, the many and varied influences on the whiskey-making process in nineteenth-century Kentucky. Comprehension of bourbon whiskey's manufacture and the landscape of its production cannot be achieved if the interwoven skein of processes and peoples that contributed to its making are separated and simplified or ignored.

One's appreciation of the business's geographic history can be refined with relevant detail that allows a better understanding of the decision making of the people involved, the workings of the processes they initiated, and the resulting patterns created by their singular and collective actions. An ecological perspective can call attention to seemingly minor events that influenced particular distillers or were compounded to affect the general industrial process. Importantly, this perspective can reveal the general and specific relationships between physical environments and cultural attributes such as technology and law. These actions took place not on a featureless plain but within a physical environment that became a tangible and haptic theater of operations, the landscape that people made and remade over the last two centuries. Nor can one argue that the development of whiskey distilling in Kentucky was a narrowly parochial process. Indeed, the industry grew from international ancestries. Distillers and millers drew from long experience in Ireland, Scotland, and France. Bourbon and rye whiskeys issued from stills invented and refined in Europe. Coopers completed apprenticeships in France and Germany. Practiced European coppersmiths migrated to Kentucky, as did yeast makers, stave cutters, and common laborers. But Kentucky's role in the industry's development was not passive. It was, in fact, a kind of "crucible of place," a select physical environment into which this broad group of skilled contributors was drawn, and from which a distinctive landscape emerged. Kentucky's distilling landscapes are the product of their own histories. Or, stated more concisely, making bourbon also made landscape.

Archives, census records, and other printed sources can be sorted and winnowed to place people within the landscapes they created and provide detailed information on some aspects of landscape construction. Although landscapes include their residents, landscape artifacts may be all that remain of anonymous people whose lives were not chronicled and whose presence and contributions would otherwise be irretrievable. Landscape artifacts allow one to "look over the shoulder" of their authors, those long-absent people who, during their time, contributed ideas and labor to make the places where they lived.[1] Kentucky's distilling landscape—its grounds and structures; its roads, rails, and river ports; its archaeological imprints—records a distinctive heritage and serves as a reference for those interested in maintaining that heritage as a prideful legacy that can be selectively harvested for symbols and images, whether to underwrite contemporary product marketing or for personal interest. Appreciation of that heritage is enhanced by thoughtful landscape study. Anyone can engage in reading landscape, either to enrich one's vocation or to enjoy as an informative avocation. It is, as Donald Meinig has written, "a humane art, unrestricted to any profession, unbounded by any field, unlimited in its challenges and pleasures."[2] Reading landscape is an important element in the "art of inquiry" and a worthy tool to employ in studying the historical ecology of place.[3]

Acknowledgments

Books and essays are, for me, not solely the product of my own devices but a summation of what my mentors and colleagues have taught me through their own ruminations, written and verbal. My interest in observing and interpreting our world, both the visible and the concealed, was fostered and guided by my teachers and advisers: Eugene C. Mather, John Fraser Hart, Fred Luckermann, John Borchert, Robert Kiste, Philip Porter, Dwight Brown, John A. Wolter, Russell Adams, Ward Barrett, John Webb, Leslie Hewes, and J. B. Jackson.

Many generous colleagues provided guidance, ideas, and insights: Ron Abler, John Adams, Charles Aiken, Arne Alenan, Daniel Arreola, Tom Bell, Stephen Birdsall, Stanley Brunn, Geoffrey Buckley, Clyde Carpenter, Harry Caudill, Grady Clay, Craig Colton, Michael Conzen, Ned Crankshaw, Ronald Eller, Nikky Finney, Philip Gersmehl, Glenn Harper, George Herring, Warren Hofstra, John Hudson, John Jakle, John Paul Jones III, Penny Kaiserlian, Kathi Kern, James Klotter, Karen Koegler, Charles Kovacik, Thomas Leinbach, Pierce Lewis, Gregory Luhan, Kent Mathewson, Thomas McIlwraith, Donald Meinig, Tyrel Moore, Keith Mountain, Edward K. Muller, Wolfgang Natter, Gary O'Dell, Robert Ostergren, Risa Palm, Kevin Patrick, Chip Peterson, Jonathan Phillips, John Pickles, Richard Pillsbury, Susan Roberts, Robert Sack, Ted Schatzki, Thomas Schlereth, Keith Sculle, James Shortridge, Gerald L. Smith, Jonathan M. Smith, Paul Starrs, John Stephenson, Susan Stonich, George F. Thompson, Stanley Trimble, Monica Udvardy, Richard Ulack, Dell Upton, Richard Walker, Jon L. Walstrom, Charles Walters, Joseph Wood, Nan Woodruff, William Wyckoff, Don Yaeger, Wilbur Zelinsky, and Chester Zoeller.

I am indebted to library staff colleagues for their generous and insightful assistance: William Marshall, Sarah Dorpinghaus, Matt Harris, Gordon Hogg, Terry Kopana, James Birchfield, Tina Brooks, and Kate Black, University of Kentucky Library and Library Special Collections, Lexington; Gwen Curtis and Sarah Watson, Science and Engineering Library, University of Kentucky; Faith Harders and Lalana Powell, Hunter M. Adams College of Design Library, University of Kentucky; B. J. Gooch, Transylvania University Library Special Collections, Lexington; Amy Hanaford Purcell, University of Louisville Library Special Collections; Johna L. Ebling and Aaron Rosenblum, Filson Historical Society, Louisville; Nancy Richey, Western Kentucky University Library Special Collections, Bowling Green; Erin Campbell and Devhra Bennett Jones, Lloyd Library and Museum, Cincinnati; Francis Alba, Center for Research Libraries, Chicago; and Kenneth Butcher, National Institute of Standards and Technology, Washington, DC.

Colleagues Nancy O'Malley, former assistant director of the William S. Webb Museum of Anthropology, University of Kentucky, and Professor Richard Schein, Department of Geography and Arts & Sciences, University of Kentucky, contributed enlightenment and enriching critique over the years. Master cartographer Richard Gilbreath, director of the Gyula Pauer Center for Cartography and GIS, University of Kentucky, crafted the maps, graphs, and tables, and Jeffrey E. Levy, Geographic Information Systems Program coordinator, Department of Geography, University of Kentucky, assisted with the line drawings and photographic work.

I also wish to extend my deep appreciation to friends and colleagues for their assistance and wise counsel: Professor David Hamilton, Department of History, University of Kentucky; Professor Mary Gilmartin, Department of Geography, National University of Ireland, Maynooth; Emeritus Professor William Nolan, Department of Geography, University College, Dublin, Ireland; Hal Craven, Skerries, Ireland; Seamus Daugherty, Draperstown, Northern Ireland; Mark Brown and Meredith Moody, Buffalo Trace Distillery, Frankfort, KY; Bill Samuels Jr., Maker's Mark Distillery, Loretto, KY; Drew Kulsveen, Willett Distillery, Bardstown, KY; Kevin Didio, Bulleit Frontier Whiskey Experience, Bulleit Distillery, Shelbyville, KY; Dwayne Kozlowski, Bulleit Distillery, Shelbyville; John Pogue, Old Pogue Distillery, Maysville, KY; Amanda Craft, Phillip Miles, and Samuel Hamilton, Barton Distillery, Bardstown, KY; Kentucky Cooperage, Independent Stave Company, Lebanon, KY; Vendome Copper & Brass Works, Louisville, KY; Mary Ellyn Hamilton, Oscar Getz Museum of Whiskey History, Bardstown, KY; Mackenzie Morris, Saffell House, Lawrenceburg, KY; Adam Johnson, Kentucky Distillers' Association, Frankfort; Samantha Davis and Tim Tharpe, Kentucky Transportation Cabinet, Frankfort; Nicolas Laracuente, archaeologist, Kentucky Heritage Council, Frankfort; and Marty Perry, Kelli Carmean, Margaret Rogers, Kim McBride, James Montgomery, Julie Riesenweber, and Natalie Wilkerson, Kentucky Heritage Council's Kentucky Historic Preservation Review Board, Frankfort.

And I wish to express my deep gratitude to Leila Salisbury, former director of the University Press of Kentucky; Anne Dean Dotson, senior acquisitions editor; Natalie O'Neal, acquisitions assistant; Ila McEntire, senior editing supervisor; and copyeditor Linda Lotz for their encouragement and guidance in seeing this project and its companion volume, *Bourbon's Backroads: A Journey through Kentucky's Distilling Landscape,* through to completion.

Appendix

Weights and Measures Circa 1882

Bushel Equivalents

The following weights constitute a bushel:

Barley, 47 pounds
Malted barley, 34 pounds
Bluegrass seed, 14 pounds
Bran, 20 pounds
Corn—on the cob, 70 pounds
Corn—shelled, 56 pounds
Cornmeal, 50 pounds
Clover seed, 60 pounds
Rye, 56 pounds
Salt, 50 pounds
Fine salt, 55 pounds
Timothy seed, 45 pounds
Wheat, 60 pounds
Distiller's bushel, 56 pounds, regardless of type of grain
Stone coal (including all mined coals), 76 pounds

Other Weights

Liquids, 42 gallons (160 liters) per barrel, historic measure; 53 gallons (200 liters), contemporary American Standard Barrel (ASB) measure
Water, 8.34 pounds per gallon
Ethyl alcohol, 6.79 pounds per gallon at 190 proof and 60°F
White oak whiskey barrel (standard 53 gallons circa 1960), about 110 pounds empty; about 520 pounds full

Note that barreled whiskey may evaporate through the wood at a rate of roughly 2 percent per year, based on warehouse temperature and humidity, stave wood moisture content, proof, and other factors.

Sources: C. E. Bowman, *Annual Report of the Kentucky Bureau of Agriculture, Horticulture, and Statistics* (Frankfort, KY: S. I. M. Major, 1882), 290–91; Herman F. Willkie and Joseph A. Prochaska, *Fundamentals of Distillery Practice* (Louisville, KY: Joseph E. Seagram & Sons, 1943), 180.

Notes

Introduction

Epigraphs: Edmunds V. Bunkše, *Geography and the Art of Life* (Baltimore: Johns Hopkins University Press, 2004), 25; William H. Perrin, ed., *History of Bourbon, Scott, Harrison and Nicholas Counties, Kentucky* (Chicago: O. L. Baskin, 1882), 65; Bernard De Voto, *The Hour: A Cocktail Manifesto* (1948; reprint, Portland, OR: Tin House Books, 2010), 15.

1. The *Oxford English Dictionary* records the etymology of the term *whisky* or *whiskey* as short for *whiskybae* and the Gaelic *uisqebeathay,* or "water of life." The term referred to a spirituous liquor originally distilled in Ireland and Scotland from malted barley, with or without unmalted barley and other grains. In nineteenth-century trade usage, Scotch whisky and Irish whiskey were distinguished by their different spellings. *Whisky* became the usual spelling in Britain, while *whiskey* was commonly used in the United States. Consider, for example, Mark Twain's observations about the spirit's role in the Anglo settlement of America's interior in his book *Life on the Mississippi.* "How solemn and beautiful is the thought," he wrote, "that the earliest pioneer of civilization, the van-leader of civilization, is never the steamboat, never the railroad, never the newspaper, never the Sabbath-school, never the missionary— but always whiskey." A literal reading of Twain's choice of spellings might be interpreted as referring to the Irish and American traditions of usage. But this is pressing etymologic conventions too far. As a generic term, there is no general concurrence in its spelling, and the term was and is routinely spelled either way. Maker's Mark Distillery makes and sells bourbon whisky, and Brown-Forman purveys Early Times bourbon whisky. Most other bourbon distillers offer whiskey to their customers, including Brown-Forman's Old Forester whiskey. This book follows the American convention and uses the spelling *whiskey* unless the alternative spelling appears in a quotation. See *Oxford English Dictionary,* 2d ed., vol. 20 (Oxford: Clarendon Press, 1989), 254; Mark Twain, *Life on the Mississippi* (New York: Harper & Brothers, 1903), 428; Henry G. Crowgey, *Kentucky Bourbon: The Early Years of Whiskeymaking* (1971; reprint, Lexington: University Press of Kentucky, 2008), xiv; Michael R. Veach, *Kentucky Bourbon Whiskey: An American Heritage* (Lexington: University Press of Kentucky, 2013), 13.

2. See "Bourbon Facts," Kentucky Distillers' Association, http://kybourbon.com/bourbon_culture-2/key_bourbon_facts/.

3. See Sally Van Winkle Campbell, *But Always Fine Bourbon* (Frankfort, KY: Old Rip Van Winkle Distillery, 1999); Carla Harris Carlton, *Barrel Strength Bourbon: The Explosive Growth of America's Whiskey* (Birmingham, AL: Clerisy Press, 2017); Gerald Carson, *The Social History of Bourbon* (1963; reprint, Lexington: University Press of Kentucky, 2010); Sam K. Cecil, *Bourbon: The Evolution of Kentucky Whiskey* (New York: Turner Publishing, 2010); Charles K. Cowdery, *Bourbon Straight: The Uncut and Unfiltered Story of American Whiskey* (Chicago: Charles K. Cowdery, 2004); Crowgey, *Kentucky Bourbon;* Dane Huckelbridge, *A History of the American Spirit* (New York: William Morrow, 2014); Frank Kane, *Anatomy of the Whiskey Business* (Manhasset, NY: Lake House Press, 1965); Fred Minnick, *Bourbon: The Rise, Fall, and*

Rebirth of an American Whiskey (Minneapolis: Voyageur Press, 2016); F. Paul Pacult, *American Still Life: The Jim Beam Story and the Making of the World's #1 Bourbon* (Hoboken, NJ: John Wiley & Sons, 2003); John Ed Pearce, *Nothing Better in the Market* (Louisville, KY: Brown-Forman Distillers, 1970); Gary Regan and Mardee Haidin Regan, *The Book of Bourbon and Other Fine American Whiskeys* (Shelburne, VT: Chapters Publishing, 1995); Susan Reigler, *Kentucky Bourbon Country: The Essential Travel Guide* (Lexington: University Press of Kentucky, 2013); Richard Taylor, *The Great Crossing: A Historic Journey to Buffalo Trace Distillery* (Frankfort, KY: Buffalo Trace Distillery, 2002); Veach, *Kentucky Bourbon Whiskey*; Chester Zoeller, *Bourbon in Kentucky: A History of Distilleries in Kentucky*, 2d ed. (Louisville, KY: Butler Books, 2010); Chester Zoeller, *Kentucky Bourbon Barons: Legendary Distillers from the Golden Age of Whiskey Making* (Louisville, KY: Butler Books, 2014).

4. "Lexington Porter & Fine Ale Brewery," *Kentucky Gazette* 5, no. 15 (1814): 4.

5. Merle Patchett, "Historical Geographies of Apprenticeship: Rethinking and Retracing Craft Conveyance over Time and Place," *Journal of Historical Geography* 55 (2017): 31–32.

6. This claim about the first use of machine mashing is unclear because other distillers were using steam power by the eve of the Civil War. W. H. Perrin, J. H. Battle, and G. C. Kniffin, *Kentucky: A History of the State*, 4th ed. (Louisville, KY: F. A. Battey, 1887), 834–35; *An Atlas of Nelson and Spencer Counties, Kentucky* (Philadelphia: D. J. Lake, 1882), 27; Cecil, *Bourbon*, 69–70; Zoeller, *Bourbon in Kentucky*, 150.

7. John F. Quinn, *Father Mathew's Crusade: Temperance in Nineteenth-Century Ireland and Irish America* (Amherst: University of Massachusetts Press, 2002), 1–9; John A. Krout, *The Origins of Prohibition* (New York: Russell & Russell, 1925), 179–81.

8. Tim Ingold, *Making: Anthropology, Archaeology, Art and Architecture* (London: Routledge, 2013), 1. See also Daniel Miller, "The Power of Making," in *The Power of Making*, ed. Daniel Miller (London: V & A Publishing, 2011), 14–27; Chantel Carr and Chris Gibson, "Geographies of Making: Rethinking Materials and Skills for Volatile Futures," *Progress in Human Geography* 40, no. 3 (2016): 297–315; Chris Gibson, "Material Inheritances: How Place, Materiality, and Labor Process Underpin the Path-Dependent Evolution of Contemporary Craft Production," *Economic Geography* 92, no. 1 (2016): 61–86; Dydia DeLyser and Paul Greenstein, "The Devotions of Restoration: Materiality, Enthusiasm, and Making Three 'Indian Motocycles' Like New," *Annals of the Association of American Geographers* 107, no. 6 (2017): 1461–78.

9. Cf. Brooke Hindle, "The Exhilaration of Early American Technology," in *Early American Technology: Making and Doing Things from the Colonial Era to 1850*, ed. Judith A. McGraw (Chapel Hill: University of North Carolina Press, 1994), 63.

10. Ingold, *Making*, 107.

11. Dell Upton, "Seen, Unseen, and Scene," in *Understanding Ordinary Landscapes*, ed. Paul Groth and Todd W. Bressi (New Haven, CT: Yale University Press, 1997), 174–79.

12. Ingold, *Making*, 1–4.

13. Ibid., 7.

14. Ibid., 48.

15. Hindle, "Exhilaration of Early American Technology," 63.

16. John Wylie, *Landscape* (London: Routledge, 2007), 139.

17. Richard H. Schein, "Cultural Landscapes," in *Research Methods in Geography*, ed. Basil Gomez and John Paul Jones III (West Sussex, UK: Wiley-Blackwell, 2010), 222. One may also conceive of landscape as a form of "material space." See Theodore R. Schatzki, *Social Change in a Material World* (London: Routledge, 2019), 110.

18. Carl O. Sauer, "Foreword to Historical Geography," *Annals of the Association of American Geographers* 31, no. 1 (1941): 1–24.

19. William H. Marquardt and Carole L. Crumley, "Theoretical Issues in the Analysis of Spatial Patterning," in *Regional Dynamics: Burgundian Landscapes in Historical Perspective,* ed. Carole L. Crumley and William H. Marquardt (San Diego, CA: Academic Press, 1987), 1.

20. Carville Earle, *Geographical Inquiry and American Historical Problems* (Stanford, CA: Stanford University Press, 1992), 6.

21. D. W. Meinig, "Symbolic Landscapes: Some Idealizations of American Communities," in *The Interpretation of Ordinary Landscapes: Geographical Essays,* ed. D. W. Meinig (New York: Oxford University Press, 1979), 165–92.

22. See David Lowenthal, *The Past Is a Foreign Country* (Cambridge: Cambridge University Press, 1985); Jack Temple Kirby, *Mockingbird Song: Ecological Landscapes of the South* (Chapel Hill: University of North Carolina Press, 2006); Richard Schein, "Normative Dimensions of Landscape," in *Everyday America: Cultural Landscape Studies after J. B. Jackson,* ed. Chris Wilson and Paul Groth (Berkeley: University of California Press, 2003), 199–218; D. W. Meinig, "Reading the Landscape: An Appreciation of W. G. Hoskins and J. B. Jackson," in Meinig, *Interpretation of Ordinary Landscapes,* 195–244; W. G. Hoskins, *The Making of the English Landscape* (London: Hodder & Stoughton, 1955). For the Churchill quote, see https://www.parliament.uk/about/living-heritage/building/palace/architecture/palace structure/churchill/.

23. Wylie, *Landscape,* 7.

24. Ibid.

25. Ibid., 159.

26. Barbara Bender, "Theorising Landscapes, and the Prehistoric Landscapes of Stonehenge," *Man,* n.s., 27, no. 4 (1992): 735–36; David Frazer, "'Fields of Radiance': The Scientific and Industrial Scenes of Joseph Wright," in *The Iconography of Landscape,* ed. Denis Cosgrove and Stephen Daniels (Cambridge: Cambridge University Press, 1988), 119–41; David Lowenthal, *Possessed by the Past: The Heritage Crusade and the Spoils of History* (New York: Free Press, 1996), 148–72.

27. Tim Ingold, "The Temporality of the Landscape," *World Archaeology* 25, no. 2 (1993): 152. See also Marisa Lazzari, "The Texture of Things: Objects, People and Landscape in Northwest Argentina," in *Archaeologies of Materiality,* ed. Lynn Meskell (Oxford: Blackwell, 2005), 126–61; Bjornar Olsen, *In Defense of Things: Archaeology and the Ontology of Objects* (New York: Altamira Press, 2010), 1–19; Timothy J. Lecain, *The Matter of History: How Things Create the Past* (Cambridge: Cambridge University Press, 2017), 67–139.

28. D. W. Meinig, "The Beholding Eye: Ten Versions of the Same Scene," in Meinig, *Interpretation of Ordinary Landscapes,* 44.

29. Bruce P. Winterhalder, "Concepts in Historical Ecology: The View from Evolutionary Ecology," in *Historical Ecology: Cultural Knowledge and Changing Landscapes,* ed. Carole L. Chumley (Santa Fe, NM: School of American Research Press, 1994), 18–19; James D. Clarkson, "Ecology and Spatial Analysis," *Annals of the Association of American Geographers* 60, no. 4 (1970): 700–716.

30. Tristram R. Kidder and William Balee, "Epilogue," in *Advances in Historical Ecology,* ed. William Balee (New York: Columbia University Press, 1998), 405.

31. Marquardt and Crumley, "Theoretical Issues in the Analysis of Spatial Patterning," 6.

32. Winterhalder, "Concepts in Historical Ecology," 17–41; Christine Meisner Rosen, "The Business-Environment Connection," *Environmental History* 10, no. 1 (2005): 77–79.

33. Richard H. Schein, "A Conceptual Framework for Interpreting an American Scene," *Annals of the Association of American Geographers* 87, no. 4 (1997): 663.

34. Betsy Hunter Bradley, *The Works: The Industrial Architecture of the United States* (New York: Oxford University Press, 1999), 3.

35. Schein, "Normative Dimensions of Landscape," 199–218.

36. Alan L. Olmstead and Paul W. Rhode, "The Red Queen and the Hard Reds: Productivity Growth in American Wheat, 1800–1940," *Journal of Economic History* 62, no. 4 (2002): 929–66.

1. Heritage and Process

Epigraph: Anthony Boucherie, *The Art of Making Whiskey* (Lexington, KY: Worsley & Smith, 1819), 5.

1. Esther Kellner, *Moonshine: Its History and Folklore* (New York: Weathervane Books, 1971), 15–17.

2. Malachy Magee, *1,000 Years of Irish Whiskey* (Dublin: O'Brien Press, 1980), 7; Alfred Barnard, *The Whiskey Distilleries of the United Kingdom* (1887; reprint, Edinburgh: Birlinn Press, 2008), 3.

3. Thomas P. Slaughter, *The Whiskey Rebellion: Frontier Epilogue to the American Revolution* (New York: Oxford University Press, 1986), 223–24.

4. Henry G. Crowgey, *Kentucky Bourbon: The Early Years of Whiskeymaking* (1971; reprint, Lexington: University Press of Kentucky, 2008), 28–29.

5. "James Crow, Whiskey Maker," *New York Sun* 65, no. 5 (1897): 5.

6. Although residential proximity does not guarantee a professional calling, distillers in Lowland Scotland, east and west of Edinburgh, were producing malt whiskey by the mid-eighteenth century. Bulzion, St. Magdalene, and Kirkliston Distilleries operated west of Edinburgh, and several distilleries southeast of Edinburgh were in production by 1825. Mr. Crow's home in Dirleton was about ten miles northeast of the Glenkinchie Distillery, which began operations in 1837. See Barnard, *Whiskey Distilleries of the United Kingdom*, 326–43.

7. Gerald Carson, *The Social History of Bourbon* (1963; reprint, Lexington: University Press of Kentucky, 2010), 47.

8. H. V. Harlan and M. L. Martini, "Problems and Results in Barley Breeding," in *U.S.D.A. Yearbook of Agriculture, 1936* (Washington, DC: GPO, 1936), 310.

9. Howard T. Walden, *Native Inheritance: The Story of Corn in America* (New York: Harper & Row, 1966), 135.

10. US Congress, House, *The Reports of Committees: H.R. 4165*, 50th Cong., 2d sess., 1888–1889, 12.

11. The American alcohol scale ranges from 0 to 200. A spirit containing 50 percent alcohol is 100 proof.

12. US Department of the Treasury, Alcohol and Tobacco Tax and Trade Bureau, Title 27, Alcohol, Tobacco Products and Firearms; Part 5, Labeling and Advertising of Distilled Spirits; Subpart C, Standards of Identity; Section 5.22, Standards of Identity for Distilled Spirits, 48–52, https://www.gpo.gov/fdsys/pkg/CFR-2011-title27-vol1/pdf/CFR-2011-title27-vol1-chapI.pdf.

13. Barry Kornstein and Jay Luckett, "The Economic and Fiscal Impacts of the Distilling Industry in Kentucky" (Urban Studies Institute, University of Louisville, KY, 2014), 34. Grain production is related to climate, but the average annual temperatures in central Kentucky and central Indiana, for example, vary by only one-half of one degree Fahrenheit.

14. John W. Alexander, *Economic Geography* (Englewood Cliffs, NJ: Prentice-Hall, 1963), 309–10.

15. US Congress, *Reports of Committees: H.R. 4165*, 55.

16. Thomas H. Appleton Jr., "'Moral Suasion Has Had Its Day': From Temperance to Prohibition in Antebellum Kentucky," in *A Mythic Land Apart: Reassessing Southerners and Their History*, ed. John D. Smith and Thomas H. Appleton Jr. (Westport, CT: Greenwood Press, 1997), 19–42.

17. Lewis Loder, *The Loder Diary, 1857–1903*, transcribed by William Conrad (Florence, KY: Boone County Schools, 1985–1988), 82.

18. Alexander, *Economic Geography*, 309–10.

2. Kentucky Distilling

Epigraph: Harrison Hall, *The Distiller* (1818; reprint, 2d rev. ed., Hayward, CA: White Mule Press, 2015), 35.

1. Gerald Carson, *The Social History of Bourbon* (1963; reprint, Lexington: University Press of Kentucky, 2010), 24–28.

2. Victor S. Clark, *History of Manufactures in the United States 1607–1860* (Washington, DC: Carnegie Institution, 1916), 480.

3. Warren R. Hofstra, *The Planting of New Virginia: Settlement and Landscape in the Shenandoah Valley* (Baltimore: Johns Hopkins University Press, 2004), 224; Priscilla Ann Evans, "Merchant Gristmills and Communities, 1820–1880: An Economic Relationship," *Missouri Historical Review* 68, no. 3 (1974): 317–26.

4. John Moylan, Ledger A, Lexington, 1792, 12, 23, 47, 49, 116, 124, 135, 216, MS042, Special Collections, University of Kentucky.

5. Hofstra, *Planting of New Virginia*, 156.

6. Clark, *History of Manufactures*, 480.

7. Kentucky's population in 1810 was 406,571, or 203 people per distillery. Distilleries produced 2,220,773 gallons of whiskey in 1810, or 5.4 gallons per capita. By 1840, the population had increased to 779,828, but the number of distilleries had declined, yielding a ratio of 877 people per distillery.

8. Brian Page and Richard Walker, "From Settlement to Fordism: The Agro-Industrial Revolution in the American Midwest," *Economic Geography* 67, no. 4 (1991): 291.

9. Clarence H. Danhof, "Agriculture," in *The Growth of the American Economy*, ed. Harold F. Williamson (New York: Prentice-Hall, 1951), 182.

10. Ibid., 133–53.

11. Ibid., 151.

12. Page and Walker, "From Settlement to Fordism," 293.

13. Arthur G. Peterson, "Flour and Grist Milling in Virginia: A Brief History," *Virginia Magazine of History and Biography* 43, no. 2 (1935): 103.

14. The county boundaries depicted in the 1810 map of Kentucky changed as the legislature created additional counties in subsequent decades.

15. Lewis Loder, *The Loder Diary, 1857–1903*, transcribed by William Conrad (Florence, KY: Boone County Schools, 1985–1988), 82. Lawrenceburg, Indiana, has retained its distilling tradition, which began with the Dunn and Ludlow works in 1807 and expanded in the 1840s with the addition of the Squibb and Rossville Union Distilleries. Seagram's bought a facility there and began producing industrial alcohol in 1941. Although the Seagram's name still appears on some of the buildings, the works subsequently became MGP (Midwest Grain Products), a large-scale third-party distiller that produces unique mash bills for customers using specialized grains; rye whiskey is one of its specialties. One of the Middle West's primary grain shippers, Consolidated Grain & Barge, operates a large terminal on the Ohio River four miles away in Aurora, Indiana. See https://www.thespiritsbusiness.com/2018/05/uncovering-third-party-spirits-operations/2/; http://www.cincinnatimagazine.com/high-spirits-blog/mgp-ingredients-lawrenceburg/.

16. Hall, *The Distiller*, 23.

17. Page and Walker, "From Settlement to Fordism," 294.

18. Ibid., 294–96.

19. Leo Rogin, *The Introduction of Farm Machinery in Its Relation to the Productivity of Labor in the Agriculture of the United States during the Nineteenth Century* (Berkeley: University of California Press, 1931), 83.

20. William T. Hutchinson, *Cyrus Hall McCormick: Seed-Time, 1809–1856* (New York: Century, 1930), 351.

21. George H. Cook and John C. Smock, *Report on the Clay Deposits of Woodbridge, South Amboy, and Other Places in New Jersey, Together with Their Uses for Fire Brick, Pottery, etc.* (Trenton, NJ: Geological Survey of New Jersey, 1878); R. E. Lamborn, C. R. Austin, and Downs Schaaf, *Shales and Surface Clays of Ohio,* Bulletin 39 (Columbus: Ohio Geological Survey, 1938).

3. Kentucky's Distilling Environment

Epigraph: W. H. Auden, "In Praise of Limestone," in *Collected Shorter Poems, 1927–1957* (London: Faber & Faber, 1966), 238–41.

1. Carville Earle, *Geographical Inquiry and American Historical Problems* (Stanford, CA: Stanford University Press, 1992), 5–7.

2. William Cronon, "A Place for Stories: Nature, History, and Narrative," *Journal of American History* 78, no. 4 (1992): 1349.

3. C. E. Bowman, *Annual Report of the Kentucky Bureau of Agriculture, Horticulture, and Statistics,* Legislative Document 7 (Frankfort, KY: S. I. M. Major, 1881), 14 (emphasis in original).

4. Eugene H. Walker, *The Deep Channel and Alluvial Deposits of the Ohio Valley in Kentucky,* US Geological Survey Water Supply Paper 1411 (Washington, DC: GPO, 1957), 5–14.

5. W. N. Palmquist and F. R. Hall, *Public and Industrial Water Supplies of the Blue Grass Region, Kentucky,* US Geological Survey Circular 299 (Washington, DC: US Geological Survey, 1953), 13.

6. "Strike Fine Spring on Elkhorn Creek," *Frankfort Weekly News and Roundabout* 31, no. 43 (1908): 1.

7. John H. Procter, *Map of Kentucky* (Frankfort: Kentucky Geological Survey, 1881).

8. E. H. Taylor Jr., *Description of the O.F.C., Carlisle, and J. S. Taylor Distilleries* (Frankfort, KY: E. H. Taylor Jr. Co., [1886]), 20. See also advertising card, "The Whiskey," Taylor-Hay Family Papers, (Franklin County, KY), 1783–1991, MssAT238d, folder 73, Filson Historical Society, Louisville, KY.

9. Taylor, *Description of the O.F.C., Carlisle, and J. S. Taylor Distilleries,* 8.

10. Ibid., 9. The Procter map labels the Ordovician-age strata as Lower Silurian. The massive and dense Tyrone limestone does indeed outcrop near the bottom of the Kentucky River gorge from Frankfort upstream to southern Fayette County. The Tyrone is some sixty feet thick, exceedingly dense, and overlain by multiple members of the thin-bedded Lexington limestone formation. Groundwater at Frankfort likely moved through several of these Lexington formations, but percolation through the Tyrone would have been a protracted process. See A. M. Miller, *Geology of Franklin County,* Kentucky Geological Survey Reports, 4th series, vol. 2, pt. 3 (Frankfort: Kentucky Geological Survey, 1914). Moreover, in 1912 geologists characterized the water drawn from the Tyrone limestone as "salty." See M. L. Fuller and F. G. Clapp, *The Underground Waters of Southwestern Ohio,* US Geological Survey Water Supply Paper 259 (Washington, DC: GPO, 1912), 23.

11. *Sanborn's Survey of the Whiskey Warehouses of Kentucky and Tennessee* (New York: Sanborn-Perris Map Company, 1894), 52.

12. "Whisky Trade," *Louisville Courier-Journal,* n.s., 38, no. 148 (1868): 1.

13. Robert Peter, *Lower Blue Lick Spring: The Quantitative Chemical Analysis of the Water in the Lower Blue Lick Spring, in Nicholas County, Kentucky* (n.p., 1850), 2–3.

14. Walker, *Deep Channel and Alluvial Deposits of the Ohio Valley in Kentucky.*

15. Chester Zoeller, *Bourbon in Kentucky: A History of Distilleries in Kentucky,* 2d ed. (Louisville, KY: Butler Books, 2010), 59–60.

16. Edward Orton, *Report of the Occurrence of Petroleum, Natural Gas and Asphalt Rock in Western Kentucky: Based on Examinations Made in 1888 and 1889,* Kentucky Geological Survey and State Museum (Frankfort, KY: E. Polk Johnson, 1891), 194.

17. Herman Stabler, *Prevention of Stream Pollution by Distillery Refuse,* US Geological Survey Water-Supply and Irrigation Paper 179 (Washington, DC: GPO, 1906), 6.

18. M. I. Rorabaugh, F. F. Schrader, and L. B. Laird, *Water Resources of the Louisville Area, Kentucky and Indiana,* US Geological Survey Circular 276 (Washington, DC: GPO, 1953), 43.

19. US Department of Health, Education, and Welfare, Public Health Service, *Public Health Service Drinking Water Standards,* Publication 956 (Washington, DC: GPO, 1962), 42–43.

20. Ralph C. Heath, *Basic Ground-Water Hydrology,* US Geological Survey Water-Supply Paper 2220, rev. (Washington, DC: GPO, 2004), 64–67.

21. Palmquist and Hall, *Public and Industrial Water Supplies of the Blue Grass Region.* See also B. R. Scanlon and J. Thrailkill, "Chemical Similarities among Physically Distinct Spring Types in a Karst Terrain," *Journal of Hydrology* 89, no. 3–4 (1987): 259–79; Alan E. Fryer, "Springs and the Origin of Bourbon," *Ground Water* 47, no. 4 (2009): 605–10.

22. Palmquist and Hall, *Public and Industrial Water Supplies of the Blue Grass Region,* 41.

23. Ibid., 48.

24. Ibid., 52.

25. Ibid., 74.

26. Edwin A. Bell, Robert W. Kellogg, and Willis K. Kulp, *Progress Report on the Ground-Water Resources of the Louisville Area Kentucky, 1949–55,* US Geological Survey Water Supply Paper 1579 (Washington, DC: GPO, 1963), 45.

27. Ibid., 37–39. Cf. Michael R. Veach, *Kentucky Bourbon Whiskey: An American Heritage* (Lexington: University Press of Kentucky, 2013), 19.

28. Bell, Kellogg, and Kulp, *Progress Report on the Ground-Water Resources of the Louisville Area,* 16.

29. Martin Mayer, "The Volatile Business," *Esquire* 52, no. 2 (1959): 104.

30. Ibid.

31. Tim Ingold, "Toward an Ecology of Materials," *Annual Review of Anthropology* 41 (2012): 433–35.

32. "The Drought," *Franklin Farmer* 3, no. 3 (1839): 1.

33. John W. Jones Papers (Bourbon County, KY), 1857–1896, 1857–1889 (bulk dates), 75M26, Special Collections, University of Kentucky.

34. Edgar M. Hoover, *The Location of Economic Activity* (New York: McGraw Hill, 1948), 83–84.

35. Carville Earle, "A Staple Interpretation of Slavery and Free Labor," *Geographical Review* 68, no. 1 (1978): 51–65.

36. Carville Earle and Ronald Hoffman, "The Foundation of the Modern Economy: Agriculture and the Costs of Labor in the United States and England, 1800–1860," *American Historical Review* 85, no. 5 (1980): 1056–60.

37. Peter Mathias, "Agriculture and the Brewing and Distilling Industries in the Eighteenth Century," *Economic History Review,* n.s., 5, no. 2 (1952): 249.

38. Harry G. Enoch, *Colonel John Holder, Boonesborough Defender & Kentucky Entrepreneur* (Morley, MO: Acclaim Press, 2009), 189–211.

39. Ibid., 208–10.

40. Harry G. Enoch, *Grimes Mill: Kentucky Landmark on Boone Creek, Fayette County* (Bowie, MD: Heritage Books, 2002), 11–14.

41. John R. Robertson, *Petitions of the Early Inhabitants of Kentucky to the General Assembly of Virginia, 1769–1792* (Louisville, KY: John P. Morton, 1914), 144–45.

42. Nancy O'Malley, "Ruddle's Mills, Bourbon County, Kentucky: Early Industry in the Kentucky Bluegrass," *Mill Stone* 5, no. 2 (2006): 21–29. Note that Isaac Ruddell's name is spelled differently from the place known as Ruddles Mills.

43. Harry G. Enoch and Nancy O'Malley, "Early Mills of Bourbon County," *Mill Stone* 4, no. 1 (2005): 25–38.

44. Samuel D. Livingston, "Land for Sale," *Kentucky Whig* 1, no. 33 (1826): 4 (emphasis in original).

45. Coleman Sellers Jr., "Oliver Evans and His Inventions," *Scientific American,* Supplement 552 (February 15, 1879): 8813–14.

46. Livingston, "Land for Sale," 4.

47. Ibid.

48. *Franklin Farmer* 3, no. 5 (1839): 40.

49. *Franklin Farmer* 2, no. 43 (1839): 343.

50. US Census Bureau, 1850 Census, table IX, 623.

51. Frank R. Cox, *Soil Survey of Daviess and Hancock Counties, Kentucky,* USDA Soil Conservation Service and Kentucky Agricultural Experiment Station (Washington, DC: GPO, 1974); Steven J. Blanford, Patrick S. Aldridge, and Robert A. Eigel, *Soil Survey of Jefferson County, Kentucky,* USDA Soil Conservation Service and Kentucky Agricultural Experiment Station (Washington, DC: GPO, 1966, reissued 1991).

52. Karl Raitz and Nancy O'Malley, "An Index of Soil Production Potential as Applied to Historical Agricultural Adaptation: A Kentucky Example," *Historical Methods* 18, no. 5 (1985): 137.

53. Cf. Adam Beatty, *Southern Agriculture* (New York: C. M. Saxon, 1843), 201–40.

54. C. E. Bowman, *Annual Report of the Kentucky Bureau of Agriculture, Horticulture, and Statistics, 1880* (Frankfort, KY: E. H. Porter, 1880), 42–45; Raitz and O'Malley, "Index of Soil Production Potential," 143.

55. Alfred J. Richardson, Rudy Forsythe, and Hubert B. Odor, *Soil Survey of Bourbon and Nicholas Counties, Kentucky,* USDA Soil Conservation Service and Kentucky Agricultural Experiment Station (Washington, DC: GPO, 1982).

56. Douglas F. B. Black, Earle R. Cressman, and William C. MacQuown Jr., *The Lexington Limestone (Middle Ordovician) of Central Kentucky,* US Geological Survey Bulletin 1224-C (Washington, DC: GPO, 1965).

57. Herman P. McDonald, David Keltner, Pamela Wood, Bruce A. Waters, and Orville J. Whitaker, *Soil Survey of Anderson and Franklin Counties, Kentucky,* USDA Soil Conservation Service and Kentucky Agricultural Experiment Station (Washington, DC: GPO, 1985); Herman P. McDonald, Raymond P. Sims, and Dan Isgrig, *Soil Survey of Jessamine and Woodford Counties, Kentucky,* USDA Soil Conservation Service and Kentucky Agricultural Experiment Station (Washington, DC: GPO, 1983).

58. Fred S. Arms, James P. Fehr, David T. Carroll, Byron L. Wilson, Herman P. McDonald, and James C. Ross, *Soil Survey of Nelson County, Kentucky,* USDA Soil Conservation Service and Kentucky Agricultural Experiment Station (Washington, DC: GPO, 1971); John A. Kelley and William H. Craddock, *Soil Survey of Marion County, Kentucky,* USDA Soil Conservation Service and Kentucky Agricultural Experiment Station (Washington, DC: GPO, 1991); William H. Craddock, *Soil Survey of Washington County, Kentucky,* USDA Soil Conservation Service and Kentucky Agricultural Experiment Station (Washington, DC: GPO, 1986).

59. Tim Ingold, *The Perception of the Environment: Essays on Livelihood, Dwelling, and Skill* (London: Routledge, 2000), 320.

4. Distilling Grain, Feeding Livestock

Epigraph: Michael Krafft, *The American Distiller or, the Theory and Practice of Distilling* (Philadelphia: Archibald Bariram, 1804; reprint, Gale, Sabin Americana Print Editions, n.d.), 70–71.

1. US Congress, House, *The Reports of Committees: H.R. 4165,* 50th Cong., 2d sess., 1888–1889, 3. Note that in addition to bourbon, which required a mash bill of at least 51 percent corn, nineteenth-century Kentucky distillers made other whiskeys that used rye, wheat, barley, or malted barley as the primary grain.

2. *Cynthiana Democrat* 1, no. 25 (1869): 3.

3. Grenville Bathe and Dorothy Bathe, *Oliver Evans: A Chronicle of Early American Engineering* (Philadelphia: Historical Society of Pennsylvania, 1935).

4. Howard T. Walden, *Native Inheritance: The Story of Corn in America* (New York: Harper & Row, 1966), 133.

5. J. F. D. Smyth, *A Tour of the United States of America,* vol. 1 (London: G. Robinson, 1784), 298–99. Smyth's calculation was based on the measure of five bushels of corn per barrel, although other measures were also used. If Smyth's count was correct, eighteenth-century planting rates were substantially lower than those in contemporary corn farming. A twenty-first-century corn farmer likely plants about 30,000 seeds per acre. Seed counts per bushel vary from 65,000 to 100,000, so a bushel of seed plants only two to three acres.

6. George Imlay, *A Topographical Description of the Western Territory of North America* 2d ed. (London: J. Debrett, 1793), 166–69.

7. C. E. Bowman, *Annual Report of the Kentucky Bureau of Agriculture, Horticulture, and Statistics, 1880* (Frankfort, KY: E. H. Porter, 1880), 51.

8. Neal O. Hammon, "Settlers, Land Jobbers, and Outliers: A Quantitative Analysis of Land Acquisition on the Kentucky Frontier," *Register of the Kentucky Historical Society* 84, no. 3 (1986): 247–48.

9. Ellen Eslinger, "Farming on the Kentucky Frontier," *Register of the Kentucky Historical Society* 107, no. 1 (2009): 4–5.

10. Ibid., 8. An acre is 43,560 square feet, or about the equivalent of an American football field minus one end zone. Clearing such a parcel of land of trees or cane or breaking up sod with the tools of the day was no small accomplishment.

11. Neal O. Hammon, "Land Acquisition on the Kentucky Frontier," *Register of the Kentucky Historical Society* 78, no. 4 (1980): 297–98.

12. Henry A. Wallace and William L. Brown, *Corn and Its Early Fathers* (East Lansing: Michigan State University Press, 1956), 34–35.

13. US Bureau of the Census, *Report of the Superintendent of the Census, December 1, 1852* [7th Census] (Washington, DC: Robert Armstrong, 1853), 60.

14. Wallace and Brown, *Corn and Its Early Fathers,* 54–63.

15. "Baden Corn," *Franklin Farmer* 2, no. 16 (1838): 42, 80.

16. Arthur M. Schlesinger, "A Dietary Interpretation of American History," *Massachusetts Historical Society Proceedings* 68 (1944–1947): 202.

17. Caroline E. MacGill, *History of Transportation in the United States before 1860* (Washington, DC: Carnegie Institution, 1917), 111.

18. W. J. Rorabaugh, *The Alcoholic Republic: An American Tradition* (New York: Oxford University Press, 1979), 117–18.

19. US Bureau of the Census, *Report of the Productions of Agriculture*, 10th Census (Washington, DC: GPO, 1883), 103.

20. US Census Bureau, *Report of the Superintendent of the Census, 1852*, 88–89, 61.

21. Donald L. Kemmerer, "The Pre–Civil War South's Leading Crop, Corn," *Agricultural History* 23, no. 4 (1949): 236; William N. Parker, "A Note on Regional Culture in the Corn Harvest," *Agricultural History* 46, no. 1 (1972): 185.

22. Henry A. Wallace, "Corn and the Midwestern Farmer," *Proceedings of the American Philosophy Society* 100, no. 5 (1956): 455.

23. R. L. Ardrey, *American Agricultural Implements* (Chicago: R. L. Ardrey, 1894), 30–37.

24. Eugene C. Brooks, *The Story of Corn and the Westward Migration* (New York: Rand McNally, 1916), 221.

25. "Harvesting of Corn," *Franklin Farmer* 1, no. 4 (1837): 25.

26. Cf. Henry Glassie, *Pattern in the Material Folk Culture of the Eastern United States* (Philadelphia: University of Pennsylvania Press, 1968), 209–12.

27. Bird Smith, "Management and Diseases of Hogs," *Franklin Farmer* 1, no. 46 (1838): 366.

28. "Drawing Corn in the Shock," *American Agriculturist* 39 (October 1880): 430; John Fraser Hart, *The Rural Landscape* (Baltimore: Johns Hopkins University Press, 1998), 243.

29. These measures are based on a model developed by John Hudson. See John C. Hudson, *Making the Corn Belt: A Geographical History of Middle-Western Agriculture* (Bloomington: Indiana University Press, 1994), 7–8.

30. John W. Jones Papers (Bourbon County, KY), 1857–1896, 1857–1889 (bulk dates), 75M26, Special Collections, University of Kentucky.

31. Geographer Sam Hilliard estimated that in the 1850s, farmers in the Deep South rarely produced sufficient corn to feed hogs, and many counties in the Coastal Plain produced insufficient corn for livestock feed because much of the total production was consumed by humans. A 120-pound hog could eat about 5 pounds of feed per day. If the feed ration was 50 percent corn, and one bushel of corn weighed 56 pounds, then one hog would eat a bushel of corn in twenty-two days. Hilliard estimated that southern farmers on the most productive land might feed their hogs as much as five bushels per year, or roughly three-quarters of a pound of corn per day. See Sam Bowers Hilliard, *Hog Meat and Hoecake: Food Supply in the Old South, 1840–1860* (Carbondale: Southern Illinois University Press, 1972), 157–58.

32. "Corn Wanted," *Lexington Observer and Reporter* 60, no. 16 (1867): 1.

33. US Census Bureau, *Report of the Productions of Agriculture*, 10th Census, xvi; US Bureau of the Census, *Report of the Statistics of Agriculture in the United States*, 11th Census (Washington, DC: GPO, 1896), 8; C. E. Bowman, *Annual Report, Kentucky Bureau of Agriculture, Horticulture, and Statistics, 1881*, Legislative Document 7 (Frankfort, KY: S. I. M. Major, 1881); William H. Perrin, ed., *History of Bourbon, Scott, Harrison and Nicholas Counties, Kentucky* (Chicago: O. L. Baskin, 1882), 66.

34. "North Dakota Corn Makes Fine Whiskey," *Bismarck Daily Tribune* 31 (August 10, 1911): 8.

35. US Census Bureau, *Report of the Superintendent of the Census, 1852*, 56.

36. *National Magazine* 2, no. 9 (1846): 801.

37. Clarence H. Danhof, *Change in Agriculture: The Northern United States, 1820–1870* (Cambridge, MA: Harvard University Press, 1969), 31.

38. Bowman, *Annual Report of the Kentucky Bureau of Agriculture, Horticulture, and Statistics, 1880*, 42–45.

39. Alan L. Olmstead and Paul W. Rhode, "The Red Queen and the Hard Reds: Productivity Growth in American Wheat, 1800–1940," *Journal of Economic History* 62, no. 4 (2002): 931, 940.

40. Bowman, *Annual Report of the Kentucky Bureau of Agriculture, Horticulture, and Statistics, 1880*, 57.

41. "Agricultural Review of 1838," *Franklin Farmer* 2, no. 30 (1839): 237.

42. Bowman, *Annual Report of the Kentucky Bureau of Agriculture, Horticulture, and Statistics, 1880*, 46.

43. *National Magazine* 2, no. 11 (1846): 1033–37.

44. William Cronon, *Nature's Metropolis: Chicago and the Great West* (New York: W. W. Norton, 1991), 104–8.

45. US Department of Agriculture, *Yearbook, 1922* (Washington, DC: GPO, 1923), 503.

46. US Census Bureau, *Report of the Productions of Agriculture*, 10th Census, 123.

47. US Census Bureau, *Report of the Superintendent of the Census, 1852*, 58.

48. "Agricultural Review of 1838," *Franklin Farmer* 2, no. 30, (1839): 237.

49. US Census Bureau, *Report of the Superintendent of the Census, 1852*, 59.

50. Twentieth-century farmers often used rye as a winter cover and green manure crop, although the production of fine-grade rye grain declined. Twenty-first-century distillers import rye from producers on the American Great Plains and in Canada and Scandinavia.

51. US Census Bureau, *Report of the Productions of Agriculture*, 10th Census, 124.

52. "Winter and Spring Pasture," *Franklin Farmer* 3, no. 4 (1839): 30; Bowman, *Annual Report of the Kentucky Bureau of Agriculture, Horticulture, and Statistics, 1881*, 126–27.

53. John C. Weaver, "Barley in the United States: A Historical Sketch," *Geographical Review* 33, no. 1 (1943): 56.

54. Ibid., 58.

55. Warren M. Parsons, *Beer and Brewing in America: An Economic Study* (New York: United Brewers Industrial Foundation, 1941), 5; John C. Weaver, "United States Malting Barley Production," *Annals of the Association of American Geographers* 34, no. 2 (1944): 98.

56. US Census Bureau, *Report of the Superintendent of the Census, 1852*, 71.

57. "Agriculture," *National Magazine* 1, no. 1 (1845): 91.

58. US Census Bureau, *Report of the Productions of Agriculture*, 10th Census, 117.

59. H. V. Harlan and M. L. Martini, "Problems and Results in Barley Breeding," *U.S.D.A. Yearbook of Agriculture, 1936* (Washington, DC: GPO, 1936), 304.

60. Weaver, "Barley in the United States," 60–65; John C. Weaver, "Climatic Relations of American Barley Production," *Geographical Review* 33, no. 4 (1943): 569–72.

61. H. V. Harlan and G. A. Wiebe, *Growing Barley for Malt and Feed*, USDA Farmers' Bulletin 1732 (Washington, DC: GPO, 1934), 1–7.

62. US Census Bureau, *Report of the Productions of Agriculture*, 10th Census, 121; H. L. Shands and A. D. Dickson, "Barley: Botany, Production, Harvesting, Processing, Utilization, and Economics," *Economic Botany* 7, no. 1 (1953): 19.

63. Shands and Dickson, "Barley," 21.

64. Weaver, "United States Malting Barley Production," 98.

65. Elkhorn Distillery Records (Scott County, KY), 1868–1872, 1997 MS419, Book 1, p. 323, Special Collections, University of Kentucky.

66. Joseph Nimmo Jr., *Report on the Internal Commerce of the United States, 1881–1882* Treasury Department, US Bureau of Statistics (Washington, DC: GPO, 1884), 438.

67. R. B. Weir, "Distilling and Agriculture: 1870–1939," *Agricultural History Review* 32, no. 1 (1984): 49–62.

68. Smith, "Management and Diseases of Hogs," 365.

69. Ibid., 366; "Traveling Capacity of Berkshire Hogs," *Franklin Farmer* 3, no. 5 (1839): 36.

70. *Franklin Farmer* 3, no. 5 (1839): 40.

71. See, for example, "Livestock Sales," *Farmers' Home Journal*, October 16, 1879, 1; "T. W. Samuels & Sons Beech Grove Farm," *Farmers' Home Journal*, January 19, 1884, 1.

72. "Commissioner's Sale of Distillery Land," *Cynthiana News* 20, no. 22 (1870): 3.

73. Tom Kimmerer, *Venerable Trees: History, Biology, and Conservation in the Bluegrass* (Lexington: University Press of Kentucky, 2015), 15.

74. Lewis Sanders, "On Cattle," *Franklin Farmer* 3, no. 21 (1840): 165–66.

75. Paul C. Henlein, *Cattle Kingdom in the Ohio Valley, 1783–1860* (Lexington: University of Kentucky Press, 1959), 7.

76. Paul C. Henlein, "Cattle Driving in the Ohio Country, 1800–1850," *Agricultural History* 28, no. 2 (1954): 84.

77. Paul C. Henlein, "Early Cattle Ranges of the Ohio Valley," *Agricultural History* 35, no. 3 (1961): 150.

78. Adam Beatty, *Southern Agriculture* (New York: C. M. Saxon, 1843), 265–69; Paul C. Henlein, "Shifting Range-Feeder Patterns in the Ohio Valley before 1860," *Agricultural History* 31, no. 1 (1957): 1–2.

79. Henlein, "Cattle Driving in the Ohio Country," 89.

80. Samuel D. Martin, "Short Horn and Devon Cattle," *Franklin Farmer* 3, no. 12 (1839): 95; Lewis Loder, *The Loder Diary, 1857–1903*, transcribed by William Conrad (Florence, KY: Boone County Schools, 1985–1988), 86.

81. Bowman, *Annual Report of the Kentucky Bureau of Agriculture, Horticulture, and Statistics, 1880*, 76, 221–23.

82. Danhof, *Change in Agriculture*, 95.

83. Lewis McKee and Linda K. Bond, *History of Anderson County: 1780–1936* (Frankfort, KY: Roberts Print Company, 1936), 151.

5. Distillery Configurations

Epigraph: Harrison Hall, *The Distiller* (1818; reprint, 2d rev. ed., Hayward, CA: White Mule Press, 2015), 21.

1. Otto A. Rothert, *A History of Muhlenberg County, Kentucky* (Louisville, KY: John P. Morton, 1913), 121. See also Edward A. Kendall, *Travels through the Northern Parts of the United States in the Years 1807–1808*, vol. 3 (New York: I. Riley, 1809), 33–34.

2. "To Be Rented on Very Moderate Terms," *Kentucky Gazette* 15, no. 794 (1801): 3.

3. Peter Mathias, "Agriculture and the Brewing and Distilling Industries in the Eighteenth Century," *Economic History Review*, n.s., 5, no. 2 (1952): 249–51, 256.

4. Ibid., 252.

5. Although the chemical action of yeast was not understood until demonstrated by Louis Pasteur in the 1850s, early distillers knew that yeast was required for proper fermentation. See Anthony Boucherie, *The Art of Making Whiskey* (Lexington, KY: Worsley & Smith, 1819); Hall, *The Distiller*, 103.

6. Betsy Hunter Bradley, *The Works: The Industrial Architecture of the United States* (New York: Oxford University Press, 1999), 16; John Stilgoe, *Common Landscape of America, 1580 to 1845* (New Haven, CT: Yale University Press, 1982), 4–5, 300–333; Fred Kniffen and Henry Glassie, "Building in Wood in the Eastern United States: A Time-Place Perspective," *Geographical Review* 56, no. 1 (1966): 40–66.

7. Harry G. Enoch, *Grimes Mill: Kentucky Landmark on Boone Creek, Fayette County* (Bowie, MD: Heritage Books, 2002), 21–48; Kevin Murphy and Stephen C. Gordo, "Guyn's Mill in Woodford County, Kentucky, as Recorded by the Historic American Engineering Record," *Millstone: Journal of the Kentucky Old Mill Association* 12, no. 2 (2013): 33–52.

8. James M. Richards, *The Functional Tradition in Early Industrial Buildings* (London: Architectural Press, 1958), 14–15.

9. Michael Krafft, *The American Distiller or, the Theory and Practice of Distilling* (Philadelphia: Archibald Bariram, 1804; reprint, Gale, Sabin Americana Print Editions, n.d.), 56–57.

10. "Charged" refers to the volume of fermented mash liquid or distiller's beer that is transferred to the still to initiate the distilling process. See William M. Ambrose, *Bottled in Bond under U.S. Government Supervision* (Lexington, KY: William M. Ambrose, 2007), 3.

11. Krafft, *American Distiller*, 8–20; Hall, *The Distiller*, 27.

12. Terry S. Reynolds, *Stronger than a Hundred Men: A History of the Vertical Waterwheel* (Baltimore: Johns Hopkins University Press, 1983), 170–74.

13. Louis C. Hunter, *A History of Industrial Power in the United States, 1780–1930*, vol. 1, *Waterpower in the Century of the Steam Engine* (Charlottesville: University Press of Virginia, 1979), 104–6.

14. John Fuller, *Art of Coppersmithing: A Practical Treatise on Working Sheet Copper into All Forms* (1894; reprint, Mendham, NJ: Astragal Press, 1993), 6–13.

15. Krafft, *American Distiller*, 33–37.

16. J. A. Nettleton, *The Manufacture of Spirit as Conducted at the Various Distilleries of the United Kingdom* (London: Marcus Ward, 1893), 19–21, 133–45. Cf. Ian Buxton and Paul S. Hughes, *The Science and Commerce of Whiskey* (Cambridge: Royal Society of Chemistry, 2014), 10.

17. Boucherie, *Art of Making Whiskey*.

18. Bradley, *The Works*, ix. See Krafft, *American Distiller*, 22, on use of the term "works" to describe the distillery structure and equipment.

19. Krafft, *American Distiller*, 152, 157; Hall, *The Distiller*, 23.

20. Boucherie, *Art of Making Whiskey*; Hall, *The Distiller*; Krafft, *American Distiller*.

21. Nathen Rosenberg, *Exploring the Black Box: Technology, Economics, and History* (Cambridge: Cambridge University Press, 1994), 3–4.

22. Tim Ingold, "Making Culture and Weaving the World," in *Matter, Materiality and Modern Culture*, ed. P. M. Graves-Brown (London: Routledge, 2000), 60–62.

23. Thomas Hubka, "Just Folks Designing: Vernacular Designers and the Generation of Form," *Journal of Architectural Education* 32, no. 3 (1979): 27–29.

24. Bradley, *The Works*, 5–7.

25. *Franklin Farmer* 2, no. 43 (1839): 343.

26. Frederick Overman, *Mechanics for the Millwright, Machinist, Engineer, Civil Engineer, Architect and Student* (Philadelphia: Lippincott, Grambo, 1851), 191–96.

27. Harlan I. Halsey, "The Choice between High-Pressure and Low-Pressure Steam Power in America in the Early Nineteenth Century," *Journal of Economic History* 41, no. 4 (1981): 730–31. Oliver Evans's steam engines were distributed primarily to locations accessible by water transport. Four of his engines reached Kentucky by 1811—one each in Louisville and Frankfort, and two in Lexington. The Lexington engines powered a paper mill and a gristmill. Grenville Bathe and Dorothy Bathe, *Oliver Evans: A Chronicle of Early American Engineering* (Philadelphia: Historical Society of Pennsylvania, 1935), 207.

28. Henry M'Murtrie, *Sketches of Louisville* (Louisville, KY: S. Penn, 1819), 55.

29. B. Drake and E. D. Mansfield, *Cincinnati in 1826* (Cincinnati, OH: Morgan, Lodge & Fisher, 1827), 59.

30. Carroll W. Pursell Jr., *Early Stationary Steam Engines in America: A Study in the Migration of a Technology* (Washington, DC: Smithsonian Institution Press, 1969), 67.

31. Levi Woodbury, "Steam-Engines: Letter from the Secretary of the Treasury," US Congress, House, 25th Cong., 3d sess., 1838, doc. 21, 325. The treasury secretary stated that the national count of

stationary steam engines was flawed because states did not fully report information or, in the case of Kentucky, neglected to file a report (ibid., 379).

32. Louis C. Hunter, *A History of Industrial Power in the United States, 1780–1930*, vol. 2, *Steam Power* (Charlottesville: University Press of Virginia, 1985), 249.

33. "Jefferson County Kentucky Land for Sale," *Louisville Daily Democrat* 9, no. 44 (1852): 1.

34. "Local Notices," *Louisville Courier-Journal*, n.s., 38, no. 145 (1868): 2.

35. "Paris Machine Shop," *Bourbon News* 2, no. 119 (1883): 4.

36. Charles H. Fitch, *Report on the Manufacture of Engines and Boilers*, US Census, 1880 (Washington, DC: GPO, 1881), 50.

37. David E. Schob, "Woodhawks & Cordwood: Steamboat Fuel on the Ohio and Mississippi Rivers, 1820–1860," *Journal of Forest History* 21, no. 3 (1977): 124–32.

38. US Congress, House, *The Reports of Committees: H.R. 4165*, 50th Cong., 2d sess., 1888–1889, 2.

39. Hunter, *History of Industrial Power in the United States*, 2:427; Pursell, *Early Stationary Steam Engines*, 132.

40. Elkhorn Distillery Records (Scott County, KY), 1868–1872, 1997 MS419, book 1, p. 179 (February 10, 1869), Special Collections, University of Kentucky; Dan Day, "Corncobs—Burning Them with Coal," *Ag Connection* (University of Missouri Extension) 12, no. 12 (2006): 1–2.

41. "United States Marshall's Sale," *Cynthiana News* 22, no. 22 (1870): 3.

42. "Commissioner's Sale of Distillery Land," *Cynthiana News* 22, no. 22 (1870): 3.

43. Ed Vasser and Jerry Sudduth, *Frankfort and Cincinnati Railroad: History and Remembrances of the Bourbon Road* (Lexington, KY: CreateSpace.com, 2013), 12–13; Hamilton County [Ohio] Agricultural Society, *Western Agriculturalist and Practical Farmer's Guide* (Cincinnati, OH: Robinson & Fairbank, 1830), x–xi, 68; Karl Raitz, "Rock Fences and Preadaptation," *Geographical Review* 85, no. 1 (1995): 54–55.

44. Brian Page and Richard Walker, "From Settlement to Fordism: The Agro-Industrial Revolution in the American Midwest," *Economic Geography* 67, no. 4 (1991): 282.

45. See, for example, "No Gates Left," *Paducah Daily Sun* 1, no. 96 (1896): 1.

46. Hunter, *History of Industrial Power in the United States*, 2:433.

47. US Congress, *Reports of Committees: H.R. 4165*, 4, 6.

48. Ibid., 50.

49. See, for example, K. Austin Kerr, *Organized for Prohibition: A New History of the Anti-Saloon League* (New Haven, CT: Yale University Press, 1985), 17–18.

50. Charles Cross, *The American Compounder or Cross' Guide for Retail Liquor Dealers* (St. Louis: Charles Cross, 1899), 60, 79.

51. Ibid., 133–34.

52. Ibid., 72–73, 115–16.

53. *Practical Druggist and Pharmaceutical Review of Reviews* 35, no. 11 (1917): 36.

54. US Congress, *Reports of Committees: H.R. 4165*, 13.

55. Ibid., 3, 12; *Congressional Record*, 46th Cong., 2d sess., 1880, vol. 10, pt. 3, 2838.

56. "Big Boost Given the Manufacturers of Straight Whiskies," *Frankfort Weekly News and Roundabout* 31, no. 43 (1908): 1.

57. Jeremiah W. Jenks, "The Development of the Whiskey Trust," *Political Science Quarterly* 4, no. 2 (1889): 299–300.

58. *Congressional Record*, 46th Cong., 2d sess., 1880, vol. 10, pt. 3, 2834–45. See also Ambrose, *Bottled in Bond*, 16.

59. Jenks, "Development of the Whiskey Trust," 301–5.

60. Ibid., 305–6; Karen Clay and Werner Troesken, "Strategic Behavior in Whiskey Distilling, 1887–1895," *Journal of Economic History* 62, no. 4 (2002): 1001.

61. "The Distillery Combine," *Bourbon News* 19, no. 10 (1899): 5; Chester Zoeller, *Bourbon in Kentucky: A History of Distilleries in Kentucky*, 2d ed. (Louisville, KY: Butler Books, 2010), 28.

62. Thomas C. Martin, *Electrical Machinery, Apparatus, and Supplies*, US Bureau of the Census Special Reports, *Manufactures 1905*, pt. 4 (Washington, DC: GPO, 1908), 627.

63. Hunter, *History of Industrial Power in the United States*, 2:433.

64. Martin, *Electrical Machinery, Apparatus, and Supplies*, 632.

65. Ibid., 218–20.

66. Malcolm MacLaren, *The Rise of the Electrical Industry during the Nineteenth Century* (Princeton, NJ: Princeton University Press, 1943), 184–86.

67. "Lexington, Kentucky," *Niles Weekly Register* 7, no. 22 (1815): 339–40.

68. Zadok Cramer, *The Navigator*, 7th ed. (Pittsburgh: Cramer, Spear & Eichbaum, 1811), 126–27.

69. Mary Verhoeff, *The Kentucky River Navigation*, Filson Club Publication 28 (Louisville, KY: John P. Morton, 1917), 69.

70. C. E. Bowman, *Annual Report of the Kentucky Bureau of Agriculture, Horticulture, and Statistics, 1881*, Legislative Document 7 (Frankfort, KY: S. I. M. Major, 1881), 264.

71. Leland R. Johnson and Charles E. Parrish, *Triumph at the Falls: The Louisville and Portland Canal* (Louisville, KY: US Corps of Engineers, 2007), 49–61.

72. Woodbury, "Steam-Engines: Letter from the Secretary of the Treasury," 9.

73. Louis C. Hunter, *Steamboats on the Western Rivers: An Economic and Technological History* (New York: Dover Publications, 1949), 569–70.

74. Bowman, *Annual Report of the Kentucky Bureau of Agriculture, Horticulture, and Statistics, 1881*, 351–52.

75. Sam K. Cecil, *Bourbon: The Evolution of Kentucky Whiskey* (New York: Turner Publishing, 2010), 165; Zoeller, *Bourbon in Kentucky*, 126.

76. "Louisville Manufacture-Alcohol Distillery," *Louisville Daily Courier* 29, no. 12 (1859): 2.

77. Carl E. Kramer, "The Evolution of the Residential Land Subdivision Process in Louisville, 1772–2008," *Register of the Kentucky Historical Society* 107, no. 1 (2009): 38–52.

78. Harland Bartholomew, *Urban Land Uses*, Harvard City Planning Studies, vol. 4 (Cambridge, MA: Harvard University Press, 1932), plate IX (between pp. 100 and 101).

79. C. E. Bowman, *Annual Report of the Kentucky Bureau of Agriculture, Horticulture, and Statistics, 1880* (Frankfort, KY: E. H. Porter, 1880), 264–66.

80. Ibid., 264.

81. Zoeller, *Bourbon in Kentucky*, 104, 105, 117.

82. Tabulated from Zoeller, *Bourbon in Kentucky*, 100–127.

83. Distilling's complementary industries in Louisville in the 1870s and early 1880s included copper-smiths (3), brass and iron foundries (17), steam engine builders (13), maltsters (4), pork and beef packers (13), barrel stave and headings manufacturers (3), and tanners and curriers (17). *Caron's Annual Directory of the City of Louisville*, vol. 1 (Louisville, KY: Bradley & Gilbert, 1871); *Kentucky State Gazetteer & Business Directory 1883–4* (Detroit, MI: R. L. Polk & A. C. Danser, 1884), 749.

84. Francis Walker, *The Statistics of the Wealth and Industry of the United States*, US Census, 1870, vol. 3 (Washington, DC: GPO, 1872), 667, table XI.

85. George H. Yater, *Two Hundred Years at the Falls of the Ohio: A History of Louisville and Jefferson County* (Louisville, KY: Heritage Corporation, 1979), 106–7.

86. Peter C. Welsh, "A Craft that Resisted Change: American Tanning Practices to 1850," *Technology and Culture* 4, no. 3 (1963): 306.

87. Walker, *Statistics of the Wealth and Industry of the United States,* 667.

88. Bowman, *Annual Report of the Kentucky Bureau of Agriculture, Horticulture, and Statistics, 1880,* 264–66.

89. Welsh, "Craft that Resisted Change," 317.

90. J. Leander Bishop, *A History of American Manufactures from 1608 to 1860,* vol. 1 (Philadelphia: Edward Young, 1866), 453.

91. *Insurance Maps of Louisville, Kentucky,* vol. 1 (New York: Sanborn-Perris Map Company, 1892), 22, 24.

92. *Globe-Republican,* August 20, 1890, 2.

93. *Insurance Maps of Louisville, Kentucky,* vol. 4 (New York: Sanborn Map Company, 1905), 131.

94. *Caron's Annual Directory of the City of Louisville,* vol. 39 (Louisville, KY: Bradley & Gilbert, 1909), 2046–54. Note that this count does not include distillers that maintained business offices on adjacent streets or whose distillery works operated in other parts of the city. The south side of the 200 block of West Main Street, for example, included three wholesalers and rectifiers. Distillers like Paul Jones, who was a rectifier and made redistilled or blended whiskeys, did not require grain mills and other equipment and facilities used by fine whiskey distilleries.

95. John C. Hudson, *Making the Corn Belt: A Geographical History of Middle-Western Agriculture* (Bloomington: Indiana University Press, 1994), 6–12.

96. J. M. Nickles, "The Geology of Cincinnati," *Journal of the Cincinnati Society of Natural History* 20, no. 2 (1902): 49.

97. Wilton N. Melhorn and John P. Kempton, "The Teays Valley Problem: A Historical Perspective," in *Geology and Hydrology of the Teays-Mahomet Bedrock Valley Systems,* Special Paper 258, ed. Wilton N. Melhorn and John P. Kempton (Boulder, CO: Geological Society of America, 1991), 5. See also Daniel Drake, "Geological Account of the Valley of the Ohio: In a Letter from Daniel Drake, M.D. to Joseph Correa de Serra," *Transactions of the American Philosophical Society* 2, no. 4 (1825): 124–39.

98. Eugene H. Walker, *Geology and Ground-Water Resources of the Covington-Newport Alluvial Area, Kentucky,* US Geological Survey Circular 240 (Washington, DC: GPO, 1953), 7.

99. *Kentucky State Gazetteer and Business Directory, 1881–1882,* vol. 3, pt. 2 (Detroit, MI: R. L. Polk & A. C. Danser, 1882), 634, 821.

100. Zoeller, *Bourbon in Kentucky,* 172–75; Cecil, *Bourbon,* 111–12, 186–87.

101. Walker, *Geology and Ground-Water Resources of the Covington-Newport Alluvial Area,* 5, 25.

102. Cf. Paul A. Tenkotte, James C. Claypool, and David E. Schroeder, eds., *Gateway City: Covington, Kentucky, 1815–2015* (Covington, KY: Clerisy Press, 2015), 177.

103. Cf. George C. Matson, *Water Resources of the Blue Grass Region Kentucky,* US Geological Survey Water Supply Paper 233 (Washington, DC: GPO, 1909), 104–6, 136–38; A. B. Gibbons, *Geologic Map of Parts of Newport and Withamsville Quadrangles, Campbell and Kenton Counties, Kentucky,* map GQ-1072, US Geological Survey (Washington, DC: GPO, 1973); Stanley J. Luft, *Geologic Map of Part of the Covington Quadrangle, Northern Kentucky,* map GQ-955, US Geological Survey (Washington, DC: GPO, 1971).

104. Walker, *Geology and Ground-Water Resources of the Covington-Newport Alluvial Area,* 6, 25.

105. *Covington, Kentucky, 1886* (New York: Sanborn Map & Publishing Company, 1886).

106. From the colonial period to the end of the nineteenth century, Americans fouled their air and water with industrial and household pollutants. Raw sewage poured from pipes into streams and rivers,

where it mixed with all forms of noxious materials dumped by industries large and small. Waterborne diseases killed thousands each year. For example, in Cincinnati, which drew its water from the Ohio River, death rates from typhoid were 70 per 100,000 people in 1880, second among large cities only to Pittsburgh's rate of 135 per 100,000. As recently as 1946, the pollution load of untreated sewage and industrial waste dumped into the Ohio River and its tributaries was so heavy that one-fourth of the river's flow during low-water periods was from sewers. Martin Melosi, *The Sanitary City: Urban Infrastructure in America from Colonial Times to the Present* (Baltimore: Johns Hopkins University Press, 2000), 138, 333. See also Chase Palmer, "Quality of the Underground Waters in the Blue Grass Region," in Matson, *Water Resources of the Blue Grass Region*, 82.

107. Nancy O'Malley, "McConnell Springs in Historical Perspective" (paper, Department of Anthropology, University of Kentucky, n.d.), 23.

108. Gary O'Dell, "Water Supply and the Early Development of Lexington, Kentucky," *Filson Club Historical Quarterly* 67, no. 4 (1993): 443, 447.

109. Ibid., 450; Richard Ulack, Karl Raitz, and Gyula Pauer, eds., *Atlas of Kentucky* (Lexington: University Press of Kentucky, 1998), 35.

110. O'Dell, "Water Supply and the Early Development of Lexington," 453.

111. William H. Perrin, ed., *History of Fayette County, Kentucky* (Chicago: O. L. Baskin, 1882), 35. A chalybeate is impregnated with appreciable quantities of salts of iron.

112. George W. Ranck, *History of Lexington, Kentucky* (Cincinnati, OH: Robert Clarke, 1872), 18–19.

113. Perrin, *History of Fayette County*, 205.

114. Ibid., 206; *Lexington, Kentucky* (New York: Sanborn Map & Publishing Company, 1885), 18.

115. Warren R. Hofstra, *The Planting of New Virginia: Settlement and Landscape in the Shenandoah Valley* (Baltimore: Johns Hopkins University Press, 2004), 118.

116. Perrin, *History of Fayette County*, 206–9; O'Malley, "McConnell Springs in Historical Perspective," 17, 48; O'Dell, "Water Supply and the Early Development of Lexington," 431–61; *Lexington, Kentucky* [1885], 18.

117. Zoeller, *Bourbon in Kentucky*, 191–92.

118. Gary O'Dell, "Hillenmeyer Spring and Boiling Spring (Spring Lake Spring)" (unpublished paper, n.d.).

119. Perrin, *History of Fayette County*, 210; *Lexington, Kentucky* (New York: Sanborn-Perris Map Company, 1891), 22.

120. William M. Andrews Jr., "Geological Controls on Plio-Pleistocene Drainage Evolution of the Kentucky River in Central Kentucky" (PhD diss., Department of Geological Sciences, University of Kentucky, 2004), 77–84.

121. Nicholas Cresswell, *The Journal of Nicholas Cresswell, 1774–1777* (New York: Dial Press, 1924), 77–79.

122. Fortesque Cuming, *Sketches of a Tour to the Western Country, through the States of Ohio and Kentucky* (Pittsburgh: Cramer, Spear & Eichbaum, 1810), 169; B. N. Griffing, *An Atlas of Franklin County, Kentucky* (Philadelphia: D. J. Lake, 1882), 18–19.

123. Cuming, *Sketches of a Tour to the Western Country*, 168.

124. *Frankfort, Kentucky* (New York: Sanborn Map & Publishing Company, 1886), 10; E. H. Taylor Jr., *Description of the O.F.C., Carlisle, and J. S. Taylor Distilleries* (Frankfort, KY: E. H. Taylor Jr. Co., [1886]), 4.

125. *Frankfort, Kentucky*, 9; Cecil, *Bourbon*, 137–38.

126. Carl E. Kramer, *Capital on the Kentucky* (Frankfort, KY: Historic Frankfort, 1986), 278–81.

127. Zoeller, *Bourbon in Kentucky*, 194–99.

128. For example, the Basil Mattingly family lived in Maryland in the 1770s before moving to Washington County, Kentucky, where son George was born in 1802. George married Nancy Johnson of Washington County in 1829, and they moved to thinly settled Daviess County in 1832. The Monarch family moved from Maryland to Washington County in 1801. M. D. Monarch married Susan Davis of Washington County, and in 1834 they moved to Daviess County. Thomas J. Monarch was born in 1836, and he married Elisa Mattingly, daughter of George Mattingly, in 1885. Thomas Monarch built a distillery on the Green River near Birk City in 1859; he later opened a new distillery at Grissom's Landing on the Ohio River, ten miles downstream from Owensboro. See Leo McDonough, *An Illustrated Historical Atlas Map of Daviess County, Ky.* (1876; reprint, Evansville, IN: Unigraphic, 1978), 49, 52.

129. Ibid., 21.

130. US Census Bureau, *1870 Agriculture Census*. Neighboring Henderson County, part of the Western Coal Field, had 101,706 acres of improved farmland and 114,083 acres of woodland, the latter accounting for about 53 percent of the land in farms.

131. Frank R. Cox, *Soil Survey of Daviess and Hancock Counties, Kentucky*, USDA Soil Conservation Service and Kentucky Agricultural Experiment Station (Washington, DC: GPO, 1974).

132. Wilbur G. Burroughs, *The Geography of the Western Coal Field*, Kentucky Geological Survey, series 6, vol. 24 (Frankfort: Kentucky Geological Survey, 1924), 55–57.

133. Roger D. Marsden, *Drainage: Kentucky, Statistics for the State and Its Counties*, US Census Bulletin 1920 (Washington, DC: GPO, 1921), 8–9.

134. Burroughs, *Geography of the Western Coal Field*, 67; John Fraser Hart, "Change in the Corn Belt," *Geographical Review* 76, no. 1 (1986): 51–72.

135. William F. Switzler, *Report on the Internal Commerce of the United States* (Washington, DC: GPO, 1888), 491.

136. T. Edgar Harvey, *Commercial History of the State of Kentucky* (Louisville, KY: Courier-Journal Printing, 1899), 180.

137. McDonough, *Illustrated Historical Atlas Map of Daviess County*.

138. Based on Cecil, *Bourbon*, 116–26; Zoeller, *Bourbon in Kentucky*, 58–69; *History of Daviess County, Kentucky* (Chicago: Inter-State Publishing, 1883), 341–45.

139. Lee A. Dew and Aloma W. Dew, *Owensboro: The City on the Yellow Banks* (Owensboro, KY: Rivendell Publications, 1988), 65–79.

140. Lewis Loder, *The Loder Diary, 1857–1903*, transcribed by William Conrad (Florence, KY: Boone County Schools, 1985–1988), 88.

141. See, for example, *Owensboro, Kentucky* (New York: Sanborn Map & Publishing Company, 1885), 8–13.

142. *History of Daviess County*, 341. The same wording appeared seven years earlier in McDonough, *Illustrated Historical Atlas Map of Daviess County*, 52.

143. *History of Daviess County*, 341.

144. Mai Treude, *Windows to the Past: A Bibliography of Minnesota County Atlases*, Publication CURA 80-3 (Minneapolis: Center for Urban and Regional Affairs, University of Minnesota, 1980), 1–8.

145. Michael P. Conzen, "Landownership Maps and County Atlases," *Agricultural History* 58, no. 2 (1984): 119–20; Michael P. Conzen, "The County Landownership Map in America: Its Commercial Development and Social Transformation, 1814–1939," *Imago Mundi* 36 (1984): 17–26; Norman J. W. Thrower, "The County Atlas of the United States," *Surveying and Mapping* 21, no. 3 (1961): 365–73.

146. Thomas J. Schlereth, *Victorian America: Transformations in Everyday Life, 1876–1915* (New York: HarperCollins, 1991), 243.

6. Technology's Tools

Epigraph: Harrison Hall, *The Distiller* (1818; reprint, 2d rev. ed., Hayward, CA: White Mule Press, 2015), 35.

1. Warren R. Hofstra, "Private Dwellings, Public Ways, and the Landscape of Early Rural Capitalism in Virginia's Shenandoah Valley," in *Gender, Class, and Shelter: Perspectives in Vernacular Architecture V*, ed. Elizabeth C. Cromley and Carter L. Hudgins (Knoxville: University of Tennessee Press, 1995), 213.

2. F. Sherwood Taylor, "The Evolution of the Still," *Annals of Science* 5, no. 3 (1945): 186.

3. Robert H. Thurston, *A Manual of the Steam-Engine* (New York: John Wiley & Sons, 1891), 2–3 (emphasis in original). Thurston was the first president of the American Society of Mechanical Engineers in 1883.

4. Everett M. Rogers and F. Floyd Shoemaker, *Communication of Innovations*, 2d ed. (New York: Free Press, 1971), 16–17; Everett M. Rogers, *Diffusion of Innovations*, 5th ed. (New York: Free Press, 2003), 279–85. See also Nathan Rosenberg, *Perspectives on Technology* (Cambridge: Cambridge University Press, 1976), 173–74.

5. Cf. Alan L. Olmstead, "The Mechanization of Reaping and Mowing in American Agriculture, 1833–1870," *Journal of Economic History* 35, no. 3 (1975): 344–47.

6. William T. Hutchinson, *Cyrus Hall McCormick: Seed-Time, 1809–1856* (New York: Century, 1930), 467.

7. Edmund Burke, *List of Patents for Inventions and Designs Issued by the United States from 1790 to 1847* (Washington, DC: J. & G. S. Gideon, 1847).

8. *Fisher's National Magazine and Industrial Record* 3, no. 2 (1846): 154–57.

9. The term "merchant-distiller" refers to those distillers who did not produce any of the grains they consumed but purchased them from local, regional, or even national markets.

10. Introduction to Alfred Barnard, *The Whisky Distilleries of the United Kingdom* (1887; reprint, Edinburgh: Birlinn Press, 2008).

11. Henry Tritton, "Patent for an Improved Apparatus for Distilling," *Annals of Philosophy* 11 (1818): 445–47; Henry Tritton, "On Distillation," *Register of the Arts and Sciences, Improvements & Discoveries* 10 (February 21, 1824): 147–50.

12. Luke Herbert, *The Engineer's and Mechanic's Encyclopedia*, vol. 1 (London: Thomas Kenny, 1836), 59–62; Charles Tovey, *British & Foreign Spirits* (London: Whittaker, 1864), 15–18.

13. Anthony Perrier, "Sir Anthony Perrier's Improved Apparatus for Distillation," *Register of the Arts and Sciences* 1 (1824): 10–12.

14. "Winter's Patent Distilling Apparatus," *Register of the Arts and Sciences* 11 (1824): 161–64; Herbert, *Engineer's and Mechanic's Encyclopedia*, 62–64.

15. "Patent Granted to Robert Stein," *Repertory of Patent Inventions* 8, no. 51 (1830): 538–40; J. A. Nettleton, *The Manufacture of Spirit as Conducted at the Various Distilleries of the United Kingdom* (London: Marcus Ward, 1893), 19; Royal Commission on Whiskey and Other Potable Spirits, *Minutes of Evidence Taken by the Royal Commission on Whiskey and Other Potable Spirits*, vol. 1 (London: Jas. Truscott & Son, 1908), 251–52, appendix B.

16. John Booth, *A Toast to Ireland: A Celebration of Traditional Irish Drinks* (Boulder, CO: Roberts Rinehart, 1995), 36–40. Some sources state that Aeneas Coffey was born in Dublin.

17. Peter Mulryan, *The Whiskeys of Ireland* (Dublin, Ireland: O'Brien Press, 2016), 40–41.

18. "Patent Granted to Aeneas Coffey," *Repertory of Patent Inventions* 11, no. 72 (1831): 190–96; Nettle-

ton, *Manufacture of Spirit,* 133–45; Royal Commission on Whiskey and Other Potable Spirits, *Minutes of Evidence,* 252, appendix B, 382. Note that distillers in Scotland refer to their product as "whisky," whereas contemporary Irish distillers prefer the spelling "whiskey."

19. "Improvement in Stills," George W. Robson, Cincinnati, OH, and Melvin T. Hughes, Paris, KY, US Patent No. 87,971, March 16, 1869.

20. "Improvement in Stills for Alcoholic Spirits," William and George W. Robson, Cincinnati, OH, US Patent No. 92,477, July 13, 1869. The patent drawing was a partially sectioned elevation arranged in one plane for illustrative purposes. The various units in the installed apparatus would not have been suspended above the floor.

21. "Process of and Apparatus for Distilling," John C. Peden, Lawrenceburg, KY, US Patent No. 349,449, September 21, 1886.

22. Ian Buxton and Paul S. Hughes, *The Science and Commerce of Whiskey* (Cambridge: Royal Society of Chemistry, 2014), 10.

23. Sam K. Cecil, *Bourbon: The Evolution of Kentucky Whiskey* (New York: Turner Publishing, 2010), 158.

24. "A New Distillery," *Cynthiana News* 19, no. 2 (1869): 2.

25. Ibid.

26. Andrew J. Oots Cooper Shop Records, 1880–1926, 1880–1910 (bulk dates), 1 FM-49, Special Collections, University of Kentucky.

27. "Pure Copper Whisky," *Louisville Daily Express* 1, no. 172 (1869): 3.

28. M. D. Leggett, *Subject-Matter Index of Patents for Inventors Issued by the United States Patent Office, 1790–1873,* vol. 1 (Washington, DC: GPO, 1874), 622–23.

29. The Peter Jepson grain dryer, patented in 1887, is an example of innovation. Jepson's dryer was specifically designed to process spent grain. The design included a vertical cylinder that contained a series of funnel-shaped hoppers and perforated disks mounted on a rotating shaft. Pipes circulated pressurized hot air onto the grain, which, when dried, exited the drying cylinder through a chute on the bottom. See "Grain-Drying Apparatus," Peter Jepson, Port Chester, NY, US Patent No. 373,140, November 15, 1887.

30. In 1904 a fire in the dryer room at the M. S. Greenbaum Distillery at Midway in Woodford County caused $25,000 to $30,000 in damage.

31. "Drier," John E. Turney, Louisville, KY, US Patent No. 740,607, October 6, 1903.

32. "Drier," John E. Turney and Geiger, Koop & Fiske, Louisville, KY, US Patent No. 774,859, November 15, 1904.

33. "Business Notes," *Engineering Record* 43, no. 9 (1901): 208; "Business Notes," *Engineering Record* 51, no. 11 (1905): 46; "Turney Drier Company," *Railroad Age* 12 (March 24, 1905): 394; "Drying Plant," *Wine and Spirit Bulletin* 19, no. 1 (1905): 41.

34. "The Louisville Grains Dryer," *Western Brewer* 30, no. 4 (1906): 192.

35. Michael R. Veach, *Kentucky Bourbon Whiskey: An American Heritage* (Lexington: University Press of Kentucky, 2013), 39.

36. A version of this hook-and-rope device is still in use in some distillery warehouses. Steel cable has replaced the rope, and lifting power is provided by an electric winch.

37. "Improvement in Hoisting-Machines," Henry Reedy, Cincinnati, OH, US Patent No. 78,829, June 9, 1868; reissued 1871.

38. See Mike Veach, http://www.bourbonenthusiast.com/forum/viewtopic.php?f=17&t=115.

39. "Improvement in Racks for Tiering Barrels," Frederick Stitzel, Louisville, KY, US Patent No. 221,945, November 25, 1879; reissued April 27, 1880.

40. At some point, the term "rack" transformed into "rick," and some distillers referred to their aging warehouses as "rick houses." Traditionally, "rick" referred to a symmetrical stack of something, such as firewood. It is possible, but doubtful, that distillers conflated the terms, given that a rack warehouse under construction could be interpreted as resembling a stack of wood.

41. Betsy Hunter Bradley, *The Works: The Industrial Architecture of the United States* (New York: Oxford University Press, 1999), 184; John R. Stilgoe, *Common Landscape of America, 1580–1845* (New Haven, CT: Yale University Press, 1982), 329; James M. Richards, *The Functional Tradition in Early Industrial Buildings* (London: Architectural Press, 1958).

42. Bradley, *The Works,* 60–65.

43. "Walsh's New Distillery," *Bourbon News* 17, no. 29 (1897): 5.

44. Lewis Loder, *The Loder Diary, 1857–1903,* transcribed by William Conrad (Florence, KY: Boone County Schools, 1985–1988), 82.

45. US Congress, House, *The Reports of Committee:. H.R. 4165,* 50th Cong., 2d sess., 1888–1889, 66.

46. Richard Peters, ed., *The Public Statutes at Large of the United States of America from the Organization of the Government in 1789, to March 3, 1845,* vol. 3 (Boston: Charles C. Little & James Brown, 1846), 43.

47. US Congress, *Reports of Committees: H.R. 4165,* 22.

48. Grant McCracken, *Culture & Consumption: New Approaches to the Symbolic Character of Consumer Goods and Activities* (Bloomington: Indiana University Press, 1988), 106.

7. Complementary Industries

Epigraph: Michael Krafft, *The American Distiller or, the Theory and Practice of Distilling* (Philadelphia: Archibald Bariram, 1804; reprint, Gale, Sabin Americana Print Editions, n.d.), 30.

1. Leander J. Bishop, *A History of American Manufactures from 1608 to 1860,* vol. 1 (Philadelphia: Edward Young, 1866), 492.

2. Henry H. Work, *Wood, Whiskey, and Wine: A History of Barrels* (London: Reaktion Books, 2014), 17.

3. Kenneth Kilby, *The Cooper and His Trade* (Fresno, CA: Linden Publishing, 1971), 65–67.

4. "Wanted Immediately," *Kentucky Gazette Extra,* January 15, 1801, 2.

5. Joseph B. Wagner, *Cooperage: A Treatise on Modern Shop Practice and Methods, from the Tree to the Finished Article* (Yonkers, NY: J. B. Wagner, 1910), 153–54.

6. Wayne D. Rasmussen, "Wood on the Farm," in *Material Culture of the Wooden Age,* ed. Brooke Hindle (Tarrytown, NY: Sleepy Hollow Press, 1981), 15–34.

7. J. R. Mosedale, "Effects of Oak Wood on the Maturation of Alcoholic Beverages with Particular Reference to Whisky," *Forestry* 68, no. 3 (1998): 211.

8. N. S. Shaler and A. R. Crandall, *Report on the Forests of Greenup, Carter, Boyd & Lawrence Counties,* Geological Survey of Kentucky, 2d series, pt. 1, vol. 1 (Frankfort, KY: Yeoman Press, 1884), 13–14.

9. Henry Hall, *Report on the Ship-Building Industry of the United States* (Washington, DC: Department of the Interior, Census Office, 1884), 191.

10. Simon Difford, "Casks," in *Difford's Guide,* 5, https://www.diffordsguide.com/encyclopedia/481.

11. "How a Barrel Is Made Up," *Paducah Evening Sun* 25, no. 92 (1909): 8.

12. Mary Verhoeff, *The Kentucky River Navigation,* Filson Club Publication 28 (Louisville, KY: John P. Morton, 1917), 188–89; Steven A. Schulman, "The Lumber Industry of the Upper Cumberland River Valley," *Tennessee Historical Quarterly* 32, no. 3 (1973): 256; "Correspondence," *Mountain Advocate* 5, no. 8 (1908): 4.

13. Lewis Loder, *The Loder Diary, 1857–1903,* transcribed by William Conrad (Florence, KY: Boone County Schools, 1985–1988), 118.

14. Schulman, "Lumber Industry of the Upper Cumberland River Valley," 258.

15. "Along the Cincinnati Southern Railroad," *Farmers' Home Journal* 27, no. 46 (1879): 1.

16. "Good Thing for Patesville," *Breckenridge News* 28, no. 7 (1903): 7.

17. "A Big Business," *Mt. Sterling Advocate* 15, no. 20 (1904): 3.

18. US Congress, House, *The Reports of Committees: H.R. 4165,* 50th Cong., 2d sess., 1888–1889, 2.

19. C. E. Bowman, *Annual Report of the Kentucky Bureau of Agriculture, Horticulture, and Statistics, 1880* (Frankfort, KY: E. H. Porter, 1880), 82. In 1901 the Kern Company of Vienna, Austria, employed 450 men in Magoffin County, Kentucky, to cut and ship more than 15 million staves. The stave cutters included 150 men from the county "and 300 . . . Slavonians from Austria." "Facts and Observations," *Mt. Sterling Advocate* 11, no. 30 (1901): 6.

20. Henry Gannett and Jasper E. Whelchel, "Lumber," in US Department of Commerce, *Twelfth Census of the United States, 1900,* pt. 3, vol. 9, *Manufacturers* (Washington, DC: GPO, 1900), 853.

21. Verhoeff, *Kentucky River Navigation,* 113.

22. Gannett and Whelchel, "Lumber," 884, 824.

23. Mosedale, "Effects of Oak Wood on the Maturation of Alcoholic Beverages," 204.

24. Work, *Wood, Whiskey, and Wine,* 126–36.

25. Verhoeff, *Kentucky River Navigation,* 195–96. See also Mosedale, "Effects of Oak Wood on the Maturation of Alcoholic Beverages," 211.

26. Kenneth L. Cope, *American Cooperage Machinery and Tools* (Mendham, NJ: Astragal Press, 2003), 6.

27. Joseph Fleischman, *The Art of Blending and Compounding Liquors and Wines* (New York: Dick & Fitzgerald, 1885), 10–12.

28. Victor S. Clark, *History of Manufactures in the United States 1607–1860* (Washington, DC: Carnegie Institution, 1916), 472.

29. Cope, *American Cooperage Machinery and Tools,* 6.

30. US Bureau of the Census, *1850 Census* (Washington, DC: Robert Armstrong, 1853), 623, table X.

31. "Our Louisville Letter," *National Coopers Journal* 26, no. 12 (1911): 11.

32. Wagner, *Cooperage,* 193.

33. "Local Matters," *Cynthiana News* 8, no. 44 (1867): 3.

34. "Weekly Roundabout," *Weekly Roundabout* 3, no. 29 (1880): 5.

35. US Congress, *Reports of Committees: H.R. 4165,* 2.

36. "Queen City Copper Works," *Cynthiana News* 19, no. 44 (1869): 4.

37. William F. Switzler, *Report on the Internal Commerce of the United States* (Washington, DC: GPO, 1888), 491, 494.

38. Louisville Board of Trade, Committee on Industrial and Commercial Improvement, *The City of Louisville and a Glimpse of Kentucky, 1887* (Louisville, KY: Courier-Journal Press, 1887), 17.

39. Switzler, *Report on the Internal Commerce of the United States,* 437.

40. "T. B. Ripy and the Anderson County Distilling Company Distilleries," *Anderson News,* Souvenir Supplement, June 1906, 51.

41. "In Neighboring Counties," *Semi-Weekly Interior Journal* 31, no. 9 (1903): 1.

42. Andrew J. Oots Cooper Shop Records (Fayette County, KY), 1880–1926, 1880–1910 (bulk dates) 1 FM-49, Special Collections, University of Kentucky. See also *Kentucky State Gazetteer and Business Directory, 1881–1882,* vol. 3, pt. 2 (Detroit, MI: R. L. Polk & A. C. Danser, 1882), 629, which lists one cooper in

Lexington in 1881–1882 (H. F. Oots). *Prather's Directory of the City of Lexington, Kentucky, August, 1895* (Lexington, KY: Jas. H. Prather, 1895), lists Andrew J. Oots, cooper, 227 W. Main; H. F. Oots, livestock dealer, 227 W. Main; Robt. L. Oots, letter carrier, 227 W. Main; Wm. E. Oots, cooper, 227 W. Main; and W. E. Oots & Company (W. E. and H. F. Oots), coopers, 239 W. Main.

43. Whiskey barrel capacity was changed after World War II from 48 to 53 gallons, or 200 liters, the largest size that would fit in the racks used in most aging warehouses.

44. See *Lexington, Kentucky* (New York: Sanborn-Perris Map Company, 1890), 9.

45. G. D. Cole, "Woodworking Plants," *Industrial Development and Manufacturers' Record* 53, no. 4 (1908): 67. See also *National Coopers' Journal* 23, no. 11 (1908): 2.

46. Chester Zoeller, *Bourbon in Kentucky: A History of Distilleries in Kentucky*, 2d ed. (Louisville, KY: Butler Books, 2010), 170.

47. Ibid., 182–83.

48. William H. Perrin, ed., *History of Bourbon, Scott, Harrison and Nicholas Counties, Kentucky* (Chicago: O. L. Baskin, 1882), 510.

49. Donna Reaich, "The Influence of Copper on Malt Whisky Character," in *Proceedings of the Fifth Aviemore Conference on Malting, Brewing and Distilling*, Aviemore, Inverness-shire, Scotland, May 25–28, 1998) 141–52; Barry Harrison, Olivier Fagnen, Frances Jack, and James Brosnan, "The Impact of Copper in Different Parts of Malt Whisky Pot Stills on New Make Spirit Composition and Aroma," *Journal of the Institute of Brewing* 117, no. 1 (2011): 106–12; "Copper," *Whisky Science* (blog), http://whiskyscience.blogspot.com/2014/10/copper.html.

50. "Copper Mines and the Copper Trade," *National Magazine* 1, no. 2 (1845): 117–23.

51. James A. Mulholland, *A History of Metals in Colonial America* (University: University of Alabama Press, 1981), 37–43.

52. Bishop, *History of American Manufactures*, 547–48.

53. Otis E. Young, "Origins of the American Copper Industry," *Journal of the Early Republic* 3, no. 2 (1983): 118–25.

54. Ibid., 127, 133. See also Henry J. Kauffman, *American Copper & Brass* (Camden, NJ: Thomas Nelson & Sons, 1968), 21–23.

55. Young, "Origins of the American Copper Industry," 136–37; Clark, *History of Manufactures in the United States*, 525; Orris C. Herfindahl, *Copper Costs and Prices: 1870–1957* (Baltimore: Johns Hopkins University Press, 1959), 70–71.

56. M. L. Quinn, "Industry and Environment in the Appalachian Copper Basin, 1890–1930," *Technology and Culture* 34, no. 3 (1993): 575–612.

57. US Bureau of the Census, *Manufactures 1905*, pt. 4, *Special Reports on Selected Industries* (Washington, DC: GPO, 1908), 162.

58. Thomas J. Schlereth, "The New York Artisan in the Early Republic: A Portrait from Graphic Evidence, 1787–1853," *Landscape, Place, & Material Culture* 20, no. 1 (1988): 3.

59. Shirley Austin referred to this human quality as "the sympathetic touch of the skilled workman." Shirley P. Austin, "Glass," in US Department of Commerce, *Twelfth Census of the United States, 1900*, pt. 3, vol. 9, *Manufactures* (Washington, DC: GPO, 1900), 978.

60. Nicolas R. Laracuente, "Archaeological Investigations of the O.F.C. Building, Buffalo Trace Distillery," (2016), 4, 11, on file at Buffalo Trace Distillery, Frankfort, KY; *Sanborn's Survey of the Whiskey Warehouses of Kentucky and Tennessee* (New York: Sanborn-Perris Map Company, 1894), 52.

61. "Fishel & Gallatine, Copper & Tin Smiths," *Kentucky Gazette* 15, no. 334 (1802): 3.

62. "They Made Moonshine Whisky," *Climax* 10, no. 28 (1896): 4; "Personal," *Climax* 4, no. 38 (1891): 3.

63. The cost of copper sheets or plates depended on thickness and weight (ounces per square foot). Twenty-ounce copper weighs twenty ounces per square foot of plate about 0.027 inch thick. Kauffman, *American Copper & Brass*, 13.

64. "Stills for Sale," *Kentucky Gazette* 24, no. 1302 (1810): 1 (emphasis in original).

65. Karl Raitz and Nancy O'Malley, *Kentucky's Frontier Highway: Historical Landscapes along the Maysville Road* (Lexington: University Press of Kentucky, 2012), 63.

66. "John Metcalfe & Bro.," *Louisville Daily Courier* 27, no. 155 (1858): 4.

67. "Louisville Manufacturers," *Louisville Daily Courier* 39, no. 46 (1859): 1. See also "Lightburn & Ward," *Louisville Daily Courier* 31, no. 73 (1860): 1.

68. Everett M. Rogers and F. Floyd Shoemaker, *Communication of Innovations*, 2d ed. (New York: Free Press, 1971), 14–15. See also Everett M. Rogers, *Diffusion of Innovations*, 5th ed. (New York: Free Press, 2003), 305–8; Joe S. Bain, *Industrial Organization* (New York: John Wiley & Sons, 1959), 58–62; Kauffman, *American Copper & Brass*, 107.

69. "Louisville Manufacturers," *Louisville Daily Courier* 39, no. 46 (1859): 1.

70. "Hoffman, Ahlers & Co.," *Kentucky Irish American* 18, no. 12 (1907): 10.

71. Vendome continues to operate on Franklin Street in Louisville, a short distance from the original Main Street location. See http://www.vendomecopper.com.

72. See, for example, "Old Hickory Whisky Now on Tap," *Crittenden Press* 27, no. 23 (1905): 2.

73. Helen McKearin and Kenneth M. Wilson, *American Bottles & Flasks and Their Ancestry* (New York: Crown, 1978), 170, 562–63.

74. Ellsworth Steele, "The Flint Glass Workers' Union in the Indiana Gas Belt and the Ohio Valley in the 1890's," *Indiana Magazine of History* 50, no. 3 (1954): 238; Quentin R. Skrabec, *Michael Owens and the Glass Industry* (Gretna, LA: Pelican Publishing, 2007), 43.

75. Sam K. Cecil, *Bourbon: The Evolution of Kentucky Whiskey* (New York: Turner Publishing, 2010), 173–74; William M. Ambrose, *Bottled in Bond under U.S. Government Supervision* (Lexington, KY: William M. Ambrose, 2007), 18.

76. Austin, "Glass," 983–84.

77. Warren C. Scoville, "Growth of the American Glass Industry to 1800," *Journal of Political Economy* 52, no. 3 (1944): 203–4; Jack K. Paquette, *The Glassmakers Revisited* (Toledo, OH: Xlibris Press, 2010), 22.

78. Skrabec, *Michael Owens and the Glass Industry*, 45.

79. Austin, "Glass," 950.

80. Warren C. Scoville, "Growth of the American Glass Industry to 1880—Continued," *Journal of Political Economy* 52, no. 4 (1944): 340–42.

81. Steele, "Flint Glass Workers' Union," 238.

82. Warren C. Scoville, *Revolution in Glassmaking: Entrepreneurship and Technological Change in the American Industry, 1880–1920* (Cambridge, MA: Harvard University Press, 1948), 154.

83. Austin, "Glass," 950.

84. Walter G. Lezius, "Geography of Glass Manufacture at Toledo, Ohio," *Economic Geography* 13, no. 4 (1937): 404. See also Ernest F. Burchard, "Glass-Sand Industry of Indiana, Kentucky, and Ohio," in *Contributions to Economic Geology 1906*, ed. S. F. Emmons and E. C. Eckel, US Geological Survey Bulletin 315, series A, Economic Geology (Washington, DC: GPO, 1907), 361–76.

85. Paquette, *Glassmakers Revisited*, 17–18.

86. "Machine for Blowing Glass," M. J. Owens, Toledo, OH, US Patent No. 548,588, October 22, 1895.

87. Edward Meigh, *The Story of the Glass Bottle* (Stoke-on-Trent, UK: C. E. Ramsden, 1972), 44–49. See

also Austin, "Glass," 973; George L. Miller and Catherine Sullivan, "Machine-Made Glass Containers and the End of Production for Mouth-Blown Bottles," *Historical Archaeology* 18, no. 2 (1984): 86.

88. "Glass-Shaping Machine," Michael J. Owens, Toledo, OH, US Patent No. 766,768, August 2, 1904.

89. Skrabec, *Michael Owens and the Glass Industry,* 223.

90. Owen R. Lovejoy, "Child Labor in the Glass Industry," *Annals of the American Academy of Political and Social Science* 27, no. 1 (1906): 43.

91. George E. Barnett, *Chapters on Machinery and Labor* (Cambridge, MA: Harvard University Press, 1926), 88–89.

92. Scoville, *Revolution in Glassmaking,* 189.

93. US Department of Labor, Bureau of Labor Statistics, "Displacement of Labor by Machinery in the Glass Industry," *Monthly Labor Review* 24, no. 4 (1927): 1–4.

94. Steele, "Flint Glass Workers' Union," 248; Paquette, *Glassmakers Revisited,* 29; Skrabec, *Michael Owens and the Glass Industry,* 212.

95. "U.S. Bottlers' Supply Co.," *Wine and Spirit Bulletin* 15, no. 10 (1901): 23.

96. Paquette, *Glassmakers Revisited,* 30.

97. Scoville, *Revolution in Glassmaking,* 345.

98. McKearin and Wilson, *American Bottles & Flasks,* 170–71; David Whitten, "Louisville Glass Factories of the 19th Century—Part 1," in *Bottles and Extras,* 2–3, http://www.fohbc.org/wp-content/uploads/2014/06/LouisvilleGlassFactories_P1_2Spring2005.pdf.

99. David Whitten, "Louisville Glass Factories of the 19th Century—Part 2," in *Bottles and Extras,* 3, https://sha.org/bottle/pdffiles/LouisvilleGlass2_Whitten.pdf.

100. Austin, "Glass," 952.

101. Ibid., 951, 958. At about the same time, a large Cincinnati glassworks licensed to produce liquor bottles with Owens machines employed about 700 people. That plant operated until the eve of Prohibition in 1919, when it closed. See Scoville, *Revolution in Glassmaking,* 255.

102. Charles H. Richardson, *The Glass Sands of Kentucky,* Kentucky Geological Survey, series 6, vol. 1 (Frankfort, KY: State Journal Company, 1920), 91–97.

103. "What Natural Gas Will Do," *Mt. Sterling Advocate* 3, no. 29 (1893): 1.

104. Austin, "Glass," 975.

105. Mike McCormick, *Terre Haute: Queen City of the Wabash* (Charleston, SC: Arcadia Press, 2005), 100. The ingredients used to manufacture English flint glass included silica and powdered flint or chert. The term was later used more generally to describe lustrous, high-quality glass made with a variety of metal oxides. In America, skilled glassworkers were known as "flints," and in the 1870s they organized the American Flint Glass Workers' Union. Steele, "Flint Glass Workers' Union," 229–50.

106. Warren C. Scoville, "Spread of Techniques: Minority Migrations and the Diffusion of Technology," *Journal of Economic History* 11, no. 4 (1951): 349.

107. "Frankfort-Modes-Glass Works," *Frankfort Roundabout* 29, no. 48 (1906): 4.

108. "Trial Start at Glass Works," *Frankfort Roundabout* 30, no. 20 (1907): 4.

109. "Glass Works under Way," *Frankfort Roundabout* 30, no. 21 (1907): 6.

110. "Glass Works Shuts Down," *Frankfort Roundabout* 30, no. 37 (1907): 6.

111. "Thorn Hill Hotel," *Frankfort Roundabout* 31, no. 3 (1907): 6.

112. "Fire out at Glass Works," *Frankfort Roundabout* 31, no. 11 (1907): 3.

113. "Glass Factory Closed Down," *Frankfort Roundabout* 31, no. 9 (1907): 5.

114. Carl E. Kramer, *Capital on the Kentucky* (Frankfort, KY: Historic Frankfort, 1986), 272.

115. Barnett, *Chapters on Machinery and Labor*, 89.

116. "Construction of Corn Shellers," John A. Whitford, Saratoga Springs, NY, US Patent No. 1946, January 23, 1841.

117. "Corn-Sheller," Augustus Adams, Sandwich, IL, US Patent No. 32,971, August 6, 1861; reissued March 20, 1866.

118. R. L. Ardrey, *American Agricultural Implements* (Chicago: R. L. Ardrey, 1894), 124–25.

119. "Local Matters," *Cynthiana News* 8, no. 44 (1867): 3.

8. Signatures of Risk

Epigraph: Paducah Sun 17, no. 264 (1905): 1.

1. W. Paul Strassmann, *Risk and Technological Innovation: American Manufacturing Methods during the Nineteenth Century* (Ithaca, NY: Cornell University Press, 1959), 11.

2. Dalit Baranoff, "Shaped by Risk: The American Fire Insurance Industry, 1790–1920," *Enterprise & Society* 6, no. 4 (2005): 563.

3. "Tremendous Rains," *Frankfort Weekly News and Roundabout* 31, no. 30 (1908): 6; "Heavy Rainfall," *Frankfort Roundabout* 18, no. 44 (1895): 4.

4. "The Flood," *Frankfort Roundabout* 6, no. 22 (1883): 2.

5. "The Story of the Flood," *Breckenridge News* 7, no. 32 (1883): 2.

6. "The Storms," *Frankfort Roundabout* 9, no. 40 (1886): 2.

7. "Electricity on the Rampage," *Frankfort Roundabout* 10, no. 44 (1887): 3.

8. "The Bad Weather," *It* 1, no. 11 (1903): 5.

9. "River News," *Louisville Daily Democrat* 8, no. 74 (1851): 3.

10. "In Town and Country," *It* 1, no. 11 (1903): 1.

11. "Did You Ever," *It* 1, no. 7 (1902): 4.

12. "Kentucky Whisky," *Wine and Spirit Bulletin* 15, no. 11 (1901): 14.

13. Andrew Ure, *The Philosophy of Manufactures* (London: Charles Knight, 1835), 32; "Times in Frankfort, No. VI," *Frankfort Roundabout* 29, no. 2 (1906): 5.

14. "Local Matters," *Cynthiana News* 12, no. 46 (1867): 3.

15. Lewis Loder, *The Loder Diary, 1857–1903,* transcribed by William Conrad (Florence, KY: Boone County Schools, 1985–1988), 1.

16. Ibid., 60.

17. Ibid., 71.

18. "Horrible Accident," *Bourbon News,* March 22, 1904, 5. A similar accident at a large distillery in Peoria, Illinois, killed three employees. "News Notes," *Interior Journal* 33, no. 12 (1905): 1.

19. "Kentucky State News," *Hickman Courier* 28, no. 45 (1893): 4.

20. Loder, *Loder Diary,* 135; "News Condensed," *Semi-Weekly Interior Journal* 17, no. 27 (1889): 2; "Badly Mashed," *Bourbon News* 23 (May 26, 1903): 5; "Boys Suffocated," *Breckenridge News* 30, no. 42 (1906): 9; "Johnson Recovered $4,000 Damages," *Citizen* 6, no. 31 (1905): 7; "Hand Mangled," *Semi-Weekly Interior Journal* 31, no. 92 (1904): 3.

21. Loder, *Loder Diary,* 144.

22. Levi Woodbury, "Steam-Engines: Letter from the Secretary of the Treasury," US Congress, House, 25th Cong., 3d sess., 1838, doc. 21, 8.

23. D. E. Salmon, *First Annual Report of the Bureau of Animal Industry, 1884,* US Department of Agriculture (Washington, DC: GPO, 1885), 27–34; "The Daily Bulletin," *Daily Evening Bulletin* 3, no. 114 (1884): 3.

24. "Washington," *Daily Evening Bulletin* 3, no. 283 (1884): 3; Alan L. Olmstead and Paul W. Rhode, *Arresting Contagion: Science, Policy, and Conflicts over Animal Disease Control* (Cambridge, MA: Harvard University Press, 2015), 42–62.

25. D. E. Salmon, *Hog Cholera: Its History, Nature, and Treatment,* US Department of Agriculture, Bureau of Animal Industry (Washington, DC: GPO, 1889), 9.

26. "Cattle Disease in England," *Kentucky Live Stock Record* 13, no. 7 (1881): 106; "Pleuro-Pneumonia, Where and When," *Chicago Tribune,* September 23, 1886, 3; "The Cattle Disease in Kentucky," *Louisville Weekly Courier* 20, no. 38 (1866): 2; "The Cattle Plague," *Farmers' Home Journal* 27, no. 24 (1879): 5.

27. "Pleuro-Pneumonia," *Chicago Tribune,* September 22, 1886, 1.

28. "Must Be Slaughtered," *Chicago Tribune,* September 24, 1886, 1.

29. D. E. Salmon, *Third Annual Report, Bureau of Animal Industry, 1886,* US Department of Agriculture (Washington, DC: GPO, 1887), 12–14.

30. Ibid., 9–10; "Monthly Crop Report," *Big Sandy News* 8, no. 15 (1889): 2.

31. Salmon, *First Annual Report, Bureau of Animal Industry,* 27; Olmstead and Rhode, *Arresting Contagion,* 64–67.

32. D. E. Salmon, *Second Annual Report, Bureau of Animal Industry, 1885,* US Department of Agriculture (Washington, DC: GPO, 1886), 9–24; Salmon, *Third Annual Report, Bureau of Animal Industry,* 9–11; "The Pleuro-Pneumonia," *Daily Evening Bulletin* 3, no. 243 (1884): 1; Olmstead and Rhode, *Arresting Contagion,* 67; "The Kentucky Legislature," *Hazel Green Herald* 2, no. 2 (1886): 2.

33. Salmon, *Third Annual Report, Bureau of Animal Industry,* 10.

34. Salmon, *First Annual Report, Bureau of Animal Industry,* 28.

35. Salmon, *Hog Cholera,* 12; Sheryl Crowell, *New Richmond* (Charleston, SC: Arcadia Publishing, 2012), 69.

36. "Agricultural," *Louisville Weekly Journal* 27, no. 51 (1857): 4.

37. "Hog Cholera," *Louisville Daily Courier* 31, no. 231 (1860): 1; "Fearful Ravages of the Hog Cholera," *Louisville Daily Express,* n.s., 1, no. 174 (1869): 4.

38. William A. Penn, *Kentucky Rebel Town: The Civil War Battles of Cynthiana and Harrison County* (Lexington: University Press of Kentucky, 2016), 57.

39. Loder, *Loder Diary,* 93.

40. "Farm and Home," *Bourbon News* 1, no. 4 (1882): 3; "News Paragraphs," *Climax* 3, no. 23 (1889): 5; "Kentucky Newslets," *Spout Spring Times* 4, no. 15 (1899): 1.

41. "Farm and Trade Items," *Semi-Weekly Interior Journal* 16, no. 102 (1889): 1.

42. "City and Vicinity," *Semi-Weekly Interior Journal* 22, no. 36 (1894): 3; "Crab Orchard," *Semi-Weekly Interior Journal* 23, no. 18 (1895): 1.

43. Salmon, *Hog Cholera,* 7–12; Olmstead and Rhode, *Arresting Contagion,* 145–48.

44. "Fires," *Bourbon News* 1, no. 6 (1882): 3.

45. *The Chronicle Fire Tables for 1897* (New York: Chronicle Company, 1897), 300.

46. US Internal Revenue Service, *Report of the Commissioner of Internal Revenue, Fiscal Year Ended June 30, 1900* (Washington, DC: GPO, 1900), 146.

47. "Fire at the G. G. White Distillery," *Bourbon News* 18, no. 41 (1898): 2; "Small Fire Saturday," *Frankfort Roundabout* 24, no. 30 (1901): 5.

48. Sara E. Wermiel, *The Fireproof Building: Technology and Public Safety in the Nineteenth-Century American City* (Baltimore: Johns Hopkins University Press, 2000), 2–3.

49. "The Fire Last Night," *Louisville Evening Express*, n.s., 1, no. 95 (1869): 1.

50. In 1884, 62 percent of all fires reported in Kentucky were set by incendiaries, the highest rate in the nation and more than twice the national average of 30 percent. Kentucky's high arson rate may have been related to the long-standing southern tradition of autumn woods-burning; more likely, it was related, at least in part, to strong temperance proclivities. *The Chronicle Fire Tables for 1885* (New York: Chronicle Company, 1885), 9; *Chronicle Fire Tables for 1897*, 361; Stephen J. Pyne, *Fire in America: A Cultural History of Wildland and Rural Fire* (Princeton, NJ: Princeton University Press, 1982), 142–43.

51. "Kentucky State News," *Hickman Courier* 29, no. 6 (1893): 4. For a report on arson at a distillery in Henderson, Kentucky, see "A $200,000 Fire," *Hopkinsville Kentuckian* 18, no. 56 (1896): 1.

52. "Another Outrage," *Mt. Sterling Advocate* 17, no. 31 (1908): 3.

53. "Distillery Burned, Blood Hounds Sent For," *Mt. Sterling Advocate* 15, no. 16 (1904): 6.

54. Wermiel, *Fireproof Building*, 49.

55. "News Condensed," *Semi-Weekly Interior Journal* 18, no. 75 (1890): 2.

56. "Fire at Silver Creek," *Richmond (KY) Climax* 28, no. 37 (1902): 2.

57. "The Fire," *Frankfort Roundabout* 7, no. 35 (1884): 2.

58. "Fire Wednesday Night," *Frankfort Roundabout* 13, no. 3 (1890): 4.

59. "Frankfort Roundabout," *Frankfort Roundabout* 7, no. 32 (1884): 2.

60. "The Telegraph Line," *Frankfort Roundabout* 7, no. 49 (1884): 2.

61. Betsy Hunter Bradley, *The Works: The Industrial Architecture of the United States* (New York: Oxford University Press, 1999), 112.

62. "Don't Fear Fire Now," *Frankfort Roundabout* 15, no. 31 (1892): 4. The newly constructed Bulleit Distillery opened in rural Shelby County in 2017. The fire-suppression system includes twin diesel engines running high-pressure pumps that draw water from a 200,000-gallon pond and a nearby reservoir. A pressurized sprinkler system protects the building interiors, and exterior fire hydrants connect to large hydrant-mounted nozzles that can spray the building and grain storage tank exteriors. There are four single-story aging warehouses, each with a capacity exceeding 55,000 barrels. The warehouse exterior walls are set atop four- to five-foot poured concrete "dike" walls that form a fully enclosed perimeter around the building's base. In the event of a fire and collapse of the barrels stacked inside, the concrete walls and floors would act as a trough to contain all the burning whiskey within the building's base, preventing it from flowing downslope and endangering other structures. Interview with Dwayne Kozlowski, Bulleit Distillery, Shelbyville, KY, June 7, 2017.

63. "Distillers Insurance," *Paducah Sun* 9, no. 260 (1902): 2.

64. "Petition of Andrew Jackson for Relief from the Whiskey Tax," February 12, 1803, Record of the US House of Representatives, 1789–2015, Petitions and Memorials, 1793–1911, National Archives Catalog.

65. See, for example, US Congress, House, *A Bill for the Benefit of W. L. Berry, of Kentucky*, H.R. 439, 42d Cong., 1st sess., April 20, 1871; US Congress, House, *A Bill to Remit the Internal-Revenue Tax upon Five Hundred and Forty-Nine Barrels of Whisky . . . State of Kentucky*, H.R. 699, 42d Cong., 2d sess., December 18, 1871.

66. "Fire at Old Crow Distillery," *Frankfort Roundabout* 30, no. 52 (1907): 1.

67. "The Commonwealth," *Climax* 4, no. 4 (1890): 6; Chester Zoeller, *Bourbon in Kentucky: A History of Distilleries in Kentucky*, 2d ed. (Louisville, KY: Butler Books, 2010), 118–19.

68. "Whisky Hot," *New Ulm Weekly Review* 13, no. 34 (1890): 2; "News Condensed," *Semi-Weekly Interior Journal* 18, no. 49 (1890): 2.

69. "Government Heavy Loser," *Winchester News* 1, no. 37 (1908): 2.

70. "The Fire King," *Daily Public Ledger*, April 8, 1893, 4.

71. "Lancaster, Garrard County," *Semi-Weekly Interior Journal* 22, no. 37 (1894): 1.

72. "Moved Warehouse," *Frankfort Roundabout* 15, no. 47 (1904): 7.

73. Francis C. Moore, *Fire Insurance and How to Build* (New York: Baker & Taylor, 1903), 63.

74. Baranoff, "Shaped by Risk," 569; Peter L. Bernstein, *Against the Gods: The Remarkable Story of Risk* (New York: John Wiley, 1996), 203–4.

75. Frank H. Knight, *Risk, Uncertainty and Profit* (Boston: Houghton Mifflin, 1921; reprint, Mansfield Centre, CT: Martino Publishing, 2014), 247.

76. Edward V. French, *The Development of the Prevention Idea* (Boston: privately printed for Arkwright Mutual Fire Insurance Company, 1912), 110.

77. "Insurance Company of North America," *Fortune* 15, no. 2 (1937): 166.

78. Moore, *Fire Insurance and How to Build,* 356–57.

79. Ibid., 70.

80. Ibid., 457.

81. Ibid., 445.

82. "Fires," *Bourbon News* 1, no. 6 (1882): 3.

83. A similar distillery in Willisburg operated by B. F. Colvin in 1892 was also assessed at a rate of $7. Rates on distillery buildings were higher than on warehouses. Distilleries with external boiler houses at least thirty feet away received a 30-cent rate reduction. S. L. Rice, *Established Rates of Kentucky Distilleries, Bonded and Free Warehouses, Cattle and Pens* (Indianapolis: Rough Notes, 1892), 44.

84. "The Recent Advance in the Rates Causes Feeling among Distillers," *Citizen* 7, no. 21 (1905): 7.

85. Walter W. Ristow, "United States Fire Insurance and Underwriters Maps 1852–1968," *Quarterly Journal of the Library of Congress* 25, no. 3 (1968): 198.

86. *Sanborn's Surveys of the Whiskey Warehouses of Kentucky and Tennessee* (New York: Sanborn-Perris Map Company, 1894).

87. *Sanborn's Surveys of the Distilleries and Warehouses of Kentucky and Tennessee* (New York: Sanborn Map Company, 1910).

88. Zoeller, *Bourbon in Kentucky,* 209–10.

89. "All Sorts and Sizes," *Semi-Weekly Interior Journal* 16, no. 283 (1887): 4.

90. "Fire Destroys Midway Distillery," *Frankfort Weekly News* 31, no. 48 (1908): 1.

91. "Burned Whisky," *Frankfort Weekly News and Roundabout* 31, no. 49 (1908): 8.

92. "Current Events," *Adair County News* 11, no. 43 (1908): 6.

93. *Sanborn's Surveys of the Distilleries and Warehouses of Kentucky and Tennessee,* 67.

94. Richard Taylor, *The Great Crossing: A Historic Journey to Buffalo Trace Distillery* (Frankfort, KY: Buffalo Trace Distillery, 2002), 67–70.

95. *Sanborn's Surveys of the Distilleries and Warehouses of Kentucky and Tennessee,* 73.

96. US Congress, House, *The Reports of Committees: H.R. 4165,* 50th Cong., 2d sess., 1888–1889, 39.

97. Ibid., 41.

98. Ibid., 46.

9. By-products

Epigraphs: Michael Krafft, *The American Distiller or, the Theory and Practice of Distilling* (Philadelphia: Archibald Bariram, 1804; reprint, Gale, Sabin Americana Print Editions, n.d.), 152; "Stamping Ground," *Frankfort Roundabout* 30, no. 21 (1907): 5.

1. Angela O'Keeney, *Looking Back on Ballinascreen* (Draperstown, Ireland: Ballinascreen Historical Society, 1989), 41.

2. Herman Stabler, *Prevention of Stream Pollution by Distillery Refuse*, US Geological Survey Water Supply and Irrigation Paper 179 (Washington, DC: GPO, 1906), table 11, 26.

3. "Cheap Stock Feed for Sale," *Kentucky Sentinel* 1, no. 19 (1868): 4.

4. Nineteenth-century farmers in County Donegal, Ireland, distilled illegal poteen (or poitín), a whiskey made from grains or potatoes. They often kept pigs, to which they fed the waste products. See E. Estyn Evans, "Some Survivals of the Irish Openfield System," *Geography* 24, no. 1 (1939): 32.

5. "Two Big Concerns," *Bourbon News* 24, no. 104 (1904): 5.

6. "Local Matters," *Cynthiana News* 14, no. 19 (1868): 3.

7. "Cheap Feed," *Crittenden Press* 23, no. 32 (1902): 7.

8. "Distillery Slops for Sale," *Daily Evening Bulletin* 6, no. 26 (1886): 2.

9. "Hog Raising Is the Poor Farmer's Fortune and the Rich Farmer's Protection," *Hartford Herald* 42, no. 17 (1916): 7.

10. US Congress, House, *The Reports of Committees: H.R. 4165*, 50th Cong., 2d sess., 1888–1889, 41.

11. "Personal," *Daily Public Ledger* 1 (December 5, 1892): 1.

12. "Concerning Farmers," *Climax* 4, no. 48 (1891): 3.

13. "Jefferson County Kentucky Land," *Louisville Daily Democrat* 9, no. 24 (1852): 1.

14. "A Talk about the Prices of Cattle," *Climax* 11, no. 50 (1889): 1.

15. "Farm and Trade Items," *Semi-Weekly Interior Journal* 23, no. 26 (1895): 2.

16. "Cincinnati Market," *Climax* 28, no. 52 (1902): 3.

17. "Farm and Trade Items," *Semi-Weekly Interior Journal* 17, no. 78 (1889): 3.

18. "Farm and Trade Items," *Semi-Weekly Interior Journal* 23, no. 37 (1895): 2; "Farm and Trade Items," *Semi-Weekly Interior Journal* 23, no. 8 (1895): 1.

19. "Garrard County Department," *Semi-Weekly Interior Journal* 16, no. 231 (1887): 2.

20. "Farm & Trade Items," *Semi-Weekly Interior Journal* 19, no. 14 (1891): 1; "Farm and Trade Items," *Semi-Weekly Interior Journal* 16, no. 18 (1888): 2.

21. "Buena Vista," *Central Record* 13, no. 37 (1902): 2.

22. "Land, Stock and Crop," *Mount Vernon Signal* 17, no. 17 (1904): 1; Chester Zoeller, *Bourbon in Kentucky: A History of Distilleries in Kentucky*, 2d ed. (Louisville, KY: Butler Books, 2010), 234.

23. "Brief News Items," *Bourbon News* 23 (December 30, 1902): 5.

24. "Live Stock, Crop, etc." *Bourbon News* 25 (March 4, 1904): 4; Zoeller, *Bourbon in Kentucky*, 168.

25. "Enough to Put the Town 'on the Hog,'" *It* 1, no. 4 (1902): 3.

26. "Land, Stock and Crop," *Semi-Weekly Interior Journal* 16, no. 219 (1887): 3; "Farm and Trade Items," *Semi-Weekly Interior Journal* 23, no. 8 (1895): 1.

27. L. F. Allen, "Berkshire Pigs," *American Agriculturalist* 3 (1844): 175; John Lossing, "Berkshire Pigs," *American Farmer* 6, no. 9 (1840): 66–67.

28. "New York Cattle Market," *Louisville Daily Courier* 28, no. 29 (1859): 4; "New York Cattle Market," *Louisville Daily Courier* 31, no. 212 (1860): 4; "New York Cattle Market," *Louisville Daily Courier* 32, no. 49 (1861): 4.

29. "Kentucky Gleanings," *Citizen* 11, no. 49 (1910): 7.

30. William J. Novak, *The People's Welfare: Law & Regulation in Nineteenth-Century America* (Chapel Hill: University of North Carolina Press, 1996), 227.

31. George F. Longsdorf, ed., *Current Law: A Complete Encyclopedia of New Law* (St. Paul, MN: Keefe-Davidson, 1904), 1062.

32. Irving Browne, ed., *The American Reports*, vol. 50 (San Francisco: Bancroft-Whitney, 1912), 167–71;

Christine Meisner Rosen, "'Knowing' Industrial Pollution: Nuisance Law and the Power of Tradition in a Time of Rapid Economic Change, 1840–1864," *Environmental History* 8, no. 4 (2003): 565–97.

33. W. W. Thum, *Supplement to 1909 Kentucky Statutes* (Cincinnati, OH: W. H. Anderson, 1915), 546.

34. Novak, *People's Welfare,* 217–18.

35. "Trade Notes and Personals," *Wine and Spirit Bulletin* 15, no. 10 (1901): 20; Zoeller, *Bourbon in Kentucky,* 191.

36. "Fine Bourbon County Stock Farm for Sale," *Bourbon News,* August 16, 1904, 9.

37. "Commonwealth v. Kentucky Distilleries & Warehouse Co., et al.," *Southwestern Reporter* 159 (September 3–November 19, 1913): 571.

38. "After Them," *Mt. Sterling Advocate* 11, no. 44 (1901): 2.

39. Large-scale hog producers using factory-type management techniques in North Carolina, Iowa, Illinois, Indiana, and Minnesota began to employ surface lagoon waste storage systems in the 1960s. See Jackie W. D. Robbins, David H. Howells, and George J. Kriz, "Stream Pollution from Animal Production Units," *Journal* (Water Pollution Control Federation) 44, no. 8 (1972): 1536–44.

40. "Warning," *Mt. Sterling Advocate* 12, no. 41 (1902): 2.

41. Zoeller, *Bourbon in Kentucky,* 204.

42. "A New Enterprise," *Mt. Sterling Advocate* 13, no. 18 (1902): 3.

43. "A Good Sale," *Mt. Sterling Advocate* 17, no. 2 (1907): 8.

44. "Peacock Distilling Co. v. Commonwealth," *Kentucky Law Reporter* 25, pt. 2 (1904): 1778–81. The $1,500 fine would be equivalent to roughly $36,000 in today's currency. In Franklin County, the sheriff offered a reward for information leading to an indictment of the party guilty of poisoning Elkhorn Creek for miles above Frankfort, killing aquatic life. Distilleries and oil refineries were suspected, but the cause was not determined. "Unknown Poison," *Frankfort Weekly News and Roundabout* 31, no. 44 (1908): 5.

45. "Weekly Roundabout," *Weekly Roundabout* 5, no. 3 (1881): 1.

46. "Millville," *Frankfort Roundabout* 22, no. 48 (1899): 6.

47. "Agreement," E. H. Taylor Jr. & Sons Inc. and Simon Weil, October 27, 1903, Taylor-Hay Family Papers (Franklin County, KY), 1783–1991, MssAT238d, folder 76, Filson Historical Society, Louisville, KY.

48. "Agreement," E. H. Taylor Jr. & Sons. Inc. and Simon Weil, November 23, 1904, ibid.

49. "Agreement," E. H. Taylor Jr. & Sons Inc. and Simon Weil, September 30, 1906, ibid.

50. "Big Warehouses etc.," *Frankfort Roundabout* 29, no. 10 (1906): 5.

51. J. N. McCormack, *Biennial Report 1910 and 1911,* Bulletin of the State Board of Health of Kentucky (Bowling Green, KY, 1912), 10, 108–14, 129–31.

52. "Kentucky Slop Feeding," *Hartford Republican* 23, no. 28 (1911): 3.

53. Martin Melosi, *The Sanitary City: Urban Infrastructure in America from Colonial Times to the Present* (Baltimore: Johns Hopkins University Press, 2000), 19.

54. "Feeds for Pigs," in *Fifteenth Annual Report of the Kentucky Agricultural Experiment Station* (Lexington: State College of Kentucky, 1902), 157.

55. "Distillery-Milk Report, V," *Science* 10, no. 243 (1887): 157–58.

56. "Distillery-Milk Report, II," *Science* 9, no. 228 (1887): 579–81.

57. *Kentucky Statutes,* 2d ed. (Louisville, KY: Courier-Journal Printing, 1899), sec. 1274, 550.

58. R. M. Allen, "Distillery and Brewery By-products," *Kentucky Farmer and Breeder* 1, no. 25 (1904): 2.

59. Kentucky Agricultural Experiment Station, *Twentieth Annual Report* (Lexington: State College of Kentucky, 1907), xix; Robert M. Allen, *Milk Supply of Kentucky—Louisville,* Bulletin 134, Agricultural Experiment Station of the State College of Kentucky (Lexington, KY, 1908).

60. Allen, "Distillery and Brewery By-products," 4; W. A. Henry and F. B. Morrison, *Feeds and Feeding: A Handbook for the Student and Stockman,* 16th ed. (Madison, WI: Henry-Morrison, 1916), 188.

61. Henry and Morrison, *Feeds and Feeding,* 188, 465, 657.

10. Connections

Epigraph: D. W. Meinig, *The Shaping of America: A Geographical Perspective on 500 Years of History,* vol. 2, *Continental America 1800–1867* (New Haven, CT: Yale University Press, 1993), 251–52.

1. Warren R. Hofstra and Karl Raitz, eds., *The Great Valley Road of Virginia: Shenandoah Landscapes from Prehistory to the Present* (Charlottesville: University of Virginia Press, 2010), 3–4.

2. Caroline MacGill, *History of Transportation in the United States before 1860* (Washington, DC: Carnegie Institution, 1917), 100.

3. "Boats for Sale," *Kentucky Gazette* 22, no. 1220 (1809), 1; James D. Birchfield, "Porter Clay, 'a Very Excellent Cabinetmaker,' Part One: Biographic Account," *Journal of Early Southern Decorative Arts* 36 (2015), http://www.mesdajournal.org. See also Ethel H. Bjerkoe, *The Cabinetmakers of America* (Garden City, NY: Doubleday, 1957), 65.

4. Henry Hall, *Report on the Ship-Building Industry of the United States* (Washington, DC: Department of the Interior, Census Office, 1884), 174; Louis C. Hunter, *Studies in the Economic History of the Ohio Valley* (New York: DaCapo Press, 1973), 6–23.

5. Lewis Loder, *The Loder Diary, 1857–1903,* transcribed by William Conrad (Florence, KY: Boone County Schools, 1985–1988), 111.

6. Ibid., 123.

7. Hall, *Report on the Ship-Building Industry,* 177.

8. MacGill, *History of Transportation,* 78.

9. Ibid., 80, 87.

10. Ibid., 112.

11. Roseann R. Hogan, "Buffaloes in the Corn: James Wade's Account of Pioneer Kentucky," *Register of the Kentucky Historical Society* 89, no. 1 (1991): 21.

12. Victor S. Clark, *History of Manufactures in the United States 1607–1860* (Washington, DC: Carnegie Institution, 1916), 337–38.

13. Zadok Cramer, *The Navigator,* 7th ed. (Pittsburgh: Cramer, Spear & Eichbaum, 1811), 62; Mac-Gill, *History of Transportation,* 110.

14. John H. White Jr., *The American Railroad Freight Car: From the Wood-Car Era to the Coming of Steel* (Baltimore: Johns Hopkins University Press, 1993), 30, 172–75.

15. Theodore G. Gronert, "Trade in the Blue-Grass Region, 1810–1820," *Mississippi Valley Historical Review* 5, no. 3 (1918): 317–18.

16. "In Convention," *Kentucky Gazette* 3, no. 1 (1789): 2.

17. Karl Raitz and Nancy O' Malley, "Local-Scale Turnpike Roads in Nineteenth-Century Kentucky," *Journal of Historical Geography* 33, no. 1 (2007): 5–6.

18. Karl Raitz and Nancy O'Malley, *Kentucky's Frontier Highway: Historical Landscapes along the Maysville Road* (Lexington: University Press of Kentucky, 2012), 66.

19. Ibid., 70.

20. "Bad Accident," *Frankfort Roundabout* 24, no. 22 (1901): 5.

21. The Kentucky legislature met annually until 1850, and every other year thereafter. The graph "Kentucky Turnpikes Chartered" employs bar spacing on the x-axis to reflect this change. Note that

the primary graph of state turnpike charters is in absolute numbers; the inset graph of distilling county charters represents the percentage of all turnpikes chartered.

22. The urban distilling centers in Jefferson, Fayette, Kenton, and Campbell Counties are not included in the county sample, in part because many distilleries there were located in towns, and although turnpike construction in those counties was active, it was focused on farm-to-market roads rather than distillery access roads.

23. See, for example, *Acts of the Kentucky General Assembly*, vol. 2, chap. 1737 (Frankfort, KY, 1871), 296–97.

24. Ibid., chap. 1736, 294–98.

25. *Acts of the Kentucky General Assembly*, vol. 1, chap. 372 (Frankfort, KY, 1870), 368–69.

26. *Acts of the Kentucky General Assembly*, vol. 1, chap. 751 (Frankfort, KY, 1884), 1357.

27. Distiller Richard Cummins was born in County Carlow, Ireland, in 1830. His grave site is in the St. Vincent de Paul Catholic Cemetery south of New Hope on Coon Hollow Road, near members of the Masterson family.

28. Raitz and O'Malley, *Kentucky's Frontier Highway*, 68–70.

29. John R. Borchert, "American Metropolitan Evolution," *Geographical Review* 57, no. 3 (1967): 302–3.

30. Zadok Cramer, *The Navigator*, 8th ed. (Pittsburgh: Cramer, Spear & Eichbaum, 1814), 360.

31. James Hall, *The West: Its Commerce and Navigation* (Cincinnati, OH: H. W. Derby, 1848), 81.

32. Cramer, *Navigator* (8th ed.), 31; Louis C. Hunter, "The Invention of the Western Steamboat," *Journal of Economic History* 3, no. 2 (1943): 201–20.

33. Louis C. Hunter, *Steamboats on the Western Rivers: An Economic and Technological History* (New York: Dover Publications, 1949), 28–31; James Mak and Gary M. Walton, "Steamboats and the Great Productivity Surge in River Transportation," *Journal of Economic History* 32, no. 3 (1972): 620.

34. Clark, *History of Manufactures*, 471.

35. Henry G. Crowgey, *Kentucky Bourbon: The Early Years of Whiskeymaking* (1971; reprint, Lexington: University Press of Kentucky, 2008), 78–79.

36. William F. Switzler, *Report on the Internal Commerce of the United States* (Washington, DC: GPO, 1888), 195–229.

37. MacGill, *History of Transportation*, 467.

38. Maury Klein, *History of the Louisville & Nashville Railroad* (New York: Macmillan, 1972; reprint, Lexington: University Press of Kentucky, 2003), 5–17; Charles B. Kuhlmann, "Processing Agricultural Products in the Pre-Railroad Age," in *The Growth of the American Economy*, ed. Harold F. Williamson (New York: Prentice-Hall, 1951), 167.

39. US Bureau of the Census, *Report of the Superintendent of the Census, December 1, 1852* (Washington, DC: Robert Armstrong, 1853), 101–2.

40. D. R. Haggard, *Report of the Board of Internal Improvement*, Kentucky General Assembly, November 1851 (Frankfort, KY: A. G. Hodges, 1851), 740.

41. Clark, *History of Manufactures*, 338.

42. William T. Hutchinson, *Cyrus Hall McCormick: Seed-Time, 1809–1856* (New York: Century, 1930), 204–5, 251.

43. James E. Vance, *Capturing the Horizon: The Historical Geography of Transportation* (New York: Harper & Row, 1986), 265–327; James E. Vance, *The North American Railroad: Its Origin, Evolution, and Geography* (Baltimore: Johns Hopkins University Press, 1995), 103–91; George Rogers Taylor, *The Transportation Revolution, 1815–1860* (New York: Harper & Row, 1951), 85.

44. William Cronon, *Nature's Metropolis: Chicago and the Great West* (New York: W. W. Norton, 1991),

109–11; Guy A. Lee, "The Historical Significance of the Chicago Grain Elevator System," *Agricultural History* 11, no. 1 (1937): 16–32.

45. Edward K. Muller, "Selective Urban Growth in the Middle Ohio Valley, 1800–1860," *Geographical Review* 66, no. 2 (1976): 185–87.

46. H. E. Mead, *Kentucky and Tennessee: A Complete Guide to Their Railroads* (Louisville, KY: H. E. Mead, 1867); Joseph Nimmo Jr., *Report on the Internal Commerce of the United States, 1881–1882*, Treasury Department, US Bureau of Statistics (Washington, DC: GPO, 1884), 7.

47. Klein, *History of the Louisville & Nashville Railroad*, 6–7.

48. Sam K. Cecil, *Bourbon: The Evolution of Kentucky Whiskey* (New York: Turner Publishing, 2010), 230; Chester Zoeller, *Bourbon in Kentucky: A History of Distilleries in Kentucky*, 2d ed. (Louisville, KY: Butler Books, 2010), 142.

49. Cecil, *Bourbon*, 227; Zoeller, *Bourbon in Kentucky*, 147.

50. Cecil, *Bourbon*, 229; Zoeller, *Bourbon in Kentucky*, 150.

51. *Sanborn's Surveys of the Distilleries and Warehouses of Kentucky and Tennessee* (New York: Sanborn Map Company, 1910), 36.

52. Cecil, *Bourbon*, 234; Zoeller, *Bourbon in Kentucky*, 153.

53. Klein, *History of the Louisville & Nashville Railroad*, 9.

54. Cecil, *Bourbon*, 13.

55. Ibid., 193; Zoeller, *Bourbon in Kentucky*, 132.

56. Cecil, *Bourbon*, 142; Zoeller, *Bourbon in Kentucky*, 169.

57. Zoeller, *Bourbon in Kentucky*, 168; "Among the Independent Distilleries," *Wine and Spirit Bulletin* 17, no. 9 (1903): 23.

58. Zoeller, *Bourbon in Kentucky*, 166.

59. "The Shorthorn Record," *Farmers' Home Journal* 27, no. 49 (1879): 6; "Great Sales of Shorthorns!" *Farmers' Home Journal* 27, no. 29 (1879): 8; "Local Matters," *Cynthiana News* 14, no. 29 (1868): 3.

60. *Kentucky State Gazetteer & Business Directory for 1879–1880* (Detroit, MI: R. L. Polk, 1880).

61. *Sanborn's Surveys of the Distilleries and Warehouses of Kentucky and Tennessee*, 61; Zoeller, *Bourbon in Kentucky*, 181.

62. *Sanborn's Surveys of the Distilleries and Warehouses of Kentucky and Tennessee*, 8.

63. Ibid., 64–65.

64. Cecil, *Bourbon*, 285–86; Zoeller, *Bourbon in Kentucky*, 214.

65. "To Build Branch Railroad," *Frankfort Roundabout* 30, no. 22 (1907): 7.

66. "Opens Office—Hard at Work," *Frankfort Roundabout* 30, no. 40 (1907): 5.

67. Ed Vasser and Jerry Sudduth, *Frankfort and Cincinnati Railroad: History and Remembrances of the Bourbon Road* (Lexington, KY: CreateSpace.com, 2013), 30.

68. Ibid., 6, 10, 20, 30–32; Zoeller, *Bourbon in Kentucky*, 197–98; Cecil, *Bourbon*, 278–79.

69. Louisville's nineteenth-century railroad system comprised several lines that merged, changed names, and built and abandoned lines and yards. This general description highlights only the primary lines running in 1905. *Louisville, Kentucky*, vol. 1 (New York: Sanborn Map Company, 1905), 7–8.

70. *Sanborn's Surveys of the Distilleries and Warehouses of Kentucky and Tennessee*, 31; Zoller, *Bourbon in Kentucky*, 108.

71. The standard US measure for shell corn and rye is 56 pounds per bushel. Rye and corn were sold and transported in two-bushel (112-pound) and two-and-a-half-bushel (140-pound) sacks. See William Parry, *Federal and State Laws Relating to Weights and Measures*, 3d ed. Department of Commerce, Miscel-

laneous Publication 20 (Washington, DC: GPO, 1926), 48. Processing 1,200 bushels of grain each day would have produced more than 50,000 gallons of slop that had to be used or disposed of in some way.

72. *Sanborn's Surveys of the Distilleries and Warehouses of Kentucky and Tennessee,* 27; Zoeller, *Bourbon in Kentucky,* 104.

73. *Sanborn's Surveys of the Distilleries and Warehouses of Kentucky and Tennessee,* 19; Zoeller, *Bourbon in Kentucky,* 105.

74. Borchert, "American Metropolitan Evolution," 304–5.

75. Elkhorn Distillery Records (Scott County, KY), 1868–1872, 1997MS419, Book 1, Special Collections, University of Kentucky.

76. US Congress, House, *The Reports of Committees: H.R. 4165,* 50th Cong., 2d sess., 1888–1889, 15.

77. Ibid., 2.

78. Ibid., 12.

79. Charles Cross, *The American Compounder or Cross' Guide for Retail Liquor Dealers* (St. Louis: Charles Cross, 1899), 111; US Congress, *Reports of Committees: H.R. 4165,* 55.

80. US Congress, *Reports of Committees: H.R. 4165,* 55.

11. Making It Work

Epigraph: Bertolt Brecht, "Questions from a Worker Who Reads," in *Bertolt Brecht Poems 1913–1956,* ed. John Willett and Ralph Manheim (New York: Methuen, 1976), 252.

1. Cf. Stewart Brand, *How Buildings Learn: What Happens after They're Built* (New York: Penguin Books, 1994), 2.

2. Tim Ingold, *Making: Anthropology, Archaeology, Art and Architecture* (London: Routledge, 2013), 48.

3. Merle Patchett, "Historical Geographies of Apprenticeship: Rethinking and Retracing Craft Conveyance over Time and Place," *Journal of Historical Geography* 55 (2017): 33.

4. Clarence H. Danhof, *Change in Agriculture: The Northern United States, 1820–1870* (Cambridge, MA: Harvard University Press, 1969), 74–76.

5. US Bureau of the Census, *1850 Census* (Washington, DC: Robert Armstrong, 1853), 623–24, table IX.

6. Cf. Paul W. Gates, *The Farmer's Age: Agriculture 1815–1860* (Armonk, NY: M. E. Sharpe, 1960), 154.

7. Marion B. Lucas, *A History of Blacks in Kentucky: From Slavery to Segregation, 1760–1891,* 2d ed. (Frankfort: Kentucky Historical Society, 2003), 3. See also Thomas D. Clark, *A History of Kentucky* (Ashland, KY: Jesse Stuart Foundation, 1992), 193–94. Clark's contention that producing grain, tobacco, and hemp involved complicated tasks that required close supervision is not sustained by historical evidence or contemporary experience. Performing each of these tasks once or twice was sufficient to learn them. When working large acreages, the magnitude of the job required mental tenacity and physical strength and endurance, which are quite different from "knowing how to do it."

8. James F. Hopkins, *A History of the Hemp Industry in Kentucky* (Lexington: University Press of Kentucky, 1951), 24–25.

9. Carville Earle and Ronald Hoffman, "The Foundation of the Modern Economy: Agriculture and the Costs of Labor in the United States and England, 1800–1860," *American Historical Review* 85, no. 5 (1980): 1059.

10. William H. English, *Conquest of the Country Northwest of the River Ohio, 1778–1783,* vol. 1 (Indianapolis: Bowen-Merrill, 1896), 83.

11. William H. Perrin, ed., *History of Bourbon, Scott, Harrison and Nicholas Counties, Kentucky* (Chicago: O. L. Baskin, 1882), 59.

12. Register report, second generation, family of Henry Spears and Regina Froman, 2, 5, http://freepages.genealogy.rootsweb.ancestry.com/katy/keller/sources. It is possible that Noah Spears's land in Greene County was north of Xenia near Tawawa Springs, the site of a pre–Civil War resort frequented in the summer by southern planters. When the resort closed, the site became a refuge for black freedmen and runaway slaves moving north along the Underground Railroad. The AME Church bought the property in 1856 and established Wilberforce University on the site.

13. Industrial slavery in central Kentucky included hemp manufacturing and salt boiling. See Robert S. Starobin, *Industrial Slavery in the Old South* (New York: Oxford University Press, 1970), 17–18, 24–25.

14. Lewis Loder, *The Loder Diary, 1857–1903*, transcribed by William Conrad (Florence, KY: Boone County Schools, 1985–1988), 75; Matthew E. Becher, "The Distillery at Petersburg, Kentucky: Snyder's Old Rye Whiskey," http://bcplfusion.bcpl.org/Repository/Petersburg_Distillery_History_complete.pdf.

15. Loder, *Loder Diary*, 75.

16. The Petersburg Distillery was also known as the Boone County Distillery and by some seventy other names. See Chester Zoeller, *Bourbon in Kentucky: A History of Distilleries in Kentucky*, 2d ed. (Louisville, KY: Butler Books, 2010), 161.

17. Becher, "Distillery at Petersburg, Kentucky."

18. Keith C. Barton, "'Good Cooks and Washers': Slave Hiring, Domestic Labor, and the Market in Bourbon County, Kentucky," *Journal of American History* 84, no. 2 (1997): 448.

19. "Bond & Lillard Distillery," *Anderson News*, Souvenir Supplement, June 1906, 34.

20. US Census Bureau, 1870 Census, Anderson County.

21. The award is not included on the official list and cannot be confirmed. The exhibition's international jury gave only two awards for whiskey—one to Hannis Distilling of Philadelphia, and the other to Sattler & Company of Baltimore. Both winners were rye whiskeys. See US Centennial Commission, *List of Awards, International Exhibition, 1876, at Philadelphia* (Philadelphia: S. T. Souder, 1876), 101.

22. "Cedar Brook," *Anderson News*, Souvenir Supplement, June 1906, 36. See also Sam K. Cecil, *Bourbon: The Evolution of Kentucky Whiskey* (New York: Turner Publishing, 2010), 94.

23. *The Nelson County Record—Illustrated Historical and Industrial Supplement*, reprinted in Dixie Hibbs, *Before Prohibition: Distilleries in Nelson County Kentucky, 1880–1920* (New Hope, KY: St. Martin de Porres Print Shop, 2012), 57. See also the biographical statement at http://www.findagrave.com/cgi-bin/fg.cgi?page=gr&GRid=88934626.

24. Hibbs, *Before Prohibition*, 43.

25. Philip Bonfort, *Bonfort's Wine & Liquor Trade Directory for the United States* (New York: Philip Bonfort, 1875), 144.

26. Exhibition judges gave the top award for fermented liquors, the Medal of Merit, to John Gibson's Son & Company of Pittsburgh, Pennsylvania, for its rye whiskey. Robert H. Thurston, *Reports of the Commissioners of the United States to the International Exhibition, Vienna, 1873*, vol. 1 (Washington, DC: GPO, 1876), 169–70, 403. See also Hibbs, *Before Prohibition*, 43–44.

27. H. J. Davenport, "The Extent and the Significance of the Unearned Increment," *American Economic Review* 1, no. 2 (1911): 322–32; William H. Dawson, *The Unearned Increment: Reaping without Sowing* (London: Swan Sonnenschein, 1890), 22–39.

28. Barton, "'Good Cooks and Washers,'" 436.

29. Jonathan D. Martin, *Divided Mastery: Slave Hiring in the American South* (Cambridge, MA: Harvard University Press, 2004), 109.

30. Lucas, *History of Blacks in Kentucky*, 101–7.

31. Emma Jane Cleaveland Journal, in Lyne-Smith Family Papers (Woodford County, KY), 1820–1932, 1997 MS387, box 3, Special Collections, University of Kentucky. See also Clement Eaton, "Slave-Hiring in the Upper South: A Step toward Freedom," *Mississippi Valley Historical Review* 46, no. 4 (1960): 663.

32. John W. Jones Papers (Bourbon County, KY), 1857–1896, 1857–1889 (bulk dates), 75M26, Special Collections, University of Kentucky.

33. Earle and Hoffman, "Foundation of the Modern Economy," 1056.

34. Perrin, *History of Bourbon, Scott, Harrison and Nicholas Counties*, 66–67.

35. More than thirty resettlement villages for freedmen developed across Kentucky's Inner Bluegrass counties in the late 1860s and 1870s, and many others were created in the central and western Pennyroyal. Peter C. Smith and Karl B. Raitz, "Negro Hamlets and Agricultural Estates in Kentucky's Inner Bluegrass," *Geographical Review* 64, no. 2 (1974): 217–234. See also Sallie L. Powell, "African American Hamlets," in *The Kentucky African American Encyclopedia*, ed. Gerald L. Smith, Karen C. McDaniel, and John A. Hardin (Lexington: University Press of Kentucky, 2015), 5.

36. US Census Bureau, 1920 Census, Anderson County.

37. "Personals," *Interior Journal* 33, no. 61 (1905): 3.

38. US Census Bureau, 1900 Census, Nelson County.

39. "Frankfort Roundabout," *Frankfort Roundabout* 7, no. 23 (1884): 3.

40. "News Paragraphs," *Climax* 3, no. 23 (1889): 5.

41. "Here and There," *It* 1, no. 9 (1903): 1. On January 8, 1903, the personals section of Lawrenceburg's *It* newspaper noted that a Miss Hill of Shelbyville had been employed as a stenographer at the Clover Bottom Distillery at Tyrone.

42. US Census Bureau, 1910 Census, Nelson County.

43. Karl Raitz and Nancy O'Malley, *Kentucky's Frontier Highway: Historical Landscapes along the Maysville Road* (Lexington: University Press of Kentucky, 2012), 155–56.

44. "Athertonville," *Adair County News* 10, no. 14 (1907): 6.

45. The Thomas Lincoln farm was Abraham Lincoln's birthplace.

46. Clerk of Court, Larue County, Hodgenville, KY, *Larue County Deed Record Book 6*, 348.

47. Ibid., *Book 7*, 27–28.

48. Ibid., *Book 6*, 581.

49. Ibid., *Book 12*, 446.

50. Ibid., 208.

51. Zoeller, *Bourbon in Kentucky*, 128.

52. *Sanborn's Surveys of the Distilleries and Warehouses of Kentucky and Tennessee* (New York: Sanborn Map Company, 1910), 41, 42.

53. Zoeller, *Bourbon in Kentucky*, 128.

54. *Larue County Deed Record Book 13*, 148, 153.

55. Ibid., 149.

56. Ibid., 155.

57. George C. Buchanan, *Fine Whisky Facts* (Louisville, KY: George Buchanan, 1892), 20.

58. *Sanborn's Surveys of the Whiskey Warehouses of Kentucky and Tennessee* (New York: Sanborn-Perris Map Company, 1894), 27.

59. Cecil, *Bourbon*, 188. One source listed aggregate grain consumption here at 1,800 bushels per day in 1886. See J. M. Elstner, *The Industries of Louisville, Kentucky, and New Albany, Indiana* (Louisville, KY: J. M. Elstner, 1886), 93.

60. "A Cattle Shortage," *Daily Public Ledger* 7 (February 28, 1898): 4.

61. "Hodgenville Herald," *Central Record* 15, no. 7 (1904): 4.

62. "Land, Stock and Crop," *Mount Vernon Signal* 22, no. 14 (1908): 2.

63. "Stock, Crop and Farm Notes," *Bourbon News* 29, no. 35 (1909): 3.

64. Leifur Magnusson, *Housing by Employers in the United States*, US Department of Labor, Bureau of Labor Statistics Bulletin 263 (Washington, DC: GPO, 1920), 1–20.

65. Elstner, *Industries of Louisville, Kentucky, and New Albany, Indiana*, 93.

66. US Census Bureau, 1880 Census, Larue County. Census enumerators made their tabulations during the late spring or early summer, after the distilleries had shut down for the season.

67. Elstner, *Industries of Louisville, Kentucky, and New Albany, Indiana*, 93.

68. "Athertonville, KY," *Adair County News* 10, no. 14 (1907): 6.

69. "Athertonville, KY," *Adair County News* 6, no. 18 (1903): 3.

70. "From Larue County," *Adair County News* 11, no. 45 (1908): 5.

71. US Census Bureau, 1900 Census, Larue County.

72. *Larue County Deed Record Book 14*, 304.

73. Ibid., *Book 12, 19, 185, 352, 407*.

74. Ibid., *Book 14, 252*.

75. Ibid., *Book 13, 294*.

76. Ibid., *Book 14, 189*.

77. See, for example, ibid., *Book 16, 458; Book 18, 244; Book 20, 137, 389; Book 21, 456*.

78. Cecil, *Bourbon*, 188; Zoeller, *Bourbon in Kentucky*, 127.

79. *An Atlas of Nelson and Spencer Counties, Kentucky* (Philadelphia: D. J. Lake, 1882), 27.

80. Cecil, *Bourbon*, 230–31.

81. Clerk of Court, Nelson County, Bardstown, KY, *Nelson County Deed Record Book 26*, 415.

82. Ibid., *Book 28, 366*.

83. Ibid., *Book 34, 111*.

84. Ibid., *Book 35, 609*.

85. Cecil, *Bourbon*, 231.

86. "The Distilleries," *Hartford Herald* 9, no. 21 (1883): 2; "Notes of Current Events," *Semi-Weekly Interior Journal* 10, no. 510 (1883): 2.

87. *Sanborn's Surveys of the Distilleries and Warehouses of Kentucky and Tennessee*, 43.

88. "Equality Echoes," *Hartford Herald* 9, no. 13 (1883): 3.

89. *Nelson County Deed Record Book 39, 6; Book 41, 122; Book 44, 144; Book 45, 357, 441, 493; Book 49, 259; Book 50, 372; Book 53, 198*.

90. The street name "Sherley Avenue" is misspelled on extant street signs as "Shirley Avenue."

91. *Nelson County Deed Record Book 50, 372*.

92. "T. B. Ripy and the Anderson County Distilling Company Distilleries," *Anderson News*, Souvenir Supplement, June 1906, 51.

93. Clerk of Court, Anderson County, Lawrenceburg, KY, *Anderson County Deed Record Book M, 1:456*.

94. As a convenient reference, an acre is 43,560 square feet, or roughly the size of an American football field minus one end zone.

95. "Lancaster, Garrard County," *Semi-Weekly Interior Journal* 16, no. 91 (1888): 2; "Farm and Trade Items," *Semi-Weekly Interior Journal* 19, no. 14 (1891): 1; "Farm and Trade Items," *Semi-Weekly Interior Journal* 20, no. 35 (1892): 3.

96. "Weekly Roundabout," *Frankfort Roundabout* 3, no. 51 (1880): 3.

97. "The Frankfort Roundabout," *Frankfort Roundabout* 9, no. 13 (1885): 1; "Personals," *Frankfort Roundabout* 10, no. 28 (1887): 2.

98. "A New Departure," *Mt. Sterling Advocate* 5, no. 36 (1895): 7; "Local Jottings," *Adair County News* 4, no. 14 (1901): 3.

99. "To Tyrone and Return," *Frankfort Roundabout* 27, no. 50 (1904): 1; "Delightful Excursion," *Frankfort Roundabout* 29, no. 40 (1906): 3.

100. *Anderson County Deed Record Book N*, 48, 412.

101. Ibid., *Book Q*, 156, 356, 423; *Book R*, 433; *Book S*, 197, 217.

102. Ibid., *Book P*, 328.

103. Ibid., *Book Q*, 156; *Book S*, 217.

104. According to the *Anderson News*, Tyrone was incorporated in 1879, but official incorporation papers filed with the clerk of the court were dated December 7, 1881. See *Anderson County Deed Record Book Q*, 469.

105. *Anderson County Deed Record Book Q*, 469; "Tyrone," *It* 1, no. 15 (1903): 11.

106. Cecil, *Bourbon*, 93.

107. *Anderson County Deed Record Book S*, 398.

108. "Tyrone," *Anderson News*, Souvenir Supplement, June 1906, 52.

109. "Tyrone," *It* 1, no. 15 (1903): 11.

110. "Little Jottings by the Way," *Frankfort Roundabout* 29, no. 42 (1906): 1.

111. Zoller, *Bourbon in Kentucky*, 219.

112. "The Commonwealth," *Big Sandy News* 4, no. 45 (1889): 1; "The Great Bridge at Tyrone," *Climax* 3, no. 11 (1889): 3.

113. "Largest Sale of Whisky on Record," *Climax* 28, no. 44 (1902): 2.

114. *Anderson News*, Souvenir Supplement, June 1906, 51; Cecil, *Bourbon*, 93; Zoeller, *Bourbon in Kentucky*, 222.

115. Julie Riesenweber and Karen Hudson, eds., *Kentucky's Bluegrass Region* (Frankfort: Kentucky Heritage Council, 1990), 86–88.

116. E. A. Hewitt and G. W. Hewitt, *Topographical Map of the Counties of Bourbon, Fayette, Clark, Jessamine and Woodford, Kentucky* (New York: Smith, Gallup, 1861).

117. Cecil, *Bourbon*, 285–88; Zoeller, *Bourbon in Kentucky*, 212–14.

118. "The L & N Railroad Company," *Frankfort Roundabout* 7, no. 9 (1882): 1.

119. "Work Begins," *Frankfort Roundabout* 29, no. 40 (1906): 8.

120. "Kentucky Highland Railroad," *Frankfort Weekly News and Roundabout* 31, no. 33 (1908): 3.

121. Carl E. Kramer, *Capital on the Kentucky* (Frankfort, KY: Historic Frankfort, 1986), 214–15.

122. Addison, Kentucky, on the Ohio River in northern Breckinridge County, is the site of L. D. Addison's Old Breckenridge Distillery, which was in operation in the early 1900s. Mr. Addison also ran a general store in Holt, a railroad stop about half a mile southwest of the river. The US Corps of Engineers began building a lock and dam at Addison in 1919; it also built ten substantial houses on the bluff above the dam. The distillery made whiskey as well as apple brandy—Addison bought apples by the ton. Although the houses built by the Corps of Engineers resemble a "company town," they likely had no relationship to the distillery, which had probably stopped operations by the time dam construction began. See "Old Breckenridge Distillery," *Wine and Spirit Bulletin* 18, no. 1 (1904): 41; Roy Hampton, "Ohio River Lock and Dam No. 45," Addison, KY, Kentucky Historic Resources Individual Survey Form BC 12 (Frankfort: Kentucky Heritage Council, 1999).

123. Cf. Ralf Weber, "The Myth of Meaningful Forms: Comparing the Forms of Indigenous and Classical Architecture," *Traditional Dwellings and Settlements Review* 2, no. 2 (1991): 66, 68.

124. A. J. Downing, *The Architecture of the Country House* (New York: D. Appleton, 1850), 25. See also Judith K. Major, *To Live in the New World: A. J. Downing and American Landscape Gardening* (Cambridge, MA: MIT Press, 1997), 30.

125. Edward N. Kaufman, "Architectural Representation in Victorian England," *Journal of the Society of Architectural Historians* 46, no. 1 (1987): 30.

126. Dell Upton, *Another City: Urban Life and Urban Spaces in the New American Republic* (New Haven, CT: Yale University Press, 2008), 9.

127. James E. Vance, *This Scene of Man: The Role and Structure of the City in the Geography of Western Civilization* (New York: Harper's College Press, 1977), 393.

128. Richard L. Bushman, *The Refinement of America* (New York: Vintage Books, 1993), 250–55.

129. Russell Lynes, *The Taste-Makers* (New York: Harper & Brothers, 1955), 128.

130. The term "bay" designates a front wall window opening. The Spears house had five windows across the width of the front façade, with the central window space on the first floor occupied by the front door.

131. C. M. Wooley, "Jacob Spears Distillery," Paris, KY, National Register of Historic Places Registration Form (Frankfort: Kentucky Heritage Council, 1982).

132. Robert M. Polsgrove, "T. Jeremiah Beam House," Clermont, KY, National Register of Historic Places Registration Form (Frankfort: Kentucky Heritage Council, 1987).

133. Perrin, *History of Bourbon, Scott, Harrison and Nicholas Counties,* 664.

134. Thelma Taylor, "Monticello," Cynthiana, KY, National Register of Historic Places Registration Form (Frankfort: Kentucky Heritage Council, 1974).

135. Glenda Thacker, "Le Vega Clements House," National Register of Historic Places Registration Form (Frankfort: Kentucky Heritage Council, 1986).

136. Lewis McKee and Linda K. Bond, *History of Anderson County: 1780–1936* (Frankfort, KY: Roberts Print Company, 1936), 210.

137. Charlotte Schneider, "Dowling House," Lawrenceburg, KY, National Register of Historic Places Registration Form (Frankfort: Kentucky Heritage Council, 1979).

138. Typescript copies of *Anderson News,* June 7, 1888, Ripy Family Papers (Anderson County, KY), 1888–1963, 1F66M-725, Special Collections, University of Kentucky.

139. "T. B. Ripy and the Anderson County Distilling Company Distilleries," *Anderson News,* Souvenir Supplement, June 1906, 51.

140. "T. B. Ripy House," *Anderson News,* November 28, 2012, 8.

141. Carolyn Brooks, "Country Estates of River Road," Louisville, KY, National Register of Historic Places Registration Form (Frankfort: Kentucky Heritage Council, 1999).

142. *Anderson News,* Souvenir Supplement, June 1906, 9, 10, 15, 16–18, 23, 34, 37, 38–39, 53.

143. Clarence Cook, "New York Daguerreotyped," *Putnam's Monthly* 3, no. 15 (1854): 241, 247, cited in Jay E. Cantor, "A Monument of Trade: A. T. Stewart and the Rise of the Millionaire's Mansion in New York," *Winterthur Portfolio* 10 (1975): 173.

144. Kaufman, "Architectural Representation in Victorian England," 33.

12. External Control and Landscape

Epigraph: Bertolt Brecht, "Song of the Cut-Price Poets," in *Bertolt Brecht Poems, 1913–1956,* ed. John Willett and Ralph Manheim (New York: Methuen, 1976), 162.

1. W. J. Rorabaugh, *The Alcoholic Republic: An American Tradition* (New York: Oxford University

Press, 1979), 53; Thomas P. Slaughter, *The Whiskey Rebellion: Frontier Epilogue to the American Revolution* (New York: Oxford University Press, 1986), 143–57; William Hogeland, *The Whiskey Rebellion: George Washington, Alexander Hamilton, and the Frontier Rebels Who Challenged America's Newfound Sovereignty* (New York: Scribner, 2006), 51–70.

2. Chapter 108, article X, "License Tax," in J. Barbour and John D. Carroll, *The Kentucky Revised Statutes: Containing All General Laws, Including Those Passed at Session of 1894* (Louisville, KY: Courier-Journal Printing, 1894), 1348.

3. "Reduced the Valuation," *Frankfort Roundabout* 24, no. 17 (1901): 8.

4. "Dodging the Tax," *Climax* 9, no. 7 (1895): 2.

5. Dell Upton, "Seen, Unseen, and Scene," in *Understanding Ordinary Landscapes,* ed. Paul Groth and Todd W. Bressi (New Haven, CT: Yale University Press, 1997), 174–75.

6. Karl B. Raitz, "The Governmental Institutionalization of Tobacco Acreage in Wisconsin," *Professional Geographer* 23, no. 2 (1971): 123–26.

7. August F. Fehlandt, *A Century of Drink Reform in the United States* (Cincinnati, OH: Jennings & Graham, 1904), 189.

8. Ibid., 150.

9. "An Act Imposing Taxes on Distilled Spirits and Tobacco, and for Other Purposes," 40th Cong., 2d sess., July 20, 1868, in *The Statutes at Large, Treaties, and Proclamations, of the United States of America from December 1867, to March 1869,* vol. 15, ed. George P. Sanger (Boston: Little, Brown, 1869), 125.

10. A proof gallon was a standard federal measure for tax purposes. Liquid spirits were taxed on the proportion of alcohol they contained rather than basic volume. A proof gallon is a gallon of spirits that is 50 percent alcohol (100 proof) at 60°F. A gallon of whiskey barreled at 125 proof would equal 1.25 proof gallons.

11. Ibid.

12. Ibid., 130.

13. Ibid., 125–26.

14. Ibid., 130.

15. Ibid., 151.

16. US Congress, House, *The Reports of Committees: H.R. 4165,* 50th Cong., 2d sess., 1888–1889, 24–25.

17. "Internal Tax Bill," *Congressional Globe,* US House of Representatives, 40th Cong., 2d sess., 1868, 3397–538.

18. K. Austin Kerr, *Organized for Prohibition: A New History of the Anti-Saloon League* (New Haven, CT: Yale University Press, 1985), 38–39.

19. "Internal Tax Bill," 3399.

20. Ibid., 3397.

21. Theodore E. Burton, *John Sherman* (Boston: Houghton Mifflin, 1908), 197.

22. The distilleries north of the Ohio River, such as those at Peoria, made whiskey primarily to be sold immediately to rectifiers. See John Atherton's testimony in US Congress, *Reports of Committees: H.R. 4165,* 19–20.

23. "Internal Tax Bill," 3405.

24. Ibid., 3480.

25. Ibid.

26. Ibid., 3485.

27. *Journal of the Executive Proceedings of the Senate of the United States of America from April 12, 1869 to April 22, 1869, Inclusive,* 41st Cong., special sess. (1869), vol. 17 (Washington, DC: GPO, 1869), 213.

28. "Internal Tax Bill," 3397. See also comments by Representatives Omar Conger (R-MI) and William D. Kelley (R-PA), in *Congressional Record*, 46th Cong., 2d sess., 1880, vol. 10, pt. 3, 2837, 2840.

29. "Internal Tax Bill," 3538.

30. *Congressional Globe*, 40th Cong., 2d sess., 1868, pt. 4, 3753.

31. Ibid. The bonded warehouse system was first used in England in 1803 to control imported goods and ensure that duties were paid before the goods' distribution and use. In America, an early bonded warehouse system allowed speculators, commission merchants, distillers, and other businesses to house unsold whiskey, tobacco, and other goods until the taxes or duties were paid. Government and privately owned warehouses were often located in seaport cities such as New York, Boston, Philadelphia, and New Orleans, well removed from distillery premises. See B. W. Arnold Jr., *History of the Tobacco Industry in Virginia, from 1860 to 1894*, Johns Hopkins University Studies in Historical and Political Science, 15th series (Baltimore: Johns Hopkins University Press, 1897), 62. Distillery aging warehouses could be bonded—that is, sealed and secured by appropriate government agents. According to the tax law of 1868, distillers who stored whiskey in distillery warehouses were required to "give bond" conditional on the withdrawal and tax payment within one year from the date the whiskey was barreled and stored. In 1897 the Bottled-in-Bond Act specified that to qualify as pure and unadulterated, whiskey had to meet several conditions, including being stored and aged in federally bonded warehouses under government supervision for at least four years.

32. Burton, *John Sherman*, 196–97.

33. *Congressional Record*, 45th Cong., 2d sess., 1878, vol. 7, pt. 2, 1662–63.

34. *Congressional Record*, 46th Cong., 2d sess., 1880, vol. 10, pt. 3, 2834–45. The allowance was named for John G. Carlisle, Democratic senator from Kentucky. Carlisle was conservative and pro-business. He served as Speaker of the House from 1883 to 1889 and was secretary of the treasury under President Grover Cleveland from 1893 to 1897.

35. Representative Albert S. Willis (D-KY) complained that this legislation was passed by a Democratically controlled Congress. *Congressional Record*, 46th Cong., 2d sess., 1880, vol. 10, pt. 3, 2842. Republicans published, as part of their 1880 national platform, statements decrying the loss of tax revenue and the potential for fraud. See Republican Congressional Committee, *The Republican Campaign Text Book for 1880* (Washington, DC: Republican Congressional Committee, 1880), xxx, 111–13.

36. US Commissioner of Internal Revenue, *Report of the Commissioner of Internal Revenue, 1880*, Treasury Department Document 53 (Washington, DC: GPO, 1880), xxxi–xxxii.

37. Ibid., lxxxviii.

38. US Congress, *Reports of Committees: H.R. 4165*, 4.

39. US Commissioner of Internal Revenue, *Report of the Commissioner of Internal Revenue, 1880*, 6.

40. US Congress, *Reports of Committees: H.R. 4165*, 9.

41. US Congress, House, *The Reports of Committees: H.R. 276*, 53d Cong., 2d sess., 1893–1894 (Washington, DC: GPO, 1894), 2.

42. US Congress, Senate, *Tariff Laws of 1890 and 1894. Report 707, Part 2*, 53d Cong., 2d sess., 1894, 156–68.

43. Gerald Carson, *The Social History of Bourbon* (1963; reprint, Lexington: University Press of Kentucky, 2010), 97.

44. US Internal Revenue Service, *Historical Study: IRS Historical Fact Book; a Chronology 1646–1992* (Chamblee, GA, 2012), 44, accessed through governmentattic.org.

45. "An Act Imposing Taxes on Distilled Spirits and Tobacco," 146.

46. Ibid., 141.

47. Ibid., 134.

48. "Internal Tax Bill," 3401.

49. "An Act Imposing Taxes on Distilled Spirits and Tobacco," 135.

50. Ibid., 136–37.

51. Upton, "Seen, Unseen, and Scene," 174–75; Richard Walker, "Unseen and Disbelieved: A Political Economist among Cultural Geographers," in Groth and Bressi, *Understanding Ordinary Landscapes*, 171.

52. "An Act Imposing Taxes on Distilled Spirits and Tobacco," 127.

53. Ibid., 142.

54. *Journal of the Executive Proceedings of the Senate of the United States of America from December 4, 1865 to March 3, 1867, Inclusive*, 39th Cong., 1st sess. (1866), vol. 14 (Washington, DC: GPO, 1867), 509–10.

55. H.R. 2007, 41st Cong., 2d sess., May 1870, 13–15.

56. "An Act Imposing Taxes on Distilled Spirits and Tobacco," 128.

57. US Department of Internal Revenue, *Regulations and Instructions Concerning the Tax on Distilled Spirits under the Revised Statutes of the United States and Subsequent Acts*, series 7, no. 7 (Washington, DC: GPO, 1877), 12.

58. "An Act Imposing Taxes on Distilled Spirits and Tobacco," 131; US Department of Internal Revenue, *Regulations and Instructions Concerning the Tax on Distilled Spirits*, 9, 12.

59. US Department of Internal Revenue, *Government Drawings, Registered Distillery No. 47, Internal Revenue Bonded Warehouse No. 36, Tax Paid Bottling House No. 17, Bluegrass Distillery, Gethsemane, Kentucky* (n.d.), in Oscar Getz Museum, Bardstown, KY.

60. "An Act Imposing Taxes on Distilled Spirits and Tobacco," 130.

61. US Department of Internal Revenue, *Regulations and Instructions Concerning the Tax on Distilled Spirits*, 10.

62. "An Act Imposing Taxes on Distilled Spirits and Tobacco," 131.

63. US Department of Internal Revenue, *Regulations and Instructions Concerning the Tax on Distilled Spirits*, 10.

64. "An Act Imposing Taxes on Distilled Spirits and Tobacco," 130–31; US Department of Internal Revenue, *Regulations and Instructions Concerning the Tax on Distilled Spirits*, 23. Cf. Michael R. Veach, *Kentucky Bourbon Whiskey: An American Heritage* (Lexington: University Press of Kentucky, 2013), 66.

65. "An Act Imposing Taxes on Distilled Spirits and Tobacco," 132, 139–40.

66. Ibid., 127–28.

67. Ibid., 132–33.

68. US Department of Internal Revenue, *Regulations and Instructions Concerning the Tax on Distilled Spirits*, 15.

69. "An Act Imposing Taxes on Distilled Spirits and Tobacco," 133.

70. Ibid., 134.

71. Ibid.; US Department of Internal Revenue, *Regulations and Instructions Concerning the Tax on Distilled Spirits*, 15.

72. US Department of Internal Revenue, *Regulations and Instructions Concerning the Tax on Distilled Spirits*, 32–33.

73. Nicholas K. Bromley, *Law, Space, and the Geographies of Power* (New York: Guilford Press, 1994), xiii.

13. Temperance Troubles

Epigraph: Paul E. Johnson, *A Shopkeeper's Millennium: Society and Revivals in Rochester, New York, 1815–1837*, rev. 1st ed. (New York: Hill & Wang, 2004), 80.

1. John A. Krout, *The Origins of Prohibition* (New York: Russell & Russell, 1925), 234–35.

2. Michael L. Dorn, "(In)temperate Zones: Daniel Drake's Medico-Moral Geographies of Urban Life in the Trans-Appalachian West," *Journal of the History of Medicine and Allied Sciences* 55, no. 3 (2000): 275.

3. W. J. Rorabaugh, *The Alcoholic Republic: An American Tradition* (New York: Oxford University Press, 1979), 83–90.

4. August F. Fehlandt, *A Century of Drink Reform in the United States* (Cincinnati, OH: Jennings & Graham, 1904).

5. Samuel M. Wilson, *History of Kentucky*, vol. 2 (Chicago: S. J. Clarke, 1928), 286.

6. Krout, *Origins of Prohibition*, 144. See also *Centennial Temperance Volume: A Memorial of the International Temperance Conference, Philadelphia, June, 1876* (New York: National Temperance Society, 1877), 441.

7. Ian R. Tyrrell, "Drink and Temperance in the Antebellum South: An Overview and Interpretation," *Journal of Southern History* 48, no. 4 (1982): 485–86; Thomas H. Appleton Jr., "'Moral Suasion Has Had Its Day': From Temperance to Prohibition in Antebellum Kentucky," in *A Mythic Land Apart: Reassessing Southerners and Their History*, ed. John D. Smith and Thomas H. Appleton Jr. (Westport, CT: Greenwood Press, 1997), 23–24.

8. John Kobler, *Ardent Spirits: The Rise and Fall of Prohibition* (New York: Putnam, 1973), 56.

9. K. Austin Kerr, *Organized for Prohibition: A New History of the Anti-Saloon League* (New Haven, CT: Yale University Press, 1985), 35–36.

10. "The Temperance Cause," *Semi-Weekly Interior Journal* 17, no. 40 (1889): 1; Raymond B. Fosdick and Albert L. Scott, *Toward Liquor Control* (New York: Harper & Brothers, 1933), 1–2.

11. "Whisky Trade," *Louisville Courier-Journal* 38, no. 148 (1868): 1; Thomas H. Appleton Jr., "'Like Banquo's Ghost': The Emergence of the Prohibition Issue in Kentucky Politics" (PhD diss., Department of History, University of Kentucky, 1981), 3–4, 10. See also John Ed Pearce, *Nothing Better in the Market* (Louisville, KY: Brown-Forman Distillers, 1970), 40.

12. "Gospel in a Mash Room," *Climax* 25, no. 19 (1898): 4; "Kentucky State News," *Hickman Courier* 32, no. 40 (1897): 1.

13. Thomas J. Schlereth, *Victorian America: Transformations in Everyday Life, 1876–1915* (New York: HarperCollins, 1991), 197.

14. "W.C.T.U. Notes," *Frankfort Roundabout* 8, no. 23 (1885): 1.

15. "Father Mathew," *Kentucky Irish American* 10, no. 11 (1903): 1.

16. John F. Quinn, *Father Mathew's Crusade: Temperance in Nineteenth-Century Ireland and Irish America* (Amherst: University of Massachusetts Press, 2002), 166–67.

17. "Lecture To-night at City Hall," *Hickman Courier* 36, no. 13 (1901): 8.

18. *Caron's Annual Directory of the City of Louisville*, vol. 11 (Louisville, KY: Bradley & Gilbert, 1881), 46.

19. "A.O.U.W.," *Evening Bulletin* 16, no. 159 (1897): 3.

20. "Set Heads to Win," *Breckenridge News* 30, no. 30 (1906): 1.

21. "Home Jottings," *Interior Journal* 4, no. 20 (1875): 3.

22. "Daily Evening Bulletin," *Evening Bulletin* 8, no. 153 (1889): 3; "Program Prohibition State Convention, Music Hall, Louisville," *Mt. Sterling Advocate* 8, no. 41 (1903): 2.

23. "Temperance Hall Demolished," *Record* 15, no. 15 (1913): 3; "Temperance Hall," *Record* 15, no. 16 (1913): 2. Ironically, the small stream paralleling North Main Street in Greenville is named Whiskey Run.

24. Paul Connerton, *How Societies Remember* (Cambridge: Cambridge University Press, 1989), 21–23; Jacques Le Goff, *History and Memory* (New York: Columbia University Press, 1992), 94–96.

25. See, for example, "Francis Murphy," *Frankfort Roundabout* 7, no. 46 (1884): 1; "Ruggles Camp Meeting," *Evening Bulletin* 1, no. 236 (1882): 3; "Ruggles Camp Meeting," *Evening Bulletin* 1, no. 237 (1882): 3; "Editorial Correspondence," *Weekly Roundabout* 3, no. 44 (1880): 1. Religious camp meetings had been the "hubs of western religion" since the early nineteenth century. See John Ellis, "The Holy 'Knock-'em-Down': Methodism Remodels for the Ohio Valley, 1790s–1820s," *Ohio Valley History* 18, no. 2 (2018): 14.

26. "Has Buried Her Hatchet," *Breckenridge News* 28, no. 13 (1903): 3. When interviewed in Jessamine County, Kentucky, in 1903, Carrie Nation said she had given up her violent crusade against saloons and "would devote her time and energy to trying to suppress the manufacture of liquor, as she had found she could never accomplish her end by saloon smashing."

27. US Congress, House, *Journal of the House of Representatives of the United States,* 43d Cong., 1st sess., 1874, vol. 74 (Washington, DC: GPO, 1873), 299, accessed from Library of Congress, *A Century of Lawmaking for a New Nation: U.S. Congressional Documents and Debates, 1774–1875,* https://memory.loc.gov/cgi-bin/am.

28. Wilson, *History of Kentucky,* 286–87.

29. "Constitutional," *Mt. Sterling Advocate* 16, no. 43 (1907): 6.

30. Nollie Olin Taft, *History of State Revenue and Taxation in Kentucky* (Nashville, TN: George Peabody College for Teachers, 1931), 158.

31. Ibid., 92–100; chapter 108, article X, "License Tax," in J. Barbour and John D. Carroll, *The Kentucky Revised Statutes: Containing All General Laws, Including Those Passed at Session of 1894* (Louisville, KY: Courier-Journal Printing, 1894), 1348.

32. Taft, *History of State Revenue and Taxation in Kentucky,* 68, 110.

33. US Congress, House, *The Reports of Committees: H.R. 4165,* 50th Cong., 2d sess., 1888–1889, 30–31.

34. Ibid., 32. See also John Atherton's comments before a Cincinnati convention of the NPA in 1887 in Walter W. Spooner, ed., *The Political Prohibitionist for 1888: A Handbook for the Aggressive Temperance People of the United States* (New York: Funk & Wagnalls, 1888); Fehlandt, *Century of Drink Reform,* 182.

35. Thomas H. Appleton Jr., "Prohibition and Politics in Kentucky: The Gubernatorial Campaign and Election of 1915," *Register of the Kentucky Historical Society* 75, no. 1 (1977): 28.

36. Not coincidently, white traditionally symbolized good, purity, and cleanliness, whereas black was often associated with evil, aggression, and heavy-handed power.

37. Appleton, "'Like Banquo's Ghost,'" vi.

38. Ibid., 38; "A Temperance Church," *Interior Journal* 5, no. 41 (1876): 1.

39. Wilson, *History of Kentucky,* 287; Laura H. Hendrix, *Citizen's Guide to the Kentucky Constitution,* Research Report 137 (Frankfort, KY: Legislative Research Commission, 2013), 137. Kentucky's prohibition amendment was not repealed until 1935, some two years after the Twenty-First Amendment was ratified, ending national Prohibition.

40. E. S. C., "Constitutional Law. Interstate Commerce. Original Package Doctrine," *Virginia Law Review* 21, no. 4 (1935): 438–39.

41. Ibid., 434–35.

42. Krout, *Origins of Prohibition,* 26.

43. US Congress, *Reports of Committees: H.R. 4165,* 16.

44. Dorn, "(In)temperate Zones," 258.

45. Charles Jewett, "The Medical Uses of Alcohol," in *Centennial Temperance Volume*, 258.

46. Cited in William T. Hutchinson, *Cyrus Hall McCormick: Harvest, 1856–1884* (New York: D. Appleton-Century, 1935), 300.

47. "The Verdict Unanimous," *Frankfort Roundabout* 9, no. 42 (1886): 1.

48. "Relief from Malarial Poison," *Frankfort Roundabout* 8, no. 34 (1885): 1. Electric Bitters were also said to "end bone scraping" by physicians seeking to heal open sores.

49. Arthur J. Cramp, *Nostrums and Quackery*, vol. 2 (Chicago: American Medical Association, 1921), 555–60.

50. National Prohibition Act of 1919, Public Law 66–66, 41 Stat. 305, sec. 7, 311, http://legisworks.org/congress/66/publaw-66.pdf. Sacramental wine was also exempted from Prohibition. See also Jacob M. Appel, "'Physicians Are Not Bootleggers': The Short, Peculiar Life of the Medicinal Alcohol Movement," *Bulletin of the History of Medicine* 82, no. 2 (2008): 356.

51. "Kentucky May Be 'Dry,'" *Breckenridge News* 31, no. 37 (1907): 6.

14. Making and Selling Whiskey at the Henry McKenna Distillery

Epigraph: H. McKenna, circular, Fairfield, KY, 1885, in Marcella McKenna Distillery Collection (Nelson County, KY), 1824–1945, 1858–1942 (bulk dates), 1 F65M-593, Special Collections, University of Kentucky.

1. Ralph H. Brown, *Historical Geography of the United States* (New York: Harcourt, Brace & World, 1948), 3–6; Tim Ingold, *Making: Anthropology, Archaeology, Art and Architecture* (London: Routledge, 2013), 2–13; Theodore R. Schatzki, *Social Change in a Material World* (London: Routledge, 2019), 131–32.

2. Chapters 14 and 15 are based on records in the Marcella McKenna Distillery Collection at the University of Kentucky (hereafter cited as MMDC). The records in the ledgers and account books are recorded on six reels of microfilm. All ledgers cited are part of the MMDC.

3. Sarah Smith, *Historic Nelson County, Its Towns and People* (Bardstown, KY: GBA/Delmar, 1983), 156. We do not know the reason for Henry McKenna's migration to America. His emigration path was similar to that of others from the Draperstown area, who walked to Derry and then took a boat to Liverpool. From that English port, the thirty-two-year-old McKenna boarded the transatlantic ship *Tempest* to New York City. His departure some eight years before the beginning of the Great Famine was likely not motivated by high rates of pauperism or mortality. The northern and northeastern parts of Ireland were marked by stretches of productive land—albeit often in large, inaccessible holdings—and had a diversified economy, and the residents were comparatively well off during this period. It is possible that the attraction of employment opportunities and potential landownership were sufficient to prompt immigration to America. See S. H. Cousens, "The Regional Pattern of Emigration during the Great Irish Famine, 1846–51," *Transactions and Papers of the Institute of British Geographers* 28 (1960): 119–34.

4. Angela O'Keeney, *Looking Back on Ballinascreen* (Draperstown, Ireland: Ballinascreen Historical Society, 1989), 40.

5. Clerk of Court, Nelson County, Bardstown, KY, *Nelson County Deed Record Book 29*, 100; O'Keeney, *Looking Back on Ballinascreen*, 40.

6. H. McKenna to Kiefaber Delicacy Company, Dayton, OH, January 18, 1907, MMDC, reel 4.

7. Harry H. Kroll, *Bluegrass, Belles, and Bourbon: A Pictorial History of Whiskey in Kentucky* (New York: A. S. Barnes, 1967), 27.

8. Marcella McKenna, "Dan McKenna," 1964, MMDC, reel 1. Stafford McKenna was Marcella McKenna's father.

9. James S. McKenna, "H. McKenna, Incorporated in Kentucky," 1933, MMDC, reel 2. It is dif-

ficult to discern who entered the distillery's personal accounts, although there is some consistency in handwriting from the 1850s to the 1870s. Some entries include brief notes relating to Daniel, James, or Stafford McKenna.

10. Marcella McKenna, "An Aspect of Industry in Early Kentucky—To End of Civil War," presented to the Nelson County Historical Society, July 5, 1966, typescript copy in Oscar Getz Museum of Whiskey, Bardstown, KY.

11. "H. McKenna Distiller," *Wine and Spirit Bulletin* 18, no. 6 (1904): 41.

12. *Sanborn's Surveys of the Distilleries and Warehouses of Kentucky and Tennessee* (New York: Sanborn Map Company, 1910), 32.

13. Ledger 1, 55.

14. Ledger 2, 117.

15. Ledger 2, 383.

16. Ledger 2, 369.

17. Ledger 2, 392.

18. The W. B. Belknap Hardware Company in Louisville carried several varieties of galvanized barbed wire in 1886. See J. M. Elstner, *The Industries of Louisville, Kentucky, and New Albany, Indiana* (Louisville, KY: J. M. Elstner, 1886), 95–96.

19. Ledger 2, 410.

20. Ledger 3, 55.

21. Ledger 2, 68; Ledger 3, 185, 247.

22. Ledger 2, 65.

23. Ledger 2, 209.

24. See Ciaran Buckley and Chris Ward, *Strong Farmer: The Memoirs of Joe Ward* (Dublin, Ireland: Liberties Press, 2007), 65–73; Kevin Kenny, *The American Irish: A History* (Harlow, UK: Longman, 2000), 47–48. Some strong farmers had access to fertile lands, a rarity in western and northwestern Ireland. Many others succeeded by intensively managing and improving marginal property through brush and rock removal, field fallowing in alternate years, and heavy manuring. They raised wheat when it was in high demand and switched to cattle and sheep when grain prices fell. Their homes were often comparatively large and finely furnished and appointed; they burned coal in their hearths, whereas poor tenants and laborers burned bog turf. Strong farmers often hired full-time or seasonal laborers, housed them on the farm, and gave them access to small plots on which to grow potatoes. On the eve of the Great Famine in 1841, Ireland had about 128,000 strong farmers, some of whom owned or controlled more than 100 acres of land, as well as some 900,000 landless laborers, most of whom worked for strong farmers and a daily wage. Consider how Ireland's Nobel Laureate poet William Butler Yeats described the emotions an aspiring strong farmer might experience when considering an idealized townland, an impossibly Edenic country, in the folkloric poem "The Happy Townland":

> There's many a strong farmer
> Whose heart would break in two,
> If he could see the townland
> That we are riding to;
> Boughs have their fruit and blossom
> At all times of the year:
> Rivers are running over
> With red beer and brown beer.

W. B. Yeats, *Collected Poems* (London: Macmillan Collector's Library, 2016), 133–34. See also James Pethica, "Contextualising the Lyric Moment: Yeats's 'The Happy Townland' and the Abandoned Play *The Country of the Young*," *Yeats Annual No. 10*, ed. Warwick Gould (1993): 65–91.

25. "Dan McKenna" statement, 138–39, MMDC, reel 2.

26. MMDC, reel 6, contains three ledgers, each with more than 500 pages, that list customer accounts. Ledger 3 (1874–1876) lists 351 different customer accounts.

27. Ledger 1, 19, 21.

28. Ledger 1, 606–7.

29. Ledger 1, 579.

30. See, for example, the charge of one quart of whiskey by N. Pitt on the account of Bew Miller, December 4, 1876, in Ledger 3, 124.

31. H. P. Rickman, ed., *Pattern & Meaning in History: Wilhelm Dilthey, Thoughts on History & Society* (New York: Harper & Row, 1961), 24–33; Theodore R. Schatzki, *The Site of the Social: A Philosophical Account of the Constitution of Social Life and Change* (University Park: Pennsylvania State University Press, 2002), 99–100.

32. O'Keeney, *Looking Back on Ballinascreen*, 39–40.

33. Ibid. A similar "primitive" log-distilling process is described by Gerald Carson, *The Social History of Bourbon* (1963; reprint, Lexington: University Press of Kentucky, 2010), 45.

34. US Department of Internal Revenue, *Regulations and Instructions Concerning the Tax on Distilled Spirits under the Revised Statutes of the United States and Subsequent Acts*, series 7, no. 7 (Washington, DC: GPO, 1877), 21.

35. Ledger 1, 123.

36. MMDC, reel 2.

37. Ledger 1, 48.

38. Ledger 1, 30.

39. *An Atlas of Nelson and Spencer Counties, Kentucky* (Philadelphia: D. J. Lake, 1882), 26.

40. "Louisville Dots," *Bonfort's Wine and Spirit Circular* 20, no. 11 (1883): 213.

41. "Louisville Dots," *Bonfort's Wine and Spirit Circular* 21, no. 5 (1883): 106.

42. "Louisville Dots," *Bonfort's Wine and Spirit Circular* 33, no. 2 (1889): 52.

43. Dean Brothers to H. McKenna, June 28, 1892, MMDC, reel 4.

44. N. P. Bowsher to H. McKenna, December 22, 1897, MMDC, reel 4.

45. The standard dimension of a cord of wood is 128 cubic feet; when stacked, it measures four feet wide by four feet high by eight feet long. Nonstandard measurements include the "face cord," which is cut in lengths convenient to burn in a stove (about sixteen inches) and stacked four feet high by eight feet long. Henry McKenna always specified his wood purchases and sales as cords, not face cords.

46. Ledger 1, 130.

47. Ledger 1, 205.

48. Ledger 1, 187.

49. Ledger 1, 368.

50. Ledger 1, 394.

51. Ledger 2, 388.

52. Ledger 2, 368.

53. Ledger 2, 377.

54. Ledger 2, 430, 494.

55. Ledger 3, 103.

56. Kentucky General Assembly, *Journal of the Senate of the Commonwealth of Kentucky,* appendix to the Senate Journal, Report of the Board of Internal Improvement (Frankfort, KY: A. G. Hodges, 1837), 59–61.

57. The straight-line distance between Fairfield and Monterey is about forty-five miles. Ransdell & Birchett to H. McKenna, January 4, 1884, MMDC, reel 4.

58. Ledger 1, 304.

59. George F. Swain, "Statistics of Water Power Employed in Manufacturing in the United States," *Publications of the American Statistical Association* 1, no. 1 (1888): 11–12. See also Peter Temin, "Steam and Waterpower in the Early Nineteenth Century," *Journal of Economic History* 26, no. 2 (1966): 196.

60. Ledger 2, 441.

61. Ledger 2, 495.

62. Clay McShane and Joel A. Tarr, *The Horse in the City: Living Machines in the Nineteenth Century* (Baltimore: Johns Hopkins University Press, 2007), 9–11; R. J. Moore-Colyer, "Aspects of the Trade in British Pedigree Draught Horses with the United States and Canada, c. 1850–1920," *Agricultural History Review* 48, no. 1 (2000): 45–56.

63. "The Turf. Field and Farm," *Evening Bulletin* 8, no. 87 (1889): 2.

64. "Livestock," *Semi-Weekly Interior Journal* 21, no. 9 (1893): 3.

65. "Percheron-Morgan Cross," *Semi-Weekly Interior Journal* 20, no. 66 (1892): 3. See also Ann Norton Green, *Horses at Work: Harnessing Power in Industrial America* (Cambridge, MA: Harvard University Press, 2008), 84–88.

66. Ledger 2, 63.

67. Ledger 1, 279.

68. Ledger 1, 65.

69. Ledger 1, 49.

70. Ledger 1, 177.

71. Ledger 2, 402.

72. Ledger 1, 348.

73. Ledger 2, 37.

74. Ledger 2, 39.

75. Ledger 2, 62.

76. US Census Bureau, 1860 Census, Nelson County.

77. Ledger 1, 438.

78. Ledger 2, 411.

79. Ledger 3, 159.

80. Ledger 3, 247.

81. Ledger 2, 405, 407; Ledger 3, 37.

82. *An Atlas of Henry and Shelby Counties, Kentucky* (Philadelphia: D. J. Lake, 1882); Elmer G. Sulzer, *Ghost Railroads of Kentucky* (Bloomington: Indiana University Press, 1967), 173.

83. US Post Office Department, L & N Railroad Archive, box 34:2, contract no. 1672, Shelbyville to Bloomfield, September 26, 1888, Special Collections, Ekstrom Library, University of Louisville; J. B. Hoeing, *Preliminary Map of Kentucky, 1891: Railroad Systems* (New York: Kentucky Geological Survey, 1889).

84. P. C. Kelleher to H. McKenna, September 12, 1890, MMDC, reel 4.

85. O'Keeney, *Looking Back on Ballinascreen,* 41.

86. Ibid., 40.

87. Ledger 2, 120.

88. Ledger 2, 154.

89. Ledger 2, 271.

90. Ledger 2, 120.

91. Ledger 2, 154.

92. Ledger 2, 271, 292.

93. Ledger 2, 292.

94. Ledger 2, 351.

95. Ledger 2, 321.

96. Ledger 2, 351.

97. Ibid.

98. Ledger 2, 383.

99. Ledger 3, 19.

100. Ledger 3, 19, 54.

101. Ledger 3, 54. Several improved threshing machines were available in the 1870s, and the McKenna records do not specify which type Sweeney purchased.

102. Ledger 3, 104.

103. Ledger 3, 142.

104. Ledger 3, 143.

105. Ledger 3, 204.

106. Ledger 3, 290.

107. Ledger 3, 328.

108. Ledger 3, 328–29.

109. Ledger 3, 329, 342–43.

110. Ledger 3, 354.

111. Ledger 3, 377.

112. Ledger 3, 381.

113. O'Keeney, *Looking Back on Ballinascreen*, 41. One can reconstruct a portion of Allen "Coleman" Bixler's employment record from the US census. In 1910 he was proprietor of a distillery in Frankfort. By 1920, he had moved to Louisville, where he worked as a steamfitter; Prohibition had eliminated his distilling work. In 1930, before repeal, he was back in Fairfield, working as a building contractor. In 1940 he was again working as a distiller and still living in Fairfield.

114. Ledger 1, 33, 257, 309, 395.

115. Ledger 1, 531; Ledger 2, 172.

116. Ledger 3, 147.

117. Ledger 2, 36.

118. US Census Bureau, 1870 Census, Nelson County, KY.

119. Wheat bran is the outer layer of the seed coat and contains fiber. The inner layer of the seed coat is referred to as shorts and contains starch.

120. Ledger 2, 314.

121. Ledger 2, 382, 24, 477. A fence panel was usually about sixteen feet long.

122. Ledger 3, 72.

123. Ledger 3, 72–73.

124. Ledger 3, 96–97. Although there is no record of the transaction, McKenna apparently bought his cooper shop back from Blanford.

125. Ledger 3, 108.

126. Ledger 3, 109, 167, 130.

127. Ledger 3, 130–31.

128. Ledger 3, 167, 243.

129. Ledger 3, 249.

130. Ledger 3, 266–67.

131. Ledger 3, 267.

132. Ledger 3, 304, 318.

133. US Census Bureau, 1900 Census, Spencer County, KY.

134. Edwin Hergesheimer, *Map Showing the Distribution of the Slave Population of the Southern States of the United States, 1860* (Washington, DC: US Census Office, 1861). See also John J. Zaborney, *Slaves for Hire: Renting Enslaved Laborers in Antebellum Virginia* (Baton Rouge: Louisiana State University Press, 2012), 9–27; Harold D. Tallant, *Evil Necessity: Slavery and Political Culture in Antebellum Kentucky* (Lexington: University Press of Kentucky, 2003), 8–9; Robert S. Starobin, *Industrial Slavery in the Old South* (New York: Oxford University Press, 1970), 12–18; Jonathan Martin, *Divided Mastery: Slave Hiring in the American South* (Cambridge, MA: Harvard University Press, 2004), 75–104. The distribution of Kentucky's distilling industry generally coincided with the areas of highest agricultural production, although many Pennyroyal distilleries were established after 1870 and produced spirits in limited amounts. Chester Zoeller, *Bourbon in Kentucky: A History of Distilleries in Kentucky,* 2d ed. (Louisville, KY: Butler Books, 2010), 49–88.

135. Ledger 1, 159.

136. Ledger 1, 174.

137. Ledger 1, 610, 575.

138. Ledger 1, 106.

139. Ledger 2, 13. Henry McKenna also spelled Stephen Wickam's name as Wickum.

140. Ledger 1, 30.

141. Ledger 1, 70, 232.

142. Ledger 3, 319.

143. Ledger 2, 121.

144. Ledger 2, 162.

145. Ledger 2, 199.

146. Ledger 2, 38.

147. Ledger 2, 68.

148. Ledger 1, 349–51.

149. Ledger 1, 316.

150. Ledger 1, 165; Ledger 2, 358.

151. Ledger 3, 63.

152. Ledger 2, 368.

153. Ledger 2, 246, 389.

154. Ledger 3, 147.

155. Ledger 3, 205, 246.

156. Ledger 3, 158.

157. Ledger 1, 369.

158. Ledger 1, 205.

159. Ledger 1, 123.

160. Ledger 1, 235.

161. Ledger 1, 151; US Census Bureau, 1860 Slave Census, Nelson County, KY.

162. "Old McKenna Distillery Reopens," *Louisville Courier-Journal*, February 18, 1934, sec. 6, 3.

163. Ledger 1, 54.

164. Ledger 1, 158.

165. Ledger 1, 55; Ledger 3, 360.

166. Ledger 2, 65.

167. Ledger 1, 53; Ledger 2, 327.

168. Ledger 2, 380–81.

169. Ledger 2, 495; Ledger 3, 152.

170. Ledger 2, 430, 494.

171. J. C. Drake to Dan McKenna, May 12, 1896, MMDC, reel 4.

172. Ledger 1, 131.

173. Ledger 1, 62, 176.

174. Ledger 2, 56.

175. Ledger 1, 315; Ledger 3, 247.

176. See, for example, Ledger 2, 404–5.

177. Ledger 3, 107, 145.

178. Ledger 3, 161.

179. *Atlas of Nelson and Spencer Counties*, 26.

180. Ledger 3, 230, 321.

181. Ledger 2, 317. B. B. Wootten was also a slaveholder in 1860, owning a twenty-seven-year-old woman and her three children. US Census Bureau, 1860 Slave Census, Nelson County, KY.

182. Ledger 2, 180; Ledger 3, 205, 159. Alfred Bodine operated a large farm near Fairfield and owned sixteen slaves in 1860, seven males and nine females. It is possible that some of the African Americans living with McKenna in the 1870s were descendants of the Bodine slaves.

183. Ledger 3, 205.

184. Ledger 3, 291.

185. Ledger 1, 230; Ledger 3, 248.

186. Ledger 2, 12; Ledger 3, 178, 212, 295.

187. Ledger 3, 184.

188. Ledger 1, 534.

189. Ledger 1, 325.

190. Ledger 1, 114.

191. Ledger 1, 230.

192. Ledger 1, 270.

193. Ledger 1, 271.

194. Ledger 1, 63.

195. Ledger 1, 267, 278.

196. E. Harrison Cawker, "Kentucky Flour Mills Listed in the 1880 Edition of Cawker's Flour Mill Directory," *Millstone: Journal of the Kentucky Old Mill Association* 12, no. 2 (2013): 7–12.

197. Karl Raitz and Nancy O'Malley, *Kentucky's Frontier Highway: Historical Landscapes along the Maysville Road* (Lexington: University Press of Kentucky, 2012), 28–29.

198. "Fire at Fairfield Almost Wipes out Town," *Bourbon News* 37 (November 9, 1917): 4. The Fairfield fire started in the Masonic Temple and spread to the McKenna Hotel, two residences, a harness shop, an electric light plant, and a large general store. An explosion, thought to be caused by a burning coal-oil

tank, damaged ten stores and the bank. The loss was estimated at $30,000. See also O'Keeney, *Looking Back on Ballinascreen*, 42.

199. Karl Raitz and John Paul Jones III, "The City Hotel as Landscape Artifact and Community Symbol," *Journal of Cultural Geography* 9 (Fall/Winter 1988): 23.

200. Clerk of Court, Nelson County, Bardstown, KY, *Nelson County Deed Record Book 29*, 100.

201. Ledger 1, 31.

202. Ledger 1, 403. McKenna recorded Mrs. Simpson's name as Elizabeth (her middle name) or as Faney or Frances (her first name) in different account entries. Ledger 1, 535.

203. Ledger 2, 55, 82.

204. Ledger 2, 114, 164.

205. Thomas J. Schlereth, *Victorian America: Transformations in Everyday Life, 1876–1915* (New York: HarperCollins, 1991), 142–45.

206. Alfred Barnard, *The Whiskey Distilleries of the United Kingdom* (1887; reprint, Edinburgh: Birlinn Press, 2008), 438–43.

207. Ballinascreen Parish and St. Eugene's Roman Catholic Church are on Moydamlaght Road in Moneyneaney, about two miles northwest of Draperstown, in southern County Derry (also known as County Londonderry). The church cemetery includes several McKenna grave plots, two marked by engraved white marble Celtic crosses. The cemetery also contains the burial plots of fifteen McGuigan family members; McGuigan was Elizabeth McKenna's maiden name. See Patrick Kelly and Graham Mawhinney, *Ballinascreen Gravestone Inscriptions* (Ballinascreen, Ulster: Ballinascreen Historical Society, 2009).

208. McKenna, "An Aspect of Industry in Early Kentucky," 4.

209. Mary Louise O'Donnell, *Ireland's Harp: The Shaping of Irish Identity c. 1770–1880* (Dublin, Ireland: University College Dublin Press, 2014), 1–7.

210. Ibid., 43.

211. "For Sixty Years," *Kentucky Irish American* 16, no. 11 (1906): 9.

15. The McKenna Family Distillery

Epigraph: Thomas Brown to H. McKenna, August 1, 1896, reel 4, Marcella McKenna Distillery Collection (Nelson County, KY), 1824–1945, 1858–1942 (bulk dates), 1 F65M-593, Special Collections, University of Kentucky.

1. J. M. Elstner, *The Industries of Louisville, Kentucky, and New Albany, Indiana* (Louisville, KY: J. M. Elstner, 1886), 194–95.

2. Clerk of Court, Nelson County, Bardstown, KY, *Nelson County Deed Record Book 45*, 332.

3. "Got a Friend?" *Kentucky Irish American* 13, no. 51 (1903): 2.

4. Compiled from the Marcella McKenna Distillery Collection (hereafter cited as MMDC), reel 4. The tabulation was based on order letters written to the distillery and is not comprehensive. Repeat orders were common, but customers were counted only once to represent the general distribution of orders.

5. John Beckwith & Son to H. McKenna, January 28, 1893. Unless otherwise specified, all correspondence cited is from MMDC, reel 4.

6. W. J. Queen to H. McKenna, May 1884.

7. Everitt Osburn to H. McKenna, December 28, 1889.

8. S. Molly O'Casey to Daniel McKenna, August 18, 1892.

9. Adolph Wesseling to H. McKenna, August 13, 1890.

10. W. F. Boden to H. McKenna, November 15, 1892.

11. C. H. Clason to H. McKenna, March 29, 1892.

12. F. S. Clank to Daniel McKenna, December 14, 1892.

13. V. H. Robinson to H. McKenna, October 29, 1892.

14. The Burdetts may have lived in Lancaster, Kentucky, in Garrard County, in 1870. B. M. Burdett to H. McKenna, December 14, 1892.

15. Mrs. B. W. Burdett to H. McKenna, December 4, 1893.

16. Alfred Mullin to H. McKenna, November 4, 1884.

17. John Moran to H. McKenna, May 15, 1892.

18. Quinn, Woodland & Company to H. McKenna, October 26, 1888.

19. Eugene Hagan to H. McKenna, September 4, October 15, 1890.

20. US Census Bureau, 1890 Census, Nelson County, KY.

21. William C. P. Muir to H. McKenna, March 22, 1891.

22. Boekhoff & Mack to H. McKenna, August 5, 1889.

23. Thomas D. Samuel to H. McKenna, January 14, 1890; George W. Robertson to H. McKenna, January 27, 1892.

24. S. O'Bryan to H. McKenna, December 8, 1902.

25. Chester Zoeller, *Bourbon in Kentucky: A History of Distilleries in Kentucky*, 2d ed. (Louisville, KY: Butler Books, 2010), 129.

26. J. A. Elliott to H. McKenna, May 8, 1895.

27. A. C. Paul to H. McKenna, July 20, October 29, 1900.

28. New Mexico did not become a state until 1912.

29. N. E. Crenshaw to H. McKenna, January 20, 1902.

30. W. L. Caldwell to H. McKenna, September 19, 1888.

31. Merit Street to H. McKenna, July 5, 1886.

32. S. Web Sandifer to H. McKenna, August 6, 1890.

33. S. B. Hopkins to H. McKenna, November 4, 1893.

34. J. McCormick to H. McKenna, March 25, 1887.

35. Summit, Mississippi, was on the Illinois Central Railroad line, about 110 miles north of New Orleans and 8 miles north of Magnolia, Mississippi. John E. Penn to H. McKenna, August 14, 1890.

36. R. S. Campbell to James McKenna, September 24, 1892 (emphasis in original).

37. W. T. Martin to Henry Mckinney [*sic*], April 24, 1886.

38. W. T. Robb to H. McKenna, April 24, 1889.

39. H. B. Daugherty to H. McKenna, December 30, 1896; W. M. Davis to H. McKenna, December 30, 1896.

40. Fred Wooten to Dan McKenny [*sic*], March 3, 1887.

41. George H. Graves to Mr. McKenney [*sic*], February 16, 1889; W. F. Shaner to H. McKenna, December 29, 1885.

42. Henry B. Adsit to H. McKenna, June 28, 1893.

43. Bush, Morse & Company to H. McKenna, February 17, 1888.

44. H. Y. Yates to H. McKenna, December 14, 1907.

45. McKenna's logo places the harp in profile. The harp's forepillar, or bow, was fashioned in the shape of the Maiden of Eire, creating an image akin to the carved figureheads that graced the prows of sailing ships from the sixteenth to nineteenth centuries.

46. "Dan McKenna," statement, 1964, 138–39, MMDC, reel 2.

47. Dr. Dudley Reynolds to H. McKenna, August 1, 1888.

48. Gary A. O'Dell, "At the Starting Post: Racing Venues and the Origins of Thoroughbred Racing in Kentucky, 1783–1865," *Register of the Kentucky Historical Society* 116, no. 1 (2018): 65–67.

49. Elstner, *Industries of Louisville, Kentucky, and New Albany, Indiana,* 126–27; "Crab Orchard Springs," *Semi-Weekly Interior Journal* 16, no. 70 (1888): 8; *Semi-Weekly Interior Journal* 16, no. 43 (1888): 2.

50. M. M. Wingfield to H. McKenna, March 24, 1887.

51. W. A. Alvey to Stafford McKenna, December 17, 1896.

52. J. M. Archuleta to H. McKenna, January 5, 1893.

53. Everett Osbourn to James McKenna, October 10, 1887.

54. Felix Brannigan to H. McKenna, January 14, 1892.

55. W. A. Sullivan to H. McKenna, October 21, 1886.

56. E. L. Edwards to H. McKenna, June 12, 1906.

57. A. T. Gerard to H. McKenna, March 5, 1902. J. W. McCullough was a former internal revenue agent who bought the Green River Distillery in 1888 and moved it to the Louisville, Henderson & St. Louis Railroad tracks. "Green River" was one of McCullough's brands. Zoeller, *Bourbon in Kentucky,* 63.

58. "It Pays," *Evening Bulletin* 9, no. 186 (1890): 2.

59. "Be Sure to Call for McKenna Whisky," *Kentucky Irish American* 42, no. 11 (1919): 8.

60. See, for example, "Hickey's for McKenna Whisky," *Kentucky Irish American* 4, no. 20 (1900): 4; "Millersburg," *Bourbon News* 21 (November 1, 1901): 8.

61. "Professor Dudley Sharpe Reynolds," in L. A. Williams, *History of the Ohio Falls Cities and Their Counties,* vol. 1 (Cleveland, OH: L. A. Williams, 1882), 452–545.

62. G. E. Matthews to H. McKenna, June 6, 1886.

63. D. M. Raymond to H. McKenna, June 23, 1886.

64. William Cotton to H. McKenna, August 15, 1892.

65. Andrew Hepburn, *Great Resorts of North America* (New York: Doubleday, 1965), 3–15.

66. C. W. Cullen to H. McKenna, May 21, 1896.

67. Sister M. Austin to H. McKenna, March 1, 1893.

68. J. A. Smith to H. McKenna, October 1, 1886.

69. R. F. Vaughan to H. McKenna, April 26, 1887 (emphasis in original).

70. Dr. W. A. Statan to H. McKenna, September 1, 1887.

71. Dr. R. D. Robinson to H. McKenna, August 21, 1888.

72. Dr. H. V. Donovan to H. McKenna, December 24, 1895.

73. Hiram Burgess to H. McKenna, January 23, 1892.

74. John L. Cardin to H. McKenna, December 4, 1893.

75. T. H. Sisk to M. M. [*sic*] McKenna, March 1, 1884. See also P. C. Wideman to H. McKenna, April 16, 1890.

76. D. T. Davenport to H. McKenna, March 16, 1893.

77. W. Furr to H. McKenna, June 20, 1906.

78. H. McKenna to E. H. Walker, October 30, 1917.

79. Cooper Whiteside to H. McKenna, October 23, 1917.

80. G. D. Batcheldor to H. McKenna, March 10, 1892. See also W. Sappington (Blandinsville, IL) to H. McKenna, August 1884; W. H. Newman (Quitman, GA) to H. McKenna, July 2, 1886; H. P. Smith (WaKeeney, KS) to H. McKenna, September 18, 1886; W. T. Seymour (Seneca Falls, NY) to H. McKenna, October 22, 1886; W. S. Skellenger (Jackson, MS) to H. McKenna, March 14, 1887; W. V. Sprigg (Glendale,

KY) to H. McKenna, March 17, 1887; M. M. Wingfield (Ellettsville, IN) to H. McKenna, March 24, 1887; J. W. Norris (Palmyra, MO) to H. McKenna, October 6, 1893.

81. John Marbach to H. McKenna, August 8, 1892.

82. T. T. McCandle (Rodney, MS) to H. McKenna, n.d.

83. H. D. Daniel to H. McKenna, February 3, 1888.

84. John B. Eddins to H. McKenna, February 9, 1888.

85. N. S. Bridge to Stafford McKenna, August 2, 1893.

86. O. B. Bush to H. McKenna, April 19, 1895.

87. US Army Corps of Engineers, Mississippi River Commission, *Geological Investigation of Mississippi River Activity, Memphis, Tennessee, to Mouth of Arkansas River,* Technical Memorandum 3-288 (Vicksburg, MS: Waterways Experiment Station, 1949), 49, plate 13.

88. Cox Brothers to H. McKenna, February 8, 1902.

89. Correspondingly, the distillers sent barrel number 5272, valued at more than $126. Druehl & Franken to H. McKenna, January 19, 1892.

90. W. T. Seymour to H. McKenna, October 22, 1886.

91. W. S. Langley to H. McKenna, January 31, 1888.

92. J. M. Clayton to H. McKenna, August 31, 1892.

93. Davis & Forman to H. McKenna, December 20, 1893.

94. Lynch to Stafford McKenna, December 18, 1907.

95. Charles Price to H. McKenna, February 15, 1901.

96. Cited in Walter W. Spooner, ed., *The Political Prohibitionist for 1888: A Handbook for the Aggressive Temperance People of the United States* (New York: Funk & Wagnalls, 1888), 14.

97. George H. Gibson to James McKenna, April 12, 1888.

98. A. L. Lyman to H. McKenna, April 11, 1889, January 22, 1890. See also F. M. Newkirk (Strawn, KS) to H. McKenna, May 17, 1892; Clark & Rogers Druggists (Syracuse, KS) to H. McKenna, September 25, 1886; J. Cleurry (Brookville, KS) to H. McKenna, February 1, 1888; C. D. Arnold (Leoti, KS) to H. McKenna, February 12, 1889; W. J. Wadleigh (Topeka, KS) to H. McKenney [*sic*], January 7, 1890.

99. C. H. Fenell to H. McKenna, November 13, 1888.

100. W. J. Bennett to H. McKenna, April 23, 1903.

101. C. J. Donovan to H. McKenna, October 18, 1895.

102. W. J. Doummy to H. McKenna, May 13, 1908.

103. Willie B. Hicks to H. McKenna, May 18, 1888.

104. S. W. Offett to H. McKenna, August 26, 1885. The Winstead Distillery opened in 1880. See Zoeller, *Bourbon in Kentucky,* 73.

105. W. J. Bennett to H. McKenna, April 23, 1903 (emphasis in original).

106. LeRoy Davidson to Messrs McKenna, May 29, 1906; H. McKenna to LeRoy Davidson, n.d. (emphasis in original).

107. Paul Aaron and David Musto, "Temperance and Prohibition in America: A Historical Overview," in *Alcohol and Public Policy: Beyond the Shadow of Prohibition,* ed. Mark H. Moore and Dean R. Gerstein (Washington, DC: National Academy Press, 1981), 148.

108. D. D. McColl to H. McKenna, November 30, December 21, 1893.

109. US Congress, House, *An Act to Regulate Trade and Intercourse with the Indian Tribes, and to Preserve Peace on the Frontiers,* 23d Cong., 1st sess., 1834, 729–35.

110. Ibid., 732–33.

111. T. J. Morgan, *Fifty-Eighth Annual Report of the Commissioner of Indian Affairs* (Washington, DC: GPO, 1889), 286.

112. R. J. Huston to H. McKenna, September 9, 1889 (emphasis in original).

113. J. H. Wright to H. McKenna, August 18, 1904. The note was only partially legible.

114. J. W. Kirsch to Dan McKenny [*sic*], October 19, 1904.

115. See, for example, M. McKenna to Vollmayer, September 11, 1917, regarding compliance with government regulations about the maximum size of mail-order shipping containers; and Herbert Jackson to the National Association of Mail Order Liquor Dealers Incorporated, November 27, 1917, which stated that the Reed Amendment prohibited all spirits shipments to dry points in local-option states and that district attorneys would seek indictments in all such cases. For a summary of the laws pertaining to liquor manufacturing and transportation, see Aute Lee Carr, "Liquor and the Constitution," *Law & Contemporary Problems* 7, no. 4 (1940): 709–16, accessed through HeinOnline, January 13, 2017.

116. J. W. Bradley to H. McKenna, September 28, 1896 (emphasis in original).

117. J. E. O'Conner to H. McKenna, July 24, 1917, February 6 and 12, 1918.

118. H. L. Bridge to Stafford McKenna, August 2, 1893; B. W. Burdett to H. McKenna, December 4, 1893.

119. John C. Drewry to H. McKenna, October 2, 1903.

16. Building James Stone's Elkhorn Distillery

Epigraph: James M. Stone, January 13, 1869, Elkhorn Distillery Records (Scott County, KY), 1868–1872, 1997 MS419, Special Collections, University of Kentucky.

1. Elkhorn Distillery Records (Scott County, KY), 1868–1872, 1997 MS419, Special Collections, University of Kentucky (hereafter cited as EDR). The three surviving account books have been recorded on microfilm and can also be accessed through the Kentucky Digital Library at https://nyx.uky.edu/fa/findingaid/?id=xt76dj58dk4s#fa-heading-contents-of-the-collection. Book 1 begins in 1868 and was apparently written by James Stone and office clerks; it includes copies of correspondence and a few account summaries. Stone's crabbed handwriting is a challenge to read, and some letters are illegible. Most letters address day-to-day distillery operations, grain brokers, freight shipments, internal revenue agents, and wholesale and retail customers. Books 2 and 3 were written largely by James H. Shropshire, Stone's business partner. Book 3 starts in November 1871. Entries dated from late April 1870 to late November 1871 are missing, suggesting that there was another book. Based on the correspondence in the three books, the distillery's primary business connections were in Georgetown, Lexington, Frankfort, Louisville, Cincinnati, St. Louis, Chicago, New York City, Troy (a city in New York and a major industrial center of the day), Philadelphia, New Orleans, and Galveston. These letter books; US censuses of population, agriculture, and slaveholders; and other sources constitute a diverse and lucid record of distillery operations, although some important information is absent. Several personnel are identified only by their vocations, such as the millers who operated the distillery's gristmill and the distiller who made the whiskey. Most agents, brokers, and contractors are identified by specific accounts, but day laborers who conducted the distillery's supporting operations are unnamed and unknown. What we know of their work is through generic references.

2. Tom Kimmerer, *Venerable Trees: History, Biology, and Conservation in the Bluegrass* (Lexington: University Press of Kentucky, 2015), 92–97. The term "savanna" is sometimes used to describe the pre-settlement plant communities found in the Inner Bluegrass, but savannas were generally maintained by

periodic fires, which likely did not occur in central Kentucky. See Alice L. Heikens, "Savanna, Barrens, and Glade Communities of the Ozark Plateaus Province," in *Savannas, Barrens, and Rock Outcrop Plant Communities of North America,* ed. Roger C. Anderson, James S. Fralish, and Jerry M. Baskin (New York: Cambridge University Press, 1999), 220.

3. EDR, Book 1, 441.

4. Clerk of Court, Scott County, Georgetown, KY, *Scott County Deed Record Book 4,* 322.

5. Chester Zoeller, *Bourbon in Kentucky: A History of Distilleries in Kentucky,* 2d ed. (Louisville, KY: Butler Books, 2010), 209; EDR, Book 1, 125.

6. "Whisky Trade," *Louisville Courier-Journal* 38, no. 148 (1868): 1.

7. See, for example, EDR, Book 1, 308.

8. EDR, Book 2, 223.

9. Zoeller, *Bourbon in Kentucky,* 209.

10. Jas. H. Prather, *Prather's Lexington City Directory, 1875–1876* (Lexington: Jas. H. Prather, 1876), 209.

11. EDR, Book 1, 90.

12. EDR, Book 2, 33.

13. Andrew Morrison, *The Industries of Cincinnati: Manufacturing Establishments and Business Houses* (Cincinnati, OH: Aldine Printing, 1886), 187.

14. EDR, Book 1, 42, 45, 57.

15. EDR, Book 1, 72.

16. EDR, Book 1, 43, 57. During the Civil War, William McComas served as an artist and topographical engineer in U. S. Grant's army.

17. EDR, Book 1, 94, 113.

18. The William Readings corn sheller received US Patent No. 9,120A. EDR, Book 1, 167.

19. EDR, Book 1, 2.

20. EDR, Book 2, 8–9.

21. EDR, Book 2, 159.

22. EDR, Book 1, 5, 55.

23. EDR, Book 1, 30, 33.

24. EDR, Book 1, 41, 47, 131.

25. EDR, Book 1, 29, 66.

26. EDR, Book 1, 135, 182.

27. EDR, Book 1, 106, 186, 364.

28. EDR, Book 1, 10.

29. EDR, Book 1, 134, 206.

30. EDR, Book 1, 248.

31. Stone & Shropshire may have used copper sulfate as a cleaning agent to kill algae and fungi in mashing and fermenting vats. See EDR, Book 1, 256.

32. EDR, Book 1, 327.

33. EDR, Book 1, 308.

34. EDR, Book 1, 193, 199, 200, 201, 207.

35. J. M. Elstner, *The Industries of Louisville, Kentucky, and New Albany, Indiana* (Louisville, KY: J. M. Elstner, 1886), 95.

36. EDR, Book 1, 47, 159, 171, 355.

37. EDR, Book 1, 54.

38. EDR, Book 1, 123, 183.

39. EDR, Book 2, 400.

40. EDR, Book 1, 428, 448, 451.

41. EDR, Book 2, 388, 393.

42. EDR, Book 1, 256. Diana Twede, "The Cask Age: The Technology and History of Wooden Barrels," *Packaging Technology and Science* 18, no. 5 (2005): 260–61, www.interscience.wiley.com.

43. EDR, Book 1, 140.

44. EDR, Book 1, 234.

45. EDR, Book 1, 181, 187.

46. EDR, Book 1, 94, 177.

47. EDR, Book 1, 81, 83.

48. EDR, Book 1, 132, 133, 141.

49. EDR, Book 2, 387.

50. EDR, Book 1, 161, 305, 406.

51. EDR, Book 2, 31.

52. EDR, Book 1, 235.

53. EDR, Book 1, 431.

54. EDR, Book 2, 159. It is likely that in addition to the temperature differential between the ground floor and the upper floor, there was a variance in humidity—the second floor likely had lower humidity than the first floor. Low humidity tends to increase alcohol content in aging barrels because the evaporation rate of water is higher than that of alcohol. In high-humidity conditions, proof declines because more alcohol than water evaporates, given that alcohol evaporation is independent of humidity. See J. R. Mosedale, "Effects of Oak Wood on the Maturation of Alcoholic Beverages with Particular Reference to Whisky," *Forestry* 68, no. 3 (1998): 223.

55. EDR, Book 2, 159, 166, 169.

56. John H. White Jr., *The American Railroad Freight Car: From the Wood-Car Era to the Coming of Steel* (Baltimore: Johns Hopkins University Press, 1993), 177–78.

57. EDR, Book 1, 67 (emphasis in original).

58. EDR, Book 1, 69.

59. EDR, Book 1, 78. While the Elkhorn Distillery records do not reveal the exact manner in which railroad employees might have unloaded coal or grain "on the ground," archaeological research at Buffalo Trace Distillery in Frankfort suggests that freight agents or distillers may have constructed "pads" or platforms of fieldstone slabs that would have been somewhat more agreeable for temporary storage space than simply dumping cargo on the bare earth. See Nicolas R. Laracuente and V. Camille Westmont, "Results of Volunteer Archaeological Excavations at Riverside, Buffalo Trace Distillery, Frankfort, Kentucky," 2012 (on file at Buffalo Trace Distillery, Frankfort, KY), 42–44.

60. A shipment of new whiskey barrels sent by railroad to Stone & Shropshire from a cooperage in Vanceburg, Kentucky, went missing in early November 1868 and was never found. EDR, Book 1, 50.

61. EDR, Book 1, 11.

62. EDR, Book 2, 5.

63. EDR, Book 1, 118, 127, 144, 146.

64. EDR, Book 1, 155, 160.

65. EDR, Book 1, 165, 166, 176.

66. The actual distance between the depot and Georgetown was about five miles. EDR, Book 1, 177–78.

67. EDR, Book 1, 177, 281.

68. The dark brown soils in Shelby and Henry Counties are part of the Lowell-Nicholson-Shelbyville soils series, which is derived from Drake formation limestones and calcareous shales and windblown dust or loess. Orville J. Whitaker and Robert A. Eigel, *Soil Survey of Henry and Trimble Counties, Kentucky*, USDA Soil Conservation Service and Kentucky Agricultural Experiment Station (Washington, DC: GPO, 1988).

69. Christiansburg Depot was on the Louisville, Cincinnati & Lexington Railroad, south of Christiansburg and northeast of Shelbyville in Shelby County. The depot was about four miles from the site of Diageo's new $115 million Bulleit Distillery that began operations in 2017.

70. *An Atlas of Henry and Shelby Counties, Kentucky* (Philadelphia: D. J. Lake, 1882), 34–35.

71. Zoeller, *Bourbon in Kentucky*, 99.

72. EDR, Book 1, 8–9.

73. EDR, Book 1, 14.

74. EDR, Book 1, 6–7.

75. EDR, Book 1, 8–9.

76. EDR, Book 1, 60, 73, 74, 154.

77. EDR, Book 1, 60.

78. EDR, Book 1, 63. Corn harvested at 24 percent moisture, for example, can be stored in a bin for only about six days at 70°F before it starts to degrade.

79. EDR, Book 1, 381. Stone apparently entered his debits and credits out of order.

80. EDR, Book 1, 334, 335, 346, 393. See Alexander Maydwell, *Maydwell's Lexington City Directory for 1867* (Cincinnati, OH: Miami Printing & Publishing, 1867), 52, which carried a full-page advertisement for Woolfolk, Craig & Gay that described the firm as follows: "General Commission Merchants, Dealers in Grain, Produce & Salt, Nos 66 and 68 West Main Street, Lexington, Ky." D. F. Wolf & J. G. Yellman also placed a full-page advertisement in the directory, which read: "Lexington Brewery and Malt House, Wolf & Yellman, Brewers & Maltsters and dealers in hops, constantly on hand in wood & glass, XX Pale, Cream, & Stock Ales, Porter, Brown Stout, Lager & Table Beer. Distillers can be supplied at all times with a superior article of Barley, Malt & Hops. All orders filled with promptness & dispatch. Depot North East Corner Mill & Water" (ibid., 74).

81. EDR, Book 1, 111, 129, 140.

82. EDR, Book 1, 60.

83. EDR, Book 1, 308.

84. EDR, Book 1, 324, 330.

85. Michael Krafft, *The American Distiller or, the Theory and Practice of Distilling* (Philadelphia: Archibald Bariram, 1804; reprint, Gale, Sabin Americana Print Editions, n.d.), 117; Harrison Hall, *The Distiller* (1818; reprint, 2d rev. ed., Hayward, CA: White Mule Press, 2015), 111.

86. Herman F. Willkie and Joseph A. Prochaska, *Fundamentals of Distillery Practice* (Louisville, KY: Joseph E. Seagram & Sons, 1943), 57.

87. EDR, Book 1, 58, 104.

88. EDR, Book 1, 334, 346, 357.

89. Apologies to James Lipton for the title of this section. See Lipton's *An Exaltation of Larks* (New York: Penguin Books, 1991). Following this "logic," would the appropriate collective noun for whiskey makers be a "mash of distillers"?

90. EDR, Book 1, 12, 294.

91. James F. Hopkins, *A History of the Hemp Industry in Kentucky* (Lexington: University Press of Kentucky, 1951), 112–51.

92. EDR, Book 1, 71.

93. "Monetary & Commercial," *Louisville Evening Express,* n.s., 1, no. 99 (1869): 4. Cf. "Monetary & Commercial," *Louisville Evening Express,* n.s., 1, no. 37 (1869): 4.

94. William Cronon, *Nature's Metropolis: Chicago and the Great West* (New York: W. W. Norton, 1991), 104–15.

95. EDR, Book 1, 6–7.

96. EDR, Book 1, 152.

97. EDR, Book 1, 12, 106, 139.

98. EDR, Book 1, 152.

99. EDR, Book 1, 190.

100. EDR, Book 1, 208, 210.

101. EDR, Book 1, 211, 212.

102. EDR, Book 1, 207.

103. EDR, Book 1, 273, 327, 376, 377, 378.

104. EDR, Book 1, 149, 270.

105. EDR, Book 1, 135, 145, 271.

106. EDR, Book 1, 245, 246, 333.

107. EDR, Book 1, 298.

108. EDR, Book 1, 245.

109. EDR, Book 1, 304.

110. EDR, Book 1, 272.

111. EDR, Book 1, 318 (partly illegible).

112. EDR, Book 1, 129, 167.

113. "Corn Sheller," Wm. Reading, US Patent No. 9,120, July 13, 1852.

114. EDR, Book 1, 49.

115. Although the record does not permit a tabulation of all coal oil purchases, it is likely that Stone & Shropshire were using about five gallons of kerosene each week. EDR, Book 1, 64.

116. EDR, Book 1, 71, 115, 142, 327.

117. EDR, Book 1, 73, 373.

118. White, *American Railroad Freight Car,* 184–86.

119. EDR, Book 1, 98.

120. EDR, Book 1, 124.

121. EDR, Book 1, 222.

122. EDR, Book 1, 164.

123. EDR, Book 1, 164, 222, 228, 237, 373.

124. EDR, Book 1, 179.

125. *Sanborn's Surveys of the Whiskey Warehouses of Kentucky and Tennessee* (New York: Sanborn-Perris Map Company, 1894).

126. Dan Day, "Corn Cobs—Burning Them with Coal," *Ag Connection* (University of Missouri Extension) 12, no. 12 (2006): 1–2.

127. White, *American Railroad Freight Car,* 172–76. By the 1860s, livestock could be shipped in general-purpose wooden railcars, which were fully roofed and had ventilated doors on either end, or in dedicated livestock cars with slatted side walls.

128. EDR, Book 1, 88.

129. EDR, Book 1, 76, 92.

130. EDR, Book 1, 94, 356, 371, 372.

17. The Elkhorn Distillery's Demise

Epigraph: James H. Shropshire to James M. Stone, December 4, 1869, Elkhorn Distillery Records (Scott County, KY), 1868–1872, 1997 MS419, Special Collections, University of Kentucky.

1. Elkhorn Distillery Records (hereafter EDR) (Scott County, KY), 1868–1872, 1997 MS419, Special Collections, University of Kentucky, Book 2, 304.

2. "Finance and Trade," *Louisville Daily Express,* n.s., 1, no. 221 (1869): 1.

3. EDR, Book 1, 288.

4. EDR, Book 1, 370.

5. EDR, Book 2, 384, 462, 494.

6. EDR, Book 1, 61, 75.

7. EDR, Book 1, 307.

8. "In the Louisville District . . . ," *Evening Bulletin* 12, no. 258 (1893): 2.

9. See, for example, "Praise the Lord," *Semi-Weekly Interior Journal* 10, no. 510, n.s., no. 47 (1882): 2; "Mt. Vernon, Rockcastle County," *Semi-Weekly Interior Journal* 14, no. 164 (1886): 2. Each month, the Elkhorn Distillery paid the federal collector gauging fees of $175 to $180. EDR, Book 1, 220, 396.

10. EDR, Book 2, 328. Andrew M. January and Benjamin W. Wood owned a cotton mill in Maysville that had been in operation since the 1830s; it manufactured cordage, twine, and carpet yarn. January and Wood made sizable loans to Stone & Shropshire.

11. In February 1872 Shropshire discovered a federal gauger's error in measuring alcohol proof in more than 100 gallons out of a total purchase of 2,074 gallons. He reported the error, saving the customer $165. EDR, Book 3, 125, 127.

12. EDR, Book 2, 170.

13. EDR, Book 2, 384.

14. EDR, Book 2, 373.

15. EDR, Book 2, 478.

16. Adolph Brandeis's son Louis was appointed an associate justice of the US Supreme Court in 1916.

17. "River Intelligence, Port of Louisville," *Louisville Evening Express,* n.s., 1, no. 102 (1869): 4.

18. EDR, Book 1, 155.

19. EDR, Book 1, 288.

20. EDR, Book 1, 291.

21. Chester Zoeller, *Bourbon in Kentucky: A History of Distilleries in Kentucky,* 2d ed. (Louisville, KY: Butler Books, 2010), 117.

22. EDR, Book 2, 343.

23. EDR, Book 2, 355.

24. EDR, Book 1, 293.

25. EDR, Book 1, 302, 304.

26. EDR, Book 1, 236, 319, 326.

27. EDR, Book, 2, 40.

28. EDR, Book 2, 159.

29. EDR, Book 2, 35.

30. EDR, Book 2, 40–41.

31. EDR, Book 2, 44, 45.

32. EDR, Book 1, 276, 288, 299.

33. EDR, Book 1, 148.

34. EDR, Book 1, 158.

35. EDR, Book 2, 33.

36. EDR, Book 2, 21.

37. EDR, Book 1, 321, 324.

38. EDR, Book 1, 334, 340.

39. EDR, Book 2, 334.

40. EDR, Book 2, 348.

41. EDR, Book 1, 364.

42. EDR, Book 1, 385 (emphasis in original).

43. EDR, Book 1, 347, 352, 384.

44. EDR, Book 1, 393.

45. William H. Perrin, ed., *History of Bourbon, Scott, Harrison and Nicholas Counties, Kentucky* (Chicago: O. L. Baskin, 1882), 471–72.

46. EDR, Book 2, 175–77.

47. J. M. Elstner, *The Industries of Louisville, Kentucky, and New Albany, Indiana* (Louisville, KY: J. M. Elstner, 1886), 150.

48. EDR, Book 2, 185.

49. The collateral behind a warehouse receipt consisted of specific barrels of whiskey stored in a government-approved warehouse. Its value was based partly on speculation that the whiskey was aging properly and increasing in value. Warehouse receipts were negotiable and were regarded as sound investments if the whiskey was properly insured. Banks would often lend money on a receipt—up to 80 percent of its value.

50. EDR, Book 2, 187.

51. EDR, Book 2, 193, 194, 197.

52. EDR, Book 2, 203.

53. EDR, Book 2, 204.

54. EDR, Book 2, 206.

55. EDR, Book 2, 215.

56. EDR, Book 2, 244.

57. EDR, Book 2, 284.

58. EDR, Book 2, 284.

59. EDR, Book 2, 346.

60. EDR, Book 2, 219.

61. EDR, Book 2, 230.

62. EDR, Book 2, 248–49.

63. EDR, Book 2, 252. Note the distinction between freight price per barrel and price per 100 pounds used by other freighters.

64. Freight agent E. P. Wilson later worked for the Louisville, Paducah, & Southwestern Railroad.

65. EDR, Book 2, 264–65. Forty gallons of whiskey, or roughly the amount contained in one barrel, weighed about 295 pounds; the empty white oak barrel weighed 90 to 100 pounds. At the rate of 35 cents per 100 pounds, the Louisville freight rate would have been roughly $1.40 per barrel. In contrast, the Cincinnati rate to St. Louis was only 80 cents per barrel, or 57 percent of the Louisville rate.

66. EDR, Book 2, 257.

67. EDR, Book 2, 262.

68. In present-day value, $20,000 is roughly equivalent to $340,000.

69. EDR, Book 2, 304.

70. EDR, Book 2, 304.

71. EDR, Book 2, 322.

72. EDR, Book 2, 291.

73. EDR, Book 2, 287.

74. EDR, Book 2, 379, 393.

75. EDR, Book 2, 414.

76. EDR, Book 2, 328, 338.

77. EDR, Book 2, 345, 350.

78. EDR, Book 2, 371.

79. EDR, Book 2, 362.

80. EDR, Book 2, 370.

81. EDR, Book 2, 400.

82. EDR, Book 2, 399.

83. EDR, Book 2, 402.

84. EDR, Book 2, 403.

85. EDR, Book 2, 411.

86. EDR, Book 2, 424.

87. EDR, Book 2, 424.

88. EDR, Book 2, 433.

89. EDR, Book 2, 437, 443.

90. EDR, Book 2, 445.

91. EDR, Book 2, 447.

92. EDR, Book 2, 450.

93. EDR, Book 2, 454.

94. EDR, Book 2, 472.

95. EDR, Book 2, 457.

96. EDR, Book 2, 456.

97. EDR, Book 2, 460 (emphasis in original).

98. EDR, Book 2, 461 (emphasis in original).

99. EDR, Book 2, 467.

100. EDR, Book 2, 483.

101. EDR, Book 2, 487.

102. EDR, Book 2, 488, 489.

103. EDR, Book 2, 494.

104. This section is based on EDR, Book 3, which includes approximately 150 letters, telegrams, and financial statements written by James H. Shropshire. The distillery records contain little direct information about James Stone's activities during this period, but a good deal can be inferred from Shropshire's letters to Stone and his son Charles. Stone apparently borrowed a considerable amount of money from the partnership and could not or would not pay it back, which soured the relationship between the two men and contributed to the eventual dissolution of the firm.

105. EDR, Book 3, 27. Stone & Shropshire's financial dealings with wholesalers were, for the most part, honestly and ethically executed. James Shropshire offered the J. D. & M. Williams firm of Boston

140 barrels of Elkhorn's June 1869 bourbon at $1.55 per gallon cash, apparently unaware that the Williams firm had been the subject of a fraud investigation by the US Treasury Department in 1866. The government charged that J. D. & M. Williams had, for two decades, imported champagne and sherry from France at reduced invoice prices to avoid paying a 40 percent ad valorem tax. See J. Z. Goodrich, *Exposition of the J. D. & M. Williams Fraud* (Boston: Rockwell & Rollins, 1866), 2–10. Some companies tried to avoid paying the fees specified in sales contracts, or they alleged that the barrels shipped to them were only partially full—also termed "short" or "out"—a common condition, given leakage, evaporation, or inaccurate measurement and gauging at the distillery. For example, in November 1871 wholesaler G. I. Birch & Company of Albany, New York, refused to make full payment on a ten-barrel order, claiming that the barrels were short.

106. William M. Ambrose, *Bottled in Bond under U.S. Government Supervision* (Lexington, KY: William M. Ambrose, 2007), 25–26.

107. EDR, Book 3, 51, 78.

108. EDR, Book 1, 444.

109. EDR, Book 3, 8.

110. "Debt Statement for January, 1871," *Commercial and Financial Chronicle* 12, no. 289 (1871): 8.

111. Newcomb, Buchanan & Company, "Circular," *Commercial and Financial Chronicle* 15, no. 367 (1872): 4.

112. EDR, Book 3, 151.

113. EDR, Book 1, 320.

114. EDR, Book 1, 373.

115. Zoeller, *Bourbon in Kentucky*, 209; Clerk of Court, Scott County, Georgetown, KY, *Scott County Deed Record Book 14*, 453–54. After James Stone ceased distilling operations, he may have sold the works to Robert Dedman, a farmer from the Midway area in Woodford County.

116. "Kentucky State News," *Hickman Courier* 28, no. 39 (1893): 2; Richard Taylor, *The Great Crossing: A Historic Journey to Buffalo Trace Distillery* (Frankfort, KY: Buffalo Trace Distillery, 2002), 56–57, 62–66; Zoeller, *Bourbon in Kentucky*, 213.

117. Cf. Simon Patten and Thomas Hanfrey, "The Cultural Keystone Concept: Insights from Ecological Anthropology," *Human Ecology* 37, no. 4 (2009): 491–500.

18. Naming and Branding

Epigraph: Barry Lopez, *Crossing Open Ground* (New York: Vintage Books, 1989), 64–65.

1. Sidney J. Levy, *The Theory of the Brand* (Wilmette, IL: DecaBooks, 2016), 5; Thomas J. Schlereth, *Victorian America: Transformations in Everyday Life, 1876–1915* (New York: HarperCollins, 1991), 162–63.

2. Andy Pike, "Geographies of Brands and Branding," *Progress in Human Geography* 33, no. 5 (2009): 625–28.

3. "Commissioners Sale of Distillery & Land," *Cynthiana News*, August 4, 1870, 3.

4. Paul Duguid, "Developing the Brand: The Case of Alcohol, 1800–1880," *Enterprise & Society* 4, no. 3 (2003): 414.

5. Frank H. Foster and Robert L. Shook, *Patents, Copyrights, & Trademarks*, 2d ed. (New York: John Wiley & Sons, 1993), 166–67.

6. "Yellowstone Trade Mark Application," Taylor & Williams Inc., Louisville, KY, trademark registration no. 0280199, February 10, 1931.

7. F. Paul Pacult, *American Still Life: The Jim Beam Story and the Making of the World's #1 Bourbon* (Hoboken, NJ: John Wiley & Sons, 2003), 106.

8. "Old Tub," Jim Beam Brands, Deerfield, IL, trademark registration no. 4293249, February 19, 2013.

9. "Knob Creek Trade Mark Application," Jim Beam Company, Louisville, KY, trademark registration no. 2156225, May 12, 1998.

10. "Old Forester Trade Mark Application," Brown-Forman Company, Louisville, KY, trademark registration no. 0044366, July 4, 1905.

11. "Geo. G. Brown Est'd Old Forester 1870 First Bottled Bourbon, Old Forester Est'd 1870 Kentucky Straight Bourbon Whisky," Early Times Distillers Company, Louisville, KY, trademark registration no. 4392866, August 27, 2013. In 2018 Brown-Forman brought back a "dead" bourbon brand name, "King of Kentucky," which originated in 1881. The distiller first acquired the brand in 1936 and allowed it to go out of production in 1968. The brand was assigned to a fourteen-year-old single-barrel bourbon with a production of less than 1,000 bottles.

12. Foster and Shook, *Patents, Copyrights, & Trademarks*, 165. Cf. Susan Strasser, *Satisfaction Guaranteed: The Making of the American Mass Market* (New York: Pantheon Books, 1989), 37.

13. "H. McKenna Trade-Mark Application," James S. McKenna, Fairfield, KY, trademark 324,832, February 20, 1905; "H. McKenna Trade-Mark Application," James S. McKenna, Fairfield, KY, trademark 344,114, February 20, 1905, copies in Marcella McKenna Distillery Collection, reel 2.

14. "Reapplication of H. McKenna, Incorporated, Trade-Mark," James S. McKenna, Fairfield, KY, trademark 344,114, December 31, 1934, copy in Marcella McKenna Distillery Collection, reel 2.

15. From Conway P. Coe, Commissioner of Patents, Washington, DC, re: trademark 344,114, March 16, 1937, copy in Marcella McKenna Distillery Collection, reel 2.

16. "McKenna," William H. Hollander, Wyatt, Tarrant & Combs, Louisville, KY, registration no. 324,832, December 7, 1994; http://assignments.uspto.gov/assignments/q?db=tm&reel=0667&frame=0890. In 2018 H. McKenna's single-barrel, ten-year-old bourbon won Best Bourbon honors at the San Francisco World Spirits Competition.

17. "Elijah Culpepper," newspaper clipping, January 10, 1894, Taylor-Hay Family Papers (Franklin County, KY), 1783–1991, folder 66, MssAT238d, Filson Historical Society, Louisville, KY (hereafter cited as THFP).

18. "James Crow, Whiskey Maker," *Sun* 65, no. 5 (1897): 5.

19. William Mida, *Mida's Compendium of Information for the Liquor Interests* (Chicago: Criterion Publishing, 1899), 176–78. Cf. Chester Zoeller, *Bourbon in Kentucky: A History of Distilleries in Kentucky*, 2d ed. (Louisville, KY: Butler Books, 2010), 212.

20. "Elijah Culpepper," newspaper clipping, January 10, 1894, THFP; *Sanborn's Surveys of the Whiskey Warehouses of Kentucky and Tennessee* (New York: Sanborn-Perris Map Company, 1894), 61; Zoeller, *Bourbon in Kentucky*, 189.

21. Mida, *Mida's Compendium of Information*, 177–78.

22. George C. Buchanan, *Fine Whisky Facts* (Louisville, KY: George Buchanan, 1892), 97–100.

23. "Pepper Enjoined," *Frankfort Roundabout* 14, no. 45 (1891): 4.

24. "Old Taylor," Thos. A. Clark, National Distillers Products Corporation, trademark 507,794, March 22, 1949.

25. Letterhead, May 1, 1899, folder 71, THFP. In addition to the image of the bottle and label on the left, the right side of the letterhead included in large letters "Old Taylor the Premier Kentucky Distillery" and the large red-ink signature of E. H. Taylor and Sons. In the center, a word block read "Remem-

ber the Signature! It Guarantees the only Genuine 'Taylor' Whiskey." A doll-like figure of a small man dressed in blue trousers, a vest, and a cutaway coat covered in stars and stripes sat atop the bottle. This figure may have been intended to represent a colonel or Mr. Taylor, who was often referred to as Colonel Taylor. See letter from T. M. Gilmore, *Bonfort's Wine and Spirit Circular,* to J. Swigert Taylor, June 4, 1909, folder 82, THFP.

26. Letter from E. H. Taylor, Jr., & Sons to Dabney S. Taylor Jr., April 29, 1910, folder 83, THFP.

27. Wm. McKee Duncan, Atty., "Petition of Appellants for Modification and Extension of Opinion," E. H. Taylor, Jr., & Sons Appellants vs. Marion E. Taylor, etc. Appellees, Court of Appeals of Kentucky, April 15, 1905, 3, folder 78, THFP; Richard Taylor, *The Great Crossing: A Historic Journey to Buffalo Trace Distillery* (Frankfort, KY: Buffalo Trace Distillery, 2002), 88.

28. "Won Big Victory," *Frankfort Roundabout* 28, no. 29 (1905): 1.

29. "Suit for Infringement of Trade-Mark," *Frankfort Roundabout* 33, no. 12 (1906): 4; "Won Their Suit," *Frankfort Roundabout* 30, no. 37 (1907): 1; letter from T. M. Gilmore, *Bonfort's Wine and Spirit Circular,* to E. H. Taylor, Jr., & Sons, March 26, 1913, folder 86, THFP.

30. Andy Pike, "Placing Brands and Branding: A Scio-Spatial Biography of Newcastle Brown Ale," *Transactions of the Institute of British Geographers* 36, no. 2 (2011): 208.

31. Mida, *Mida's Compendium of Information,* 181–82.

32. "Whisky Trade in Bourbon," *Kentucky Tribune* 14, no. 23 (1857): 2.

33. E. H. Taylor Jr., *Description of the O.F.C., Carlisle, and J. S. Taylor Distilleries* (Frankfort, KY: E. H. Taylor Jr. Co., [1886]).

34. Leo McDonough, *An Illustrated Historical Atlas Map of Daviess County, Ky.* (1876; reprint, Evansville, IN: Unigraphic, 1978), 52.

35. Dennis Rook, comp., *Brands, Consumers, Symbols & Research: Sidney J. Levy on Marketing* (Thousand Oaks, CA: Sage Publications, 1999), 244–46.

36. Pike, "Placing Brands and Branding," 210.

37. US Congress, House, *The Reports of Committees: H.R. 4165,* 50th Cong., 2d sess., 1888–1889, 1.

38. Ibid., 12.

39. Ibid., 25.

40. T. J. Jackson Lears, "The Rise of American Advertising," *Wilson Quarterly* 7, no. 5 (1983): 159.

41. US Congress, *Reports of Committees: H.R. 4165,* 50.

42. Ibid.

43. Cf. Duguid, "Developing the Brand," 405.

44. Ibid., 407.

45. Linda A. Fisher, ed., *Joseph J. Mersman: The Whiskey Merchant's Diary; an Urban Life in the Emerging Midwest* (Athens: Ohio University Press, 2007), 2–3.

46. Duguid, "Developing the Brand," 415.

47. "Application for Renewal of Registration of Trademark," Old Sunny Brook Company, National Distillers and Chemical Corporation, New York, NY, trademark 710,690, January 31, 1961; Zoeller, *Bourbon in Kentucky,* 102–3.

48. Pamela W. Laird, *Advertising Progress: American Business and the Rise of Consumer Marketing* (Baltimore: Johns Hopkins University Press, 1998), 236; Daniel Pope, *The Making of Modern Advertising* (New York: Basic Books, 1983), 198.

49. See, for example, Oscar Getz, *Whiskey: An American Pictorial History* (New York: David McKay, 1978), 124–40.

50. Levy, *Theory of the Brand,* 47–49.

51. Walter Dill Scott, *The Psychology of Advertising in Theory and Practice* (Boston: Small, Maynard, 1921), 335–57.

52. Laird, *Advertising Progress*, 252, 270–71.

53. "'Old Taylor' Films Seen by Hundreds," *Bourbon News* 32 (September 18, 1914): 1.

54. It is impossible to know how many customers took the time to read whiskey bottle labels, but the fact that distillers incorporated historical information suggests that they believed it to be important. A more cynical view is that the detailed label information was really targeted at competitors and constituted a kind of claim to status—a declaration of entitlement embedded in a trademark. See "Application for Renewal of Registration of Trademark," Old Sunny Brook Company, National Distillers and Chemical Corporation, New York, NY, trademark 710,690, January 31, 1961. Note that the awards list at the 1904 Louisiana Purchase Exposition included more than 33,000 grand prizes and gold, silver, and bronze medals. Jack Daniels Distillery in Tennessee also claimed that its whiskey won a gold medal.

55. Al Young, *Four Roses: The Return of a Whiskey Legend* (Louisville, KY: Butler Books, 2010), 45.

56. Ibid., 40–41; V. N. Balasubramanyam and M. A. Salisu, "Brands and the Alcoholic Drinks Industry," in *Adding Value: Brands and Marketing in Food and Drink,* ed. Geoffrey Jones and Nicholas J. Morgan (New York: Routledge, 1994), 61.

57. Julian L. Watkins, *The 100 Greatest Advertisements* (New York: Dover Publications, 1959), 150–51; Stephen Fox, *The Mirror Makers: A History of American Advertising and Its Creators* (New York: William Morrow, 1984), 140.

58. Lears, "Rise of American Advertising," 163.

59. James M. Ferguson, "Advertising and Liquor," *Journal of Business* 40, no. 4 (1967): 425–26. National Distillers renewed the trademark registration for Old Sunny Brook (registration no. 710,690) in 1961 and as Sunny Brook (registration no. 710,690) in 1980.

60. Balasubramanyam and Salisu, "Brands and the Alcoholic Drinks Industry," 70.

61. Grant McCracken, *Culture & Consumption: New Approaches to the Symbolic Character of Consumer Goods and Activities* (Bloomington: Indiana University Press, 1988), 32–43; Neil McKendrick, John Brewer, and J. H. Plumb, *The Birth of a Consumer Society: The Commercialization of Eighteenth-Century England* (Bloomington: Indiana University Press, 1982), 21–24, 28–30.

62. Maoz Azaryahu and Kenneth E. Foote, "Historical Space as Narrative Medium: On the Configuration of Spatial Narratives of Time at Historical Sites," *GeoJournal* 73, no. 3 (2008): 179–94.

63. Michael B. Beverland, "Crafting Brand Authenticity: The Case of Luxury Wines," *Journal of Management Studies* 42, no. 5 (2005): 1003–4.

64. Michael B. Beverland, "The 'Real Thing': Branding Authenticity in the Luxury Wine Trade," *Journal of Business Research* 59, no. 2 (2006): 251.

65. Mark Gottdiener, *The Theming of America: Dreams, Visions, and Commercial Spaces* (Boulder, CO: Westview Press, 1997); Scott A. Lukas, ed., *The Themed Space: Locating Culture, Nation, and Self* (Lanham, MD: Lexington Books, 2007).

66. Jim Murray, *Jim Murray's Whisky Bible, 2017* (Litchborough, UK: Dram Good Books, 2017).

67. Balasubramanyam and Salisu, "Brands and the Alcoholic Drinks Industry," 70–71.

68. Joe S. Bain, *Industrial Organization* (New York: John Wiley & Sons, 1959), 229.

69. Ferguson, "Advertising and Liquor," 429–32; Pacult, *American Still Life,* 118–19.

70. Charles F. McGovern, *Sold American: Consumption and Citizenship, 1890–1945* (Chapel Hill: University of North Carolina Press, 2006), 73–74; Ferguson, "Advertising and Liquor," 414–16; Bain, *Industrial Organization,* 229–30. Some whiskey manufacturers, Sunny Brook among them, diversified their advertisements in the 1960s and 1970s, featuring fine paintings by artists such as Frederic Remington.

The ads were often overtly masculine portrayals of the Old West showcasing the US Cavalry and other stereotypical images. The appeal may have been to both historical sensibilities and the dynamism of action. And the ads may have attracted the attention of those who appreciated fine art.

71. Balasubramanyam and Salisu, "Brands and the Alcoholic Drinks Industry," 61, 69.

72. Bain, *Industrial Organization*, 229.

73. Balasubramanyam and Salisu, "Brands and the Alcoholic Drinks Industry," 63, 68.

74. Sally Van Winkle Campbell, *But Always Fine Bourbon: Pappy Van Winkle and the Story of Old Fitzgerald* (Frankfort, KY: Old Rip Van Winkle Distillery, 1999), 124–25.

75. Bill Samuels Jr., *My Autobiography* (Louisville, KY: Saber Publishing–Butler Books, 2009).

76. Douglas B. Holt, *How Brands Become Icons: The Principles of Cultural Branding* (Boston: Harvard Business School Press, 2004), 3–12.

77. Ann Brigham, "Behind-the-Scenes Spaces: Promoting Production in a Landscape of Consumption," in Lukas, *Themed Space*, 207.

78. Several distillers have active research and experimentation programs. Some companies have created product variations by adding honey or cinnamon, for example. And, since the late 1980s, master distiller Harlen Wheatley of Buffalo Trace Distillery has conducted more than 1,500 barrel experiments for the distillery's Experimental Whiskey Collection. See Fred Minnick, "Barreling Ahead: Whiskey-Makers Break Cherished Traditions to Create New Flavors," *Scientific American* 14 (March 2013), https://www.scientificamerican.com/article/whiskey-makers-break-tradition-to-make-new-flavors/?print=true.

79. Elliott West, "Selling the Myth: Western Images in Advertising," *Magazine of Western History* 46, no. 2 (1996): 44–45.

80. Lukas, *Themed Space*, 1.

81. Balasubramanyam and Salisu, "Brands and the Alcoholic Drinks Industry," 74–75.

82. Ibid., 59–75; John Urry and Jonas Larsen, *The Tourist Gaze 3.0* (Los Angeles: Sage, 2011), 120–23.

83. See, for example, Cayuga Lake Wine Trail—America's First, www.cayugawinetrail.com; FingerLakesWineCountry.com.

84. Donovan D. Rypkema, *Historic Preservation and the Economy of the Commonwealth: Kentucky's Past at Work for Kentucky's Future* (Frankfort: Kentucky Heritage Council & Commonwealth Preservation Advocates, 1997), 12.

85. Certic Inc., *Economic Impact of Kentucky's Travel and Tourism Industry—2016 and 2017* (Frankfort: Kentucky Tourism, Arts & Heritage Cabinet, 2018).

86. Rypkema, *Historic Preservation and the Economy of the Commonwealth*, 12.

87. David E. Nye, *American Technological Sublime* (Cambridge, MA: MIT Press, 1996), 25.

88. Ibid., 114.

89. John R. Stilgoe, *Landscape and Images* (Charlottesville: University of Virginia Press, 2005), 130–32; John R. Stilgoe, *Metropolitan Corridor: Railroads and the American Scene* (New Haven, CT: Yale University Press, 1983), 78.

90. Brian Goodall, "Industrial Heritage and Tourism," *Built Environment* 19, no. 2 (1993): 94.

91. Pat Yale, *From Tourist Attractions to Heritage Tourism*, 3d ed. (Huntingdon, UK: ELM Publications, 2004), 7–14.

92. Goodall, "Industrial Heritage and Tourism," 99–100.

93. Ibid., 100; Thomas J. Schlereth, *Cultural History and Material Culture: Everyday Life, Landscapes, Museums* (Charlottesville, VA: University Press of Virginia, 1990), 396–405.

94. Nuala C. Johnson, "Where Geography and History Meet: Heritage Tourism and the Big House

in Ireland," *Annals of the Association of American Geographers* 86, no. 3 (1996): 551; Dean MacCannell, *Empty Meeting Grounds: The Tourist Papers* (London: Routledge, 1992), 1.

95. http://kybourbon.com/bourbon_culture-2/key_bourbon_facts/. In 2019 the Kentucky governor's office announced that Log Still Distilling planned to invest $12 million to revitalize a historical distillery site near Gethsemane, in Nelson County. See https://www.lanereport.com/114188/2019/06/startup-distillery-with-historic-roots-to-locate-in-nelson-county-with-12m-investment/.

96.   http://www.kentuckytourism.com/industry/development_incentives/devincentiveprogram.aspx.

97. "Kentucky Highland Railroad," *Frankfort Weekly News and Roundabout* 31, no. 35 (1908): 3; "Go out from the City Each Day," *Frankfort Roundabout* 30, no. 31 (1907): 8.

98. "Kentucky Elks' Reunion Association Hold Meeting at Frankfort," *Bourbon News* 30, no. 68 (1910): 7.

99. "Frankfort Plans Elaborate Entertainment for Kentucky Hosts at the Inauguration," *Adair County News* 19, no. 6 (1915): 1, 3.

100. R. E. Hughes and C. C. Owsley, *Kentucky the Beautiful*, Louisville & Nashville Railroad Passenger Department (Chicago: Corbitt Railway Printing, 1901), 39–41.

101. Ibid., 39. John Ritchie was born in Scotland in 1752 and migrated to Nelson County with Benjamin Linn, becoming one of the area's first settlers. He built and operated a cedar log distillery near Linn's station in 1785. See Sarah B. Smith, *Historic Nelson County, Its Towns and People* (Bardstown, KY: GBA/Delmar, 1983), 101.

102. Hughes and Owsley, *Kentucky the Beautiful*, 41; "The First Distillery Erected in Kentucky," *Hartford Herald* 40, no. 35 (1914): 7.

103. Urry and Larsen, *Tourist Gaze 3.0*, 11.

104. Jon Goss, "The 'Magic of the Mall': An Analysis of Form, Function, and Meaning in the Contemporary Retail Built Environment," *Annals of the Association of American Geographers* 83, no. 1 (1993): 19; H. P. Rickman, ed., *Pattern & Meaning in History: Wilhelm Dilthey, Thoughts on History & Society* (New York: Harper & Row, 1961), 38–42.

105. Balasubramanyam and Salisu, "Brands and the Alcoholic Drinks Industry," 76.

106. Kentucky's twenty-first-century distilling industry has an extensive following on social media that supports both production and tourism. Distilleries maintain their own websites, as do industry trade organizations such as the Kentucky Distillers' Association. See kybourbon.com. The *Bourbon Review*, published quarterly in hard-copy and digital online formats, and the *Whiskey Advocate* are professionally produced enthusiasts' magazines that offer industry news, product critiques, evaluations of production strategies, and special events calendars. See http://whiskyadvocate.com/. In addition, numerous blogs post analyses of industry events. Brian Haara's *Sipp'n Corn* presents a discussion of important legal cases, past and present, that relate to bourbon production, trademarks, marketing, and other industry matters; see http://sippncorn.blogspot.com. Tom Fischer's *BourbonBlog*, at http://www.bourbonblog.com/media-coverage/, provides a wide range of industry news, as does Chuck Cowdery's blog at http://chuckcowdery.blogspot.com/. Mike Veach's *BourbonVeach*, at https://bourbonveach.com/, is oriented toward industrial history, and *StraightBourbon*, at https://www.straightbourbon.com /community/, bills itself as the Internet's foremost source of curated bourbon discussion. For in-depth tasting reviews and useful historical notes, see the *Copper Tot* at http://www.cooperedtot.com/.

107. John Brinckerhoff Jackson, "The Accessible Landscape," in *Landscape in Sight: Looking at America* (New Haven, CT: Yale University Press, 1997), 68–77.

108. http://kybourbontrail.com/history/.

109. Map, *Kentucky Bourbon Trail, 2017* (Frankfort, KY: Kentucky Distillers' Association, 2017). In 2017 the Bourbon Trail's ten participating industrial distilleries were Angels Envy, Bulleit, Evan Williams, Four Roses, Heaven Hill, Jim Beam, Maker's Mark, Town Branch, Wild Turkey, and Woodford Reserve. Since 2009, the Sazerac-owned distilleries Buffalo Trace (Frankfort), Barton (Bardstown), and Glenmore (Owensboro) have not been KDA members and do not participate in KDA promotional programs such as the Bourbon Trail tour. Given the rapid change in the number of distilleries that are undergoing reclamation or retrofitting and the start-up of new craft distilleries, the total number of distilleries in the state, and those that subscribe to KDA membership, changes periodically.

110. *Kentucky Revised Statutes* 243.0305, http://abc.ky.gov/.

111. Marichela Sepe, "Places and Perceptions in Contemporary City," *Urban Design International* 18 (2013): 111–13.

112. Gottdiener, *Theming of America*, 209.

113. Americus Reed II, Mark R. Forehand, Stefano Puntoni, and Luk Warlop, "Identity-Based Consumer Behavior," *International Journal of Research in Marketing* 29, no. 4 (2012): 310–15.

114. Kentucky Transportation Cabinet, *Limited Supplemental Guide Signs,* http://transportation. ky.gov/Sign-Programs-and-Standards/Pages/Limited_Supplemental_Guide_Signs.aspx.

In 2012 Kentucky's small artisan distilleries organized the Kentucky Bourbon Trail Craft Tour as an adjunct to the KDA Bourbon Trail tour. Thirteen distilleries participate, including New Riff Distillery, Newport; Boone County Distillery, near Florence; Hartfield & Company Distillery, Paris; Limestone Branch Distillery, Lebanon; Old Pogue Distillery, Maysville; Wilderness Trail Distillery, Danville; and Willett Distillery, Bardstown. Some new and revived craft whiskey distilleries are also members of the KDA, including Barrel House Distilling Company, Lexington; Corsair Artisan Distillery, Bowling Green; Hartfield & Company, Paris; Kentucky Peerless Distilling Company, Louisville; Limestone Branch Distillery, Lebanon; MB Roland Distillery, Pembroke; New Riff Distillery, Newport; Old Pogue Distillery, Maysville; Wilderness Trace Distillery, Danville; and Willett Distillery, Bardstown. Craft distillers that are not KDA members and are not on the Kentucky Bourbon Trail Craft Tour include Bluegrass Distillers, Lexington; Boone County Distilling Company, Florence; Boundary Oak Distillery, Radcliff; Bourbon 30 Spirits, Georgetown; Casey Jones Distillery, Hopkinsville; Glenns Creek Distillery, Frankfort; Grease Monkey Distillery, Louisville; Jeptha Creed Distillery, Shelbyville; Kentucky Artisan Distillery, Crestwood; Kentucky Mist, Whitesburg; Michter's Distillery, Louisville; Olde Towne Distillery, Harrodsburg; Paducah Distilled Spirits, Paducah; Route 52 Moonshine, Irvine; Royal Springs Branch Distillery, Georgetown; Second Sight Distillery, Ludlow; Wade Lyn Ranch Distilling, Waynesburg; and Whiskey Thief Distilling, Frankfort.

Bourbon Country, a Louisville-based organization, sponsors visitors' centers across central Kentucky, as well as an Urban Bourbon Tour of Louisville that features stops at distilleries, museums, restaurants, and hotels. See http://www.bourboncountry.com/things-to-do/urban-bourbon -trail/index.aspx. Kentucky Cooperage, owned by the Independent Stave Company, operates whiskey barrel cooperages in Lebanon, Kentucky, and Lebanon, Missouri, and in wine-producing areas such as California, France, South Africa, and Australia. The Lebanon, Kentucky, cooperage stands amidst acres of white oak staves stacked to air-dry and offers tours that demonstrate the barrel-making process.

115. Colin M. Hall, "The Definition and Analysis of Hallmark Tourist Events," *GeoJournal* 19, no. 3 (1989): 264.

116. Mihalis Kavaratzis and G. J. Ashworth, "City Branding: An Effective Assertion of Identity or a Transitory Marketing Trick," *Tijdschrift voor Economische en Sociale Geografie* 96, no. 5 (2005): 506–9.

19. A Reconstructed Past Lives in the Present

Epigraph: Kevin Didio, manager, Kentucky Visitor Experience, Bulleit Frontier Whiskey Experience, Bulleit Distillery, Shelbyville, KY, June 7, 2017.

1. A Resolution Designating September 2007 as "National Bourbon Heritage Month," US Senate Res. 294, 110th Cong., https://www.congress.gov/bill/110th-congress/senate-resolution/294.

2. Cf. Richard Walker, "Unseen and Disbelieved: A Political Economist among Cultural Geographers," in *Understanding Ordinary Landscapes,* ed. Paul Groth and Todd W. Bressi (New Haven, CT: Yale University Press, 1997), 171.

3. David Lowenthal, *The Heritage Crusade and the Spoils of History* (Cambridge: Cambridge University Press, 1998), 14–17.

4. David E. Nye, *Narratives and Spaces: Technology and the Construction of American Culture* (New York: Columbia University Press, 1997), 179.

5. Ibid., 183.

6. Jorn Rusen, "Historical Narration: Foundation, Types, Reason," *History and Theory* 26, no. 4 (1987): 88.

7. The concern here is not with a static, externally imposed authenticity but with a narrative that those in the distilling industry regard as a truthful expression of their past. See Lesley Head, "Cultural Landscapes," in *The Oxford Handbook of Material Culture Studies,* ed. Dan Hicks and Mary C. Beaudry (Oxford: Oxford University Press, 2010), 427–39.

8. Carville Earle, *Geographical Inquiry and American Historical Problems* (Stanford, CA: Stanford University Press, 1992), 5–11.

9. Raphael Samuel, *Theatres of Memory: Past and Present in Contemporary Culture* (London: Verso, 2012), 8.

10. Henry Glassie, "Tradition," *Journal of American Folklore* 108, no. 430 (1995): 395.

11. Ibid., 396, 405.

12. Richard A. Peterson, "In Search of Authenticity," *Journal of Management Studies* 42, no. 5 (2005): 1084.

13. David Lowenthal, *Possessed by the Past: The Heritage Crusade and the Spoils of History* (New York: Free Press, 1996), 1–20. See also Eric Laurier, "Replication and Restoration: Ways of Making Maritime Heritage," *Journal of Material Culture* 3, no. 1 (1988): 24–25.

14. David Lowenthal, *The Past Is a Foreign Country Revisited* (Cambridge: Cambridge University Press, 2015), 595; Laurier, "Replication and Restoration," 25.

15. David Lowenthal, *The Past Is a Foreign Country* (Cambridge: Cambridge University Press, 1985), 336–48.

16. Robert Ulin, "Invention and Representation as Cultural Capital: Southwest French Winegrowing History," *American Anthropologist* 97, no. 3 (1995): 519–20.

17. Peterson, "In Search of Authenticity," 1083–84.

18. Ibid., 1088.

19. Lowenthal, *Possessed by the Past,* ix; Samuel, *Theatres of Memory,* 274–87.

20. Lowenthal, *Possessed by the Past,* 98.

21. R. Gerald Alvey, *Kentucky Bluegrass Country* (Jackson: University Press of Mississippi, 1992), 237–39.

22. Paul Coomes and Barry Kornstein, "The Economic and Fiscal Impacts of the Distilling Industry in Kentucky," paper prepared for the Kentucky Distillers' Association (Louisville, KY: Urban Studies Institute, University of Louisville, 2019); Janet Kelly, Barry Kornstein, and Ryan Marshall, "The Eco-

nomic and Fiscal Impacts of the Distilling Industry in Kentucky," paper prepared for the Kentucky Distillers' Association (Louisville, KY: Urban Studies Institute, University of Louisville, 2017).

23. D. W. Meinig, "The Beholding Eye: Ten Versions of the Same Scene," in *The Interpretation of Ordinary Landscapes: Geographical Essays,* ed. D. W. Meinig (New York: Oxford University Press, 1979), 44.

24. Peirce Lewis, "The Future of the Past: Our Clouded Vision of Historic Preservation," *Pioneer America* 7, no. 2 (1975): 3–9.

25. Ibid., 9–20.

26. John Urry and Jonas Larsen, *The Tourist Gaze 3.0* (Los Angeles: Sage, 2011), 11.

27. National Historic Preservation Act of 1966, Public Law 102-575, https://www.nps.gov/history/local-law/nhpa1966.htm.

28. http://www.heritage.ky.gov/.

29. National Park Service, "Historic Preservation Tax Incentives," Technical Preservation Services, 2012, www.nps.gov/tps.

See, for example, J. Daniel Pezzoni, "George T. Stagg Distillery (Buffalo Trace Distillery)," Frankfort, KY, National Register of Historic Places registration form (Frankfort: Kentucky Heritage Council, 2001). Other registered distilleries include Burk's Distillery or Maker's Mark Distillery, Loretto, Marion County, 1974; Van Meter Distillery, Winchester, Clark County, 1976; Warehouse D, White Mills Distillery Company, Louisville, Jefferson County, 1978; Jacob Spears Distillery, Paris, Bourbon County, 1982; McCracken Distillery and Mill, Tyrone, Anderson County, 1983; Old Prentice Distillery or Four Roses Distillery, Lawrenceburg, Anderson County, 1986; T. W. Samuels Distillery Historic District, Deatsville, Nelson County, 1987; Labrot & Graham's Old Oscar Pepper Distillery, McCracken Pike, Woodford County, 1995; J. T. S. Brown & Son's Complex, Louisville, Jefferson County, 1998; James E. Pepper Distillery, Lexington, Fayette County, 2008; and Old Taylor Distillery Historic District, Millville, Woodford County, 2016.

30. Registered homes and other distilling-related structures include Monticello, Cynthiana, Harrison County, 1974; Dowling House, Lawrenceburg, Anderson County, 1979; Bernheim Distillery Bottling Plant, Louisville, Jefferson County, 1983; Boone County Distillery Superintendent's House and Guest House, Petersburg, Boone County, 1986; T. Jeremiah Beam House, Clermont, Bullitt County, 1987; Country Estates of River Road, Louisville, Jefferson County, 1999; Whiskey Row Historic District, Louisville, Jefferson County, 2010; Ritte's East Historic District, Covington, Kenton County, 2013; and Nelson Distillery Warehouse, Anderson-Nelson Distillery, Louisville, Jefferson County, 2014.

31. A National Register "property" may be one building or a district with dozens of structures.

32. National Park Service, "National Historic Landmarks Program" 2017, www.nps.gov/nhl.

33. Kentucky's registered properties are exceeded in number only by those in Ohio, Massachusetts, and New York.

34. Dean MacCannell, *The Tourist: A New Theory of the Leisure Class* (Berkeley: University of California Press, 2013), 100–102.

35. Donovan D. Rypkema, *Historic Preservation and the Economy of the Commonwealth: Kentucky's Past at Work for Kentucky's Future* (Frankfort: Kentucky Heritage Council & Commonwealth Preservation Advocates, 1997), 1.

36. Nicolas R. Laracuente and V. Camille Westmont, "Results of Volunteer Archaeological Excavations at Riverside, Buffalo Trace Distillery, Frankfort, Kentucky," 2012, on file at Buffalo Trace Distillery, Frankfort, KY; Nicolas R. Laracuente, "Archaeological Investigations of the O.F.C. Building, Buffalo Trace Distillery," 2016, on file at Buffalo Trace Distillery, Frankfort, KY; Carolyn Brooks, "George T. Stagg Distillery," Frankfort, KY, National Historic Landmark nomination (Frankfort: Kentucky Heri-

tage Council, 2012); *Sanborn's Surveys of the Whiskey Warehouses of Kentucky and Tennessee* (New York: Sanborn-Perris Map Company, 1894), 52; *Sanborn's Surveys of the Distilleries and Warehouses of Kentucky and Tennessee* (New York: Sanborn Map Company, 1910), 45.

37. Rypkema, *Historic Preservation and the Economy of the Commonwealth*, 2.

38. Quoted in ibid., 17. See also Jean K. Wolf, "Labrot & Graham's Old Oscar Pepper Distillery," Woodford County, KY, National Historic Landmark nomination (Frankfort: Kentucky Heritage Council, 1999).

39. Rypkema, *Historic Preservation and the Economy of the Commonwealth*, 10.

40. Jayne Goddard, Brett Connors, and Amy Crossfield, "Old Taylor Distillery Historic District," Woodford County, KY, National Register of Historic Places registration form (Frankfort: Kentucky Heritage Council, 2016), 3–34; Janet Patton, "Former Woodford Distillery to Spring Back to Life," *Herald-Leader*, July 11, 2016, 1B, 6B.

41. Phil Breen, "Pogue House," Maysville, KY, National Register of Historic Places registration form (Frankfort: Kentucky Heritage Council, 2005), 1–8.

42. Kathryne M. Joseph and Craig A. Potts, "James E. Pepper Distillery," Lexington, KY, National Register of Historic Places registration form (Frankfort: Kentucky Heritage Council, 2008), 1–19.

43. Lexington-Fayette Urban County Government and Lexington Distillery District Foundation, *Lexington Distillery District Tax Increment Financing Development Area & Plan* (Lexington, KY: Lexington-Fayette Urban County Government, 2008), 4.

44. Chester Zoeller, *Bourbon in Kentucky: A History of Distilleries in Kentucky*, 2d ed. (Louisville, KY: Butler Books, 2010), 190.

45. Sam K. Cecil, *Bourbon: The Evolution of Kentucky Whiskey* (New York: Turner Publishing, 2010), 159.

46. "Exhibit of the Old Times Distillery Company at the Fair," *World's Columbian Exposition Illustrated* 3, no. 1 (1893): 213.

47. Ibid.

48. Ibid.

49. A variation on the small-scale distillery as a demonstration device is in place at the Jim Beam Distillery in Clermont, Kentucky. Visitors are introduced to the complexity of industrial distilling by first visiting a "small-batch" distillery that produces one barrel of white whiskey per day; the industrial still next door produces about forty barrels per hour. The downsized works readily demonstrates the general distilling process from raw grain to freshly distilled white whiskey flowing into undersized tail boxes.

50. In the 1980s many distillers began bottling their bourbons in decorative ceramic decanters to spur flagging sales. Jim Beam has carried that tradition forward, and its current decanter is a replica of the American Stillhouse visitors' center, which is itself a replica of the 1935 structure built by the Beam family.

Epilogue

Epigraph: John Brinckerhoff Jackson, *Discovering the Vernacular Landscape* (New Haven, CT: Yale University Press, 1984), 148.

1. Irish author Malachy Kelly offers an observation about those whose lives are commemorated not only by their gravestone epitaphs but also by the landscapes they shaped. Writing of those resting

in the Moneyneena graveyard in Ballinascreen, Ulster, he remarks: "Not all who are buried here have a monument. Quite a few have made their mark shaping the landscape and left no monument other than that." See Patrick Kelly and Graham Mawhinney, *Ballinascreen Gravestone Inscriptions* (Ballinascreen, Ulster: Ballinascreen Historical Society, 2009).

2. D. W. Meinig, "Reading the Landscape: An Appreciation of W. G. Hoskins and J. B. Jackson," in *The Interpretation of Ordinary Landscapes: Geographical Essays,* ed. D. W. Meinig (New York: Oxford University Press, 1979), 236.

3. Tim Ingold, *Making: Anthropology, Archaeology, Art and Architecture* (London: Routledge, 2013), 6–8.

Bibliography

Books, Monographs, Articles, and Reports

Aaron, Paul, and David Musto. "Temperance and Prohibition in America: A Historical Overview." In *Alcohol and Public Policy: Beyond the Shadow of Prohibition*, edited by Mark H. Moore and Dean R. Gerstein, 127–81. Washington, DC: National Academy Press, 1981.

Acts of the Kentucky General Assembly. Frankfort, KY, 1825–1920.

"Agriculture." *National Magazine* 1, no. 1 (1845): 91–94.

Alderman, Neil. "Industrial Innovation Diffusion: The Extent of Use and Disuse of Process Technologies in Engineering." *Area* 30, no. 2 (1998): 107–16.

Alexander, John W. *Economic Geography.* Englewood Cliffs, NJ: Prentice-Hall, 1963.

Allen, Frederick. "American Spirit." *American Heritage* 49, no. 3 (1998): 82–92.

Allen, L. F. "Berkshire Pigs." *American Agriculturalist* 3 (1844): 175.

Allen, Robert M. *Milk Supply of Kentucky—Louisville.* Bulletin 134, Agricultural Experiment Station of the State College of Kentucky. Lexington, KY, 1908.

Allen, T. W. "The Turnpike System in Kentucky: A Review of State Road Policy in the Nineteenth Century." *Filson Club Historical Quarterly* 28 (1954): 239–59.

Alvey, R. Gerald. *Kentucky Bluegrass Country.* Jackson: University Press of Mississippi, 1992.

Ambrose, William M. *Bottled in Bond under U.S. Government Supervision.* Lexington, KY: William M. Ambrose, 2007.

Ames, Kenneth L. "Ideologies in Stone: Meanings in Victorian Gravestones." *Journal of Popular Culture* 14, no. 4 (1981): 641–56.

"Among the Independent Distilleries." *Wine and Spirit Bulletin* 17, no. 9 (1903): 23.

Anderson, Roger C., James S. Fralish, and Jerry M. Baskin. *Savannas, Barrens, and Rock Outcrop Plant Communities of North America.* New York: Cambridge University Press, 1999.

Andrews, William M., Jr. "Geological Controls on Plio-Pleistocene Drainage Evolution of the Kentucky River in Central Kentucky." PhD diss., Department of Geological Sciences, University of Kentucky, 2004.

"Appalled by Fever." *Washington Historical Quarterly* 2, no. 2 (1908): 163–64.

Appel, Jacob M. "'Physicians Are Not Bootleggers': The Short, Peculiar Life of the Medicinal Alcohol Movement." *Bulletin of the History of Medicine* 82, no. 2 (2008): 356.

Appleton, Thomas H., Jr. "'Like Banquo's Ghost': The Emergence of the Prohibition Issue in Kentucky Politics." PhD diss., Department of History, University of Kentucky, 1981.

———. "'Moral Suasion Has Had Its Day': From Temperance to Prohibition in Antebellum Kentucky." In *A Mythic Land Apart: Reassessing Southerners and Their History*, edited by John D. Smith and Thomas H. Appleton Jr., 19–42. Westport, CT: Greenwood Press, 1997.

———. "Prohibition and Politics in Kentucky: The Gubernatorial Campaign and Election of 1915." *Register of the Kentucky Historical Society* 75, no. 1 (1977): 28–54.

Ardrey, R. L. *American Agricultural Implements.* Chicago: R. L. Ardrey, 1894.

Arms, Fred S., James P. Fehr, David T. Carroll, Byron L. Wilson, Herman P. McDonald, and James C. Ross. *Soil Survey of Nelson County, Kentucky.* USDA Soil Conservation Service and Kentucky Agricultural Experiment Station. Washington, DC: GPO, 1971.

Arnold, B. W., Jr. *History of the Tobacco Industry in Virginia from 1860 to 1894.* Johns Hopkins University Studies in Historical and Political Science, 15th series. Baltimore: Johns Hopkins University Press, 1897.

Auden W. H. *Collected Shorter Poems, 1927–1957.* London: Faber & Faber, 1966.

Austin, Shirley P. "Glass." In US Department of Commerce, *Twelfth Census of the United States, 1900,* pt. 3, vol. 9, *Manufacturers,* 948–1119. Washington, DC: GPO, 1900.

Azaryahu, Maoz, and Kenneth E. Foote. "Historical Space as Narrative Medium: On the Configuration of Spatial Narratives of Time at Historical Sites." *GeoJournal* 73, no. 3 (2008): 179–94.

Bailey, Robert G. *Ecosystem Geography: From Ecoregions to Sites.* 2d ed. New York: Springer Science, 2009.

Bain, Joe S. *Industrial Organization.* New York: John Wiley & Sons, 1959.

Baker, Oliver E. "Agricultural Regions of North America, Part II—The South." *Economic Geography* 3, no. 1 (1927): 50–86.

———. "Agricultural Regions of North America, Part III—The Middle Country Where South and North Meet." *Economic Geography* 3, no. 3 (1927): 309–39.

———. "Agricultural Regions of North America, Part IV—The Corn Belt." *Economic Geography* 3, no. 4 (1927): 447–65.

Balasubramanyam, V. N., and M. A. Salisu. "Brands and the Alcoholic Drinks Industry." In *Adding Value: Brands and Marketing in Food and Drink,* edited by Geoffrey Jones and Nicholas J. Morgan, 59–75. New York: Routledge, 1994.

Baranoff, Dalit. "Shaped by Risk: The American Fire Insurance Industry, 1790–1920." *Enterprise & Society* 6, no. 4 (2005): 561–70.

Barbour, J., and John D. Carroll. *The Kentucky Revised Statutes: Containing All General Laws, Including Those Passed at Session of 1894.* Louisville, KY: Courier-Journal Printing, 1894.

Barnard, Alfred. *The Whiskey Distilleries of the United Kingdom.* 1887. Reprint, Edinburgh: Birlinn Press, 2008.

Barnett, George E. *Chapters on Machinery and Labor.* Cambridge, MA: Harvard University Press, 1926.

Bartholomew, Harland. *Urban Land Uses.* Harvard City Planning Studies, vol. 4. Cambridge, MA: Harvard University Press, 1932.

Barton, Keith C. "'Good Cooks and Washers': Slave Hiring, Domestic Labor, and the Market in Bourbon County, Kentucky." *Journal of American History* 84, no. 2 (1997): 436–60.

Basso, Keith H. *Wisdom Sits in Places: Landscape and Language among the Western Apache.* Albuquerque: University of New Mexico Press, 1996.

Bathe, Grenville, and Dorothy Bathe. *Oliver Evans: A Chronicle of Early American Engineering.* Philadelphia: Historical Society of Pennsylvania, 1935.

Beatty, Adam. *Southern Agriculture.* New York: C. M. Saxon, 1843.

Becher, Matthew E. "The Distillery at Petersburg, Kentucky: Snyder's Old Rye Whiskey." http://bcplfusion.bcpl.org/Repository/Petersburg_Distillery_History_complete.pdf.

Bell, Edwin A., Robert W. Kellogg, and Willis K. Kulp. *Progress Report on the Ground-Water Resources of the Louisville Area Kentucky, 1949–55.* US Geological Survey Water Supply Paper 1579. Washington, DC: GPO, 1963.

Bender, Barbara. "Theorising Landscapes, and the Prehistoric Landscapes of Stonehenge." *Man,* n.s., 27, no. 4 (1992): 735–55.

Bernstein, Peter L. *Against the Gods: The Remarkable Story of Risk.* New York: John Wiley, 1996.

Beverland, Michael B. "Crafting Brand Authenticity: The Case of Luxury Wines." *Journal of Management Studies* 42, no. 5 (2005): 1003–29.

———. "The 'Real Thing': Branding Authenticity in the Luxury Wine Trade." *Journal of Business Research* 59, no. 2 (2006): 251–58.

Biggs, Lindy. *The Rational Factory: Architecture, Technology, and Work in America's Age of Mass Production.* Baltimore: Johns Hopkins University Press, 1996.

Birchfield, James D. "Porter Clay, 'a Very Excellent Cabinetmaker,' Part One: Biographic Account." *Journal of Early Southern Decorative Arts* 36 (2015). http://www.mesdajournal.org.

Bishop, J. Leander. *A History of American Manufactures from 1608 to 1860.* Vol. 1. Philadelphia: Edward Young, 1866.

Bjerkoe, Ethel H. *The Cabinetmakers of America.* Garden City, NY: Doubleday, 1957.

Black, Douglas F. B., Earle R. Cressman, and William C. Macquown Jr. *The Lexington Limestone (Middle Ordovician) of Central Kentucky.* US Geological Survey Bulletin 1224-C. Washington, DC: GPO, 1965.

Blanford, Steven J., Patrick S. Aldridge, and Robert A. Eigel. *Soil Survey of Jefferson County, Kentucky.* USDA Soil Conservation Service and Kentucky Agricultural Experiment Station. Washington, DC: GPO, 1966; reissued 1991.

Booth, John. *A Toast to Ireland: A Celebration of Traditional Irish Drinks.* Boulder, CO: Roberts Rinehart, 1995.

Borchert, John R. "American Metropolitan Evolution." *Geographical Review* 57, no. 3 (1967): 301–32.

Boucherie, Anthony. *The Art of Making Whiskey.* Lexington, KY: Worsley & Smith, 1819.

Bowman, C. E. *Annual Report of the Kentucky Bureau of Agriculture, Horticulture, and Statistics, 1880.* Frankfort, KY: E. H. Porter, 1880.

———. *Annual Report of the Kentucky Bureau of Agriculture, Horticulture, and Statistics, 1881.* Legislative Document 7. Frankfort, KY: S. I. M. Major, 1881.

Bradley, Betsy Hunter. *The Works: The Industrial Architecture of the United States.* New York: Oxford University Press, 1999.

Brand, Stewart. *How Buildings Learn: What Happens after They're Built.* New York: Penguin Books, 1994.

Brecht, Bertolt. *Bertolt Brecht Poems, 1913–1956.* Edited by John Willett and Ralph Manheim. New York: Methuen, 1976.

Brigham, Ann. "Behind-the-Scenes Spaces: Promoting Production in a Landscape of Consumption." In *The Themed Space: Locating Culture, Nation, and Self,* edited by Scott A. Lukas, 207–24. New York: Lexington Books, 2007.

Bromley, Nicholas K. *Law, Space, and the Geographies of Power.* New York: Guilford Press, 1994.

Brooks, Eugene C. *The Story of Corn and the Westward Migration.* New York: Rand McNally, 1916.

Brown, Ralph H. *Historical Geography of the United States.* New York: Harcourt, Brace & World, 1948.

Browne, Irving, ed. *The American Reports.* Vol. 50. San Francisco: Bancroft-Whitney, 1912.

Buchanan, George C. *Fine Whisky Facts.* Louisville, KY: George Buchanan, 1892.

Buckley, Ciaran, and Chris Ward. *Strong Farmer: The Memoirs of Joe Ward.* Dublin, Ireland: Liberties Press, 2007.

Bunkše, Edmunds V. *Geography and the Art of Life.* Baltimore: Johns Hopkins University Press, 2004.

Burchard, Ernest F. "Glass-Sand Industry of Indiana, Kentucky, and Ohio." In *Contributions to Economic Geology 1906,* edited by S. F. Emmons and E. C. Eckel, 361–76. US Geological Survey Bulletin 315, series A, Economic Geology. Washington, DC: GPO, 1907.

Burke, Edmund. *List of Patents for Inventions and Designs Issued by the United States from 1790 to 1847.* Washington, DC: J. & G. S. Gideon, 1847.

Burroughs, Wilbur G. *The Geography of the Western Coal Field.* Kentucky Geological Survey Reports, series 6, vol. 24. Frankfort: Kentucky Geological Survey, 1924.

Burton, Theodore E. *John Sherman.* Boston: Houghton Mifflin, 1908.

Buschena, David, Richard S. Gray, and Ethan Sverson. *Changing Structures in the Barley Production and Malting Industries of the United States and Canada.* Policy Issues Paper 8. Boseman: Trade Research Center, Montana State University, 1998.

Bushman, Richard L. *The Refinement of America.* New York: Vintage Books, 1993.

"Business Notes." *Engineering Record* 43, no. 9 (1901): 208.

"Business Notes." *Engineering Record* 51, no. 11 (1905): 46.

Butler, Frances C. "Eating the Image: The Graphic Designer and the Starving Audience." *Design Issues* 1, no.1 (1984): 27–40.

Buxton, Ian, and Paul S. Hughes. *The Science and Commerce of Whiskey.* Cambridge: Royal Society of Chemistry, 2014.

Campbell, Sally Van Winkle. *But Always Fine Bourbon: Pappy Van Winkle and the Story of Old Fitzgerald.* Frankfort, KY: Old Rip Van Winkle Distillery, 1999.

Cantor, Jay E. "A Monument of Trade: A. T. Stewart and the Rise of the Millionaire's Mansion in New York." *Winterthur Portfolio* 10 (1975): 165–97.

Carlton, Carla Harris. *Barrel Strength Bourbon: The Explosive Growth of America's Whiskey.* Birmingham, AL: Clerisy Press, 2017.

Carpenter, Lawrence E. "Animal Feeds Produced by the Modern Beverage Distilling Industry." *American Biology Teacher* 18, no. 2 (1956): 93–97.

Carr, Aute Lee. "Liquor and the Constitution." *Law & Contemporary Problems* 7, no. 4 (1940): 709–16.

Carr, Chantel, and Chris Gibson. "Geographies of Making: Rethinking Materials and Skills for Volatile Futures." *Progress in Human Geography* 40, no. 3 (2016): 297–315. https://doi.org/10.1177/0309132515578775.

Carson, Gerald. *The Social History of Bourbon.* 1963. Reprint, Lexington: University Press of Kentucky, 2010.

Cawker, E. Harrison. "Kentucky Flour Mills Listed in the 1880 Edition of Cawker's Flour Mill Directory." *Millstone: Journal of the Kentucky Old Mill Association* 12, no. 2 (2013): 7–12.

Cecil, Sam K. *Bourbon: The Evolution of Kentucky Whiskey.* New York: Turner Publishing, 2010.

Centennial Temperance Volume: A Memorial of the International Temperance Conference, Philadelphia, June, 1876. New York: National Temperance Society, 1877.

Certic Inc. *Economic Impact of Kentucky's Travel and Tourism Industry—2016 and 2017.* Frankfort: Kentucky Tourism, Arts & Heritage Cabinet, 2018.

The Chronicle Fire Tables for 1885. New York: Chronicle Company, 1885.

The Chronicle Fire Tables for 1897. New York: Chronicle Company, 1897.

Clark, Thomas D. *A History of Kentucky.* Ashland, KY: Jesse Stuart Foundation, 1992.

———. "Kentucky Logmen." *Journal of Forest History* 25, no. 3 (1981): 144–57.

Clark, Victor S. *History of Manufactures in the United States 1607–1860.* Washington, DC: Carnegie Institution, 1916.

Clarkson, James D. "Ecology and Spatial Analysis." *Annals of the Association of American Geographers* 60, no. 4 (1970): 700–716.

Clay, Karen, and Werner Troesken. "Strategic Behavior in Whiskey Distilling, 1887–1895." *Journal of Economic History* 62, no. 4 (2002): 999–1023.

Clotfelter, Elizabeth R. "The Agricultural History of Bourbon County, Kentucky Prior to 1900." Master's thesis, Department of History, University of Kentucky, 1953.

Cole, G. D. "Woodworking Plants." *Industrial Development and Manufacturers' Record* 53, no. 4 (1908): 67.

"Commonwealth v. Kentucky Distilleries & Warehouse Co., et al." *Southwestern Reporter* 159 (September 3–November 19, 1913): 570–73.

Congressional Globe. 46 vols. Washington, DC, 1834–1873.

Connerton, Paul. *How Societies Remember.* Cambridge: Cambridge University Press, 1989.

Conzen, Michael P. "The County Landownership Map in America: Its Commercial Development and Social Transformation, 1814–1939." *Imago Mundi* 36 (1984): 9–31.

———. "Landownership Maps and County Atlases." *Agricultural History* 58, no. 2 (1984): 118–22.

Cook, Clarence. "New York Daguerreotyped." *Putnam's Monthly* 3, no. 15 (1854): 233–49.

Cook, George H., and John C. Smock. *Report on the Clay Deposits of Woodbridge, South Amboy, and Other Places in New Jersey, Together with Their Uses for Fire Brick, Pottery, etc.* Trenton, NJ: Geological Survey of New Jersey, 1878.

Coomes, Paul, and Barry Kornstein. "The Economic and Fiscal Impacts of the Distilling Industry in Kentucky." Paper prepared for the Kentucky Distillers' Association. Louisville, KY: Urban Studies Institute, University of Louisville, 2019.

Cope, Kenneth L. *American Cooperage Machinery and Tools.* Mendham, NJ: Astragal Press, 2003.

"Copper Mines and the Copper Trade." *National Magazine* 1, no. 2 (1845): 116–28.

Cousens, S. H. "The Regional Pattern of Emigration during the Great Irish Famine, 1846–51." *Transactions and Papers of the Institute of British Geographers* 28 (1960): 119–34.

Cowan, Ruth Schwartz. *A Social History of American Technology.* New York: Oxford University Press, 1997.

Cowdery, Charles K. *Bourbon Straight: The Uncut and Unfiltered Story of American Whiskey.* Chicago: Charles K. Cowdery, 2004.

Cowdrey, Albert E. *This Land, This South: An Environmental History.* Rev. ed. Lexington: University Press of Kentucky, 1996.

Cox, Frank R. *Soil Survey of Daviess and Hancock Counties, Kentucky.* USDA Soil Conservation Service and Kentucky Agricultural Experiment Station. Washington, DC: GPO, 1974.

Coxe, Tench. *A Statement of the Arts and Manufactures of the United States of America, 1810.* Philadelphia: A. Cornman, 1814.

Craddock, William H. *Soil Survey of Washington County, Kentucky.* USDA Soil Conservation Service and Kentucky Agricultural Experiment Station. Washington, DC: GPO, 1986.

Cramer, Zadok. *The Navigator.* 7th and 8th eds. Pittsburgh: Cramer, Spear & Eichbaum, 1811, 1814.

Cramp, Arthur J. *Nostrums and Quackery.* Vol. 2. Chicago: American Medical Association, 1921.

Cresswell, Nicholas. *The Journal of Nicholas Cresswell, 1774–1777.* New York: Dial Press, 1924.

Cronon, William. *Nature's Metropolis: Chicago and the Great West.* New York: W. W. Norton, 1991.

———. "A Place for Stories: Nature, History, and Narrative." *Journal of American History* 78, no. 4 (1992): 1347–76.

Cross, Charles. *The American Compounder or Cross' Guide for Retail Liquor Dealers.* St. Louis: Charles Cross, 1899.

Crowell, Sheryl. *New Richmond.* Charleston, SC: Arcadia Publishing, 2012.

Crowgey, Henry G. *Kentucky Bourbon: The Early Years of Whiskeymaking.* 1971. Reprint, Lexington: University Press of Kentucky, 2008.

Cuming, Fortesque. *Sketches of a Tour to the Western Country, through the States of Ohio and Kentucky.* Pittsburgh: Cramer, Spear & Eichbaum, 1810.

Cutshall, Alden D. "Industrial Geography of the Lower Wabash Valley." *Economic Geography* 17, no. 3 (1941): 297–307.

Danhof, Clarence H. "Agriculture." In *The Growth of the American Economy*, edited by Harold F. Williamson, 133–53. New York: Prentice-Hall, 1951.

———. *Change in Agriculture: The Northern United States, 1820–1870*. Cambridge, MA: Harvard University Press, 1969.

Davenport, H. J. "The Extent and the Significance of the Unearned Increment." *American Economic Review* 1, no. 2 (1911): 322–32.

Davidson, Alex B., and M. W. Tatlock. "Treatment of Distillery Wastes." *Sewage and Industrial Wastes* 22, no. 5 (1950): 654–65.

Davidson, Bart. *Water Well and Spring Map of the Louisville 30 x 60 Minute Quadrangle, Kentucky*. Lexington: Kentucky Geological Survey, 2002.

Davis, Darrell Haug. *Geography of the Blue Grass Region of Kentucky*. Kentucky Geological Survey Reports, series 6, vol. 23. Frankfort: Kentucky Geological Survey, 1927.

Dawson, A. H. "The Great Increase in Barley Growing in Scotland." *Geography* 65, no. 3 (1980): 213–17.

Dawson, William H. *The Unearned Increment: Reaping without Sowing*. London: Swan Sonnenschein, 1890.

Day, Dan. "Corn Cobs—Burning Them with Coal." *Ag Connection* (University of Missouri Extension) 12, no. 12 (2006): 1–2.

"Debt Statement for January, 1871." *Commercial and Financial Chronicle* 12, no. 289 (1871): 8.

DeLyser, Dydia, and Paul Greenstein. "The Devotions of Restoration: Materiality, Enthusiasm, and Making Three 'Indian Motocycles' Like New." *Annals of the Association of American Geographers* 107, no. 6 (2017): 1461–78.

Denzin, Norman K. "On Semiotics and Symbolic Interactionism." *Symbolic Interaction* 10, no. 1 (1987): 1–19.

De Voto, Bernard. *The Hour: A Cocktail Manifesto*. Boston: Houghton Mifflin, 1948. Reprint, Portland, OR: Tin House Books, 2010.

Dew, Lee A., and Aloma W. Dew. *Owensboro: The City on the Yellow Banks*. Owensboro, KY: Rivendell Publications, 1988.

Dicken, Peter. *Global Shift: Mapping the Changing Contours of the World Economy*. 6th ed. New York: Guilford Press, 2011.

Difford, Simon. "Casks." In *Difford's Guide*. https://www.diffordsguide.com/encyclopedia/481.

"Distillery-Milk Report, II." *Science* 9, no. 228 (1887): 579–81.

"Distillery-Milk Report, V." *Science* 10, no. 243 (1887): 157–58.

Dorn, Michael L. "(In)temperate Zones: Daniel Drake's Medico-Moral Geographies of Urban Life in the Trans-Appalachian West." *Journal of the History of Medicine and Allied Sciences* 55, no. 3 (2000): 256–91.

Downard, William L. *Dictionary of the History of American Brewing and Distilling Industries*. Westport, CT: Greenwood Press, 1980.

Downing, A. J. *The Architecture of the Country House*. New York: D. Appleton, 1850.

Drake, B., and E. D. Mansfield. *Cincinnati in 1826*. Cincinnati, OH: Morgan, Lodge & Fisher, 1827.

Drake, Daniel. "Geological Account of the Valley of the Ohio: In a Letter from Daniel Drake, M.D. to Joseph Correa de Serra." *Transactions of the American Philosophical Society* 2, no. 4 (1825): 124–39.

"Drawing Corn in the Shock." *American Agriculturist* 39 (October 1880): 430.

"Drying Plant." *Wine and Spirit Bulletin* 19, no. 1 (1905): 41.

Duguid, Paul. "Developing the Brand: The Case of Alcohol, 1800–1880." *Enterprise & Society* 4, no. 3 (2003): 405–41.

Dunn, Cecil G. "The Industrial Utilization of Yeasts." *Economic Botany* 12, no. 2 (1958): 145–63.

Earle, Carville. *Geographical Inquiry and American Historical Problems.* Stanford, CA: Stanford University Press, 1992.

——. "A Staple Interpretation of Slavery and Free Labor." *Geographical Review* 68, no. 1 (1978): 51–65.

Earle, Carville, and Ronald Hoffman. "The Foundation of the Modern Economy: Agriculture and the Costs of Labor in the United States and England, 1800–1860." *American Historical Review* 85, no. 5 (1980): 1055–94.

East, Ernest E. "The Distillers' and Cattle Feeders' Trust, 1887–1895." *Journal of the Illinois State Historical Society* 45, no. 2 (1952): 101–23.

Eaton, Clement. "Slave-Hiring in the Upper South: A Step toward Freedom." *Mississippi Valley Historical Review* 46, no. 4 (1960): 663–78.

Ellis, John. "The Holy 'Knock-'em-Down': Methodism Remodels for the Ohio Valley, 1790s–1820s." *Ohio Valley History* 18, no. 2 (2018): 3–24.

Elstner, J. M. *The Industries of Louisville, Kentucky, and New Albany, Indiana.* Louisville, KY: J. M. Elstner, 1886.

English, William H. *Conquest of the Country Northwest of the River Ohio, 1778–1783.* Vol. 1. Indianapolis: Bowen-Merrill, 1896.

Enoch, Harry G. *Colonel John Holder, Boonesborough Defender & Kentucky Entrepreneur.* Morley, MO: Acclaim Press, 2009.

——. *Grimes Mill: Kentucky Landmark on Boone Creek, Fayette County.* Bowie, MD: Heritage Books, 2002.

Enoch, Harry G., and Nancy O'Malley. "Early Mills of Bourbon County." *Mill Stone* 4, no. 1 (2005): 25–38.

E. S. C. "Constitutional Law. Interstate Commerce. Original Package Doctrine." *Virginia Law Review* 21, no. 4 (1935): 433–42.

Eslinger, Ellen. "Farming on the Kentucky Frontier." *Register of the Kentucky Historical Society* 107, no. 1 (2009): 3–32.

Evans, E. Estyn. "Some Survivals of the Irish Openfield System." *Geography* 24, no. 1 (1939): 24–36.

Evans, Priscilla Ann. "Merchant Gristmills and Communities, 1820–1880: An Economic Relationship." *Missouri Historical Review* 68, no. 3 (1974): 317–26.

Fehlandt, August F. *A Century of Drink Reform in the United States.* Cincinnati, OH: Jennings & Graham, 1904.

Ferguson, James M. "Advertising and Liquor." *Journal of Business* 40, no. 4 (1967): 414–34.

Fifteenth Annual Report of the Kentucky Agricultural Experiment Station. Lexington: State College of Kentucky, 1902.

Fisher, Linda A., ed. *Joseph J. Mersman: The Whiskey Merchant's Diary; an Urban Life in the Emerging Midwest.* Athens: Ohio University Press, 2007.

Fisher's National Magazine and Industrial Record 3, no. 2 (1846): 154–57.

Fishlow, Albert. "Antebellum Interregional Trade Reconsidered." *American Economic Review* 54, no. 3 (1964): 352–64.

Fitch, Charles H. *Report on the Manufacture of Engines and Boilers.* US Census, 1880. Washington, DC: GPO, 1881.

Fleischman, Joseph. *The Art of Blending and Compounding Liquors and Wines.* New York: Dick & Fitzgerald, 1885.

Fosdick, Raymond B., and Albert L. Scott. *Toward Liquor Control.* New York: Harper & Brothers, 1933.

Foster, Frank H., and Robert L. Shook. *Patents, Copyrights, & Trademarks.* 2d ed. New York: John Wiley & Sons, 1993.

Fox, Stephen. *The Mirror Makers: A History of American Advertising and Its Creators.* New York: William Morrow, 1984.

Frazer, David. "'Fields of Radiance': The Scientific and Industrial Scenes of Joseph Wright." In *The Iconography of Landscape,* edited by Denis Cosgrove and Stephen Daniels, 119–41. Cambridge: Cambridge University Press, 1988.

French, Edward V. *The Development of the Prevention Idea.* Boston: privately printed for Arkwright Mutual Fire Insurance Company, 1912.

Fryer, Alan E. "Springs and the Origin of Bourbon." *Ground Water* 47, no. 4 (2009): 605–10.

Fuller, John. *Art of Coppersmithing: A Practical Treatise on Working Sheet Copper into All Forms.* 1894. Reprint, Mendham, NJ: Astragal Press, 1993.

Fuller, M. L., and F. G. Clapp. *The Underground Waters of Southwestern Ohio.* US Geological Survey Water Supply Paper 259. Washington, DC: GPO, 1912.

Gannett, Henry, and Jasper E. Whelchel. "Lumber." In US Department of Commerce, *Twelfth Census of the United States, 1900,* pt. 3, vol. 9, *Manufacturers,* 805–97. Washington, DC: GPO, 1900.

Garber, John H. "Alcoholic Liquors." In US Department of Commerce, *Twelfth Census of the United States, 1900,* pt. 3, vol. 9, *Manufacturers,* 595–635. Washington, DC: GPO, 1900.

Gaston, Kay Baker. "Robertson County Distilleries, 1796–1909." *Tennessee Historical Quarterly* 43, no. 1 (1984): 49–67.

Gates, Paul W. *The Farmer's Age: Agriculture, 1815–1860.* Armonk, NY: M. E. Sharpe, 1960.

George, Peter. *The Emergence of Industrial America: Strategic Factors in American Economic Growth since 1870.* Albany: State University of New York Press, 1882.

Getz, Oscar. *Whiskey: An American Pictorial History.* New York: David McKay, 1978.

Gibson, Chris. "Material Inheritances: How Place, Materiality, and Labor Process Underpin the Path-Dependent Evolution of Contemporary Craft Production." *Economic Geography* 92, no. 1 (2016): 61–86. https://doi.org/10.1080/00130095.2015.1092211.

Glassie, Henry. *Pattern in the Material Folk Culture of the Eastern United States.* Philadelphia: University of Pennsylvania Press, 1968.

——. "Tradition." *Journal of American Folklore* 108, no. 430 (1995): 395–412.

Goodall, Brian. "Industrial Heritage and Tourism." *Built Environment* 19, no. 2 (1993): 92–104.

Goodrich, J. Z. *Exposition of the J. D. & M. Williams Fraud.* Boston: Rockwell & Rollins, 1866.

Goss, Jon. "The Built Environment and Social Theory: Towards an Architectural Geography." *Professional Geographer* 40, no. 4 (1988): 392–403.

——. "The 'Magic of the Mall': An Analysis of Form, Function, and Meaning in the Contemporary Retail Built Environment." *Annals of the Association of American Geographers* 83, no. 1 (1993): 18–47.

Gottdiener, Mark. *The Theming of America: Dreams, Visions, and Commercial Spaces.* Boulder, CO: Westview Press, 1997.

Gould, Peter. *Spatial Diffusion.* Resource Paper 4. Washington, DC: Association of American Geographers, 1969.

Gray, Henry H. "Archaeological Sedimentology of Overbank Silt Deposits on the Floodplain of the Ohio River near Louisville, Kentucky." *Journal of Archaeological Science* 11 (1984): 421–32.

Gray, Lewis C. *History of Agriculture in the Southern United States to 1860.* 2 vols. Carnegie Institution of Washington Publication 430. New York: Peter Smith, 1941.

Green, Ann Norton. *Horses at Work: Harnessing Power in Industrial America.* Cambridge, MA: Harvard University Press, 2008.

Gronert, Theodore G. "Trade in the Blue-Grass Region, 1810–1820." *Mississippi Valley Historical Review* 5, no. 3 (1918): 313–23.

Groth, Paul, and Todd W. Bressi, eds. *Understanding Ordinary Landscapes.* New Haven, CT: Yale University Press, 1997.

"H. McKenna Distiller." *Wine and Spirit Bulletin* 18, no. 6 (1904): 41.

Haggard, D. R. *Report of the Board of Internal Improvement.* Kentucky General Assembly, November 1851. Frankfort, KY: A. G. Hodges, 1851.

Hale, H. M. *Wood Used for Tight Cooperage Stock in 1905.* USDA Forest Service Circular 53. Washington, DC: GPO, 1907.

Hall, Colin M. "The Definition and Analysis of Hallmark Tourist Events." *GeoJournal* 19, no. 3 (1989): 263–68.

Hall, Harrison. *The Distiller.* 1818. Reprint, 2d rev. ed., Hayward, CA: White Mule Press, 2015.

Hall, Henry. *Report on the Ship-Building Industry of the United States.* Washington, DC: Department of the Interior, Census Office, 1884.

Hall, James. *The West: Its Commerce and Navigation.* Cincinnati, OH: H. W. Derby, 1848.

Halsey, Harlan I. "The Choice between High-Pressure and Low-Pressure Steam Power in America in the Early Nineteenth Century." *Journal of Economic History* 41, no. 4 (1981): 723–44.

Hamann, Fred. "Stiller's Argot." *American Speech* 21, no. 3 (1946): 193–95.

Hamilton County [Ohio] Agricultural Society. *Western Agriculturalist and Practical Farmer's Guide.* Cincinnati, OH: Robinson & Fairbank, 1830.

Hammon, Neal O. "Land Acquisition on the Kentucky Frontier." *Register of the Kentucky Historical Society* 78, no. 4 (1980): 297–321.

———. "Settlers, Land Jobbers, and Outliers: A Quantitative Analysis of Land Acquisition on the Kentucky Frontier." *Register of the Kentucky Historical Society* 84, no. 3 (1986): 241–62.

Harlan, H. V., and M. L. Martini. "Problems and Results in Barley Breeding." In *U.S.D.A. Yearbook of Agriculture, 1936,* 303–46. Washington, DC: GPO, 1936.

Harlan, H. V., and G. A. Wiebe. *Growing Barley for Malt and Feed.* US Department of Agriculture Farmers' Bulletin 1732. Washington, DC: GPO, 1934.

Harrison, Barry, Olivier Fagnen, Frances Jack, and James Brosnan. "The Impact of Copper in Different Parts of Malt Whisky Pot Stills on New Make Spirit Composition and Aroma." *Journal of the Institute of Brewing* 117, no. 1 (2011): 106–12.

Harrison, Lowell H., and James C. Klotter. *A New History of Kentucky.* Lexington: University Press of Kentucky, 1997.

Hart, John Fraser. "Change in the Corn Belt." *Geographical Review* 76, no. 1 (1986): 51–72.

———. *The Rural Landscape.* Baltimore: Johns Hopkins University Press, 1998.

Harvey, T. Edgar. *Commercial History of the State of Kentucky.* Louisville, KY: Courier-Journal Printing, 1899.

Hazen, Margaret Hindle, and Robert M. Hazen. *Keepers of the Flame: The Role of Fire in American Culture, 1775–1925.* Princeton, NJ: Princeton University Press, 1992.

Head, Lesley. "Cultural Landscapes." In *The Oxford Handbook of Material Culture Studies,* edited by Dan Hicks and Mary C. Beaudry, 427–39. Oxford: Oxford University Press, 2010.

Heath, Ralph C. *Basic Ground-Water Hydrology.* US Geological Survey Water Supply Paper 2220, rev. Washington, DC: GPO, 2004.

Heikens, Alice L. "Savanna, Barrens, and Glade Communities of the Ozark Plateaus Province." In *Sa-*

vannas, Barrens, and Rock Outcrop Plant Communities of North America, edited by Roger C. Anderson, James S. Fralish, and Jerry M. Baskin, 220–30. New York: Cambridge University Press, 1999.

Hendrix, Laura H. *Citizen's Guide to the Kentucky Constitution*. Research Report 137. Frankfort, KY: Legislative Research Commission, 2013.

Henlein, Paul C. "Cattle Driving in the Ohio Country, 1800–1850." *Agricultural History* 28, no. 2 (1954): 83–95.

———. *Cattle Kingdom in the Ohio Valley, 1783–1860*. Lexington: University of Kentucky Press, 1959.

———. "Early Cattle Ranges of the Ohio Valley." *Agricultural History* 35, no. 3 (1961): 150–54.

———. "Shifting Range-Feeder Patterns in the Ohio Valley before 1860." *Agricultural History* 31, no. 1 (1957): 1–12.

Henry, W. A., and F. B. Morrison. *Feeds and Feeding: A Handbook for the Student and Stockman*. 16th ed. Madison, WI: Henry-Morrison, 1916.

Hepburn, Andrew. *Great Resorts of North America*. New York: Doubleday, 1965.

Herbert, Luke. *The Engineer's and Mechanic's Encyclopedia*. Vol. 1. London: Thomas Kenny, 1836.

Herfindahl, Orris C. *Copper Costs and Prices: 1870–1957*. Baltimore: Johns Hopkins University Press, 1959.

Hibbs, Dixie. *Before Prohibition: Distilleries in Nelson County, Kentucky, 1880–1920*. New Hope, KY: St. Martin de Porres Print Shop, 2012. Includes reprinted sections of *The Nelson County Record—Illustrated Historical and Industrial Supplement*. Bardstown, KY, 1896.

Hilliard, Sam Bowers. *Hog Meat and Hoecake: Food Supply in the Old South, 1840–1860*. Carbondale: Southern Illinois University Press, 1972.

Hindle, Brooke. "The Exhilaration of Early American Technology." In *Early American Technology: Making and Doing Things from the Colonial Era to 1850*, edited by Judith A. McGraw, 40–67. Chapel Hill: University of North Carolina Press, 1994.

Hindle, Brooke, and Steven Lubar. *Engines of Change: The American Industrial Revolution, 1790–1860*. Washington, DC: Smithsonian Institution Press, 1986.

Historical Cerrillos. Cerrillos, NM: Cerrillos Historical Society, n.d.

History of Daviess County, Kentucky. Chicago: Inter-State Publishing, 1883.

Hofstra, Warren R. *The Planting of New Virginia: Settlement and Landscape in the Shenandoah Valley*. Baltimore: Johns Hopkins University Press, 2004.

———. "Private Dwellings, Public Ways, and the Landscape of Early Rural Capitalism in Virginia's Shenandoah Valley." In *Gender, Class, and Shelter: Perspectives in Vernacular Architecture V*, edited by Elizabeth C. Cromley and Carter L. Hudgins, 211–24. Knoxville: University of Tennessee Press, 1995.

Hofstra, Warren R., and Karl Raitz, eds. *The Great Valley Road of Virginia: Shenandoah Landscapes from Prehistory to the Present*. Charlottesville: University of Virginia Press, 2010.

Hogan, Roseann R. "Buffaloes in the Corn: James Wade's Account of Pioneer Kentucky." *Register of the Kentucky Historical Society* 89, no. 1 (1991): 1–31.

Hogeland, William. *The Whiskey Rebellion: George Washington, Alexander Hamilton, and the Frontier Rebels Who Challenged America's Newfound Sovereignty*. New York: Scribner, 2006.

Holt, Douglas B. *How Brands Become Icons: The Principles of Cultural Branding*. Boston: Harvard Business School Press, 2004.

Hoover, Edgar M. *The Location of Economic Activity*. New York: McGraw Hill, 1948.

Hopkins, James F. *A History of the Hemp Industry in Kentucky*. Lexington: University Press of Kentucky, 1951.

Horning, Audrey J. "Myth, Migration, and Material Culture: Archaeology and the Ulster Influence on Appalachia." *Historical Archaeology* 36, no. 4 (2002): 129–49.

Hoskins, W. G. *The Making of the English Landscape.* London: Hodder & Stoughton, 1955.

Howlett, Leon. *The Kentucky Bourbon Experience: A Visual Tour of Kentucky's Bourbon Distilleries.* Morley, MO: Acclaim Press, 2012.

Hubka, Thomas. "Just Folks Designing: Vernacular Designers and the Generation of Form." *Journal of Architectural Education* 32, no. 3 (1979): 27–29.

Huckelbridge, Dane. *A History of the American Spirit.* New York: William Morrow, 2014.

Hudson, John C. *Making the Corn Belt: A Geographical History of Middle-Western Agriculture.* Bloomington: Indiana University Press, 1994.

Hughes, R. E., and C. C. Owsley. *Kentucky the Beautiful.* Louisville & Nashville Railroad Passenger Department. Chicago: Corbitt Railway Printing, 1901.

Hughes, Thomas P. *Networks of Power: Electrification in Western Society, 1880–1930.* Baltimore: Johns Hopkins University Press, 1983.

Hunter, Louis C. *A History of Industrial Power in the United States, 1780–1930.* Vol. 1, *Waterpower in the Century of the Steam Engine.* Charlottesville: University Press of Virginia, 1979.

———. *A History of Industrial Power in the United States, 1780–1930.* Vol. 2, *Steam Power.* Charlottesville: University Press of Virginia, 1985.

———. "The Invention of the Western Steamboat." *Journal of Economic History* 3, no. 2 (1943): 201–20.

———. *Steamboats on the Western Rivers: An Economic and Technological History.* New York: Dover Publications, 1949.

———. *Studies in the Economic History of the Ohio Valley.* New York: DaCapo Press, 1973.

Hutchinson, William T. *Cyrus Hall McCormick: Harvest, 1856–1884.* New York: D. Appleton–Century, 1935.

———. *Cyrus Hall McCormick: Seed-Time, 1809–1856.* New York: Century, 1930.

Imlay, George. *A Topographical Description of the Western Territory of North America.* 2d ed. London: J. Debrett, 1793.

Ingold, Tim. *Making: Anthropology, Archaeology, Art and Architecture.* London: Routledge, 2013.

———. "Making Culture and Weaving the World." In *Matter, Materiality and Modern Culture,* edited by P. M. Graves-Brown, 50–71. London: Routledge, 2000.

———. *The Perception of the Environment: Essays on Livelihood, Dwelling, and Skill.* London: Routledge, 2000.

———. "The Temporality of the Landscape." *World Archaeology* 25, no. 2 (1993): 152–74.

———. "Toward an Ecology of Materials." *Annual Review of Anthropology* 41 (2012): 433–35.

Innis, Harold A. "On the Economic Significance of Culture." *Journal of Economic History* 4 (1944): 80–97.

"Insurance Company of North America." *Fortune* 15, no. 2 (1937), 94–100, 164, 166, 168, 170, 173, 174, 176, 178, 184, 186.

Internal Revenue Record and Customs Journal 11, no. 13 (March 26, 1994).

Jackson, John Brinckerhoff. *Discovering the Vernacular Landscape.* New Haven, CT: Yale University Press, 1984.

———. *Landscape in Sight: Looking for America.* Edited by Helen Lefkowitz Horowitz. New Haven, CT: Yale University Press, 1997.

———. *A Sense of Place, a Sense of Time.* New Haven, CT: Yale University Press, 1994.

Jenks, Jeremiah W. "The Development of the Whiskey Trust." *Political Science Quarterly* 4, no. 2 (1889): 296–319.

Jewett, Charles. "The Medical Uses of Alcohol." In *Centennial Temperance Volume: A Memorial of the International Temperance Conference, Philadelphia, June, 1876,* 258–79. New York: National Temperance Society, 1877.

Johnson, Leland R., and Charles E. Parrish. *Triumph at the Falls: The Louisville and Portland Canal.* Louisville, KY: US Corps of Engineers, 2007.

Johnson, Nuala C. "Where Geography and History Meet: Heritage Tourism and the Big House in Ireland." *Annals of the Association of American Geographers* 86, no. 3 (1996): 551–66.

Johnson, Paul E. *A Shopkeeper's Millennium: Society and Revivals in Rochester, New York, 1815–1837.* Rev. 1st ed. New York: Hill & Wang, 2004.

Johnson, R. W. "Land Use in the Bluegrass Basins." *Economic Geography* 16, no. 3 (1940): 315–35.

Journal of the Executive Proceeding of the Senate of the United States of America from December 4, 1865 to March 3, 1867, Inclusive. 39th Cong., 1st sess. (1866), vol. 14. Washington, DC: GPO, 1867.

Journal of the Executive Proceedings of the Senate of the United States of America from April 12, 1869 to April 22, 1869, Inclusive. 41st Cong., special sess. (1869), vol. 17. Washington, DC: GPO, 1869.

Kane, Frank. *Anatomy of the Whiskey Business.* Manhasset, NY: Lake House Press, 1965.

Kauffman, Henry J. *American Copper & Brass.* Camden, NJ: Thomas Nelson & Sons, 1968.

Kaufman, Edward N. "Architectural Representation in Victorian England." *Journal of the Society of Architectural Historians* 46, no. 1 (1987): 30–38.

Kavaratzis, Mihalis, and G. J. Ashworth. "City Branding: An Effective Assertion of Identity or a Transitory Marketing Trick." *Tijdschrift voor Economische en Sociale Geografie* 96, no. 5 (2005): 506–14.

Kelley, John A., and William H. Craddock. *Soil Survey of Marion County, Kentucky.* USDA Soil Conservation Service and Kentucky Agricultural Experiment Station. Washington, DC: GPO, 1991.

Kellner, Esther. *Moonshine: Its History and Folklore.* New York: Weathervane Books, 1971.

Kelly, Janet, Barry Kornstein, and Ryan Marshall. "The Economic and Fiscal Impacts of the Distilling Industry in Kentucky." Paper prepared for the Kentucky Distillers' Association. Louisville, KY: Urban Studies Institute, University of Louisville, 2017.

Kelly, Patrick, and Graham Mawhinney. *Ballinascreen Gravestone Inscriptions.* Ballinascreen, Ulster: Ballinascreen Historical Society, 2009.

Kemmerer, Donald L. "The Pre–Civil War South's Leading Crop, Corn." *Agricultural History* 23, no. 4 (1949): 236–39.

Kendall, Edward A. *Travels through the Northern Parts of the United States in the Years 1807–1808.* Vol. 3. New York: I. Riley, 1809.

Kennedy, William H. "An Introduction to the Problem: A Resume of the Federal Liquor Laws and Their Application to the Industry." *Proceedings of the Annual Conference on Taxation under the Auspices of the National Tax Association* 46 (1953): 8–13.

Kenny, Kevin. *The American Irish: A History.* Harlow, UK: Longman, 2000.

Kentucky Agricultural Experiment Station. *Twentieth Annual Report of the Kentucky Agricultural Experiment Station.* Lexington: State College of Kentucky, 1907.

Kentucky General Assembly. *Journal of the Senate of the Commonwealth of Kentucky.* Frankfort, KY: A. G. Hodges, 1837.

Kentucky Revised Statutes. http://abc.ky.gov/.

Kentucky Statutes. 2d ed. Louisville, KY: Courier-Journal Printing, 1899.

Kentucky Transportation Cabinet. *Limited Supplemental Guide Signs.* http://transportation.ky.gov/Sign-Programs-and-Standards/Pages/Limited_Supplemental_Guide_Signs.aspx.

Kerr, K. Austin. *Organized for Prohibition: A New History of the Anti-Saloon League.* New Haven, CT: Yale University Press, 1985.

Kidder, Tristram R., and William Balee. "Epilogue." In *Advances in Historical Ecology,* edited by William Balee, 405–10. New York: Columbia University Press, 1998.

Kilby, Kenneth. *The Cooper and His Trade.* Fresno, CA: Linden Publishing, 1971.

———. *Coopers and Coopering.* Princes Risborough, UK: Shire Publications, 2004.

Kimmerer, Tom. *Venerable Trees: History, Biology, and Conservation in the Bluegrass.* Lexington: University Press of Kentucky, 2015.

Kirby, Jack Temple. *Mockingbird Song: Ecological Landscapes of the South.* Chapel Hill: University of North Carolina Press, 2006.

Kleber, John E., ed. *The Encyclopedia of Louisville.* Lexington: University Press of Kentucky, 2001.

———. *The Kentucky Encyclopedia.* Lexington: University Press of Kentucky, 1992.

Klein, Maury. *History of the Louisville & Nashville Railroad.* New York: Macmillan, 1972. Reprint, Lexington: University Press of Kentucky, 2003.

Kniffen, Fred, and Henry Glassie. "Building in Wood in the Eastern United States: A Time-Place Perspective." *Geographical Review* 56, no. 1 (1966): 40–66.

Knight, Frank H. *Risk, Uncertainty and Profit.* Boston: Houghton Mifflin, 1921. Reprint, Mansfield Centre, CT: Martino Publishing, 2014.

Kobler, John. *Ardent Spirits: The Rise and Fall of Prohibition.* New York: Putnam, 1973.

Kornstein, Barry, and Jay Luckett. "The Economic and Fiscal Impacts of the Distilling Industry in Kentucky." Paper, Urban Studies Institute, University of Louisville, Louisville, KY, 2014.

Krafft, Michael. *The American Distiller or, the Theory and Practice of Distilling.* Philadelphia: Archibald Bariram, 1804. Reprint, Gale, Sabin Americana Print Editions, n.d.

Kragh, Helge. *An Introduction to the Historiography of Science.* Cambridge: Cambridge University Press, 1987.

Kramer, Carl E. *Capital on the Kentucky.* Frankfort, KY: Historic Frankfort, 1986.

———. "The Evolution of the Residential Land Subdivision Process in Louisville, 1772–2008." *Register of the Kentucky Historical Society* 107, no. 1 (2009): 33–81.

Kroll, Harry H. *Bluegrass, Belles, and Bourbon: A Pictorial History of Whiskey in Kentucky.* New York: A. S. Barnes, 1967.

Krout, John A. *The Origins of Prohibition.* New York: Russell & Russell, 1925.

Kuhlmann, Charles B. "Processing Agricultural Products in the Pre-Railroad Age." In *The Growth of the American Economy,* edited by Harold F. Williamson, 154–71. New York: Prentice-Hall, 1951.

Laird, Pamela W. *Advertising Progress: American Business and the Rise of Consumer Marketing.* Baltimore: Johns Hopkins University Press, 1998.

Lamborn, R. E., C. R. Austin, and Downs Schaaf. *Shales and Surface Clays of Ohio.* Bulletin 39. Columbus: Ohio Geological Survey, 1938.

Laracuente, Nicolas R. "Archaeological Investigations of the O.F.C. Building, Buffalo Trace Distillery." 2016. On file at Buffalo Trace Distillery, Frankfort, KY.

Laracuente, Nicolas R., and V. Camille Westmont. "Results of Volunteer Archaeological Excavations at Riverside, Buffalo Trace Distillery, Frankfort, Kentucky." 2012. On file at Buffalo Trace Distillery, Frankfort, KY.

Laurier, Eric. "Replication and Restoration: Ways of Making Maritime Heritage." *Journal of Material Culture* 3, no. 1 (1988): 21–50.

Lazzari, Marisa. "The Texture of Things: Objects, People and Landscape in Northwest Argentina." In *Archaeologies of Materiality,* edited by Lynn Meskell, 126–61. Oxford: Blackwell, 2005.

Lears, T. J. Jackson. "The Rise of American Advertising." *Wilson Quarterly* 7, no. 5 (1983): 156–67.

Lecain, Timothy J. *The Matter of History: How Things Create the Past.* Cambridge: Cambridge University Press, 2017.

Lee, Guy A. "The Historical Significance of the Chicago Grain Elevator System." *Agricultural History* 11, no. 1 (1937): 16–32.

Leggett, M. D. *Subject-Matter Index of Patents for Inventors Issued by the United States Patent Office, 1790–1873*. Vol. 1. Washington, DC: GPO, 1874.

Le Goff, Jacques. *History and Memory*. New York: Columbia University Press, 1992.

Levy, Sidney J. *The Theory of the Brand*. Wilmette, IL: DecaBooks, 2016.

Lewis, Peirce. "The Future of the Past: Our Clouded Vision of Historic Preservation." *Pioneer America* 7, no. 2 (1975): 1–20.

Lexington-Fayette Urban County Government and Lexington Distillery District Foundation. *Lexington Distillery District Tax Increment Financing Development Area & Plan*. Lexington, KY: Lexington-Fayette Urban County Government, 2008.

Lezius, Walter G. "Geography of Glass Manufacture at Toledo, Ohio." *Economic Geography* 13, no. 4 (1937): 402–12.

Lipton, James. *An Exaltation of Larks*. New York: Penguin Books, 1991.

Longsdorf, George F., ed. *Current Law: A Complete Encyclopedia of New Law*. St. Paul, MN: Keefe-Davidson, 1904.

Lopez, Barry. *Crossing Open Ground*. New York: Vintage Books, 1989.

Lossing, John. "Berkshire Pigs." *American Farmer* 6, no. 9 (1840): 66–67.

Louisville Board of Trade, Committee on Industrial and Commercial Improvement. *The City of Louisville and a Glimpse of Kentucky, 1887*. Louisville, KY: Courier-Journal Press, 1887.

"Louisville Dots." *Bonfort's Wine and Spirit Circular* 20, no. 11 (1883): 213.

"Louisville Dots." *Bonfort's Wine and Spirit Circular* 21, no. 5 (1883): 106.

"Louisville Dots." *Bonfort's Wine and Spirit Circular* 33, no. 2 (1889): 52.

"The Louisville Grains Dryer." *Western Brewer* 30, no. 4 (1906): 192.

Lovejoy, Owen R. "Child Labor in the Glass Industry." *Annals of the American Academy of Political and Social Science* 27, no. 1 (1906): 42–53.

Lowenthal, David. *The Heritage Crusade and the Spoils of History*. Cambridge: Cambridge University Press, 1998.

——. *The Past Is a Foreign Country*. Cambridge: Cambridge University Press, 1985.

——. *The Past Is a Foreign Country Revisited*. Cambridge: Cambridge University Press, 2015.

——. "Past Time, Present Place: Landscape and Memory." *Geographical Review* 65, no. 1 (1975): 1–36.

——. *Possessed by the Past: The Heritage Crusade and the Spoils of History*. New York: Free Press, 1996.

Lucas, Marion B. *A History of Blacks in Kentucky: From Slavery to Segregation, 1760–1891*. 2d ed. Frankfort: Kentucky Historical Society, 2003.

Lukas, Scott A., ed. *The Themed Space: Locating Culture, Nation, and Self*. Lanham, MD: Lexington Books, 2007.

Lynes, Russell. *The Taste-Makers*. New York: Harper & Brothers, 1955.

MacCannell, Dean. *Empty Meeting Grounds: The Tourist Papers*. London: Routledge, 1992.

——. *The Tourist: A New Theory of the Leisure Class*. Berkeley: University of California Press, 2013.

MacGill, Caroline E. *History of Transportation in the United States before 1860*. Washington, DC: Carnegie Institution, 1917.

MacLaren, Malcolm. *The Rise of the Electrical Industry during the Nineteenth Century*. Princeton, NJ: Princeton University Press, 1943.

Magee, Malachy. *1000 Years of Irish Whiskey*. Dublin, Ireland: O'Brien Press, 1980.

Magnusson, Leifur. *Housing by Employers in the United States*. US Department of Labor, Bureau of Labor Statistics Bulletin 263. Washington, DC: GPO, 1920.

Major, Judith K. *To Live in the New World: A. J. Downing and American Landscape Gardening*. Cambridge, MA: MIT Press, 1997.

Mak, James, and Gary M. Walton. "Steamboats and the Great Productivity Surge in River Transportation." *Journal of Economic History* 32, no. 3 (1972): 619–40.

Mangelsdorf, Paul C. *Corn: Its Origin, Evolution and Improvement.* Cambridge, MA: Belknap Press, 1974.

Marquardt, William H., and Carole L. Crumley. "Theoretical Issues in the Analysis of Spatial Patterning." In *Regional Dynamics: Burgundian Landscapes in Historical Perspective,* edited by Carole L. Crumley and William H. Marquardt, 1–18. San Diego, CA: Academic Press, 1987.

Marsden, Roger D. *Drainage: Kentucky, Statistics for the State and Its Counties.* US Census Bulletin 1920. Washington, DC: GPO, 1921.

Martin, Jonathan D. *Divided Mastery: Slave Hiring in the American South.* Cambridge, MA: Harvard University Press, 2004.

Martin, Samuel D. "Short Horn and Devon Cattle." *Franklin Farmer* 3, no. 12 (1839): 95.

Martin, Thomas C. *Electrical Machinery, Apparatus, and Supplies.* US Bureau of the Census Special Reports, *Manufactures 1905,* pt. 4. Washington, DC: GPO, 1908.

Mathias, Peter. "Agriculture and the Brewing and Distilling Industries in the Eighteenth Century." *Economic History Review,* n.s., 5, no. 2 (1952): 249–57.

Matson, George C. *Water Resources of the Blue Grass Region of Kentucky.* US Geological Survey Water Supply Paper 233. Washington, DC: GPO, 1909.

Mayer, Martin. "The Volatile Business." *Esquire* 52, no. 2 (1959): 102–10, 112, 114–18.

McCormack, J. N. *Biennial Report 1910 and 1911.* Bulletin of the State Board of Health of Kentucky. Bowling Green, KY, 1912.

McCormick, Mike. *Terre Haute: Queen City of the Wabash.* Charleston, SC: Arcadia Press, 2005.

McCracken, Grant. *Culture & Consumption: New Approaches to the Symbolic Character of Consumer Goods and Activities.* Bloomington: Indiana University Press, 1988.

McDonald, Herman P., David Keltner, Pamela Wood, Bruce A. Waters, and Orville J. Whitaker. *Soil Survey of Anderson and Franklin Counties, Kentucky.* USDA Soil Conservation Service and Kentucky Agricultural Experiment Station. Washington, DC: GPO, 1985.

McDonald, Herman P., Raymond P. Sims, and Dan Isgrig. *Soil Survey of Jessamine and Woodford Counties, Kentucky.* USDA Soil Conservation Service and Kentucky Agricultural Experiment Station. Washington, DC: GPO, 1983.

McGovern, Charles F. *Sold American: Consumption and Citizenship, 1890–1945.* Chapel Hill: University of North Carolina Press, 2006.

McKearin, Helen, and Kenneth M. Wilson. *American Bottles & Flasks and Their Ancestry.* New York: Crown, 1978.

McKee, Lewis, and Linda K. Bond. *History of Anderson County: 1780–1936.* Frankfort, KY: Roberts Print Company, 1936.

McKendrick, Neil, John Brewer, and J. H. Plumb. *The Birth of a Consumer Society: The Commercialization of Eighteenth-Century England.* Bloomington: Indiana University Press, 1982.

McShane, Clay, and Joel A. Tarr. *The Horse in the City: Living Machines in the Nineteenth Century.* Baltimore: Johns Hopkins University Press, 2007.

Mead, H. E. *Kentucky and Tennessee: A Complete Guide to Their Railroads.* Louisville, KY: H. E. Mead, 1867.

Meigh, Edward. *The Story of the Glass Bottle.* Stoke-on-Trent, UK: C. E. Ramsden, 1972.

Meinig, D. W. "The Beholding Eye: Ten Versions of the Same Scene." In *The Interpretation of Ordinary Landscapes: Geographical Essays,* edited by D. W. Meinig, 33–48. New York: Oxford University Press, 1979.

———. "Reading the Landscape: An Appreciation of W. G. Hoskins and J. B. Jackson." In *The Interpreta-

tion of Ordinary Landscapes: Geographical Essays, edited by D. W. Meinig, 195–244. New York: Oxford University Press, 1979.

———. *The Shaping of America: A Geographical Perspective on 500 Years of History*. Vol. 2, *Continental America 1800–1867*. New Haven, CT: Yale University Press, 1993.

———. "Symbolic Landscapes: Some Idealizations of American Communities." In *The Interpretation of Ordinary Landscapes: Geographical Essays*, edited by D. W. Meinig, 165–92. New York: Oxford University Press, 1979.

Melhorn, Wilton N., and John P. Kempton. "The Teays Valley Problem: A Historical Perspective." In *Geology and Hydrology of the Teays-Mahomet Bedrock Valley Systems*, edited by Wilton N. Melhorn and John P. Kempton, 3–8. Special Paper 258. Boulder, CO: Geological Society of America, 1991.

Melosi, Martin. *The Sanitary City: Urban Infrastructure in America from Colonial Times to the Present*. Baltimore: Johns Hopkins University Press, 2000.

Mida, William. *Mida's Compendium of Information for the Liquor Interests*. Chicago: Criterion Publishing, 1899.

Miller, A. M. *Geology of Franklin County*. Kentucky Geological Survey Reports, 4th series, vol. 2, pt. 3. Frankfort: Kentucky Geological Survey, 1914.

Miller, Daniel. "The Power of Making." In *The Power of Making*, edited by Daniel Miller, 14–27. London: V & A Publishing, 2011.

Miller, George L., and Catherine Sullivan. "Machine-Made Glass Containers and the End of Production for Mouth-Blown Bottles." *Historical Archaeology* 18, no. 2 (1984): 83–96.

Minnick, Fred. "Barreling Ahead: Whiskey-Makers Break Cherished Traditions to Create New Flavors." *Scientific American* 14 (March 2013). https://www.scientificamerican.com/article/whiskey-makers-break-tradition-to-make-new-flavors/?print=true.

———. *Bourbon: The Rise, Fall, and Rebirth of an American Whiskey*. Minneapolis: Voyageur Press, 2016.

M'Murtrie, Henry. *Sketches of Louisville*. Louisville, KY: S. Penn, 1819.

Moore, Francis C. *Fire Insurance and How to Build*. New York: Baker & Taylor, 1903.

Moore-Colyer, R. J. "Aspects of the Trade in British Pedigree Draught Horses with the United States and Canada, c. 1850–1920." *Agricultural History Review* 48, no. 1 (2000): 42–59.

Morgan, T. J. *Fifty-Eighth Annual Report of the Commissioner of Indian Affairs*. Washington, DC: GPO, 1889.

Morrison, Andrew. *The Industries of Cincinnati: Manufacturing Establishments and Business Houses*. Cincinnati, OH: Aldine Printing, 1886.

Mosedale, J. R. "Effects of Oak Wood on the Maturation of Alcoholic Beverages with Particular Reference to Whisky." *Forestry* 68, no. 3 (1998): 203–30. http://forestry.oxfordjournals.org.

Mugerauer, Robert. *Interpreting Environments: Tradition, Deconstruction, Hermeneutics*. Austin: University of Texas Press, 1995.

Mulholland, James A. *A History of Metals in Colonial America*. University: University of Alabama Press, 1981.

Muller, Edward K. "Selective Urban Growth in the Middle Ohio Valley, 1800–1860." *Geographical Review* 66, no. 2 (1976): 178–99.

Mulryan, Peter. *The Whiskeys of Ireland*. Dublin, Ireland: O'Brien Press, 2016.

Munslow, Alun. *Narrative and History*. New York: Palgrave Macmillan, 2007.

Murphy, Kevin, and Stephen C. Gordo. "Guyn's Mill in Woodford County, Kentucky, as Recorded by the Historic American Engineering Record." *Millstone: Journal of the Kentucky Old Mill Association* 12, no. 2 (2013): 33–52.

Murray, Jim. *Jim Murray's Whisky Bible, 2017*. Litchborough, UK: Dram Good Books, 2017.

National Distillers. *National Distillers: An American History.* N.p.: Low Down Printer, 1985.

National Historic Preservation Act of 1966. Public Law 102-575. https://www.nps.gov/history/local-law/nhpa1966.htm.

National Park Service. "Historic Preservation Tax Incentives." Technical Preservation Services, 2012. www.nps.gov/tps.

———. "National Historic Landmarks Program." 2017. www.nps.gov/nhl.

Nettleton, J. A. *The Manufacture of Spirit as Conducted at the Various Distilleries of the United Kingdom.* London: Marcus Ward, 1893.

Newcomb, Buchanan & Company. "Circular." *Commercial and Financial Chronicle* 15, no. 367 (1872): 4.

Nickles, J. M. "The Geology of Cincinnati." *Journal of the Cincinnati Society of Natural History* 20, no. 2 (1902): 49–100.

Nimmo, Joseph, Jr. *Report on the Internal Commerce of the United States, 1881–1882.* Treasury Department, US Bureau of Statistics. Washington, DC: GPO, 1884.

Novak, William J. *The People's Welfare: Law & Regulation in Nineteenth-Century America.* Chapel Hill: University of North Carolina Press, 1996.

Nye, David E. *American Technological Sublime.* Cambridge, MA: MIT Press, 1996.

———. *Narratives and Spaces: Technology and the Construction of American Culture.* New York: Columbia University Press, 1997.

O'Dell, Gary A. "At the Starting Post: Racing Venues and the Origins of Thoroughbred Racing in Kentucky, 1783–1865." *Register of the Kentucky Historical Society* 116, no. 1 (2018): 29–78.

———. "Hillenmeyer Spring and Boiling Spring (Spring Lake Spring)." Unpublished paper, n.d.

———. "Water Supply and the Early Development of Lexington, Kentucky." *Filson Club Historical Quarterly* 67, no. 4 (1993): 431–61.

O'Donnell, Mary Louise. *Ireland's Harp: The Shaping of Irish Identity c. 1770–1880.* Dublin, Ireland: University College Dublin Press, 2014.

O'Keeney, Angela. *Looking Back on Ballinascreen.* Draperstown, Ireland: Ballinascreen Historical Society, 1989.

"Old Breckenridge Distillery." *Wine and Spirit Bulletin* 18, no. 1 (1904): 41.

Olmstead, Alan L. "The Mechanization of Reaping and Mowing in American Agriculture, 1833–1870." *Journal of Economic History* 35, no. 3 (1975): 327–52.

Olmstead, Alan L., and Paul W. Rhode. *Arresting Contagion: Science, Policy, and Conflicts over Animal Disease Control.* Cambridge, MA: Harvard University Press, 2015.

———. "The Red Queen and the Hard Reds: Productivity Growth in American Wheat, 1800–1940." *Journal of Economic History* 62, no. 4 (2002): 929–66.

Olsen, Bjornar. *In Defense of Things: Archaeology and the Ontology of Objects.* New York: Altamira Press, 2010.

O'Malley, Nancy. "McConnell Springs in Historical Perspective." Paper, Department of Anthropology, University of Kentucky, n.d.

———. "Ruddle's Mills, Bourbon County, Kentucky: Early Industry in the Kentucky Bluegrass." *Mill Stone* 5, no. 2 (2006): 21–29.

"On Distillation." *Register of the Arts and Sciences, Improvements & Discoveries* 10 (February 21, 1824): 147–50.

Orton, Edward. *Report of the Occurrence of Petroleum, Natural Gas and Asphalt Rock in Western Kentucky: Based on Examinations Made in 1888 and 1889.* Kentucky Geological Survey and State Museum. Frankfort, KY: E. Polk Johnson, 1891.

Overman, Frederick. *Mechanics for the Millwright, Machinist, Engineer, Civil Engineer, Architect and Student.* Philadelphia: Lippincott, Grambo, 1851.

Oxford English Dictionary. 2d ed. Oxford: Clarendon Press, 1989.

Pacult, F. Paul. *American Still Life: The Jim Beam Story and the Making of the World's #1 Bourbon.* Hoboken, NJ: John Wiley & Sons, 2003.

Page, Brian, and Richard Walker. "From Settlement to Fordism: The Agro-Industrial Revolution in the American Midwest." *Economic Geography* 67, no. 4 (1991): 281–315.

Palmer, Chase. "Quantity of the Underground Waters in the Blue Grass Region." In *Water Resources of the Blue Grass Region of Kentucky,* by George C. Matson. US Geological Survey Water Supply Paper 233. Washington, DC: GPO, 1909.

Palmquist, W. N., and F. R. Hall. *Public and Industrial Water Supplies of the Blue Grass Region, Kentucky.* US Geological Survey Circular 299. Washington, DC: US Geological Survey, 1953.

Paquette, Jack K. *The Glassmakers Revisited.* Toledo, OH: Xlibris Press, 2010.

Parker, William N. "A Note on Regional Culture in the Corn Harvest." *Agricultural History* 46, no. 1 (1972): 181–90.

Parry, William. *Federal and State Laws Relating to Weights and Measures.* 3d ed. Department of Commerce, Miscellaneous Publication 20. Washington, DC: GPO, 1926.

Parsons, Warren M. *Beer and Brewing in America: An Economic Study.* New York: United Brewers Industrial Foundation, 1941.

Patchett, Merle. "Historical Geographies of Apprenticeship: Rethinking and Retracing Craft Conveyance over Time and Place." *Journal of Historical Geography* 55 (2017): 30–43. https://doi.org/10.1016/j.jhg.2016.11.006.

"Patent Granted to Aeneas Coffey." *Repertory of Patent Inventions* 11, no. 72 (1831): 190–96.

"Patent Granted to Robert Stein." *Repertory of Patent Inventions* 8, no. 51 (1830): 538–40.

Patten, Simon, and Thomas Hanfrey. "The Cultural Keystone Concept: Insights from Ecological Anthropology." *Human Ecology* 37, no. 4 (2009): 491–500.

"Peacock Distilling Co. v. Commonwealth." *Kentucky Law Reporter* 25, pt. 2 (1904): 1778–81.

Pearce, John Ed. *Nothing Better in the Market.* Louisville, KY: Brown-Forman Distillers, 1970.

Penn, William A. *Kentucky Rebel Town: The Civil War Battles of Cynthiana and Harrison County.* Lexington: University Press of Kentucky, 2016.

Perrier, Anthony. "Sir Anthony Perrier's Improved Apparatus for Distillation." *Register of the Arts and Sciences* 1, (1824): 10–12.

Perrin, W. H., J. H. Battle, and G. C. Kniffin. *Kentucky: A History of the State.* 4th ed. Louisville, KY: F. A. Battey, 1887.

Perrin, William H., ed. *History of Bourbon, Scott, Harrison and Nicholas Counties, Kentucky.* Chicago: O. L. Baskin, 1882.

———. *History of Fayette County, Kentucky.* Chicago: O. L. Baskin, 1882.

Peter, Robert. *Lower Blue Lick Spring: The Quantitative Chemical Analysis of the Water in the Lower Blue Lick Spring, in Nicholas County, Kentucky.* N.p., 1850.

Peters, Richard, ed. *The Public Statutes at Large of the United States of America from the Organization of the Government in 1789, to March 3, 1845.* Vol. 3. Boston: Charles C. Little & James Brown, 1846.

Peterson, Arthur G. "Flour and Grist Milling in Virginia: A Brief History." *Virginia Magazine of History and Biography* 43, no. 2 (1935): 97–108.

Peterson, Richard A. "In Search of Authenticity." *Journal of Management Studies* 42, no. 5 (2005): 1083–98.

Pethica, James. "Contextualising the Lyric Moment: Yeats's 'The Happy Townland' and the Abandoned Play *The Country of the Young.*" *Yeats Annual No. 10,* ed. Warwick Gould (1993): 65–91.

"Petition of Andrew Jackson for Relief from the Whiskey Tax." February 12, 1803. Record of the US

House of Representatives, 1789–2015, Petitions and Memorials, 1793–1911. National Archives Catalog.

Pezzoni, J. Daniel. *An Evaluation of Kentucky's Historic Distilleries*. Frankfort: Kentucky Heritage Council, 2000.

Phillips, U. B. "Transportation in the Ante-bellum South: An Economic Analysis." *Quarterly Journal of Economics* 19 (1905): 434–51.

Pike, Andy. "Geographies of Brands and Branding." *Progress in Human Geography* 33, no. 5 (2009): 619–45.

———. "Placing Brands and Branding: A Scio-Spatial Biography of Newcastle Brown Ale." *Transactions of the Institute of British Geographers* 36, no. 2 (2011): 206–22.

Pope, Daniel. *The Making of Modern Advertising*. New York: Basic Books, 1983.

Powell, Sallie L. "African American Hamlets." In *The Kentucky African American Encyclopedia*, edited by Gerald L. Smith, Karen C. McDaniel, and John A. Hardin, 5–6. Lexington: University Press of Kentucky, 2015.

Power, Richard L. *Planting Corn Belt Culture: The Impress of the Upland Southerner and Yankee in the Old Northwest*. Indianapolis: Indiana Historical Society, 1953.

Pratt, Joseph H. "The Lumber and Forest-Products Industry of the South." *Annals of the American Academy of Political and Social Science* 153 (1931): 63–75.

Purcell, Aaron D. "Bourbon to Bullets: Louisville's Distilling Industry during World War II." *Register of the Kentucky Historical Society* 96, no. 2 (1998): 61–87.

Pursell, Carroll W., Jr. *Early Stationary Steam Engines in America: A Study in the Migration of a Technology*. Washington, DC: Smithsonian Institution Press, 1969.

Pyne, Stephen J. *Fire in America: A Cultural History of Wildland and Rural Fire*. Princeton, NJ: Princeton University Press, 1982.

Quinn, John F. *Father Mathew's Crusade: Temperance in Nineteenth-Century Ireland and Irish America*. Amherst: University of Massachusetts Press, 2002.

Quinn, M. L. "Industry and Environment in the Appalachian Copper Basin, 1890–1930." *Technology and Culture* 34, no. 3 (1993): 575–612.

Raitz, Karl B. "The Government Institutionalization of Tobacco Acreage in Wisconsin." *Professional Geographer* 23, no. 2 (1971): 123–26.

———. "Rock Fences and Preadaptation." *Geographical Review* 85, no. 1 (1995): 50–62.

Raitz, Karl, and John Paul Jones III. "The City Hotel as Landscape Artifact and Community Symbol." *Journal of Cultural Geography* 9 (Fall/Winter 1988): 17–36.

Raitz, Karl, and Nancy O'Malley. "An Index of Soil Production Potential as Applied to Historical Agricultural Adaptation: A Kentucky Example." *Historical Methods* 18, no. 5 (1985): 137–45.

———. *Kentucky's Frontier Highway: Historical Landscapes along the Maysville Road*. Lexington: University Press of Kentucky, 2012.

———. "Local-Scale Turnpike Roads in Nineteenth-Century Kentucky." *Journal of Historical Geography* 33, no. 1 (2007): 1–23.

———. "The Nineteenth-Century Evolution of Local-Scale Roads in Kentucky's Bluegrass." *Geographical Review* 94, no. 4 (2004): 415–39.

Ranck, George W. *History of Lexington, Kentucky*. Cincinnati, OH: Robert Clarke, 1872.

Rasmussen, Wayne D. "Wood on the Farm." In *Material Culture of the Wooden Age*, edited by Brooke Hindle, 15–34. Tarrytown, NY: Sleepy Hollow Press, 1981.

Reaich, Donna. "The Influence of Copper on Malt Whisky Character." In *Proceedings of the Fifth Aviemore Conference on Malting, Brewing and Distilling*, 141–52. Aviemore, Inverness-shire, Scotland, May 25–28, 1998.

Reed, Americus, II, Mark R. Forehand, Stefano Puntoni, and Luk Warlop. "Identity-Based Consumer Behavior." *International Journal of Research in Marketing* 29, no. 4 (2012): 310–21.

Regan, Gary, and Mardee Haidin Regan. *The Book of Bourbon and Other Fine American Whiskeys.* Shelburne, VT: Chapters Publishing, 1995.

Reigler, Susan. *Kentucky Bourbon: The Essential Travel Guide.* Lexington: University Press of Kentucky, 2013.

Republican Congressional Committee. *The Republican Campaign Text Book for 1880.* Washington, DC: Republican Congressional Committee, 1880.

A Resolution Designating September 2007 as "National Bourbon Heritage Month." S. Res. 294, 110th Cong. https://www.congress.gov/bill/110th-congress/senate-resolution/294.

Reynolds, Terry S. *Stronger than a Hundred Men: A History of the Vertical Waterwheel.* Baltimore: Johns Hopkins University Press, 1983.

Rice, Otis K. "Importations of Cattle into Kentucky, 1785–1860." *Register of the Kentucky Historical Society* 49 (1951): 35–47.

Rice, S. L. *Established Rates of Kentucky Distilleries, Bonded and Free Warehouses, Cattle and Pens.* Indianapolis: Rough Notes, 1892.

Richards, James M. *The Functional Tradition in Early Industrial Buildings.* London: Architectural Press, 1958.

Richardson, Alfred J., Rudy Forsythe, and Hubert B. Odor. *Soil Survey of Bourbon and Nicholas Counties, Kentucky.* USDA Soil Conservation Service and Kentucky Agricultural Experiment Station. Washington, DC: GPO, 1982.

Richardson, Charles H. *The Glass Sands of Kentucky.* Kentucky Geological Survey, series 6, vol. 1. Frankfort, KY: State Journal Company, 1920.

Rickman, H. P., ed. *Pattern & Meaning in History: Wilhelm Dilthey, Thoughts on History & Society.* New York: Harper & Row, 1961.

Riesenweber, Julie, and Karen Hudson, eds. *Kentucky's Bluegrass Region.* Frankfort: Kentucky Heritage Council, 1990.

Ristow, Walter W. "United States Fire Insurance and Underwriters Maps 1852–1968." *Quarterly Journal of the Library of Congress* 25, no. 3 (1968): 194–218.

Robbins, Jackie W. D., David H. Howells, and George J. Kriz. "Stream Pollution from Animal Production Units." *Journal* (Water Pollution Control Federation) 44, no. 8 (1972): 1536–44.

Robertson, John R. *Petitions of the Early Inhabitants of Kentucky to the General Assembly of Virginia, 1769–1792.* Louisville, KY: John P. Morton, 1914.

Rogers, Everett M. *Diffusion of Innovations.* 5th ed. New York: Free Press, 2003.

Rogers, Everett M., and F. Floyd Shoemaker. *Communication of Innovations.* 2d ed. New York: Free Press, 1971.

Rogin, Leo. *The Introduction of Farm Machinery in Its Relation to the Productivity of Labor in the Agriculture of the United States during the Nineteenth Century.* Berkeley: University of California Press, 1931.

Rook, Dennis, comp. *Brands, Consumers, Symbols & Research: Sidney J. Levy on Marketing.* Thousand Oaks, CA: Sage Publications, 1999.

Rorabaugh, M. I., F. F. Schrader, and L. B. Laird. *Water Resources of the Louisville Area, Kentucky and Indiana.* US Geological Survey Circular 276. Washington, DC: GPO, 1953.

Rorabaugh, W. J. *The Alcoholic Republic: An American Tradition.* New York: Oxford University Press, 1979.

Rosen, Christine Meisner. "The Business-Environment Connection." *Environmental History* 10, no. 1 (2005): 77–79.

——. "'Knowing' Industrial Pollution: Nuisance Law and the Power of Tradition in a Time of Rapid Economic Change, 1840–1864." *Environmental History* 8, no. 4 (2003): 565–97.

Rosenberg, Nathan. *Exploring the Black Box: Technology, Economics, and History.* Cambridge: Cambridge University Press, 1994.

——. *Perspectives on Technology.* Cambridge: Cambridge University Press, 1976.

Rosenbloom, Morris V. *The Liquor Industry: A Survey of Its History, Manufacture, Problems of Control and Importance.* Braddock, PA: Ruffsdale Distilling Company, 1935.

Rothert, Otto A. *A History of Muhlenberg County, Kentucky.* Louisville, KY: John P. Morton, 1913.

Royal Commission on Whiskey and Other Potable Spirits. *Minutes of Evidence Taken by the Royal Commission on Whiskey and Other Potable Spirits.* Vol. 1. London: Jas. Truscott & Son, 1908.

Rusen, Jorn. "Historical Narration: Foundation, Types, Reason." *History and Theory* 26, no. 4 (1987): 87–97.

Russell, Inge, and Graham Stewart. *Whiskey: Technology, Production and Marketing.* 2d ed. Oxford: Academic Press, 2014.

Rypkema, Donovan D. *Historic Preservation and the Economy of the Commonwealth: Kentucky's Past at Work for Kentucky's Future.* Frankfort: Kentucky Heritage Council & Commonwealth Preservation Advocates, 1997.

Salmon, D. E. *First Annual Report of the Bureau of Animal Industry, 1884.* US Department of Agriculture. Washington, DC: GPO, 1885.

——. *Hog Cholera: Its History, Nature, and Treatment.* US Department of Agriculture, Bureau of Animal Industry. Washington, DC: GPO, 1889.

——. *Second Annual Report, Bureau of Animal Industry, 1885.* US Department of Agriculture. Washington, DC: GPO, 1886.

——. *Third Annual Report, Bureau of Animal Industry, 1886.* US Department of Agriculture. Washington, DC: GPO, 1887.

Samuel, Raphael. *Theatres of Memory: Past and Present in Contemporary Culture.* London: Verso, 2012.

Samuels, Bill, Jr. *My Autobiography.* Louisville, KY: Saber Publishing–Butler Books, 2009.

Sanders, Alvin H. *Short-Horn Cattle: Historical Sketches, Memoirs and Records.* 2d ed. Chicago: Sanders Publishing, 1901.

Sanders, Lewis. "On Cattle." *Franklin Farmer* 3, no. 21 (1840): 165.

Sanger, George P., ed. *The Statutes at Large, Treaties, and Proclamations, of the United States of America from December 5, 1859, to March 3, 1863.* Vol. 12. Boston: Little, Brown, 1863.

——. *The Statutes at Large, Treaties, and Proclamations, of the United States of America from December 1867, to March 1869.* Vol. 15. Boston: Little, Brown, 1869.

Sauer, Carl O. "Foreword to Historical Geography." *Annals of the Association of American Geographers* 31, no. 1 (1941): 1–24.

Scanlon, B. R. "Physical Controls on Hydrochemical Variability in the Inner Bluegrass Karst Region of Central Kentucky." *Ground Water* 27, no. 5 (1989): 639–46.

Scanlon, B. R., and J. Thrailkill. "Chemical Similarities among Physically Distinct Spring Types in a Karst Terrain." *Journal of Hydrology* 89, no. 3–4 (1987): 259–79.

Schatzki, Theodore R. *The Site of the Social: A Philosophical Account of the Constitution of Social Life and Change.* University Park: Pennsylvania State University Press, 2002.

——. *Social Change in a Material World.* London: Routledge, 2019.

Schein, Richard H. "A Conceptual Framework for Interpreting an American Scene." *Annals of the Association of American Geographers* 87, no. 4 (1997): 660–80.

——. "Cultural Landscapes." In *Research Methods in Geography*, edited by Basil Gomez and John Paul Jones III, 222–39. West Sussex, UK: Wiley-Blackwell, 2010.

——. "Normative Dimensions of Landscape." In *Everyday America: Cultural Landscape Studies after J. B. Jackson*, edited by Chris Wilson and Paul Groth, 199–218. Berkeley: University of California Press, 2003.

Schlereth, Thomas J. *Cultural History and Material Culture: Everyday Life, Landscapes, Museums.* Charlottesville: University Press of Virginia, 1990.

——. "The New York Artisan in the Early Republic: A Portrait from Graphic Evidence, 1787–1853." *Landscape, Place, & Material Culture* 20, no. 1 (1988): 1–31.

——. *Victorian America: Transformations in Everyday Life, 1876–1915.* New York: HarperCollins, 1991.

Schlesinger, Arthur M. "A Dietary Interpretation of American History." *Massachusetts Historical Society Proceedings* 68 (1944–1947): 199–227.

Schob, David E. "Woodhawks & Cordwood: Steamboat Fuel on the Ohio and Mississippi Rivers, 1820–1860." *Journal of Forest History* 21, no. 3 (1977): 124–32.

Schulman, Steven A. "The Lumber Industry of the Upper Cumberland River Valley." *Tennessee Historical Quarterly* 32, no. 3 (1973): 255–64.

Scott, Walter Dill. *The Psychology of Advertising in Theory and Practice.* Boston: Small, Maynard, 1921.

Scovell, M. A. "Report of the Director of the Kentucky Agricultural Experiment Station to the Governor of Kentucky, on the Enforcement of the State Pure Food Law." In *Twentieth Annual Report of the Kentucky Agricultural Experiment Station*, xvi–xix. Lexington: State College of Kentucky, 1907.

Scoville, Warren C. "Growth of the American Glass Industry to 1800." *Journal of Political Economy* 52, no. 3 (1944): 193–216.

——. "Growth of the American Glass Industry to 1800—Continued." *Journal of Political Economy* 52, no. 4 (1944): 340–55.

——. *Revolution in Glassmaking: Entrepreneurship and Technological Change in the American Industry, 1880–1920.* Cambridge, MA: Harvard University Press, 1948.

——. "Spread of Techniques: Minority Migrations and the Diffusion of Technology." *Journal of Economic History* 11, no. 4 (1951): 347–60.

Sellers, Coleman, Jr. "Oliver Evans and His Inventions." *Scientific American*, Supplement 552 (February 15, 1879): 8813–14.

Sepe, Marichela. "Places and Perceptions in Contemporary City." *Urban Design International* 18 (2013): 111–13.

Shaler, N. S., and A. R. Crandall. *Report on the Forests of Greenup, Carter, Boyd & Lawrence Counties.* Geological Survey of Kentucky, 2d series, vol. 1, pt. 1. Frankfort, KY: Yeoman Press, 1884.

Shands, H. L., and A. D. Dickson. "Barley: Botany, Production, Harvesting, Processing, Utilization and Economics." *Economic Botany* 7, no. 1 (1953): 3–26.

Skrabec, Quentin R. *Michael Owens and the Glass Industry.* Gretna, LA: Pelican Publishing, 2007.

Slaton, Amy E. *Reinforced Concrete and the Modernization of American Building, 1900–1930.* Baltimore: Johns Hopkins University Press, 2001.

Slaughter, Thomas P. *The Whiskey Rebellion: Frontier Epilogue to the American Revolution.* New York: Oxford University Press, 1986.

Smith, Merritt Roe, and Gregory Clancey, eds. *Major Problems in the History of American Technology.* Boston: Houghton Mifflin, 1998.

Smith, Peter C., and Karl B. Raitz. "Negro Hamlets and Agricultural Estates in Kentucky's Inner Bluegrass." *Geographical Review* 64, no. 2 (1974): 217–234.

Smith, Sarah B. *Historic Nelson County, Its Towns and People.* Bardstown, KY: GBA/Delmar, 1983.

Smyth, J. F. D. *A Tour of the United States of America.* Vol. 1. London: G. Robinson, 1784.

Spooner, Walter W., ed. *The Political Prohibitionist for 1888: A Handbook for the Aggressive Temperance People of the United States.* New York: Funk & Wagnalls, 1888.

Stabler, Herman. *Prevention of Stream Pollution by Distillery Refuse.* US Geological Survey Water Supply and Irrigation Paper 179. Washington, DC: GPO, 1906.

Starobin, Robert S. *Industrial Slavery in the Old South.* New York: Oxford University Press, 1970.

Steele, Ellsworth. "The Flint Glass Workers' Union in the Indiana Gas Belt and the Ohio Valley in the 1890's." *Indiana Magazine of History* 50, no. 3 (1954): 229–50.

Stewart, Bruce E. "'This Country Improves in Cultivation, Wickedness, Mills, and Still': Distilling and Drinking in Antebellum Western North Carolina." *North Carolina Historical Review* 83, no. 4 (2006): 447–78.

Stilgoe, John R. *Common Landscape of America, 1580 to 1845.* New Haven, CT: Yale University Press, 1982.

———. *Landscape and Images.* Charlottesville: University of Virginia Press, 2005.

———. *Metropolitan Corridor: Railroads and the American Scene.* New Haven, CT: Yale University Press, 1983.

Storck, John, and Walter Dorwin Teague. *Flour for Man's Bread: A History of Milling.* Minneapolis: University of Minnesota Press, 1952.

Storrie, Margaret C. "The Scotch Whiskey Industry." *Transactions and Papers of the Institute of British Geographers* 31 (1962): 97–114.

Strasser, Susan. *Satisfaction Guaranteed: The Making of the American Mass Market.* New York: Pantheon Books, 1989.

Strassman, W. Paul. *Risk and Technological Innovation: American Manufacturing Methods during the Nineteenth Century.* Ithaca, NY: Cornell University Press, 1959.

Sulzer, Elmer G. *Ghost Railroads of Kentucky.* Bloomington: Indiana University Press, 1967.

Swain, George F. "Statistics of Water Power Employed in Manufacturing in the United States." *Publications of the American Statistical Association* 1, no. 1 (1888): 5–44.

Switzler, William F. *Report on the Internal Commerce of the United States.* Washington, DC: GPO, 1888.

Taft, Nollie Olin. *History of State Revenue and Taxation in Kentucky.* Nashville, TN: George Peabody College for Teachers, 1931.

Tallant, Harold D. *Evil Necessity: Slavery and Political Culture in Antebellum Kentucky.* Lexington: University Press of Kentucky, 2003.

Taylor, E. H., Jr. *Description of the O.F.C., Carlisle, and J. S. Taylor Distilleries.* Frankfort, KY: E. H. Taylor Jr. Co., [1886].

Taylor, F. Sherwood. "The Evolution of the Still." *Annals of Science* 5, no. 3 (1945): 185–202.

Taylor, George Rogers. *The Transportation Revolution, 1815–1860.* New York: Harper & Row, 1951.

Taylor, Richard. *The Great Crossing: A Historic Journey to Buffalo Trace Distillery.* Frankfort, KY: Buffalo Trace Distillery, 2002.

Temin, Peter. "Steam and Waterpower in the Early Nineteenth Century." *Journal of Economic History* 26, no. 2 (1966): 187–205.

Tenkotte, Paul A., James C. Claypool, and David E. Schroeder, eds. *Gateway City: Covington, Kentucky, 1815–2015.* Covington, KY: Clerisy Press, 2015.

Thrower, Norman J. W. "The County Atlas of the United States." *Surveying and Mapping* 21, no. 3 (1961): 365–73.

Thum W. W. *Supplement to 1909 Kentucky Statutes.* Cincinnati, OH: W. H. Anderson, 1915.

Thurston, Robert H. *A Manual of the Steam-Engine.* New York: John Wiley & Sons, 1891.

———. *Reports of the Commissioners of the United States to the International Exhibition, Vienna, 1873.* Vol. 1, *Introduction; Executive Commission; Agriculture.* Washington, DC: GPO, 1876.

Tovey, Charles. *British & Foreign Spirits.* London: Whittaker, 1864.

"Trade Notes and Personals." *Wine and Spirit Bulletin* 15, no. 10 (October 1, 1901): 20.

Treude, Mai. *Windows to the Past: A Bibliography of Minnesota County Atlases.* Publication CURA 80-3. Minneapolis: Center for Urban and Regional Affairs, University of Minnesota, 1980.

Tritton, Henry. "On Distillation," *Register of the Arts and Sciences, Improvements & Discoveries* 10 (February 21, 1824): 147–50.

———. "Patent for an Improved Apparatus for Distilling." *Annals of Philosophy* 11 (1818): 445–47.

Troesken, Werner. "Exclusive Dealing and the Whiskey Trust, 1890–1895." *Journal of Economic History* 58, no. 3 (1998): 755–78.

"Turney Drier Company." *Railroad Age* 12 (March 24, 1905): 394.

Twain, Mark. *Life on the Mississippi* New York: Harper & Brothers, 1903.

Twede, Diana. "The Cask Age: The Technology and History of Wooden Barrels." *Packaging Technology and Science* 18 (2005): 253–64. www.interscience.wiley.com.

Tyrrell, Ian R. "Drink and Temperance in the Antebellum South: An Overview and Interpretation." *Journal of Southern History* 48, no. 4 (1982): 485–510.

Ulin, Robert C. "Invention and Representation as Cultural Capital: Southwest French Winegrowing History." *American Anthropologist* 97, no. 3 (1995): 519–27.

The United States Internal Revenue and Tariff Law (Passed July 13, 1870) Together with the Act Imposing Taxes on Distilled Spirits and Tobacco, and for Other Purposes (Approved July 20, 1868) and Such Other Acts or Parts of Acts Relating to Internal Revenue as Are Now in Effect. Compiled by Horace E. Dresser. New York: Harper & Brothers, 1870.

Upton, Dell. *Another City: Urban Life and Urban Spaces in the New American Republic.* New Haven, CT: Yale University Press, 2008.

———. "Seen, Unseen, and Scene." In *Understanding Ordinary Landscapes,* edited by Paul Groth and Todd W. Bressi, 174–79. New Haven, CT: Yale University Press, 1997.

Urban, Raymond, and Richard Mancke. "Federal Regulation of Whiskey Labelling: From the Repeal of Prohibition to the Present." *Journal of Law and Economics* 15, no. 2 (1972): 411–26.

Ure, Andrew. *The Philosophy of Manufactures.* London: Charles Knight, 1835.

Urry, John, and Jonas Larsen. *The Tourist Gaze 3.0.* Los Angeles: Sage, 2011.

US Army Corps of Engineers, Mississippi River Commission. *Geological Investigation of Mississippi River Activity, Memphis, Tennessee, to Mouth of Arkansas River.* Technical Memorandum 3-288. Vicksburg, MS: Waterways Experiment Station, 1949.

"U.S. Bottlers' Supply Co." *Wine and Spirit Bulletin* 15, no. 10 (1901): 23.

US Bureau of Animal Industry. *Third Annual Report, 1886.* Washington, DC: GPO, 1887.

US Bureau of the Census. *1850 Census.* Washington, DC: Robert Armstrong, 1853.

———. *Manufactures 1905.* Pt. 4, *Special Reports on Selected Industries.* Washington, DC: GPO, 1908.

———. *Report of the Productions of Agriculture.* 10th Census. Washington, DC: GPO, 1883.

———. *Report of the Statistics of Agriculture in the United States.* 11th Census. Washington, DC: GPO, 1896.

———. *Report of the Superintendent of the Census, December 1, 1852.* Washington, DC: Robert Armstrong, 1853.

US Centennial Commission. *List of Awards, International Exhibition, 1876, at Philadelphia.* Philadelphia: S. T. Souder, 1876.

US Commissioner of Internal Revenue. *Report of the Commissioner of Internal Revenue, 1880.* Treasury Department Document 53. Washington, DC: GPO, 1880.

———. *Report of the Commissioner of Internal Revenue, Fiscal Year Ended June 30, 1887.* Treasury Department Document 1031, 2d ed. Washington, DC: GPO, 1887.

———. *Report of the Commissioner of Internal Revenue, Fiscal Year Ended June 30, 1900.* Washington, DC: GPO, 1900.

US Congress, House. *An Act to Regulate Trade and Intercourse with the Indian Tribes, and to Preserve Peace on the Frontiers.* 23d Cong., 1st sess., 1834.

———. *Journal of the House of Representatives of the United States.* 43d Cong., 1st sess., 1874, vol. 74. Washington, DC: GPO, 1873.

———. *The Reports of Committees: H.R. 276.* 53d Cong., 2d sess., 1893–1894.

———. *The Reports of Committees: H.R. 4165.* 50th Cong., 2d sess., 1888–1889.

US Congress, Senate. *Tariff Laws of 1890 and 1894: Report 707, Part 2.* 53d Cong., 2d sess., 1894.

US Department of Agriculture. *Yearbook, 1922.* Washington, DC: GPO, 1923.

US Department of Health, Education, and Welfare, Public Health Service. *Public Health Service Drinking Water Standards.* Publication 956. Washington, DC: GPO, 1962.

US Department of Internal Revenue. *Government Drawings, Registered Distillery No. 47, Internal Revenue Bonded Warehouse No. 36, Tax Paid Bottling House No. 17, Bluegrass Distillery, Gethsemane, Kentucky.* N.p., n.d.

———. *Historical Study: IRS Historical Fact Book; a Chronology 1646–1992.* Chamblee, GA, 2012. Accessed through governmentattic.org.

———. *Regulations and Instructions Concerning the Tax on Distilled Spirits under the Revised Statutes of the United States and Subsequent Acts.* Series 7, no. 7. Washington, DC: GPO, 1877.

US Department of Labor, Bureau of Labor Statistics. "Displacement of Labor by Machinery in the Glass Industry." *Monthly Labor Review* 24, no. 4 (1927): 1–13.

US Department of the Treasury. *Annual Report of the Secretary of the Treasury on the State of the Finances for the Year 1889.* Treasury Department Document 1244, 4th ed. Washington, DC: GPO, 1889.

US Department of the Treasury, Alcohol and Tobacco Tax and Trade Bureau. Title 27, Alcohol, Tobacco Products and Firearms; Part 5, Labeling and Advertising of Distilled Spirits; Subpart C, Standards of Identity; Section 5.22, Standards of Identity for Distilled Spirits. https://www.gpo.gov/fdsys/pkg/CFR-2011-title27-vol1/pdf/CFR-2011-title27-vol1-chapI.pdf.

Vance, James E. *Capturing the Horizon: The Historical Geography of Transportation.* New York: Harper & Row, 1986.

———. *The North American Railroad: Its Origin, Evolution, and Geography.* Baltimore: Johns Hopkins University Press, 1995.

———. *This Scene of Man: The Role and Structure of the City in the Geography of Western Civilization.* New York: Harper's College Press, 1977.

Vasser, Ed, and Jerry Sudduth. *Frankfort and Cincinnati Railroad: History and Remembrances of the Bourbon Road.* Lexington, KY: CreateSpace.com, 2013.

Veach, Michael R. *Kentucky Bourbon Whiskey: An American Heritage.* Lexington: University Press of Kentucky, 2013.

Verhoeff, Mary. *The Kentucky River Navigation.* Filson Club Publication 28. Louisville, KY: John P. Morton, 1917.

Wagner, Joseph B. *Cooperage: A Treatise on Modern Shop Practice and Methods, from the Tree to the Finished Article.* Yonkers, NY: J. B. Wagner, 1910.

Walden, Howard T. *Native Inheritance: The Story of Corn in America.* New York: Harper & Row, 1966.

Walker, Eugene H. *The Deep Channel and Alluvial Deposits of the Ohio Valley in Kentucky.* US Geological Survey Water Supply Paper 1411. Washington, DC: GPO, 1957.

———. *Geology and Ground-Water Resources of the Covington-Newport Alluvial Area, Kentucky.* US Geological Survey Circular 240. Washington, DC: GPO, 1953.

Walker, Francis. *The Statistics of the Wealth and Industry of the United States.* US Census, 1870, vol. 3. Washington, DC: GPO, 1872.

Walker, Richard. "Unseen and Disbelieved: A Political Economist among Cultural Geographers." In *Understanding Ordinary Landscapes,* edited by Paul Groth and Todd W. Bressi, 162–73. New Haven, CT: Yale University Press, 1997.

Wallace, Henry A. "Corn and the Midwestern Farmer." *Proceedings of the American Philosophy Society* 100, no. 5 (1956): 455–66.

Wallace, Henry A., and William L. Brown. *Corn and Its Early Fathers.* East Lansing: Michigan State University Press, 1956.

Watkins, Julian L. *The 100 Greatest Advertisements.* New York: Dover Publications, 1959.

Weaver, John C. "Barley in the United States: A Historical Sketch." *Geographical Review* 33, no. 1 (1943): 56–73.

———. "Climatic Relations of American Barley Production." *Geographical Review* 33, no. 4 (1943): 569–88.

———. "United States Malting Barley Production." *Annals of the Association of American Geographers* 34, no. 2 (1944): 97–131.

Weber, Ralf. "The Myth of Meaningful Forms: Comparing the Forms of Indigenous and Classical Architecture." *Traditional Dwellings and Settlements Review* 2, no. 2 (1991): 65–73.

Weir, R. B. "Distilling and Agriculture: 1870–1939." *Agricultural History Review* 32, no. 1 (1984): 49–62.

Welsh, Peter C. "A Craft that Resisted Change: American Tanning Practices to 1850." *Technology and Culture* 4, no. 3 (1963): 299–317.

Wermiel, Sara E. *The Fireproof Building: Technology and Public Safety in the Nineteenth-Century American City.* Baltimore: Johns Hopkins University Press, 2000.

West, Elliott. "Selling the Myth: Western Images in Advertising." *Magazine of Western History* 46, no. 2 (1996): 36–49.

Whitaker, Orville J., and Robert A. Eigel. *Soil Survey of Henry and Trimble Counties, Kentucky.* USDA Soil Conservation Service and Kentucky Agricultural Experiment Station. Washington, DC: GPO, 1988.

White, John H., Jr. *The American Railroad Freight Car: From the Wood-Car Era to the Coming of Steel.* Baltimore: Johns Hopkins University Press, 1993.

Whitten, David. "Louisville Glass Factories of the 19th Century—Part 1." In *Bottles and Extras,* 1–6. http://www.fohbc.org/wp-content/uploads/2014/06/LouisvilleGlassFactories_P1_2Spring2005.pdf.

———. "Louisville Glass Factories of the 19th Century—Part 2." In *Bottles and Extras,* 2–4. https://sha.org/bottle/pdffiles/LouisvilleGlass2_Whitten.pdf.

Williams, L. A. *History of the Ohio Falls Cities and Their Counties.* Vol. 1. Cleveland, OH: L. A. Williams, 1882.

Willkie, Frederick, and Harrison C. Blankmeyer. *An Outline for Industry.* Springfield, IL: Charles C. Thomas, 1944.

Willkie, Herman F., and Joseph A. Prochaska. *Fundamentals of Distillery Practice.* Louisville, KY: Joseph E. Seagram & Sons, 1943.

Wilson, Chris, and Paul Groth, eds. *Everyday America: Cultural Landscape Studies after J. B. Jackson.* Berkeley: University of California Press, 2003.

Wilson, Samuel M. *History of Kentucky.* Vol. 2. Chicago: S. J. Clarke, 1928.

Winterhalder, Bruce. "Concepts in Historical Ecology: The View form Evolutionary Ecology." In *Historical Ecology: Cultural Knowledge and Changing Landscapes,* edited by Carole L. Crumley, 17–41. Santa Fe, NM: School of American Research Press, 1994.

"Winter's Patent Distilling Apparatus." *Register of the Arts and Sciences* 11 (1824): 161–64.

Woodbury, Levi. "Steam-Engines: Letter from the Secretary of the Treasury." US Congress, House, 25th Cong., 3d sess., 1838, doc. 21.

Work, Henry H. *Wood, Whiskey, and Wine: A History of Barrels.* London: Reaktion Books, 2014.

Wright, Helen. "Insurance Mapping and Industrial Archeology." *Journal of the Society for Industrial Archeology* 9, no. 1 (1983): 1–18.

Wylie, John. *Landscape.* London: Routledge, 2007.

Yale, Pat. *From Tourist Attractions to Heritage Tourism.* 3d ed. Huntingdon, UK: ELM Publications, 2004.

Yater, George H. *Two Hundred Years at the Falls of the Ohio: A History of Louisville and Jefferson County.* Louisville, KY: Heritage Corporation, 1979.

Yeats, W. B. *Collected Poems.* London: Macmillan Collector's Library, 2016.

Young, Al. *Four Roses: The Return of a Whiskey Legend.* Louisville, KY: Butler Books, 2010.

Young, Otis E. "Origins of the American Copper Industry." *Journal of the Early Republic* 3, no. 2 (1983): 117–37.

Zaborney, John J. *Slaves for Hire: Renting Enslaved Laborers in Antebellum Virginia.* Baton Rouge: Louisiana State University Press, 2012.

Zarnowitz, Victor. *Business Cycles: Theory, History, Indicators and Forecasting.* Chicago: University of Chicago Press, 1996.

Zoeller, Chester. *Bourbon in Kentucky: A History of Distilleries in Kentucky.* 2d ed. Louisville, KY: Butler Books, 2010.

———. *Kentucky Bourbon Barons: Legendary Distillers from the Golden Age of Whiskey Making.* Louisville, KY: Butler Books, 2014.

Zube, Ervin H., ed. *Landscapes: Selected Writings of J. B. Jackson.* Amherst: University of Massachusetts Press, 1970.

Newspapers

Adair County News (Columbia, KY)

Anderson News (Lawrenceburg, KY)

Big Sandy News (Louisa, KY)

Bismarck (ND) Daily Tribune

Bourbon News (Paris, KY)

Breckenridge (KY) News

Central Record (Lancaster, KY)

Chicago Tribune

The Citizen (Berea, KY)

The Climax (Richmond, KY)

Crittenden Press (Marion, KY)

Cynthiana (KY) Democrat

Cynthiana (KY) News

Daily Commonwealth (Frankfort, KY)

Daily Evening Bulletin (Maysville, KY)

Daily Public Ledger (Maysville, KY)

Evening Bulletin (Maysville, KY)

Farmers' Home Journal (Louisville, KY)

Frankfort (KY) Roundabout

Frankfort (KY) Weekly News and Roundabout
Franklin Farmer (Frankfort, KY)
Hartford (KY) Herald
Hartford (KY) Republican
Hazel Green (KY) Herald
Herald-Leader (Lexington, KY)
Hickman (KY) Courier
Hopkinsville Kentuckian
Interior Journal (Stanford, KY)
It (Lawrenceburg, KY)
Kentucky Farmer and Breeder (Lexington, KY)
Kentucky Gazette (Lexington, KY)
Kentucky Irish American (Louisville, KY)
Kentucky Live Stock Record (Lexington, KY)
Kentucky Sentinel (Mt. Sterling, KY)
Kentucky Tribune (Danville, KY)
Kentucky Whig (Lexington, KY)
Lexington (KY) Observer and Reporter
Louisville (KY) Courier-Journal
Louisville (KY) Daily Courier
Louisville (KY) Daily Democrat
Louisville (KY) Daily Express
Louisville (KY) Evening Express
Louisville (KY) Weekly Courier
Louisville (KY) Weekly Journal
Mount Vernon (KY) Signal
Mountain Advocate (Barbourville, KY)
Mt. Sterling (KY) Advocate
National Coopers Journal
New Ulm (MN) Weekly Review
New York Sun
Niles Weekly Register
Paducah (KY) Daily Sun
Paducah (KY) Evening Sun
Paducah (KY) Sun
The Record (Greenville, KY)
Richmond (KY) Climax
Semi-Weekly Interior Journal (Stanford, KY)
Spout Spring (KY) Times
Weekly Roundabout (Frankfort, KY)
Winchester (KY) News

Atlases, Maps, and Directories

Atlas of Bourbon, Clark, Fayette and Woodford Counties, Kentucky. Philadelphia: D. G. Beers, 1877.
An Atlas of Henry and Shelby Counties, Kentucky Philadelphia: D. J. Lake, 1882.

Atlas of Jefferson and Oldham Counties, Kentucky. Philadelphia: D. G. Beers & J. Lanagan, 1879.

An Atlas of Nelson and Spencer Counties, Kentucky. Philadelphia: D. J. Lake, 1882.

Bonfort, Philip. *Bonfort's Wine & Liquor Trade Directory for the United States* New York: Philip Bonfort, 1875.

Caron's Annual Directory of the City of Louisville. Vol. 1. Louisville, KY: Bradley & Gilbert, 1871.

Caron's Annual Directory of the City of Louisville. Vol. 11. Louisville, KY: Bradley & Gilbert, 1881.

Caron's Annual Directory of the City of Louisville. Vol. 39. Louisville, KY: Bradley & Gilbert, 1909.

Covington, Kentucky, 1886. New York: Sanborn Map & Publishing Company, 1886.

Frankfort, Kentucky. New York: Sanborn Map & Publishing Company, 1886.

Gibbons, A. B. *Geologic Map of Parts of Newport and Withamsville Quadrangles, Campbell and Kenton Counties, Kentucky.* Map GQ-1072, US Geological Survey. Washington, DC: GPO, 1973.

Griffing, B. N. *An Atlas of Franklin County, Kentucky.* Philadelphia: D. J. Lake, 1882.

Hergesheimer, Edwin. *Map Showing the Distribution of the Slave Population of the Southern States of the United States, 1860"* Washington, DC: US Census Office, 1861.

Hewitt, E. A., and G. W. Hewitt. *Topographical Map of the Counties of Bourbon, Fayette, Clark, Jessamine and Woodford, Kentucky.* New York: Smith, Gallup, 1861.

Hoeing, J. B. *Preliminary Map of Kentucky, 1891: Railroad Systems.* New York: Kentucky Geological Survey, 1889.

Insurance Maps of Louisville, Kentucky. Vol. 1. New York: Sanborn-Perris Map Company, 1892.

Insurance Maps of Louisville, Kentucky. Vol. 4. New York: Sanborn Map Company, 1905.

Kentucky Bourbon Trail, 2017. Frankfort: Kentucky Distillers' Association, 2017.

Kentucky State Gazetteer & Business Directory for 1879–1880. Detroit, MI: R. L. Polk, 1880.

Kentucky State Gazetteer and Business Directory, 1881–1882. Detroit, MI: R. L. Polk & A. C. Danser, 1882.

Kentucky State Gazetteer & Business Directory, 1883–4. Detroit, MI: R. L. Polk & A. C. Danser, 1884.

Lexington, Kentucky. New York: Sanborn Map & Publishing Company, 1885.

Lexington, Kentucky. New York: Sanborn-Perris Map Company, 1890.

Lexington, Kentucky. New York: Sanborn-Perris Map Company, 1891.

Louisville, Kentucky. Vol. 1. New York: Sanborn Map Company, 1905.

Luft, Stanley J. *Geologic Map of Part of the Covington Quadrangle, Northern Kentucky.* Map GQ-955, US Geological Survey. Washington, DC: GPO, 1971.

Map of Harrison County, Kentucky. Philadelphia: D. G. Beers, 1877.

Map of Marion and Washington Counties, Kentucky. Philadelphia: D. G. Beers, 1877.

Maydwell, Alexander. *Maydwell's Lexington City Directory for 1867.* Cincinnati, OH: Miami Printing & Publishing, 1867.

McDonough, Leo. *An Illustrated Historical Atlas Map of Daviess County, Ky. 1876.* Reprint, Evansville, IN: Unigraphic, 1978.

Owensboro, Kentucky. New York: Sanborn Map & Publishing Company, 1885.

Prather, Jas. H. *Prather's Directory of the City of Lexington, Kentucky, August, 1895.* Lexington, KY: Jas. H. Prather, 1895.

———. *Prather's Lexington City Directory, 1875–1876.* Lexington, KY: Jas. H. Prather, 1875–1876.

Procter, John H. *Map of Kentucky.* Frankfort: Kentucky Geological Survey, 1881.

Railroad Map of Kentucky Issued by Authority of the Railroad Commission. Frankfort, KY: Interstate Publishing Company, 1912.

Sanborn's Surveys of the Distilleries and Warehouses of Kentucky and Tennessee. New York: Sanborn Map Company, 1910.

Sanborn's Surveys of the Whiskey Warehouses of Kentucky and Tennessee. New York: Sanborn-Perris Map Company, 1894.

Ulack, Richard, Karl Raitz, and Gyula Pauer, eds. *Atlas of Kentucky.* Lexington: University Press of Kentucky, 1998.

National Register of Historic Places (NRHP) Registration Forms and National Historic Landmark (NHL) Nominations

ACW-AOJ-CMT. "Van Meter Distillery," Winchester, KY. NRHP. Frankfort: Kentucky Heritage Council, 1976.

Boldrick, Sam S. "Burk's Distillery or Maker's Mark Distillery," Loretto, KY. NRHP. Frankfort: Kentucky Heritage Council, 1974.

Breen, Phil. "Pogue House," Maysville, KY. NRHP. Frankfort: Kentucky Heritage Council, 2005.

Brooks, Carolyn. "Country Estates of River Road," Louisville, KY. NRHP. Frankfort: Kentucky Heritage Council, 1999.

———. "George T. Stagg Distillery," Frankfort, KY. NHL. Frankfort: Kentucky Heritage Council, 2012.

Goddard, Jayne, Brett Connors, and Amy Crossfield. "Old Taylor Distillery Historic District," Woodford County, KY. NRHP. Frankfort: Kentucky Heritage Council, 2016.

Hall, David H. "T. W. Samuels Distillery Historic District," Deatsville, KY. NRHP. Frankfort: Kentucky Heritage Council, 1987.

Hampton, Roy. "Ohio River Lock and Dam No. 45," Addison, KY. Kentucky Historic Resources Individual Survey Form BC 12. Frankfort: Kentucky Heritage Council, 1999.

Hedgepeth, Marty. "Bernheim Distillery Bottling Plant," Louisville, KY. NRHP. Frankfort: Kentucky Heritage Council, 1983.

Joseph, Kathryne M., and Craig A. Potts. "James E. Pepper Distillery," Lexington, KY. NRHP. Frankfort: Kentucky Heritage Council, 2008.

Kinsman, Mary Jean. "Warehouse D, White Mills Distillery Company," Louisville, KY. NRHP. Frankfort: Kentucky Heritage Council, 1978.

Kramer, Carl E. "J. T. S. Brown & Son's Complex," Louisville, KY. NRHP. Frankfort: Kentucky Heritage Council, 1998.

Norwood, Clare, and Lisa Gillham. "Ritte's East Historic District," Covington, KY. NRHP. Frankfort: Kentucky Heritage Council, 2013.

Pezzoni, J. Daniel. "George T. Stagg Distillery (Buffalo Trace Distillery)," Frankfort, KY. NRHP. Frankfort: Kentucky Heritage Council, 2001.

Pierson, Joe. "Whiskey Row Historic District," Louisville, KY. NRHP. Frankfort: Kentucky Heritage Council, 2010.

Polsgrove, Robert M. "Old Prentice Distillery (Four Roses Distillery)," Lawrenceburg, KY. NRHP. Frankfort: Kentucky Heritage Council, 1986.

———. "T. Jeremiah Beam House," Clermont, KY. NRHP. Frankfort: Kentucky Heritage Council, 1987.

Schneider, Charlotte. "Dowling House," Lawrenceburg, KY. NRHP. Frankfort: Kentucky Heritage Council, 1979.

Taylor, Thelma. "Monticello," Cynthiana, KY. NRHP. Frankfort: Kentucky Heritage Council, 1974.

Thacker, Glenda. "Le Vega Clements House." NRHP. Frankfort: Kentucky Heritage Council, 1986.

Whisman, Eric. "Nelson Distillery Warehouse, Anderson-Nelson Distillery," Louisville, KY. NRHP. Frankfort: Kentucky Heritage Council, 2014.

Wolf, Jean K. "Labrot & Graham's Old Oscar Pepper Distillery," Woodford County, KY. NHL. Frankfort: Kentucky Heritage Council, 1999.

Wooley, C. M. "Jacob Spears Distillery," Paris, KY. NRHP. Frankfort: Kentucky Heritage Council, 1982.

———. "McCrackin Distillery and Mill," Tyrone, KY. NRHP. Frankfort: Kentucky Heritage Council, 1983.

Patents and Trademarks

"Application for Renewal of Registration of Trademark." Old Sunny Brook Company, National Distillers and Chemical Corporation, New York, NY. Trademark 710,690, January 31, 1961.

"Construction of Corn Shellers." John A. Whitford, Saratoga Springs, NY. US Patent No. 1,946, January 23, 1841.

"Corn-Sheller." Augustus Adams, Sandwich, IL. US Patent No. 32,971, August 6, 1861. Reissued March 20, 1866.

"Corn Sheller." Wm. Reading, Washington, DC. US Patent No. 9,120, July 13, 1852.

"Drier." John E. Turney, Louisville, KY. US Patent No. 740,607, October 6, 1903.

"Drier." John E. Turney and Geiger, Koop & Fiske, Louisville, KY. US Patent No. 774,859, November 15, 1904.

"Feed-Mill." Nelson P. Bowsher, South Bend, IN. US Patent No. 528,070, October 23, 1894.

"Geo. G. Brown Est'd Old Forester 1870 First Bottled Bourbon Old Forester Est'd 1870 Kentucky Straight Bourbon Whisky." Early Times Distillers Company, Louisville, KY. Trademark registration no. 4392866, August 27, 2013.

"Glass-Shaping Machine." Michael J. Owens, Toledo, OH. US Patent No. 766,768, August 2, 1904.

"Grain-Drying Apparatus." Peter Jepson. Port Chester, NY. US Patent No. 373,140, November 15, 1887.

"H. McKenna Trade-Mark Application." James S. McKenna, Fairfield, KY. Trademark 324,832, February 20, 1905. Copy in Marcella McKenna Distillery Collection, Special Collections, University of Kentucky.

"H. McKenna Trade-Mark Application." James S. McKenna, Fairfield, KY. Trademark 344,114, February 20, 1905. Copy in Marcella McKenna Distillery Collection, Special Collections, University of Kentucky.

"Improvement in Hoisting-Machines." Henry Reedy, Cincinnati, OH. US Patent No. 78,829, June 9, 1868. Reissued 1871.

"Improvement in Racks for Tiering Barrels." Frederick Stitzel, Louisville, KY. US Patent No. 221,945, November 25, 1879. Reissued April 27, 1880.

Improvement in Stills." George W. Robson, Cincinnati, OH, and Melvin T. Hughes, Paris, KY. US Patent No. 87,971, March 16, 1869.

"Improvement in Stills for Alcoholic Spirits." William and George W. Robson, Cincinnati, OH. US Patent No. 92,477, July 13, 1869.

"Knob Creek Trade Mark Application." Jim Beam Company, Louisville, KY. Trademark registration no. 2156225, May 12, 1998.

"Machine for Blowing Glass." Michael J. Owens, Toledo, OH. US Patent No. 548,588, October 22, 1895.

"McKenna." William H. Hollander, Wyatt, Tarrant & Combs, Louisville, KY. Registration no. 324,832, December 7, 1994. http://assignments.uspto.gov/assignments /q?db=tm&reel=0667&frame=0890.

"Old Forester Trade Mark Application." Brown-Forman Company, Louisville, KY. Trademark registration no. 0044366, July 4, 1905.

"Old Taylor." Thos. A. Clark, National Distillers Products Corporation. Trademark 507,794, March 22, 1949.

"Old Tub." Jim Beam Brands, Deerfield, IL. Trademark registration no. 4293249, February 19, 2013.

"Process of and Apparatus for Distilling." John C. Peden, Lawrenceburg, KY. US Patent No. 349,449, September 21, 1886.

"Reapplication of H. McKenna, Incorporated, Trade-Mark." James S. McKenna, Fairfield, KY. Trade-mark 344,114, December 31, 1934. Copy in Marcella McKenna Distillery Collection, Special Collections, University of Kentucky.

"Yellowstone Trade Mark Application." Taylor & Williams Inc., Louisville, KY. Trademark registration no. 0280199, February 10, 1931.

Kentucky County Records

Clerk of Court, Anderson County, Lawrenceburg, KY. *Anderson County Deed Record Books M, N, P, Q, R, S.*

Clerk of Court, Larue County, Hodgenville, KY. *Larue County Deed Record Books 6, 7, 12–16, 18.*

Clerk of Court, Nelson County, Bardstown, KY. *Nelson County Deed Record Books 26, 28, 34, 35, 39, 41, 44, 45, 49, 50, 53.*

Clerk of Court, Scott County, Georgetown, KY. *Scott County Deed Record Books 4, 14.*

Manuscript Collections

Cleaveland, Emma Jane. Journal, 1820–1932. In Lyne-Smith Family Papers (Woodford County, KY). 1997 MS387. Special Collections, University of Kentucky.

Coyte, H. W. "History of Kentucky Distilleries," 1940–1987. Manuscript. University Archives and Records Center, University of Louisville, Louisville, KY.

Daviess County Distilling Company Records, 1850–1940, Medley, Meschendorf, Owensboro. Microfilm, 1F67M-621. Special Collections, University of Kentucky.

Elkhorn Distillery Records (Scott County, KY), 1868–1872. 1997 MS419. Special Collections, University of Kentucky.

Jones, John W. Papers (Bourbon County, KY), 1857–1896. 1857–1889 (bulk dates), 75M26. Special Collections, University of Kentucky.

Loder, Lewis. *The Loder Diary, 1857–1903.* Transcribed by William Conrad. Florence, KY: Boone County Schools, 1985–1988. Microfilm. http://bcpl.org/cbc/doku.php/lewis_loder.

McKenna, James S. "H. McKenna, Incorporated in Kentucky." 1933. Marcella McKenna Distillery Collection, Special Collections, University of Kentucky.

McKenna, Marcella. "An Aspect of Industry in Early Kentucky—To End of Civil War." Presented to the Nelson County Historical Society, July 5, 1966. Typescript copy in Oscar Getz Museum, Bardstown, KY.

———. "Dan McKenna." 1964. In Marcella McKenna Distillery Collection, Special Collections, University of Kentucky.

McKenna, Marcella. Distillery Collection (Nelson County, KY), 1824–1945. 1858–1942 (bulk dates), 1 F65M-593. Special Collections, University of Kentucky.

McKenna, Marcella. Papers. Oscar Getz Museum, Bardstown, KY.

Moylan, John. Ledger A, Lexington (Fayette County, KY), 1792. MS042. Special Collections, University of Kentucky.

Oots, Andrew J. Cooper Shop Records (Fayette County, KY), 1880–1926. 1880–1910 (bulk dates), 1 FM-49. Special Collections, University of Kentucky.

Ripy Family Papers (Anderson County, KY), 1888–1963. 1 F66M-725. Special Collections, University of Kentucky.

Taylor-Hay Family Papers, (Franklin County, KY), 1783–1991. MssAT238d, Filson Historical Society, Louisville, KY.
US Post Office Department. L & N Railroad Archive. Special Collections, Ekstrom Library, University of Louisville.

Websites

General Sources

http://athena.uky.edu/cgi/i/image/image-idx?cc=beasanic;page=browse;key=beasanic_lo;size=50;c=beasanic;back=back1456695601;value=l
http://babel.hathitrust.org/cgi/pt?id=coo.31924015431400;view=1up;seq=7.
http://banjosbourbonbarrelsandbluegrass.tumblr.com/post/68069901962/1838–1903-lexington-city-directories-available
http://chroniclingamerica.loc.gov
http://kdl.kyvl.org

US Census Tables and Reports

http://www.agcensus.usda.gov/Publications/Historical_Publications/
https://www.census.gov/history/pdf/manufactures1810–1890.pdf
https://www.census.gov/prod/www/decennial.html
http://www.nrcs.usda.gov/wps/portal/nrcs/surveylist/soils/survey/state/?stateId=KY

Historic Structures and Landscapes

http://www.citylab.com/housing/2014/10/the-accidental-revelations-of-sanborn-maps/381262/   (Sanborn maps)
http://www.heritage.ky.gov/
https://www.irelandxo.com/ireland/derry/ballynascreen/message-board/moneyneena-drumderg-mckenna
http://www.kentucky.com/2012/06/12/2221667_tom-eblen-family-saves-historic.html?rh=1
https://www.nps.gov/history/local-law/nhpa1966.htm
http://www.nps.gov/nr/research/data_downloads.htm#spreadsheets

Whiskey Subjects

http://kybourbon.com/bourbon_culture-2/key_bourbon_facts/
http://www.kybourbon.com/images/uploads/economic_impact_2014.pdf
http://www.bourbonenthusiast.com
http://www.cincinnatimagazine.com/high-spirits-blog/mgp-ingredients-lawrenceburg/
http://www.findagrave.com/cgi-bin/fg.cgi?page=gr&GRid=88934626
http://www.gibbsbrothers.com/about.htm
http://gobourbon.com/using-oak-barrels-to-age-whiskey/
http://www.kentuckybarrels.com/barrels.html
http://www.kentuckytourism.com/industry/development_incentives/devincentiveprogram.aspx
http://www.kyhistory.com/cdm/search/searchterm/distillery/mode/all/order/descri/ad/desc/page/2
https://www.lanereport.com/114188/2019/06/startup-distillery-with-historic-roots-to-locate-in-nelson-county-with-12m-investment/

http://lexhistory.org/wikilex/bourbon-and-distilleries
http://lexhistory.org/wikilex/henry-clay-distillery-rd-5
http://mcginnisbourbonbarrels.com/aboutus.html
http://nunncenter.org/bourbon/tag/w-b-saffell-distillery/
http://pdfpiw.uspto.gov/
http://www.pre-pro.com
http://www.rootsweb.ancestry.com/~flbbm/heritage/cooper/barrelmaking.htm
http://www.rootsweb.ancestry.com/~kyfrankl/SWelchgenealogy.htm
http://www.sha.org/bottle/References.htm
http://sippncorn.blogspot.com
https://www.thespiritsbusiness.com
http://whiskyscience.blogspot.com/2014/10/copper.html

Maps and Atlases

http://www.arcgis.com/home/webmap/viewer.html?useExisting=1
http://geonames.usgs.gov/apex/f?p=262:1:7064394388960
http://www.uky.edu/Libraries/forms/kylandowners.pdf

Kentucky-Based Traditional Trades

Brown-Forman Cooperage, Louisville. http://bourbontrailtours.com/Brown-Forman-Cooperage-Tour.html.

Buzick Construction, Tom Blincoe (president). Kentucky's only remaining building contractor for warehouse construction. http://www.teambuzick.com/.

Copper & Kings, Butchertown, Louisville. http://www.copperandkings.com/.

Independent Stave Company, Lebanon. Part of the Bourbon Trail tour. http://www.independentstavecompany.com/tours.

Kelvin Cooperage, Louisville. Wine and craft distillery barrels. http://kelvincooperage.com/.

Robinson Stave and Cooperage, East Bernstadt. http://www.robinsonstavecooperage.com/.

Scotland Barrel Recyclers, Shepherdsville. http://speysidecooperageky.com/.

Vendome Copper and Brass Works, Louisville. http://www.vendomecopper.com/; https://www.youtube.com/watch?v=jhGiutzV438&list=PLhNST-gqaeMMWbh_sCDoZGgLrjpJ9IOQd&index=3.

Index

Italic page numbers refer to illustrations.

A. Adams & Sons Company, *157, 158*
A. Brandis & Son, 102
A. B. & W. L. Crabb, 390
accidents, 162. *See also* fires
Adam, Edward, 118
Adams, Augustus, *157, 158*
Adams, Samuel, 17
Adams corn sheller, *157, 158*
advertising: development of advertising strategies, 453–54; by the Henry McKenna Distillery, 351–69; promoting tradition, 454–56; transition to modern branding and advertising, 451–59
African Americans: community account holders of the Henry McKenna Distillery, 328–29; distillery and distilling-related employment, 232, 233, 234, 235–36, 243; farm laborers for Henry McKenna, 303; laborers at the Elkhorn Distillery, 381, 396; in the Henry McKenna household, 299, *300, 301*; in the James Stone household, *372, 373*. *See also* slavery
aging. *See* barrel aging
Agricultural Adjustment Act, 266
agriculture: costs of transporting whiskey versus corn, 194–95; ecological context of distilling and, 36, 37; grains and distilling, 56–70 (*see also* grains); labor requirements and labor force, 221, 222–23; and land reclamation in Daviess County, 110; livestock, 70–72 (*see also* livestock); mechanization and, 156–58; multimodal transportation networks, 218–19; railroad transportation and the shipping of grain, 204–8; river transportation and, 192–93; seasonality and, 46–47; soils and physiography, 50–55; upland sites, 50; weather and climate, 45–46

A. H. Heilman & Company, 449
Alanant-O-Wamiowee, 108
alcohol: distilling process and, 19; sold and consumed as a medicinal, 292–94; spirits consumption in America, 1800–1900, 285; views of in colonial and frontier America, 283
aldehydes, 19
alembics, 19
alfisols, 50
Allen, J. A., 320
Allen, Robert M., 188–89
Allen-Bradley Distilling Company, 172
Alvey, W. J., 353
A. Mayfield and Company, 241, 451
American Distiller, The (Krafft), 393
American Plan, 196
American Temperance Society, 284
Ancient Age–Schenley Distillery, 216
Ancient Order of United Workmen, 287
Anderson County: distilleries, 89, 143, 144, 145, 213, 489; distiller–slave owners, 226, 227; distillery production trends in the 1800s, 30; distilling-related employment, 232–33; environmental risks to distillers, 161; physiography and soils, 52, *53;* railroads, 213; sites of mills and distilleries, 48; slop dumping, animal waste, and stream pollution, 187; spent grains and livestock feeding, 183; turnpikes, 198. *See also* Lawrenceburg, KY; Tyrone, KY
Anderson County Distillery No. 418, 141
Anderson Distillery, 408
Andrew J. Oots & Sons, 122, 141–45
animal diseases, 163–66
Anti-Saloon League, 284
Appalachian Plateau, 50
Appleton, William, 130